ELECTROMAGNETIC SCATTERING

Academic Press Rapid Manuscript Reproduction

ELECTROMAGNETIC SCATTERING

EDITED BY

PIERGIORGIO L. E. USLENGHI

Communications Laboratory
Department of Information Engineering
University of Illinois at Chicago Circle
Chicago, Illinois

ACADEMIC PRESS **New York San Francisco London** **1978**
A Subsidiary of Harcourt Brace Jovanovich, Publishers

ACADEMIC PRESS, INC.
111 Fifth Avenue, New York, New York 10003

United Kingdom Edition published by
ACADEMIC PRESS, INC. (LONDON) LTD.
24/28 Oval Road, London NW1 7DX

Library of Congress Cataloging in Publication Data

Main entry under title:

Electromagnetic scattering.

1. Electromagnetic waves—Scattering.
I. Uslenghi, P. L. E.
QC665.S3E43 530.1'41 78-15929
ISBN 0-12-709650-7

PRINTED IN THE UNITED STATES OF AMERICA

CONTENTS

LIST OF CONTRIBUTORS

Numbers in parentheses indicate the pages on which authors' contributions begin.

Carl E. Baum (471, 571), Air Force Weapons Laboratory, Kirtland Air Force Base, Albuquerque, New Mexico 80117

C. Leonard Bennett (393), Sperry Research Center, Sudbury, Massachusetts 01776

Shu-Kong Chang (359), Department of Electrical Engineering and Computer Sciences, University of California at Berkeley, Berkeley, California 94720

Charles L. Dolph (503), Department of Mathematics, The University of Michigan, Ann Arbor, Michigan 48109

Leopold B. Felsen (29), Department of Electrical Engineering, Polytechnic Institute of New York, Brooklyn, New York 11201

Roger F. Harrington (429), Department of Electrical and Computer Engineering, Syracuse University, Syracuse, New York 13210

J. Richard Huynen (653), Engineering Consultant, Los Altos Hills, California 94022

Ralph E. Kleinman (1), Department of Mathematics, University of Delaware, Newark, Delaware 19711

S. W. Lee (67), Electromagnetics Laboratory, Department of Electrical Engineering, University of Illinois at Urbana-Champaign, Urbana, Illinois 61801

Robert J. Mailloux (713), Deputy for Electronic Technology (RADC), Hanscom Air Force Base, Bedford, Massachusetts 01731

Joseph R. Mautz (429), Department of Electrical and Computer Engineering, Syracuse University, Syracuse, New York 13210

Kenneth K. Mei (359), Department of Electrical Engineering and Computer Sciences, University of California at Berkeley, Berkeley, California 94720

E. K. Miller (315), Lawrence Livermore Laboratory, Livermore, California 94550

R. Mittra (121), Electromagnetics Laboratory, Department of Electrical Engineering, University of Illinois at Urbana-Champaign, Urbana, Illinois 61801

Michael E. Morgan (359), Department of Electrical Engineering and Computer Sciences, University of California at Berkeley, Berkeley, California 94720

A. J. Poggio (315), Lawrence Livermore Laboratory, Livermore, California 94550

Y. Rahmat-Samii (121), Electromagnetics Laboratory, Department of Electrical Engineering, University of Illinois at Urbana-Champaign, Urbana, Illinois 61801

Richard A. Scott (503), Department of Applied Mechanics and Engineering Science, The University of Michigan, Ann Arbor, Michigan 48109

Thomas B. A. Senior (185), Radiation Laboratory, Department of Electrical and Computer Engineering, The University of Michigan, Ann Arbor, Michigan 48109

Victor Twersky (221), Department of Mathematics, University of Illinois at Chicago Circle, Chicago, Illinois 60680

James R. Wait (253), Cooperative Institute for Research in Environmental Sciences, University of Colorado, Boulder, Colorado 80309

Vaughan H. Weston (289), Division of Mathematical Sciences, Purdue University, West Lafayette, Indiana 47907

PREFACE

The idea for this book first occurred to me during the planning of a National Conference on Electromagnetic Scattering, which took place during the summer of 1976 and brought together scientists from government, universities, and industry, to assess the state of the art and to outline the most important areas of future research in electromagnetic scattering and its applications, as well as the most promising ways of performing such research during the next decade. At that conference I approached a number of prominent researchers with the request to contribute a chapter to a book elucidating methods, state of the art, applications, and future research in electromagnetic scattering. Most of them responded enthusiastically and worked very hard during the past 18 months to produce the 17 chapters that comprise this volume. The result is a book that has several unique features. It is a collection of minimonographs of long-lasting value to applied mathematicians and physicists, as well as to electrical engineers who are engaged in the design or use of antennas and radars for both civilian and military applications, or in the applications of electromagnetic scattering theory to a variety of fields (geophysical prospecting, environmental pollution, biomedicine, etc.). Since the authors have wisely refrained from delving into too many specific problems and have confined themselves to illustrating general methods, this book will also be very useful to graduate students and all others who are not yet specialists in scattering theory but desire to acquire a general view of this subject before dealing with the myriad of specific problems that are discussed in professional journals.

The book consists of four groups of chapters. The first group (Chapters 1–8) deals with analytical methods for the solution of electromagnetic scattering problems. A review of low-frequency scattering is followed by three chapters on high-frequency techniques devoted, respectively, to recent developments in asymptotic methods, to scattering by edges, and to the spectral theory of diffraction. Chapter 5 is devoted to some new exact solutions of boundary value problems by function-theoretic methods. Chapters 6 and 7 present the general theories of multiple scattering for periodic and random distributions, and the applications to scattering by grids and meshes. The ever-growing importance of inverse scattering and the difficult analytical problems associated with it are addressed in Chapter 8. These eight chapters provide the reader with an up-to-date, comprehensive treatment of the most successful analytical tools presently employed in direct and inverse scattering research.

The obvious limitations of analytical approaches to scattering problems have been partially overcome in recent years by numerical methods, especially by the method of moments and by finite methods. These techniques and their limitations are discussed in Chapters 9 and 10. The particular problems that arise in the numerical treatment of transients are faced in Chapter 11, whereas Chapter 12 is devoted to applications of numerical techniques to penetration of electromagnetic fields through apertures. This second group of four chapters gives the reader the necessary tools for numerically solving a wide variety of scattering and radiation problems with the aid of digital computers.

A subject that has attracted increased attention by the scientific community in recent years is the study of natural and man-made electromagnetic pulses (lightning, nuclear explosions, geophysical prospecting, wideband radar, etc.). The necessity to screen electronic circuits and systems from electromagnetic pulses has led to significant advances in the understanding of electromagnetic interference (Chapter 13). General theoretical methods to study pulse propagation and scattering have also been developed; among these, perhaps the most promising and successful is the singularity expansion method, whose mathematical foundations are laid in Chapter 14 for electromagnetic as well as elastic wave propagation and whose engineering applications are discussed in Chapter 15.

The last two chapters are, respectively, concerned with a theory of radar targets from a phenomenological viewpoint and with modern trends in arrays research.

The chapter authors and I gratefully acknowledge the financial support that the Air Force Office of Scientific Research, Air Force Systems Command, United States Air Force, provided under Grant AFOSR-76-2888. Additional financial support was provided by the University of Illinois at Chicago Circle. I am indebted to Dr. Robert N. Buchal of AFOSR/NM for his guidance and encouragement, to those who typed the manuscript, and to the publishers for their assistance.

LOW FREQUENCY ELECTROMAGNETIC SCATTERING

Ralph E. Kleinman
Department of Mathematics
University of Delaware

I. INTRODUCTION

More than a decade has passed since the publication of a review of analytic methods and results in Rayleigh scattering [13]. The intervening years have seen considerable progress in this general area and the present paper attempts to record the current status of low frequency methods in electromagnetic scattering.

Undoubtedly the most dramatic advances have been in numerical solutions of scattering problems. However the limitation to low frequencies in numerical work is a pragmatic one, based on machine imposed constraints. It is not an essential mathematical restriction such as that imposed by finite radius of convergence of a power series, the criteria which formed the basis for the definition of the Rayleigh region adopted in the earlier cited review. For this reason, also because the subject of numerical methods is treated elsewhere in this volume, no attempt will be made to describe developments in this very important area.

Other notable omissions include two dimensional and scalar three dimensional problems. The present focus on electromagnetics explains the omission of scalar three dimensional problems, even though they often serve as prototypes for the vector, electromagnetic, case and they were discussed in the 1965 study. However,

[1] This work was supported by the U.S. Air Force Office of Scientific Research under Grant 74-2634.

ISBN 0-12-709650-7

only an arbitrary decision to accede to the pressures of time explains the absence of two dimensional problems, which, though scalar in character, have electromagnetic interpretations.

Two dimensional problems were somewhat cavalierly dismissed in the earlier study though the existence of sensible two dimensional problems was acknowledged. A partial explanation for this lies in the fact that in two dimensions, the functions of interest, be they field quantities, surface currents, far field coefficients or scattering cross sections, are not analytic function of the wave number, k, at $k=0$, whereas in three dimensions they are analytic, hence are developable in convergent power series. Because $\omega = kc$, where ω is frequency and c is the velocity of propagation, low frequency and small k approximations are equivalent. Since such convergent power series underlay the definition of the Rayleigh region, two dimensional problems were awkward in that they did not fall in the Rayleigh region so defined.

Recent work has suggested a different way to characterize the low frequency regime which will be equally applicable in all dimensions. The idea rests on integral equation formulations for the unknown field quantities. In cases where the unknown field satisfies an integral equation of the second kind which obtains when the scatterer has no infinitesimally thin edges then the kernel will depend on wave number and the equation will have the form

$$u = K_k u + f_k \tag{1}$$

where u is the desired field quantity, f_k is known and K_k is a compact operator valued function of k, a boundary or volume integral operator. In scalar problems K_k may be a boundary integral operator with kernel equal to the normal derivative of the free space Green's function or fundamental solution (a double layer distribution); in the Born approximation K_k is a volume integral operator with kernel merely the free space Green's function; in electromagnetic scattering by perfect conductors K_k is a vector valued boundary integral operator with a kernel which is the gradient of the free space Green's function (see Equation 43). The spectral radius of K_k is the radius of the smallest circle which contains all points in the spectrum of K_k (values of λ for which $\lambda I-K_k$ is not invertible). Any λ larger than the spectral radius is in the resolvent set hence

$$\lambda u = K_k u + f_k \tag{2}$$

is soluble. Moreover if the spectral radius is less than one, not only is $\lambda = 1$ (which corresponds to the case of interest, Equation 1) in the resolvent set but the solution is given by the convergent Neumann series, $u = \sum_{n=0}^{\infty} K_k^n f_k$. In practice, the only examples where such convergence has been demonstrated are in the Rayleigh region. The classic solution of boundary value problems in potential theory has been obtained as the Neumann series solution of

$$u = K_0 u + f_0$$

using these considerations of the spectral radius by [22] and [38] and extensions to the scalar Helmholtz equation have been carried out by [1] and [21]. The electromagnetic case for perfect conductors has been treated by [7] and for diaphanous scatterers, the Born approximation [10] is precisely a Neumann series solution of Equation (1).

Of course, it is well known that while integral equation formulations of scattering problems may have unique solutions, the formulation itself is not unique. That is, there are many (infinitely many) different integral equations, all of which will have the same solutions. Thus while Equation (1) correctly characterizes a class of integral equations which arise in electromagnetics there may be many different possible choices of the kernel K_k. This ambiguity is accentuated if the compactness requirement on K_k is removed so that multiples of the identity may be included. One way of removing the ambiguity has been found [18] so that out of some class of possible kernels, K_k is chosen to have the smallest spectral radius. This suggests a new definition:

the Rayleigh region is that range of k for which the spectral radius of K_k is less than one. This definition has the advantage of not only including the usual cases, convergent power series expansions in three dimensions, but also two dimensional cases. The disadvantage lies in the ambiguity remaining in the choice of K_k.

There are however additional advantages to this definition. The problem of estimating the radius of convergence of a low frequency expansion in any particular case was recognized as severe. To estimate the spectral radius of a complicated integral operator is also a problem but is more tractable in that there is available considerable machinery for estimating the largest eigenvalue of an operator which in many cases is all that is needed for an estimate of the spectral radius. Moreover this way of characterizing the low frequency regime may provide new weapons to attack the persistant problem of finding ways to extend analytical methods from low frequencies into the resonance region.

This functional analytic approach to electromagnetics is perhaps one of the most promising developments in low frequency scattering. However at the present time, there remains much to be done. While outgrowths of this approach may dominate a review of low frequency methods a decade from now, results of the last decade are the principal concern of this paper.

The physical problem under consideration is the scattering of a time harmonic incident electromagnetic wave by a closed bounded, homogeneous object in three dimensions. Precise conditions on the target geometry as well as some basic integral representations are given in Section II. Section III is intended as a handbook of low frequency results including explicit reduction of the determination of coefficients in a low frequency expansion to problems in classical potential theory as well as some iterative methods for actually constructing the solutions. Here again we distinguish between methods which reduce a problem to other, simpler, problems which still must be solved, e.g., the Rayleigh-Stevenson method (III.A.1), and methods which actually produce solutions, e.g., low frequency iteration (III.A.2,3). Section III.A. considers perfectly reflecting targets via the Rayleigh-Stevenson method as well as two iterative methods, one involving volume integrals, one with only surface integrals. In addition results of the application of Polarizability tensors to the description of the Rayleigh term in the far field coefficient for plane wave scattering is presented. Section III.B is concerned with penetrable homogeneous scatterers and is devoted primarily to the Rayleigh-Stevenson method. The results presented have been rederived and are hopefully more usable than Stevenson's original presentation. Iterative methods and the applicability of Polarizability tensors are also discussed.

Any effort such as this is bound to be incomplete. For example, not specifically included is the perturbation procedure described in [25] and developed further [27]. There the scattering problem is reduced to a sequence of integral equations of the first kind, each equation having the same kernel, namely that kernel which arises in the static (k=0) case. Thus this method is analogous to the Rayleigh-Stevenson approach. Despite the fact that there remain some mathematical questions of existence and uniqueness some interesting results have been obtained by this method [11, 9, 2].

Additional references to a wide class of related problems may be found in the triennial URSI Reviews of Radio Science [24, 5]. Finally, the interested reader is urged to consult a forthcoming article [37], which, though not available during the preparation of this summary, will undoubtedly be a most useful reference. Results for particular shapes are given in [40,41].

II. NOTATION AND REPRESENTATION THEOREMS

Let S be a closed piecewise Lyapunoff surface which divides $\mathbb{R}^3$ into simply connected disjoint domains, an interior S_{in} (bounded) and an exterior S_{ex} (unbounded). A Lyapunoff surface has a continuous unique normal [23] and by piecewise Lyapunoff is meant a finite union of surface segments each of which is part of a Lyapunoff surface. In addition S is required to satisfy a modified cone condition [1], i.e., there exist positive quantities α and h such that each point of S is the common vertex of a pair of otherwise nonintersecting right circular cones of vertex angle α and height h, one lying entirely in S_{in} and one in S_{ex}. This permits S to have corners, wedge-like edges and conical points but not infinitesmally thin edges or spines. Denote by $\hat{n}_p$ the unit normal to S at the point p pointing from S into S_{ex}. The point p will have coordinates (x,y,z) or (x_1,x_2,x_3) with respect to a rectangular Cartesian system with origin in S_{in} and spherical polar coordinates (r,θ,ϕ). The position vector will also be denoted p, whereas the unit vector in the direction of p will be denoted by $\hat{p}$. Subscripts will be added only if there is ambiguity as to the point in question, e.g., r_p, θ_p etc. and $p = x_p\hat{x} + y_p\hat{y} + z_p\hat{z}$. The geometry is pictured in Fig. 1. The exterior domain, S_{ex}, is characterized by constitutive parameters ε_0,μ_0 denoting permitivity and permeability and is assumed to have zero conductivity whereas S_{in} is characterized by constants ε,μ,σ where the conductivity σ may be infinite. The total electromagnetic field $(\bar{E},\bar{H})$ will consist of the sum of incident $(\bar{E}^{inc},\bar{H}^{inc})$ and scattered $(\bar{E}^S,\bar{H}^S)$ terms where the incident field is assumed to originate in S_{ex}. All fields are assumed to have a harmonic time dependence $e^{-i\omega t}$, not explicitly exhibited. The frequency ω and constitutive parameters are related to wave numbers appropriate to S_{ex} and S_{in} by

$$k = \omega\sqrt{\varepsilon_0\mu_0} \quad , \quad k_i = \omega\sqrt{\left(\varepsilon + \frac{i\sigma}{\omega}\right)\mu} \ . \tag{3}$$

The appropriate time harmonic Maxwell equations are

$$\text{a)} \left.\begin{aligned} \nabla \times \bar{E}^S &= i\omega\mu_0\bar{H}^S \\ \nabla \times \bar{H}^S &= -i\omega\varepsilon_0\bar{E}^S \end{aligned}\right\} p \,\varepsilon\, S_{ex} \qquad \text{b)} \left.\begin{aligned} \nabla \times \bar{E} &= i\omega\mu\bar{H} \\ \nabla \times \bar{H} &= (-i\omega\varepsilon+\sigma)\bar{E} \end{aligned}\right\} p \,\varepsilon\, S_{in} \tag{4}$$

where it is noted that the incident field satisfies (4a) almost everywhere in $\mathbb{R}^3$. In addition the scattered field satisfies the radiation condition [39]

$$\hat{r}\times\nabla\times\bar{E}^S + ik\bar{E}^S = \hat{r}\times\nabla\times\bar{H}^S + ik\bar{H}^S = o\left(\frac{1}{r}\right) \text{ uniformly in } \hat{r}. \quad (5)$$

While the field quantities are infinitely differentiable except at source points in S_{ex} and S_{in}, some field components will be discontinuous at S. Denoting by subscripts the different limiting values of the field quantities as S is approached from the interior and exterior, the following conditions hold for $p \varepsilon S$:

a)
$$\hat{n}_p \times \bar{E}_{ex}(p) = \hat{n}_p \times \bar{E}_{in}(p) \qquad \hat{n}\cdot\mu_0\bar{H}_{ex} = \hat{n}\cdot\mu\bar{H}_{in}$$
$$\hat{n}\times\bar{H}_{ex} = \hat{n}\times\bar{H}_{in} \qquad \hat{n}\cdot\varepsilon_0\bar{E}_{ex} = \hat{n}\cdot\left(\varepsilon + \frac{i\sigma}{\omega}\right)\bar{E}_{in} \quad (6)$$

while, if $\sigma \to \infty$,

b) $\hat{n}\times\bar{E}_{ex} = 0 \qquad \hat{n}\cdot\bar{H}_{ex} = 0$

Integral representations of electromagnetic fields are available in standard texts (e.g., [36, 10, 26] and a comprehensive tabulation is given [29]. However conditions on the surface and differentiability requirements on the fields are often either overlooked or overstringent. For example, [39] requires the electromagnetic fields to be differentiable in the closure of the domain of interest and while with Maxwell's equations it is easy to establish differentiability to any order at interior points the restriction on the boundary is unnecessarily severe. To obtain the desired form of the representations first introduce the following measure of solid angle subtended by S at p [23, 1]

$$\sigma_3(p) = -\lim_{\varepsilon\to 0}\int_{\partial B_\varepsilon(p)\cap S_{ex}} \frac{\partial}{\partial n_q}\frac{1}{|p-q|}\,dS_q \quad (7)$$

where $\partial B_\varepsilon(p)$ is the boundary of a ball of radius ε with center at p. If S is Lyapunoff

$$\sigma_3(p) = 4\pi,\ p\ \varepsilon\ S_{ex}\ ;\quad 2\pi,\ p\ \varepsilon\ S\ ;\quad 0,\ p\ \varepsilon\ S_{in} \quad (8)$$

and while this evaluation of σ_3 remains valid in S_{ex} and S_{in} if S is piecewise Lyapunoff, $\sigma_3 = 2\pi$ only at smooth points of piece wise Lyapunoff surfaces (non boundary points of the constituent Lyapunoff segments). With this notation the following representations hold for all $p \in \mathbb{R}^3$. If $(\bar{E},\bar{H})$ satisfy (4b) in S_{in} and are continuous on S (piecewise Lyapunoff) then

$$\int_S \{v_i(p,q)(i\omega\varepsilon-\sigma)\hat{n}_q \times \bar{E}_{in}(q) - \nabla_q v_i \hat{n}_q \cdot \bar{H}_{in}(q) + \nabla_q v_i \times (\hat{n}_q \times \bar{H}_{in}(q))\}dS_q = \frac{\sigma_3(p)-4\pi}{4\pi}\bar{H}(p) \tag{9}$$

$$\int_S \{-v_i i\omega\mu\hat{n}_q \times \bar{H}_{in}(q) - \nabla_q v_i \hat{n}_q \cdot \bar{E}_{in}(q) + \nabla_q v_i \times (\hat{n}_q \times \bar{E}_{in}(q))\}dS_q = \frac{\sigma_3(p)-4\pi}{4\pi}\bar{E}(p) \tag{10}$$

where $$v_i(p,q) = -\frac{e^{ik_i|p-q|}}{4\pi|p-q|},$$

$$|p-q| = \sqrt{(x_p - x_q)^2 + (y_p - y_q)^2 + (z_p - z_q)^2} \;;$$

if $(\bar{E}^S,\bar{H}^S)$ satisfy (4a) and (5) in S_{ex} and are continuous on S then

$$\int_S \{vi\omega\varepsilon_0\hat{n}_q \times \bar{E}^S_{ex}(q) - \nabla_q v\hat{n}_q \cdot \bar{H}^S_{ex}(q) + \nabla_q v \times (\hat{n}_q \times \bar{H}^S_{ex}(q)\}dS_q = \frac{\sigma_3(p)}{4\pi}\bar{H}^S(p) \tag{11}$$

$$\int_S \{-vi\omega\mu_0\hat{n}_q \times \bar{H}^S_{ex}(q) - \nabla_q v\hat{n}_q \cdot \bar{E}^S_{ex}(q) + \nabla_q v \times (\hat{n}_q \times \bar{E}^S_{ex}(q)\}dS_q = \frac{\sigma_3(p)}{4\pi}\bar{E}^S(p) \tag{12}$$

where $v = -\frac{e^{ik|p-q|}}{4\pi|p-q|}$. If $p \in S$ then the right sides of (9) and (10) represent $\bar{H}_{in}$ and $\bar{E}_{in}$ whereas the right hand sides of (11) and (12) represent $\bar{H}^S_{ex}$ and $\bar{E}^S_{ex}$. It is convenient to adopt the convention that $E^S(p)$, $p \in S$ always is taken to mean the limit of E^S from S_{ex} so that the subscript on E^S_{ex} may be omitted. The incident field, either plane wave, point source in S_{ex}, or superposition of such sources, satisfies (4a) in S_{in} and is continuous across S hence satisfies

$$\int_S \{vi\omega\varepsilon_0\hat{n}_q \times \bar{E}^{inc}(q) - \nabla_q v\hat{n}_q \cdot H^{inc}(q) + \nabla_q v \times (\hat{n}_q \times \bar{H}^{inc}(q))\}dS_q = \frac{\sigma_3(p)-4\pi}{4\pi}\bar{H}^{inc}(p) \tag{13}$$

$$\int_S \{-vi\omega\mu_0\hat{n}_q \times \bar{H}^{inc}(q) - \nabla_q v\hat{n}_q \cdot \bar{E}^{inc}(q) + \nabla_q v \times (\hat{n}_q \times \bar{E}^{inc}(q))\}dS_q = \frac{\sigma_3(p)-4\pi}{4\pi}\bar{E}^{inc}(p) \tag{14}$$

for all $p \in \mathbb{R}^3$.

Expressions for the far field could be obtained using the asymptotic form of v in (11) and (12) however a representation exhibiting the k dependence more explicitly, which proves more useful for low frequency purposes, [15]

$$\bar{E}^S(p) = -\frac{e^{ikr_p}}{4\pi r_p}k^2\{\hat{p}\times\hat{p}\times\int_S \bar{q}\, e^{-ik\hat{p}\cdot\bar{q}}\, Z\hat{n}_q\cdot\hat{p}\times\bar{H}^S(q)+\hat{n}_q\cdot\bar{E}^S(q))dS_q$$
$$+\hat{p}\times\int_S \bar{q}\, e^{ik\hat{p}\cdot\bar{q}}(-\hat{n}_q\cdot\hat{r}_p\times\bar{E}^S(q) + Z\hat{n}_q\cdot\bar{H}^S(q))dS_q\} + o\left(\frac{1}{r_p}\right) \tag{15a}$$

$$\bar{H}^S(p) = \hat{p}\times\bar{E}^S(p) + o\left(\frac{1}{r_p}\right) \tag{15b}$$

where

$$Z = \sqrt{\frac{\mu_0}{\varepsilon_0}}\ ,\ \left(Y = \frac{1}{Z}\right) \tag{16}$$

is the free space impedance, (Y is the free space admittance). An alternate expression in terms of the field in S_{in} (if it exists) is [14]

$$\bar{E}^S(p) = \frac{e^{ikr_p}}{4\pi r_p}k^2\left(1-\frac{\varepsilon+\frac{i\sigma}{\omega}}{\varepsilon_0}\right)\hat{p}\times\hat{p}\times\int_{S_{in}} e^{-ikp\cdot\bar{q}}\,\bar{E}(q)dv_q$$
$$+ Z\left(1-\frac{\mu}{\mu_0}\right)\hat{p}\times\int_{S_{in}} e^{-ik\hat{p}\cdot\hat{q}}\,\bar{H}(q)dv_q\} + o\left(\frac{1}{r_p}\right). \tag{17}$$

In the limiting case, $k = 0$, an even more useful form of (15) is [17]

$$\bar{E}^S(p) = -\frac{e^{ikr}}{4\pi r} k^2\{\hat{p}\times\hat{p}\times\int_S \bar{q}(\hat{n}_q\cdot\bar{E}_0^S(q)) + \frac{1}{2}\bar{r}_q\times(\hat{n}_q\times\bar{E}_0^S(q))\}dS_q$$

$$+ Z\hat{p}\times\int_S [\bar{q}\ \hat{n}_q\cdot\bar{H}_0^S(q) + \frac{1}{2}\bar{q}\times(\hat{n}_q\times\bar{H}_0^S(q))]dS_q\} + o\left(\frac{1}{r}\right). \quad (18)$$

With $\bar{E}_0^S = \nabla u_0^e$ and $\bar{H}_0^S = \nabla v_0^e$ (a consequence of 4a when $\omega = 0$) this may also be written as

$$\bar{E}^S(p) = \frac{e^{ikr}}{4\pi r} k^2\{\hat{p}\times\hat{p}\times\int_S\left(\hat{n}_q u_0^e(q) - \bar{r}_q\frac{\partial\mu_0^e}{\partial n_q}\right)dS_q$$

$$+ Z\hat{p}\times\int_S\left(\hat{n}_q v_0^e(q) - \bar{q}\frac{\partial v_0^e}{\partial n_q}\right)dS_q\} + o\left(\frac{1}{r}\right) \quad (19)$$

III. LOW FREQUENCY SOLUTIONS

A. Perfect Conductivity

The problem here is that of finding $(\bar{E}^S, \bar{H}^S)$ satisfying 4a, 5 and 6b for some given $(\bar{E}^{inc}, \bar{H}^{inc})$.

1. *Raleigh-Stevenson Method*

Lord Rayleigh [31,32] proposed reducing the problem, in the zero frequency limit, to problems in potential theory whose solutions were the low frequency approximations of the near field. The far field can then be obtained using (15) although Rayleigh used a different argument. Stevenson [33] extended this approach to obtain higher order terms in the near field and some ambiguities in this approach were removed by Kleinman [16]. The essence of this approach involves expansions of incident and scattered fields:

$$\bar{E}^{inc}(p) = \sum_{n=0}^{\infty} (ik)^n \bar{E}_n^{inc}(p),\ \bar{H}^{inc}(p) = \sum_{n=0}^{\infty} (ik)^n \bar{H}_n^{inc}(p),$$

$$\bar{E}^S(p) = \sum_{n=0}^{\infty} (ik)^n \bar{E}_n^S(p),\ \bar{H}^S(p) = \sum_{n=0}^{\infty} (ik)^n \bar{H}_n^S(p). \quad (20)$$

While $\bar{E}_n^{inc}$ and $\bar{H}_n^{inc}$ are presumed known, $\bar{E}_n^S$ and $\bar{H}_n^S$ are not and the determination of these coefficients is reduced to problems in potential theory. The zeroth order (Rayleigh) terms are found to be

$$\bar{E}_0^S(p) = \nabla_p u_0^e(p) \qquad \bar{H}_0^S(p) = \nabla v_0^e(p) \tag{21}$$

where u_0^e and v_0^e are solutions of the following potential problems.

$$\begin{aligned} &\nabla_p^2 u_0^e(p) = 0, \quad p \in S_{ex} && \nabla_p^2 v_0^e(p) = 0, \quad p \in S_{ex} \\ &\hat{n}_p \times \nabla_p u_0^e = -\hat{n}_p \times \bar{E}_0^{inc}, \quad p \in S && \frac{\partial v_0^e}{\partial n} = -\hat{n} \cdot \bar{H}_0^{inc}, \quad p \in S \\ &\int_S \frac{\partial u_0^e}{\partial n}\, dS = 0 && v_0^e = 0\left(\frac{1}{r}\right) \\ &u_0^e = 0\left(\frac{1}{r}\right), \quad \frac{\partial u_0^e}{\partial r} = o\left(\frac{1}{r}\right) && \frac{\partial v_0^e}{\partial r} = o\left(\frac{1}{r}\right) \end{aligned} \tag{22}$$

whereas higher order terms are cast as potential problems in which the preceeding terms are considered known:

$$\bar{E}_n^S(p) = \bar{F}_n^e(p) + \nabla u_n^e(p) \qquad \bar{H}_n^S(p) = \bar{G}_n^e(p) - \bar{g}_n(p) + \nabla v_n^e(p) \tag{23}$$

where

$$\begin{aligned} \bar{F}_n^e(p) = {} & \frac{1}{4\pi} \sum_{m=0}^{n} \int_S \frac{\{Zn_q \times \bar{H}_{m-1}^S(q) - \nabla_p \times [\hat{n}_q \times \bar{E}_m^{inc}(q)]\}dS_q}{(n-m)!\ |p-q|^{m+1-n}} \\ & - \frac{1}{4\pi} \sum_{m=0}^{n-1} \nabla_p \int_S \frac{\hat{n}_q \cdot \bar{E}_m^S(q)\, dS_q}{(n-m)!\ |p-q|^{m+1-n}} \end{aligned} \tag{24}$$

$$\begin{aligned} \bar{G}_n^e(p) = {} & \frac{1}{4\pi} \sum_{m=0}^{n} \int_S \frac{\{\nabla_p [\hat{n}_q \cdot \bar{H}_m^{inc}(q)] - Y\hat{n}_q \times \bar{E}_{m-1}^S(q)\}dS_q}{(n-m)!\ |p\ q|^{m+1-n}} \\ & + \frac{1}{4\pi} \sum_{m=0}^{n-1} \nabla_p \times \int_S \frac{\hat{n}_q \times \bar{H}_m^S(q)}{(n-m)!\ |p-q|^{m+1-n}}\, dS_q \end{aligned} \tag{25}$$

$$\bar{g}_n = \frac{Y}{4\pi} \nabla_p \times \left\{ \int_{S_{ex}} \frac{\nabla_q u_{n-1}^e(q)\, dv_q}{|p-q|} + \int_{S_{in}} \frac{\nabla_q \phi_n^i(q)\, dv_q}{|p-q|} \right\} \tag{26}$$

and u_n^e, ϕ_n^i, v_n^e are solutions of potential problems;

$$
\text{(a)}\quad
\begin{aligned}
&\nabla^2 u_n^e = 0 \qquad p \in S_{ex} \\
&\hat{n} \times \nabla u_n^e = -\hat{n} \times \bar{F}_n^e - \hat{n} \times \bar{E}_n^{inc} \qquad p \in S \\
&\int_S \frac{\partial u_n^e}{\partial n}\, dS = -\int_S \hat{n} \cdot \bar{F}_n^e \, dS
\end{aligned}
\qquad
\text{(b)}\quad
\left\{
\begin{aligned}
&\nabla^2 \phi_n^i = 0 \qquad p \in S_{in} \\
&\frac{\partial \phi_n^i}{\partial n} \quad \frac{\partial u_{n-1}^e}{\partial n} \qquad p \in S
\end{aligned}
\right.
$$

$$
u_n^e = O\left(\frac{1}{r}\right) \quad \frac{\partial u_n^e}{\partial r} = o\left(\frac{1}{r}\right)
$$

$$
\text{(c)}\quad
\left\{
\begin{aligned}
&\nabla^2 v_n^e = 0 \quad p \in S_{ex} \quad ; \quad \frac{\partial v_n^e}{\partial n} = -\hat{n} \cdot (\bar{g}_n - \bar{G}_n^e - \bar{H}_n^{inc}) \quad p \in S \; ; \\
&v_n^e = O\left(\frac{1}{r}\right), \quad \frac{\partial v_n^e}{\partial r} = o\left(\frac{1}{r}\right).
\end{aligned}
\right.
\tag{27}
$$

Note that $\bar{g}_n$ hence ϕ_n^i must be found before v_n^e may be determined. The free space impedance and admittance, Z and Y, are defined in (16). Equations 23-26 present explicit expressions for the expansion coefficients which are valid for $p \in S_{ex}$. The values for $p \in S$ are obtained as the continuous limit of these quantities as $p \to S$ from S_{ex}.

With n near field terms one may insert them in equation 15 to find a like number of terms in the low frequency expansion of the far field coefficient.

It is clear that the procedure is lengthy and requires the solution of three potential problem at each stage beyond the first, two exterior and one interior.

2. *Low Frequency Iteration - Volume Integrals*

The Rayleigh-Stevenson procedure has been systematized in terms of the Green's functions for Laplace's equation so that the terms may be obtained iteratively [4]. Actually, low frequency expansions are found for modified fields $\bar{e}$ and $\bar{h}$ where

$$
\bar{E}^S(p) = e^{ikr_p}\, \bar{e}(p) \quad , \quad \bar{H}^S(p) = e^{ikr_p}\, \bar{h}(p). \tag{28}
$$

With the expansions

$$\bar{e}(p) = \sum_{n=0}^{\infty} (ik)^n \bar{e}_n(p) \quad , \quad \bar{h}(p) = \sum_{n=0}^{\infty} (ik)^n \bar{h}_n(p) \tag{29}$$

the coefficients $\bar{e}_n$ and $\bar{h}_n$ are expressed recursively as

$$\bar{e}_0(p) = \bar{g}_0(p)$$

$$\bar{e}_{n+1}(p) = -\int_{S_{ex}} \Big\{ [Z\bar{h}_n(q) - q \times \bar{e}_n(q)] \cdot \bar{\bar{H}}(p,q) + \hat{q} \cdot \bar{e}_n(p) \nabla_p \left[G^e(p,q) + \frac{U(p) - U(p)U(q)}{4\pi C} \right] \Big\} dv_q + \bar{g}_{n+1}(p), \tag{30}$$

$$\bar{h}_0(p) = \bar{f}_0(p)$$

$$\bar{h}_{n+1}(p) = \int_{S_{ex}} \{ [Y\bar{e}_n(p) + \hat{q} \times \bar{h}_n(p)] \cdot \bar{\bar{E}}(p,q) - \hat{q} \cdot \bar{h}_n(q) \nabla_p N^e(p,q) \} dv_q + \bar{f}_{n+1}(p); \tag{31}$$

where

$$\bar{f}_n(p) = \sum_{m=0}^{n} \frac{1}{(n-m)!} \nabla_p \int_S (-r_q)^{n-m} \bar{H}_m^{inc}(q) N^e(p,q) dS_q \tag{32}$$

$$\bar{g}_n(p) = -\sum_{m=0}^{n} \frac{1}{(n-m)!} \int_S (-r_q)^{n-m} [\hat{n}_q \times \bar{E}_m^{inc}(q)] \cdot \bar{\bar{H}}(p,q) dS_q; \tag{33}$$

$G^e(p,q)$ and $N^e(p,q)$ are the exterior Dirichlet and Neumann Green's functions for S, for Laplace's equation, i.e.,

$$G^e(p,q) = -\frac{1}{4\pi|p-q|} + g(p,q), \quad N^e(p,q) = -\frac{1}{4\pi|p-q|} + n(p,q)$$

$$\nabla_p^2 g = \nabla_p^2 n = 0 \ , \ p,q \in S_{ex} \ ; \quad g = n = O\left(\frac{1}{r_p}\right), \ \frac{\partial g}{\partial r_p} = \frac{\partial n}{\partial r_p} = o\left(\frac{1}{r_p}\right)$$

$$G^e(p,q) = 0, \ \frac{\partial N^e(p,q)}{\partial n_p} = 0 \quad p \in S, \ q \in S_{ex} \ ; \tag{34}$$

$U(p)$ is the conductor potential for S, i.e.,

$$U(p) = \int_S \frac{\partial}{\partial n_q} G^e(p,q) dS_q, \quad p \in S_{ex} \tag{35}$$

C is the capacity of S, i.e.,

$$C = \frac{1}{4\pi} \int_S \frac{\partial}{\partial n_q} U(q) dS_q \ ; \tag{36}$$

$\bar{\bar{E}}(p,q)$ and $\bar{\bar{H}}(p,q)$ are dyadic potential Green's functions

$$\bar{\bar{E}}(p,q) = \nabla_q \times \left(- \frac{I}{4\pi|p-q|} \right) +$$

$$+ \sum_{j=1}^{3} \left[- \frac{1}{4\pi} \nabla_q \times \int_S \frac{\hat{x}_j \cdot \nabla_p N^e(p_S, p)}{|p_S - q|} \hat{n}_{p_S} dS_{p_S} + \nabla_q G_j(p,q) \right] \hat{x}_j \tag{37}$$

$$\bar{\bar{H}}(p,q) = \nabla_q \times \left(- \frac{I}{4\pi|p-q|} \right)$$

$$+ \sum_{j=1}^{3} \left[- \frac{1}{4\pi} \nabla_q \times \int_S \frac{\hat{x}_j \cdot \nabla_p N^i(p_S, p)}{|p_S - q|} \hat{n}_{p_S} dS_{p_S} + \nabla_q N_j(p,q) \right] \hat{x}_j \; ; \tag{38}$$

I is the identity dyadic, G_j and N_j are exterior potential functions defined as the solutions of the following problems for $j = 1, 2, 3$,

$$\nabla_q^2 G_j(p,q) = 0 \quad , \quad p,q \in S_{ex} \quad ; \quad G_j = O\left(\frac{1}{r_q}\right), \; \frac{\partial G}{\partial r_q} = o\left(\frac{1}{r_q}\right) \; ;$$

$$\hat{n}_q \times \nabla_q G_j(p,q) = \frac{1}{4\pi} \hat{n}_q \times \nabla_q \times \left(\frac{\hat{x}_j}{|p-q|} + \int_S \frac{\hat{x}_j \cdot \nabla_p N^e(p_S, p)}{|p_S - q|} \hat{n}_{p_S} dS_{p_S} \right),$$

$$p \in S_{ex}, \; q \in S;$$

$$\int_S \hat{n}_q \cdot \nabla_q G_j(p,q) S_q = 0, \qquad p \in S_{ex} \, , \tag{39}$$

$$\nabla_q^2 N_j(p,q) = 0 \quad , \quad p,q \in S_{ex} \quad ; \quad N_j = O\left(\frac{1}{r_q}\right), \; \frac{\partial N_j}{\partial r_q} = o\left(\frac{1}{r_q}\right)$$

$$\hat{n}_q \cdot \nabla_q N_j(p,q) = \hat{n}_q \cdot \nabla_q \times \left(\frac{\hat{x}_j}{4\pi|p-q|} + \frac{1}{4\pi} \int_S \frac{\hat{x}_j \cdot \nabla_p N^i(p_S, p)}{|p_S - q|} \hat{n}_{p_S} dS_{p_S} \right),$$

$$p \in S_{ex} \quad q \in S \tag{40}$$

and $N^i(p,q)$ is an interior potential in the first variable, viz,

$$\nabla_{p_1}^2 N^i(p_1, p) = 0, \qquad p_1 \in S_{in}, \qquad p \in S_{ex},$$

$$\hat{n}_{p_1} \cdot \nabla_{p_1} N^i(p_1, p) = -\hat{n}_{p_1} \cdot \nabla_{p_1} \left(G^e(p_1, p) - \frac{U(p_1) U(p)}{4\pi C} \right), \; p_1 \in S, \; p \in S_{ex}. \tag{41}$$

Unit coordinate vectors are denoted by $\hat{x}_j$, $j = 1, 2, 3$. This technique is still cumbersome, involves volume integrals over the infinite domain S_{ex} and requires knowledge of the appropriate potential Green's functions which in a sense is an even stronger requirement than that implicit in the Rayleigh-Stevenson approach which demands the solution of a sequence of potential problems.

3. *Low Frequency Iteration - Boundary Integrals*

A considerable improvement is contained in the work of Gray [7] which provides an iterative method of solution involving only surface integrals without requiring any knowledge of the Green's functions for the potential problem. The method at present has been proven to be valid only for Lyapunoff rather than piecewise Lyapunoff surfaces. The scheme is based on the magnetic field integral equation or Maue formulation which is easily obtained by adding equations 11 and 13, using the boundary conditions 6b and remembering that $\sigma_3(p) = 2\pi \;\; p \in S$ for S Lyapunoff, i.e.,

$$\frac{\sigma_3(p)}{4\pi}\bar{H}(p) = \int_S \nabla_q v(p,q) \times (\hat{n}_q \times \bar{H}(q))\,dS_q + \bar{H}^{inc}(p) \tag{42}$$

The unknown tangential field $\hat{n}_q \times \bar{H}$ is found by considering the tangential components of (42) when $p \in S$, which is a pure integral equation,

$$\hat{n}_p \times H(p) = 2\hat{n}_p \times \int_S \nabla_q v \times (\hat{n}_q \times \bar{H}(q))\,dS_q + 2\hat{n}_p \times \bar{H}^{inc}(p). \tag{43}$$

Gray has shown that, provided $H^{inc}(p)$ is sufficiently differentiable (in the scattering problem with no sources on S, $\bar{H}^{inc}(p)$ is infinitely differentiable on S which is by any estimate sufficient) and k is sufficiently small, the sequence of iterates

$$\hat{n}_p \times \bar{H}^{(0)}(p) = 2\hat{n}_p \times \bar{H}^{inc}(p),$$

$$\hat{n}_p \times \bar{H}^{(N)} = 2\hat{n}_p \times \int_S \nabla_q v(\hat{n}_q \times \bar{H}^{(N-1)}(q))\,dS_q + 2\hat{n}_p \times \bar{H}^{inc} \tag{44}$$

converges to the desired solution. Values of the field at points off the surface are obtained with equation 42 and in the far field with 15 and 16. This is clearly a low frequency method since convergence is proven only for small k. However, N iterations do not produce the exact first N terms in equation 20. If the result after N iterations is developed in a power series in k then each coefficient will approach its exact value as $N \to \infty$.

4. *Polarizability Tensors and Far Field Scattering*

When the incident field is a plane wave and only the first term in the low frequency expansion is taken into account, substantial simplification of the far field coefficient is possible using a method of [12]. If

$$\bar{E}^{inc} = \hat{a}e^{ik\hat{k}\cdot p} \quad \bar{H}^{inc} = Y\hat{b}e^{ik\hat{k}\cdot p} \tag{45}$$

where $(\hat{a},\hat{b},\hat{k})$ form an orthonormal set then the boundary conditions in (22) become

$$u_0^e = \hat{a}\cdot\bar{p}+c \qquad \frac{\partial v_0^e}{\partial n} = -Y\hat{n}\cdot\hat{b} \tag{46}$$

where the constant c is determined through $\int_S \frac{\partial u_0^e}{\partial n_q}\, dS_q = 0$. Introducing potentials ϕ_i and ψ_i, $i = 1, 2, 3$ where

$$\nabla^2\phi_i = \nabla^2\psi_i = 0,\ p\in S_{ex},\quad \phi_i = \psi_i = 0\left(\frac{1}{r}\right)\ \frac{\partial\phi_i}{\partial r} = \frac{\partial\psi_i}{\partial r} = o\left(\frac{1}{r}\right)$$

$$\phi_i = x_i + c_i\,,\quad \frac{\partial\psi_i}{\partial n} = \hat{n}\cdot\hat{x}_i,\ p\in S\,,\ \int_S \frac{\partial\phi_i}{\partial n}\, dS = 0; \tag{47}$$

where x_i are the rectangular coordinate variables, it is easily seen that

$$u_0^e = -\sum_{i=1}^{3}(\hat{a}\cdot\hat{x}_i)\phi_i,\quad v_0^e = -Y\sum_{i=1}^{3}(\hat{b}\cdot\hat{x}_i)\psi_i \tag{48}$$

and, with (15a) and the boundary conditions,

$$\bar{E}^S(p) = -\frac{e^{ikr}}{4\pi r}k^2\{\hat{p}\times\hat{p}\times(\bar{\bar{P}}\cdot\hat{a}) - \hat{r}\times(\bar{\bar{M}}\cdot\hat{b})\} + o\left(\frac{1}{r}\right) \tag{49}$$

where the 3×3 polarizability tensors $\bar{\bar{P}}$ and $\bar{\bar{M}}$ have elements

$$p_{ij} = V\delta_{ij} - \int_S x_{iq}\hat{n}_q\cdot\nabla\phi_j(q)\, dS_q \tag{50a}$$

$$m_{ij} = V\delta_{ij} - \int_S \hat{n}_q\cdot\hat{x}_{iq}\psi_j(q)\, dS_q \tag{50b}$$

where V is the volume of S_{in} and δ_{ij} is the Kronecker delta. These tensors are symmetric ($p_{ij} = p_{ji}$, $m_{ij} = m_{ji}$) and independent of the incident filed. Thus the determination of the first far field term is reduced to finding at most twelve constants which depend solely on the geometry. These tensor elements are constrained by inequalities [12,28]. Moreover when the surface S is axially symmetric the number of independent elements is reduced to three in which case the far field expression in equation 49 becomes

$$\bar{E}^S(p) = -\frac{e^{ikr}}{4\pi r} k^2 \Big\{ p_{11}\hat{p}\times\hat{p}\times\hat{a} + (p_{33} - p_{11})(\hat{a}\cdot\hat{z})\hat{p}\times\hat{p}\times\hat{z} - M_{11}\hat{p}\times\hat{b} - \left(\frac{p_{11}}{2} - m_{11}\right)(\hat{b}\cdot\hat{z})\hat{p}\times\hat{z}\Big\} + o\left(\frac{1}{r}\right) . \tag{51}$$

The tensor elements are expressed in (50) in terms of solutions of potential problems. While no explicit solution is known for general S, The problem is readily amenable to numerical methods and evaluation of the tensor elements for a variety of axially symmetric configurations were given in [20]. Simplifications for special directions of incidence and observation (e.g., nose on incidence, backscattering, etc.) as well as explicit expressions for radar cross section are given in [12,19].

B. Penetrable Homogeneous Scatterer

When S is the boundary of a homogeneous scatterer with constant constitutive parameters ε, μ, σ differing from those of the surrounding medium in which an incident field originates, the problem is that of finding $(\bar{E},\bar{H})$ in S_{ex} and S_{in} satisfying Maxwell's equations (4a,b), the transition conditions (6a) at S and the radiation condition *(5)*. This problem was reduced to a potential problem by Lord Rayleigh [31,32] in the limiting case of zero frequency and the method was extended to nonzero ω in [33,34]. The latter treatment is not wholly unambiguous and while it was clarified in [14], this work is not now readily available therefore an expanded summary of this approach is presented here.

In the case of infinite conductivity, the field quantities were expanded in powers of k, however in the present case the propagation constants differ in S_{in} and S_{ex} therefore it is more convenient to expand the field quantities everywhere in powers of $i\omega$, where ω is the frequency common in both regions. Thus we write

$$\bar{E}^{inc} = \sum_{n=0}^{\infty} (i\omega)^n \bar{E}_n^{inc}, \quad \bar{E}^S = \sum_{n=0}^{\infty} (i\omega)^n \bar{E}_n^S, \quad \bar{E} = \sum_{n=0}^{\infty} (i\omega)^n \bar{E}_n \tag{52}$$

and similar expansions for the magnetic field. The goal is to reduce the problem of determining $(\bar{E}_n, \bar{H}_n)$ to problems in potential theory, assuming these coefficients are known for $0, 1, \ldots, n-1$. Substitution of these expansions in Maxwell's equation (4) and equating like powers of $i\omega$ leads to a set of recursive differential equations:

$$\text{a)} \left.\begin{aligned} \nabla\times\bar{E}_n^S &= \mu_0 \bar{H}_{n-1}^S \\ \nabla\times\bar{H}_n^S &= -\varepsilon_0 \bar{E}_{n-1}^S \end{aligned}\right\} p \in S_{ex}, \quad \text{b)} \left.\begin{aligned} \nabla\times\bar{E}_n &= \mu\bar{H}_{n-1} \\ \nabla\times\bar{H}_n &= -\varepsilon\bar{E}_{n-1} + \sigma\bar{E}_n \end{aligned}\right\} p \in S_{in} \tag{53}$$

The coefficients in an expansion of the incident field $(\bar{E}^{inc}, \bar{H}^{inc})$ will satisfy equation 53a almost everywhere in S_{ex}. A similar procedure involving the transition conditions (6a) leads to

$$\left.\begin{aligned} &\text{a)}\ \hat{n}\times(\bar{E}_n^{inc} + \bar{E}_n^S) = \hat{n}\times\bar{E}_n \qquad \text{c)}\ \mu_0 \hat{n}\cdot(\bar{H}_n^{inc} + \bar{H}_n^S) = \mu\hat{n}\cdot\bar{H}_n \\ &\text{b)}\ \hat{n}\times(\bar{H}_n^{inc} + \bar{H}_n^S) = \hat{n}\times\bar{H}_n \qquad \text{d)}\ \varepsilon_0 \hat{n}\cdot(\bar{E}_{n-1}^{inc} + \bar{E}_{n-1}^S) = \varepsilon\hat{n}\cdot\bar{E}_{n-1} + \sigma\hat{n}\cdot\bar{E}_n \end{aligned}\right\} p \varepsilon S. \tag{54}$$

Here we denote by $(\bar{E}_n, \bar{H}_n)$ the coefficients of the expansion of $\bar{E}$ and $\bar{H}$ in S_{in} <u>and</u> their limiting values as $p \to S$ from S_{in}. In these equations it is to be understood that coefficients with negative subscripts are identically zero, e.g., $\bar{E}_{-1}^S \equiv 0$, etc. In addition all coefficients are divergence free, e.g., $\nabla\cdot\bar{E}_n = 0$, etc. in their appropriate domains and consequently

$$\int_S \hat{n} \cdot \bar{E}_n dS \equiv 0 \tag{55}$$

and similar identities hold for all other coefficients. The first step is to find a particular solution of

$$\nabla \times \bar{E}_n = \mu \bar{H}_{n-1} , \quad p \in S_{in} \tag{56}$$

Following the method of Stevenson [35], define

$$\bar{F}_n^{in} := \frac{1}{4\pi} \nabla_p \times \left\{ \int_{S_{ex}} \frac{\nabla_q \phi_n(q)}{|p-q|} \, dv_q + \mu \int_{S_{in}} \frac{\bar{H}_{n-1}(q)}{|p-q|} \, dv_q \right\} \tag{57}$$

where ϕ_n is the exterior potential function satisfying

$$\nabla^2 \phi_n = 0, \; p \in S_{ex} \; ; \; \frac{\partial \phi_n}{\partial n} = \mu \hat{n} \cdot \bar{H}_{n-1}, \; p \in S \; ; \; \phi_n = O\left(\frac{1}{r}\right) , \; \frac{\partial \phi_n}{\partial r} = o\left(\frac{1}{r}\right) \tag{58}$$

This is the first potential problem that must be solved. With $\bar{F}_n^{in}$ so defined it follows by direct calculation that

$$\nabla \times \bar{F}_n^{in} = \mu \bar{H}_{n-1} . \tag{59}$$

and therefore

$$\nabla \times (\bar{E}_n - \bar{F}_n^{in}) = 0, \quad p \in S_{in} . \tag{60}$$

From this one concludes that $\bar{E}_n - F_n^{in}$ is the gradient of a scalar function and, since $\bar{E}_n$ and $\bar{F}_n^{in}$ are divergence free, this scalar is a potential function, i.e.,

$$\bar{E}_n = \bar{F}_n^{in} + \nabla u_n^{in} \tag{61}$$

where

$$\nabla^2 u_n^{in} = 0, \quad p \in S_{in} . \tag{62}$$

Here one must distinguish between two cases.

a) $\underline{\sigma \neq 0}$. If the conductivity is nonzero, the boundary condition 54d together with (61) implies that

$$\frac{\partial u_n^{in}}{\partial n} = n^i \left(\frac{\varepsilon}{\sigma} \bar{E}_{n-1} - \frac{\varepsilon_0}{\sigma} \bar{E}_{n-1}^{inc} - \frac{\varepsilon \bar{E}_{n-1}^{S}}{\sigma} - \bar{F}_n^{in} \right) , \quad p \in S . \tag{63}$$

With the normal derivative thus specified, the interior Neumann potential problem posed by (62) and (63) constitutes the second potential problem that must be solved. It has a solution (though not a unique one since any constant may be added) because the requisite compatibility condition

$$\int_S \frac{\partial u_n^{in}}{\partial n} dS = 0 \tag{64}$$

may be shown to hold with the data specified in (63). If this problem is solved, and therefore $\bar{E}_n$ is determined through equation (61), the next step is to find $\bar{E}_n^S$. For this purpose, expand the quantities in equation (12) in powers of $i\omega$, use equation (3) and equate like powers of $i\omega$ to obtain

$$\frac{\sigma_3(p)}{4\pi} \bar{E}_n^S(p) = \frac{1}{4\pi}\int_S \sum_{m=0}^{n} \frac{(\varepsilon_0\mu_0)^{\frac{n-m}{2}}}{(n-m)!} \{\mu_0 |p-q|^{n-m-1} \hat{n}_q \times \bar{H}_{m-1}^S(q)$$

$$- \nabla_q |p-q|^{n-m-1} \times (\hat{n}_q \times E_m^S(q)) + \nabla_q |p-q|^{n-m-1} \hat{n}_q \cdot \bar{E}_m^S(q)\} dS_q. \tag{65}$$

For $p \in S_{ex}$, this may be written, with equation (8) and (54a), as

$$\bar{E}_n^S(p) = \bar{F}_n^{ex} + \nabla u_n^{ex} \tag{66}$$

where

$$\bar{F}_n^{ex} = \frac{1}{4\pi}\int_S \sum_{m=0}^{n} \frac{(\varepsilon_0\mu_0)^{\frac{n-m}{2}}}{(n-m)!} \{\mu_0 |p-q|^{n-m-1} \hat{n}_q \times \bar{H}_{m-1}^S(q)$$

$$- \nabla_q |p-q|^{n-m-1} \times \hat{n}_q \times (\bar{E}_m(q) - \bar{E}_m^{inc}(q))\} dS_q$$

$$+ \frac{1}{4\pi}\int_S \sum_{m=0}^{n-1} \frac{(\varepsilon_0\mu_0)^{\frac{n-m}{2}}}{(n-m)!} \hat{n}_q \cdot \bar{E}_m^S(q) \nabla_q |p-q|^{n-m-1} dS_q \tag{67}$$

and

$$\nabla_p u_n^{ex} = \frac{1}{4\pi}\int_S \hat{n}_q \cdot \bar{E}_n^S(q) \nabla_q \frac{1}{|p-q|} dS_q. \tag{68}$$

Under the assumptions, $\bar{F}_n^{ex}$ is known but u_n^{ex} is unknown since $\hat{n} \cdot \bar{E}_n^S$ has not been found. However it follows from equation (68) that u_n^{ex} is an exterior potential function, regular at infinity and, with (54a) and (55), the problem of finding u_n^{ex} is reduced to the well-posed potential problem:

$$\nabla^2 u_n^{ex} = 0, \quad p \in S_{ex}, \quad u_n^{ex} \text{ regular at infinity,}$$

$$\hat{n} \times \nabla u_n^{ex} = \hat{n} \times \bar{E}_n - \hat{n} \times \bar{E}_n^{inc} - \hat{n} \times \bar{F}_n^{ex}, \quad p \in S \quad \text{, and} \tag{69}$$

$\int_S \frac{\partial u_n^{ex}}{\partial n} dS = -\int_S \hat{n} \cdot \bar{F}_n^{ex} dS$. This is the third potential problem that must be solved and its solution together with equation (66) completes the determination of $\bar{E}_n$ and $\bar{E}_n^S$ in the case where $\sigma \neq 0$.

b) $\underline{\sigma = 0}$. This case does not permit the complete specification of the interior field $\bar{E}_n$ before finding $\bar{E}_n^S$. Rather they must be found simultaneously as the solution of a transition (coupled interior and exterior) potential problem unless $\varepsilon = \varepsilon_0$. Recall that $\bar{E}_n$ is written in terms of an interior potential function in equation (61) however the boundary condition (54d) no longer yields an explicit expression for $\frac{\partial u_n^{in}}{\partial n}$ in terms of known quantities. Instead it is coupled with the values of $\bar{E}_n^S$ on S. For this reason we proceed to consider $\bar{E}_n^S$ in the same way as in case a. That is, after expanding the scattered field in powers of $i\omega$, equation (65) is obtained just before. Now however $\hat{n} \times \bar{E}_n^S$ is not known hence the decomposition of the terms in (65) is carried out slightly differently. For $p \in S_{ex}$, it may be shown with (53b) and standard vector manipulations that

$$\frac{1}{4\pi} \int_S \nabla_q \frac{1}{|p-q|} \times (\hat{n}_q \times E_n(q)) dS_q =$$
$$\frac{\mu}{4\pi} \nabla_p \times \int_{S_{in}} \frac{\bar{H}_{n-1}}{|p-q|} dv_q - \int_S \hat{n}_q \cdot \bar{E}_n(q) \nabla_q \frac{1}{|p-q|} dS_q . \tag{70}$$

This result allows (65) to be written, for $p \in S_{ex}$, as

$$\bar{E}_n^S(p) = \bar{f}_n^{ex} + \nabla \tilde{u}_n^{ex} \tag{71}$$

where

$$\bar{f}_n^{ex} = \frac{1}{4\pi} \int_S \sum_{m=0}^{n-1} \frac{(\varepsilon_0\mu_0)^{\frac{n-m}{2}}}{(n-m)!} \{\mu_0 |p-q|^{n-m-1} \hat{n}_q \times \bar{H}_{m-1}^S(q)$$
$$+ (\hat{n}_q \times \bar{E}_m^S) \times \nabla_q |p-q|^{n-m-1} + \hat{n}_q \cdot \bar{E}_n^S(q) \nabla_q |p-q|^{n-m-1}\} dS_q$$
$$+ \frac{\mu_0}{4\pi} \int_S \frac{\hat{n}_q \times \bar{H}_{n-1}^S(q)}{|p-q|} dS_q - \frac{1}{4\pi} \int_S (\hat{n}_q \times \bar{E}_n^{inc}) \times \nabla_q \frac{1}{|p-q|} dS_q$$
$$+ \frac{\mu}{4\pi} \nabla_p \times \int_{S_{in}} \frac{\bar{H}_{n-1}}{|p-q|} dV_q \tag{72}$$

and

$$\tilde{u}_n^{ex} = \frac{1}{4\pi} \int_S \frac{\hat{n}_q \cdot (\bar{E}_n(q) - \bar{E}_n^S(q))}{|p-q|} \, dS_q . \tag{73}$$

Under the assumptions, that all coefficients are known up through index $n-1$, $\bar{f}_n^{ex}$ is determined. On the other hand $\tilde{u}_n^{ex}$ is only known if $\varepsilon = \varepsilon_0$. In that case equation (54d) implies that

$$\hat{u}_n^{ex} = \frac{1}{4\pi} \int_S \frac{\hat{n}_q \cdot \bar{E}_n^{inc}}{|p-q|} \, dS_q . \tag{74}$$

In this case ($\varepsilon = \varepsilon_0$, $\sigma = 0$) it is appropriate to find $\bar{E}_n^S$ first and then employ (61) and (54d) to determine $\bar{E}_n$. Explicitly $\bar{E}_n = \bar{F}_n^{in} + \nabla u_n^{in}$ where $\bar{F}_n^{in}$ is given by (57) and $\nabla^2 u_n^{in} = 0$, $\dfrac{\partial u_n^{in}}{\partial n} = \hat{n} \cdot \bar{E}_n^{inc} + \hat{n} \cdot \bar{E}_n^S - \hat{n} \cdot \bar{F}_n^{in}$. This interior Neumann problem has a solution, though as before, the solution is not unique.

When $\varepsilon \neq \varepsilon_0$, $\bar{E}_n$ and $\bar{E}_n^S$ or equivalently u_n^{in} and $\tilde{u}_n^{ex}$ must be found concurrently. Explicitly

$$\bar{E}_n = \bar{F}_n^{in} + \nabla u_n^{in} \qquad \bar{E}_n^S = \bar{f}_n^{ex} + \nabla \tilde{u}_n^{ex} \tag{75}$$

where $\bar{F}_n^{in}$ and $\bar{f}_n^{ex}$ are given by (57) and (72) respectively and

$$\nabla^2 u_n^{in} = 0, \quad p \in S_{in}, \quad \nabla^2 \tilde{u}_n^{ex} = 0, \; p \in S_{ex} \tag{76}$$

$\tilde{u}_n^{ex}$ is regular at infinity and, from (54a,d),

$$\left.\begin{array}{ll} \text{a)} & \hat{n} \times \nabla (\tilde{u}_n^{ex} - u_n^{in}) = \hat{n} \times (\bar{F}_n^{in} - \bar{f}_n^{ex} - \bar{E}_n^{inc}) \\ \text{b)} & \varepsilon_0 \dfrac{\partial \tilde{u}_n^{ex}}{\partial n} - \varepsilon \dfrac{\partial u_n^{in}}{\partial n} = \varepsilon \hat{n} \cdot \bar{F}_n^{in} - \varepsilon_0 \hat{n} \cdot \bar{f}_n^{ex} - \varepsilon_0 \hat{n} \cdot \bar{E}_n^{inc} \end{array}\right\} p \in S. \tag{77}$$

Any ambiguity arising from the form of (77a) is removed by the conditions

$$\int_S \hat{n} \cdot \bar{E}_n \, dS_q = \int_S \hat{n} \cdot \bar{E}_n^S \, dS_q = 0. \tag{78}$$

Turning next to the task of determining $\bar{H}_n$ and $\bar{H}_n^S$ it is noted that the order in which the electric and magnetic field coefficients are considered is arbitrary if $\sigma = 0$ however the electric field terms must be found first if $\sigma \neq 0$. This is due to the term $\sigma\bar{E}_n$ on the right hand side of (53b) which vanishes if $\sigma = 0$ but is present otherwise. Finding $\bar{H}_n$ and $\bar{H}_n^S$ begins in the same way as the procedure for finding $\bar{E}_n$, $\bar{E}_n^S$. A particular solution of

$$\nabla \times \bar{H}_n = -\varepsilon\bar{E}_{n-1} + \sigma\bar{E}_n, \quad p \in S_{in} \tag{79}$$

is given by

$$\bar{G}_n^{in} = \frac{1}{4\pi} \nabla_p \times \left\{ \iint_{S_{ex}} \frac{\nabla_q \psi_n}{|p-q|} dv_q + \int_{S_{in}} \frac{(-\varepsilon\bar{E}_{n-1} + \sigma\bar{E}_n)}{|p-q|} dv_q \right\} \tag{80}$$

where ψ_n is the exterior potential function satisfying

$$\nabla^2\psi_n = 0, \quad p \in S_{ex}; \quad \frac{\partial\psi_n}{\partial n} = \hat{n} \cdot (-\varepsilon\bar{E}_{n-1} + \sigma\bar{E}_n), \quad p \in S;$$

$$\psi_n = O\left(\frac{1}{r}\right), \quad \frac{\partial\psi_n}{\partial n} = o\left(\frac{1}{r}\right) \tag{81}$$

A direct calculation verifies that

$$\nabla^2 \bar{G}_n^{in} = -\varepsilon\bar{E}_{n-1} + \sigma\bar{E}_n, \quad p \in S_{in} \tag{82}$$

therefore

$$\nabla \times (\bar{H}_n - \bar{G}_n^{in}) = 0, \quad p \in S_{in}. \tag{83}$$

This in turn implies that

$$\bar{H}_n = \bar{G}_n^{in} + \nabla v_n^{in} \tag{84}$$

where, since $\nabla \cdot \bar{H}_n = \nabla \cdot \bar{G}_n^{in} = 0$,

$$\nabla^2 v_n^{in} = 0, \quad p \in S_{in}. \tag{85}$$

Next examine the exterior field by expanding the quantities in equation (11) and equating coefficients of powers of $i\omega$. This yields

$$\frac{\sigma_3(p)}{4\pi} \bar{H}_n^S(p) = \frac{1}{4\pi} \int_S \sum_{m=0}^{m} \frac{(\varepsilon_0\mu_0)^{\frac{n-m}{2}}}{(n-m)!} \{-\varepsilon_0 \hat{n}_q \times \bar{E}_{m-1}^S |p-q|^{n-m-1}$$

$$+ \hat{n}_q \cdot \bar{H}_m^S \nabla_q |p-q|^{n-m-1} + (\hat{n}_q \times \bar{H}_m^S) \times \nabla_q |p-q|^{n-m-1}\} dS_q . \tag{86}$$

For $p \in S_{ex}$, this may be written as

$$\bar{H}_n^S(p) = \bar{G}_n^{ex} + \nabla v_n^{ex} \tag{87}$$

where

$$\bar{G}_n^{ex} = \int_S \frac{1}{4\pi} \sum_{m=0}^{n-1} \frac{(\varepsilon_0\mu_0)^{\frac{n-m}{2}}}{(n-m)!} \{-\varepsilon_0 \hat{n}_q \times \bar{E}_{m-1}^S |p-q|^{n-m-1}$$

$$+ \hat{n}_q \cdot \bar{H}_m^S \nabla_q |p-q|^{n-m-1} + (\hat{n}_q \times \bar{H}_m^S) \times \nabla_q |p-q|^{n-m-1}\} dS_q \tag{88}$$

$$-\frac{1}{4\pi}\int_S \varepsilon_0 \frac{\hat{n}_q \times E_{n-1}^S}{|p-q|} dS_q - \frac{\nabla_p}{4\pi} \times \int_S \frac{\hat{n}_q \times \bar{H}_n^{inc}}{|p-q|} dS_q + \frac{\nabla_p}{4\pi} \times \int_{S_{in}} \frac{(-\varepsilon E_{n-1} + \sigma E_n)}{|p-q|} dv_q$$

and

$$v_n^{ex} = \frac{1}{4\pi} \int_S \frac{\hat{n}_q \cdot (\bar{H}_n - \bar{H}_n^S)}{|p-q|} dS_q. \tag{89}$$

The decomposition into a known term, $\bar{G}_n^{ex}$, and an unknown exterior potential function, v_n^{ex}, is made possible with the following relation, comparable to (70),

$$\frac{1}{4\pi} \nabla_p \times \int_S \frac{\hat{n}_q \times \bar{H}_n}{|p-q|} dS_q = \frac{\nabla_p}{4\pi} \times \int_{S_{in}} \frac{(-\varepsilon \bar{E}_{n-1} + \sigma \bar{E}_n)}{|p-q|} dv_q + \frac{\nabla_p}{4\pi} \int_S \frac{\hat{n}_q \cdot \bar{H}_n}{|p-q|} dS_q. \tag{90}$$

It should be noted that when $\mu = \mu_0$, v_n^{ex} is explicitly given by

$$v_n^{ex} = \frac{1}{4\pi} \int_S \frac{\hat{n}_q \cdot \bar{H}_n^{inc}}{|p-q|} dS_q, \tag{91}$$

in which case $\bar{H}_n^S$ is completely known and $\bar{H}_n$ is found with (84) where the interior potential function v_n^{in} satisfies the Neumann data,

$$\frac{\partial v_n^{in}}{\partial n} = -\hat{n} \cdot \bar{G}_n^{in} + \hat{n} \cdot (\bar{H}_n^{inc} + \bar{H}_n^S). \tag{92}$$

When $\mu \neq \mu_0$ then $\bar{H}_n$ and $\bar{H}_n^S$ or equivalently, v_n^{in} and v_n^{ex}, must be found concurrently. Explicitly $\bar{H}_n = \bar{G}_n^{in} + \nabla v_n^{in}$, $p \in S_{in}$; $\bar{H}_n^S = \bar{G}_n^{ex} + \nabla v_n^{ex}$, $p \in S_{ex}$, v_n^{ex} is regular at infinity and, from (54b,c)

$$\left.\begin{aligned} &\text{a)}\quad \hat{n}\times\nabla(v_n^{ex} - v_n^{in}) = \hat{n}\times(\bar{G}_n^{in} - \bar{G}_n^{ex} - \bar{H}_n^{inc}) \\ &\text{b)}\quad \mu_0\frac{\partial v_n^{ex}}{\partial n} - \mu\frac{\partial v_n^{in}}{\partial n} = \mu\hat{n}\cdot\bar{G}_n^{in} - \mu_0\hat{n}\cdot\bar{G}_n^{ex} - \mu_0\hat{n}\cdot\bar{H}_n^{inc} \end{aligned}\right\} p\varepsilon S. \tag{93}$$

As before (93a) does not unambiguously define the difference $v_n^{ex} = v_n^{in}$ on S but the conditions

$$\int_S \hat{n}\cdot H_n dS = \int_S \hat{n}\cdot\bar{H}_n^S\, dS = 0 \tag{94}$$

remove this ambiguity. This completes the reduction of the determination of the coefficients $\bar{E}_n$, $\bar{E}_n^S$, $\bar{H}_n$, $\bar{H}_n^S$ to a sequence of potential problems. In general there are six potential functions to be determined for each order and this number is reduced by one if $\mu = \mu_0$ or $\varepsilon = \varepsilon_0$ and $\sigma = 0$.

This process is clearly a lengthy one and not readily suited for determining high order low frequency approximations. A systematic iterative procedure involving Green's functions for Laplace's equation, comparable to that found in the perfectly conducting case (equations 13 and 14 et seq) has been developed by Asvestas [3] however the expressions are almost prohibitively cumbersome.

When the incident field is a plane wave and attention is restricted to the first term in the low frequency expansion of the far field, the simplified expression in terms of polarizability tensors (III.A.4) may still be used. The expression for the far field given for the perfectly conducting case, equation (49) viz

$$\bar{E}^S(p) = \frac{-e^{ikr}}{4\pi r} k^2\{\hat{p}\times\hat{p}\times(\bar{\bar{P}}\cdot\hat{a}) - \hat{p}\times(\bar{\bar{M}}\cdot\hat{b})\} + o\left(\frac{1}{r}\right)$$

is still valid however the definitions of the polarizability tensor elements must be modified. In the case when $\sigma = 0$, the tensor elements are [30]

$$\text{a)}\quad P_{ij} = X_{ij}\left(\frac{\varepsilon}{\varepsilon_0}\right), \qquad \text{b)}\quad M_{ij} = -X_{ij}\left(\frac{\mu}{\mu_0}\right) \tag{95}$$

where

$$X_{ij}(\tau) = (1-\tau)\int_S \hat{n}\cdot\hat{x}_{iq}\,\Phi_j(q)\,dS_q \tag{96}$$

and $\Phi_j(q)$ are the boundary values of potential functions which may be characterized in terms of transition problems for Laplace's equation:

$$\text{a)}\quad \Phi_j = \Phi_j^S - x_j, \quad p \in S_{ex} \quad ; \quad \text{d)}\quad \Phi_j^S = \Psi_j + x_j + c_j \; ; \; \frac{\partial \phi_j^S}{\partial n} = \tau \frac{\partial \psi_j}{\partial n} + \hat{n}\cdot\hat{x}_j$$

$$\text{b)}\quad \nabla^2 \Psi_j = 0, \; p \in S_{in}, \; \nabla^2 \Phi_j^S = 0 \quad p \in S_{ex} \quad ; \qquad \text{e)}\quad \int_S \frac{\partial \Phi_j^S}{\partial n_q}(q)\, dS_q = 0,$$

$$\text{c)}\quad \Phi_j^S \text{ regular at infinity} \quad ; \tag{97}$$

or as solutions of the integral equation

$$\Phi_j(p) = -\frac{2}{1+\tau} x_{jp} + \frac{1-\tau}{1+\tau} \frac{1}{2\pi} \int_S \Phi_j(q) \frac{\partial}{\partial n_q} \frac{1}{|p-q|}\, dS_q . \tag{98}$$

Note that if $\varepsilon = \varepsilon_0$ then $p_{ij} \equiv 0$ whereas if $\mu = \mu_0$ then $M_{ij} \equiv 0$.

If $\sigma \neq 0$ then p_{ij} reverts to the definition (50a) whereas M_{ij} is still given by (95b). Thus if $\sigma \neq 0$ and $\mu = \mu_0$ there is no difference between the first terms of the far field coefficients for the perfectly conducting and penetrable cases.

These expressions were used to obtain numerical results for scattering by a dielectric cube by Herrick and Senior [8].

IV. REFERENCES

1. Ahner, J.F. and R.E. Kleinman, The Exterior Neumann Problem for the Helmholtz Equation, *Arch. Ration. Mech. & Anal. 52 (1)*, 26-43 (1973).
2. Alawneh, A.D. and R.P. Kanwal, Singularity Methods in Mathematical Physics, *SIAM Rev. 19(3)*, 437-471 (1977).
3. Asvestas, J.A., Low Frequency Scattering by Imperfectly Conducting Obstacles, AFOSR Scientific Report TR-72-0807, University of Delaware, Math Dept. Report 1794-4 (1972).
4. Asvestas, J.S. and R.E. Kleinman, Low Frequency Scattering by Perfectly Conducting Obstacles, *J. Math. Phys. 12(5)*, 795-811 (1971).
5. Bowhill, S.A. (Ed.), "Review of Radio Science 1972-74," International Union of Radio Science (URSI), Brussels, Belgium (1975).
6. Courant, R. and D. Hilbert, "Methods of Mathematical Physics, Vol. II," Interscience, New York (1962).
7. Gray, G., Low Frequency Iterative Solution of Integral Equations in Electromagnetic Scattering Theory, Ph.D. Dissertation, University of Delaware, Math. Dept. (1977).

8. Herrick, D.F. and T.B.A. Senior, The Dipole Moments of a Dielectric Cube, *IEEE Trans. Antennas & Propag. 25(4)*, 590-592 (1977).
9. Jain, D.L. and R.P. Kanwal, An Integral Equation Perturbation Technique in Applied Mathematics II, *Appl. Anal. 4,* 297-329 (1975).
10. Jones, D.S., "The Theory of Electromagnetism," MacMillan Co., (Pergamon Press), New York (1964).
11. Kanwal, R.P., An Integral Equation Perturbation Technique in Applied Mathematics, *J. Math. & Mech. 19(7)*, 625-656 (1970).
12. Keller, J.B., R.E. Kleinman and T.B.A. Senior, Dipole Moments in Rayleigh Scattering, *J. Inst. Math. & Appl. 9,* 14-22 (1972).
13. Kleinman, R.E., The Rayleigh Region, *Proc. IEEE 53(8)*, 848-856 (1965).
14. Kleinman, R.E., Low Frequency Methods in Classical Scattering Theory, Report NB 18, Lab. of Electromagnetic Theory, Tech. Univ. of Denmark (1966).
15. Kleinman, R.E., Far Field Scattering at Low Frequencies, *Appl. Sci. Res. 18,* 1-8 (1967).
16. Kleinman, R.E., "Low Frequency Solutions of Electromagnetic Scattering Problems, Electromagnetic Wave Theory (Delft Symposium)," Pergamon Press, Oxford, New York, 891-905 (1967).
17. Kleinman, R.E., Dipole Moments and Near Field Potentials, *Appl. Sci. Res. 27,* 335-340 (1973).
18. Kleinman, R.E., Iterative Solutions of Boundary Value Problems, "Lecture Notes in Mathematics #561: Function Theoretic Methods for Partial Differential Equations," Springer, New York (1976).
19. Kleinman, R.E. and T.B.A. Senior, Rayleigh Scattering Cross Sections, *Radio Sci. 7(10)*, 937-942 (1972).
20. Kleinman, R.E. and T.B.A. Senior, Low Frequency Scattering by Space Objects, *IEEE Trans. Aerosp. & Electron. Syst. 11 (4)*, 672-675 (1975).
21. Kleinman, R.E. and W. Wendland, On Neumann's Method for the Exterior Neumann Problem for the Helmholtz Equation, *J. Math. Anal. & Appl. 57(1)*, 170-202 (1977).
22. Král, J., The Fredholm Method in Potential Theory, *Trans. Am. Math. Soc. 125,* 511-547 (1966).
23. Mikhlin, S.G., "Mathematical Physics, An Advanced Course," North Holland Publishing Co., Amsterdam-London (1970).
24. Minnis, C.M. and Y. Bogitch (Eds.), "Review of Radio Science 1969-72," International Union of Radio Science (URSI), Brussels, Belgium (1972).
25. Morse, P.M. and H. Feshbach, "Methods of Theoretical Physics," McGraw-Hill, New York (1953).
26. Müller, C., "Foundations of the Mathematical Theory of Electromagnetic Waves," Springer-Verlag, Berlin (1969).
27. Noble, B., Integral Equation Perturbation Methods in Low Frequency Diffraction, in "Electromagnetic Waves" (R. Langer, Ed.), The University of Wisconsin Press (1962).

28. Payne, L.E., Isoperimetric Inequalities and Their Applications, *SIAM Rev. 9(3)*, 453-488 (1967).
29. Poggio, A.J. and E.K. Miller, Integral Equation Solutions of Three-dimensional Scattering Problems, in "Computer Techniques for Electromagnetics" (R. Mittra, Ed.), Pergamon Press, Oxford (1973).
30. Senior, T.B.A., Low Frequency Scattering by a Dielectric Body, *Radio Sci. 11,* 477-482 (1976).
31. Lord Rayleigh, On the Incidence of Aerial and Electric Waves upon Small Obstacles in the Form of Ellipsoids or Elliptic Cylinders, and On the Passage of Electric Waves through a Circular Aperture in a Conducting Screen, *Philos. Mag. 44,* 28-52 (1897).
32. Lord Rayleigh, On the Passage of Waves through Apertures in Plane Screens, *Philos. Mag. 43,* 259-272 (1897).
33. Stevenson, A.F., Solutions of Electromagnetic Scattering Problems as Power Series in the Ratio (Dimension of Scatterer Wavelength), *J. Appl. Phys. 24,* 1134-1142 (1953).
34. Stevenson, A.F., Electromagnetic Scattering by an Ellipsoid in the Third Approximation, *J. Appl. Phys. 24(9)*, 1143-1151 (1953).
35. Stevenson, A.F., Note on the Existence and Determination of a Vector Potential, *Q. Appl. Math. XII(2)*, 194-197 (1954).
36. Stratton, J.A., "Electromagnetic Theory," McGraw-Hill, New York, 465 (1941).
37. Van Bladel, J., Low Frequency Asymptotic Techniques, in "Modern Topics in Electromagnetics and Antennas," (R. Mittra, Ed.), Peter Peregrinus, Ltd., Stevenage, England (1977).
38. Wendland, W., Die Behandlung von Randwertaufgaben im R_3 mit Hilfe von Einfach und Doppelschichtpotentialen, *Numer. Math. 11*, 380-404 (1968).
39. Wilcox, C.H., An Expansion Theorem for Electromagnetic Fields, *Comm. Pure & Appl. Math IX,* 115-134 (1956).
40. Bowman, J.J., T.B.A. Senior and P.L.E. Uslenghi, "Electromagnetic and Acoustic Scattering by Simple Shapes," North Holland Publishing Co., Amsterdam (1969).
41. Ruck, G.T., D.E. Barrick, W.D. Stuart and C.K. Krichbaum, "Radar Cross Section Handbook," Plenum Press, New York-London (1970).

ELECTROMAGNETIC SCATTERING

ASYMPTOTIC METHODS IN HIGH-FREQUENCY PROPAGATION AND SCATTERING

Leopold B. Felsen
Department of Electrical Engineering
Polytechnic Institute of New York

I. INTRODUCTION

Asymptotic methods for the analysis of high-frequency electromagnetic propagation and diffraction have been thoroughly exploited during the past three decades and have reached a high level of sophistication. Of more recent vintage are asymptotic techniques for transient propagation in nondispersive and dispersive media, and also the inclusion of the effects of losses within an asymptotic framework. The physical foundation for these asymptotic procedures is the *local* character of wave processes in the relevant parameter regime. For propagation and diffraction in an otherwise unbounded medium, the basic constituents are local plane waves in the time-harmonic regime and wave packets in the transient regime. When the domain of interest is confined by extended boundaries or by stratification that can trap and guide the field energy, the formulation may be modified by conversion of multiply reflected local plane waves into guided modes. Alternatively, the local fields may be cast directly into modal form and then tracked from one portion of the guiding region to another. When the physical configuration has elements of periodicity, the asymptotic description can be phrased so as to incorporate these features into the local field representation. From these considerations, it may be surmised that the asymptotic analysis of high-frequency time-harmonic or of transient wave phenomena can be performed in different ways, depending on the initial choice of the local environment and hence of the local wave types. This important

ISBN 0-12-709650-7

aspect lends present asymptotic theory much of its vitality.

Asymptotic techniques in propagation and diffraction have been reviewed in several recent publications [1-4]. Moreover, the contributions to this symposium include applications that need not be repeated here. Therefore the emphasis in the present paper is on developments that were not included in these earlier surveys or elsewhere at this Symposium. One of these is the extension of asymptotic theory to local fields with complex phase, thereby incorporating into the class of treatable wave phenomena those effects, such as dissipation and evanescence, which cause fields to decay exponentially. Evanescent fields in particular are of current interest because they include as a special case the highly collimated Gaussian beams which are utilized in integrated optics, optical fiber, and microwave acoustics applications. Evanescent fields also describe leaky waves, fields on the dark side of caustics, penetration of totally reflected fields and other important wave phenomena. Various techniques for the asymptotic treatment of evanescent waves have been explored, including evanescent wave tracking, complex source points, complex rays and the like. Some problems arising in this connection are discussed in this paper.

Ray methods for radiation and diffraction in waveguides or ducts have also been the subject of recent interest. While most of the contributions to date have been for parallel plane geometry, the theory has now been extended to relatively arbitrary waveguide configurations. These generalizations are discussed below and applied in particular to the study of open optical resonators, which are employed in connection with the generation of optical frequency fields by laser sources. Other applications of asymptotic techniques are summarized more succinctly, and the presentation is concluded with some final remarks concerning the present status of asymptotic methods.

II. EVANESCENT FIELDS

A. Complex Source Points and Gaussian Beams

It has recently been noted [5,6] that, by assigning complex values to the source coordinate locations of oscillating isotropic point or line source radiators in free space, one may generate a highly collimated field that behaves in the vicinity of its maximum like a three-dimensional or two-dimensional Gaussian beam. This implies that the complex-source-point substitution converts point or line source Green's functions for propagation and diffraction in various environments into field solutions for incident Gaussian beams. Thus, the whole arsenal of rigorous and asymptotic diffraction solutions yields, without further effort, the field response for beam excitation, provided that one can perform the analytic continuation [7]. We discuss here some problems that can arise in this context. For validity of the beam solution, one must verify that convergence of a particular Green's function representation is retained. The convergence requirement imposes restrictions that generally depend on the type of representation employed. A few examples will illustrate what is involved.

1. *Free Space*

We recall first the properties of the closed form free-space Green's function G_f when the source point $r' = (x',y',z')$ is made complex. Let

$$\underset{\sim}{r}' \rightarrow \underset{\sim}{r}'_b = \underset{\sim}{r}_o + i\underset{\sim}{b}, \quad \underset{\sim}{r}_o = (x_o, y_o, z_o), \quad \underset{\sim}{b} = (b_x, b_y, b_z) \quad (1)$$

where $\underset{\sim}{r}_o$ and $\underset{\sim}{b}$ are real, denote the complex substitution. Then

$$G_f = \frac{\exp(ik\bar{r})}{4\pi\bar{r}}, \quad \bar{r} = [(x-x')^2 + (y-y')^2 + (z-z')^2]^{\frac{1}{2}} \quad (2)$$

becomes

$$G_{fb} = \frac{\exp(ik\bar{r}_b)}{4\pi\bar{r}_b}, \quad \bar{r} = [(x-x'_b)^2 + (y-Y'_b)^2 + (z-z'_b)^2]^{\frac{1}{2}} \quad (3)$$

$$Re\, \bar{r}_b \geq 0$$

which remains an exact solution of the wave equation. The subscript b denotes quantities subjected to the transformation in (1). Taking $\underset{\sim}{r}_o = 0$, $b_x = b_y = 0$, for the present and $b_z = b > 0$, one has in the far zone $r >> kb^2$,

$$\bar{r}_b \sim r - ib\cos\theta, \quad r = (x^2 + y^2 + z^2)^{\frac{1}{2}}, \quad \cos\theta = \frac{z}{r}, \tag{4}$$

and in the paraxial region $\rho^2 << z^2 + b^2$ near the z axis,

$$\bar{r}_b \sim z - ib + \frac{\rho^2}{2(z - ib)} = z - ib + \frac{\rho^2(z + ib)}{2(z^2 + b^2)}, \; \rho^2 = x^2 + y^2. \tag{5}$$

Thus, the far zone field has a radiation pattern

$$f(\theta) = \exp(kb\cos\theta), \tag{6}$$

which decays strongly away from its maximum $\exp(kb)$ on the z axis $(\theta = 0)$. The paraxial field everywhere decays from (5) according to the Gaussian variation $\exp[-kb\rho^2/2(z^2 + b^2)]$. The substitution $\underset{\sim}{r}' = (0, 0, ib)$ therefore converts the spherical wave in (2) into the rotationally symmetric beam in (3), which becomes highly collimated near the positive z axis when kb is large. When (1) is used in its entirety, the phase center (waist) of the beam is shifted to $\underset{\sim}{r}_o$ and the beam axis (trajectory of the field maxima) lies along the direction of $\underset{\sim}{b}$. The disk of radius b transverse to the beam axis may be regarded as an equivalent source distribution for the beam field (Fig. 1 (a)).

In two dimensions, where $\underset{\sim}{r}' = (y', z')$ and

$$G_f = \frac{i}{4} H_o^{(1)}(k\bar{r}) \sim \frac{\exp(ik\bar{r} + i\pi/4)}{2(2\pi k\bar{r})^{\frac{1}{2}}},$$
$$\bar{r} = [(y - y')^2 + (z - z')^2]^{\frac{1}{2}}, \tag{7}$$

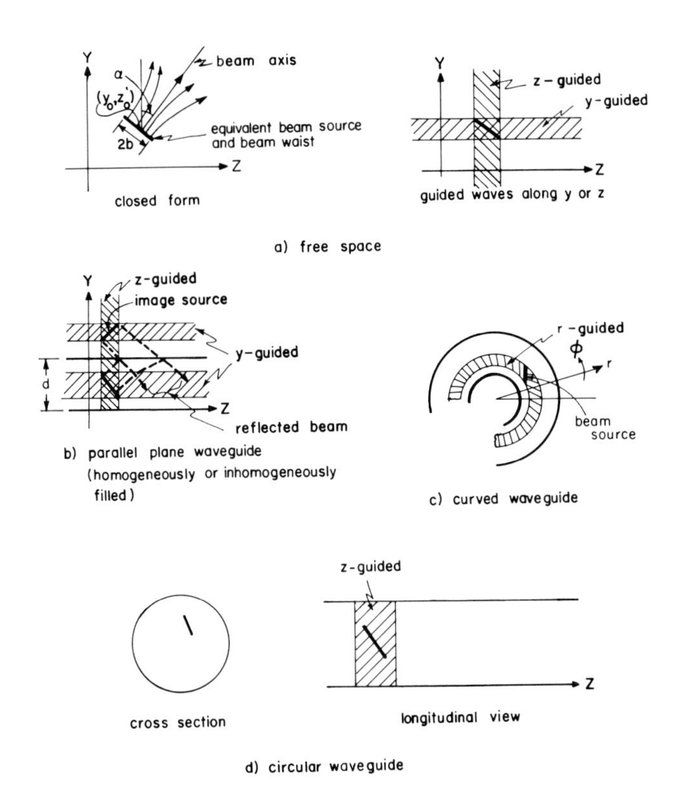

Fig. 1. Restrictions imposed when complex-source-point substitution is made in ordinary Green's functions. Excluded domains are shown only in the first figure.

one arrives at the same conclusions as above for an x-independent sheet beam. Finally, by making the substitution (1) in the dyadic Green's function for a vector dipole source, one may generate a vector beam.

While the replacement of $\bar{r}$ by $\bar{r}_b$, with $Re\ \bar{r}_b > 0$, guarantees that the corresponding G_{fb} satisfies the wave equation and radiation condition, this is not equally evident when Green's function representations, rather than closed form solutions are employed. Just how the domain of validity of $\bar{r} \rightarrow \bar{r}_b$ depends on the field representation is best illustrated by some typical cases.

2. *Plane Stratified Media*

a. *Isotropic case.* When a line source is located at (y',z') exterior to an isotropic half space that may be stratified along the y direction and occupies the region $y > 0$, the incident and reflected fields in the region $y > 0$ may be represented in terms of the generic plane wave spectral representation [8]:

$$I^{\pm} = \int_{-\infty}^{\infty} f^{\pm}(\eta) \exp[ik\psi^{\pm}(\eta)]d\eta \tag{8}$$

$$\psi^{\pm}(\eta) = \eta(z - z') \pm \kappa(y \mp y'), \quad \kappa = (1 - \eta^2)^{\frac{1}{2}} \tag{8a}$$

where the upper sign refers to the incident field and the lower sign to the reflected field; for the incident field, it is required that $y > y'$. This representation is in terms of guided waves along y. $f^{\pm}(\eta)$ is a spectral amplitude function that does not behave exponentially (but may possess pole and branch point singularities), while κ is defined so that $\kappa = +1$ when $\eta = 0$ and $Im\ \kappa > 0$ when κ is non-real; k is the wavenumber in the half space $y < 0$. To generate a beam whose center lies at $(0, y_o)$ and whose axis is inclined through an angle α with the positive y axis, one chooses

$$z' = ib \sin\alpha, \quad y' = y_o + ib \cos\alpha, \quad y_o < 0, \quad b > 0. \tag{9}$$

Then the exponential in (8), whose behavior for large $|\eta|$ governs the convergence of the integral, becomes

$$\psi^{\pm} \to \eta z - i\eta b \sin\alpha \pm i|\eta|(y \mp y_o) + |\eta|b \cos\alpha, \quad |\eta| \to \infty. \tag{10}$$

Thus, $\exp(ik\psi_b^{\pm})$ decays as $|\eta| \to \infty$ provided that

$$|y \mp y_o| < b \sin\alpha. \tag{11}$$

For the incident field, this restriction implies that the observation points (y,z) must be removed from the beam center at $z = 0$, $y = y_o$, by at least the projected distance $y > y_o = b \sin\alpha$. For the reflected field at all observation points $y < 0$, the condition $|y_o| > b \sin\alpha$ is adequate. Thus, the integral in (8), with (9), may be employed to represent the incident and reflected beam fields provided that the restriction in (11) is obeyed. When applied to the incident field, whose closed form solution G_{fb} (see (3), etc.), with (9), is nonrestricted, one observes that unless the beam axis is oriented perpendicular to the z coordinate, which serves as the reference for the spectral decomposition, the range of admissible observation points is limited, with the limitation becoming stronger as the inclination of the beam axis with respect to the spectral coordinate increases (Fig. 1(a)).

b. Anisotropic case. When the medium is anisotropic, the y-guided spectral representations in (8) are modified in that $\kappa = \kappa(\eta)$ is no longer given by $(1 - \eta^2)^{\frac{1}{2}}$; moreover, different functions κ_i and κ_r generally characterize the incident and reflected fields, respectively. Proceeding as for the isotropic case and assuming that κ is imaginary for large enough $|\eta|$ (i.e., the refractive index surface $\kappa = \kappa(\eta)$ is described by a closed curve), one finds that condition (11) remains applicable provided that the right-hand side is multiplied by a positive constant which depends on the form of $\kappa(\eta)$. Thus, when the line source field is changed by complex source point displacement, the integral representation remains valid at observation points whose coordinates are sufficiently far removed from the beam waist. It may be verified that the resulting field describes a Gaussian beam provided that the sign of b is chosen in a manner that takes into account the curvature of the refractive index surface at the point corresponding to the direction of the beam axis [9].

3. Waveguides

a. Parallel Plane waveguide. When the source point displacement (see (1)) with $b_x = 0$, $b_y = b\cos\alpha$, $b_z = b\sin\alpha$, $b > 0$) is applied to the z-guided mode representation of the two-dimensional Dirichlet Green's function for a homogeneously filled parallel plane waveguide with perfectly conducting boundaries $y = 0$ and $y = d$, one finds that for $z > z_o$ [10],

$$G(\underset{\sim}{r},\underset{\sim}{r}_b') = \frac{1}{2d} \sum_{m=1}^{\infty} \sin\left(\frac{m\pi y}{d}\right) \frac{\exp[ik\tau_m(z - z_o)]}{k\tau_m} \cdot \qquad (12)$$

$$\cdot \{\exp[im\pi y_o/d - kb \sin(\theta_m - \alpha)] - \exp[-im\pi y_o/d + kb \sin(\theta_m + \alpha)]\}$$

where $\theta_m = \cos^{-1}\tau_m = \sin(m\pi/kd)$ is the propagatiaon angle, measured counteclockwise from the waveguide axis z, of the constituent plane waves that synthesize the mth mode. Since $\tau_m \to im\pi/kd$ for large m, the series converges for $z > z_0 + b \cos\alpha$, thereby confirming the condition in (11) (noting that (12) involves a z-guided and (11) a y-guided representation). One may remark that the analytically continued Green's function in (12) is the same as would be obtained when the beam is regarded as a non-isotropic source located at (y_o,z_o), with far zone radiation pattern $f(\theta) = \exp[kb|\sin(\alpha - \theta)|]$ decaying away from the beam axis $\theta = \pi/2 - \alpha$ (see (6)); this equivalence is confirmed by comparing (12) with (39). Evidently, the beam source excites most strongly those modes whose propagation angles $\pm\theta_m$ are closest to the beam axis. When the waveguide contains an inhomogeneous medium with refractive index $n(y)$, the formulation in (39), with $f(\theta) = exp [kn(y_o)b|\sin(\alpha - \theta)|]$, may be used to infer that the convergence condition remains unchanged provided that $n(y)$ varies slowly over an interval on the scale of b [10].

If the waveguide Green's function had been expressed in terms of a plane wave spectrum involving a Fourier integral in z and a one-dimensional Green's function g_y in the y domain (y-guided representation), the analytically continued solution would be invalid inside a strip of width $2b \sin\alpha$ perpendicular to the y axis as in Fig. 1(b).

The parallel plane waveguide Green's function, and hence its analytically continued form for the beam problem, can also be expressed as a series of image sources in free space. This representation implies direct tracking of the incident beam as it progresses via wall reflections down the waveguide. It is evident that since each of the image sources is subject to the constraints $z > z_o + b\cos\alpha$ or $y > \bar{y}_0 + b\sin\alpha$ referred to the image location $(\bar{y}_o, z_o)$, only the real source at (y_o, z_o) imposes a restriction on the field observed in the waveguide region $0 \leq y < d$. When the y-guided formulation is decomposed into traveling waves by image expansion of g_y, one may show likewise that the excluded region in Fig. 1(b) applies only to the incident beam but not to the multiply reflected beams. Thus, the restrictions on $G(\underset{\sim}{r}, \underset{\sim}{r}'_b)$ arise only from the actual beam source and not from multiply reflected beam contributions.

b. Curved waveguide. The region between two concentric perfectly conducting cylinders may serve as a two-dimensional model for a curved waveguide, in which (axially independent) propagation takes place in the radial or azimuthal directions. If the Green's function is represented in terms of constituents that, in the asymptotic regime, represent r-guided traveling waves which are reflected repeatedly at the waveguide walls, one may examine the integral representation for each such constituent when $\underset{\sim}{r}'$ is changed into $\underset{\sim}{r}'_b$. Assuming that $kr_o >> 1$, where (r_o, ϕ_o) denotes the beam source location, one finds [11] that the representation for the incident beam is not valid inside an annular strip that contains the equivalent beam source distribution (see Fig. 1(c)). No such restriction is found to encumber the multiply reflected contributions, thereby confirming again that convergence problems for beam Green's function representations arise solely from the source-generated incident beam.

The preceding discussion shows that guided mode representations of the analytically continued waveguide Green's functions generally have an excluded domain of observation points that occupies a strip containing the equivalent beam source distribution and oriented perpendiularly to the assumed guiding direction. On the other hand, when the field is resolved into multiply reflected beams, only the incident portion is subject to this restriction. The distinction between the two cases arises because both the incident and reflected beams are expressed in terms of the same waveguide modes in the guided mode expansions, whereas each of the multiply reflected beams has its own representation in the beam tracking formulation.

It should also be remarked that the excluded domains coincide with the appropriately projected beam source distributions only when the modal fields in the waveguide behave like local plane waves over the extent of the source region. Loosely speaking, this is the range of applicability of the WKB approximations for the mode functions (see (30a) and (33)). When these approximations are not applicable (for example, near the axis of a circular waveguide (see Fig. 1d), the excluded domain may have to be enlarged [12].

4. *Wedge*

The wedge diffraction problem will serve as an example of how convergence for the complex source point field can be secured by contour deformation. Whe a line source is located at the point $\underset{\sim}{r}' = (y', z') = (r', \phi')$ in the presence of a wedge whose axis $r = 0$ coincides with the x axis and whose sides are the intersecting half planes at $\phi = 0$ and $\phi = \beta$, respectively, the scalar Green's function for observation points in the geometrical shadow region has the generic form [ref. 8, Sec. 6.3].

$$G_d(\underset{\sim}{r}, \underset{\sim}{r}') = \int_{i\infty}^{-i\infty} H_0^{(1)}(k\tau)\, V(\phi, \phi', w)\, dw, \quad \phi, \phi' \le \beta, \tag{13}$$

where the subscript d denotes the diffraction field in the shadow zone. Here,

$$\tau(w) = (r^2 + r'^2 + 2r\, r' \cos w)^{\frac{1}{2}}, \quad \text{Re}\, \tau > 0 \tag{13a}$$

and V is a function that is well behaved at $|Imw| \to \infty$ but has pole singularities on the real w axis; the detailed structure of

V depends on the boundary conditions satisfied by the field on the wedge surface. As before, $r = (y^2 + z^2)^{\frac{1}{2}}$ and $\phi = \tan^{-1}(y/z)$ are polar coordinates. The exterior wedge angle β is assumed to be greater than $3\pi/2$. It is also assumed that the source angle ϕ' is such as to admit observation angles ϕ satisfying $|\phi - \phi'| > \pi$; this domain constitutes the geometrical shadow region. Since τ is real when r' is real, the asymptotic behavior of the Hankel function (see (7)) assures that the integral is convergent. In the illuminated region $|\phi - \phi'| < \pi$, the field in (13) must be augmented by geometric-optical incident and possible reflected field contributions, which arise from poles of V that cross the $Im\, w$ axis as $|\phi - \phi'|$ decreases. For a "perfectly absorbing" wedge, there are no reflected field contributions. In that event, the function V has the form [Ref. 8, Sec. 6.4]

$$V(\phi,\phi',w) = \frac{1}{8\pi}\left(\frac{1}{\pi - |\phi-\phi'| - w} + \frac{1}{\pi + |\phi-\phi'| + w}\right), \tag{14}$$

and a field solution G valid at arbitrary observation points is obtained by adding to (13) the incident field G_i in the illuminated region:

$$G(\underset{\sim}{r},\underset{\sim}{r}') = G_i(\underset{\sim}{r},\underset{\sim}{r}') + G_d(\underset{\sim}{r},\underset{\sim}{r}') , \tag{15}$$

$$G_i(\underset{\sim}{r},\underset{\sim}{r}') = \frac{i}{4} H_o^{(1)}(k\bar{r})\, U(\pi - |\phi - \phi'|) , \tag{15a}$$

where $U(x)$ equals unity or zero when $x > 0$ or $x < 0$, respectively, and $\bar{r}$ is defined in (7).

To convert (13) into a solution that describes a beam field incident on the edge, we assume that

$$z' = z_o , \quad y' = -\bar{y}_o + ib , \quad \bar{y}_o > 0 , \quad b > 0 . \tag{16a}$$

These complex coordinates, when employed in (15a) (without the unit step function), provide an incident beam whose waist is centered at $(-\bar{y}_o, z_o)$ and whose axis is parallel to the y axis, with propagation taking place along the positive y direction. The polar coordinates corresponding to (16a) are

$$r_b' = [(\bar{y}_o - ib)^2 + z_o^2]^{\frac{1}{2}}, \quad \phi_b' = \tan^{-1}\left(-\frac{\bar{y}_o + ib}{z_o}\right), \tag{16b}$$

with r_b' defined to be positive, and $\phi_b' > \pi$, when $b = 0$. To determine whether (13) remains valid as b increases from zero, it is adequate to examine τ for large values of $w_i = Im\, w$, i.e.,

$$\tau \sim (2rr_b' \cosh w_i)^{\frac{1}{2}} \sim r^{\frac{1}{2}} \exp(|w_i|/2)|r_b'|^{\frac{1}{2}} \exp(-i\varphi/2), \tag{17}$$

where $2\varphi = \tan^{-1}[2b\bar{y}_o(\bar{y}_o^2 + z_o^2 - b^2)^{-1}]$ (note that $\varphi = \tan^{-1}(b/\bar{y}_o)$ when $z = 0$). Since $\varphi > 0$ when $b \neq 0$, it follows that $Im\tau < 0$ and, hence, $H_o^{(1)}(k\tau)$ diverges strongly as $|w_i| \to \infty$. Thus, the analytic continuation from $b = 0$ to $b > 0$ cannot be performed in (13).

To remedy this situation, it is suggestive to seek alternative integration paths whereon $Im\tau > 0$. Writing $w = w_r + iw_i$, where $w_r = Re w$, one finds as $w_i \to \pm\infty$,

$$\tau \sim (2rr_b' \cos w)^{\frac{1}{2}} \sim r^{\frac{1}{2}} \exp(|w_i|/2)|r_b'|^{\frac{1}{2}} \exp(\mp iw_r/2 - i\varphi/2). \tag{18}$$

Thus, $Im\tau > 0$ if

$$w_r < -\varphi \text{ as } w_i \to \infty, \quad w_r > \varphi \text{ as } w_i \to -\infty. \tag{19}$$

By displacing the endpoints of the integration path in (13) so that they terminate in the regions specified by (19) (this is possible since $Im\tau > 0$ in sectors $-\pi < w_r < 0$, $w_i > 0$ and $0 < w_r < \pi$, $w_i < 0$, when r' is real), one may subsequently carry out the analytic continuation from $b = 0$ to $b > 0$ along the deformed path P. If the integration path along the imaginary axis in (13) is replaced by P, the resulting function $G_d(\underset{\sim}{r}, \underset{\sim}{r}_b')$ represents the diffraction field in the region $\phi < \pi/2$ due to an incident beam whose axis lies along the line $z = z_o$ and whose waist is located at $y = -\bar{y}_o$.

The modified integral (13) may be evaluated asymptotically by the saddle point method when kr and (*or*) $|kr_b'|$ are large; $H_o^{(1)}(k\tau)$ is then approximated as in (7). The saddle point $w_s = 0$, obtained as a solution of $d\tau/dw = 0$, is not affected by $b \neq 0$. Since the integration path can be deformed into the steepest descent path through $w = 0$, the saddle point approximation of the complex-source-point field is the same as the analytically continued saddle point approximation of the real-source-point field G_d [7]:

$$G_d(\underset{\sim}{r}, \underset{\sim}{r}_b') \sim -\frac{1}{4\pi} V(\phi, \phi_b', 0) \frac{\exp(ikr + i\pi/4)}{2(2\pi kr)^{\frac{1}{2}}} \frac{\exp(ikr_b' + i\pi/4)}{2(2\pi kr_b')^{\frac{1}{2}}}, \tag{20}$$

where V for the perfectly absorbing wedge is given explicitly in (14). The last factor in (20) represents the value of the incident beam field at the edge; when the edge lies in the paraxial region, the Gaussian illumination at the edge may be inferred by proceeding as in (5). The remaining factors in (20) represent the diffraction field at any (r,ϕ) exterior to transition regions due to a unit amplitude plane wave incident along the complex angle ϕ_b'. This fact confirms the continued applicability of the local character of edge diffraction even when the incidence angle is complex. However, the character of the transition regions is markedly affected by the complex incidence angle [13].

5. *Paraxial Approximations*

While the complex-source-point method generates rigorous field solutions for beam excitation from those for point source or line source excitation, the connection between the physically transparent ray-optical asymptotic fields for the latter and the complex-source-point fields is generally obscured by the analytic continuation. An exception occurs when observation points are restricted to the paraxial region near the axis of the incident and reflected beams. In that event, the same physical parameters, which are employed for tracking a real ray bundle from the source point to the observation point, appear also in the beam tracking formulas. For example, when a line-source excited ray bundle is reflected from a circular cylindrical boundary as in Fig. 2(a) (this configuration is relevant for ray and beam tracking inside an optical fiber), the paraxial field near the central ray is given by [11] (for plane stratified media see [10]):

$$G(\underset{\sim}{r},\underset{\sim}{r}') \sim \frac{i}{4}\sqrt{\frac{2}{\pi k}}\left|\frac{L_{fo}}{R}\right|^{\frac{1}{2}} \frac{1}{L_o^{\frac{1}{2}}} \exp(ik\psi - \frac{i\pi}{2}\sigma_f - \frac{i\pi}{4}) \quad (21)$$

where the wall reflection coefficient is assumed to equal (+1) and the phase function ψ is

$$\psi = 2sL_a - L_o + L + \frac{d^2}{2R}, \quad R = L - L_{fo}. \quad (22)$$

The lengths L_o, L_{fo} and L_a, measured from the perpendicular bisector of the central ray whose orientation is fixed by the angle γ_a, are defined in Fig. 2(a). The integer s counts the number of reflections, and σ_f counts the number of times that the central ray passes through a ray tube focus at L_{fo}. R is the radius of curvature of the wavefront, d is the perpendicular distance from the central ray and

$$L_{fo} = L_a L_o (L_a - 2sL_o)^{-1} \quad (23)$$

When the complex-source-point method is applied to convert the ray bundle into a paraxial Gaussian beam, one may show that this implies replacement of L_o by $L_o + ib$ [11]. Then

$$G_b \sim \frac{i}{4}\sqrt{\frac{2\pi}{k}}\left(\frac{L_{fb}}{L - L_{fb}}\right)^{\frac{1}{2}} \frac{1}{(L_o + ib)^{\frac{1}{2}}} \exp(ik\psi_b)e^{-i\pi/4} \quad (24)$$

where $L_{fb} = Re\, L_{fb} + i\, Im\, L_{fb}$, with

$$Re\ L_{fb} = \frac{L_a L_o (L_a - 2s\ L_o) - 2sb^2}{(L_a - 2s\ L_o)^2 + 4s^2 b^2} \tag{25a}$$

$$Im\ L_{fb} = \frac{bL_a^2}{(L_a - 2s\ L_o)^2 + 4s^2 b^2} \tag{25b}$$

Also,

$$\psi_b = 2s\ L_a - (L_o + ib) + L + \frac{d^2 (L - Re\ L_{fb} + i\ Im\ L_{fb})}{2[(L - Re\ L_{fb})^2 + (Im\ L_{fb})^2} \tag{26}$$

The square roots are defined so that $G_b \to G$ in (21) when $b \to 0$. From (26), the beam width minima occur at $L = Re\ L_{fb}$, which locations are displaced from paraxial ray tube focus $L = L_{fo}$ (see Fig. 2(b)).

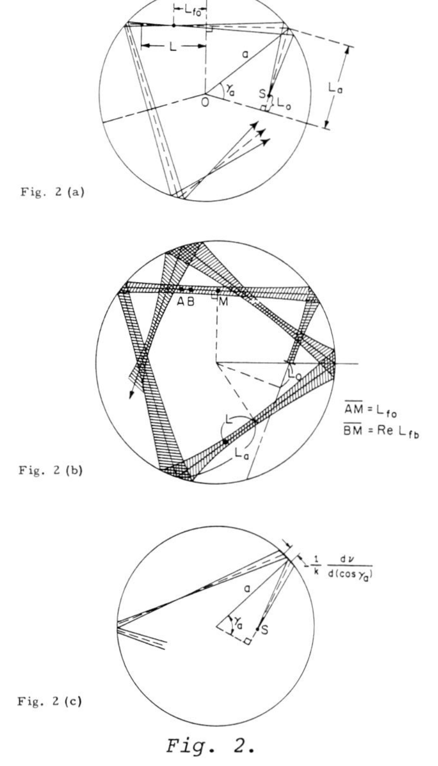

Fig. 2.

6. *Lateral Shifts of Reflected Rays and Beams*

When a boundary is characterized by an incidence-angle-dependent reflection coefficient $\Gamma(\theta) = \exp[i\nu(\theta)]$, a totally reflected ray does not emerge from the point of impact of the incident ray but from a location shifted by $-(i/k)\, d\nu/d(\cos\theta)$, where θ is the angle of incidence measured from the boundary. Reflected fields computed along the conventional non-shifted geometric optical ray path differ negligibly from those obtained via the shifted path, when the number of reflections is small. However, in layered or other guiding regions, where many reflections may occur, discrepancies between the two formulations arise. It may be shown [10] that the shifted paths must be employed when multiply reflected rays are converted into guided modes; otherwise, the conversion does not yield the correct modal dispersion equation. The same statement applies when the incident field is a Gaussian beam.

To account for a circular cylindrical boundary characterized by $\Gamma(\gamma_a)$, the phase function ψ in (22) contains the additional term $-(s/k)[d\nu/d(\cos\gamma_a)]\cos\gamma_a$, and the amplitude is augmented by the factor $\exp[is\nu(\gamma_a)]$. The shifted trajectory is shown in Fig. 2(c).

B. Evanescent Wave Tracking

Alternative to the complex-source-point method for Gaussian beams is the more general method of evanescent wave tracking, which proceeds in real coordinate space and constitutes an extension of the conventional geometrical theory of diffraction (GTD) for non-evanescent fields. The asymptotic expansion, on which the GTD is based, has the form:

$$u(\underset{\sim}{r}) \sim A(\underset{\sim}{r}) \exp[ik\psi(\underset{\sim}{r})] + \exp[ik\psi(\underset{\sim}{r})] \sum_{n=1}^{\infty} A_n(\underset{\sim}{r})/(ik)^n \qquad (27)$$

where the wavenumber k is the large parameter and the phase ψ is real. Substituting (1) into the wave equation and equating to zero the coefficients of k^n, $n = 0,1,2,...$ in the resulting series, one obtains the eikonal and transport equations for determination of the phase ψ and the amplitude A of the geometric optical fields as well as equations for the higher order amplitudes A_n, $n \geq 1$. Because ψ is real, $A \exp(ik\psi)$ represents a homogeneous local plane wave field.

By allowing ψ to be complex, this procedure can be extended to include inhomogeneous local plane waves which are required for the asymptotic representation of evanescent fields. As noted in Sec. I, one accommodates thereby wave fields such as Gaussian beams, leaky waves, surface waves, waves on the dark side of caustics, etc., which were excluded by the restriction to real ψ. The theory for evanescent wave tracking has been developed in references [14,15] and applied to evanescent plane wave and Gaussian beam diffraction by a circular cylinder, to guided mode propagation on graded index dielectric slabs and fibers, and to propagation along curved dielectric waveguides. For a summary review of these developments, the reader is referred to references [16] and [17].

C. Complex Rays

Rays are conventionally defined as those trajectories whereon the wavenumbers associated with a plane wave field remain constant. When the wavenumbers are complex, as for evanescent fields or for fields in dissipative media, real-space trajectories are generally inadequate. There, the systematic extension of ray methods from fields with real phase to fields with complex phase requires the use of a complex coordinate space. The implications of complex rays are well understood [18,19], and their utilization unifies, at least conceptually, the asymptotic treatment of this general class of wave problems. However, procedures for actual field tracking from an initial surface to an observation point via an interposed medium or a scattering object remain to be further explored. Problems arising in the context have been noted in references [15] and [19].

III. RADIATION AND DIFFRACTION IN WAVEGUIDES OR DUCTS

A. Background

Asymptotic methods have been applied to waveguides or ducts which can support guided modes. When such configurations are excited by a primary radiating source or by secondary induced sources generated by discontinuities and obstacles, it is often desirable to convert a poorly convergent ray-optical formulation involving direct rays and rays multiply reflected at the boundaries (or refracted due to medium inhomogeneities) into modal fields. While the relation between rays and modes was recognized and employed some time ago in connection with wave propagation in the earth ionosphere waveguide[1], a systematic study extending these concepts to diffracted rays and the ray-optical evaluation of scattering matrix or equivalent network representations for waveguide discontinuities was first carried out within the last decade [20]. This investigation of reflection from an open-ended parallel plane waveguide (Fig. 3) provided lowest order modal reflection coefficients based on singly diffracted rays, and improved values incorporating also multiple diffraction between the two edges. The calculation of multiple diffraction is complicated by the fact that the interaction path lies along the shadow boundary of the geometrically reflected field. The first attempt to deal with this problem by postulating an equivalent isotropic line source to represent an n-th order diffracted field emanating from one edge and traveling to the other edge [20] led to remark-

---

[1]Note, however, the importance of incorporating lateral ray shifts when the boundaries have an incidence-angle-dependent reflection coefficient (see Sec. II.A.6).

ably accurate numerical values for the modal reflection coefficients. However, it was shown subsequently [21] that the multiple diffraction fields so generated do not agree with the asymptotic expansion of the exact solution. When the multiple diffraction process is analyzed by the uniform asymptotic theory of edge diffraction, which avoids the isotropic line source postulate, the defect in [20] is removed [22]. This important demonstration of the validity of the ray method when applied properly to a complicated problem of multiple interaction between canonical (single-edge scattering) constituents has significantly enlarged the class of scattering configurations for which ray techniques have been found to supply correct asymptotic solutions.

The investigations in parallel plane waveguides [23,24] have been generalized to include multiwave media [25], inhomogeneous media [26], and waveguides with non-planar boundaries [27].[2] In conjunction with the latter, it has been necessary to provide a ray optical description of modal propagations in non-uniform guiding regions. The strict definition of modal fields requires the waveguide boundaries to conform to constant coordinate surfaces in a separable coordinate system, since only then can the longitudinal and transverse field dependance be separated one from the other. When the boundaries are curved or non-parallel, the asymptotic modal fields are generated by modal caustics which belong either to the family of curves descriptive of the waveguide walls or to the family orthogonal thereto. In the latter event, a portion of the waveguide is devoid of propagating fields and is penetrated only by evanescent waves (see Fig. 4). By introducing the concept of local separability, one may remove the restriction to separable boundaries and accommodate more general configurations.

---

[2] Some multiple scattering problems in these configurations require the application of uniform asymptotic methods; this has not yet been done.

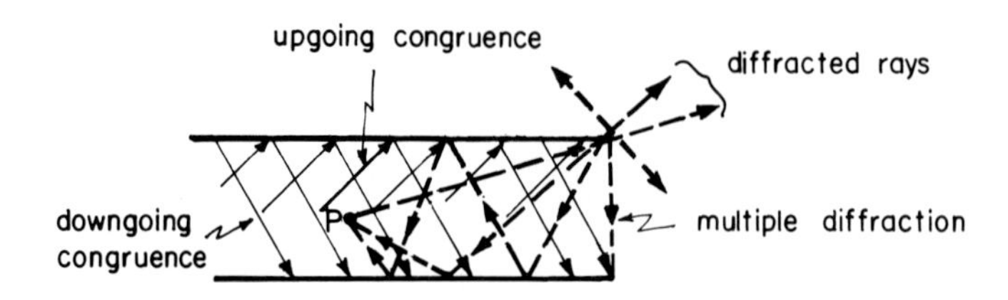

Fig. 3. Ray methods applied to radiation and reflection from an open-ended waveguide. The incident modal field is decomposed into upgoing and downgoing ray congruences. The edge diffracted rays, shown dashed, reach an observation point P by direct and multiply reflected paths. Corresponding rays from the lower edge are not shown. The multiply reflected fields are converted into waveguide modes by Poisson summation. The modal reflection and coupling coefficients are based on singly diffracted rays and on rays experiencing multiple diffraction between the two edges.

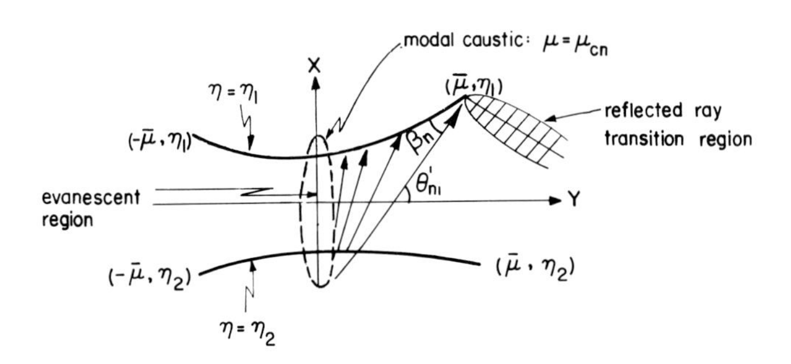

Fig. 4. Modal ray congruence in non-uniform homogeneously filled waveguide. Only the upgoing congruence is shown. The caustic would lie exterior to the waveguide region. When the waveguide is truncated at $\mu = \pm\bar{\mu}$, the resulting open-ended configuration serves as a model for an (unstable) optical resonator for high power laser systems. The waveguide walls then become the reflecting mirrors whose large axial separation far exceeds the mirror width. Edge effects in this resonator can be calculated from (39). If the edge at $(\bar{\mu}, \eta_1) \to (\mu', \eta')$ is regarded as an equivalent non-isotropic source, then θ'_{n1} represents the angle appearing in (36) and β_n represents the angle appearing in (40). If the waveguide medium is inhomogeneous, the same considerations apply but the modal rays follow curved trajectories.

B. Formulation for Inhomogeneous but Separable Geometries

Basic to the ray-optical treatment of radiation and diffraction in a waveguide environment is the ability to determine the excitation of guided modes by a non-isotropic radiating source since such a source can be employed to represent the far zone diffraction pattern of an edge discontinuity or other scattering center located in the guiding region. The solution can be constructed from the corresponding one for an isotropic source, i.e., the waveguide Green's function. We assume a two-dimensional geometry that may comprise curvilinear boundaries and an inhomogeneous interior but subject to the restriction that the wave equation is separable in this configuration. If the boundaries are located at $\eta = \eta_1$ and $\eta = \eta_2$ in an orthogonal (μ,η) curvilinear coordinate system and the source is placed at $\underset{\sim}{r}' = (\mu', \eta')$, the two-dimensional scalar Dirichlet Green's function can be represented as [Ref. 8, Sec. 3.3]

$$G(\underset{\sim}{r},\underset{\sim}{r}') = \sum_n \frac{\Phi_n(\eta)\Phi_n(\eta')}{N_n^2} g_n(\mu,\mu') , \quad \underset{\sim}{r} = (\mu,\eta), \tag{28}$$

where $\Phi_n(\eta)$ are the eigenfunctions satisfying the boundary conditions $\Phi_n(\eta_{1,2}) = 0$ on the waveguide walls and N_n is their normalization constant. The one-dimensional Green's function $g_n(\mu,\mu')$, descriptive of propagation along the waveguide axis, is synthesized in terms of traveling wave solutions $F_{n1}(\mu)$ and $F_{n2}(\mu)$ of the source-free μ-domain.

To synthesize the field $u(\mu,\eta)$ due to a non-isotropic source with radiation pattern $f(\theta)$, we assume that the source is comprised of currents $J(\mu,\eta)$ throughout an area A centered at (μ',η'). Then

$$u(\mu,\eta) = \sum_n \frac{\Phi_n(\eta)}{N_n^2} \int_A J(\hat{\mu},\hat{\eta})\Phi_n(\hat{\eta})\, g_n(\mu,\hat{\mu})d\hat{A} \tag{29}$$

where $d\hat{A} = h_1 h_2 d\hat{\mu} d\hat{\eta}$ is an area element and $h_{1,2}$ are metric coefficients. Let the standing wave functions $\Phi_n(\eta)$ be expressed in terms of constituents $\Phi_{n1}(\eta)$ and $\Phi_{n2}(\eta)$ which, asymptotically, describe traveling waves:

$$\Phi_n(\eta) = \Phi_{n1}(\eta) + \Phi_{n2}(\eta), \tag{30}$$

where $\Phi_{n1}(\eta)$ and $\Phi_{n2}(\eta)$ behave asymptotically according to the WKB forms

$$\Phi_{n1,2}(\eta) \sim \mp \exp[\mp ik \int_{\eta}^{\eta_1} \psi_n(\xi)d\xi]/2i[\psi_n(\eta)]^{\frac{1}{2}} \tag{30a}$$

with $\psi_n(\eta)$ representing the local plane wave propagation coefficient in the η domain. Then (29) becomes

$$u = \sum_n \frac{\Phi_n(\eta)}{N_n^2} (B_{n1} + B_{n2}) \tag{31}$$

where

$$B_{n1,2} = \int_A J(\hat{\mu},\hat{\eta})\; \Phi_{n1,2}(\hat{\eta}) g_n(\mu,\hat{\mu}) d\hat{A} . \tag{32}$$

If the waveguide cross section and interior medium vary slowly over the source region A, and if the source region is removed from the modal caustics (if any) whereon $\psi_n(\eta) \to 0$, then one may employ (30a) and expand the modal functions about η':

$$\Phi_{n1,2}(\hat{\eta}) \sim \Phi_{n1,2}(\eta')\exp[\pm ik\psi_n(\eta')(\hat{\eta} - \eta')] \tag{33}$$

Similarly, using the WKB form for $g_n(\mu,\mu')$, one has

$$g_n(\mu,\hat{\mu}) \sim g_n(\mu,\mu') \exp[ik\varphi_n(\mu')(\hat{\mu} - \mu')] \tag{34}$$

where $\varphi_n(\mu)$ is the local plane wave propagation coefficient in the μ domain. Thus, (32) becomes

$$B_{n1,2} \sim f(\pi + \theta'_{n1,2})\; \Phi_{n1,2}(\eta') g_n(\mu,\mu') \tag{35}$$

where

$$f(\pi + \theta'_{n1,2}) \equiv \int_A J(\hat{\mu},\hat{\eta})\exp[\pm ik\psi_n(\eta')(\hat{\eta} - \eta') + ik\varphi_n(\mu')(\hat{\mu} - \mu')]d\hat{A} \tag{36}$$

The integral in (36) may be shown to express the free space far zone radiation pattern of the assumed current distribution,

$$f(\pi + \theta) = \int_A J(\hat{x},\hat{y})\exp[-ik(\hat{x} - x')\sin(\pi + \theta) - ik(\hat{y} - y')\cos(\pi + \theta)]d\hat{x}\,d\hat{y} \tag{37}$$

when transformed into the (μ,η) coordinates. The observation angle θ is measured from the positive y axis toward the positive x axis (Fig. 4). When applied to (36), there appear the angles $\theta'_{n1,\,2}$, the characteristic propagation angles of the modal plane wave constituents in (30a) at the source location (μ',η'). The characteristic propagation angles are defined by

$$\sin\theta_{n1,2} = \underset{\sim}{x}_o \cdot \nabla[\int^{\eta} \psi_n(\xi)d\xi \pm \int^{\mu} \varphi_n(\xi)d\xi], \tag{38}$$

where $\underset{\sim}{x}_o$ is a unit vector in the x direction. Thus, the excitation strength of the nth mode depends on the strength of the radiation pattern of the source in the directions of the modal rays at the source location. When (35) is substituted into (31), one has the final result

$$u \sim \sum_n \frac{f(\pi + \theta'_{n1})\Phi_{n1}(\eta') + f(\pi + \theta'_{n2})\Phi_{n2}(\eta')}{N_n^2}\,\Phi_n(\eta)g_n(\mu,\mu'), \tag{39}$$

which could also have been derived from purely ray-optical considerations by converting the multiply reflected ray contributions into guided modes [26]. This formula is basic for the ray-optical treatment of scattering by edge discontinuities or other localized scattering centers in a relatively arbitrary waveguide environment. Note that the mode functions and longitudinal

Green's function have here been retained in their rigorous form, with asymptotic considerations applied only to the calculation of the modal amplitudes. The traveling wave functions $\Phi_{n1,2}$ can be defined in terms of a superposition of the standing wave functions $\Phi_n(\eta)$ and their derivatives $d\Phi_n(\eta)/d\eta$ so as to satisfy the asymptotic behavior in (30a).

The result in (39) has been used for calculation of modal excitation, reflection, and coupling in a variety of applications ranging from waveguide discontinuities in conventional waveguides [24] to scattering in the earth-ionosphere waveguide [28]. Because the procedure utilizes rigorous diffraction coefficients in the evaluation of $f(\theta)$, even the primary diffraction calculations for sharp-edged apertures or obstacles are an improvement over results commonly derived by the Kirchhoff approximation. By analytic continuation to complex modal angles, the expression in (39) can be employed for leaky modes, evanescent modes, or modes with dissipation losses. An interesting recent application is to the propagation and diffraction of Gaussian beams, which simulate closely the output fields of laser oscillators. Since a Gaussian beam has a far zone pattern as in (6), insertion of that pattern function into (39) determines the coupling from an incident beam in a layered medium to the guided modes. These considerations are relevant to integrated optical systems.[3]

C. Application to Unstable Optical Resonators

The injection of quasi-optic waveguide considerations into the analysis of open optical resonators for laser oscillators has provided another timely application. Here, the region containing the active lasing material (usually encased in a large tubular configuration) is terminated at both ends by reflecting mirrors,

[3] Alternatively the Gaussian beam problem can be tackled by complex displacement of the coordinate location of an isotropic point radiator; see Sec. II.A.3.

between which the resonant oscillations are established. Depending on the shape of the mirrors, these oscillations can be stable or unstable; in the former case, the fields are confined to the axial region and reach the edges only by evanescent coupling (Fig. 5) while in the latter case, the edges are strongly illuminated (Fig. 4). The mirrors are very large compared to the optical wavelength and their axial separation in turn is very large compared to their width. Thus, the waveguide so formed supports a huge number of guided modes. What makes the waveguide method tractable is the fact that the modes of interest are those near cutoff since their modal rays propagate essentially parallel to the resonator axis (x-axis); modes with ray congruences strongly inclined with respect to the resonator axis escape from the resonator and hence suffer high losses. Furthermore, it is found that although the edge couples all of the waveguide modes, two modes and sometimes a single mode are adequate to describe the resonant behavior [27,29].

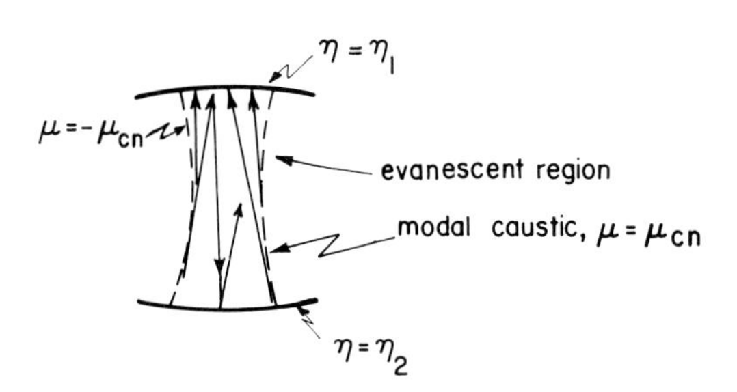

Fig. 5. Stable optical resonator.

To establish the resonance properties by the waveguide approach, one applies the transverse resonance procedure familiar from microwave network theory. To this end, it is necessary to have available the modal reflection coefficient from the caustic and from the mirror edges. The former can be calculated by uniform asymptotic methods (note that the caustic in Fig. 4 is generally so narrow that the two branches cannot be regarded as isolated [27]), while the latter follows by the technique described above. We consider the unstable configuration in Fig. 4, with symmetrical mirrors $\eta_2 = \eta_1$. With a suitably normalized incident field in the j-th mode given by $\Phi_j(\eta)F_{j1}(\mu)$, the reflected field in the n-th mode is $\Gamma_{jn}\Phi_n(\eta)F_{n2}(\mu)$, where the reflection or coupling coefficient Γ_{jn} has the form [27]:

$$\Gamma_{jn} \propto \frac{V(\beta_n,\beta_j)[1 + (-1)^{n+j}]}{[\psi_n(\eta_1)\psi_j(\eta_1)\varphi_n(\bar{\mu})\,\varphi_j(\bar{\mu})]^{\frac{1}{2}}} \qquad (40)$$

Here, β_n is the angle between the tangent plane at the edge and the ray in the n-th mode congruence passing through the edge (Fig. 4), while $V(\beta_n,\beta_j)$ is the diffraction coefficient for a perfectly conducting half plane illuminated by a unit strength plane wave incident at the angle β_j:

$$V(\beta_n,\ \beta_j) = -\sec\frac{\beta_n - \beta_j}{2} + \sec\frac{\beta_n + \beta_j}{2}, \qquad \beta_n \pm \beta_j \not\approx \pi. \qquad (41)$$

The coupling coefficient in (40) accounts only for single diffraction at the edges and is valid only when the reflected ray transition region in Fig. 4 does not intercept the opposite edge. This implies sufficiently large mirror curvature. Because of the large separation of the mirrors, multiple interaction between the edges can be ignored under these circumstances. As the mirror curvature is decreased so that the transition region intercepts the opposite edge, the mirror configuration approaches the parallel plane case. Multiple interaction now becomes important and can be accounted for approximately by local parallel plane modeling of the region near the edges and use of the interaction function for an open-ended parallel plane waveguide [30]. Since it can be shown that the local parallel plane formulation and the single diffraction formulation in (40) have overlapping regions of applicability, the modal coupling coefficients are available for arbitrary mirror curvatures [27].

When the resonant mode losses (i.e., the complex eigenvalues) as a function of mirror size and shape are calculated by insertion of the caustic and edge reflection coefficients into the transverse resonance relation, one obtains remarkably good agreement with data from numerical solution of the integral equation for the currents induced on the mirror surfaces [31]. This provides justification for the use of the ray-optical method in analyzing optical cavities by the waveguide approach.

IV. GUIDED MODES

Asymptotic methods in a free space environment are based on the description of the propagation process in terms of local plane wave fields (see (27)). When field solutions are sought in guiding regions such as dielectric layers or inhomogeneous ducts, it may be advantageous to proceed from an asymptotic expansion incorporating local guided modes. By a guided mode ansatz, one accounts compactly for the local field dependence in the layer cross section, whose height is measured by y, and restricts the ray trajectories essentially to curves in the (x,z) plane followed by the local modal fields. This is in contrast to the ray-optical description of guided modes in terms of caustic-generated ray congruences that are reflected self-consistently at the layer boundaries so as to satisfy the imposed boundary conditions (see Fig. 4; similar considerations of self-consistency apply when guiding takes place by refraction in a continuously stratified medium).

The guided mode expansion takes the form [32,33]

$$u(\underset{\sim}{r}) \sim \sin[\gamma(x,z)y]\, e^{ik\sigma(x,z)} \left[A(x,z) + \sum_{n=1}^{\infty} A_n(\underset{\sim}{r})/(ik)^n\right] \qquad (42)$$

where $y = \pm H(x,z)$ defines the waveguide boundaries on which a surface impedance condition may be specified. Insertion of (42) into the wave equation and boundary conditions yields, on proceeding as noted in connection with (27), a local modal resonance equation for γ, a modal eikonal equation for σ, and transport equations for the amplitude coefficients.

When the guiding region is a straight duct with continuous transverse stratification, the transverse exponential decay of the modal fields can be incorporated into a different ansatz that utilizes the theory of evanescent waves [34]. This ansatz resembles that in (27), with ψ complex, and with the imposition of the following modal constraints: linear phase variation along the propagation direction, planar phase fronts, and modal amplitudes that depend only on the transverse coordinates. Under these conditions, it is found that the transport equations can be integrated explicitly and generate the exact amplitude coefficients to any order n. This assertion has been verified [34] by comparison of the evanescent wave solution with the asymptotic expansion of the exact solution of the wave equation for canonical duct profiles such as the parabolic or the hyperbolic tangent. An advantage of the complex phase formulation is that it avoids the need for treatment of caustics which arises when ψ is real.

The dependence of a guided mode on a preferred propagation direction can also be exhibited by separation of a longitudinally dependent phase from the modal field, with the remaining wave function then satisfying a parabolic equation.

V. QUASI-PERIODIC STRUCTURES

A new application of ray optical techniques has been to the analysis of periodic and quasi-periodic structures on curved surfaces, with emphasis on conformal antenna arrays. Here, the local environment that describes propagation on, radiation from, or scattering by such a structure involves not the smooth surface, which is perturbed by discrete scattering centers, but rather the collective effects of the periodically loaded surface. Consequently, an incident plane wave is reflected not only along the specular direction but also, if the interelement spacing exceeds a half wavelength, along directions corresponding to the higher order grating (or periodic structure) modes. Creeping waves along such a surface have propagation speeds and decay rates modified by periodicity. If the element spacing is large enough to admit higher order grating rays, then the creeping waves do not only shed energy along the forward tangential direction away from the surface but also along the characteristic angles of the grating rays.

These considerations are illustrated in Fig. 6 for the case of radiation from a single slit element in a finite two-dimensional array of quasi-periodic slits on a curved surface. The canonical problems required for solution by the ray optical technique are as follows: a) determination of the propagation along, and radiation from, a single element on a periodically loaded plane surface (see Fig. 7a); b) determination of the properties of the creeping waves (surface rays) when the surface is deformed into a circular cylinder (Fig. 7b); c) determination of the effects of the array edge, i.e. scattering from the end of a planar semi-infinite periodic array; this involves cylindrical wave radiation into space as well as reflection into the surface ray (Fig. 7c). The composite contributions to the far zone radiated field are schematized in Fig. 8.

The ray optical model for propagation along, and radiation from, the conformal slit array as depicted in Figs. 6-8 has been remarkably successful in providing an interpretation of phenomena that are otherwise obscured when the scattering process is regarded as multiple interaction between individual elements. Details may be found in references [35,36].

VI. CONCAVE SURFACES

Canonical problems still remain within the framework of the conventional GTD, and they continue to arise in connection with

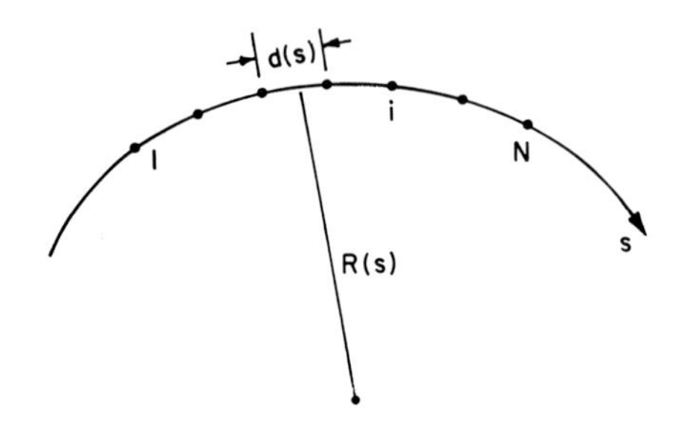

Fig. 6. Quasi-periodic N element array with variable spacing d(s) on a convex perfectly conducting surface with variable radius of curvature R(s). Each dot represents a narrow slit with its feed structure from below the surface.

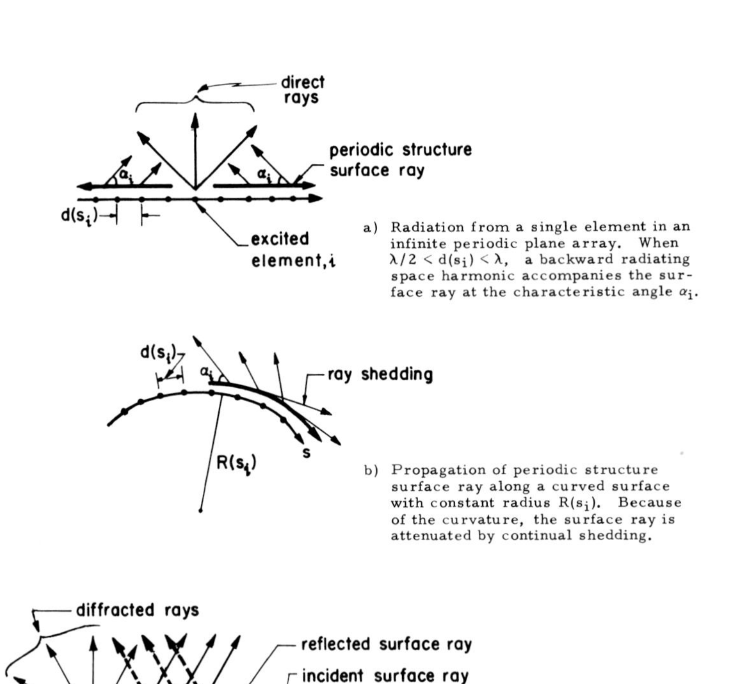

a) Radiation from a single element in an infinite periodic plane array. When $\lambda/2 < d(s_i) < \lambda$, a backward radiating space harmonic accompanies the surface ray at the characteristic angle α_i.

b) Propagation of periodic structure surface ray along a curved surface with constant radius $R(s_i)$. Because of the curvature, the surface ray is attenuated by continual shedding.

c) Scattering and reflection by the edge of the array when a surface ray is incident.

Fig. 7. Canonical problems for radiation from ith element in Fig. 1. The parameters for the local environment near the excited element are denoted by subscript 1.

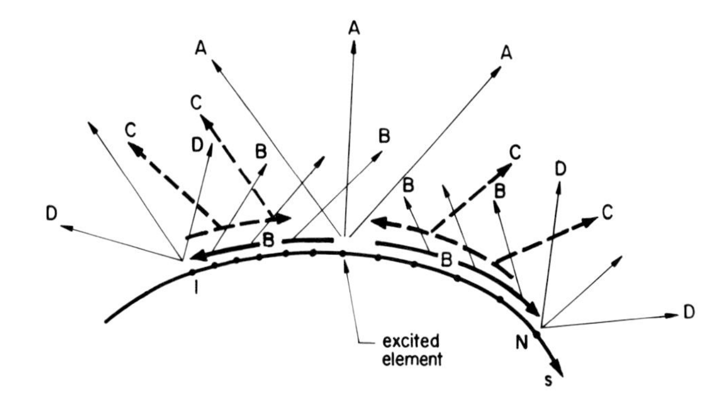

Fig. 8. Ray contributions to the far zone field of a single excited element. A - direct rays; B - source excited surface rays; C - edge reflected surface rays; D - edge diffracted rays. Additional contributions from species C and D due to multiple reflection and diffraction between the two array edges are not shown. Transition regions surround the horizon of the excited element and the directions along which rays A and B or C and D are parallel.

newly developed asymptotic field representations. One of the more interesting of these involves concave boundary shapes. While the excitation of, and the propagation along, convex surfaces has been thoroughly explored in the technical literature, much less has been done for concave shapes. Nevertheless, concave configurations are of importance for a variety of applications including ground wave propagation over terrain with smooth hills and depressions, scattering from reflectors and similar open structures, scattering from enclosures whose interior is accessible by apertures or entry ports, mutual coupling in dome-shaped conformal antenna arrays, and others. Of special interest, and most difficult to analyze, are the fields observed on or near the surface when the source (acutal, or induced as for edge scattering) is also located on or near the surface.

The most fundamental difference between the excitation of

concave and convex surfaces of large (compared to the wavelength) radius of curvature is the absence in the former of a geometrical shadow region, from which the source is invisible. This circumstance gives rise not only to a more intricate geometric-optical field comprising multiply reflected rays (Fig. 9) but also, in an alternative guided wave description, to the presence of whispering gallery modes which cling to the surface (Fig. 10) and, in the absence of dissipation, experience no attenuation. Their counterparts on a convex surface, the creeping waves, lose energy by tangential shedding along the propagation path. The problem is complicated further by the fact that ray optics is incapable of providing the field solution since the caustics of multiply reflected rays pile up near the boundary and thus invalidate the geometric-optical field evaluation there. A field representation in terms of whispering gallery modes only (with inclusion of a continuous mode spectrum for some surfaces), while valid, is inconvenient for calculation for large separation of source and observation points on a large-radius surface since the number of modes required can be substantial. These problems do not arise on a convex surface where the distant field in the shadow region is represented compactly by the dominant creeping wave.

To gain a better physical as well as quantitative understanding of these aspects of wave propagation on a concave surface, intensive studies have been carried out on the simplest prototype configuration, the interior of a perfectly reflecting circular cylinder excited by an axial line source [37-39]. A peculiarity of the cylindrical geometry is the presence, in addition to the whispering gallery modes, of a continuous guided mode spectrum which arises because of spurious reflections from the radial coordinate origin. Elimination of these spurious contributions leads to an asymptotic field representation in terms of an integral which can be manipulated so as to exhibit ray-optical contributions, whispering gallery mode contributions, a mixture of these, or a formulation containing a reduced canonical integral

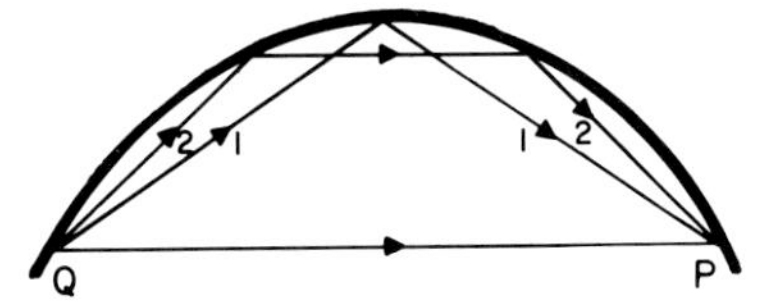

Fig. 9. Direct and multiply reflected rays.

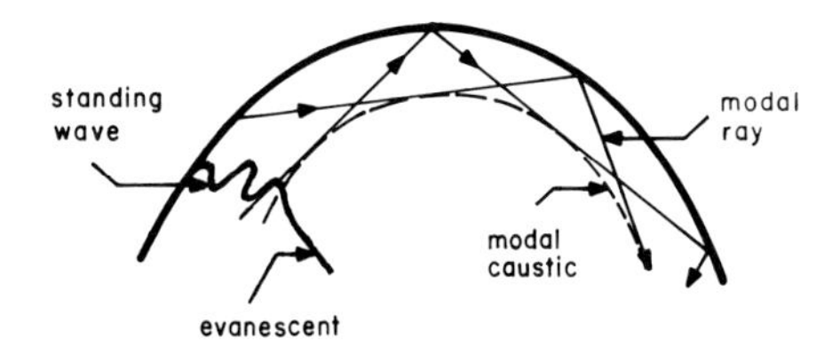

Fig. 10. Whispering gallery mode: modal field, modal rays and caustic.

analogous to the Fock integral for convex surfaces. The most effective choice depends on the parameters of the problem. The results so obtained can be generalized to apply to arbitrary concave surface shapes provided that the radius of curvature changes slowly over a wavelength interval. The validity of these variable-curvature solutions has been verified by comparison with exact calculations performed for a parabolic contour [40]. Results have also been obtained for surface impedance boundary conditions and for the transition from flat surfaces to concave surfaces with very large radius of curvature [41].

VII. TRANSIENT FIELDS

Non-time harmonic fields are playing an increasingly important role in applications involving electromagnetic and acoustic propagation and diffraction. Transient field phenomena in various environments are of interest: in nondispersive media, in dispersive media such as the ionosphere, lossy ground, etc., and in guiding configurations where dispersion enters because of mode guiding effects. No attempt is made here to review progress in this area since several surveys have recently been published [2,42]. It shall merely be noted that emphasis is being placed on direct asymptotic treatment of the time-dependent problem rather than on the conventional, and more cumbersome, route of Fourier or Laplace inversion of time-harmonic results.

VIII. SUMMARY

A principal objective of this review of asymptotic methods in propagation and scattering has been to convey that this subject area remains vital despite the fact that some phases have reached a high level of sophistication. New problems arise from various physical configurations that are of interest in emerging technologies or have been of long concern in traditional applications. The challenges involve the specification of the local wave phenomena that describe the fields in the relevant parameter range most succinctly and thereby lead to a rapidly convergent field expansion. For example, by including periodic effects in the local description, one eliminates the need for performing a multiple diffraction analysis between periodically spaced scattering elements, however at the expense of requiring a more complicated local plane wave field. By employing local evanescent waves with complex phase, one may avoid consideration of modal caustics in ducted propagation. By utilizing a mixture of ray and whispering gallery mode fields, one may describe more efficiently the excitation of concave surfaces, etc. It may be anticipated that the need for solution of new problem areas will give rise to new asymptotic field representations. For wave problems in complicated "non-canonical" environments, asymptotic techniques often afford the only tractable approach.

Closely related to various asymptotic expansions based on local field types is the development of uniform representations in parameter regimes where the basic local wave types are inadequate. A well known example is the failure of the local plane wave ansatz in transition regions surrounding foci, caustics and shadow boundaries. This difficulty can be removed by modifying the ansatz so as to incorporate in it the transition behavior. For the caustics, this requires use of Airy functions and for the shadow boundary behind a sharp-edged object the use of Fresnel integrals. Corresponding uniform expansions are required in regions of failure of other asymptotic descriptions based on fields that differ from the simple local plane wave type. The generation of a reservoir of uniform asymptotic representations for the various basic expansions poses challenging problems that have yet to be addressed.

While asymptotic field representations involve series expansions of successively decreasing terms ordered according to a suitabily defined large parameter, principal emphasis is centered on the leading term. Consideration of the usually much more complicated higher order terms is important for establishing the overall quality of the approximation but their inclusion for quantitative improvement of the leading term results is often of doubtful value. Therefore, careful judgement should be exercised before embarking on the calculation of higher order terms in an expansion. It may be much more fruitful to extract the dominant field behavior as expressed by the leading term from the overall solution and use numerical techniques for evaluation of the correction field. The interplay of analytical asymptotic and numerical procedures appears especially promising and may be expected to receive a good deal of attention.

A most encouraging aspect has been the passage of asymptotic methodology from the applied mathematician, who has been responsible for its development, to the user involved with practical problems. Ray methods as incorporated into the geometrical theory of diffraction are becoming standard fare for radar cross section and propagation specialists. Ray methods in a more general sense, applied to local wave types that differ from the simple plane waves in the geometrical theory of diffraction, may likewise be expected to find their way to those dealing with applications. As the interface between theoretician and user is further developed, one may look forward to a fertile era in asymptotic theory wherein the motivation for developing new methods and solutions derives from the need for practical answers.

IX. REFERENCES

1. Borovikov, V.A., and B. Ye. Kinber, *Proc. IEEE 62,* 1416-1437 (1974).
2. Kravtsov, Yu. A., L.A. Ostrovsky and N.S. Stepanov, *Proc. IEEE 62,* 1492-1510 (1974).
3. Knott, E.F., and T.B.A. Senior, *Proc. IEEE 62,* 1468-1474 (1974).
4. Deschamps, G.A., *Proc. IEEE 60,* 1022-1035 (1972).
5. Deschamps, G.A., *Electron. Lett. 7,* 684-685 (1971).
6. Arnaud, J.A., *Appl. Opt. 8,* 1909-1917 (1969).
7. Felsen, L.B. "Symposia Matematica", Istituto Nazionale di Alta Matematica, Vol. XVIII, pp. 40-56. Academic Press, New York and London (1976).
8. Felsen, L.B. and N. Marcuvitz, "Radiation and Scattering of Waves," Sec. 5.2. Prentice Hall, Englewood Cliffs, New Jersey (1973).
9. Shin, S. Y. and L.B. Felsen, *Appl. Phys. 5,* 239-250 (1974).
10. Felsen, L.B. and S.Y. Shin, *IEEE Trans. Microwave Theory & Tech.* Special Issue on Integrated Optics and Optical Waveguides, *MTT-23,* 150-161 (1975).
11. Felsen, L.B. and S.Y. Shin, Gaussian Beam Tracking in an Optical Fiber, *IEEE/AP-S and USNC/URSI Symposium,* Boulder, Colorado (October 1975).
12. Felsen, L.B. and S.Y. Shin, Guided Mode Excitation in an Optical Fiber by a Gaussian Beam, to be submitted.
13. Bertoni, H., A. Green and L.B. Felsen, Evanescent Plane Wave Shadowing by an Edge, published in *J. Opt. Soc. Am.*
14. Choudhary, S., and L.B. Felsen, *IEEE Trans. Antennas & Propag. AP-21,* 827-842 (1973).
15. Choudhary, S., and L.B. Felsen, *Proc. IEEE,* Special Issue on Rays and Beams, *62,* 1530-1541 (1974).
16. Felsen, L.B., *J. Opt. Soc. Am. 66,* 751-760 (1976).
17. Felsen, L.B. and S. Choudhary, *Nouv. Rev. Opt. 6,* 297-301, (1975).
18. Felsen, L.B., *Philips Res. Rep. 30,* 187-195 (1975).
19. Wang, W.Y.D. and G. Deschamps, *Proc. IEEE 62,* 1541-1551 (1974).
20. Yee, H.Y., L.B. Felsen and J.B. Keller, *SIAM. J. Appl. Math. 16,* 268-300 (1968).
21. Bowman, J.J., *SIAM. J. Appl. Math. 18,* 818-829 (1970).

22. Boersma, J., *Proc. IEEE 62,* 1475-1481 (1974).
23. Yee, H.Y. and L.B. Felsen, *IEEE Trans. Microwave Theory & Tech. 17,* 73-85 (1969).
24. Yee, H.Y. and L.B. Felsen, *IEEE Trans. Microwave Theory & Tech. 17,* 671-683 (1969).
25. Chen, L.W. and L.B. Felsen, *SIAM. J. Appl. Math. 27,* 138-158 (1974).
26. Batorsky, D.V. and L.B. Felsen, *Radio Sci. 6,* 911-923, (1971).
27. Felsen, L.B. and C. Santana, Ray Optical Calculation of Edge Diffraction in Unstable Resonators, to be published in *IEEE Trans. Microwave Theory & Tech.*
28. Batorsky, D.V. and L.B. Felsen, *Radio Sci. 8,* 547-557 (1973).
29. Chen, L.W. and L.B. Felsen, *IEEE J. Quantum Electron. QE-9,* 1102-1113 (1973).
30. Weinstein, L.A., "The Theory of Diffraction and the Factorization Method," Golem Press, Boulder, Colorado (1969). Appendix B and Problem 1.6.
31. Sanderson, R.L. and W. Streifer, *Appl. Opt. 8,* 2129-2136 (1969).
32. Rulf, B., *J. Eng. Math. 4,* 261-271 (1970).
33. Ahluwalia, D.S., J.B. Keller and B. J. Matkowsky, *J. Acoust. Soc. Am. 55,* 7-12, (1974).
34. Choudhary, S. and L.B. Felsen, Asymptotic Method for Ducted Propagation, to be published this year in the *J. Acoust. Soc. Am.*
35. Shapira, J., L.B. Felsen and A. Hessel, *IEEE Trans. Antennas & Propag. AP-22,* 49-63, (1974).
36. Shapira, J., L.B. Felsen and A. Hessel, *Proc. IEEE,* Special Issue on Rays and Beams, *62,* 1482-1492 (1974).
37. Kinber, B. Ye., *Radiotekh. & Elektron. 6,* 1273-1283 (1961).
38. Babich, V.M. and V.S. Buldyrev, "Asymptotic Methods of Short Wave Diffraction," Nauka, Moscow (1972). Chapter II, Sec. 4.
39. Wasylkiwskyj, W., *IEEE Trans. Antennas & Propag. AP-23,* 480-492 (1975).
40. Buldyrev, V.M. and A.I. Lanin, Asymptotic Formulas for a Wave Propagating along a Concave Surface. Limits of their Applicability, *Radiotekh. Elektron. 20* (1975).
41. Felsen, L.B. and A. Green, Excitation of Concave Surfaces, *International IEEE/AP-S Symposium,* Amherst, Mass. (October 1976).
42. Felsen, L.B. (editor), "Transient Electomagnetic Fields, Topics in Applied Physics, Vol. 10, Springer Verlag, New York (1976).

UNIFORM ASYMPTOTIC THEORY OF ELECTROMAGNETIC EDGE DIFFRACTION: A REVIEW[1]

S. W. Lee
Department of Electrical Engineering
University of Illinois at Urbana-Champaign

This paper studies the application of ray techniques to edge diffraction of a high-frequency electromagnetic wave. We summarize the theory of geometrical optics, the geometrical theory of diffraction, and the uniform asymptotic theory. Special emphasis is given to the presentation of final results so that they are general, unambiguous, and convenient for applications.

I. INTRODUCTION

This paper is concerned with the asymptotic solution of electromagnetic edge diffraction problems at high frequency (wavenumber $k \to \infty$). A powerful method for attacking such problems is the ray method introduced by Keller in 1957, commonly known as the geometrical theory of diffraction (GTD) [1 - 3]. As in improvement over classical geometrical optics (GO), GTD has been widely used, and has produced many significant results; but it does have a few limitations. From an application viewpoint, a most serious limitation is the infinite field predicted by GTD on the shadow boundaries of incident and reflected fields (denoted hereafter by $SB^{i,r}$). This difficulty can be explained in terms of a boundary layer arising in the asymptotic solution of a differential equation [4]. A boundary layer is a small region in a parameter space

[1]This work was supported by NSF Grant ENG-73-08218

ISBN 0-12-709650-7

where the solution varies rapidly under a small change in the parameters. In an edge diffraction problem, boundary layers exist around $SB^{i,r}$, where the field solution is very sensitive to the observation point $\vec{r}$. The "thickness" of the layers depends on the wavenumber k. As $k \to \infty$, the thickness tends to zero and the layers collapse on $SB^{i,r}$. Keller's theory is based on an expansion valid outside the layers, called outer expansion. Therefore it cannot be applied to calculate the field next to $SB^{i,r}$. The inner expansion for diffraction problems valid at $SB^{i,r}$ (or at caustics) has been studied by Buchal and Keller [5] for scalar waves. Thus, following their approach, one would generally obtain several different expressions for the field solutions: one valid outside the boundary layers and others inside.

To overcome the difficulty at $SB^{i,r}$, a different approach called the uniform asymptotic theory of edge diffraction (UAT) was recently developed. Starting from a new *Ansatz* that involves Fresnel integrals, UAT yields a single field expression that is uniformly valid across $SB^{i,r}$, thus avoiding separate expansions of the boundary layer technique altogether. When the diffracting edge is that of an (infinitely) thin screen, UAT further improves GTD in the following two aspects: (i) There exists a systematic method in UAT for calculating all the higher-order terms in the field solution beyond terms of order $k^{-1/2}$ (relative to the incident field). (ii) Excluding caustics, the total field solution of UAT is uniformly valid for all points in space, including the edge where the UAT solution satisfied the edge condition. The theoretical developements of UAT are reported in [6-10]. Some applications of UAT are given in [11-18]. It is found that the UAT solution agrees with the high-frequency asymptotic expansion of the exact solution whenever the latter is available. Several test problems have been studied for the purpose of comparing (correlating) UAT with other high-frequency techniques. The comparison of UAT with the technique used in [19] is given in [12], [13]; [20] in [14]; [21] in [15]; [22] in [10]; [23-26] in [17]; [27] in

[28]; and [29] in [30].

It should be emphasized that UAT is not a totally new theory on high-frequency edge diffraction. It is an extension (improvement) of GTD, in the same manner as GTD is an extension of classical geometrical optics (GO). In fact, to construct a UAT solution, one must first obtain the solutions by GO and GTD.

In this paper, we attempt to give a comprehensive review of UAT, with special emphasis on the summary of final results useful for general applications. Mathematical derivations and manipulations are normally omitted. We try to collect all important formulas in GO, GTD, and UAT, and present them in a unified manner. The paper consists of three major parts. In the first part (sections II and III), some general mathematical background and the asymptotic solution of Maxwell's equations are reviewed. Next, in Sections IV through VI, we build up the theory of UAT from GO and GTD, with a concluding remark in Section VII. The Appendix (not included in this book; it may be obtained from the author) contains a list of some 300 references on ray techniques in edge diffraction problems.

Several conventions are used throughout this paper. They are listed below: (i) The time factor is $\exp(-i\omega t)$ and is suppressed. (ii) The total (electric) field solution is denoted by $\vec{E}^t$. According to GO, GTD, or UAT, $\vec{E}^t$ always consists of two parts: $\vec{E}^{ti}$, the part associated with the incienet field $\vec{E}^i$, and $\vec{E}^{tr}$, associated with the reflected field $\vec{E}^r$. Since there exists a perfect symmetry between $\vec{E}^{ti}$ and $\vec{E}^{tr}$, only $\vec{E}^{ti}$ is studied in detail. (iii) $A^{i,r} = \mp B^{i,r}$ means $A^i = -B^i$ and $A^r = +B^r$.

II. MATHEMATICAL BACKGROUND

A. Asymptotic Expansion

The behavior of a function $f(k)$ for large values of k may conveneintly be described by its asymptotic expansion. In the simplest case, such an asymptotic expansion is a formal power series

$$\sum_{m=0}^{\infty} a_m k^{-m} = a_0 + a_1 k^{-1} + a_2 k^{-2} + \ldots + a_m k^{-m} + \ldots \tag{1}$$

in inverse powers of k. Let its partial sums be denoted by

$$f_M(k) = \sum_{m=0}^{M} a_m k^{-m} \tag{2}$$

Then the series (1) is said to be the *asymptotic expansion* of $f(k)$ if for any M

$$\lim_{k\to\infty} k^M [f(k) - f_M(k)] = 0; \tag{3}$$

the usual notation is

$$f(k) \sim \sum_{m=0}^{\infty} a_m k^{-m} \quad , \quad k \to \infty \; . \tag{4}$$

Using the symbol o (small oh) and O (big oh), (3) may be rewritten as

$$f(k) = \sum_{m=0}^{M} a_m k^{-m} + o(k^{-M}) \quad , \quad k \to \infty \; , \tag{5a}$$

and equivalently

$$f(k) = \sum_{m=0}^{M} a_m k^{-m} + O(k^{-M-1}) \quad , \quad k \to \infty \; . \tag{5b}$$

More generally, the function $f(k)$ may have the asymptotic expansion

$$f(k) \sim \sum_{m=0}^{\infty} a_m \phi_m(k) \quad , \quad k \to \infty \; , \tag{6a}$$

which means that for any M,

$$\lim_{k\to\infty} [f(k) - f_M(k)]/\phi_M(k) = 0 \; , \tag{6b}$$

or equivalently,

$$f(k) = \sum_{m=0}^{M} a_m \phi_m(k) + o[\phi_M(k)] \quad , \quad k \to \infty \tag{6c}$$

$$f(k) = \sum_{m=0}^{M} a_m \phi_m(k) + O[\phi_{M+1}(k)] \quad , \quad k \to \infty \ . \tag{6d}$$

It is understood that $\phi_m(k)$ in (6) is a given function subject to $\phi_{m+1}(k)/\phi_m(k) \to 0$ as $k \to \infty$. From the viewpoint of application, we list below several common properties of asymptotic expansions.

First, for a *fixed* M, the difference between $f(k)$ and $f_M(k)$ can be made arbitrarily small, provided that k is chosen large enough. This is simply a restatement of (3). The reader is undoubtedly familiar with the fact that, in many asymptotic expansions, the first few terms of the series often give surprisingly accurate results even in nonasymptotic regimes. This is so despite the fact that the series in (4) is often divergent.

Second, for a *fixed* k, the difference between $f(k)$ and $f_M(k)$ *cannot* be made arbitrarily small by increasing M. Unfortunately, in almost all the practical problems in electromagnetic theory, k is fixed. In such cases, the higher-order terms of an asymptotic expansion are useful only in the following sense:

(i) For a fixed k, the error of using $f_M(k)$ for $f(k)$ is often bounded by the magnitude of the $(M+2)$th term, the first term neglected.

(ii) For a fixed k, the magnitudes of successive terms in (4) usually first decrease until, say, the $(Q+2)$th term. Starting from the $(Q+3)$th term, the magnitudes become larger and larger, and the series diverge.

For an asymptotic expansion with the above two characteristics, it is obvious that, for a fixed k, there exists a "best" asymptotic approximation of $f(k)$, which is $f_Q(k)$.

Thirdly, for a given function $f(k)$, the asymptotic (power series) expansion in (4) in unique. A useful consequence of this property is that, if $f(k) = 0$, each coefficient a_m in (4) must be zero.

B. FRESNEL INTEGRAL

A special function that will be used frequently in this paper is the Fresnel integral $F(x)$ defined by

$$F(x) = \frac{e^{-i\pi/4}}{\sqrt{\pi}} \int_x^\infty e^{it^2} dt \quad , \quad \text{for real } x. \tag{7}$$

We will list below several useful properties of the Fresnel integral:

(i) Symmetry relation

$$F(x) + F(-x) = 1. \tag{8}$$

(ii) Differentiation

$$\frac{d}{dx} F(x) = \frac{1}{\sqrt{\pi}} \exp\left[i\left(x^2 + \frac{3\pi}{4}\right)\right]. \tag{9}$$

(iii) Series expansion

$$F(x) = \frac{1}{2} - \frac{e^{-i\pi/4}}{\sqrt{\pi}} x \sum_{n=0}^{\infty} \frac{(ix^2)^n}{n!\,(2n+1)} . \tag{10}$$

(iv) Asymptotic expansion

$$F(x) \sim \Theta(-x) + \hat{F}(x) \quad , \quad |x| \to \infty , \tag{11}$$

where

$$\Theta(y) = \begin{cases} 1, & \text{if } y > 0 \\ 0, & \text{if } y < 0 \end{cases} , \tag{12}$$

$$\hat{F}(x) = e^{i(x^2+\pi/4)} \frac{1}{2\pi x} \sum_{n=0}^{\infty} \Gamma(n+\tfrac{1}{2})\,(ix^2)^{-n} \tag{13}$$

$$\Gamma\left(n+\frac{1}{2}\right) = \sqrt{\pi}\left(\frac{1}{2}\right)\left(\frac{3}{2}\right) \cdots \left(\frac{2n-1}{2}\right) . \tag{14}$$

(v) Special values

$$F(-\infty) = 1 \quad , \quad F(0) = \frac{1}{2} \quad , \quad F(+\infty) = 0. \tag{15}$$

(vi) Integral representation

$$F(x) = \frac{e^{ix^2}}{2\pi i}\int_{-\infty}^{\infty} \frac{e^{-t^2}}{t - xe^{i\pi/4}}\, dt \quad , \quad x > 0 \tag{16}$$

$$= \frac{1}{\pi} e^{i(x^2-\pi/4)} \; x\int_{0}^{\infty} \frac{e^{-t^2}}{t^2 - ix^2}\, dt \quad , \quad x > 0 \tag{17}$$

(vii) A slightly different Fresnel integral $f(x)$ is used in many papers of UAT [6-9], [11-14], [17]. It is related to the present $F(x)$ by the relation

$$f(x) = e^{-ix^2} F(-x). \tag{18}$$

The trajectory of $F(x)$ in its complex plane using x as the parameter is presented in Fig. 1. This curve is known as the Cornu spiral (after A. Cornu).

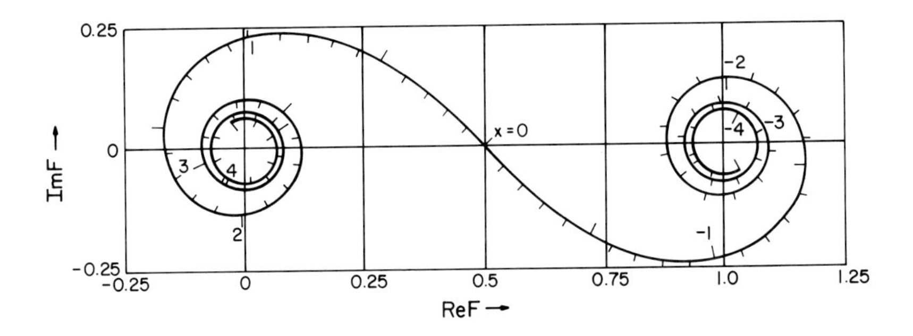

Fig. 1. Cornu spiral for the Fresnal integral F(x).

C. CURVES AND SURFACES

Elementary knowledge concerning curves and surfaces is necessary for the application of ray techniques. It can be learned from any standard textbook on differential geometry. Some important results on curves and surfaces are summarized below for convenient reference [41].

Curve. In electromagnetic edge diffraction, the edge is

assumed to be a smooth curve Γ. We may represent Γ by a parametric equation

$$\Gamma: \quad \vec{r}(\eta) = [x(\eta), y(\eta), z(\eta)] \quad , \quad \eta_1 < \eta < \eta_2 \ , \tag{21}$$

where the parameter η is the arclength of Γ. The (unit) tangent, normal, and binormal of Γ are given by

$$\hat{t} = \frac{d\vec{r}}{d\eta} \ , \tag{22}$$

$$\hat{n} = \frac{d^2\vec{r}}{d\eta^2} \Bigg/ \left| \frac{d^2\vec{r}}{d\eta^2} \right| \tag{23}$$

$$\hat{b} = \hat{t} \times \hat{n} \tag{24}$$

The curvature and torsion of Γ are given by

$$\kappa = \left| \frac{d^2\vec{r}}{d\eta^2} \right| \tag{25}$$

$$\tau = \frac{d\vec{r}}{d\eta} \cdot \frac{d^2\vec{r}}{d\eta^2} \times \frac{d^3\vec{r}}{d\eta^3} \Bigg/ \left| \frac{d^2\vec{r}}{d\eta^2} \right|^2 \tag{26}$$

Note that κ is always positive. The variation of $(\hat{t}, \hat{n}, \hat{b})$ along the curve is governed by the well-known Frenet formulas.

Surface: In electromagnetic reflection from a screen, the screen is assumed to be a smooth surface Σ, represented by a parametric equation with parameters (u,v):

$$\Sigma: \quad \vec{r}(u,v) = [x(u,v), y(u,v), z(u,v)] \quad , \quad u_1 < u < u_2 \text{ and } v_1 < v < v_2 \ . \tag{27}$$

The surface normal at (u,v) is defined by

$$\hat{N} = \mu \frac{\vec{r}_u \times \vec{r}_v}{|\vec{r}_u \times \vec{r}_v|} \tag{28}$$

where $\mu = \pm 1$. In reflection problems, we often choose μ such that $\hat{N}$ points toward the source. The subscript u of $\vec{r}_u$, for example, means the partial derivative of $\vec{r}$ with respect to u.

At any point (other than the umbilic) on Σ, there exists a pair of orthogonal directions such that the radius of curvature of the normal sections of Σ in these directions assumes the maximum and minimum values. These two directions denoted by $(\hat{a}_1,\hat{a}_2)$ are the principal directions, and the two extreme values (R_1,R_2) are the principal radii of curvature. To calculate (R_1,R_2) for a surface Σ as prescribed in (27), the following parameters are needed:

$$E = \vec{r}_u \cdot \vec{r}_u , \quad F = \vec{r}_u \cdot \vec{r}_v , \quad G = \vec{r}_v \cdot \vec{r}_v , \tag{29a}$$

$$e = \vec{r}_{uu} \cdot (\vec{r}_u \times \vec{r}_v)\Delta, \quad f = \vec{r}_{uv} \cdot (\vec{r}_u \times \vec{r}_v)\Delta, \quad g = \vec{r}_{vv} \cdot (\vec{r}_u \times \vec{r}_v)\Delta \tag{29b}$$

$$\Delta = \mu(EG - F^2)^{-1/2} . \tag{29c}$$

Then (R_1,R_2) are the roots of the following equation

$$\frac{1}{R^2} - 2\kappa_M \frac{1}{R} + \kappa_G = 0 , \tag{30}$$

where κ_M is the mean curvature

$$\kappa_M = \frac{1}{2}\left(\frac{1}{R_1} + \frac{1}{R_2}\right) = \frac{1}{2}\,\Delta^2\,(gE - 2fF + eG), \tag{31}$$

and κ_G is the Gaussian curvature

$$\kappa_G = \frac{1}{R_1 R_2} = \Delta^2\,(eg - f^2). \tag{32}$$

The sign of $R_1\,(R_2)$, as computed from the above formula, is positive if the corresponding normal section of Σ bends toward $\hat{N}$, and is negative if it bends away from $\hat{N}$. When Σ is a sphere of radius a with $\hat{N}$ pointing away from its center, we have $R_1 = R_2 = -a$.

Let us consider the approximation of Σ in the neighborhood of a point O on Σ. Introduce a rectangular coordinate system with base vectors $(\hat{x}_1,\hat{x}_2,\hat{x}_3 = \hat{z})$ at O such that $\hat{z} = \hat{N}$. Then a typical point (x_1,x_2,z) on Σ in the neighborhood of O can be represented by

$$\Sigma: \quad z = \frac{1}{2}\begin{bmatrix} x_1 \\ x_2 \end{bmatrix} \cdot \bar{\bar{Q}} \begin{bmatrix} x_1 \\ x_2 \end{bmatrix} + O\left(x_{1,2}^3\right) , \tag{33}$$

where $0(x_{1,2}^3)$ is a short notation for $0(x_1^\mu x_2^\nu)$ with $\mu + \nu = 3$. If $(\hat{x}_1,\hat{x}_2)$ coincide with the principal directions $(\hat{a}_1,\hat{a}_2)$, the curvature matrix $\bar{\bar{Q}}$ is diagonal, and is given by

$$\bar{\bar{Q}} = \begin{bmatrix} 1/R_1 & 0 \\ 0 & 1/R_2 \end{bmatrix}, \tag{34}$$

where (R_1,R_2) are the principal radii of curvature at 0. If $(\hat{x}_1,\hat{x}_2)$ make an angle ψ with $(\hat{a}_1,\hat{a}_2)$, then $\bar{\bar{Q}}$ becomes

$$\bar{\bar{Q}} = \begin{bmatrix} \cos\psi & \sin\psi \\ -\sin\psi & \cos\psi \end{bmatrix}^T \begin{bmatrix} 1/R_1 & 0 \\ 0 & 1/R_2 \end{bmatrix} \begin{bmatrix} \cos\psi & \sin\psi \\ -\sin\psi & \cos\psi \end{bmatrix}, \tag{35}$$

where T is the transpose operator.

In reflection problems, the quadratic approximation in (33) is also used to describe a wavefront W. In the latter case, we choose $\hat{z} = -\hat{N}$, instead of $+\hat{N}$. Then the factor (1/2) in (33) should be replaced by (-1/2) in the corresponding equation for W.

III. ASYMPTOTIC SOLUTION OF MAXWELL'S EQUATIONS

A. Direct and Indirect Asymptotics

The study of electromagnetic diffraction phenomena requires the solution of an appropriate boundary value problem for Maxwell's equations. Only relatively few electromagnetic diffraction problems permit an explicit exact solution, and even then, such a solution is often too complex to be useful, say for engineering applications. Thus from a practical viewpoint, it is important to develop approximate (analytical or numerical) techniques that can be used for different occasions. Throughout this work we are concerned with (high-frequency) *asymptotic methods* which apply to the diffraction of high-frequency electromagnetic fields, that is, fields with a wavelength that is small compared to either a significant dimension of the diffracting object, or to the distance between source and observation point. In the literature, several asymptotic methods have been developed for the solution of Maxwell's euqations. The application of these methods can be either direct or indirect, depending on the stage in the solution process where the asymptotic method comes in.

In the indirect approach, one first has to determine the exact solution of the diffraction problem. Then the asymptotic method is applied to this exact solution, thus yielding the *asymptotic expansion* (in the sense of Section II.A) *of the exact solution.* As an example, let the exact solution be given by an integral representation of the form

$$\vec{E}(\vec{r}) = \int_{-\infty}^{\infty} d\alpha \int_{-\infty}^{\infty} d\beta\, \vec{A}(\alpha,\beta,kz)e^{ik(\alpha x+\beta y)} \quad ; \tag{36}$$

solutions of this type may arise, e.g., in problems with a separable geometry. Frequently, either the above integral cannot be evaluated explicitly, or when it can, the result is too complex to be useful. At a high frequency, i.e., $k \to \infty$, the integral (36) can be treated by well-known asymptotic methods such as the saddle-point method or the method of stationary phase, thus leading to the asymptotic expansion

$$\vec{E}(\vec{r}) \sim \sum_{\nu} (ik)^{-\nu}\, \vec{E}_{\nu}(\vec{r},k) \quad , \quad k \to \infty \ , \tag{37}$$

where $\{\nu\}$ is a set of integers or fractional numbers, and $\{\vec{E}_{\nu}\}$ is, in general, a function of k which is bounded as $k \to \infty$. The present result (37) is the asymptotic expansion of the exact solution of the diffraction problem at hand. Studies on the indirect application of asymptotic methods are well documented in books on electromagnetic theory [31], [32].

In the direct approach, the asymptotic method is applied directly to each of Maxwell's equation at the beginning of the problem, instead of to the solution. Since it is no longer necessary to first determine the exact solution, the direct approach is invariably simpler, and, more importantly, it can be adopted to a much broader class of problems. It is the direct application of asymptotic methods to electromagnetic edge diffraction problems that will be studied in this paper.

Guided by our experience with edge diffraction problems, an "educated" conjecture is that the solution of Maxwell's equations can be represented by an asymptotic series of the form[2]

$$\vec{E}(\vec{r}) \sim k^{\tau} e^{iks(\vec{r})} \sum_{m=0}^{\infty} (ik)^{-m} \vec{e}_m(\vec{r}) \quad , \quad k \to \infty , \tag{38a}$$

$$\vec{H}(\vec{r}) \sim k^{\tau} e^{iks(\vec{r})} \sum_{m=0}^{\infty} (ik)^{-m} \vec{h}_m(\vec{r}) \quad , \quad k \to \infty . \tag{38b}$$

Here the *amplitudes* $\{\vec{e}_m, \vec{h}_m\}$ and the *phase function* $s(\vec{r})$ are the functions of the space variable $\vec{r}$, and are independent of k. When the incident field is assumed to be of order k^0, the exponent τ in (38) takes a value between -1 and 0, depending on the nature of the field represented by the series, e.g., $\tau = 0$ for a geometrical-optics field, $\tau = -\frac{1}{2}$ for an edge-diffracted field. In a given problem, the complete solution for the total field may be a superposition of several asymptotic series with possibly different τ.

In free space the source-free Maxwell's equations (with a time dependence $e^{-i\omega t}$ suppressed) take the form

$$\nabla \times \vec{E} = ikZ\vec{H} \quad , \quad \nabla \cdot \vec{E} = 0 \tag{39a}$$

$$\nabla \times \vec{H} = -ikZ^{-1}\vec{E} \quad , \quad \nabla \cdot \vec{H} = 0 \tag{39b}$$

where $k = \omega(\varepsilon\mu)^{1/2}$, $Z = (\mu/\varepsilon)^{-1/2}$, ε and μ being the constitutive parameters of free space. For the present purpose we replace (39) by the equivalent set of equations

$$\nabla^2\vec{E} + k^2\vec{E} = 0 \tag{40a}$$

---

[2] The use of the asymptotic series in the form of (38) for solving Maxwell's equations was first suggested by R. K. Luneburg in his mimeographed notes on the "Mathematical Theory of Optics" issued by Brown University in 1944; see also M. Kline and I. W. Kay, *Electromagnetic Theory and Geometrical Optics*, Interscience, New York, 1965. The series (38) is sometimes called the Luneburg-Kline expansion.

$$\nabla \cdot \vec{E} = 0 \tag{40b}$$

$$\vec{H} = (ik)^{-1}Z^{-1}\nabla \times \vec{E} \,. \tag{40c}$$

Then by formal substitution of the asymptotic series (38) into (40), we are led to the following equations for the phase s and the amplitudes $\{\vec{e}_m, \vec{h}_m\}$ [3], [7]:

$$(\nabla s)^2 = 1 \qquad \text{(eikonal equation)} \tag{41}$$

$$2(\nabla s \cdot \nabla)\vec{e}_m + \vec{e}_m \nabla^2 s = -\nabla^2 \vec{e}_{m-1} \qquad \text{(transport equation)} \tag{42}$$

$$\nabla s \cdot \vec{e}_m = -\nabla \cdot \vec{e}_{m-1} \qquad \text{(Gauss' law)} \tag{43}$$

$$\vec{h}_m = Z^{-1}[\nabla s \times \vec{e}_m + \nabla \times \vec{e}_{m-1}] \tag{44}$$

where $m = 0,1,2,\ldots,$ and $\vec{e}_{-1} = 0$ by definition. If the phase s and the amplitudes $\{\vec{e}_m, \vec{h}_m\}$ do satisfy (41)-(44), then we call (38) an *asymptotic solution* of Maxwell's equations. Let the partial sums of (38a) be denoted by

$$\vec{E}_M(\vec{r}) = k^{\tau} e^{iks(\vec{r})} \sum_{m=0}^{M} (ik)^{-m} \vec{e}_m(\vec{r}) \;; \tag{45}$$

then it is easily seen that

$$\nabla^2 \vec{E}_M + k^2 \vec{E}_M = O(k^{\tau-M}) \;,\; \nabla \cdot \vec{E}_M = O(k^{\tau-M}) \;,\; k \to \infty \;, \tag{46}$$

for any M. Alternatively, the latter relations may serve as a definition of the asymptotic solution of Maxwell's equations. Notice the difference between the concepts of asymptotic solution and asymptotic expansion: it is obvious from (46) that

$$\nabla^2(\vec{E} - \vec{E}_M) + k^2(\vec{E} - \vec{E}_M) = O(k^{\tau-M}) \;,\; \nabla \cdot (\vec{E} - \vec{E}_M) = O(k^{\tau-M}) \;,\; k \to \infty \tag{47}$$

whereas in the case of an asymptotic expansion (37),

$$\vec{E} - \vec{E}_M = O(k^{\tau-M-1}) \;,\quad k \to \infty \,. \tag{48}$$

No general proof has yet been given that the asymptotic solution is identical with the asymptotic expansion of the exact solution of the diffraction problem at hand. Nevertheless, the agreement found at various special problems provides strong evidence of the validity of the present asymptotic method.

The solution to equations (41)-(44) is elaborated in the next subsections.

B. Continuation of Phase

The surface $s(\vec{r})$ = constant defines a *wavefront*. The curves orthogonal to all wavefronts are called *rays*. As a consequence of (41), the rays in free space are straight lines in the direction of ∇s. By use of these concepts, the solution of the eikonal equation (41) proceeds as follows: As an initial condition, let s have the constant value s_0 on a given surface W_0 (initial wavefront). Then the rays are found as the two-parameter family of straight lines orthogonal to W_0. To label a ray, we may introduce two parameters (α,β). Along a given ray (fixed α and β), let σ denote the distance from W_0, measured positively in the direction of wave propagation. Then (α,β,σ) form *ray coordinates* (Fig. 2);

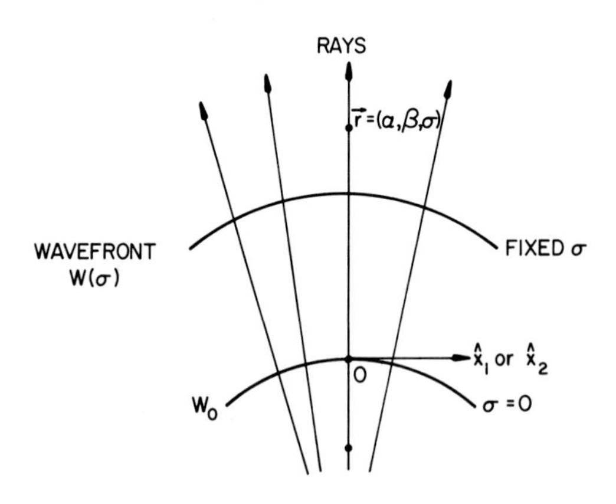

Fig. 2. Rays and wavefronts, ray coordinates (α,β,σ) and principal directions $(\hat{x}_1,\hat{x}_2)$.

in general, these coordinates are curvilinear. Along each ray the phase s is now given by

$$s(\vec{r}) = s_0 + \sigma. \tag{49}$$

This provides the solution of (41).

The wavefronts $s(\vec{r})$ = constant or σ = constant form a family of parallel surfaces. Consider a typical point O on the initial wavefront W_0. Introduce a rectangular coordinate system (x_1, x_2, σ) with origin at 0 and unit vectors $(\hat{x}_1, \hat{x}_2)$ pointing in the principal directions of W_0 at O (Fig. 2). Let the corresponding principal radii of curvature be denoted by (R_1, R_2); here, $R_1(R_2)$ is taken positive if the rays lying in the plane $x_2 = 0$ $(x_1 = 0)$ are divergent, and negative if the rays are convergent. Then proceeding along the ray through O, the principal directions of the wavefront $W(\sigma)$ through $(0,0,\sigma)$ remain parallel to $(\hat{x}_1, \hat{x}_2)$ and the principal radii of curvature are given by $(R_1 + \sigma,\ R_2 + \sigma)$. Thus the curvature matrix of $W(\sigma)$ takes the form

$$\bar{\bar{Q}}(\sigma) = \begin{bmatrix} \dfrac{1}{R_1 + \sigma} & 0 \\ 0 & \dfrac{1}{R_2 + \sigma} \end{bmatrix} \tag{50}$$

with respect to the (x_1, x_2, σ) coordinate system. In the case of general rectangular coordinates (x_1, x_2, σ) with $(\hat{x}_1, \hat{x}_2)$ making an angle ψ with the principal directions, the matrix $\bar{\bar{Q}}$ may be found from (35).

For later use we determine the Jacobian of the transformation from ray coordinates (α, β, σ) to rectangular coordinates $\vec{r} = (x, y, z)$, viz.,

$$j(\sigma) = j(\alpha, \beta, \sigma) = \frac{\partial(x, y, z)}{\partial(\alpha, \beta, \sigma)} = \frac{\partial \vec{r}}{\partial \alpha} \times \frac{\partial \vec{r}}{\partial \beta} \cdot \frac{\partial \vec{r}}{\partial \sigma}. \tag{51}$$

In the calculation of the amplitudes $\{\vec{e}_m\}$, the ratio $[j(\sigma_0)/j(\sigma)]^{1/2}$ for two points σ_0 and σ on a given ray enters. It can be shown [3], [8], [33], that this ratio depends on the curvature of the wavefront, namely

$$\left[\frac{j(\sigma_0)}{j(\sigma)}\right]^{1/2} = \left[\frac{\det \bar{\bar{Q}}(\sigma)}{\det \bar{\bar{Q}}(\sigma_0)}\right]^{1/2} = \frac{(R_1+\sigma_0)^{1/2}(R_2+\sigma_0)^{1/2}}{(R_1+\sigma)^{1/2}(R_2+\sigma)^{1/2}} \tag{52}$$

where (R_1, R_2) are the principal radii of curvature of the wavefront W_0 at $\sigma = 0$ as before. We adopt the following convention for the square roots in (52): $(R_{1,2}+\sigma)^{1/2}$ takes positive real, positive imaginary, or zero values.

The Jacobian $j(\sigma)$ along a given ray has an interesting geometrical interpretation. Consider a narrow pencil of rays with the given ray as its axial ray. Then the cross section of the pencil $da(\sigma)$ depends on σ (Fig. 3), and may be calculated from the well-known formula [41]

$$da = |j(\sigma)| \; d\alpha \; d\beta \tag{53}$$

Thus, apart from a phase factor, the Jacobian ratio in (52) is equal to the square root of the cross-section ratio of a pencil at two points on a ray.

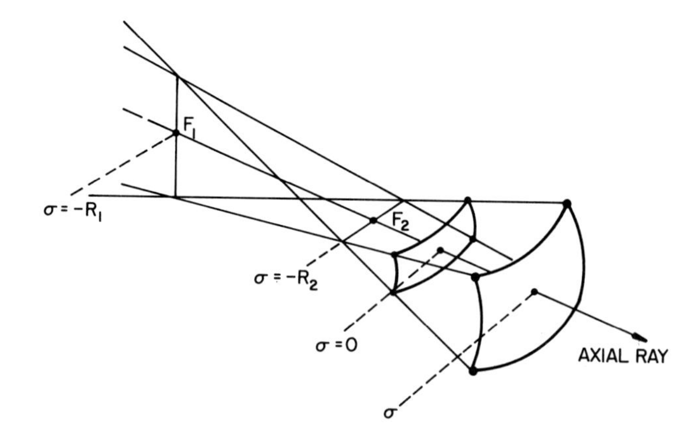

Fig. 3. Variation of cross section of a pencil. In this sketch both R_1 *and* R_2 *are positive (diverging pencil).*

C. Continuation of Field Amplitudes

Along a given ray, the transport equations in (42) reduce to ordinary differential equations. The solutions of the latter are found to be [7],[8]

$$\vec{e}_0(\sigma) = \vec{e}_0(\sigma_0)\left[\frac{j(\sigma_0)}{j(\sigma)}\right]^{1/2} \tag{54a}$$

$$\vec{e}_m(\sigma) = \vec{e}_m(\sigma_0)\left[\frac{j(\sigma_0)}{j(\sigma)}\right]^{1/2} - \frac{1}{2}\int_{\sigma_0}^{\sigma}\left[\frac{j(\sigma')}{j(\sigma)}\right]^{1/2} \nabla^2 \vec{e}_{m-1}(\sigma')\, d\sigma' ,$$

$$m = 1,2,\ldots \tag{54b}$$

The result in (54) enables one to continue $\{\vec{e}_m\}$ along a given ray. For the zeroth-order amplitude $\vec{e}_0$, it is only necessary to know one initial value of $\vec{e}_0$ at a reference point σ_0 in order to carry out this continuation. For a higher-order amplitude $\vec{e}_m$ $(m > 0)$, however, more information is needed. In addition to the initial value $\vec{e}_m(\sigma_0)$, $\nabla^2\vec{e}_{m-1}(\sigma')$ must be known for all σ' in the range $\sigma_0 < \sigma' < \sigma$. Thus, the continuation of the higher-order amplitudes is in general quite tedious.

The amplitudes $\{\vec{e}_m\}$ should also satisfy Gauss' law (43). It has been proven [7] that, whenever the solution (54) satisfies (43) at one point on a ray, then it does so along the whole ray. Because of this property, it is sufficient to enforce (43) at the reference point σ_0. In fact, (43) then comes down to an additional condition on the initial values $\vec{e}_m(\sigma_0)$.

Several remarks about the solution of $\vec{e}_0$ in (54a) are in order: (i) The direction of $\vec{e}_0$ (the polarization of the zeroth-order field) remains constant along a given ray. In addition, it is found from (43), (44) that

$$\nabla s \cdot \vec{e}_0 = 0 \quad , \quad \vec{h}_0 = Z^{-1}\nabla s \times \vec{e}_0 , \tag{55}$$

i.e., $(\vec{e}_0,\vec{h}_0,\nabla s)$ form a right-handed orthogonal system of vectors.

Thus the leading term of the asymptotic solution (38) describes a local plane-wave field. (ii) At a caustic point where $\sigma = -R_1$ or $-R_2$, $\vec{e}_0(\sigma)$ as calculated from (54a) and (52) becomes infinite, hence (54a) fails. This is a well-known difficulty of ray techniques. (iii) Tracing along a ray in the direction of wave propagation, the phase of $\vec{e}_0(\sigma)$ remains constant until a caustic point is encountered. It is well known [38] that the phase of $\vec{e}_0(\sigma)$ jumps by the amount of $(-\pi/2)$ when crossing the caustic point. This fact is accounted for through the properly chosen square root convention associated with (52).

The solution (54) is not valid when the reference point $\sigma_0 = 0$ is a caustic point. In such a case, (54) should be replaced by the alternative form [6]

$$\vec{e}_m(\sigma) = \frac{\vec{\delta}_m}{[j(\sigma)]^{1/2}} - \frac{1}{2} \int_0^\sigma \left[\frac{j(\sigma')}{j(\sigma)}\right]^{1/2} \nabla^2 \vec{e}_{m-1}(\sigma')\, d\sigma' \tag{54c}$$
$$m = 0,1,2,\ldots \quad .$$

Here the dash on the integral sign denotes the "finite part" of a divergent integral.[3] The (new) initial value $\vec{\delta}_m$ has the following interpretation. Taking limits as $\sigma \to 0$ in (54c) gives

$$\vec{\delta}_m = \lim_{\sigma \to 0} \vec{e}_m(\sigma)[j(\sigma)]^{1/2}. \tag{56}$$

At the caustic point $\sigma = 0$, $j(\sigma = 0)$ becomes zero whereas $\vec{e}_m(\sigma)$ tends to infinity as $\sigma \to 0$, in such a way that $\vec{\delta}_m$ is finite.

IV. GEOMETRICAL OPTICS (GO)

A. Formulation of the Problem

A basic problem studied in this paper is the diffraction of an incident field $\vec{E}^i$ by a perfectly conducting screen Σ. We

---

[3] Let $f(\varepsilon) = \int_\varepsilon^b g(x)dx$ have an asymptotic expansion in (perhaps fractional) powers of ε as $\varepsilon \to 0$. The coefficient of $\varepsilon^0 = 1$ in the expansion is called the "finite part" of the integral.

assume that $\vec{E}^i$ has the following high-frequency asymptotic expansion:

$$\vec{E}^i(\vec{r}) \sim e^{iks^i(\vec{r})} \sum_{m=0}^{\infty} (ik)^{-m} \vec{e}_m^{\,i}(\vec{r}) \quad , \quad k \to \infty . \tag{57}$$

The problem is to determine the asymptotic solution of the total field $\vec{E}^t$ everywhere. According to the ascending degrees of sophistication and accuracy, such a solution can be determined by any one of the following theories: (i) GO, (ii) GTD, and (iii) UAT. In this section, we will start out with the simplest one: GO. In fact, the solution obtained here is essential for the construction of the two more elaborate theories later.

According to GO, the presence of Σ blocks a part of the incident field and casts a shadow behind Σ. Let us introduce a shadow indicator $\varepsilon^i(\vec{r})$ such that

$$\varepsilon^i(\vec{r}) = \begin{cases} +1, & \text{if } \vec{r} \text{ is in the shadow region of } \vec{E}^i \\ -1, & \text{if } \vec{r} \text{ is in the lit region of } \vec{E}^i \end{cases} . \tag{58}$$

Then, in the presence of Σ, $\vec{E}^i$ in (57) is modified to become $\Theta(-\varepsilon^i)\vec{E}^i$, where the unit step function Θ is defined in (12). The surface which separates the shadow and lit regions is called the incident shadow boundary SB^i.

In addition to the incident field, GO also predicts the existence of a reflected field $\vec{E}^r$ such that

$$\vec{E}^t(\vec{r}) = \Theta(-\varepsilon^i)\vec{E}^i(\vec{r}) + \Theta(-\varepsilon^r)\vec{E}^r(\vec{r}). \tag{59}$$

The two terms on the right-hand side of (59) are also known as the geometrical-optics field. When $\vec{E}^i$ is given in (57), $\vec{E}^r$ can be expressed in a similar form:

$$\vec{E}^r(\vec{r}) \sim e^{iks^r(\vec{r})} \sum_{m=0}^{\infty} (ik)^{-m} \vec{e}_m^{\,r}(\vec{r}) \quad , \quad k \to \infty \tag{60}$$

where the phase function $s^r(\vec{r})$ and the amplitudes $\{\vec{e}_m^{\,r}\}$ are to be determined in the next subsection. The shadow indicator of the reflected wave ε^r in (59) is defined in the same manner as ε^i in (58) except that the superscript "i" should be replaced by "r". The reflected field $\vec{E}^r$ in (60), as it stands, is mathematically defined for all $\vec{r}$ (in the lit as well as shadow regions of the reflected field), despite the fact that only $\vec{E}^r$ in its lit region (where $\varepsilon^r = -1$) is physically meaningful and contributes to $\vec{E}^t$ in accordance with (59). The precise definition of $\vec{E}^r$ in its shadow region is of no concern at this moment. This subject will be discussed further in Section VI.

Two further conditions are to be enforced for the determination of $\vec{E}^r$ in its lit region. First, the usual boundary condition on the perfectly conducting screen Σ requires that the tangential components of the total electric field be zero:

$$\hat{N} \times (\vec{E}^i + \vec{E}^r) = 0 \quad , \quad \vec{r} \text{ on } \Sigma \ , \tag{61}$$

where $\hat{N}$ is the outward normal[4] of Σ. The second condition is that the total field satisfy Gauss' law:

$$\nabla \cdot (\vec{E}^i + \vec{E}^r) = 0 \ , \tag{62}$$

for all $\vec{r}$ in the common lit region of $\vec{E}^i$ and $\vec{E}^r$. It can be shown [7] that the satisfaction of Gauss' law at one point on a ray implies its satisfaction at all other points on the same ray. Because of this property, it is only necessary to enforce (62) at points on the screen Σ, as they are common points on incident and reflected rays. Thus, (62) may be replaced by

$$\nabla \cdot (\vec{E}^i + \vec{E}^r) = 0 \quad , \quad \vec{r} \text{ on } \Sigma \ . \tag{63}$$

The two conditions in (61) and (63) should be enforced on the fields in (57) and (60) for the solution of the reflected field.

---

[4] "Outward" normal means the normal pointing toward the source of $\vec{E}^i$.

B. Reflected Field

As a preparation, consider the reflection from a typical point O on the screen Σ (point of reflection). Three different sets of orthonormal base vectors at O are introduced:

(i) $(\hat{x}_1^i, \hat{x}_2^i, \hat{x}_3^i = \hat{z}^i)$ associated with the incident ray through O; $(\hat{x}_1^i, \hat{x}_2^i)$ coincide with the principal directions of the incident wavefront at O; $\hat{z}^i$ is in the direction of the incident ray through O (Fig. 4).

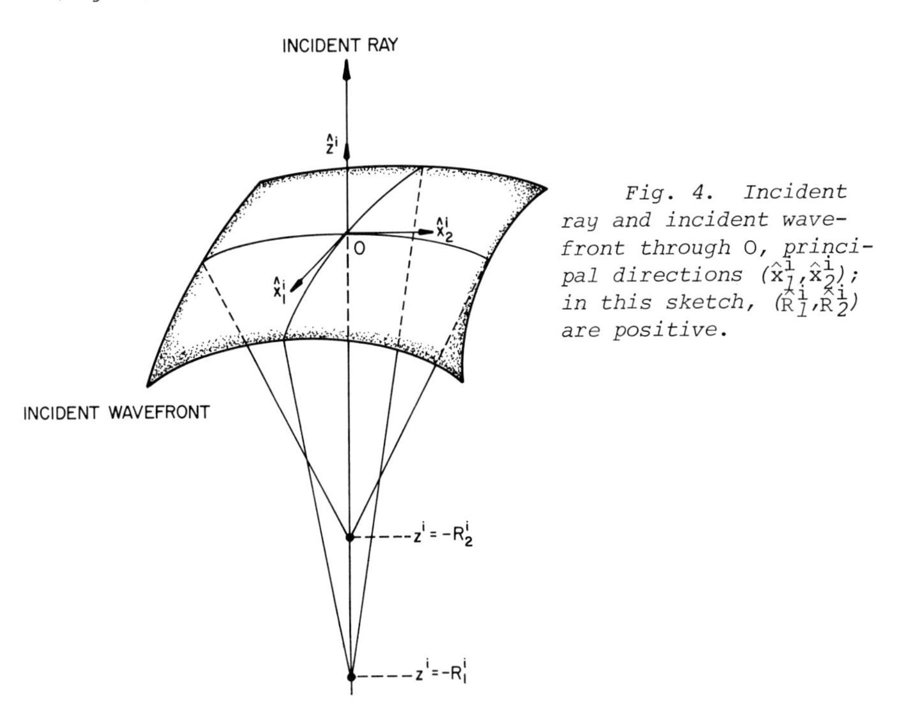

Fig. 4. Incident ray and incident wavefront through O, principal directions $(\hat{x}_1^i, \hat{x}_2^i)$; *in this sketch,* (R_1^i, R_2^i) *are positive.*

(ii) $(\hat{x}_1^r, \hat{x}_2^r, \hat{x}_3^r = \hat{z}^r)$ associated with the reflected ray through O; $(\hat{x}_1^r, \hat{x}_2^r)$ coincide with the principal directions of the reflected wavefront at O; $\hat{z}^r$ is in the direction of the reflected ray through O.

(iii) $(\hat{x}_1, \hat{x}_2, \hat{x}_3 = \hat{z})$ associated with the screen Σ; $(\hat{x}_1, \hat{x}_2)$ coincide with the principal directions of Σ at O; $\hat{z} = \hat{N}$ where $\hat{N}$ is the outward normal of Σ at O (Fig. 5).

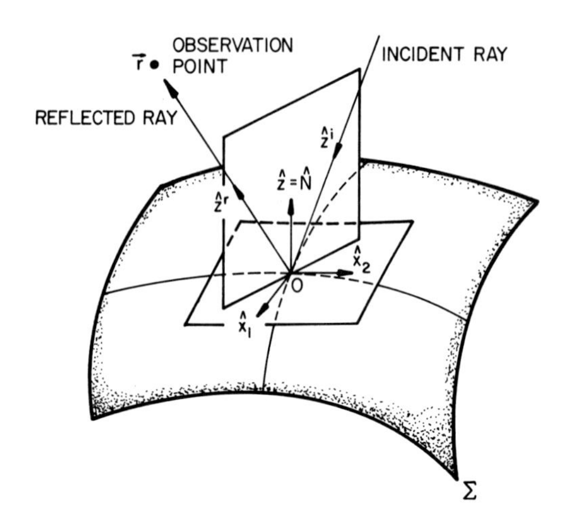

Fig. 5. Reflection from a smooth surface Σ.

With respect to the base vectors (i) and (ii), respectively, the curvature matrices of the incident and reflected wavefronts at O take the form (see (34))

$$\bar{\bar{Q}}^{i,r} = \begin{bmatrix} \frac{1}{R_1^{i,r}} & 0 \\ 0 & \frac{1}{R_2^{i,r}} \end{bmatrix}, \tag{64}$$

where $(R_1^{i,r}, R_2^{i,r})$ are the principal radii of curvature. Likewise, at O the screen Σ has the curvature matrix

$$\bar{\bar{Q}} = \begin{bmatrix} \frac{1}{R_1} & 0 \\ 0 & \frac{1}{R_2} \end{bmatrix}, \tag{65}$$

with respect to the base (iii); (R_1, R_2) denote the principal radii of curvature of Σ at O. (The sign convention is described after (32).) Finally, along the incident (reflected) ray through O, the incident (reflected) field can be resolved into transverse and longitudinal components with respect to the ray. Thus the amplitudes $\{\vec{e}_m^{\,i,r}\}$ of the incident and reflected fields are represented by

$$\vec{e}_m^{\,i,r}(\vec{r}) = \hat{x}_1^{i,r} e_{m1}^{i,r} + \hat{x}_2^{i,r} e_{m2}^{i,r} + \hat{z}^{i,r} e_{m3}^{i,r} \; ; \tag{66}$$

here, $(e_{m1}^{i,r}, e_{m2}^{i,r})$ are the transverse components, and $e_{m3}^{i,r}$ are the longitudinal components of $\vec{e}_m^{\,i,r}$ with respect to the direction of wave propagation.

For the incident field $\vec{E}^i$ as given in (57), we list below the steps and formulas for determining the reflected field $\vec{E}^r$ in (60) [33-36].

(i) For a given observation point $\vec{r}$, determine the point of reflection O by using Snell's law

$$\hat{z}^r = \hat{z}^i - 2(\hat{z} \cdot \hat{z}^i)\hat{z} \; . \tag{67}$$

The phase $s^r(\vec{r})$ on the reflected ray is then given by

$$s^r(\vec{r}) = s^i(O) + \sigma \; , \tag{68}$$

where $s^i(O)$ denotes the incident phase at O and σ is the distance to O along the ray.

(ii) Calculate the parameters

$$p_{mn}^{i,r} = \hat{x}_m^{i,r} \cdot \hat{x}_n \quad , \quad m,n = 1,2,3 \; . \tag{69}$$

There may be a difficulty in calculating $\{p_{mn}^r\}$, as the principal directions $(\hat{x}_1^r, \hat{x}_2^r)$ of the reflected wavefront are yet unknown. In many simple problems, $(\hat{x}_1^r, \hat{x}_2^r)$ can be determined by inspection. For such cases, $\{p_{mn}^r\}$ can be calculated. Otherwise, the method described in (iv) below must be followed.

(iii) Determine the curvature matrix $\bar{\bar{Q}}^r$ at O from the relation

$$(\bar{\bar{P}}^i)^T \bar{\bar{Q}}^i \bar{\bar{P}}^i + p_{33}^i \bar{\bar{Q}} = (\bar{\bar{P}}^r)^T \bar{\bar{Q}}^r \bar{\bar{P}}^r + p_{33}^r \bar{\bar{Q}} \; . \tag{70}$$

In an abbreviated form (70) is written as

$$(\bar{\bar{P}}^i)^T \bar{\bar{Q}}^i \bar{\bar{P}}^i + p_{33}^i \bar{\bar{Q}} = \{i \to r\} \tag{71}$$

where $\{i \to r\}$ means that all terms on the left of the equality sign are repeated with the superscript "i" changed into "r". The matrices $\bar{\bar{P}}^{i,r}$ in (71) are defined by

$$\bar{\bar{P}}^{i,r} = \begin{bmatrix} p_{11}^{i,r} & p_{12}^{i,r} \\ p_{21}^{i,r} & p_{22}^{i,r} \end{bmatrix}. \tag{72}$$

The notation T in (71) means the transpose operator. Since $(\hat{x}_1^r, \hat{x}_2^r)$ are principal directions of the reflected pencil, $\bar{\bar{Q}}^r$ calculated from (71) is diagonal, and is in the form of (64). Then, (R_1^r, R_2^r) can be immediately determined.

(iv) Suppose that in step (ii), $(\hat{x}_1^r, \hat{x}_2^r)$ cannot be immediately recognized. Then for the purpose of calculating $\bar{\bar{Q}}^r$, two *arbitrary* orthogonal directions $(\hat{\underline{x}}_1^r, \hat{\underline{x}}_2^r)$ transverse to $\hat{z}^r$ may be chosen. The relation in (71) remains valid when $(\hat{\underline{x}}_1^r, \hat{\underline{x}}_2^r)$ are employed in place of $(\hat{x}_1^r, \hat{x}_2^r)$. When $(\hat{\underline{x}}_1^r, \hat{\underline{x}}_2^r)$ are used, $\bar{\bar{Q}}^r$ calculated from (71) is, in general, not diagonal, and it should have the form of (35). Following the standard procedure, $\bar{\bar{Q}}^r$ should be next diagonalized, $(\hat{x}_1^r, \hat{x}_2^r)$ determined, and $\{p_{mn}^r\}$ in (69) calculated. In later manipulations, $\{p_{mn}^r\}$ are needed.

(v) The initial values of $\{\vec{e}_m^r\}$ at O are determined in succession. For $m = 0$, the components of $\vec{e}_0^r$ defined in the form of (66) can be solved from the relation

$$\vec{e}_0^r(O) = -\vec{e}_0^i(O) + 2[\vec{e}_0^i(O) \cdot \hat{N}]\hat{N} \tag{73a}$$

$$\vec{h}_0^r(O) = +\vec{h}_0^i(O) - 2[\vec{h}_0^i(O) \cdot \hat{N}]\hat{N} \tag{73b}$$

For $m > 0$, the corresponding relation reads

$$\begin{bmatrix} p_{11}^i & p_{21}^i & p_{31}^i \\ p_{12}^i & p_{22}^i & p_{32}^i \\ p_{13}^i p_{33}^i & p_{23}^i p_{33}^i & p_{33}^i p_{33}^i \end{bmatrix} \begin{bmatrix} e_{m,1}^i \\ e_{m,2}^i \\ e_{m,3}^i \end{bmatrix} + \begin{bmatrix} 0 \\ 0 \\ \nabla \cdot \vec{e}_{m-1}^i \end{bmatrix} = (-1)\{i \to r\}, \quad m = 1,2,3,\ldots . \tag{74}$$

From (74), one can solve for the value of $\vec{e}_m^{\,r}$ at O, in terms of $\vec{e}_m^{\,i}$, $\vec{e}_{m-1}^{\,i,r}$ and the derivatives of $\vec{e}_{m-1}^{\,i,r}$ at O. For a given incident field, $\vec{e}_m^{\,i}$ and its derivatives are known. Then $\vec{e}_{m-1}^{\,r}$ is found from the previous calculation. The only less explicit part in (74) is the divergence of $\vec{e}_{m-1}^{\,r}$. For $m = 1$, an explicit formula for calculating $\nabla \cdot \vec{e}_0^{\,r}$ is given in [34].

(vi) With the initial values of $\{\vec{e}_m^{\,r}\}$ at O as determined in (v), the continuation of $\{\vec{e}_m^{\,r}\}$ along the reflected ray is governed by (54). For $m = 0$, the latter formula reads

$$\vec{e}_0^{\,r}(\vec{r}) = \left[\left(1 + \frac{\sigma}{R_1^r}\right)^{1/2}\left(1 + \frac{\sigma}{R_2^r}\right)^{1/2}\right]^{-1} \vec{e}_0^{\,r}(O) , \tag{75}$$

where $(\cdot)^{1/2}$ takes positive real, positive imaginary, or zero values, (R_1^r, R_2^r) are determined in (iii), and σ measures the distance to O.

V. GEOMETRICAL THEORY OF DIFFRACTION (GTD)

A. Diffracted Field

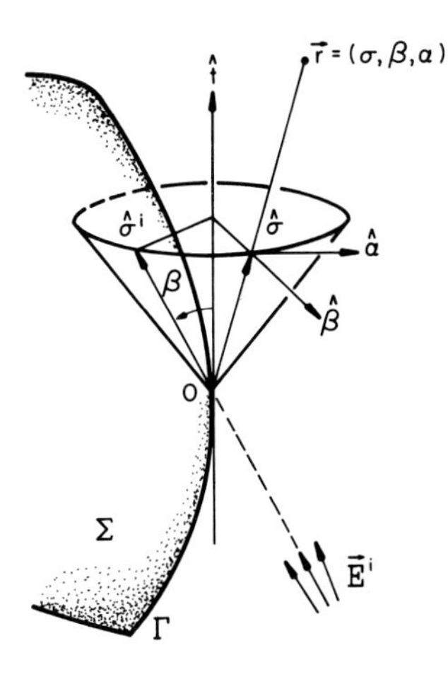

Fig. 6. Edge diffraction geometry showing the diffracted ray in the direction $\hat{\sigma}$, through the observation point $\vec{r}$. The spherical coordinates of $\vec{r}$ are (σ, β, α), with base vectors $(\hat{\sigma}, \hat{\beta}, \hat{\alpha})$.

Consider a typical edge diffraction problem sketched in Fig. 6:

a perfectly conducting screen Σ with edge Γ (a zero-angled wedge) is illuminated by an incident field $\vec{E}^i$ as given in (57). The asymptotic solution of the total field $\vec{E}^t$ is to be determined. According to GO discussed in Section IV, $\vec{E}^t$ is approximated by the geometrical-optics field $\vec{E}^g$ in (59) which reads

$$\vec{E}^g(\vec{r}) = \Theta(-\varepsilon^i)\vec{E}^i(\vec{r}) + \Theta(-\varepsilon^r)\vec{E}^r(\vec{r}) \tag{76a}$$

$$= \Theta(-\varepsilon^i)\vec{E}^i(\vec{r}) + \{i \to r\} \ . \tag{76b}$$

In (76b), we have used the notation $\{i \to r\}$ to emphasize the symmetry between the contributions of the incident and reflected fields. The result (76) predicts a zero-field in the shadow region and therefore fails to account for the diffraction phenomenon. An improvement over GO is provided by Keller's GTD [1-3], which does account for diffraction. According to the latter theory, $\vec{E}^t$ contains an additional term $\vec{E}^d$, called the diffracted field, such that

$$\text{GTD:} \quad \vec{E}^t(\vec{r}) = \vec{E}^g(\vec{r}) + \vec{E}^d(\vec{r}) \tag{77}$$

where

$$\begin{aligned} \vec{E}^d(\vec{r}) &= \vec{E}^{di}(\vec{r}) + \vec{E}^{dr}(\vec{r}) \\ &\sim k^{-1/2} e^{iks^d(\vec{r})} \sum_{m=0}^{\infty} (ik)^{-m} \vec{e}_m^{\,di}(\vec{r}) + \{i \to r\} \quad , \quad k \to \infty \ . \end{aligned} \tag{78}$$

Some preliminary remarks about $\vec{E}^d$ in (78) can be made: (i) $\vec{E}^d$ is of order $k^{-1/2}$, and is asymptotically smaller than $\vec{E}^g$, which is of order k^0. (ii) In general, $\vec{E}^d$ is nonzero everywhere. Hence, there is no shadow region of the diffracted field. (iii) As $\vec{E}^g$ in (76b), $\vec{E}^d$ consists of two symmetrical parts: $\vec{E}^{di}$ associated with the incident field and $\vec{E}^{dr}$ with the reflected field. These two parts have the same phase function s^d, but different amplitudes $\{\vec{e}_m^{\,di}\}$ and $\{\vec{e}_m^{\,dr}\}$. (iv) Since $\vec{E}^d$ should be an asymptotic solution of Maxwell's equations, s^d and $\{\vec{e}_m^{\,d}\}$ should satisfy the equations (41)-(43). Their solution is given in the next subsection.

B. Diffracted Phase and Amplitudes

The surface $s^d(\vec{r})$ = constant defines a diffracted wavefront which is orthogonal to a congruence of diffracted rays. All diffracted rays emanate from the edge Γ of the screen Σ, and may be constructed as follows. Consider an incident ray in the direction $\hat{\sigma}^i$ which meets the edge Γ at some point O (point of diffraction), and makes an angle β^i with the tangent $\hat{t}$ of Γ (Fig. 6). Such a ray gives rise to a family of diffracted rays which emanate from O and make an angle β with $\hat{t}$. In other words, for each edge point illuminated by $\vec{E}^i$, the diffracted rays generate a cone of semi-angle β with vertex at the edge point and axis tangent to the edge. The fact that

$$\beta = \beta^i \tag{79}$$

is known as the *law of edge diffraction*, the counterpart of Snell's law for reflection.

Consider a particular diffracted ray in the direction $\hat{\sigma}$ and passing through an observation point $\vec{r}$ (Fig. 6). The spherical coordinates of $\vec{r}$ are (σ,β,α) where

$$\sigma = \text{distance measured from } O \text{ to } \vec{r}, \tag{80a}$$

$$\beta = \text{polar angle measured from } \hat{t} \text{ in the range } (0,\pi), \tag{80b}$$

$$\alpha = \text{azimuthal angle in the plane perpendicular to } \hat{t}. \tag{80c}$$

The range, the positive direction, and the zero-value of α can be arbitrarily chosen, as all our formulas will be given in a form that is independent of these choices. Then the phase of the diffracted field at $\vec{r}$ is given by

$$s^d(\vec{r}) = s^i(O) + \sigma \,, \tag{81}$$

where $s^i(O)$ is the value of $s^i(\vec{r})$ at O.

Proceeding along the diffracted ray through O and $\vec{r}$, the principal directions of the diffracted wavefront remain the same. It can be shown [6] that these principal directions are given by

$$\hat{\sigma}_1 = \hat{\sigma}_2 \times \hat{\sigma} = -\hat{\beta} \ , \tag{82a}$$

$$\hat{\sigma}_2 = \frac{1}{\sin \beta} \hat{\sigma} \times \hat{t} = -\hat{\alpha} \ . \tag{82b}$$

Thus, $\hat{\sigma}_1$ lies in the plane through $\hat{\sigma}$ and $\hat{t}$, and $\hat{\sigma}_2$ is normal to this plane (Fig. 7). For the diffracted wavefront passing through O, the two principal radii of curvature in the directions of $\hat{\sigma}_1$ and $\hat{\sigma}_2$ are denoted by R_1 and R_2, respectively. Several ways exist for determining (R_1, R_2) [1],[6],[22],[33],[37].

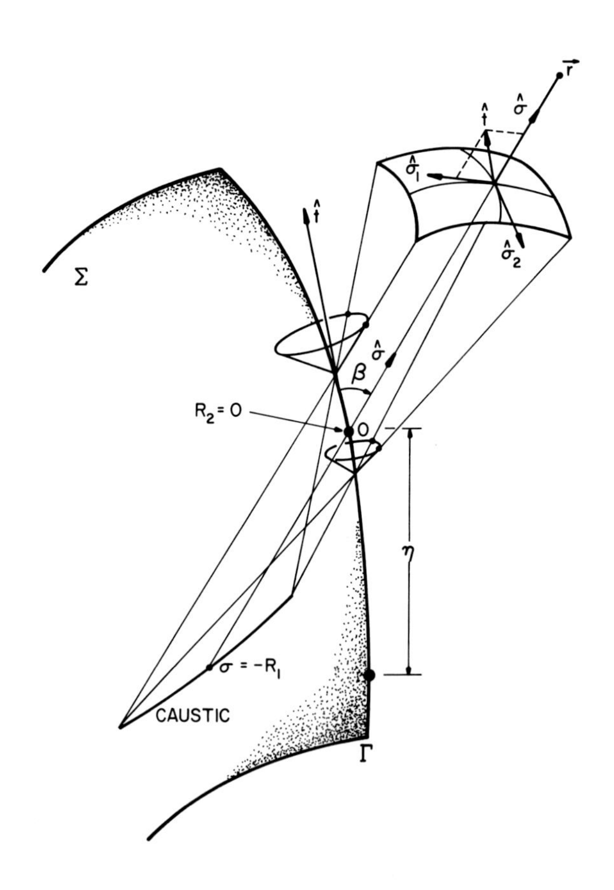

Fig. 7. Diffracted rays and diffracted wavefront, principal directions $(\hat{\sigma}_1, \hat{\sigma}_2)$. In this sketch, R_1 is positive (diverging pencil).

One way involves the calculation of the Jacobian as defined

in (51). As noted earlier, a typical point on a diffracted ray is described by the spherical coordinates (σ, β, α). For a given incident field, β is a function of η, where η is the arc length along the edge Γ measured positively in the direction of $\hat{t}$ (Fig. 7). Then (η, α, σ) form the diffracted-ray coordinates. The Jacobian of the transformation from these ray coordinates to rectangular coordinates according to (51) is found to be [6]

$$j(\sigma) = \sigma \sin^2\beta \left(1 + \frac{\sigma}{R_1}\right) , \tag{83}$$

where

$$R_1 = - \frac{\sin^2\beta}{\kappa\hat{\sigma} \cdot \hat{n} + (d\beta/d\eta) \sin \beta} . \tag{84}$$

In (84), κ is the curvature and $\hat{n}$ is the unit normal of Γ at O ($\kappa \geq 0$ and $\hat{n}$ points toward the center of curvature). From (83) and (52), it is immediately recognized that R_1 in (84) is indeed one of the two principal radii of curvature of the diffracted wavefront at σ, and the other radius is

$$R_2 = 0 , \tag{85}$$

which implies that the edge Γ is a caustic line of the diffracted field.

An alternative approach to calculating R_1 is based on the matching of the incident and diffracted phase functions s^i and s^d over Γ [33],[37]. The resulting formula reads

$$R_1 = \left[\frac{1}{R_0^i} + \frac{\kappa}{\sin^2\beta} (\hat{\sigma}^i - \hat{\sigma}) \cdot \hat{n}\right]^{-1} . \tag{86}$$

Here R_0^i is the radius of curvature of the incident wavefront at O in the plane through $\hat{t}$ and $\hat{\sigma}^i$; in terms of the principal radii of curvature (R_1^i, R_2^i) and the principal directions $(\hat{x}_1^i, \hat{x}_2^i)$ of the incident wavefront at O, R_0^i is given by

$$\frac{1}{R_0^i} = \frac{\cos^2\Omega^i}{R_1^i} + \frac{\sin^2\Omega^i}{R_2^i} , \tag{87}$$

where Ω^i is the angle between $\hat{x}_1^i$ and the projection of $\hat{t}$ on the plane through $\hat{x}_1^i$ and $\hat{x}_2^i$ (Fig. 8). The two formulas in (84) and (86) are equivalent.

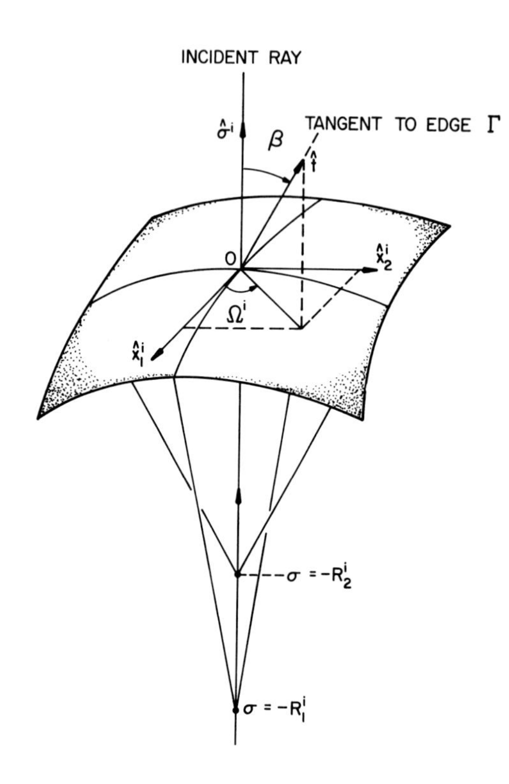

Fig. 8. Incident ray and incident wavefront through the edge point O, principal directions ($\hat{x}_1^i, \hat{x}_2^i$), principal radii of curvature (R_1^i, R_2^i).

The continuation of the diffracted field amplitudes $\{\vec{e}_m^{\,d}\}$ along a given ray is governed by (54). We choose the point of diffraction O as the reference point. Since O is also a caustic point, (54c) should be used, namely

$$\vec{e}_m^{\,d}(\vec{r}) = \vec{e}_m^{\,di}(\vec{r}) + \vec{e}_m^{\,dr}(\vec{r}) = \left[\sigma^{1/2}\left(1+\frac{\sigma}{R_1}\right)^{1/2}\right]^{-1} \vec{\delta}_m^{\,i}$$

$$- \frac{1}{2}\int_0^{\sigma} \frac{(\sigma')^{1/2}(\sigma'+R_1)^{1/2}}{(\sigma)^{1/2}(\sigma+R_1)^{1/2}} \nabla^2 \vec{e}_{m-1}^{\,di}(\vec{r}\,')\, d\sigma' + \{i \to r\} \tag{88}$$

where $\vec{r}$ and $\vec{r}'$ have coordinates (σ,β,α) and (σ',β,α), respectively. The initial values $\{\vec{\delta}_m^{i,r}\}$ are yet to be determined.

C. Zeroth-Order Diffracted Field

In general the initial values $\{\vec{\delta}_m^{i,r}\}$ depend on the amplitudes $\{\vec{e}_m^{i,r}\}$ and their (first- and higher-order) derivatives at the point of diffraction O. In GTD, only $\vec{\delta}_0^{i,r}$ can be determined by comparison of (78) and (88) with the solution of a canonical problem, namely, Sommerfeld's half-plane diffraction problem. This procedure finds its "justification" in the fact that high-frequency diffraction is a local phenomenon, and locally the screen can be approximated by a half-plane. Once $\vec{\delta}_0^{i,r}$ are found and used in (88) and (78), the dominant term of the diffracted field expansion at $\vec{r} = (\sigma,\beta,\alpha)$ is determined. The final result is

$$\vec{E}^d(\vec{r}) = g(k\sigma)\frac{1}{\sqrt{1+(\sigma/R_1)}}\frac{\chi^i}{\sin\beta}\left[e^{iks^i(0)}\ \mathrm{rot}\ \vec{e}_0^{\,i}(0)\right] + \{i \rightarrow r\} + O(k^{-3/2}) \ . \tag{89}$$

The various factors in (89) are explained below.

Cylindrical wave factor:

$$g(k\sigma) = \frac{1}{2\sqrt{2\pi k\sigma}}\ exp\left[i\left(k\sigma + \frac{\pi}{4}\right)\right] , \tag{90}$$

which may be identified with the leading term of the asymptotic expansion of the two-dimensional Green's function $(i/4)H_0^{(1)}(k\sigma)$ of the wave equation.

Divergence factor:

$$\mathrm{DF} = \frac{1}{\sqrt{1+(\sigma/R_1)}} , \tag{91}$$

where R_1 is given by (84) or (86). As explained in Section II.B, the square root in (91) takes positive real, positive imaginary, or zero values (DF is positive real, negative imaginary, or infinite). Notice that the product of g and DF behaves as $\sigma^{-1/2}$ when $\sigma \to 0$ (as a cylindrical wave), and as σ^{-1} when $\sigma \to \infty$ (as a spherical wave).

Phase of $\vec{E}^d$: $s^i(O)$ is the phase of the incident field at O. Thus, the phase s^d of the diffracted field in (89) is given by $s^i(O) + \sigma$, in accordance with (81).

Rot operator: The operator rot in (89) denotes the rotation about $\hat{t}$ that brings $\hat{\sigma}^i$ onto $\hat{\sigma}$ (Fig. 6), i.e.,

$$\text{rot } \hat{\sigma}^i = \hat{\sigma}. \tag{92}$$

Alternatively, the vector rot $\vec{e}_0^{\,i}$ can be expressed as follows by use of the spherical coordinates (σ,β,α). Along the incident and the diffracted ray through O we introduce unit vectors $(\hat{\beta}^i,\hat{\alpha}^i)$ and $(\hat{\beta},\hat{\alpha})$, respectively, pointing in the direction of increasing (β,α) and therefore orthogonal to the pertaining ray (Fig. 6). Then the amplitude $\vec{e}_0^{\,i}$ along the incident ray may be resolved into two transverse components:

$$\vec{e}_0^{\,i}(\vec{r}) = \hat{\beta}^i e_{0\beta}^i(\vec{r}) + \hat{\alpha}^i e_{0\alpha}^i(\vec{r}) , \tag{93}$$

where $(e_{0\beta}^i, e_{0\alpha}^i)$ are the (polar, azimuthal) components of $\vec{e}_0^{\,i}$. Remember that the longitudinal component of $\vec{e}_0^{\,i}$ (in the direction of the ray) vanishes, as found in (55). Under the rot operator, $(\hat{\beta}^i,\hat{\alpha}^i)$ pass into $(\hat{\beta},\hat{\alpha})$, hence we have

$$\text{rot } \vec{e}^{\,i}(O) = \hat{\beta} e_{0\beta}^i(O) + \hat{\alpha} e_{0\alpha}^i(O) . \tag{94}$$

It is easily recognized from (93) and (94) that rot $\vec{e}_0^{\,i}$ is independent of the sense of rotation (clockwise or counterclockwise), and of the positive direction of α. As another alternative, the vector rot $\vec{e}_0^{\,i}$ can be expressed as

$$\text{rot } \vec{e}_0^{\,i}(O) = \frac{1}{\sin^2\beta} \{[\vec{e}_0^{\,i}(O) \cdot \hat{t}]\,[\hat{\sigma} \times (\hat{t} \times \hat{\sigma})] + [\vec{e}_0^{\,i}(O) \cdot (\hat{t} \times \hat{\sigma}^i)]\,(\hat{t} \times \hat{\sigma})\} \tag{95a}$$

$$= \frac{1}{\sin^2\beta} \{[\vec{e}_0^{\,i}(O) \cdot \hat{t}]\,[\hat{\sigma} \times (\hat{t} \times \hat{\sigma})] + Z\,[\vec{h}^i(O) \cdot \hat{t}]\,(\hat{t} \times \hat{\sigma})\} \tag{95b}$$

where (95b) follows from (95a) and (55). When (95b) is used in (89), we note that the diffracted field $\vec{E}^{di}(\vec{r})$ is proportional to the tangential components of $(\vec{E}^i, \vec{H}^i)$ at the point of diffraction.

Reflected part of $\vec{E}^d$: At the point of diffraction O, the reflected field $\vec{E}^r$ is simply related to $\vec{E}^i$. In particular, we have

$$s^r(O) = s^i(O) \quad , \quad e^r_{0\beta}(O) = -e^i_{0\beta}(O) \quad , \quad e^r_{0\alpha}(O) = e^i_{0\alpha}(O). \tag{96}$$

Then the two parts in (89) may be combined to become

$$\vec{E}^d(\vec{r}) = g(k\sigma)\, \frac{1}{\sqrt{1 + (\sigma/R_1)}}\, \frac{1}{\sin\beta}\, e^{iks^i(O)}\left[\hat{\beta} D^s e^i_{0\beta}(O) + \hat{\alpha} D^h e^i_{0\alpha}(O)\right]$$
$$+\, O(k^{-3/2}) \tag{97}$$

where

$$D^{s,h} = \chi^i \mp \chi^r \tag{98}$$

are often known as the sound-soft and sound-hard scalar diffraction coefficients, respectively.

Diffraction coefficient: For the present case of an (infinitely) thin screen, $\chi^{i,r}$ are given by

$$\chi^{i,r} = \csc\frac{\psi^{i,r}}{2} \qquad \text{(for thin screen)}. \tag{99}$$

The angles $\psi^{i,r}$ are defined below. Referring to Figs. 6 and 9, we consider the following three rays through the point of diffraction O: an incident ray in the direction $\hat{\sigma}^i$, a reflected ray in the direction $\hat{\sigma}^r$, and a diffracted ray in the direction $\hat{\sigma}$. By

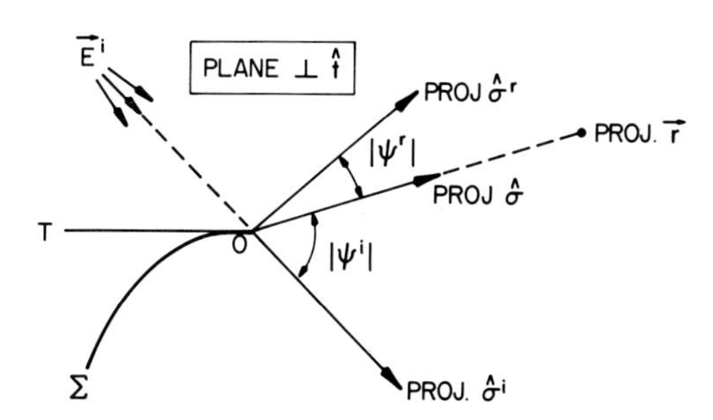

Fig. 9. Projection of Fig. 6 on a plane perpendicular to $\hat{t}$. *T is the tangent plane of* Σ *at O.*

Snell's law, $\hat{\sigma}^i$ and $\hat{\sigma}^r$ are symmetrically situated on both sides of the tangent plane T of Σ at O, while the diffracted ray passes through the observation point $\vec{r}$. The angles $\psi^{i,r}$ have magnitudes $|\psi^{i,r}|$ equal to the angles of the continuous rotation about $\hat{t}$ that brings $\hat{\sigma}^{i,r}$ onto $\hat{\sigma}$ without crossing T. These magnitudes are most easily visualized by projecting the vectors $(\hat{\sigma}^{i,r},\hat{\sigma})$ on the plane through O perpendicular to $\hat{t}$ (see Fig. 9). The signs of $\psi^{i,r}$ are chosen such that

$$\operatorname{sgn} \psi^{i,r} = \varepsilon^{i,r}(\vec{r}) \ , \tag{100}$$

where the shadow indicators $\varepsilon^{i,r}$ are defined in (58). We emphasize that the above definitions of $\chi^{i,r}$ are independent of the choice of coordinates.

In summary, the total field solution according to GTD for the diffraction problem sketched in Fig. 6 is given in (77). The dominant term of the diffracted field $\vec{E}^d$ may be calculated from (89) or (97). The corresponding diffracted magnetic field $\vec{H}^d$ can be calculated by means of (44) or

$$\vec{H}^d(\vec{r}) = Z^{-1}\hat{\sigma} \times \vec{E}^d(\vec{r}) + O(k^{-3/2}). \tag{101}$$

It can also be shown that (89) and (97) remain valid when the electric fields $(\vec{E}^d,\vec{e}^i)$ are replaced by the magnetic fields $(\vec{H}^d,\vec{h}^i)$. There is no systematic method in GTD for calculating the higher-order terms in the expansion of $(\vec{E}^d,\vec{H}^d)$ beyond the term of order $k^{-1/2}$.

D. Extension to Diffraction by a Wedge

With a simple modification, the GTD result for diffraction by a thin screen can also be applied in the case of diffraction by a wedge. The geometry of the wedge is shown in Fig. 10. The wedge is composed of two smooth, curved, perfectly conducting surfaces Σ_1 and Σ_2, which intersect along the curved edge Γ. At the point of diffraction O, the unit tangent and normal vectors to Γ are

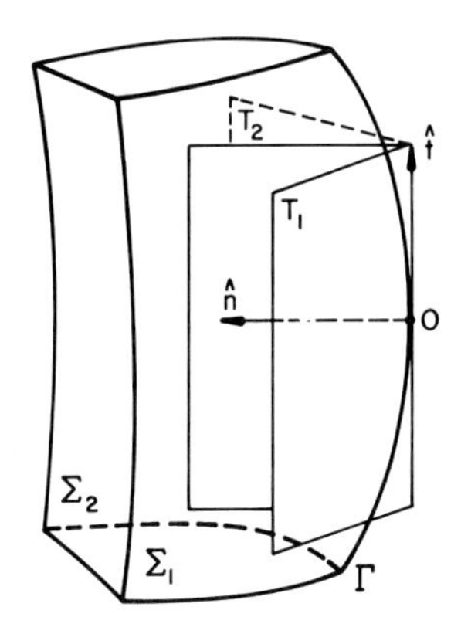

Fig. 10. Conducting curved wedge with faces Σ_1 and Σ_2; at the point O, $\hat{t}$ is tangent and $\hat{n}$ normal to the edge Γ; (T_1, T_2) are tangent planes to (Σ_1, Σ_2).

denoted by $\hat{t}$ and $\hat{n}$, respectively. The curvature κ of Γ at O is (by definition) a positive number. The half-planes T_1 and T_2 tangent to O at Σ_1 and Σ_2, respectively, form a planar wedge which approximates the curved wedge in the vicinity of O. The exterior angle between T_1 and T_2 is $m\pi$ with $1 \leq m \leq 2$.

Consider now the diffraction problem for the wedge when illuminated by the incident field $\vec{E}^i$ in (57). The total field solution, accroding to GTD, is again given in (77). The geometrical optics field $\vec{E}^g$ consists of $\vec{E}^i$, and, in general, two reflected fields: $\vec{E}_1^r$ from Σ_1 and $\vec{E}_2^r$ from Σ_2 (Fig. 11). Corresponding to $\vec{E}_1^r$, the shadow indicator $\varepsilon_1^r(\vec{r}) = -1$ if $\vec{r}$ is in the lit region of $\vec{E}_1^r$, and $\varepsilon_1^r(\vec{r}) = +1$ if $\vec{r}$ is in the complementary shadow region. A similar definition holds for ε_2^r with respect to $\vec{E}_2^r$. Then, in place of (76), the geometrical optics field in the wedge problem

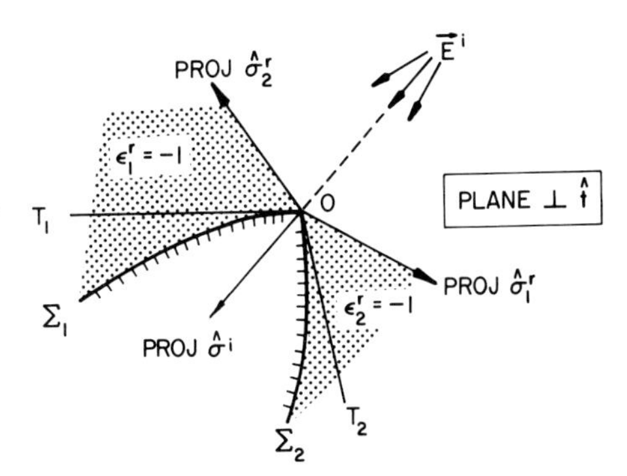

Fig. 11. Reflection by a curved wedge. If $\hat{\sigma}^i$ falls inside the two tangent planes of the wedge in Fig. 10, there are two reflected fields: $\vec{E}_1^r$ from Σ_1 and $\vec{E}_2^r$ from Σ_2.

is given by

$$\vec{E}^{g}(\vec{r}) = \Theta(-\varepsilon^{i})\vec{E}^{i}(\vec{r}) + \Theta(-\varepsilon_1^{r})\vec{E}_1^{r}(\vec{r}) + \Theta(-\varepsilon_2^{r})\vec{E}_2^{r}(\vec{r}) \ . \tag{102}$$

Note that $\vec{E}_1^r$ and $\vec{E}_2^r$ may be calculated separately according to the method outlined in Section IV.

The diffraction field $\vec{E}^d$ in the wedge problem may be calculated from (89) and (96) as before. The only modification necessary is that the diffraction coefficient in (99) be generalized to become

$$\chi^{i,r} = \frac{\frac{2}{m}\sin\frac{\pi}{m}}{\cos\frac{\pi}{m} - \cos\frac{\pi+\psi^{i,r}}{m}} \qquad \text{(for wedge)}. \tag{103}$$

The angle ψ^i has a magnitude equal to the angle of the continuous rotation about $\hat{t}$ that brings $\hat{\sigma}^i$ onto $\hat{\sigma}$ without crossing *both* T_1 and T_2, and sgn $\psi^i = \varepsilon^i(\vec{r})$ as in (79). In the general case that both faces of the wedge are illuminated (Fig. 11), the rotation that brings $\hat{\sigma}^i$ onto $\hat{\sigma}$ may cross either T_1 or T_2 (not both). The two values of ψ^i lead to the same value of χ^i. The angle ψ^r in (103) may be determined by the following rule: $|\psi^r|$ is the angle of the continuous rotation about $\hat{t}$ that brings $\hat{\sigma}_p^r$ on $\hat{\sigma}$ without crossing T_1 or T_2, and sgn $\psi^r = \varepsilon_p^r(\vec{r})$; here $p = 1$ or 2. Hence, when both faces of the wedge are illuminated, we find two values of ψ^r. Again, these two values of ψ^r lead to the same value of χ^r. Two examples for calculating $\psi^{i,r}$ are given in Fig. 12, where

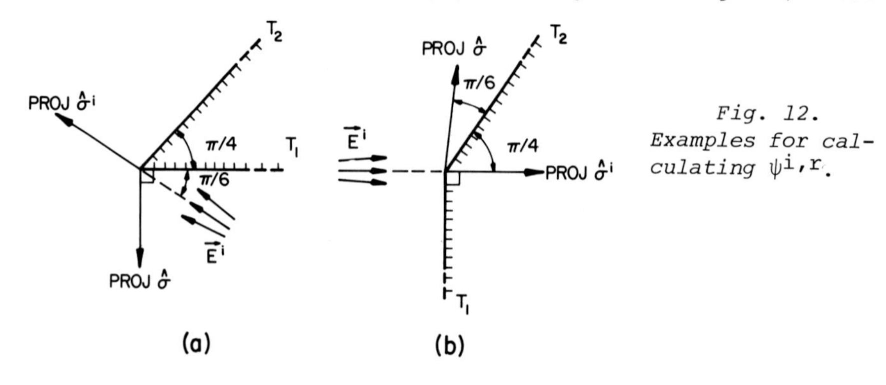

Fig. 12. Examples for calculating $\psi^{i,r}$.

the values of $\psi^{i,r}$ are found to be

(a) $\psi^i = -(2\pi/3)$, $\psi^r = -(\pi/3)$

(b) $\psi^i = -(5\pi/12)$ or $-(19\pi/12)$, $\psi^r = -(\pi/12)$ or $+(7\pi/12)$.

As a further example, let the directions $\hat{\sigma}^i$ and $\hat{\sigma}$ be described by the azimuthal angles ϕ^i and ϕ in the plane through O perpendicular to $\hat{t}$ (Fig. 13). Both angles are measured clockwise from the tangent plane T_1 of Σ_1, which is the *illuminated* face of the wedge.

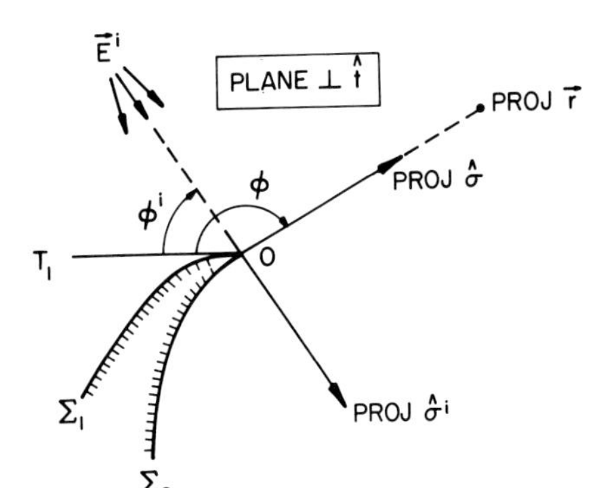

Fig. 13. Definitions of ϕ^i *and* ϕ.

(If both faces are illuminated, Σ_1 can be either one of them.) Thus ϕ^i and ϕ have ranges

$$0 < \phi^i < \pi \quad , \quad 0 < \phi < m\pi .$$

Then it can easily be shown that

$$\psi^i = \phi - (\pi + \phi^i) \quad , \quad \psi^r = \phi - (\pi - \phi^i) . \tag{104}$$

Substituting (104) into (103), an alternative form of $\chi^{i,r}$ is obtained:

$$\chi^{i,r} = \frac{\frac{2}{m} \sin \frac{\pi}{m}}{\cos \frac{\pi}{m} - \cos \frac{\phi \mp \phi^i}{m}} \qquad \text{(for wedge)} \tag{105}$$

Several properties of the diffraction coefficient χ^i in (103) are listed below:

(i) For the special case $m\pi = 2\pi$, i.e., the wedge becomes a thin screen, (103) reduces to (99) as expected.

(ii) $\chi^i(\psi^i)$ has a period $2m\pi$, and has even symmetry with respect to $\psi^i = -\pi$, i.e.,

$$\chi^i(-\pi + \alpha) = \chi^i(-\pi - \alpha). \tag{107}$$

for any α.

(iii) For small values of ψ^i,

$$\chi^i = \frac{2}{\psi^i} - \frac{1}{m} \cot \frac{\pi}{m} + O(\psi^i) . \tag{108}$$

Thus, χ^i becomes infinite as $\psi^i \to 0$.

(iv) sgn $\chi^i = \varepsilon^i(\vec{r})$, and sgn $\chi^r = \varepsilon_1^r(\vec{r})\varepsilon_2^r(\vec{r})$.

(v) χ^i depends on the deviation of $\hat{\sigma}^i$ from $\hat{\sigma}$ (Fig. 9), but is independent of their directions relative to the wedge (as long as $\hat{\sigma}$ remains in the lit or shadow region determined by $\hat{\sigma}^i$ and the wedge).

Exactly the same comments apply to χ^r as a function of ψ^r.

E. Discussion of GTD

In the present section the diffraction of a high-frequency electromagnetic field by a curved wedge (Fig. 10) has been treated by Keller's GTD. Starting from an incident field $\vec{E}^i$ as given in (57), the GTD solution for the total field $\vec{E}^t$ is given by (77) and (102), while the dominant term of the diffracted field $\vec{E}^d$ may be calculated from (89) or (97). The solution, according to GTD, involves only elementary functions and can easily be evaluated numerically.

Keller's GTD has been applied to a wide variety of high-frequency edge diffraction problems, such as diffraction by an aperture in a screen, scattering by a finite body with a sharp edge, and many others. Its solutions yield excellent agreement with experimental results. They also agree perfectly with the (high-frequency) asymptotic expansion of the few exact solutions that are known. Although GTD is an asymptotic method valid for $k \to \infty$, many examples have shown that its solutions may be remarkably accurate even when the dimension of the diffracting object is in the order of a wavelength.

Returning to the wedge diffraction problem, its GTD solution does have a number of limitations as listed below.

(i) Across the shadow boundaries $SB^{i,r}$ (where $\psi^{i,r} = 0$), the geometrical-optics field $\vec{E}^g$ in (102) is discontinuous. Furthermore, because of (108), the diffracted field $\vec{E}^d$ in (89) becomes infinite on $SB^{i,r}$. Thus the GTD solution is not valid in the neighborhoods of $SB^{i,r}$ (a more precise definition of the transition regions around $SB^{i,r}$ is given in [18] and [40]).

(ii) The diffracted field $\vec{E}^d$ becomes infinite at the edge Γ (where $\sigma = 0$) in a manner that violates the edge condition. Therefore, GTD does not predict the correct field in the neighborhood of Γ (near field).

(iii) When $R_1 < 0$, $\vec{E}^d$ becomes infinite at the caustic $\sigma = -R_1 = |R_1|$ of the diffracted wave. The total field $\vec{E}^t$ also becomes infinite at caustics of the geometrical-optics field.

(iv) The diffraction coefficients $\chi^{i,r}$ are derived by a comparison with the asymptotic expansion of the exact solution of a canonical problem, and they do not arise as an integral part of the theory.

(v) There is no systematic method for determining the higher-order terms beyond the term of order $k^{-1/2}$ in the expansion (78) for the diffracted field $\vec{E}^d$.

(vi) A rigorous proof of the asymptotic nature of the formal expansion (78) for $\vec{E}^d$ has not been given.

VI. UNIFORM ASYMPTOTIC THEORY OF EDGE DIFFRACTION (UAT)

A. *Ansatz* of UAT

Consider again the diffraction of an incident field $\vec{E}^i$ in (57) by a perfectly conducting screen Σ with edge Γ, as sketched in Fig. 6. The problem is to determine the asymptotic solution of the total field $(\vec{E}^t, \vec{H}^t)$. According to UAT [6]-[8], $\vec{E}^t$ is given by

$$\text{UAT:} \quad \vec{E}^t(\vec{r}) = \vec{E}^G(\vec{r}) + \vec{E}^d(\vec{r}) \tag{109}$$

which should be compared with the GTD solution in (77). The second term $\vec{E}^d$ in (109) is unchanged and again given by (78), whereas the first term $\vec{E}^G$ (with a capital G) is new and replaces the geometrical-optics field $\vec{E}^g$ in (76). It is the *Ansatz* of UAT that $\vec{E}^G$ is given by [6],[7]

$$\vec{E}^G(\vec{r}) = [F(k^{1/2}\xi^i) - \hat{F}(k^{1/2}\xi^i)]\vec{E}^i(\vec{r}) + \{i \to r\} . \quad (110)$$

The notations in (110) are explained below. F is the Fresnel integral defined in (7), and $\hat{F}$ is the asymptotic series in (13) associated with F. ξ^i is defined by

$$\xi^i(\vec{r}) = \varepsilon^i(\vec{r})\left|\sqrt{s^d(\vec{r}) - s^i(\vec{r})}\right| . \quad (111)$$

Note that $(\xi^i)^2$ measures the excessive ray path from the source to the observation point via the edge Γ, hence, ξ^i is known as the *detour parameter* of the incident field. $\xi^i = 0$ along the incident shadow boundary SB^i, and $\xi^i > 0$ ($\xi^i < 0$) in the shadow (lit) region of the incident field. The other detour parameter ξ^r of the reflected field is again defined by (111) after replacing superscript "i" by "r".

The same solution in (109), (78) and (110) holds for the total magnetic field $\vec{H}^t$ after replacing $(\vec{E}^\ell, \vec{e}^\ell_m)$ by $(\vec{H}^\ell, \vec{h}^\ell_m)$ with $\ell = t, i, r$, and d. Furthermore, $\vec{e}^\ell_m$ and $\vec{h}^\ell_m$ are related through (44).

The UAT solution for $(\vec{E}^t, \vec{H}^t)$ is now completely determined except for $\{\vec{e}^d_m\}$ of the diffracted field in (78). The latter amplitudes satisfy the usual transport equations, whose solutions are given in (88) with yet unknown initial values $\{\vec{\delta}^{i,r}_m\}$. Hence, our problem now reduces to determining $\{\vec{\delta}^{i,r}_m\}$.

B. Determination of $\vec{\delta}^{i,r}_m$

Let us concentrate on the determination of $\vec{\delta}^i_m$, as its final solution may be used for $\vec{\delta}^r_m$ after replacing superscript "i" by "r". We follow the approach given in [7]. First rewrite $\vec{E}^t$ in a different form. The substitution of (110) and (13) into (109) leads to

$$\vec{E}^t(\vec{r}) \sim F(k^{1/2}\xi^i)e^{iks^i(\vec{r})} \sum_{m=0}^{\infty} (ik)^{-m} \vec{e}_m^{\,i}(\vec{r}) + k^{-1/2}e^{iks^d(\vec{r})} \sum_{m=0}^{\infty} (ik)^{-m} \vec{w}_m^{\,i}(\vec{r}) + \{i \to r\} \quad , \quad k \to \infty \tag{112a}$$

where

$$\vec{w}_m^{\,i}(\vec{r}) = \vec{e}_m^{\,di}(\vec{r}) - \frac{e^{i\pi/4}}{2\pi} \sum_{n=0}^{m} \Gamma(n+\frac{1}{2})(\xi^i)^{-2n-1} \vec{e}_{m-n}^{\,i}(\vec{r}). \tag{112b}$$

Similarly, the total magnetic field may be expressed as

$$\vec{H}^t(\vec{r}) \sim F(k^{1/2}\xi^i)e^{iks^i(\vec{r})} \sum_{m=0}^{\infty} (ik)^{-m} \vec{h}_m^{\,i}(\vec{r}) + k^{-1/2}e^{iks^d(\vec{r})} \sum_{m=0}^{\infty} (ik)^{-m} \vec{v}_m^{\,i}(\vec{r}) + \{i \to r\} \quad , \quad k \to \infty \tag{113a}$$

where

$$\vec{v}_m^{\,i}(\vec{r}) = \vec{h}_m^{\,di}(\vec{r}) - \frac{e^{i\pi/4}}{2\pi} \sum_{n=0}^{m} \Gamma(n+\frac{1}{2})(\xi^i)^{-2n-1} \vec{h}_{m-n}^{\,i}(\vec{r}). \tag{113b}$$

The amplitudes $\{\vec{h}_m^{\,di}\}$ are related to $\{\vec{e}_m^{\,di}\}$ by the relation in (44).

The initial values $\{\vec{\delta}_m^{\,i}\}$ are determined by the following three conditions:

(i) The *Gauss Law*, $\nabla \cdot \vec{E} = 0$, when applied to the incident part of the diffracted field in (78), yields

$$\hat{\sigma} \cdot \vec{e}_m^{\,di} + \nabla \cdot \vec{e}_{m-1}^{\,di} = 0 \tag{114}$$

where $\hat{\sigma} = \nabla s^d$ is the direction of the diffracted ray (Fig. 6). It is proved that whenever (114) is satisfied at one point on a diffracted ray, it is satisfied along the whole ray. Thus, it is sufficient to enforce (114) at the point of diffraction O on the edge Γ (Fig. 6). However, it is also shown that the left-hand side of (114) is $O(\sigma^{-1/2})$ when $\sigma \to 0$, and therefore becomes singular at O. An equivalent of (114) at O may be expressed by

$$\lim_{\sigma\to 0} \sigma^{1/2}\left[\hat{\sigma} \cdot \vec{e}^{\,di}_{m} + \nabla \cdot \vec{e}^{\,di}_{m-1}\right] = 0. \tag{115}$$

(ii) The *edge condition* prescribes the behavior of the total electromagnetic field $(\vec{E}^t, \vec{H}^t)$ in the vicinity of the diffracting edge. More specifically, let d denote distance to the edge. Then the edge condition requires that [39]

$$\vec{E}^t \cdot \hat{t} = O(d^{1/2}) \quad , \quad \vec{H}^t \cdot \hat{t} = O(1) \quad , \quad d \to 0 \tag{116}$$

while all other components of $(\vec{E}^t, \vec{H}^t)$ are $O(d^{-1/2})$ when $d \to 0$. In (116), $\hat{t}$ is the unit tangent to the edge. Consider now $(\vec{E}^t, \vec{H}^t)$ given in (112) and (113). The constituents containing the Fresnel integrals are bounded and finite near the edge. Thus, the edge condition should be applied to the terms of the second series in (112) and (113), yielding

$$\vec{w}^{\,i}_{m} \cdot \hat{t} = O(d^{1/2}) \quad , \quad \vec{v}^{\,i}_{m} \cdot \hat{t} = O(1) \quad , \quad d \to 0 \tag{117}$$

and all other components of $\{\vec{w}^{\,i}_{m}, \vec{v}^{\,i}_{m}\}$ are $O(d^{-1/2})$ when $d \to 0$. In order to enforce these conditions, one has to expand $\vec{w}^{\,i}_{m}$ and $\vec{v}^{\,i}_{m}$ for small values of σ. It turns out that these expansions take the form

$$\vec{w}^{\,i}_{m}(\vec{r}) = \vec{A}^{\,i}_{m}\sigma^{-1/2} + O(\sigma^{1/2}) \quad , \quad \sigma \to 0 \tag{118a}$$

$$\vec{v}^{\,i}_{m}(\vec{r}) = \vec{B}^{\,i}_{m}\sigma^{-1/2} + O(\sigma^{1/2}) \quad , \quad \sigma \to 0 . \tag{118b}$$

It follows that the edge condition is completely satisfied provided that $\vec{A}_{m} \cdot \hat{t} = 0$ and $\vec{B}^{\,i}_{m} \cdot \hat{t} = 0$, or, more conveniently,

$$\lim_{\sigma\to 0} \vec{w}^{\,i}_{m} \cdot \hat{t} = 0 \tag{119}$$

$$\lim_{\sigma\to 0} \vec{v}^{\,i}_{m} \cdot \hat{t} = 0 \tag{120}$$

Summarising, the unknown initial values $\{\vec{\delta}^{\,i}_{m}\}$ in (88) for the diffracted field are to be determined by the three conditions in

(115), (119), and (120). Detailed steps for determining $\vec{\delta}_0^i$ and $\vec{\delta}_1^i$ can be found in [7]. We present only final results below.

Solution of $\vec{\delta}_0^i$ is given by

$$\vec{\delta}_0^i = \frac{e^{i\pi/4}}{2\sqrt{2\pi}} \frac{\chi^i}{\sin\beta} \operatorname{rot} \vec{e}_0^i(O) . \tag{121}$$

When (121) is used in (88) and (78), we obtain the leading term of the diffracted field $\vec{E}^d$ of UAT. It is identical to that of $\vec{E}^d$ in GTD given in (89). Recall that $\vec{\delta}_0^i$ is obtained in GTD by a comparison with the solution of a canonical problem (Section V.C). In UAT, such a procedure is unnecessary, as $\vec{\delta}_0^i$ arises as an integral part of the theory. To recapitulate the results obtained so far, the total field in UAT, up to and including terms of $O(k^{-1/2})$, is given by (109), (110), and (89). Thus, to apply UAT, one has to calculate the diffracted field $\vec{E}^d$ according to GTD.

Solution of $\vec{\delta}_1^i$ is given by

$$\vec{\delta}_1^i = \frac{1}{\sin^2\beta} [(C^i \cos\beta - \vec{A}^i \cdot \hat{t})\hat{t} + (\vec{A}^i \cdot \hat{t}\cos\beta - C^i)\hat{\sigma} - (\vec{B}^i \cdot \hat{t})(\hat{t}\times\hat{\sigma})] . \tag{122}$$

The three parameters $\vec{A}^i$, $\vec{B}^i$, and C^i in (122) are very complicated, and their explicit values are given in [7]. When (122) is used in (88), we have the solution of the first-order diffracted field. For two-dimensional problems, more explicit solutions are given in [12],[14],[17].

Solution of $\vec{\delta}_m^i$ for a general m can be determined in a similar way from (115), (119), and (120). No explicit result, however, has been worked out yet.

C. Discussion of UAT

For the diffraction by a thin conducting screen (Fig. 6), the total electric field according to UAT has its asymptotic solution given in (109), (110), and (78). The amplitudes $\{\vec{e}_m^d\}$ of the diffracted field are described in (88) in terms of the initial values $\{\vec{\delta}_m^{i,r}\}$. The first two initial values have been explicitly found and are presented in (121) and (122). The higher order $\{\delta_m^{i,r}\}$ can be determined from the three conditions in (115), (119), and (120). Some discussions of the UAT solution are listed below.

(i) It is believed that $\vec{E}^t$ in (109) is finite and continuous across $SB^{i,r}$. This has been rigorously proved (a) for terms up to and including order $k^{-1/2}$ [7], and (b) for all terms when $\vec{E}^t$ is replaced by a scalar field [8].

(ii) $\vec{E}^t$ in (109) satisfies the edge condition by construction. Hence, it can be used for the calculation of the near field as well as the far field.

(iii) When the observation point $\vec{r}$ is away from $SB^{i,r}$ and edge Γ, one has

$$k^{1/2}|\xi^{i,r}| \to \infty . \tag{123}$$

Then the asymptotic formula of F in (11) may be employed in (110), which leads to an important conclusion

$$\vec{E}^G(\vec{r}) \sim \Theta(-\varepsilon^i)\vec{E}^i + \{i \to r\} ,$$

or (big G becomes small g)

$$\vec{E}^G(\vec{r}) \to \vec{E}^g \quad , \quad k^{1/2}|\xi^{i,r}| \to \infty . \tag{124}$$

Therefore, away from $SB^{i,r}$ and Γ, the total field $\vec{E}^t$ of UAT in (109) reduces to that of GTD in (77). Thus, UAT and GTD agree as long as the observation point is not near the shadow boundaries or the edge (or whenever $k^{1/2}|\xi^{i,r}|$ is large).

(iv) In Section IV, the reflected field $\vec{E}^r$ is defined only in its lit region. In order for (102) to be defined everywhere, we must continue $\vec{E}^r$ into its shadow region. This continuation is discussed in [6].

(v) For the two components in (109), $\vec{E}^G$ satisfies the boundary conditions on the screen Σ, but $\vec{E}^d$ in general does not (except for the special case when Σ is planar). To correct the latter deficiency, additional terms corresponding to secondary (multiply) reflected fields and/or creeping waves must be introduced in (109).

(vi) UAT does not overcome the difficulty at caustics, where infinitely large fields are predicated by UAT.

(vii) The asymptotic nature of the solution in (109) has not been rigorously established.

We note that the difficulties stated in (v) through (vii) also exist in GTD.

D. Extension to Diffraction by a Wedge

It is suggested in [9],[10] that the UAT formulation in (109) can be also used when the thin screen in Fig. 6 is replaced by the wedge in Fig. 10. In the latter case, the first term in (109) becomes

$$\vec{E}^G(\vec{r}) = \left[F\left(k^{1/2}\xi^i\right) - \hat{F}\left(k^{1/2}\xi^i\right)\right]\vec{E}^i(\vec{r}) + \left[F\left(k^{1/2}\xi_1^r\right) - \hat{F}\left(k^{1/2}\xi_1^r\right)\right]\vec{E}_1^r(\vec{r})$$
$$+ \left[\left(F\ k^{1/2}\xi_2^r\right) - \hat{F}\left(k^{1/2}\xi_2^r\right)\right]\vec{E}_2^r(\vec{r}) \ . \tag{125}$$

As described in Section V.D, there are in general two reflected fields, $\vec{E}_1^r$ and $\vec{E}_2^r$, one from each face of the wedge (Fig. 11). That explains the two reflected terms in (125). For the special case that only face Σ_1 is illuminated (Fig. 12a), $k^{1/2}\xi_2^r \to \infty$ for every point in the exterior of the wedge. Then (125) reduces asymptotically to

$$\vec{E}^G(\vec{r}) \sim \left[F\left(k^{1/2}\xi^i\right) - \hat{F}\left(k^{1/2}\xi^i\right)\right]\vec{E}^i(\vec{r}) + \left[F\left(k^{1/2}\xi_1^r\right) - \hat{F}\left(k^{1/2}\xi_1^r\right)\right]\vec{E}_1^r(\vec{r})$$
$$k \to \infty \tag{126}$$

The second term $\vec{E}^d$ in (109) for the present wedge problem is identical to the diffracted field in GTD given in (78). The leading term of $\vec{E}^d$ is described in Section V.C with the diffraction coefficients $\chi^{i,r}$ defined in (103).

Discussions of the UAT solution for the wedge are listed below.

(i) It has been shown [9],[10] that (109) is finite and continuous across $SB^{i,r}$ for the terms up to and including order $k^{-1/2}$. Presumably this property holds true for all orders of k.

(ii) Unlike its counterpart for a thin screen, (109) is not valid when the observation point $\vec{r}$ is near the edge of the wedge. Thus, (109) may be regarded as an outer expansion with respect to the edge boundary layer. For a scalar wave, an inner expansion valid in the neighborhood of the edge is given in [9]. By matching the inner and outer expansions in a common region, the diffraction coefficients $\chi^{i,r}$ in (103) are derived. The same procedure, in principle, may be used to determine the higher-order terms in the diffracted field $\vec{E}^d$ in (78) for the wedge problem. Details remain to be worked out.

(iii) Discussions (iii) to (vii) in Section VI.C apply also to the present wedge problem.

E. Fields on Shadow Boundaries

A main contribution of UAT is to modify GTD for fields near $SB^{i,r}$, where the detour parameters $\xi^{i,r}$ are small. Exactly on $SB^{i,r}$, both $\vec{E}^G$ and $\vec{E}^d$ in (109) become infinite in such a manner that their singular parts cancel each other, and hence $\vec{E}^t$ is finite and continuous. We will now give an explicit expression for the leading terms of $\vec{E}^t$ when the observation point $\vec{r}$ is on SB^i.

Consider the wedge in Fig. 10, which is illuminated by the incident field in (57). At an observation point $\vec{r}$ described by the spherical coordinates (σ,β,α) and locating exactly on SB^i $(\vec{r} = \hat{\sigma}^i\sigma)$, the total field determined from UAT in (109) is [10]

$$\vec{E}^t(\vec{r}=\hat{\sigma}^i\sigma) = \vec{A} + \vec{B} + \vec{C} + \vec{D} + \vec{G} + O(k^{-1}) \quad , \quad k \to \infty \tag{127}$$

The five terms in (127) are explained next:

$$\vec{A} = \frac{1}{2}\vec{E}^i(\vec{r}) \tag{128}$$

is the most dominant term of order k^0, while other terms in (127) are of order $k^{-1/2}$. The term $\vec{B}$ is a contribution from $\vec{E}^G$:

$$\vec{B} = -2g(k\sigma)\,\frac{1}{\sin\beta}\,\frac{\sqrt{1+(\sigma/R_1^i)}\,\sqrt{1+(\sigma/R_2^i)}}{\sqrt{1+(\sigma/R_1)}}\,e^{iks^i(O)} \cdot \{\hat{\beta}\,\frac{\partial}{\partial\alpha}\,[\hat{\beta}\cdot\vec{e}_0^{\,i}(\vec{r})] + \hat{\alpha}\,\frac{\partial}{\partial\alpha}\,[\hat{\alpha}\cdot\vec{e}_0^{\,i}(\vec{r})]\} \,. \tag{129}$$

Here $g(\cdot)$ is defined in (90), and the square roots are taken according to the convention specified in connection with (52). $s^i(O)$ is the incident phase at the point of diffraction O. The positive reference direction of α in (129) is chosen toward the shadow side of $\vec{E}^i$ in the exterior region of the wedge. The factor $\{\cdot\}$ in (129) describes the azimuthal angular variation of the incident field at the observation point. The term $\vec{C}$ in (127) is a contribution from $\vec{E}^d$:

$$\vec{C} = -g(k\sigma)\,\frac{1}{\sin\beta}\,\frac{1}{\sqrt{1+(\sigma/R_1)}}\,K\vec{E}^i(O) \tag{130a}$$

where

$$K = \frac{\kappa\sigma R^i}{\sigma + R_0^i}\,\frac{\hat{\sigma}^i\times\hat{t}\cdot\hat{n}}{\sin^2\beta} + \frac{1}{m}\cot\frac{\pi}{m}\,, \tag{130b}$$

κ is the curvature of the edge (see (84)), and R_0^i is defined in (87). (In [10], an angle ω^i is introduced for calculating K in (130b), and it gives correct results only if the direction of $\hat{t}$ is chosen such that along SB^i the vector $\hat{t}\times\hat{\sigma}^i$ points into the shadow region of $\vec{E}^i$. However, when (130b) is used, no such restriction on the direction of $\hat{t}$ is necessary.) The term $\vec{D}$ in (127) is the contribution from the reflected field. Assuming SB^i is not close to SB^r, $\vec{D}$ is given by

$$\vec{D} = \Theta(-\varepsilon^r)\vec{E}^r(\vec{r}) + g(k\sigma)\,\frac{1}{\sqrt{1+(\sigma/R_1)}}\,\frac{\chi^r}{\sin\beta}\ \mathrm{rot}\ \vec{E}^r(O) \tag{131}$$

where χ^r is given in (103). The term $\vec{G}$ in (127) is a contribution of $\vec{E}^G$:

$$\vec{G} = -g(k\sigma)\,\frac{1}{\sqrt{1+(\sigma/R_1)}}\,\frac{G^i}{\sin\beta}\,\vec{E}^i(O) \tag{132a}$$

where

$$G^i = 2\sigma^2 \sin\beta\,\frac{(1+\sigma/R_1^i)(1+\sigma/R_2^i)}{(1+\sigma/R_0^i)}\Bigg[\frac{1}{2\sigma}\left(\frac{1}{R_1^i+\sigma}-\frac{1}{R_2^i+\sigma}\right)\cot\beta\,\sin\Omega^i\,\cos\Omega^i$$
$$+c_{30}^i\sin^3\Omega^i - c_{21}^i\sin^2\Omega^i\cos\Omega^i + c_{12}^i\sin\Omega^i\cos^2\Omega^i - c_{03}^i\cos^3\Omega^i\Bigg] \tag{132b}$$

$$c_{mn}^i = \frac{1}{m!n!}\left[\left(\frac{\partial}{\partial x_1^i}\right)^m\left(\frac{\partial}{\partial x_2^i}\right)^n s^i(\vec{r})\right]_{\vec{r}=(\sigma,\beta,\alpha)} \tag{132c}$$

Note that $\{c_{mn}^i\}$ in (132c) depend on the third-order derivatives of the incident phase evaluated at the observation point $\vec{r}$. The coordinates (x_1^i, x_2^i) in (132) are those along the principal directions of the incident wavefront as shown in Fig. 8. If the incident wavefront is spherical, $\vec{G}$ is identically zero.

The same formula (127) applies to the magnetic field after replacing $\vec{E}$ by $\vec{H}$. When the observation point $\vec{r}$ is exactly on a reflected shadow boundary SB^r, (127) still holds after changing superscript "i" to "r".

Let us apply (127) to a simple example. Referring to Fig. 14,

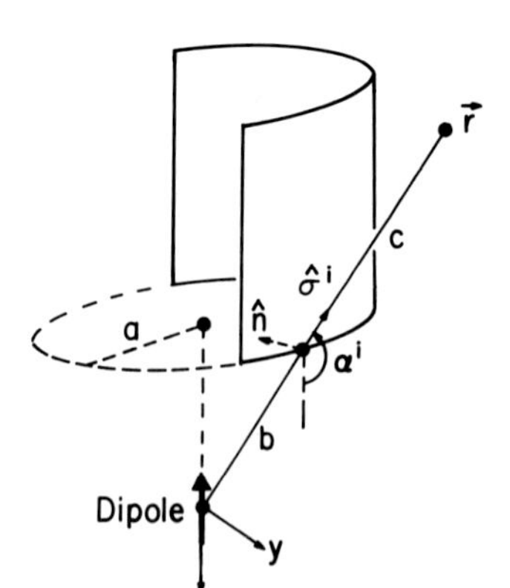

Fig. 14. Field on incident shadow boundary from point source below open-ended cylinder.

the field at $\vec{r}$ on the shadow boundary is to be determined. The incident field is that from a point source located directly below the axis of an open-ended conducting cylinder. The radiation pattern of the point source in the absence of the cylinder is assumed to be

$$\vec{E}^i(r,\theta,\phi) = \frac{e^{ikr}}{kr}\,[\hat{\theta}f_\theta(\theta,\phi) + \hat{\phi}f_\phi(\theta,\phi)] \tag{133}$$

where (r,θ,ϕ) are spherical coordinates with origin at the point source. At a point of diffraction O on the rim of the cylinder $(r = b,\ \theta = \pi/2,\ \phi = \alpha^i)$, we have

$$\beta = \pi/2,\ \sigma = c,\ \kappa = 1/a,\ R_1^i = R_2^i = b,\ R_0^i = b,\ \chi^r = \csc\alpha^i,$$

$$\hat{t} = \hat{z},\ \hat{\beta} = \hat{\theta},\ \hat{\alpha} = \hat{\phi},\ \frac{\partial}{\partial\alpha} = \frac{c}{b+c}\frac{\partial}{\partial\phi}\,.$$

Substituting the preceding results in (128)-(132), we obtain

$$\vec{A} = \frac{1}{2}\vec{E}^i = \frac{e^{ik(b+c)}}{2k(b+c)}\,[\hat{\theta}f_\theta + \hat{\phi}f_\phi] \tag{134a}$$

$$\vec{B} = g\sqrt{\frac{c}{b}}\,\frac{-2}{k(b+c)}\left[\hat{\theta}\,\frac{\partial f_\theta}{\partial\phi} + \hat{\phi}\,\frac{\partial f_\phi}{\partial\phi}\right] \tag{134b}$$

$$\vec{C} = -g\sqrt{\frac{c}{b}}\,\frac{\cos\alpha^i}{k(b+c)}\,\frac{b}{a}\,[\hat{\theta}f_\theta + \hat{\phi}f_\phi] \tag{134c}$$

$$\vec{D} = g\sqrt{\frac{c}{b}}\,\frac{\csc\alpha^i}{kc}\,[-\hat{\theta}f_\theta + \hat{\phi}f_\phi] \tag{134d}$$

$$\vec{G} = 0 \tag{134e}$$

where

$$g = [8\pi k(b+c)]^{-1/2}\,e^{ik(b+c)+i\pi/4}\,. \tag{134f}$$

F. Problems of UAT

In connection with UAT, there are several immediate problems that may be studied in the future: (i) The comparison of the results of UAT and the rigorous asymptotic solutions has been done only for simple two-dimensional geometries. Thus, more complex canonical problems should be solved, analytically or even numerically, so that the validity of the general result of UAT, such as that presented in Section VI.E, can be tested. (ii) UAT fails at caustics, and an improvement in that respect should be most useful. (iii) For the case of a wedge, UAT needs to be refined so that it becomes valid near the edge. (iv) As in all ray techniques, UAT applies in general only when the incident field is a ray field (in the form of (38)). Thus, it cannot be used to attack practical problems such as the calculation of the main beam from a parabolic reflector [35],[40].

VII. CONCLUDING REMARKS

Science progresses with the introduction of theories of ever-increasing generality and ever-finer approximations to physical phenomena. With these advances a hierarchy of theories is generated in which the most general theory reduces to a series of less general, usually earlier approximations in the limit of certain variables. In this paper, we have reported a hierarchy of electromagnetic ray theories: GO, GTD, and UAT. Undoubtedly, this hierarchy is not absolute and further refinement will be made in the future. In this connection, two points may be emphasized [40]. First, the usefulness of a theory within its domain of applicability will not be superseded by later refinements. As a tool of approximation, a theory has limits to its validity and accuracy which the user must keep in mind. Secondly, along with each increase in generality goes, usually, a greater complexity in application. From a practical viewpoint, it is necessary to develop a sense of scale, and to strike a balance between the accuracy sought in a solution and the effort in applying a theory.

VIII. ACKNOWLEDGEMENT

I would like to express my sincere gratitude to Professor

J. Boersma, from whom I learned the uniform asymptotic theory, and my admiration for his scholarship and perfectionist's attitude in doing research.

VIII. REFERENCES

1. Keller, J.B., Diffraction by an Aperture, *J. Appl. Phys. 28,* 426-444 (1957).
2. Keller, J.B., Geometrical Theory of Diffraction, *J. Opt. Soc. Am. 52,* 116-130 (1962).
3. Lewis, R.M. and J.B. Keller, Asymptotic Methods for Partial Differential Equations: The Reduced Wave Equation and Maxwell's Equations, Res. Rep. EM-194, Courant Institute of Mathematical Science, New York University, New York (1964).
4. Cole, J.D., "Perturbation Methods in Applied Mathematics," Blaisdell Publishing Company, Waltham, Massachusetts (1968).
5. Buchal, R.N. and J.B. Keller, Boundary Layer Problems in Diffraction Theory, *Comm. Pure Appl. Math 13,* 85-114 (1960).
6. Lewis, R.M. and J. Boersma, Uniform Asymptotic Theory of Edge Diffraction, *J. Math. Phys. 10,* 2291-2305 (1969).
7. Boersma, J. and P.H.M. Kersten, Uniform Asymptotic Theory of Electromagnetic Diffraction by a Plane Screen, (in Dutch), Tech. Rept., Department of Mathematics, Technical University of Eindhoven, Eindhoven (1967).
8. Ahluwalia, D.S., R.M. Lewis and J. Boersma, Uniform Asymptotic Theory of Diffraction by a Plane Screen, *SIAM J. Appl. Math. 16,* 783-807 (1968).
9. Ahluwalia, D.S., Uniform Asymptotic Theory of Diffraction by the Edge of a Three-Dimensional Body, *SIAM J. Appl. Math. 18,* 287-301 (1970).
10. Lee, S.W. and G.A. Deschamps, A Uniform Asymptotic Theory of Electromagnetic Diffraction by a Curved Wedge, *IEEE Trans. Antennas & Propag. AP-24,* 25-34 (1976).
11. Boersma, J. and M.J. van de Scheur, Diffraction of a Normally Incident, Plane, Scalar Wave through a Slit, (in Dutch), Tech. Rept., Department of Mathematics, University of Eindhoven, Eindhoven (1970).
12. Boersma, J., Ray-Optical Analysis of Reflection in an Open-Ended Parallel-Plane Waveguide. I: TM Case, *SIAM J. Appl. Math 29,* 164-195 (1975).
13. Boersma, J., Ray-Optical Analysis of Reflection in an Open-Ended Parallel-Plane Waveguide. II: TE Case, *Proc. IEEE 62,* 1475-1481 (1974).
14. Lee, S.W. and J. Boersma, Ray-Optical Analysis of Fields on Shadow Boundaries of Two Parallel Plates, *J. Math. Phys. 16,* 1746-1764 (1975).
15. Boersma, J., Diffraction by Two Parallel Half-Planes, *Q. J. Mech. & Appl. Math. 28,* 405-425 (1975).

16. Wolfe, P., Diffraction of Plane Waves by a Semicircular Strip, *Q. J. Mech. & Appl. Math. 28*, 355-371 (1975).
17. Boersma, J. and S.W. Lee, High-Frequency Diffraction of a Line-Source Field by a Half-Plane: Solutions by Ray Techniques, *IEEE Trans. Antennas & Propag. AP-25*, 171-179 (1977).
18. Menendez, R. and S.W. Lee, On the Role of Geometrical Optics Field in Aperture Diffraction, *IEEE Trans. Antennas. & Propag. AP-25*, 688-695 (1977).
19. Yee, H.Y., L.B. Felsen and J.B. Keller, Ray Theory of Reflection from the Open End of a Waveguide, *SIAM J. Appl. Math 16*, 268-300 (1968).
20. Lee, S.W., Ray Theory of Diffraction by Open-Ended Waveguides. II. Applications, *J. Math. Phys. 13*, 656-664 (1972).
21. Jones, D.S., Double Knife-Edge Diffraction and Ray Theory, *Q. J. Mech. & Appl. Math. 26*, 1-18 (1973).
22. Kouyoumjian, R.G. and P.H. Pathak, A Uniform Geometrical Theory of Diffraction for an Edge in a Perfectly Conducting Surface, *Proc. IEEE 62*, 1448-1461 (1974).
23. Mentzer, C.A., L. Peters, Jr. and R.C. Rudduck, Slope Diffraction and Its Application to Horns, *IEEE Trans. Antennas & Propag. AP-23*, 153-159 (1975).
24. Ryan, C.E., Jr. and L. Peters, Jr., Evaluation of Edge-Diffracted Fields Including Equivalent Currents for the Caustic Regions, *IEEE Trans. Antennas & Propag. AP-17*, 292-299 (1969). (See also correction: *AP-18*, 275 (1970).)
25. Knott, E.F. and T.B.A. Senior, Comparison of Three High-Frequency Diffraction Techniques, *Proc. IEEE 62*, 1468-1474 (1974).
26. Hwang, Y.M. and R.G. Kouyoumjian, papers presented at 1974 Fall URSI Meeting at Boulder, Colorado, and at 1975 URSI Meeting at Urbana-Champaign, Illinois.
27. Mittra, R., Y. Rahmat-Samii and W.L. Ko, Spectral Theory of Diffraction, *Appl. Phys. 10*, 1-13 (1976).
28. Rahmat-Samii, Y. and R. Mittra, On the Investigation of Diffracted Fields at the Shadow Boundaries of Staggered Parallel Plates-A Spectral Domain Approach, *Radio Sci. 12*, 659-670 (1977).
29. Ya, P., Ufimtsev, Method of Edge Waves in the Physical Theory of Diffraction, Air Force Systems Command, Foreign Technical Division Document ID No. FTD-HC-23-259-71 (1971) (translation from the Russian version published by Soviet Radio Publication House, Moscow, 1962).
30. Lee, S.W., Comparison of Uniform Asymptotic Theory and Ufimtsev's Theory in Electromagnetic Edge Diffraction, *IEEE Trans. Antennas & Propag. AP-25*, 162-170 (1977).
31. Jones, D.S., "The Theory of Electromagnetics," Macmillan, New York (1964).
32. Felsen, L.B. and N. Marcuvitz, "Radiation and Scattering of Waves," Prentice-Hall, Englewood Cliffs, New Jersey (1973).
33. Deschamps, G.A., Ray Techniques in Electromagnetics, *Proc. IEEE 60*, 1022-1035 (1972). (A minus sign is missing in the

definition of χ.)

34. Lee, S.W., Electromagnetic Reflection from a Conducting Surface: Geometrical Optics Solution, *IEEE Trans. Antennas & Propag. AP-23,* 184-191 (1975).
35. Keller, J.B., R.M. Lewis and B.D. Seckler, Asymptotic Solution of Some Diffraction Problems, *Comm. Pure Appl. Math. 9,* 207-265 (1956).
36. Schensted, C., Electromagnetic and Acoustic Scattering by a Semi-Infinite Body of Revolution, *J. Appl. Phys. 26,* 306-308 (1955).
37. Kouyoumjian, R.G., The Geometrical Theory of Diffraction and Its Application, "Numerical and Asymptotic Techniques in Electromagnetics," (R. Mittra, ed.), Springer-Verlag, New York, 165-215 (1975).
38. Kay, I. and J.B. Keller, Asymptotic Evaluation of the Field at a Caustic, *J. Appl. Phys. 25,* 876-883 (1954).
39. Bouwkamp, C.J., Diffraction Theory, *Rep. Prog. Phys. 17,* 35-100 (1954).
40. Menendez, R.C. and S.W. Lee, Uniform Asymptotic Theory Applied to Aperture Diffraction, University of Illinois at Urbana-Champaign, Electromagnetics Laboratory Scientific Report No. 76-9 (1976).
41. Lee, S.W., Differential Geometry for GTD Applications, University of Illinois at Urbana-Champaign, Electromagnetics Laboratory Scientific Report No. 77-21 (1977).

A SPECTRAL DOMAIN ANALYSIS OF HIGH FREQUENCY DIFFRACTION PROBLEMS[1]

R. Mittra and Y. Rahmat-Samii
Electromagnetics Laboratory
University of Illinois at Urbana-Champaign

I. INTRODUCTION

Ever since Keller's pioneering work [1] on the Geometrical Theory of Diffraction (GTD), representing an extension of Geometrical Optics (GO), there has been a considerable amount of progress made toward the development of the concepts of ray methods and their application to high-frequency diffraction problems. Onc of the principal attributes of the ray techniques is that they exploit the "local" nature of the field solution at high frequencies. The principle of the local field states that the high-frequency limit processes such as reflection and diffraction depend only on the local geometrical and electrical properties of the scatterer in the immediate neighborhood of the point of reflection and diffraction [2]. This concept is extremely useful as it allows one to construct the solution for the scattered field from a complicated geometry by isolating the so-called scattering centers, to compute their individual contributions from the knowledge of the scattering properties of associated canonical geometries, and finally to sum these contributions to generate the total field.

Although the ray methods based on Keller's theory work remarkably well for a very wide class of high-frequency diffraction

[1]The work reported in this paper was supported in part by the National Science Foundation Grant NSF-ENG-76-08305 and in part by Joint Services Electronics Program under Grant JSEP, Grant DAAB C0259.

ISBN 0-12-709650-7

problems, there are situations where the need for refining these solutions becomes clearly evident. A few examples of such situations are: (i) the observation point at which the scattered field is desired, is located at the incident or reflection shadow boundary of a diffracted edge; (ii) a second diffracting edge is located at the shadow boundary of the first edge; and (iii) near edge on incidence of a single or multiple-edged structure, e.g., a strip or an open-ended cylinder. Various approaches based on so-called uniform theories or equivalent current methods have been proposed and extensively developed for the purpose of circumventing the difficulties in Keller's theory. Typically, they attempt to preserve the spirit of Keller's ray concepts and correct it postfact with some modifications of the original formulas at the "trouble regions."

In this communication the authors develop an alternate interpretation of the high-frequency diffraction phenomenon which is based on the concepts of "spectral domain" [3]. In this method the scattered far fields are viewed as the Fourier transform of the induced current on the scatterer and are thus associated with the spectrum of the surface current distribution. Some significant advantages accrue from the use of the spectral domain concepts for the derivation of the solution of high-frequency diffraction problems. First, we show that the spectral concept yields results identical to Keller's formulas where they apply, but the formulas based on spectral concepts continue to be valid at shadow boundaries, and caustic regions, and for edge-edge interactions, etc., where Keller's formula runs into difficulty. Second, we demonstrate that the spectral formulation can be used in a very systematic fashion to produce the higher order asymptotic terms for some important cases. Finally, we show that the important task of testing and systematically improving high frequency solutions, whether derived by ray methods or spectral techniques, can be accomplished by working with the Fourier transform of the integral equation for the surface current and by using interaction or

Galerkin's method in the spectral domain.

Illustrative examples and comparison with other contemporary techniques for solving similar high-frequency diffraction problems are liberally included in the paper to show the ease of application and the systematic nature of the spectral concept and its usefulness for handling practical problems.

II. DIFFRACTION OF A PLANE WAVE BY A HALF-PLANE

The problem of plane-wave diffraction by a half-plane has been analyzed extensively in the literature since Sommerfeld's well-known solution in 1896. The reader may refer to the standard texts of Noble [4], Born and Wolf [5], Mittra and Lee [6], and others in which comprehensive reviews are found. The principal reason why the half-plane solution plays such an important role in diffraction theory is that it forms an integral part of the solution of a large class of high-frequency diffraction problems dealing with more complex bodies. In this section we re-examine this classical problem from a new angle in which the solution is constructed in the spectral domain after introducing the concept of the spectral diffraction coefficient. Only a brief discussion of the solution is presented here, mainly with the objective of laying the foundation for more complex problems to be dealt with in the following sections.

A. Basic Formulation

The geometry of a perfectly conducting half-plane located at $y = 0$, $x \leq 0$, and illuminated by a plane wave is shown in Fig. 1a. The Cartesian coordinates (x,y,z) and the cylindrical coordinates (ρ,ϕ,z) are erected at the edge of the half-plane. Angles are defined positively counter-clockwise with the range $[-\pi,\pi]$. We let the direction of propagation of the incident plane wave be normal to the edge, i.e., $\vec{k}^i \cdot \hat{z} = 0$. This assumption changes the vector nature of the three-dimensional problem to a two-

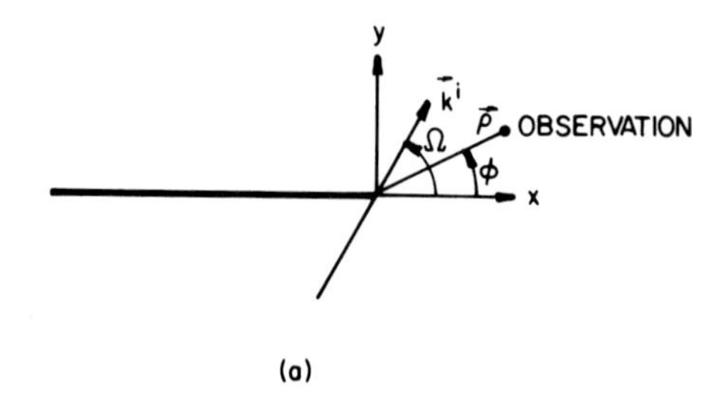

a. Diffraction of a plane wave by a half-plane, $0 \leq \Omega \leq \pi$ and $-\pi \leq \phi < \pi$.

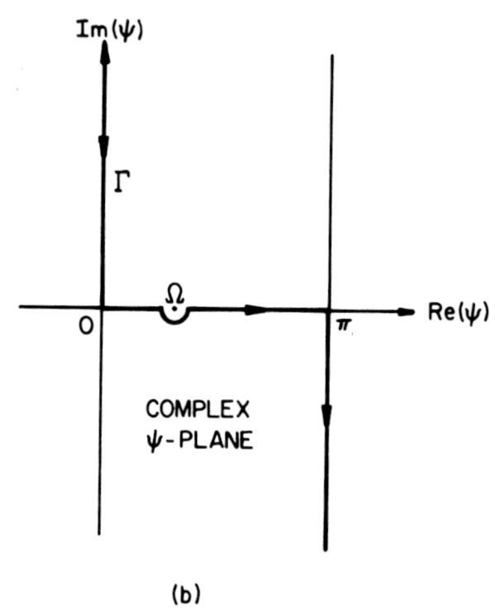

b. Integration path Γ *for integral representation (11).*

Fig. 1a and b.

dimensional scalar problem. Furthermore, the problem may be classified as in the cases of *E*-wave (nonzero field components E_z, H_x, H_y), or *H*-wave (nonzero field components H_z, E_x, E_y) by simply letting the incident field *E*-field or *H*-field be polarized alternatively along the z-axis. Unless otherwise specified, the cases of *E*- and *H*-waves are treated simultaneously, with the help of two symbols u and τ, such that

$$\text{for } E\text{-waves:} \quad u = E_z \,, \; \tau = -1 \tag{1a}$$

$$\text{for } H\text{-waves:} \quad u = H_z \,, \; \tau = +1. \tag{1b}$$

The total field u^t may be split into the incident field u^i and the scattered field u^s to give

$$u^t = u^i + u^s. \tag{2}$$

For a perfect electric conductor the total field u^t is subject to the boundary condition $u^t = 0$ or $\partial u^t/\partial y = 0$ for E-wave or H-wave cases, respectively, on the half-plane. If one defines the induced electric current on the half-plane as

$$\text{for } E\text{-wave:} \quad J_z = \left.\frac{\partial u^t}{\partial y}\right|_{0^+}^{0^-} \tag{3a}$$

$$\text{for } H\text{-wave:} \quad J_x = -u^t\Big|_{0^+}^{0^-}, \tag{3b}$$

and uses the time convention $e^{-i\omega_0 t}$, one can readily arrive at the following equations using Maxwell's equation

$$\text{for } E\text{-wave:} \quad u^s = E_z^s = i\omega_0\mu \int_{-\infty}^{0} J_z(x')g_0(k|\vec{\rho} - x'\hat{x}|)dx' \tag{4a}$$

$$\text{for } H\text{-wave:} \quad u^s = H_z^s = \frac{\partial}{\partial y}\int_{-\infty}^{0} J_x(x')g_0(k|\vec{\rho} - x'\hat{x}|)dx' \tag{4b}$$

where $k = \omega_0\sqrt{\mu\varepsilon}$, μ and ε are the permeability and permittivity of the medium, respectively, and $g_0(k\rho) = iH_0^1(k\rho)/4$ (H_0^1 is the Hankel function of first kind and zero order). The objective is to determine J_s and u^s for the half-plane illuminated by an incident plane wave. This is done by using transform techniques and employing the results given in [6].

B. Spectral Diffraction Coefficient and Total Field

Let us define the Fourier transform pair as

$$U(\alpha) = \int_{-\infty}^{\infty} u(x)e^{i\alpha x}dx = F[u(x)] \tag{5a}$$

and

$$u(x) = \frac{1}{2}\pi \int_{-\infty+i\Delta}^{\infty+i\Delta} U(\alpha)e^{-i\alpha x}d\alpha = F^{-1}[U(\alpha)] \tag{5b}$$

where Δ is a small positive number. The incident plane wave may also be written as

$$u^i = e^{i\vec{k}\cdot\vec{\rho}} = e^{i(k_x x + k_y y)} = e^{ik\rho\cos(\Omega-\phi)} \tag{6}$$

where $k_x = k \cos \Omega$, $k_y = k \sin \Omega$ and $0 \leq \Omega \leq \pi$ is the incident angle shown in Fig. 1a. Transforming (4) into the spectral (Fourier) domain and applying the well-known Wiener-Hopf construction [6], one arrives at the following

$$\text{for } E\text{-wave:} \quad F[J_z] = (i\omega_0\mu)^{-1} X(k_x,\alpha) \tag{7a}$$

$$\text{for } H\text{-wave:} \quad F[J_x] = -\gamma^{-1} X(k_x,\alpha) \tag{7b}$$

where $X(k_x,\alpha)$ is

$$X(k_x,\alpha) = 2\,\frac{\sqrt{k+\tau k_x}\,\sqrt{k+\tau\alpha}}{\alpha + k_x}\,, \tag{8}$$

and $\gamma = \sqrt{\alpha^2 - k^2}$, $\alpha = k_x$ and $Re\,\gamma \geq 0$ (see [6]). In this work, unless otherwise stated, $\sqrt{\cdot}$ and $(\cdot)^{1/2}$ are defined with their proper branch cut slightly below the negative real axis. Using the transform version of (4) and then incorporating (7), one finally obtains

$$U^s = \begin{Bmatrix} 1 \\ sgn\ (y) \end{Bmatrix} X(k_x,\alpha)\,\frac{e^{-\gamma|y|}}{2\gamma} \quad \text{for} \begin{Bmatrix} E\text{-wave} \\ H\text{-wave} \end{Bmatrix}. \tag{9}$$

Furthermore, one may notice that the following equation has been used in the construction of (9)

$$F[g_0(k\rho)] = \frac{e^{-\gamma|y|}}{2\gamma}\,. \tag{10}$$

Introducing the change of variables $x = \rho \cos \phi$, $y = \rho \sin \phi$, $k_x = k \cos \Omega$, $k_y = k \sin \Omega$, $\alpha = -k \cos \psi$ and $\gamma = -ik \sin \psi$ into (9) and substituting the result into (5b), one finally arrives at

$$u^s = \frac{i}{4\pi} \begin{Bmatrix} 1 \\ sgn\ (\phi) \end{Bmatrix} \int_\Gamma \chi(\Omega,\psi)\, e^{ik\rho \cos(\psi-|\phi|)}\, d\psi \quad \text{for} \begin{Bmatrix} E\text{-wave} \\ H\text{-wave} \end{Bmatrix}. \tag{11}$$

In the preceding equation ψ is the complex angle defined on path Γ, shown in Fig. 1b, and $\chi(\Omega,\psi)$ is

$$\chi(\Omega,\psi) = X(k \cos \Omega, -k \cos \psi) = \chi_i(\Omega,\psi) + \tau\chi_r(\Omega,\psi) \tag{12}$$

where

$$\chi_{\substack{i\\r}}(\Omega,\psi) = \mp\csc\frac{\Omega\mp\psi}{2} . \tag{13}$$

We may notice that $\chi_i(\cdot)$ and $\chi_r(\cdot)$ have the same functional form, i.e., csc $(\cdot)$. This definition of χ_i and χ_r is closely related to the definition used by Deschamps in [7]. Clearly χ_i and χ_r are infinite at $\psi = \Omega$ and $\psi = -\Omega$, respectively. These two values of ω correspond to the incident and reflection shadow boundaries appearing in the GTD technique. As a matter of fact, $\chi(\Omega,\psi)$ is precisely the angular part of Keller's diffraction coefficient, when ω is replaced by the observation angle ϕ. Although χ tends to infinity at the shadow boundaries, it does not mean that the field itself is also infinite as Keller's GTD predicts. Instead, the correct value of the field is obtained from (11), which is always bounded. To distinguish it from Keller's coefficient, which is associated with the diffracted field, we will refer to $\chi(\Omega,\psi)$ as the Spectral Diffraction Coefficient for the half-plane. This terminology is chosen since $\chi(\Omega,\psi)$ is associated with the spectrum, or equivalently, the Fourier transform, of the induced current and appears only inside the kernel of the plane wave spectrum representation for the field and not directly in the form of a factor multiplying the incident field as in the case of Keller's representation.

We may further use (4) and (7) and introduce the spectral coefficient of the physical optics field X^{po} as the Fourier transform of the physical optics induced current to arrive at

$$\text{for } E\text{-wave:} \quad X^{po}(k_x,\alpha) = \frac{2k_y}{\alpha + k_x} \tag{14a}$$

$$\text{for } H\text{-wave:} \quad X^{po}(k_x,\alpha) = \frac{2i\sqrt{\alpha^2 - k^2}}{\alpha + k_x} . \tag{14b}$$

The application of the change of variables used in (11) allows one to express (14) as

$$\chi^{po}(\Omega,\psi) = \chi_i^{po}(\Omega,\psi) + \tau\chi_r^{po}(\Omega,\psi) \quad (15)$$

where

$$\chi_r^{po}(\Omega,\psi) = \mp\text{ctn}\,\frac{\Omega \mp \psi}{2}\,. \quad (16)$$

It is worthwhile to mention that χ^f, as defined in the following equation, is bounded at the shadow boundaries

$$\chi^f(\Omega,\psi) = \chi(\Omega,\psi) - \chi^{po}(\Omega,\psi)\,. \quad (17)$$

$\chi^f(\Omega,\psi)$ could be called the fringe diffraction coefficient, and it is used for the scalar aperture diffraction problems [8].

For the problem at hand, i.e., incident plane wave, the spectral integral (11) can be expressed exactly in terms of the Fresnel integral, viz.,

$$u^s = -e^{ik\rho\cos(\Omega-\phi)}F(-\xi_i) + \tau\, e^{ik\rho\cos(\Omega+\phi)}F(\xi_r) \quad (18)$$

where the Fresnel integral F is defined as

$$F(\xi) = \frac{e^{-i\pi/4}}{\sqrt{\pi}}\int_{\xi}^{\infty} e^{it^2}\,dt \quad (19)$$

and its properties are discussed in [9] and [10]. Furthermore, ξ_i and ξ_r are

$$\xi_{\substack{i\\r}} = \mp\sqrt{2k\rho}\,\sin\frac{\Omega \mp \phi}{2}\,. \quad (20)$$

Using the analytic continuation argument, one can show that, for complex angles of incidence, (18) is still the proper solution of the diffraction problem. In this context Ω is replaced by the complex angle ω which follows the path Γ_i, $[(i\infty,0)U(0,\pi)U(\pi,-i\infty)]$, in the complex ω-plane to cover the infinite spectrum of incidence angles.

In reviewing the material presented in this section, we note that its principal contribution has been the introduction of the spectral diffraction coefficient, which, in turn, is shown to be associated with the integral representation of the scattered field. The equivalence between the GTD results and those derived from the spectral representation for observation angles not close to the shadow boundaries can be easily established by substituting the asymptotic expansion of the Fresnel integral into (18). In the next few sections, we will illustrate the broad nature of the spectral concept and its versatility of application by considering more general incident waves and complex structures than the half-plane illuminated by a plane wave.

III. DIFFRACTION OF AN ARBITRARY INCIDENT FIELD BY A HALF-PLANE

Recently there has been much interest in the problem of predicting the characteristics of antennas mounted on complex structures, such as aircraft, and in the accurate calculation of the radiation pattern of reflector antennas used for high-frequency communication. Accurate solution of these problems requires a thorough understanding of the scattering process of an arbitrary incident field impinging on a diffracting edge. Though the problem of diffraction by a half-plane due to an incident plane wave has been studied extensively in the past, it has not been analyzed in depth for the case of an arbitrary incident field. The first rigorous mathematical treatment of the problem was given by Carslaw in 1899, in which he considered the very special case of the incident field due to an isotropic line source. The latter problem was also studied by Clemmow [11], who provided much insight into the subject through the use of plane wave spectrum analysis. In his paper, Clemmow presents a good review of the existing literature up to the year 1950. A comprehensive study of the half-plane diffraction due to an incident plane wave and an isotropic line source is given in Chapter 11 of the book by Born and Wolf [5]. Khestanov [12] derived an integral representation of the field for the diffraction by an arbitrary incident field and presented a formal solution of the problem. In his construction, no details regarding the behavior of the field at the shadow boundaries were provided. Recently, Boersma and Lee [13] have employed the formalism of the Uniform Asymptotic Theory (UAT), which is based on an Ansatz developed by Ahluwalia, Lewis and Boersma [14], to obtain the asymptotic solution of the field diffracted by a half-plane due to a line source field to the order of $k^{-3/2}$ of the incident field.

Our objectives in this section are the following: (i) to present an integral representation of the incident field and study its asymptotic behavior; (ii) to derive the radiated field behavior of a multipole line source; (iii) to use spectral domain analysis (STD) in a manner discussed in the previous section and obtain a general formulation of the problem; (iv) to use asymptotic techniques and construct higher-order asymptotic expressions for the total field, including the order $k^{-5/2}$; (v) to employ numerical techniques and determine the field in the regions where asymptotic techniques cannot be used conveniently; and (vi) to compare the results with those of other asymptotic techniques and draw some unique and useful conclusions.

A. Incident Field - Asymptotic Expansion

1. General Case

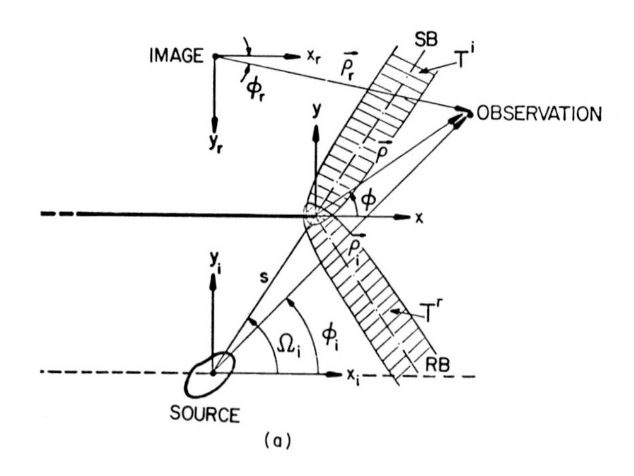

Fig. 2a. Diffraction of an arbitrary field by a half-plane.

As shown in Fig. 2a, a perfectly conductive half-plane at $x < 0$ and $y = 0$ is illuminated by an arbitrary incident field with the following spectral plane-wave representation

$$u^i(\rho_i, \phi_i) = \frac{i}{4\pi}\int_{\Gamma_i} P(\omega,k)\, e^{ik\rho_i \cos(\omega - |\phi_i|)}\, d\omega, \qquad \pi < \phi_i < \pi \tag{21}$$

where $\vec{\rho}_i = 0$ denotes the phase center for the incident field when its Fourier transform is taken along the dashed line shown in the same figure. $P(\omega,k)$ is known as the pattern function; it is assumed that this function is slowly varying in terms of a large k and has the desired differentiability property. Furthermore, the integration path Γ_i in the complex plane ω is shown in Fig. 2b

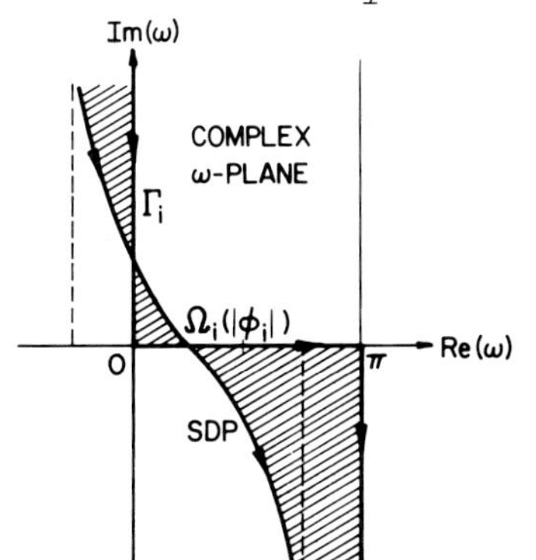

Fig. 2b. Complex ω*-plane, path* Γ_i *and steepest descent path (SDP).*

and its characteristic is discussed in [6]. As a special case, if we let $P(\omega,k) = 1$, we obtain from (21)

$$u^i = \frac{i}{4} H_0^{(1)}(k\rho_i) = \frac{i}{4\pi}\int_{\Gamma_i} e^{ik\rho_i \cos(\omega - |\phi_i|)}\, d\omega \tag{22}$$

which is proportional to the field of an isotropic line source.

For large values of k (high frequency), one is usually interested in determining the asymptotic expansion of (21). To do this, one first assumes that $P(\omega,k)$ can be expanded in terms of an asymptotic series in the following fashion

$$P(\omega,k) = \sum_{m=0}^{\infty} (ik)^{-m} P_m(\omega); \tag{23}$$

moreover, $P(\omega,k)$ does not possess any pole singularity. The path Γ_i is next deformed to the steepest descent path SDP passing through the saddle point $\omega = |\phi_i|$, as shown in Fig. 2b. On this path, the relation $Re\ [\cos\ (\omega - |\phi_i|)] = 1$ holds; thus one may introduce the following change of variable

$$t = \sqrt{2}\, e^{i\pi/4} \sin \frac{\omega - |\phi_i|}{2} \tag{24}$$

in which t is a real variable taking the domain $[-\infty,\infty]$. Substituting (24) into (21) and using the fact that t is now a real variable, one finally arrives at

$$u^i(\rho_i,\phi_i) = \frac{e^{ik\rho_i + i\pi/4}}{2\pi\sqrt{2}} \int_{-\infty}^{\infty} G(t,k)\, e^{-k\rho_i t^2}\, dt \tag{25}$$

where

$$G(t,k) = P(\omega,k) \sec \frac{\omega - |\phi_i|}{2} \tag{26a}$$

in which ω is replaced by

$$\omega = 2 \text{ Arc sin} \left(\frac{e^{-i\pi/4}}{\sqrt{2}}\, t\right) + |\phi_i|$$

$$= \mp\, i\, \ell n\, [1 + it^2 \pm (1+i)t\sqrt{1 + it^2/2}] + |\phi_i| \tag{26b}$$

$$\text{as } t \gtrless 0.$$

Using (23), the asymptotic expansion of (26a) may be written as

$$G(t,k) = \sum_{m=0}^{\infty} (ik)^{-m} G_m(t)\ ;\ G_m(t) = P_m(\omega) \sec \frac{\omega - |\phi_i|}{2} \tag{27}$$

where (26b) is also implied. The complete asymptotic expansion procedure [15] is used for the asymptotic evaluation of (25). In this procedure, one first expands $G_m(t)$ in a Taylor series

$$G_m(t) = \sum_{n=0}^{\infty} \frac{G_m^{(n)}(0)}{\Gamma(n+1)}\, t^n \tag{28}$$

where $G_m^{(n)}(0) = \left.\frac{\partial^{-n}}{\partial t^n} G_m(t)\right|_{t=0}$ and Γ is the Gamma function. Then one substitutes (28) into (26a) and (25) and uses the result

$$\int_{-\infty}^{\infty} t^n e^{-k\rho t^2} dt = \begin{cases} (k\rho)^{-(1+n)/2}\Gamma[(1+n)/2] & \text{for } n \text{ even} \\ 0 & \text{for } n \text{ odd} \end{cases} \tag{29}$$

to finally arrive at

$$u^i(\rho_i,\phi_i) = \frac{e^{ik\rho_i+i\pi/4}}{2\sqrt{2\pi k}} \sum_{m=0}^{\infty} (ik)^{-m} \sum_{n=0}^{\infty} \frac{\Gamma(n+1/2)}{\sqrt{\pi}\,\Gamma(2n+1)} e^{in\pi/2} (ik)^{-n} \cdot G_m^{(2n)}(0)\rho_i^{-n-1/2}$$

$$= \frac{e^{ik\rho_i+i\pi/4}}{2\sqrt{2\pi k}} \sum_{n=0}^{\infty} \frac{2^{-2n}}{n!} e^{in\pi/2} (ik)^{-n} G^{(2n)}(0,k)\, \rho_i^{-n-1/2}. \tag{30}$$

If we combine the order of k's in (30), we may express the incident field as

$$u^i(\rho_i,\phi_i) = \frac{e^{ik\rho_i+i\pi/4}}{2\sqrt{2\pi k}} Z^i(\rho_i,\phi_i) = \frac{e^{ik\rho_i+i\pi/4}}{2\sqrt{2\pi k}} \sum_{m=0}^{\infty} (ik)^{-m} Z_m^i(\rho_i,\phi_i) \tag{31}$$

which has a ray field characteristic [16], i.e., satisfying the Luneberg-Kline expansion. Our task is now to determine Z_m^i's in terms of P_m's. From (27) and (26b), the following relations are easily constructed:

$$\begin{cases} G_m(0) = P_m(|\phi_i|) \\ G_m^{(1)}(0) = \left(\frac{dt}{d\omega}\right)^{-1} \frac{d}{d\omega}\left[P_m(\omega)\sec\frac{\omega-|\phi_i|}{2}\right]\Bigg|_{\omega=|\phi_i|} \\ \cdot \\ \cdot \\ \cdot \\ G_m^{(n)}(0) = D^n\left[P_m(\omega)\sec\frac{\omega-|\phi_i|}{2}\right]\Bigg|_{\omega=|\phi_i|} \end{cases} \tag{32}$$

where $D = \left(\frac{dt}{d\omega}\right)^{-1}\left(\frac{d}{d\omega}\right)$. Simplifying the above relations, one obtains

$$\begin{cases} G_m^{(2)}(0) = -2i\left[\frac{d^2P_m(\omega)}{d\omega^2} + \frac{1}{4}P_m(\omega)\right]_{\omega=|\phi_i|} & \text{(33a)} \\ G_m^{(4)}(0) = -4\left[\frac{d^4P_m(\omega)}{d\omega^4} + \frac{5}{2}\frac{d^2P_m(\omega)}{d\omega^2} + \frac{9}{16}P_m(\omega)\right]_{\omega=|\phi_i|} . & \text{(33b)} \end{cases}$$

The higher-order terms may also be determined in the same fashion.

Substituting (33) into (30) and comparing the results with (31), we finally obtain

$$z_0^i(\rho_i,\phi_i) = P_0(|\phi_i|)\rho_i^{-1/2} \tag{34a}$$

$$z_1^i(\rho_i,\phi_i) = P_1(|\phi_i|)\rho_i^{-1/2} - \left[\frac{1}{2}P_0''(|\phi_i|) + \frac{1}{8}P_0(|\phi_i|)\right]\rho_i^{-3/2} \tag{34b}$$

$$z_2^i(\rho_i,\phi_i) = P_2(|\phi_i|)\rho_i^{-1/2} + \left[\frac{1}{2}P_1''(|\phi_i|) + \frac{1}{8}P_1(|\phi_i|)\right]\rho_i^{-3/2} + \left[\frac{1}{8}P_0''''(|\phi_i|) + \frac{5}{16}P_0''(|\phi_i|) + \frac{9}{128}P_0(|\phi_i|)\right]\rho_i^{-5/2} \tag{34c}$$

which completes the asymptotic evaluation of $u^i(\rho_i,\phi_i)$ up to and including the order of $k^{-5/2}$. We notice that the above procedure can be systematically followed to determine any higher-order terms in the asymptotic expansion of the incident field.

2. *Multipole Line Source*

In this section, we study the behavior of the radiated field of a multipole line source and employ the results of the previous section to construct the asymptotic expansion for the radiated field. A multipole line source, either electric or magnetic, may be defined as

$$\hat{z}J(x_i,y_i) = \hat{z}\sum_{n=0}^{N}\sum_{m=0}^{M} k^{-n-m} I_{nm}\,\delta^{(n)}(x_i)\,\delta^{(m)}(y_i) \tag{35}$$

where (n) symbolizes an nth order differentiation and I_{nm}'s are constants. The field radiated by (35) is proportional to

$$u^i(x_i,y_i) = \iint_{-\infty}^{\infty} J(x_i',y_i') \frac{i}{4} H_0^{(1)}\left[k\sqrt{(x_i - x_i')^2 + (y_i - y_i')^2}\right] dx_i' dy_i' \tag{36}$$

where the proportionality constant is $i\omega_0\mu$ for the electric current source (E-wave) and $i\omega_0\varepsilon$ for the magnetic current source (H-wave). Since

$$\int_{-\infty}^{\infty} \delta^{(n)}(x_i') e^{i\alpha x_i'} dx_i' = (-1)^n (i\alpha)^n \tag{37a}$$

and

$$\int_{-\infty}^{\infty} \delta^{(m)}(y_i') \frac{e^{-\sqrt{\alpha^2-k^2}|y_i-y_i'|}}{2\sqrt{\alpha^2 - k^2}} dy_i' = \left(\mp\sqrt{\alpha^2 - k^2}\right)^m \frac{e^{-\sqrt{\alpha^2-k^2}|y_i|}}{2\sqrt{\alpha^2 - k^2}}$$

$$\text{for } y_i \gtrless 0, \tag{37b}$$

the Fourier transform of (36) along the x_i-axis may be expressed as

$$u^i(\alpha,y_i) = \int_{-\infty}^{\infty} u^i(x_i,y_i) e^{i\alpha x_i} dx_i =$$

$$= \left[\sum_{n=0}^{N} \sum_{m=0}^{M} k^{-n-m} I_{nm} (-i\alpha)^n \left(\mp\sqrt{\alpha^2 - k^2}\right)^m\right] \cdot \frac{e^{-\sqrt{\alpha^2-k^2}|y_i|}}{2\sqrt{\alpha^2 - k^2}}$$

$$\text{for } y_i \gtrless 0. \tag{38}$$

Introducing the change of variable $\alpha = -k \cos \omega$ and $\sqrt{\alpha^2 - k^2} = -ik \sin \omega$ into (38), performing the inverse Fourier transformation, and finally comparing the result with (21), we arrive at the following interpretation for the pattern function of a multipole line source

$$P(\omega,k) = P(\omega) = \sum_{n=0}^{N} \sum_{m=0}^{M} I_{nm} e^{i(n+m)\pi/2} (\cos \omega)^n (\text{sgn } \phi_i \sin \omega)^m. \tag{39}$$

As expected, the pattern function of a multipole line source is independent of k. To construct the asymptotic expression of (36), one may simply use (31) and substitute $P_0(\omega) = P(\omega)$, defined in (39), into (34). It is evident that a closed-form expression for (36) may be obtained by substituting (35) into (36) and writing the final result in terms of the higher-order Hankel functions.

B. Diffraction by a Half-Plane

The geometry of a half-plane illuminated by an arbitrary incident field is shown in Fig. 2a. The half-plane is illuminated by an arbitrary incident field with its plane wave spectral representation given in (21) and its pattern function in (39). The total field u^t may be split into the incident field u^i and scattered field u^s as in (2). Our goal is to determine u^s with the application of the spectral diffraction coefficient. To this end, the incident field (21) may first be expressed as

$$u^i(\rho,\phi) = \frac{i}{4\pi}\int_{\Gamma_i} P(\omega)\, e^{iks\cos(\omega-\Omega)}\, e^{ik\rho\cos(\omega-\phi)}\, d\omega \qquad (40)$$

$$\text{for } \rho\sin\phi > -s\sin\Omega$$

where s and Ω are defined in Fig. 2a. If we substitute $P(\omega) = -4\pi i\delta(\omega - \Omega)$ into (40), the plane wave (6) will be recovered. Representation (40) may be interpreted as a collection of plane waves with spectral density $i(4\pi)^{-1}P(\omega)\exp[iks\cos(\omega-\Omega)]\,d\omega$ illuminating the half-plane. Comparing (40) with (6) and using (21), one may finally express the scattered field u^s as

$$u^s = \frac{i}{4\pi}\int_{\Gamma_i} d\omega\, P(\omega)\, e^{iks\cos(\omega-\Omega)} \qquad (41)$$

$$\cdot\left[\frac{i}{4\pi}\begin{Bmatrix} 1 \\ \operatorname{sgn}(\phi)\end{Bmatrix}\int_{\Gamma} \chi(\omega,\psi)\, e^{ik\rho\cos(\psi-|\phi|)}\, d\psi\right] \quad \text{for}\begin{Bmatrix} E\text{-wave} \\ H\text{-wave}\end{Bmatrix}.$$

Attempts were made to evaluate (41) asymptotically by first replacing the Γ integration by its asymptotic expression and then calculating the Γ_i integration. Due to the nonuniform nature of this procedure, proper results could not be obtained in a convenient manner.

The inner integration over Γ in (41) may be evaluated uniformly in terms of the Fresnel integral using (21) and (18). Once this is done, one may then express u^s as

$$u^s(\rho,\phi) = u_i^s(\rho,\phi) + \tau u_r^s(\rho,\phi) \tag{42a}$$

where

$$u^s_{\substack{i\\r}}(\rho,\phi) = \frac{i}{4\pi}\int_{\Gamma_i} \mp P(\omega)\; e^{iks\cos(\omega-\Omega)}\, e^{ik\rho\cos(\omega\mp\phi)}\, F(\mp\xi_{\substack{i\\r}})\; d\omega. \tag{42b}$$

In the above equation $F(\cdot)$ has already been defined in (19) and ξ's are obtained from (20) after Ω is replaced by ω, viz.,

$$\xi_{\substack{i\\r}} = \mp\sqrt{2k\rho}\,\sin\frac{\omega\mp\phi}{2}\,. \tag{42c}$$

It is evident that $u^s(\rho,\phi) = u^s(\rho,-\phi)$, and that (42b) can be interpreted as the superposition integral. Our task is now to evaluate (42b) asymptotically for different locations of the observation and source points according to the behavior of ξ_i and ξ_r. We first deform Γ_i to the steepest descent path (SDP) (see Fig. 2b) through the saddle point $\Omega = \Omega_i$ given by $Re[\cos(\omega-\Omega)] = 1$ and $Im[\cos(\omega-\Omega)] \geq 0$. It may be shown that the contribution of the closing segments of the integration path at infinity is zero. Since there is no pole in the integrand of (42b), this equation may be written as

$$u^s_{\substack{i\\r}}(\rho,\phi) = \frac{i}{4\pi}\int_{SDP} \mp P(\omega)\; e^{iks\cos(\omega-\Omega)}\, e^{ik\rho\cos(\omega\mp\phi)}\, F(\mp\xi_{\substack{i\\r}})\; d\omega. \tag{43}$$

Equation (43) is the starting point of our discussion for the asymptotic evaluation of the scattered field. Before dealing with this topic in the next few sections, let us introduce some definitions that will make our task convenient. Observation directions $\phi = \Omega$ and $\phi = -\Omega$ are called incident (SB) and reflection (RB) shadow boundaries, respectively, as shown in Fig. 2a. Regions in the vicinity of SB and RB, where ξ_i and ξ_r at $\omega = \Omega$ take some given values for different $\vec{\rho}$'s, are called transition regions and are designated by T^i and T^r, respectively.

1. Total Field for $\phi \notin T^i \cup T^r$

In this region, the Fresnel function in (43) is substituted by its asymptotic series from the following expression

$$F(\xi) = \theta[-Re(\xi e^{-i\pi/4})] + \tilde{F}(\xi) \quad \text{for } |\xi| >> 0 \tag{44a}$$

where

$$\tilde{F}(\xi) = e^{i\xi^2} \frac{e^{i\pi/4}}{2\pi\xi} \sum_{m=0}^{\infty} \Gamma(m+1/2)(i\xi^2)^{-m}, \tag{44b}$$

and θ is the unit step function. It is noticed that $\tilde{F}(\xi)$ is an odd function, i.e., $\tilde{F}(-\xi) = -\tilde{F}(\xi)$. One may then express (43) as

$$u^s_{\substack{i\\r}}(\rho,\phi) = u^g_{\substack{i\\r}}(\rho,\phi) + u^d_{\substack{i\\r}}(\rho,\phi) \tag{45a}$$

where

$$u^g_{\substack{i\\r}}(\rho,\phi) = \frac{i}{4\pi}\int_{SDP(\Omega)} \mp P(\omega)\, e^{iks\cos(\omega-\Omega)}\, e^{ik\rho\cos(\omega\mp\phi)} \cdot \theta\left[-Re\left(\mp\xi_{\substack{i\\r}}\, e^{-i\pi/4}\right)\right] d\omega \tag{45b}$$

and

$$u^d_{\substack{i\\r}}(\rho,\phi) = \frac{i\, e^{ik\rho}}{4\pi}\int_{SDP(\Omega)} P(\omega)\, e^{iks\cos(\omega-\Omega)}\, e^{-i\xi^2_{\substack{i\\r}}}\, \tilde{F}(\xi_{\substack{i\\r}})\, d\omega. \tag{45c}$$

Since $u^d_r(\rho,\phi) = u^d_i(\rho, 2\pi - \phi)$, our attention will only be focused on the asymptotic evaluation of u^d_i. On the SDP path, we introduce the change of variable

$$\omega = 2\,\text{Arcsin}\left(\frac{e^{-i\pi/4}}{\sqrt{2}}\, t\right) + \Omega = \mp i\, \ell n\left[1 + it^2 \pm (1+i)t\sqrt{1 + it^2/2}\right] + \Omega \quad \text{for } t \gtrless 0 \tag{46}$$

into (45c) to have the integration performed on the real line. We notice that $\cos(\omega - \Omega) = 1 + it^2$ and that both $\ell n\,[\cdot]$ and $\sqrt{\cdot}$ are defined with their branch cuts on the negative real axis. Following the same steps used in the previous section to generate

(30), we arrive at

$$u_i^d(\rho,\phi) = g(ks)g(k\rho)\sum_{m=0}^{\infty}(ik)^{-m}\sum_{n=0}^{\infty}\frac{\Gamma(n+1/2)}{\Gamma(2n+1)}e^{in\pi/2}(ik)^{-n}G_m^{(2n)}(0)s^{-n} \tag{47}$$

where

$$g(k\rho) = \frac{e^{ik\rho+i\pi/4}}{2\sqrt{2\pi k\rho}}, \tag{48}$$

and $G_m^{(2n)}$'s are obtained from (28) using

$$G_m(t) = -\frac{1}{\pi}\Gamma(m+1/2)(2\rho)^{-m}\csc^{2m+1}\left(\frac{\omega-\phi}{2}\right)\sec\frac{\omega-\Omega}{2}P(\omega) \tag{49}$$

such that, at the same time, ω is replaced by t from (46).

The diffracted field u_i^d may be expressed as

$$u_i^d(\rho,\phi) = u_{i1}^d(\rho,\phi) + u_{i2}^d(\rho,\phi) + u_{i3}^d(\rho,\phi) + 0(k^{-4}) \tag{50}$$

in which the r.h.s. of (50) is obtained by using (47) and simplifying the result. After performing all the necessary manipulations, i.e., using the same steps as in (32) and (34), we finally arrive at

$$u_{i1}^d(\rho,\phi) = g(k\rho)\chi_i g(ks)P(\Omega) \tag{51a}$$

$$\begin{aligned} u_{i2}^d(\rho,\phi) &= \frac{1}{4}g(k\rho)(ik\rho)^{-1}\chi_i^3 g(ks)P(\Omega) \\ &+ \frac{1}{2}g(k\rho)\left[\frac{1}{2}\chi_i^3 P(\Omega) + \cos\frac{\Omega-\phi}{2}\chi_i^2 P'(\Omega) + \chi_i P''(\Omega)\right](iks)^{-1}g(ks) \end{aligned} \tag{51b}$$

$$\begin{aligned} u_{i3}^d(\rho,\phi) &= \frac{3}{16}g(k\rho)(ik\rho)^{-2}\chi_i^5 g(ks)P(\Omega) \\ &+ \frac{1}{8}g(k\rho)(ik\rho)^{-1}\left[\left(-2\chi_i^3+3\chi_i^5\right)P(\Omega) + 3\cos\frac{\Omega-\phi}{2}\chi_i^4 P'(\Omega)\right. \\ &\left. + \chi_i^3 P''(\Omega)\right](iks)^{-1}g(ks) + \frac{1}{8}g(k\rho)\left[\frac{3}{2}\chi_i^5 P(\Omega)\right. \\ &+ \cos\frac{\Omega-\phi}{2}\left(2\chi_i^2+3\chi_i^4\right)P'(\Omega) + \left(\chi_i+3\chi_i^3\right)P''(\Omega) \\ &\left. + 2\cos\frac{\Omega-\phi}{2}\chi_i^2 P'''(\Omega) + \chi_i P''''(\Omega)\right](iks)^{-2}g(ks) \end{aligned} \tag{51c}$$

where

$$\chi_{\substack{i\\r}} = \mp\csc\frac{\Omega \mp \phi}{2}, \tag{52}$$

as defined in (13). The diffracted field u^d may then simply be constructed by using the following relation

$$u^d(\rho,\phi) = u_i^d(\rho,\phi) + \tau u_i^d(\rho,\phi) = u_i^d(\rho,\phi) + \tau u_i^d(\rho,2\pi-\phi). \tag{53}$$

The expression of u_1^d is precisely Keller's GTD solution [1]. Thus, the leading term in (53) agrees exactly with the result provided by Keller's GTD for the region where such a solution is valid. In addition, our spectral analysis also provided higher-order terms, whereas GTD is incapable of determining these terms. The expression for u_2^d agrees completely with the one obtained using the UAT formulation [13]. Furthermore, we have constructed u_3^d to show the ease in using spectral analysis for the systematic determination of higher-order terms.

To complete the asymptotic evaluation of u^t, defined in (2) one must also determine $u^g_{\substack{i\\r}}$, given in (45b). First notice that

$$s\cos(\omega-\Omega) + \rho\cos(\omega\mp\phi) = \rho_{\substack{i\\r}}\cos(\omega-\phi_{\substack{i\\r}}) \tag{54}$$

where its geometrical interpretation is shown in Fig. 2a., (ρ_i,ϕ_i) are the coordinates of the observation point seen from the source point, and (ρ_r,ϕ_r) are the coordinates of the observation point seen from the image of the source point and ϕ_r is measured positively clockwise with range $-\pi < \phi_r < \pi$. Substituting (54) into (45b), one can show that the SDP path may be deformed to the steepest descent path going through ϕ_i or ϕ_r, respectively. This deformation is valid as long as $\phi \notin T^i$ or $\phi \notin T^r$, respectively. The asymptotic evaluation of (45b) follows the same steps as used for (21) and the result takes the form (34) when P_0 is replaced by P. We therefore arrive at the following asymptotic results

$$u_i^g(\rho,\phi) = \begin{cases} -u^i(\rho_i,\phi_i) & \text{for } \Omega < \phi \notin T^i \\ 0 & \text{for } \Omega > \phi \notin T^i \end{cases} \tag{55a}$$

and

$$u_r^g(\rho,\phi) = \begin{cases} 0 & \text{for } -\Omega < \phi \notin T^r \\ u^r(\rho_r,\phi_r) & \text{for } -\Omega > \phi \notin T^r \end{cases} \tag{55b}$$

where u^r is the field radiated from the image point of the line source. Customarily, $u^i + u_i^g + \tau u_r^g$ is called the geometrical field in the GTD construction of the field.

2. Total Field at the Shadow Boundaries $\phi = \pm\Omega$

Along these boundaries, the solutions given in (51) diverge to infinity, because χ_i and χ_r take infinite values at $\phi = \Omega$ and $\phi = -\Omega$, respectively. To overcome this difficulty, we go back to (43) and rewrite it as follows

$$u_{\substack{i\\r}}^s(\rho,\pm\Omega) = \frac{i}{4\pi}\int_{SDP} \mp P(\omega)\, e^{ik(s+\rho)\cos(\omega-\Omega)}\, F\left(\sqrt{2k\rho}\,\sin\frac{\omega-\Omega}{2}\right)d\omega; \tag{56}$$

(56) denotes two equations corresponding to $u_i^s(\rho,\Omega)$ and $u_r^s(\rho,-\Omega)$, respectively. It is noticed that the evaluation of $u_i^s(\rho,-\Omega)$ and $u_r^s(\rho,\Omega)$ follows the same procedures developed in the previous section and their asymptotic expansions can be obtained from (51). Introducing the change of variable (46) into (56), one readily arrives at

$$u_{\substack{i\\r}}^s(\rho,\pm\Omega) = \frac{\sqrt{2}}{4\pi}\, e^{ik(s+\rho)+i\pi/4}\int_{-\infty}^{\infty} \mp\, p(t)\,(1+it^2/2)^{-1/2} \cdot F(e^{-i\pi/4}\sqrt{k\rho}\; t)\, e^{-k(s+\rho)t^2}\,dt \tag{57}$$

where $p(t) = P(\omega)$ after ω is replaced by t from (46). We then expand $p(t)(1+it^2/2)^{-1/2}$ in terms of the Taylor series and substitute the result into (57). This equation can now be evaluated with the help of the following relations

$$I_n = \int_{-\infty}^{\infty} F(e^{-i\pi/4}\sqrt{k\rho}t)\, e^{-k(s+\rho)t^2}\, t^n\, dt \tag{58a}$$

$$I_{2n} = \frac{1}{2}\Gamma(n+1/2)\, k^{-n-1/2}\, (s+\rho)^{-n-1/2} \tag{58b}$$

$$I_{2n+1} = nk^{-1}(s+\rho)^{-1}\, I_{2n-1} + \frac{i}{2\sqrt{\pi}}\,\Gamma(n+1/2)k^{-n-1}(s+\rho)^{-1}\, s^{-n}\,\sqrt{\rho/s}. \tag{58c}$$

Omitting all the intermediate steps, one can finally express (57) as follows [10]

$$\begin{aligned}
u_i^s(\rho,\Omega) &= -u_r^s(\rho,-\Omega)\\
&= -\frac{1}{2}g[k(s+\rho)]\left\{P(\Omega) + (ik)^{-1}\left[\frac{1}{2}P''(\Omega) + \frac{1}{8}P(\Omega)\right](s+\rho)^{-1}\right.\\
&\quad \left. + (ik)^{-2}\left[\frac{1}{8}P''''(\Omega) + \frac{5}{16}P''(\Omega) + \frac{9}{128}P(\Omega)\right](s+\rho)^{-2}\right\}\\
&\quad - g(ks)g(k\rho)\,\frac{2\rho}{\rho+s}\,P'(\Omega) - (ik)^{-1}g(ks)g(k\rho)\frac{\rho^2+3\rho s}{3s(s+\rho)^2}\\
&\quad \cdot [P'(\Omega) + P'''(\Omega)] - g(ks)g(k\rho)(ik)^{-2}\,\frac{15\rho s^2 + 10s\rho^2 + 3\rho^3}{60s^2(s+\rho)^3}\\
&\quad \cdot [4P'(\Omega) + 5P'''(\Omega) + P'''''(\Omega)] + O(k^{-7/2})\,.
\end{aligned} \tag{59}$$

The first term in the r.h.s. of (59) can be identified with the negative of one half of the incident field when it is compared with (34) for $\phi_i = \Omega > 0$, $P_0 = P$ and $P_1 = P_2 = 0$. The total field, defined in (2), may then be obtained by determining $u_r^s(\rho,\Omega) = u_i^s(\rho,2\pi-\Omega)$ from (45a) and (51). In (59) the terms up to and including the order of $k^{-5/2}$ do agree with the results given in [13] which used the UAT technique. It is noted that the dominant asymptotic term of the incident field is of the order $k^{-1/2}$.

3. *Numerical Evaluation*

In the transition regions T^i and T^r, the procedures that were employed in the previous sections for deriving the asymptotic expression of the field no longer apply. In this section we discuss the numerical evaluation of the expressions for the field, viz., (21) and (42b), for observation angles that lie in these angular regions. Since the numerical procedure is not restricted to a specific angular region, we have employed it to evaluate the diffracted field for a wider range of observation angles than just the transition regions. Though these results are not ordered asymptotically in terms of powers of k, they do provide a good basis for checking the asymptotic results derived earlier.

In their present form, (21) and (42b) are not suitable for numerical integration quadratures, because of their highly oscillatory kernels. We deform their paths to the steepest descent paths and introduce the change of variables (26b) and (46), respectively, in (21) and (42b) to arrive at

$$u^i(\rho_i,\phi_i) = \frac{e^{ik\rho_i + i\pi/4}}{2\sqrt{2}\pi}\int_0^\infty \left[P(\omega)\Big|_t + P(\omega)\Big|_{-t}\right](1+it^2/2)^{-1/2} e^{-k\rho_i t^2}\, dt \tag{60a}$$

where ω is replaced by t from (26b) and

$$u^s_{\substack{i\\r}}(\rho,\phi) = \frac{e^{ik(s+\rho)+i\pi/4}}{2\sqrt{2}\pi}\int_0^\infty \left[\mp P(\omega)\, e^{-i\xi^2_{\substack{i\\r}}} F\left(\mp\xi_{\substack{i\\r}}\right)\Big|_t \mp P(\omega)\, e^{-i\xi^2_{\substack{i\\r}}} F\left(\mp\xi_{\substack{i\\r}}\right)\Big|_{-t}\right] \cdot (1+it^2/2)^{-1/2}\, e^{-kst^2}\, dt \tag{60b}$$

where ω is replaced by t from (46). A sufficient condition for the integrand of (60b) to be bounded is that $s + \rho\cos(\Omega \mp \phi) > 0$, respectively, for u^s_i or u^s_r. Due to the fact that the integrands of (60a) and (60b) have smooth behaviors and decrease very rapidly for large values of t, conventional numerical integration techniques, e.g., Gauss quadratures, may be employed to evaluate the integrals with good accuracy. Some typical results obtained in this manner are shown in the next section, in which comparisons are made with other techniques.

C. Comparison with Other High-Frequency Techniques

Ever since the classic paper by Keller [1] on the Geometrical

Theory of Diffraction (GTD), numerous attempts have been made toward extending the domain of GTD to regions where it predicts an infinite field and is therefore invalid. Most of the efforts in this direction have been concentrated on constructing uniform formulations, such as UTD, MSD and UAT, that overcome the difficulties of GTD at the incident and reflection shadow boundaries, and at the transition regions. The Uniform Theory of Diffraction (UTD) and its Modified Slope Diffraction (MSD) version have been introduced by Kouyoumjian, Pathak and Hwang [17,18], and the Uniform Asymptotic Theory (UAT) has been developed by Ahluwalia, Lewis and Boersma [14]. The latter has been employed extensively by Boersma, Lee, Deschamps and Wolfe [13,16,19,20] to investigate a number of problems of this type. UTD, MSD and UAT are asymptotic techniques based on individual hypotheses or *Ansatz*. Typically, there is no general proof available for the validity of the completeness of these *Ansatz*, and one has to apply them to certain test cases in order to establish their accuracy. The formulations of these asymptotic theories are reduced to the classical Sommerfeld's result when the plane wave illumination on a half-plane is considered. In fact, Sommerfeld's formulation is the basic foundation of all the different *Ansatz* that have been proposed to date. For more complex situations, the validity of the various asymptotic theories is typically checked against numerical results or experimental data, since analytical results are often unavailable in closed form. Even for the problem of diffraction of a half-plane illuminated by a line source possessing isotropic or nonisotropic patterns, there is no substantial check available for establishing the accuracy of the various asymptotic theories. Specifically, at the transition regions, it is not known how well these asymptotic theories compare with each other or with the exact solution.

The object of this section is to compare the aforementioned asymptotic theories with the exact solution that has been constructed using the spectral analysis presented in this paper. The comparison has been carried out, both analytically and numerically, for a wide range of observation angles including shadow boundaries, transition regions and angles away from these regions. For completeness, different source locations and pattern functions have been investigated and extensive results have been derived. In the following, we begin by reviewing briefly the nature of the formulation of each of the uniform theories, and follow this with the presentation of the results and conclusions.

1. Formulation of Different Theories

All the uniform asymptotic theories essentially approximate (22) and (42b) in an asymptotic sense. It is interesting to note that as yet there is no proof available that justifies the use of these forms for arbitrary and complex situations. In all of these techniques the following is true for the total field

$$u^t(\rho,\phi) = v(\rho,\phi) + \tau v(\rho,2\pi - \phi) \tag{61}$$

where $v(\rho,\phi)$ takes different forms for different theories. For convenience of interpretation, all of the asymptotic theories are schematically represented in Fig. 3. The notations in this figure are explained below.

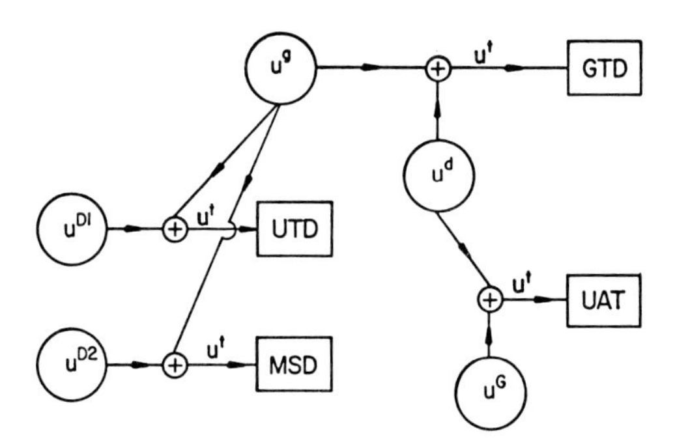

Fig. 3. Schematic diagram of inherent relationships among different asymptotic techniques.

We first give the expression of $v(\rho,\phi)$ for the GTD theory, which reads as follows

$$v(\rho,\phi) = u^g(\rho,\phi) + u^d(\rho,\phi) + O(k^{-3/2}) \tag{62a}$$

where

$$u^g(\rho,\phi) = \theta\left[\sin\frac{1}{2}(\Omega-\phi)\right]u^i(\rho_i,\phi_i) \tag{62b}$$

and

$$u^d(\rho,\phi) = -\csc\frac{\Omega-\phi}{2}\,g(k\rho)u^i(s,\Omega). \tag{62c}$$

In the preceding equations, θ is the unit step function and

$$u^i(\rho_i,\phi_i) = g(k\rho_i),P(\phi_i), \tag{63}$$

that is, the first asymptotic term of (31). It is clear that the GTD formulation fails at the shadow boundaries, i.e., $\phi = \pm\Omega$, and their vicinities.

The UTD formulation overcomes this failure of GTD at the shadow boundaries by expressing $v(\rho,\phi)$ from (61) as

$$v(\rho,\phi) = u^g(\rho,\phi) + u^{D1}(\rho,\phi) + O(k^{-3/2}) \tag{64a}$$

where u^g has already been defined in (62b) and u^{D1} is

$$u^{D1}(\rho,\phi) = 2\sqrt{\pi}\,\mathrm{sgn}\left[\sin\frac{\phi-\Omega}{2}\right]\sqrt{\frac{2k\rho s}{\rho+s}}\,g(k\rho)e^{-i(\pi/4+\xi^2)}F(\xi)u^i(s,\Omega). \tag{64b}$$

In (64b) F is the Fresnel integral (19) and

$$\xi = \left|\sin\frac{\phi-\Omega}{2}\right|\sqrt{\frac{2k\rho s}{\rho+s}} \tag{64c}$$

which goes to zero at shadow boundaries. The UTD formulation is often referred to in the literature as the Kouyoumjian-Pathak (KP) formulation.

Since UTD is only suitable for plane wave or isotropic line source types of incident field, recently a modified version of UTD that employs the notion of the slope diffraction coefficient has been introduced to generalize this procedure. This formulation, which is referred to here as MSD, modifies (64a) as

$$v(\rho,\phi) = u^g(\rho,\phi) + u^{D2}(\rho,\phi) + O(k^{-3/2}) \tag{65a}$$

where

$$u^{D2}(\rho,\phi) = u^{D1}(\rho,\phi) + \left\{ 2\sqrt{\pi}\ \mathrm{sgn}\left[\sin\frac{\phi-\Omega}{2}\right]\sqrt{2ks\rho}\ \rho(s+\rho)^{-3/2} \right.$$

$$\left. \cdot \sin(\phi-\Omega)\quad e^{-i(\pi/4+\xi^2)}F(\xi) - 2\rho(\rho+s)^{-1}\cos\frac{\phi-\Omega}{2} \right\}$$

$$\cdot\ g(k\rho)\ \left.\frac{\partial u^i}{\partial \phi^i}\right|_{s,\Omega} \quad . \tag{65b}$$

In summary, UTD and its modification MSD have been introduced to modify u^d in the GTD formulation in such a manner as to cancel the discontinuity of u^g.

Finally, the Uniform Asymptotic Theory (UAT) has been invented to circumvent the shadow boundary difficulties of the GTD by writing v^t as

$$v^t(\rho,\phi) = u^G(\rho,\phi) + u^d(\rho,\phi) + 0(k^{-3/2}) \tag{66a}$$

where u^d is the same term used in the GTD formulation, i.e., (62c). $U^G(\rho,\phi)$ has the form that exactly cancels the infinite value of u^d at the shadow boundaries. In this theory U^G takes the form

$$u^G(\rho,\phi) = [F(\zeta) - \hat{F}(\zeta)]u^i(\rho_i,\phi_i) \tag{66b}$$

where $\hat{F}$ is the first term of (44b)

$$\hat{F}(\zeta) = \frac{e^{i\pi/4+i\zeta^2}}{2\sqrt{\pi}\zeta} \tag{66c}$$

and ζ, which is called the detour function, is

$$\zeta = \mathrm{sgn}\left[\sin\frac{\phi-\Omega}{2}\right]\sqrt{k[(\rho+s)-\rho_i]}\ ;\ \rho_i = \sqrt{s^2+\rho^2+2s\rho\cos(\phi-\Omega)}\ . \tag{66d}$$

Note that at the shadow boundaries the detour ζ is zero. The higher-order asymptotic expression of UAT to the order of k^{-2} (dominant asymptotic term of u^i is $k^{-1/2}$) for the problem at hand is given by Boersma and Lee [13]. It is worth mentioning that in all of the above asymptotic techniques the field incident on the diffracting edge must be a ray field (local plane wave). Some difficulties related to the application of incident fields that do not fulfill this criterion are discussed in Section IV.

2. Numerical Results and Discussions

In the previous sections, some results based on analytical and numerical asymptotic evaluations of (21) and (42b) were presented for three situations: point of observation exactly at the shadow boundaries; inside the transfer regions; and outside the transition regions. Additional results for this problem of diffraction of a nonisotropic line source have been given in [13]. A comparative study of the various solutions allows one to deduce certain conclusions discussed below. First, spectral formulation (STD) can be used in a very systematic and straightforward fashion to determine any number of higher-order terms in the asymptotic expansion of the total field. In fact, by using this approach we have computed the asymptotic expression of the total field up to the order of k^{-3}. Second, UAT formulation, in its general version, can also furnish the higher-order asymptotic expression of the total field. However, in this method the higher-order terms in the diffracted field u^d must be generated through an explicit application of the edge condition. Results up to the order of k^{-2} have been reported in [13], and they agree perfectly with the STD results. Third, UTD provides the correct asymptotic expression of the field up to the order of k^{-1}, but only for an isotropic line source. Furthermore, UTD formulation does not provide higher-order asymptotic terms in a direct fashion. Finally, MSD, which is the modified version of UTD, allows one to determine the asymptotic values of the field for the nonisotropic line source up to the order of k^{-1}. Again, this formulation does not provide higher-order terms in a systematic manner.

All of the above observations have been made at the shadow boundaries ($\xi = \zeta = 0$) and far away from the transition regions ($\xi \gg 0, \zeta \gg 0$) where the field can be expressed in a closed form.

For the sake of completeness we have also investigated the behavior of the fields predicted by the various theories in the neighborhood of the transition regions. For this purpose the field integrals in STD, i.e. eqs. (60a) and (60b), have been evaluated using an efficient numerical algorithm discussed in Section III.B.3. Note that the computed results obtained in this manner contain all of the higher-order terms. Numerical results have also been obtained, up to the range of higher-order terms available, using GTD, UTD, MSD and UAT formulations, as given in (62a), (64a), (65a) and (66a), respectively. In order to get better insight, we have compared both the total field u^t and the scattered field u^s, defined in (2). Some typical results are shown and discussed in the following.

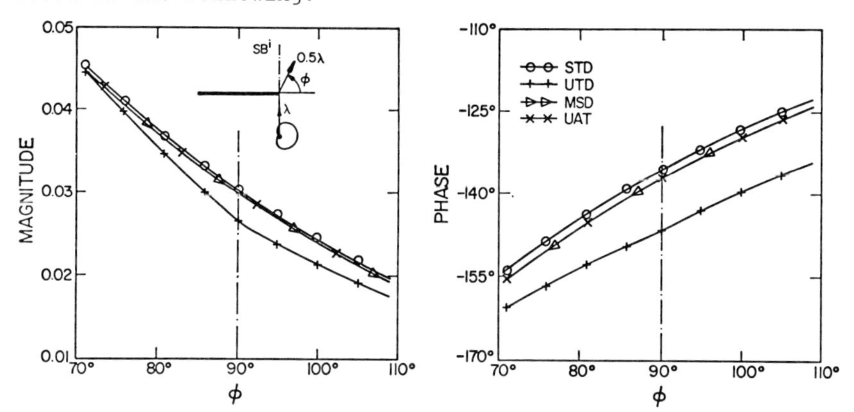

Fig. 4. Total field u^t *diffracted by a half-plane illuminated by a nonisotropic source* (E-*wave) with pattern function of (60) and* $|u^i(1,\pi/2)| = .0796$.

Fig. 4 shows the total field diffracted by a half-plane illuminated by a nonisotropic line source (*E*-wave) with the following pattern function obtained from (39)

$$P(\omega) = \cos\omega + \text{sgn}\,(\phi_i)\,\sin\omega. \tag{67}$$

It should be noted that when (67) is used in (60b), ϕ_i is replaced by Ω and hence sgn $(\Omega) = +1$. Results are shown for different techniques and the field was sampled in a 40° angular region about the incident shadow boundary (SB). The SB direction is shown in the figure, and as expected, UTD formulation fails to provide the correct result due to the lack of proper slope discontinuity compensation. Since the incident field is well-behaved in the entire region, we have also compared the scattered field u^s for different techniques in Fig. 5. Exactly at the shadow boundaries, we

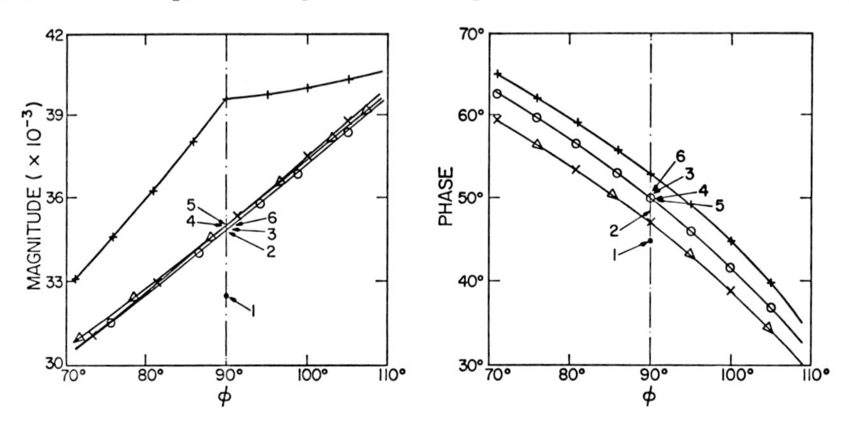

Fig. 5. Scattered field u^s *from a half-plane illuminated by a nonisotropic line source (geometry shown in Fig. 4). Points 1-6 are the number of terms in asymptotic expansion of* u^s *given in Column 3 of Table 1.*

can employ our higher-order asymptotic results, derived in (59) and (51), to determine the scattered field. The results of these calculations have been listed in Table 1 for different orders of k. In the same table the result of evaluation of (60b) has also been given. Though this evaluation implicitly contains all of the higher-order terms of k, the agreement with the results given in the table for terms beyond k^{-1} is remarkable. This is indeed a strong indication of the usefulness of asymptotic techniques for convenient computation of the diffracted field. We have also plotted the results tabulated in the third column of Table 1 in Fig. 5 with a view to exhibiting the behavior of the asymptotic solution. The number of terms used in the asymptotic evaluation is indicated by numerals inserted alongside the points representing the evaluation of the diffracted field at the shadow boundary.

TABLE 1

Asymptotic terms of the scattered field u^s *evaluated at the incident shadow boundary (SB), shown in Figs. 5 and 6. First column is the number of terms. Second and fourth columns are the numerical value of each term. Third and fifth columns are the summation of terms. The result of the numerical evaluation of (60b) at SB for the nonisotropic case is* $.034920e^{i50.11°}$.

n	Nonisotropic Source, Eq. (67)		Isotropic Source	
	$O(k^{-n/2})$	$\sum_1^n$	$O(k^{-n/2})$	$\sum_1^n$
1	$.032487e^{i45°}$	$.032487e^{i45°}$	$.032487e^{i45°}$	$.032487e^{i45°}$
2	$.002986e^{i90°}$	$.034663e^{i48.49°}$	$.008956e^{i90°}$	$.039333e^{i54.26°}$
3	$.001293e^{i135°}$	$.034766e^{i50.62°}$	$.000431e^{-i45°}$	$.039266e^{i53.64°}$
4	$.000356$	$.034993e^{i50.17°}$	$.001069$	$.039909e^{i52.41°}$
5	$.000043e^{i90°}$	$.035036e^{i50.16°}$	$.000026e^{i135°}$	$.039883e^{i52.41°}$
6	$.000184e^{i270°}$	$.034893e^{i50.66°}$	$.000326e^{-i90°}$	$.039625e^{i52.12°}$

Fig. 6 displays the behavior of the scattered field calculated using the aforementioned techniques in the transition region of a half-plane illuminated by an isotropic line source (*E*-wave). As expected, UTD and MSD are the same and all the

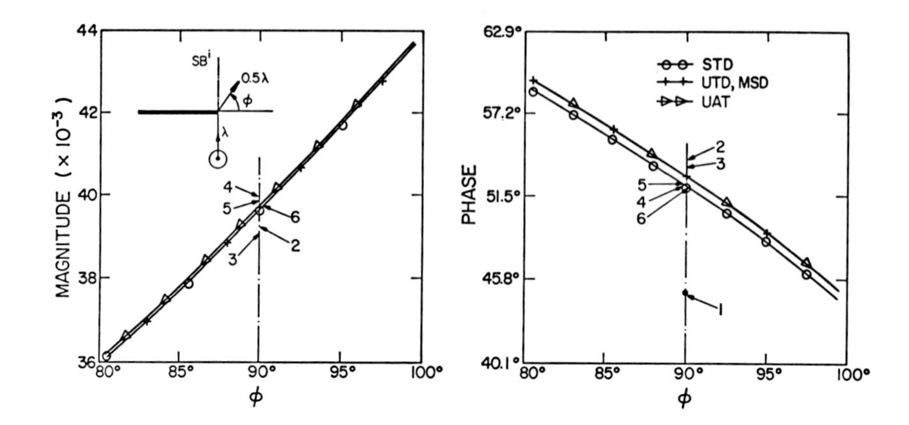

Fig. 6. Scattered field u^s *from a half-plane illuminated by an isotropic line source (E-wave) with* $|u^i(1,\pi/2)| = .0796$. *Points 1-6 are the number of terms in asymptotic expansion of* u^s *given in Column 5 of Table 1.*

techniques are in very good agreement. For completeness, the asymptotic expressions have been tabulated in Table 1, and are shown in Fig. 6. Fig. 7 shows the plot of the scattered field from a half-plane illuminated by an isotropic line source (*E*-wave) coplanar with the half-plane. For this geometry the incident and reflected shadow boundaries coincide with the half-plane itself.

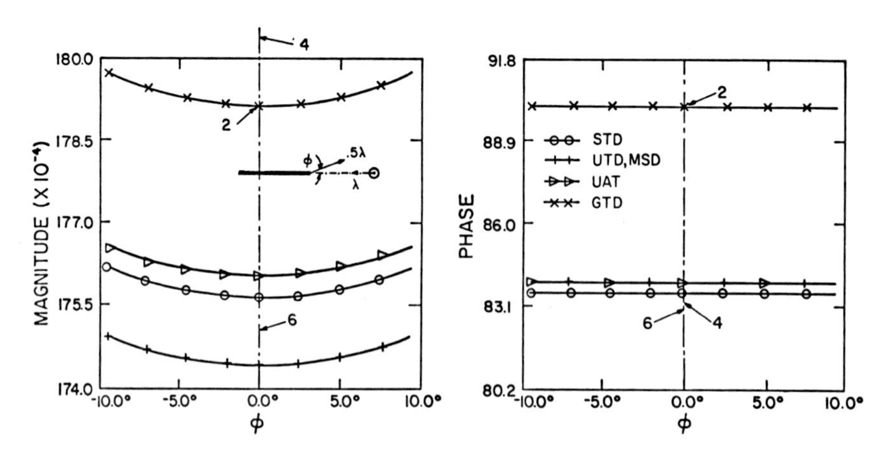

Fig. 7. Scattered field u^s *from a half-plane illuminated by an isotropic line source (E-wave) with* $|u^i(1,\pi/2)| = 0.0796$. *Points 2, 4 and 6 are the asymptotic evaluation at* $\phi = 0°$.

The scattered field was sampled in an angular region $-10° < \phi < 10°$ and results for different techniques GTD are shown. For the sake of completeness the asymptotic results computed from (62a) are also shown in Fig. 7 for the observation angle $\phi = 0$, and the oscillatory nature of this asymptotic solution is evident from the figure.

To further illustrate the comparative nature of the results derived from different theories, we employ the formulations of UAT, UTD and MSD to obtain some three-dimensional representations of the total field. Fig. 8, containing the geometry of the problem, is the three-dimensional plot of the total field $|u^t|$ diffracted from a half-plane illuminated by an isotropic line source (E-wave). To obtain the total field u^t, we employ the formulations of UAT and UTD, and find that essentially identical plots are obtained, although numerical values of these curves differ only slightly from each other. For instance, at the observation point

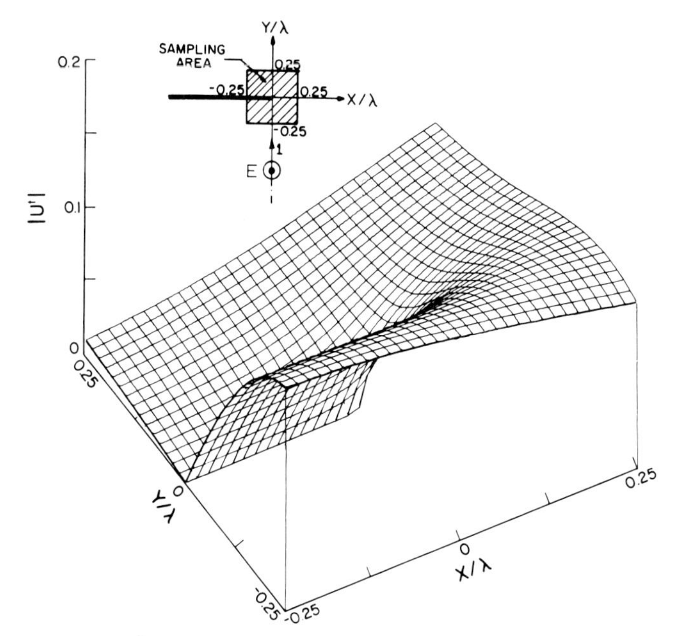

Fig. 8. $|u^t|$ *from UAT/UTD formulation.* E-*wave illumination of an isotropic source with* $|u^i(1,\pi/2)| = .08$.

$x = y = .25\lambda$, UAT and UTD formulations give $|u^t| = .05913$ and $|u^t| = .05918$, respectively. Fig. 9 is the counterpart of Fig. 8 for the *H*-wave case. As expected, the *H*-field is discontinuous at the half-plane. For additional results and three-dimensional plots the reader is referred to [10].

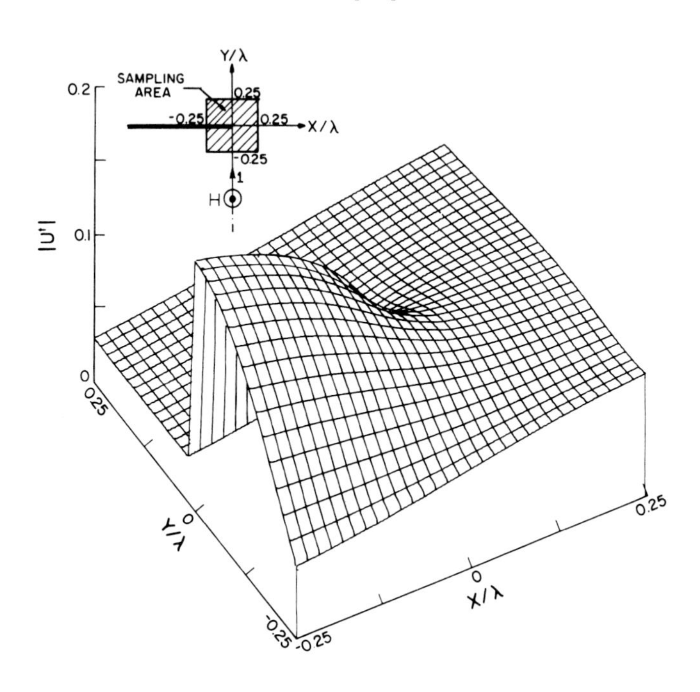

Fig. 9. $|u^t|$ *from UAT/UTD formulation.* H-*wave illumination of an isotropic source with* $|u^i(1,\pi/2)| = .08$.

IV. MULTIPLE-EDGE DIFFRACTION

The multiple-edge diffraction problem arises in a number of practical situations and is considerably more complicated than its single-edge counterpart. The complexity of this problem stems from the fact that the field, which impinges on the second edge after diffracting from the first, can no longer be treated, in general, as a locally uniform plane wave. It has been recognized that for the problem of diffraction by a nonplane-wave source, e.g., one with a pattern fucntion as in Section III, the result for the scattered field at the shadow boundaries is not just a simple modification of the plane-wave solution multiplied by the pattern function, but involves the derivative of the pattern function as well. One might assume that the problem of multiple-edge diffraction can be solved by a simple extension of the pattern fucntion illumination of a single edge. That this is not the case, however, has been pointed out by Jones [21], who has employed the Wiener-Hopf technique to construct the rigorous high-frequency expansion of the exact solution for the diffracted field by staggered parallel plates. Although Jones [21] did not explicitly obtain the results for the field at the shadow boundaries, he did examine the uniform expression he derived in some detail and concluded that the Uniform Asymptotic Theory (UAT) could not be directly (mechanically) applied at the shadow boundaries. Lee and Boersma [19] and Boersma [22] have carried out rather elaborate investigations of this problem and have shown that the UAT can still be used for this situation.

In this section, the problem of diffraction of a plane wave incident on two staggered parallel half-planes (multiple-edge) is investigated and an analysis is presented with has the following features: First, a general representation of the field after successive diffraction by two half-planes is given in terms of a double complex integral. The analysis is carried out in the Fourier transform domain and is based on the application of the results given in Sections II and III. Second, asymptotic techniques are employed to determine the total field to the order of k^{-1} in a rigorous, yet straightforward fashion for some cases of practical interest. These special cases are (see Fig. 10): (a) the edge of the top half-plane lies at or away from the shadow boundary of the bottom half-plane, i.e., the directly illuminated one; (b) the observation point lies at or away from the shadow boundary of the top half plane. In all of these cases the final results are expressed in a compact and useful form.

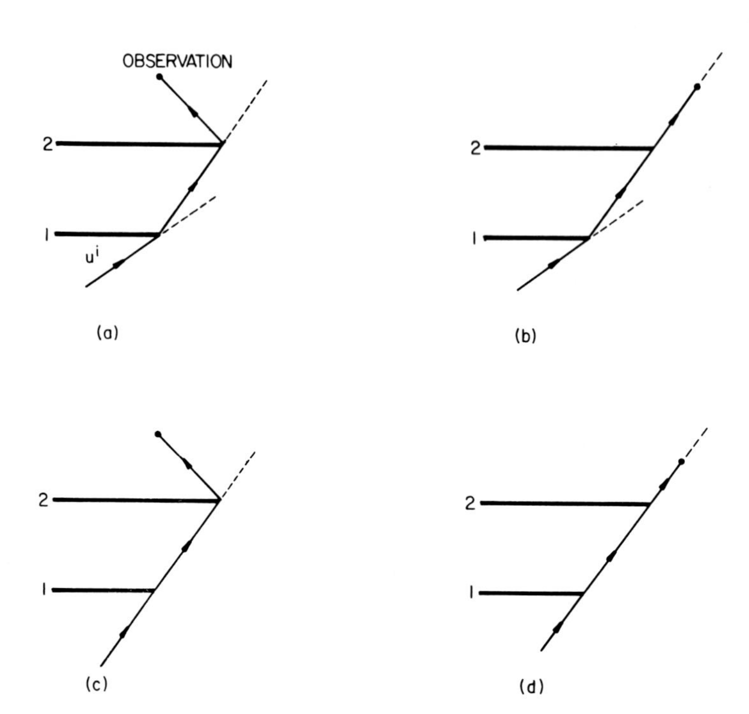

Fig. 10. Relative positions of the edge of the top half-plane and the observation point with respect to the shadow boundaries.

A. Basic Formulation

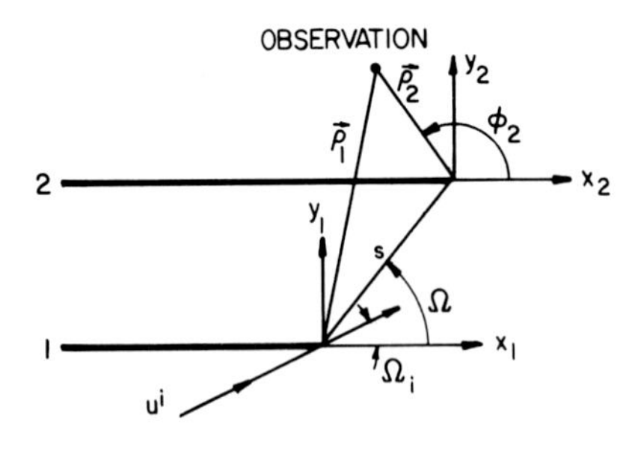

Fig. 11. Diffraction of a plane wave by staggered parallel plates, $\Omega > 0$ and $\Omega_i > 0$.

The geometry of a pair of staggered, perfectly conducting, parallel half-planes illuminated by a plane wave is shown in Fig. 11. The bottom and top planes are labeled 1 and 2, respectively. The edges are separated by distance s, and the angle between the line joining the edges and the x-axis is Ω. Let the incident plane wave with unit amplitude and incident angle Ω_i take the following form at point $\vec{\rho}_1$

$$u^i(\rho_1, \phi_1) = e^{ik\rho_1 \cos(\Omega_i - \phi_1)} . \tag{68}$$

In this work, we neglect the interaction between planes 1 and 2, because it is assumed that s/λ is a large number. The total field diffracted by plane 1, u_1^t, may now be expressed as

$$u_1^t = u^i + u_1^s \tag{69}$$

where u_1^s is the scattered field due to the induced current in plane 1. From (11), the scattered field u_1^s can be written as

$$u_1^s = \frac{i}{4\pi} \begin{Bmatrix} 1 \\ \operatorname{sgn}(\phi_1) \end{Bmatrix} \int_{\Gamma_1} \chi(\Omega_i, \omega) \, e^{ik\rho_1 \cos(\omega - |\phi_1|)} \, d\omega \quad \text{for} \begin{Bmatrix} E\text{-wave} \\ H\text{-wave} \end{Bmatrix} \tag{70}$$

where χ is defined in (12) and path Γ_1 is shown in Fig. 12a. In order to construct (70), ψ is replaced by ω in (11) for reasons which will become clear later.

Next, we consider u_1^t as the incident field impinging on the top half-plane 2. The total field diffracted by this half-plane, i.e., u_2^t, may be split into

$$u_2^t = u_1^t + u_2^s \tag{71}$$

where u_2^s is the scattered field. Since u_1^t, given in (69), consists of two parts, we may use the superposition argument and designate u_{22}^s and u_{21}^s as the scattered fields due to u^i and u_1^s, respectively. u_2^s can then be written as

$$u_2^s = u_{22}^s + u_{21}^s . \tag{72}$$

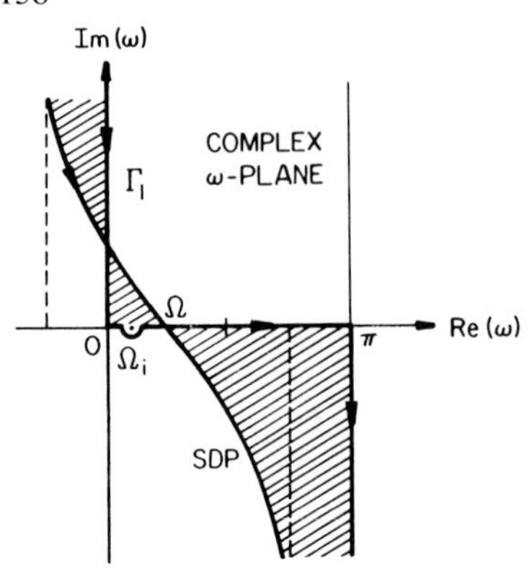

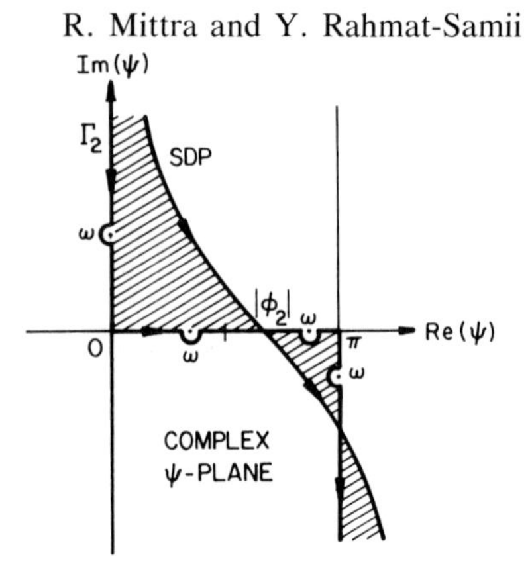

Fig. 12a and b. (a) Contour Γ_1 for (70) and steepest descent path SDP for (77). (b) Contour Γ_2 for (73).

The task is to determine u_2^t, which now consists of three parts, viz. u_1^t, u_{22}^s and u_{21}^s. Fields u_1^t and u_{22}^s can be constructed from expressions given in Section II as they are the total and scattered fields, respectively, from the half-planes 1 and 2 illuminated by the incident plane wave (68). These fields take uniform forms similar to (18) and their asymptotic expressions can readily be derived.

In determining u_{21}^s, we compare (70) with (21) and employ equations similar to (40) and (41), arriving at

$$u_{21}^s = \frac{i}{4\pi}\int_{\Gamma_1} d\omega\ \chi(\Omega_i,\omega)\ e^{iks\cos(\omega-\Omega)} \cdot \left[\frac{i}{4\pi}\begin{Bmatrix} 1 \\ \operatorname{sgn}(\phi_2)\end{Bmatrix}\int_{\Gamma_2}\chi(\omega,\psi)\,e^{ik\rho_2\cos(\phi-|\phi_2|)}\,d\psi\right], \quad \text{for } \begin{Bmatrix} E\text{-wave} \\ H\text{-wave}\end{Bmatrix} \tag{73}$$

where ϕ_1 is replaced by $\Omega > 0$ in (70) and ρ_2 and ϕ_2 are defined in Fig. 12. Comparison of (73) with (41) reveals that $\chi(\Omega_i,\omega)$ can be interpreted as the pattern function of a fictitious line source located at the edge of half-plane 1. In contrast to $P(\omega)$, which was assumed to be regular in (41), $\chi(\Omega_i,\omega)$ is singular at $\omega = \pm\Omega_i$ and, therefore, special attention must be devoted in the asymptotic evaluation of (73). Attempts were made to first approximate the inner ψ-integral in terms of its asymptotic expansion and then to evaluate the ω-integration. Due to the nonuniform nature of this procedure, proper results could not be obtained for all the cases. Since the inner ψ-integral can be performed uniformly in terms of the Fresnel integral, we may now express (73) as

$$u^s_{21} = T_1 + \tau T_2 + T_3 + \tau T_4 \tag{74}$$

where

$$\begin{Bmatrix} T_{1} \\ {}_{2} \\ T_{3} \\ {}_{4} \end{Bmatrix} = \frac{i}{4\pi}\int_{\Gamma_1} \mp \begin{Bmatrix} \chi_i(\Omega_i,\omega) \\ \tau\chi_r(\Omega_i,\omega) \end{Bmatrix} e^{iks\cos(\omega-\Omega)}\, e^{ik\rho_2\cos(\omega\mp\phi_2)}\, F(\mp\xi_{\substack{i\\r}})\,d\omega, \tag{75}$$

and, furthermore, from (13) and (20), we have

$$\chi_{\substack{i\\r}}(\Omega_i,\omega) = \mp\csc\frac{\Omega_i \mp \omega}{2} \tag{76a}$$

$$\xi_{\substack{i\\r}} = \mp\sqrt{2k\rho_2}\,\sin\frac{\omega\mp\phi_2}{2} \tag{76b}$$

It is noticed that on path Γ_1, χ_r is well-behaved, whereas χ_i possesses a pole singularity at $\omega = \Omega_i$. Therefore, asymptotic evaluation of T_3 and T_4 follows the same steps used to evaluate (42b). In the following sections, we devote our attention to the asymptotic evaluation of T_1 and T_2, which, in conclusion, allows one to determine u^t_2 from (71).

B. Total Field When $\Omega \gg \Omega_i$ and $\phi_2 \gg \Omega$ (Fig. 10a)

For this case, where the pole $\omega = \Omega_i$ and the saddle point $\omega = \Omega$ are separated, we may deform the path Γ_1 into the saddle path SDP, shown in Fig. 12a, taking into account the residue contribution of the pole $\omega = \Omega_i$, and arrive at

$$T_{\substack{1\\2}} = \frac{i}{4\pi}\int_{\mathrm{SDP}} \mp\, \chi_i(\Omega_i,\omega)\, e^{iks\cos(\omega-\Omega)}\, e^{ik\rho_2\cos(\omega\mp\phi_2)}\, F(\mp\xi_{\substack{i\\r}})\, d\omega$$
$$\pm F\left(\sqrt{2k\rho_2}\,\sin\frac{\Omega_i\mp\phi_2}{2}\right) e^{iks\cos(\Omega_i-\Omega)}\, e^{ik\rho_2\cos(\Omega_i\mp\phi_2)} \tag{77}$$

Since at $\omega = \Omega$ the relation $\xi_{\substack{i\\r}} \neq 0$ holds, one can then employ the results of Section III.B.1 and determine the asymptotic value of (77). The same procedure can also be used to evaluate T_2 and T_3, which completes the derivation of u_{21}^s, defined in (74). Using the procedure described in connection with (71) and (72), one can also determine the asymptotic values of u_1^t and u_{22}^s. The final result for the total field u_2^t is finally given as

$$u_2^t = \chi(\Omega_i,\Omega)g(ks)\chi(\Omega,\phi_2)g(k\rho_2) + O(k^{-3/2}) \tag{78}$$

where χ and $g(\cdot)$ are defined in (12) and (48), respectively. Equation (78) agrees completely with the result obtained using GTD formulation; furthermore, $\chi(\Omega_i,\Omega)$ and $\chi(\Omega,\phi_2)$ may be identified as Keller's diffraction coefficient at the edges of half-planes 1 and 2, respectively.

C. Total Field When $\Omega \gg \Omega_i$ and $\phi_2 = \Omega$ (Fig. 10b)

This is an interesting case, because from GTD consideration the observation angle coincides with the shadow boundary of the diffracted rays originating from half-plane 2. In this case, as in the previous case, the pole $\omega = \Omega_i$ and the saddle point $\omega = \Omega$ are separated. One can then replace T_1 and T_2 from (75) with their counterparts from (77). Since at $\omega = \Omega$ the relations $\xi_i = 0$ and $\xi_r \neq 0$ hold, evaluation of T_1 and T_3 from (77) and (75), respectively, requires special care. This is done by using the procedure described in Section III.B.2. Determination of T_2 and T_4 from (77) and (75), respectively, may be completed using the analysis of Section III.B.1. Finally, u_1^t and u_{22}^s can be constructed from the solution of the half-plane illuminated by a plane wave as discussed in Section II. Evaluating all these components and adding them up, we finally arrive at the following total field

$$\psi_2^t = \frac{1}{2}\chi(\Omega_i,\Omega)g(ks+k\rho_2) + \left[\tau\chi(\Omega_i,\Omega)\chi_r(\Omega,\Omega) - \frac{2\rho_2}{\rho_2+s}\chi'(\Omega_i,\Omega)\right]$$
$$\cdot\, g(ks)g(k\rho_2) + O(k^{-3/2}) \qquad (79)$$

where $\chi'(\Omega_i,\Omega) = \frac{\partial}{\partial\omega}\chi(\Omega_i,\omega)\Big|_\Omega$ and that $\phi_2 = \Omega$. Comparing the term containing χ' with the one containing P' in (59), we may once more interpret χ as the pattern function of a fictitious line source located at the edge of half-plane 1. Expression (79) can also be obtained using UAT or MSD formulations in two successive steps in a mechanical fashion from (66) and (65), respectively.

D. Total Field When $\Omega_i = \Omega$ and $\Omega << \phi_2$ (Fig. 10c)

This is another important case for which the edge of half-plane 1 coincides with the direction of the incident shadow boundary of the incident field. In contrast to the last two previous cases, the pole $\omega = \Omega_i$ and the saddle point $\omega = \Omega$ coincide for the situation under investigation in this section. Hence, when deforming Γ_1 into SDP for T_1 and T_2, defined in (75), the integration passes into a principal-value integral along SDP plus half the residue contribution, namely,

$$T_{\substack{1\\2}} = \frac{i}{4\pi}\,⨍_{SDP} \mp \chi_i(\Omega,\omega)\, e^{iks\cos(\omega-\Omega)}\, e^{ik\rho_2\cos(\omega\mp\phi_2)} F(\mp\xi_{\substack{i\\r}})\, d\omega$$
$$\pm \frac{1}{2} F\left(\sqrt{2k\rho_2}\,\sin\frac{\Omega\mp\phi_2}{2}\right) e^{ik\rho_2\cos(\Omega\mp\phi_2)}\, e^{iks}. \qquad (80)$$

Although at the saddle point where $\omega = \Omega$ the relation $\xi_{\substack{i\\r}} \neq 0$ holds, the procedure developed in Section III.B.2 cannot be used for (80) as χ_i is singular, in contrast to P, which was well-behaved in (56).

The integrand of (80) is expanded in terms of the Laurent series as

$$\mp\chi_i(\Omega,\omega)\, e^{ik\rho_2\cos(\omega\mp\phi_2)} F(\mp\xi_{\substack{i\\r}}) = \frac{\mp 1}{\omega-\Omega} A_{\substack{i\\r}} \mp B_{\substack{i\\r}} + O(\omega-\Omega) \quad (81a)$$

where

$$A_{\substack{i\\r}} = 2e^{ik\rho_2} e^{-i\xi^2_{\substack{i\\r}}} F(\mp\xi_{\substack{i\\r}})\Big|_{\omega=\Omega} \quad (81b)$$

$$B_{\substack{i\\r}} = 2e^{ik\rho_2} \frac{\partial}{\partial\omega}\left[e^{-i\xi^2_{\substack{i\\r}}} F(\mp\xi_{\substack{i\\r}})\right]\Bigg|_{\omega=\Omega}. \quad (81c)$$

Substituting (81) into (80) and using saddle-point integration, we arrive at

$$T_{\substack{1\\2}} = \mp B_{\substack{i\\r}} g(ks) - \frac{1}{2}\chi_{\substack{i\\r}}(\Omega,\phi_2) g(k\rho_2)\, e^{iks} + \begin{Bmatrix}1/2\\0\end{Bmatrix} e^{ik\rho_2\cos(\Omega-\phi_2)} e^{iks} + O(k^{-3/2}) \quad (82)$$

where $B_{\substack{i\\r}}$ takes the following asymptotic expansion

$$B_{\substack{i\\r}} = \mp 2g(k\rho_2) \frac{\partial}{\partial\omega}\chi_{\substack{i\\r}}(\omega,\phi_2)\Big|_{\omega=\Omega} + O(k^{-3/2}).$$

Having determined T_1 and T_2, one can then evaluate T_3, T_4, u_1^t and u_{22}^s in the same manner described earlier to finally arrive at

$$u_2^t = \frac{1}{2}\chi(\Omega,\phi_2) g(k\rho_2)\, e^{iks} + [\tau\chi_r(\Omega,\Omega)\chi(\Omega,\phi_2) + 2\chi'(\Omega,\phi_2)]\, g(ks) g(k\rho_2) + O(k^{-3/2}) \quad (83)$$

where $\chi'(\Omega,\phi_2) = \frac{\partial}{\partial\omega}\chi(\omega,\phi_2)\Big|_{\omega=\Omega}$. Again, the appearance of χ' is an important observation. It must be realized that for the case under study in this section, the fictitious line source interpretation does not apply anymore, because otherwise the term χ' would not have appeared in the solution. The complexity stems from the fact that the diffracted field from half-plane 1 in its shadow boundary direction is not a local plane wave.

E. Total Field When $\Omega_i = \Omega = \phi_2$ (Fig. 10d)

This is a unique situation, as the incident direction, the observation direction, and the line connecting the edge of the half-planes are all aligned with each other. Since the pole $\omega = \Omega_i$ coincides with the saddle point $\omega = \Omega$, one can transform T_1 and T_2 from (75) into the form presented in (80). At $\omega = \Omega_i$ the relation $\xi_r \neq 0$ holds; therefore, evaluation of T_2 follows the same procedure used in Section IV.D for evaluation of T_2 from (80), i.e.,

$$T_2 = -\frac{1}{2}\chi_r(\Omega,\Omega)g(k\rho_2)\, e^{iks} + 2\chi_r'(\Omega,\Omega)g(k\rho_2)g(ks) + O(k^{-3/2}). \quad (84)$$

Similarly, evaluation of T_3, given in (75), requires formulation of the same steps employed in Section III.B.2, i.e.,

$$T_3 = \frac{-1}{2}\tau\chi_r(\Omega,\Omega)g[k(s+\rho_2)] - \frac{2\rho_2}{s+\rho_2}\tau\chi_r'(\Omega,\Omega)g(ks)g(k\rho_2). \quad (85)$$

T_4, defined in (75), may be obtained using the construction of Section III.B.1 to obtain

$$T_4 = \tau\chi_r(\Omega,\Omega)g(ks)\chi_r(\Omega,\Omega)g(k\rho_2). \quad (86)$$

Since u_1^t and u_{22}^s, defined in (71) and (72), respectively, are the solutions due to plane-wave illumination of half-planes 1 and 2, respectively, one may use (18) to arrive at

$$u_1^t = \frac{1}{2}\, e^{ik(s+\rho_2)} + \tau\chi_r(\Omega,\Omega)g[k(s+\rho)] \quad (87)$$

$$u_{22}^s = -\frac{1}{2}\, e^{ik(s+\rho_2)} + \tau\chi_r(\Omega,\Omega)\, e^{iks}\, g(k\rho_2). \quad (88)$$

To complete our evaluation of u_2^t, the only step remaining is to determine T_1 from (80). Substituting $\phi_2 = \Omega$ into (80), one may express T_1 as

$$T_1 = \frac{i}{4\pi} \int_{\text{SDP}} - \csc\left(\frac{\omega - \Omega}{2}\right) F\left(\sqrt{2k\rho_2} \sin \frac{\omega - \Omega}{2}\right) e^{ik(s+\rho_2)\cos(\omega-\Omega)} \, d\omega$$

$$+ \frac{1}{4} e^{ik(s+\rho_2)} \tag{89}$$

Introducing the change of variable

$$t = \sqrt{2}\, e^{i\pi/4} \sin \frac{\omega - \Omega}{2} \tag{90}$$

into (89), one obtains

$$T_1 = \frac{1}{4} e^{ik(s+\rho_2)} - \frac{i}{2\pi} e^{ik(s+\rho_2)} \int_{-\infty}^{\infty} Q(t) F\left(e^{-i\pi/4}\sqrt{k\rho_2}\, t\right) e^{-k(s+\rho)t^2} dt \tag{91a}$$

where

$$Q(t) = t^{-1}(1 + it^2/2)^{-1/2} = t^{-1} - \frac{i}{4} t + 0(t^2). \tag{91b}$$

Substituting (91b) into (91a), one notes that in addition to the integration of type (58a) with $n = 1$, one must also evaluate the following integral, viz.,

$$\int_{-\infty}^{\infty} F\left(e^{-i\pi/4}\sqrt{k\rho_2}\, t\right) e^{-k(s+\rho)t^2} t^{-1} \, dt = I(s). \tag{92}$$

Differentiation of $I(s)$ with respect to s yields

$$I'(s) = -k \int_{-\infty}^{\infty} F\left(e^{-i\pi/4}\sqrt{k\rho_2}\, t\right) e^{-k(s+\rho_2)t^2} t \, dt$$

$$= -\frac{i}{2} (s + \rho_2)^{-1} \sqrt{\rho_2/s} \tag{93}$$

for which the result given in (58a) for $n = 1$ is incorporated. Since $I(s) \to 0$ as $s \to \infty$, one finally obtains

$$I(s) = \frac{i}{2} \sqrt{\rho_2} \int_s^{\infty} \frac{d\sigma}{\sqrt{\sigma}\,(\sigma + \rho_2)} = i \tan^{-1}\sqrt{\frac{\rho_2}{s}} \tag{94}$$

T_1 may at last be evaluated to yield

$$T_1 = \left(\frac{1}{4} + \frac{1}{2\pi} \tan^{-1}\sqrt{\frac{\rho_2}{s}}\right) e^{ik(s+\rho_2)} - \frac{\rho_2}{2(s + \rho_2)} g(ks) g(k\rho_2). \tag{95}$$

Now we have enough information to construct u_2^t from (71), (72) and (74), thus arriving at

$$u_2^t = e^{ik(s+\rho_2)}\left[\frac{1}{4}+\frac{1}{2\pi}\tan^{-1}\sqrt{\frac{\rho_2}{s}}\right] + [g(ks+k\rho_2)+e^{iks}g(k\rho_2)]\cdot\frac{\tau}{2}\chi_r(\Omega,\Omega)$$
$$+ g(ks)g(k\rho)\left[-\frac{\rho_2}{2(s+\rho_2)}+\chi_r^2(\Omega,\Omega)+\frac{2s}{s+\rho_2}\tau\chi_r'(\Omega,\Omega)\right] + O(k^{-3/2}) \tag{96}$$

where

$$\chi_r(\Omega,\Omega) = \frac{1}{\sin\Omega}, \quad \chi_r'(\Omega,\Omega) = -\frac{\cos\Omega}{2\sin^2\Omega}. \tag{97}$$

This is an important result and cannot be obtained via simple application of ray techniques.

F. Comparison With Other Techniques

In this section, we apply the formulation of uniform techniques, discussed in Section III.C.1, to the geometry of staggered parallel plates. Results obtained by a cascading approach, in which the fields scattered by the first half-plane are directly used as the incident fields for the second half-plane, are compared with those given in the previous sections using spectral analysis.

For the situation shown in Fig. 10a, where both the edge of half-plane 2 and the observation point are outside the transition regions, GTD, UTD, MSD and UAT give the same asymptotic result, and are in complete agreement with (78). This is the simplest situation for which even the GTD formulation provides an adequate result.

In the case shown in Fig. 10b, the edge of half-plane 2 is outside the transition region and the observation point is on the shadow boundary of half-plane 2. The GTD formulation can be used to determine the field diffracted from half-plane 1, but this construction fails to provide the correct result for the field diffracted off half-plane 2 along the shadow boundary direction. It actually gives an infinite field which is physically unrealizable. At this point one might be tempted to employ UTD formulation, as in (64), since it would predict a bounded (finite) result. It can be easily verified that UTD formulation does not give the correct asymptotic result when compared with the exact asymptotic solution given in (97). In fact, UTD formulation fails to predict the term χ' in (97). Two successive applications of MSD and UAT formulations do indeed provide the correct result and demonstrate perfect agreement with (97).

For the situation shown in Fig. 10c, where the edge of half-plane 2 lies at the shadow boundary of half-plane 1 and the observation point is positioned away from the transition region, the GTD formulation obviously fails to determine the field correctly. It is then logical to consider whether or not UTD, MSD or UAT can be applied twice in succession in a cascading fashion to derive the final result. Since the incident field on the half-plane 1 is a plane wave, application of each of the three formulations gives Sommerfeld's solution for the field diffracted by half-plane 1. We can readily show that this field takes the following form at the edge of half-plane 2

$$u_1^t = \frac{1}{2} e^{iks} + \tau\chi_r(\Omega,\Omega)g(ks) + 0(k^{-3/2}). \tag{98}$$

Field (98) can now be viewed as the incident field on half-plane 2. Hence, straightforward (mechanical) application of UTD, MSD and UAT would hopefully provide the field diffracted by half-plane 2 at the observation point. After some simplification, this field takes the following asymptotic form

$$u_2^t = \frac{1}{2}\chi(\Omega,\phi_2)\, e^{iks}g(k\rho_2) + \tau\chi_r(\Omega,\Omega)g(ks)\chi(\Omega,\phi_2)g(k\rho_2) + 0(k^{-3/2}). \tag{99}$$

Comparison of (99) with the correct asymptotic expansion given in (83) reveals that (99) is not complete and misses the term corresponding to $2\chi'(\Omega,\phi_2)$. In other words, straightforward (mechanical) application of UTD, MSD and UAT does not provide the correct asymptotic result when the edge of half-plane 2 lies at the shadow boundary (transition region) of half-plane 1. This important observation was first made by Jones [21] for the UAT formulation. Here we have come to a similar important observation that the mechanical application of UTD and MSD does not provide the correct result either. Boersma [22] and Lee and Boersma [19] resolved the difficulty related to UAT by using the argument that the diffracted field (98) from half-plane 1, which varies rapidly in the transition region, does not comply with a ray field behavior. Hence, this field cannot be used directly (mechanically) as the incident field via the application of the UAT formulation. They expanded (98) in terms of an infinite summation of cylindrical and plane waves and applied the UAT formulation to each term of expansion separately. Their final uniform result agrees with Jones' uniform result and with that given in (83).

The difficulties with the straightforward (mechanical) application of UTD, MSD and UAT, as expressed in the preceding paragraph, also exist for the case shown in Fig. 10d. In this case, both the edge of half-plane 1 and the observation point lie at the shadow boundaries. It is straightforward to show that the "mechanical" application of the uniform theories gives

$$u_2^t = \frac{1}{2} e^{ik(s+\rho)} + O(k^{-1}) \tag{100}$$

for the dominant term of the total field. However, the correct dominant behavior is given first in the r.h.s. of (96), which is obviously different from (100). As a matter of fact, as $\rho \to \infty$, (96) gives $(1/2)\exp(iks + ik\rho)$ for the dominant term, i.e., differs by a factor of two from the "mechanical" result, (100), of the uniform theories. Boersma [22] and Lee and Boersma [19] were able to show that, by carefully analyzing the problem in the manner described earlier, they could obtain the complete result agreeing with (96). However, the same analysis has not yet been performed for the MSD formulation.

V. SPECTRAL DOMAIN APPROACH TO VERIFICATION AND REFINEMENT OF ASYMPTOTIC SOLUTIONS

A. Introduction

One of the most challenging problems in the solution of high-frequency scattering analyses is the establishment of the accuracy of the results and the refinement of the solution when the need for its improvement is clearly indicated. The difficulty in verifying whether the asymptotic expression, typically derived from the ray approach, does indeed solve the boundary value problem under consideration stems primarily from the fact that there is no obvious way to "built-in" the boundary conditions in solution procedures based on ray methods. Another reason is that the high-frequency solutions are often constructed for the radiated far fields, whereas the application of the boundary conditions clearly requires the near-field information. In contrast, the integral equation formulation of the scattering problem is based directly on the application of the boundary condition and, consequently, the boundary condition check is redundant for this approach. However, the conventional moment method solution of integral equations is limited strictly to the low frequency and resonance regions as the matrix size becomes unmanageably large beyond the resonance region.

In this section we will briefly outline a spectral domain method for bridging the two approaches, viz., the integral equation and asymptotic techniques. The hybrid method has the desirable feature that it not only verifies the accuracy of the ray solutions but provides a systematic means for improving the solution for a large class of problems of practical interest. This fact will be illustrated via two typical examples given in this section. Other cases have also been treated and may be found in [23] and [24].

B. Development of Spectral Domain Formulation of the Integral Equation and Its Approximate Solution

The key to combining the asymptotic solution with the integral equation formulation lies in recognizing the fact that the Fourier transform of the induced current on a scatterer is directly proportional to the scattered far field and that a good approximation to this scattered field is often available from any number of asymptotic methods, e.g., GTD or the spectral approach discussed in the preceding sections. To take advantage of these facts we choose to work with the "Fourier-transformed" or "spectral domain" version of the integral equation rather than with the conventional spatial domain counterpart of the same equation. We begin, however, with the conventional electric-field integral equation (E-equation) for a perfectly conducting scatterer:

$$(\bar{\bar{G}} * \bar{J})_t = -\bar{E}_t^i \quad , \quad \bar{\bar{G}} = \text{Green's Dyadic} \tag{101}$$

where $\bar{J}(\bar{r}')$ is the unknown induced surface current density, the subscript t signifies the tangential component of the field on the surface S of the scatterer, $\bar{E}^i$ is the incident electric field on the scatterer, and "*" symbolizes the convolution operation.

As a preamble to Fourier transforming (101), we first extend it over all space. To this end we define a truncation operator $\Theta(\bar{A})$

$$\Theta(\bar{A}) = \int \bar{A}_t \delta(r - r_s)\, dr \quad , \quad r_s \in S \tag{102}$$

where δ is the Dirac delta function. Let $\hat{\Theta}(\bar{A})$ be defined as the complementary operator

$$\hat{\Theta}(\bar{A}) = \bar{A} - \Theta(\bar{A}). \tag{103}$$

Then (101) can be rewritten as

$$\bar{\bar{G}} * \bar{J} = \hat{\Theta}(-\bar{E}^i) + \hat{\Theta}(\bar{\bar{G}} * \bar{J}) \tag{104}$$

for all space. Using the definition of the Fourier transform introduced in Section II, now extended to the general three-dimensional case, we can write (104) as

$$\tilde{\bar{\bar{G}}}\tilde{\bar{J}} = -\tilde{\bar{E}}_I + \tilde{\bar{F}} \tag{105}$$

where $\tilde{\bar{F}} = F[\hat{\Theta}(\bar{\bar{G}} * (\Theta\bar{J})]$ and $\tilde{\bar{E}}_I$ is the transform of the tangential component of the incident field truncated on S, with all transformed quantities being denoted in this section by ~ on top. Note that the convolution operator in the integral equation is transformed into an algebraic product upon Fourier transformation. Note also that (105) has two unknowns, viz., $\tilde{\bar{J}}$ and $\tilde{\bar{F}}$, which must be solved for simultaneously.

At this point one can construct the solution of (105) in at least two ways. The first of these is based on an iterative procedure that begins with an approximation to $\tilde{\bar{J}}$, the transform of the induced surface current, derived from the application of some asymptotic procedure to the scattering problem. The following equation is then employed to generate the next order solution and the procedure is repeated until convergence is achieved:

$$\tilde{\bar{J}}^{(n+1)} = \tilde{\bar{\bar{G}}}^{-1}\left[-\tilde{\bar{E}}_I + F\left(F^{-1}[\tilde{\bar{\bar{G}}}\tilde{\bar{J}}^{(n)}] - \Theta\{F^{-1}[\tilde{\bar{\bar{G}}}\tilde{\bar{J}}^{(n)}]\}\right)\right] \tag{106a}$$

and in the space domain

$$\bar{J}^{(n+1)} = \Theta\left[F^{-1}[\tilde{\bar{J}}^{(n+1)}]\right]. \tag{106b}$$

It may be shown that the combination of the second and third terms in (106a) represents an approximation to $\tilde{\bar{F}}$ derived from the nth approximation of $\tilde{\bar{J}}$, i.e, $\tilde{\bar{J}}^{(n)}$. Also, it should be realized that the inverse operator $\tilde{\bar{\bar{G}}}^{-1}$ is algebraic since $\tilde{\bar{\bar{G}}}$ itself has the same nature. The check for the boundary condition may be readily applied in this procedure since $\Theta\{F^{-1}[\tilde{\bar{\bar{G}}}\tilde{\bar{J}}^{(n)}]\}$, the last term inside the square bracket in (106a), represents the tangential E-field produced on the surface of the object by the induced current. Ideally, this field should equal the negative of the tangential component of the incident E-field $\bar{E}_t^i$. A comparison between these two fields immediately reveals the extent to which the boundary condition is satisfied on the surface of the scatterer.

A second approach to handling (105) would be to employ the Galerkin procedure in the transform domain. One may write

$$\tilde{\tilde{J}} \simeq \tilde{\tilde{J}}^{(0)} + \sum C_P \tilde{\tilde{J}}_P \tag{107}$$

where $\tilde{\tilde{J}}^{(0)}$ is the approximate solution derived from a suitable asymptotic formula for the scattered field and $\tilde{\tilde{J}}_P$ represents a set of basis functions in the transform domain. Typically, there are certain angular regions in the far field where the asymptotic solutions require refinement. One may choose to concentrate the basis functions in these regions in the transform domain. Alternatively, the $\tilde{\tilde{J}}_P$'s could be chosen as the transforms of a suitable set of basis functions in the space domain, and the location (support) of these subdomain basis functions may be selected to coincide with transition regions or corners, etc., where the canonical solution of the asymptotic solution may require refinement.

In either case, the problem of determining $\tilde{\tilde{J}}$ may be reduced to that of finding the unknown coefficients C_P such that (107) satisfies (105). The Galerkin procedure provides a way for accomplishing this, as we will soon see. This technique also has the advantage that the other unknown in (105), viz., $\tilde{\tilde{F}}$, is conveniently eliminated from this equation upon application of Galerkin's method. We demonstrate this fact in the manipulations presented below.

Substituting (107) in (105) and taking a scalar product of the resulting equation with a set of suitable series of testing fuctions $\tilde{\tilde{W}}_q$, we arrive at

$$\sum C_p \langle \tilde{\tilde{W}}_q, \tilde{\tilde{\bar{G}}} \tilde{\tilde{J}}_p \rangle = -\langle \tilde{\tilde{W}}_q, \tilde{\tilde{E}}_I \rangle + \langle \tilde{\tilde{W}}_q, \tilde{\tilde{F}} \rangle \tag{108}$$

where "<,>" is the scalar inner product. If we now choose $\tilde{\tilde{W}}_q$ to be transforms of functions which are nonzero only on the surface of the scatterer, then the scalar products $\langle \tilde{\tilde{W}}_q, \tilde{\tilde{F}} \rangle$ can be shown to vanish. To show this, one uses Parseval's theorem and transforms the scalar product of $\tilde{\tilde{W}}_q$ and $\tilde{\tilde{F}}$ in terms of a similar product of

their counterparts in the space domain. Since the inverse transforms of $\tilde{\tilde{W}}_q$ and $\tilde{\tilde{F}}$ exist in complementary regions, viz., on the surface of the scatterer and in the region complementary to this surface, respectively, one finds that their scalar product is identically zero. One can now proceed in the usual manner to solve for the coefficients C_p by solving the matrix equation represented by (108) with the second term in the r.h.s.
It is evident that the use of this method would be practical only when relatively few terms are needed in (107) to modify the available asymptotic solution; however, this is typically the situation for many problems. It should also be noted that (108) represents a direct check on the satisfaction of the boundary condition, in the sense of moments. The choice of $\tilde{\tilde{W}}_p$'s is governed by the locations on the surfaces of the scatterer where these boundary conditions are applied. Typically these will be the zones where the asymptotic solution might be inaccurate, e.g., the transition region between the lit and shadow regions.

C. Illustrative Examples

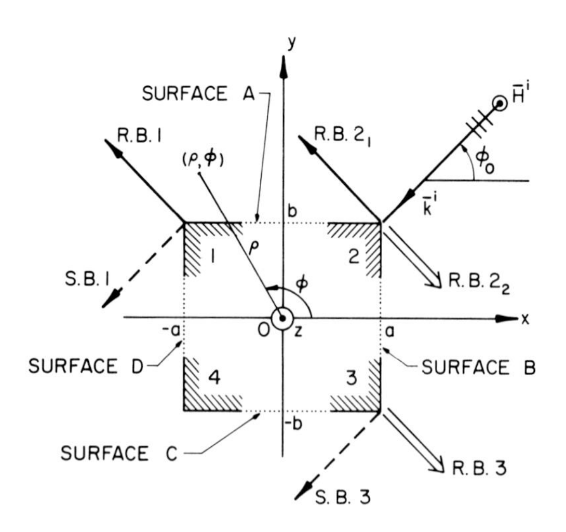

Fig. 13. For the angle of incidence ϕ_0 as shown, wedges 1, 2 and 3 are illuminated while wedge 4 is in the dark.

We will now illustrate both the iterative and Galerkin procedures by means of two examples of high-frequency scattering. For the first example we choose the problem of diffraction by a rectangular cylinder, a structure with multiple wedges. The geometry of the problem is shown in Fig. 13. We note that the wedges 1, 2 and 3 and faces A and B are illuminated by the incident wave while the faces C and D and the wedge 4 are in the shadow. Let us first consider the hemisphere above the face A. The contribution to the scattered field in this region comes from the two lit surfaces and rays diffracted by the wedges 1, 2 and 3. To derive a first approximation to the asymptotic expression for the far field scattered by the cylinder, we may use Keller's coefficients for the diffraction by the three wedges. From wedge 1 we have, for instance,

$$H^d_{z1} \sim e^{-ik(-a\cos\phi_0 + b\sin\phi_0)} \frac{e^{ik\rho}}{\sqrt{\rho}} \frac{e^{i\frac{\pi}{4}}}{\sqrt{2\pi k}} \tag{109}$$
$$\cdot \left\{ \frac{\frac{2}{3}\sin\frac{2\pi}{3}\, e^{ika\cos\phi}\, e^{-ikb\sin\phi}}{\cos\frac{2\pi}{3} - \cos\frac{2}{3}(\phi - \phi_0)} + \frac{\frac{2}{3}\sin\frac{2\pi}{3}\, e^{ika\cos\phi}\, e^{-ikb\sin\phi}}{\cos\frac{2\pi}{3} - \cos\frac{2}{3}(\phi + \phi_0)} \right\}$$

where $0 \leq \phi < \frac{\pi}{2}$ and $0 \leq \phi < 3\pi/2$. Similar expressions can be readily obtained for the wedges 2 and 3. Note that the scattered field from wedge 3 is restricted in the angular region $0 < \phi \leq \pi/2$ and $\pi \leq \phi \leq 2\pi$ for $0 \leq \phi_0 \leq \pi/2$. Hence, in the upper hemisphere only one-half of the angular range is illuminated by the scattered field from wedge 3, if Keller's formula for an <u>infinite</u> wedge is used to derive an approximation to this scattered field. Another rather important and common feature of the Keller expressions for the wedge-diffracted fields as given, for instance, by (109), is that these formulas predict fictitious infinities in the scattered fields at the shadow and reflection boundaries. One could completely eliminate these infinities by employing the spectral concepts and deriving the scattered far field from the transform of an approximation to the induced surface current, comprising the physical optics and fringe currents. This transform, being associated with a function with finite support, is always bounded and consequently free of the singular behavior present in (109).

As an alternative we may also employ one of the available uniform theories [17,14] that provide smooth and bounded transition through the reflection and shadow boundaries. It is fortuitous, however, that the aggregate contribution of the infinities from the individual wedges cancels out exactly when their contributions are superimposed. This occurs because of the unique symmetries of the geometry of the rectangular cylinder under consideration. Hence, no special care is required in this example at the transition regions. The diffracted far fields computed by using the Keller formulas for wedge diffraction are shown in Fig. 14. It is evident that the pattern is discontinuous at 0°, 90°, 180° and 270°. This behavior is attributable to the use of infinite wedge diffraction coefficients that produce a nonzero scattered field only in the region external to the wedge and, hence, produce discontinuous fields supported by induced surface currents that extend to infinity along the wedge surfaces.

To refine this far-field asymptotic solution, we may now proceed to apply the iterative procedure developed in Section V.B. To this end, we first introduce in Fourier transform the variable $\alpha = k \cos \phi$ and express the far scattered field in the hemispherical region defined by $0 < \phi < \pi$ in terms of α. We employ analytical continuation of the expressions for the wedge-diffracted fields to determine the Fourier transform in the range $|\alpha| > k$ by substituting appropriate complex values for ϕ. One of the chief advantages of using the available expressions for the asymptotic solution as a starting point for the iterative procedure is that the approximate analytical expression is convenient for estimating the scattered far field, both in the visible and invisible ranges. By Fourier inversion of the scattered E-field at infinity, we can derive the tangential E-field on a planar surface tangential to

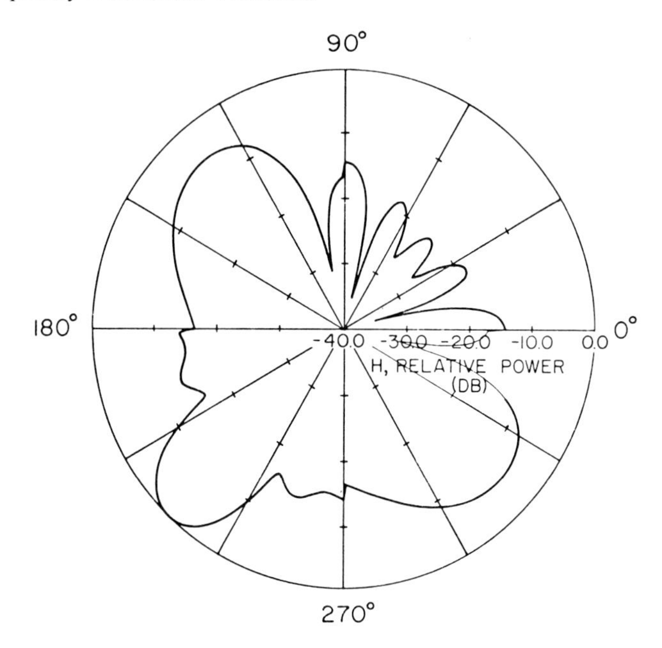

Fig. 14. GTD diffracted far-field pattern of the rectangular cylinder; $\phi_0 = \pi/4$, a = b = 1λ.

the face A. If this computed near field were to satisfy the boundary condition that the tangential E-field on surface A equal the negative of $\bar{E}^i$, and the similar situation is repeated for other faces, we would conclude that the solution so derived is an accurate one. However, we would not expect that to be true for the solution represented by (109) since, as pointed out earlier, this expression produces glitches at some angles of observation. Nevertheless, this solution provides a very good zero-order estimate for $\tilde{\tilde{F}}^{(0)}$ which is readily derived by taking the transform of the near field derived in the plane of the surface A of the scatterer, and repeating the same procedure for the other faces as well. In following this procedure, we effectively compute

$$F\left(F^{-1}[\tilde{\tilde{G}}\tilde{\tilde{J}}^{(0)}] - \Theta\{F^{-1}[\tilde{\tilde{G}}\tilde{\tilde{J}}^{(0)}]\}\right) \tag{110}$$

which is an approximation to $\tilde{\tilde{F}}$ derived by using $\tilde{\tilde{J}}^{(0)}$. The next order of approximation to $\tilde{\tilde{J}}$ is now readily obtained from (106). This quantity is Fourier inverted four times to calculate the tangential H-field on the four faces of the cylinder and the surface current on these faces, obtained from (106b), is Fourier transformed again to derive the far-field pattern.[2] The iterated far-field pattern $\tilde{\tilde{J}}^{(1)}$ is shown in Fig. 15; the disappearance of the discontinuities at 0, $\pi/2$, $3\pi/2$, etc. is immediately evident from this plot. This result has also been verified by a few other workers who have followed different procedures than those outlined here [25]. Recall, however, that the method outlined here provides a convenient "built-in" check for the satisfaction of the boundary condition and an independent check is not altogether necessary to establish the accuracy of the solution.

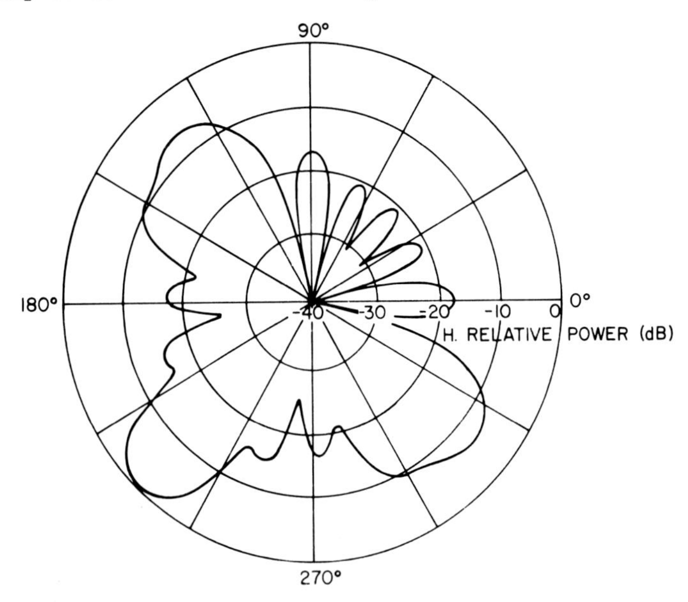

Fig. 15. Improved scattered far-field pattern of the rectangular cylinder; $\phi_0 = \pi/4$, $a = b = 1\lambda$.

The next example we discuss is that of a smooth convex surface with no wedges - a circular cylinder. One of the important attributes of this canonical geometry is that it permits

---

[2]Numerical experiments have shown that very good far-field results are obtained even when one skips the first step and uses $\tilde{\tilde{J}}^{(0)}$ itself to derive the surface current using (106b) and $\tilde{\tilde{J}}^{(1)}$ by the Fourier transformation of this current.

convenient comparison with the exact series solution available for the representation of scattered fields from this structure. The geometry of the problem is shown in Fig. 16. We consider the case of an E-polarized wave incident from $\phi = 180°$.

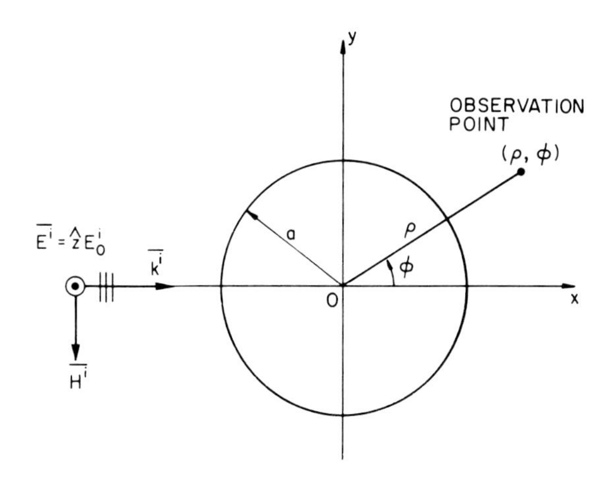

Fig. 16. Diffraction by a circular cylinder illuminated by an E-wave incident along the x-axis.

The first step in attacking the problem is to use a geometrical-optical (GO) approach [24] to derive the far-scattered field. When this is done, one obtains the dotted curve in Fig. 17, which also exhibits the exact series solution for the scattered field as a solid curve. It is evident that in the range $-60° < \phi < 60°$ the GO solution is not adequate. This is not totally surprising since it is well known that creeping-wave contributions need to be included in the scattered field expression in the shadow and transition regions. Rather than following this procedure, we will now show how the Galerkin method can be readily and conveniently applied to this problem to derive an accurate solution.

To this end we consider, as a first step, the behavior of the scattered field on a surface erected in juxtaposition to the cylinder at the point $x = a$, the farthest point away from the incident field. In the deep shadow region, say $|y| < 2$, we expect the scattered field $E_z^s = -E_z^i$ to be a very good approximation.

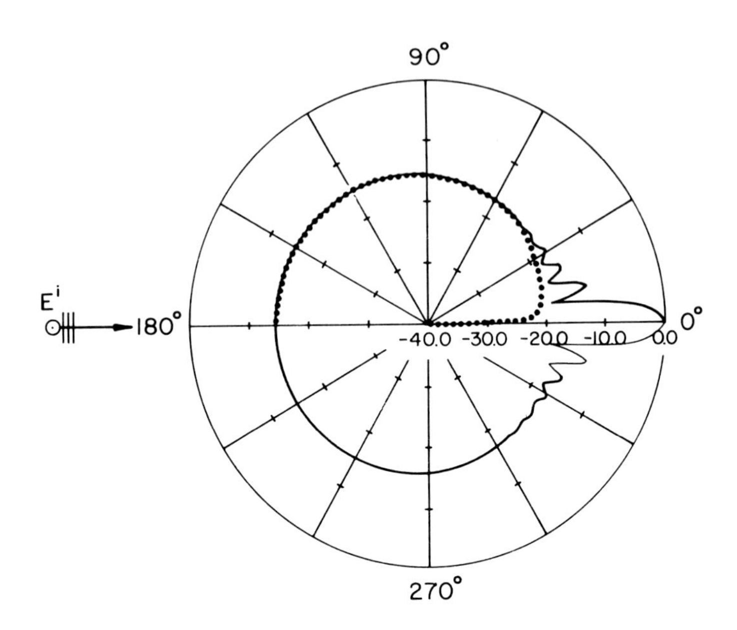

Fig. 17. Geometrical optics scattered far-field pattern in dB of a circular cylinder with radius a = *3λ. (····· = geometrical optics; ——— = exact.)*

On the other hand, when we go far onto the lit region on this surface, say for $|y| > 6$, we expect the E_z^s to be described adequately by the GO formulas. If we had a good estimate of the scattered field behavior in the transition region $2 < |y| < 6$, we would be able to get a good representation of the excess scattered field (over and above the GO field) on the entire surface at $x = a$. We should then be able to compute the field radiated in the r.h.s. of the cylinder by this excess field using the concept of Huyghen's source and use this radiated field to fill in the gap between the GO pattern and the true pattern shown in Fig. 18.

To derive the E_z^s field in the transition region, we first interpolate the magnitude of this field from E^i at $|y| = 2$, to 0 at $y = 6$ and the phase from π at $|y| = 2$, to the GO phase at $|y| = 6$. Next we introduce a set of basis functions, with undetermined coefficients, to describe the correction to the

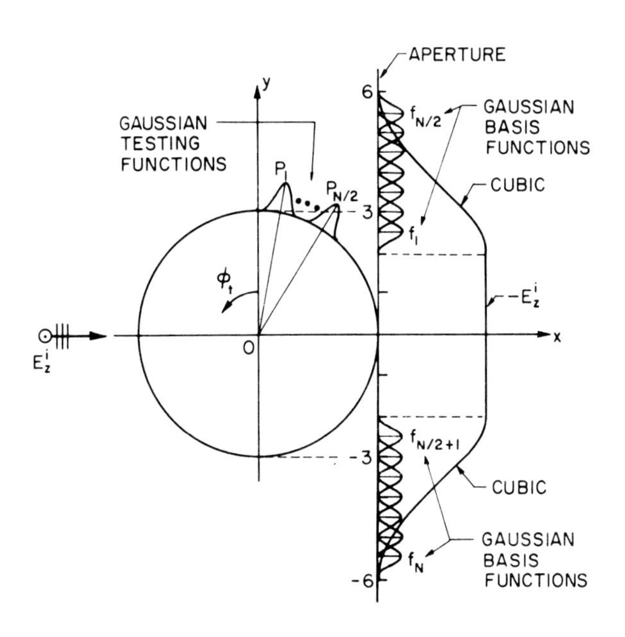

Fig. 18. Locations of the basis functions on the aperture and the testing functions on the surface of the obstacle.

interpolated E_z^s field at the plane $x = a$, $2 < |y| < 6$. To determine these coefficients, we apply the concept of Galerkin's method in the spectral domain as briefly outlined in Section V.B. In the example being considered here, the transforms of the basis functions play the role of $\tilde{\tilde{J}}_p$ in (107), and the zero-order, scattered far field $\tilde{\tilde{J}}^{(0)}$ is obtained by adding the contributions of GO and the approximate excess E_z^s field derived from the interpolation procedure just described.

The choice of the testing functions, $\tilde{\tilde{W}}_q$ in (108), is suggested by the fact that the error in the high-frequency asymptotic solution is mostly concentrated around the transition region on the surface of the cylinder, i.e., in the neighborhood of the junction between the lit and shadow regions. Thus, a suitable choice for the testing functions would be to locate them at the transition region as shown in Fig. 18, where the location of the basis functions is also shown. Note that we need not be restricted in our choice for the location of these functions by demanding that they have a common support, although this is almost always the case in the conventional moment or Galerkin methods. We may also note from Fig. 18 that the shape of the basis and testing functions are both Gaussian. Since we are dealing with transforms, this choice is not only convenient for deriving the Fourier transformations $\tilde{\tilde{J}}_p$ and $\tilde{\tilde{W}}_p$, but is also desirable from a numerical point of view because the transforms are not oscillatory as they would be for a pulse or triangular basis. This feature is important when numerically computing the scalar products $\langle \tilde{\tilde{J}}_p, \tilde{\tilde{W}}_p \rangle$, needed for the determination of the unknown coefficients C_p.

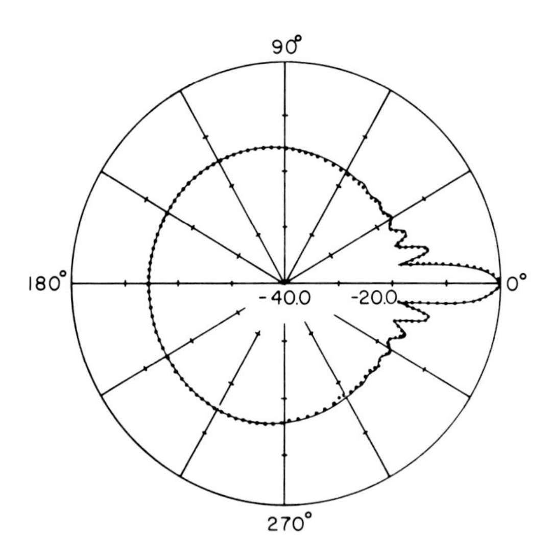

Fig. 19. Scattered far-field pattern in dB of a circular cylinder with radius a = *3λ obtained by Galerkin's method.*

Only a few unknowns C_p (three to seven) are needed to derive an accurate solution for both the radiated far field and the surface current on the cylinder. The accuracy itself can be verified by computing the tangential E-field on the surface via Fourier inversion of the hemispherical far-field pattern centered around the point to be tested. This procedure is also used to compute the surface current distribution from the knowledge of the scattered H-field at large distances. Of course an independent check is available for this problem via the exact series solution. A comparison of the Galerkin solution and the exact series solution is shown in Figs. 19 and 20 to illustrate the highly accurate nature of the Galerkin solution. In fact, the solution is surprisingly accurate, except for a slight error in the transition region, even without the Galerkin refinement, as evidenced by the dotted curve in Fig. 20 which exhibits this case.

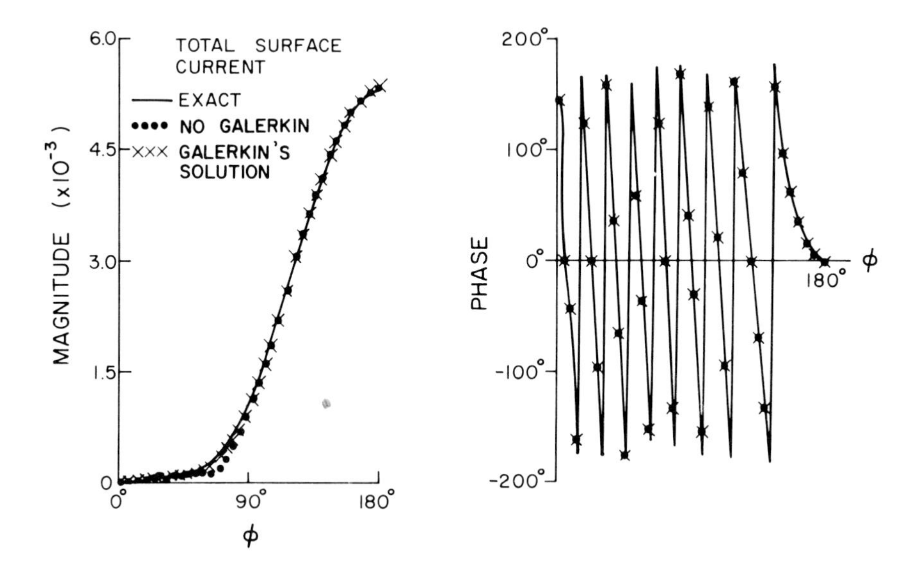

Fig. 20. Total surface current on a perfectly conducting circular cylinder with radius a = *3*λ.

This method just described could easily be employed for more complex, nonseparable shapes as long as the smooth convex nature of the surface is preserved. A generalization to even more complex geometries, e.g., combination of curved surfaces and wedges, also appears possible.

VI. REFERENCES

1. Keller, J.B., Geometrical Theory of Diffraction, *J. Opt. Soc. Am. 52,* 116-130 (1962).
2. Kouyoumjian, R.G., The Geometrical Theory of Diffraction and its Application, in "Topics in Applied Physics," (R. Mittra, Ed.), Vol. 3, Springer Verlag, New York, 165-215 (1976).
3. Mittra, R., Y. Rahmat-Samii and W.L. Ko, Spectral Theory of Diffraction, *Appl. Phys. 10,* 1-13 (1976).
4. Noble, B., "Methods Based on the Wiener-Hopf Technique," Pergamon Press, London (1958).
5. Born, M. and E. Wolf, "Principles on Optics," Macmillan, New York (1964).
6. Mittra, R. and S.W. Lee, "Analytical Techniques in the Theory of Guided Waves," Macmillan, New York (1971).
7. Deschamps, G.A., Ray Techniques in Electromagnetics, *Proc. IEEE 60,* 1022-1035 (1972).
8. Rahmat-Samii, Y. and R. Mittra, A Spectral Domain Interpretation of High-Frequency Diffraction Phenomena, *IEEE Trans. Antennas and Propag. AP-25,* 676-687 (1977).
9. Clemmow, P.C., "The Plane Wave Spectrum Representation of Electromagnetic Fields," Pergamon Press, Oxford (1966).
10. Rahmat-Samii, Y. and R. Mittra, Spectral Analysis of High Frequency Diffraction of an Arbitrary Incident Field by a Half-Plane - Comparison with Four Asymptotic Techniques, University of Illinois at Urbana-Champaign, Electromagnetics Laboratory Technical Report No. 76-10, November 1976.
11. Clemmow, P.C., A Note on the Diffraction of a Cylindrical Wave by a Perfectly Conducting Half-Plane, *Q. J. Mech. Appl. Math. III(3)*, 377-384, (1950).
12. Khestanov, R.Kh., Diffraction of Arbitrary Field at a Half-Plane, *Radio Eng. & Electron. Phys. 15,* 250-256 (1970).
13. Boersma, J. and S.W. Lee, High-Frequency Diffraction of a Line-Source Field by a Half-Plane, Part I. Solutions by Ray Techniques, *IEEE Trans. Antennas & Propag. AP-25,* 171-179 (1977).
14. Ahluwalia, D.S., R.M. Lewis and J. Boersma, Uniform Asymptotic Theory of Diffraction by a Plane Screen, *SIAM J. Appl. Math 16,* 783-807 (1968).
15. Felsen, L.B. and N. Marcuvitz, "Radiation and Scattering of Waves," Prentice Hall, Elglewood Cliffs, New Jersey (1973).
16. Lee, S.W. and G.A. Deschamps, A Uniform Asymptotic Theory of Electromagnetic Diffraction by a Curved Wedge, *IEEE Trans. Antennas & Propag. AP-24,* 25-34 (1976).
17. Kouyoumjian, R.G. and P.H. Pathak, A Uniform Geometrical Theory of Diffraction for an Edge in a Perfectly Conducting Surface, *Proc. IEEE 62,* 1448-1461 (1974).
18. Hwang, Y.M. and R.G. Kouyoumjian, A Dyadic Coefficient for an Electromagnetic Wave which is Rapidly Varying at an Edge, URSI 1974 Annual Meeting, Boulder, Colorado.
19. Lee, S.W. and J. Boersma, Ray-Optical Analysis of Fields on

Shadow Boundaries of Two Parallel Plates, *J. Math. Phys. 16*, 1746-1764 (1975).

20. Wolfe, P., Diffraction of Plane Waves by a Semicircular Strip, *Q. J. Mech. Appl. Math 28*, 355-371 (1975).
21. Jones, D.S., Double Knife-Edge Diffraction and Ray Theory, *Q. J. Mech. Appl. Math. 26*, 1-18 (1973).
22. Boersma, J., Diffraction by Two Parallel Half-Planes, *Q. J. Mech. Appl. Math. 28*, 405-426 (1975).
23. Ko, W.L. and R. Mittra, A New Approach Based on a Combination of Integral Equation and Asymptotic Techniques for Solving Electromagnetic Scattering Problems, *IEEE Trans. Antennas & Propag. AP-24*, 187-197 (1977).
24. Ko, W.L. and R. Mittra, A Method for Combining Integral Equation and Asymptotic Techniques for Solving Electromagnetic Scattering Problems, University of Illinois at Urbana-Champaign, Electromagnetics Laboratory Technical Report No. 76-6, May 1976; prepared under contracts DAHC04-74-G-0113 and N00014-75-C-0293.
25. Burnside, W.D., C.L. Yu and R.J. Marhefka, A Technique to Combine the Geometrical Theory of Diffraction and the Moment Method, *IEEE Trans. Antennas & Propag. AP-23*, 551-558 (1975).
26. Mittra, R. and Y. Rahmat-Samii, On the Investigation of Diffracted Fields at the Shadow Boundaries of Staggered Parallel Plates - A Spectral Domain Approach, *Radio Sci. 12(5)*, 659-670 (1977).
27. Mittra, R. and Y. Rahmat-Samii, Spectral Analysis of High-Frequency Diffraction of an Arbitrary Incident Field by a Half Plane - Comparison with Four Asymptotic Techniques, *Radio Sci. 13(1)*, 31-48 (1978).

SOME PROBLEMS INVOLVING IMPERFECT HALF PLANES[1]

Thomas B.A. Senior
Radiation Laboratory
The University of Michigan

As part of a continuing study of how the material of a body affects its scattering, we consider here the influence of material properties on edge diffraction. An appropriate canonical problem is the scattering of an electromagnetic wave by a half plane whose boundary condition is chosen to simulate the material in question. The half planes considered fall into two categories: impedance sheets subject to a Leontovich boundary condition on both faces, and resistive and/or conductive sheets characterized by a jump condition. In each case the material parameter (impedance, resistivity or conductivity) may be a tensor, but is assumed to have the same value at all points on a given face, corresponding to a homogeneous, anisotropic sheet. The various forms of the boundary conditions are discussed for a wave incident in a plane perpendicular to the edge and for oblique incidence when the conditions couple two components of the field. We then consider the two basic methods which are available for solving boundary value problems of this type and examine their application to the different half planes. With the advances that have taken place recently, the two methods now have the same capability, but neither works when the material is anisotropic except in those situations where the anisotropy has no effect.

[1]This work was supported by the U.S. Air Force Office of Scientific Research under Grant 72-2262.

ISBN 0-12-709650-7

I. INTRODUCTION

The scattering from any body is a function of its material properties as well as its geometry. This fact is exploited when radar absorbing materials are applied to the surface of a target to reduce its scattering, but most RAM are designed to suppress the specular contributions from a body whose electrical dimensions are large. Once these echoes have been eliminated we are left only with non-specular or diffractive sources, and though their magnitudes are much less, they are often the dominant contributors to the backscattering patterns of aerospace targets in the aspect ranges of most interest. If we are to suppress these as well, it is now necessary to examine the effect that the material properties of a body have on such non-specular scatterers as edge, creeping and traveling waves and attempt to tailor the properties of an absorber to maximize the reduction.

The last few years have seen a growing effort directed at this type of scattering. Using the integral equations for the surface fields on two-dimensional bodies subject to an impedance boundary condition, computer programs have been developed [11,12] to find the impedances which are most effective in suppressing the non-specular contribution to the backscattering from various shapes of target. Resistive (electric current) sheets placed in the vicinity of a metallic edge have also proved effective in reducing edge scattering for carefully chosen variations of the resistivities along the sheets [13]. Computer programs have been written to obtain the scattering from these types of configuration and to optimize the sheet selection: in contrast to an impedance specification of a surface, the resistivity uniquely defines the sheet, and the computer predictions have been confirmed experimentally. Program RAMVS [14] accepts many resistive and/or conductive sheets in the presence of a body and is now being used to design absorbers to suppress non-specular forms of scattering. In effect, by representing a graded RAM as a number of discrete layers or sheets, this program provides a way of specifying the material parameters of a coating that will realize the surface impedance necessary to minimize the scattering.

Although design problems of this complexity would be almost impossible to solve without the power and versatility of a digital computer, high frequency approximations such as GTD can still play a vital role in the investigation. In addition to giving insight into the type of scattering that occurs, the approximations help in pinpointing the material parameters of interest and in interpreting the computer-generated data. This is particularly true in the case of edge scattering, whether produced by the incident field directly or by a creeping or traveling wave on the surface. Edge waves are important in many applications, including the specification of antenna ground planes and the design of acoustic baffles [22]. They are major contributors to the scattering from the wings and tailplanes of an aircraft, and are almost inevitably introduced when resistive sheets are used to reduce these (and other) forms of scattering. It is therefore of interest to examine the scattering from the edge of a plate or sheet to see how it depends on the material properties, and the present paper is concerned with some aspects of the mathematical analyses involved.

To determine the diffraction coefficient of an edge it is necessary to solve an appropriate canonical problem, and a natural one to choose is a half plane whose boundary conditions are such as to simulate the material in question. It will be assumed that the primary field is a plane wave incident in a plane perpendicular to the edge (normal incidence) or obliquely (skew incidence), and though we shall devote most attention to electromagnetics, it is convenient to preface our discussions with a brief consideration of the analogous but simpler problems in acoustics. The task is then to take some of the half planes relevant to an assessment of material effects in edge scattering, and to explore the various forms of the boundary conditions and the methods which are available for the exact solution of the scattering problem.

The half planes fall into two distinct categories: impedance sheets subject to a Leontovich boundary condition, and resistive and/or conductive sheets characterized by a jump condition. In each case the material parameter (impedance, resistivity, etc.) will be allowed to be a tensor corresponding to a non-isotropic sheet with the tensor constant at all points on a given face, though different on the two faces. Note that the assumption of homogeneity excludes problems which can be solved in some particular coordinate system by virtue of a surface impedance having a specific variation matching that of the metric coefficients. Though it has proved essential for some absorber applications to have the properties of a sheet vary as a function of the distance from the edge, under most circumstances edge diffraction is still a local phenomenon determined by the properties within a small fraction of a wavelength of the edge.

In the first part of the paper we describe the various sheets and their associated boundary conditions, emphasizing those situations where the problem can be scalarized. We then turn to the methods available for their solution. There are, in effect, two basic methods. Each has a number of variations and/or extensions, and there are in addition methods which are effective in special cases and which may then provide a rather elementary approach, but one or other of the two methods is applicable in all circumstances for which solutions have been obtained so far.

II. BOUNDARY CONDITIONS

A generalization of the standard boundary conditions for soft (yielding) and hard (rigid) surfaces in acoustics is the impedance boundary condition [17]

$$\hat{n} \cdot \underline{u} + p/Z = 0 \tag{1}$$

where $\underline{u}$ is the perturbation velocity, p is the excess pressure, $\hat{n}$ is the unit outward normal to the surface and Z is the acoustic impedance. When expressed in terms of the velocity potential V for a time harmonic field, (1) becomes

$$\frac{\partial V}{\partial n} + \frac{ik}{\eta} V = 0 \tag{2}$$

where η (proportional to Z) is the complex specific acoustic impedance of the surface, and reduces to the Dirichlet boundary condition for a soft surface when $\eta = 0$ and to the Neumann boundary condition for a hard surface when $\eta = \infty$. A time dependence $e^{-i\omega t}$ has been assumed and suppressed.

The condition (2) is widely employed in studies of acoustically imperfect surfaces, for example, in the design of absorbing baffles. Its form is the same regardless of the incident field and for curved as well as planar surfaces, and though a rigorous justification may require that the radii of curvature of the surface are everywhere large compared to the wavelength, the condition has nevertheless found useful application even to edged structures such as an absorbing sheet or wedge. In the case of an absorbing half plane occupying the region $x \geq 0$ of the plane $y = 0$ in a Cartesian coordinate system, (2) implies

$$\frac{\partial V}{\partial y} \pm \frac{ik}{\eta} V = 0 \qquad (y = \pm 0) \tag{3}$$

where η may differ on the two faces; and for an absorbing wedge whose surfaces are defined as $\rho \geq 0$, $\phi = \Omega$ and $2\pi - \Omega$, in a cylindrical polar coordinate system where 2Ω is the closed angle of the wedge,

$$\frac{1}{\rho}\frac{\partial V}{\partial \phi} \pm \frac{ik}{\eta} V = 0 \qquad (\phi = \Omega,\ 2\pi - \Omega). \tag{4}$$

There are two extensions of the boundary conditions that should be mentioned. The first arises when there is a uniform main stream flow past an abosrbing wedge or half plane [20]. The result is to add a tangential derivative to (2), but since this does not affect the applicability of the methods available for the solution of the problem, we shall not discuss it further. The second extension is rather more profound. If (2) is applied on each side of a thin slice of material having η the same on both sides, then in the limit as the thickness tends to zero, addition and subtraction of the two conditions gives

$$\left[\frac{\partial V}{\partial n}\right]_-^+ + \frac{ik}{\eta}(V_+ + V_-) = 0 \quad , \quad [V]_-^+ + \frac{\eta}{ik}\left(\frac{\partial V_+}{\partial n} + \frac{\partial V_-}{\partial n}\right) = 0 \qquad (5)$$

where $\hat{n}$ is outward to the side indicated by the plus sign. In general, the eqs. (5) imply discontinuities in $\frac{\partial V}{\partial n}$ and V across the sheet, but we can conceive of situations in which one or other is continuous. This additional constraint imposed in place of one of the conditions (5) converts the boundary value problem into a transition problem. If, for example $V_+ = V_-$, then

$$V \text{ continuous} , \quad \left[\frac{\partial V}{\partial n}\right]_-^+ + \frac{2ik}{\eta} V = 0 \qquad (6)$$

and these conditions in which the pressure is equal on two sides, but creates a proportional jump discontinuity in the normal component of the fluid velocity, have been used [26] to characterize an acoustically resistive membrane. We can likewise conceive of a "conductive" membrane having

$$\frac{\partial V}{\partial n} \text{ continuous} , \quad [V]_-^+ + \frac{2\eta}{ik}\frac{\partial V}{\partial n} = 0 \qquad (7)$$

and these two membranes are complementary in the sense required for the existence of a Babinet principle [26]. For present purposes, however, an important point is the following: although (6) and (7) have obvious similarities to (5), the mathematical problems are quite distinct, and a method which is applicable for the solution of the boundary value problem for an impedance half plane may be ineffective when applied to a half plane membrane.

The electromagnetic analog of (1) is the surface impedance (or Leontovich) boundary condition

$$\underline{E} - (\hat{n} \cdot \underline{E})\hat{n} = Z\bar{\bar{\eta}} \cdot (\hat{n}_\wedge \underline{H}) \qquad (8)$$

where $\bar{\bar{\eta}}$ is a tensor surface impedance relative to the impedance Z of the free space medium above. Since (8) can be written as

$$\hat{n}_\wedge (\hat{n}_\wedge \underline{E}) = -Z\bar{\bar{\eta}} \cdot \hat{n}_\wedge \underline{H},$$

the surface magnetic current density $\underline{K}^* = -\hat{n}_\wedge \underline{E}$ is directly related to the surface electric current density $\underline{K} = \hat{n}_\wedge \underline{H}$ via

$$\hat{n}_\wedge \underline{K}^* = -Z\overline{\overline{\eta}} \cdot \underline{K}$$

implying

$$\underline{K}^* = -Z\underline{K} \cdot (\overline{\overline{\eta}}_\wedge \hat{n}), \tag{9}$$

and this is particularly convenient if Hertz vectors are used to represent the unknown scattered field.

The boundary condition (9) is intrinsically a vector one in that each component involves two components of the field. Thus, for a locally plane surface with $\hat{n} = \hat{y}$

$$E_x = \eta_1 Z H_z, \quad E_z = -\eta_3 Z H_x \tag{10}$$

where (x,y,z) are Cartesian coordinates and $\overline{\overline{\eta}} = \eta_1\hat{x}\hat{x} + \eta_3\hat{z}\hat{z}$, but in certain special cases (8) can be expressed in terms of a single field component. In particular, if the surface and the incident field are both independent of the z coordinate, the problem is two-dimensional, and for an incident field such that $\underline{E} = \hat{z}E_z$ (E polarization), (8) reduces to

$$\frac{\partial E_z}{\partial n} + \frac{ik}{\eta_3} E_z = 0, \tag{11}$$

whereas for a field having $\underline{H} = \hat{z}H_z$ (H polarization)

$$\frac{\partial H_z}{\partial n} + ik\eta_1 H_z = 0. \tag{12}$$

These are scalar conditions directly analogous to (2).

In the more general case of a two-dimensional surface with an incident field which is not E- or H-polarized (e.g. a plane wave at skew incidence) or a three-dimensional surface under any illumination, it would seem that the only field component which could satisfy a scalar boundary condition is the normal component E_n or H_n. To explore this possibility, consider a surface $u_2 =$ constant where (u_1, u_2, u_3) form a right-handed orthogonal curvilinear coordinate system with metric coefficients h_1, h_2, h_3. If now $\bar{\bar{\eta}} = \eta_1\hat{u}_1\hat{u}_1 + \eta_3\hat{u}_3\hat{u}_3$, Maxwell's equations and the divergence conditions in conjunction with (8) yield

$$\frac{\partial}{\partial u_2}(h_1h_3E_2) + ik\eta_1h_1h_2h_3E_2 = -\frac{h_3}{\eta_1}E_1\frac{\partial}{\partial u_1}(h_2\eta_1) - \frac{h_1}{h_3}E_3\frac{\partial}{\partial u_3}(h_2\eta_3) - h_2(\eta_3-\eta_1)\frac{\partial}{\partial u_3}\left(\frac{h_1}{\eta_3}E_3\right)$$

$$\frac{\partial}{\partial u_2}(h_1h_3H_2) + \frac{ik}{\eta_3}h_1h_2h_3H_2 = -\eta_3h_3H_1\frac{\partial}{\partial u_1}\left(\frac{h_2}{\eta_3}\right) - \eta_1h_1H_3\frac{\partial}{\partial u_3}\left(\frac{h_2}{\eta_1}\right) - h_2\left(\frac{1}{\eta_1}-\frac{1}{\eta_3}\right)\frac{\partial}{\partial u_3}(h_1\eta_1H_3) \tag{13}$$

and the right-hand sides are zero only if $\eta_3 = \eta_1$, i.e., the impedance is isotropic, with $h_2\eta_1$ and h_2/η_3 independent of u_1 and $h_2\eta_3$ and h_2/η_1 independent of u_3. This is not even true for a planar surface ϕ = constant in cylindrical polar coordinates, and we further comment that the normal components E_ϕ and H_ϕ do not satisfy the scalar wave equation. No scalarization has yet been found for the problem of an impedance wedge except when the conditions (11) and (12) are applicable, and thus the problem of a plane wave at skew incidence is still unsolved.

As in the acoustic case, the simplest situation is a surface in the plane $y = 0$ (say) of a Cartesian coordinate system. If the surface impedance is isotropic, i.e., $\eta_3 = \eta_1$, simple manipulation of (10) gives

$$\frac{\partial E_y}{\partial y} + ik\eta_1E_y = 0\ , \qquad \frac{\partial H_y}{\partial y} + \frac{ik}{\eta_1}H_y = 0 \tag{14}$$

in which the (scalar) surface impedance is proportional to the logarithmic normal derivatives of the normal components. The eqs. (14) are scalar boundary conditions which are the direct counterpart of the acoustic condition (3). They are valid for any type of field, but if $\eta_3 \neq \eta_1$ it follows from (13) that

$$\frac{\partial E_y}{\partial y} + ik\eta_1 E_y = \left(\frac{\eta_1}{\eta_3} - 1\right)\frac{\partial E_z}{\partial z}, \quad \frac{\partial H_y}{\partial y} + \frac{ik}{\eta_3} H_y = \left(\frac{\eta_1}{\eta_3} - 1\right)\frac{\partial H_z}{\partial z}.$$

Even for an infinitely extended planar surface in Cartesian coordinates, no scalar boundary condition has been found that is valid for an arbitrary incident field and a general anisotropic impedance, and it is extremely unlikely that one exists. This can be seen by considering an incident plane wave. The total field then consists of the incident and reflected waves, and it is easily verified that only for an isotropic impedance or for a plane wave incident in a principal direction of the impedance tensor is there a linear combination of E_y (or H_y) and its derivatives which is zero on the surface.

The above remarks are obviously applicable to an impedance half plane occupying the region $x \geq 0$ of the plane $y = 0$. The most general case is that in which the impedance is anisotropic and differs on the two sides of the half plane, and the boundary conditions can then be written as

$$\begin{aligned} E_x &= \eta_1 Z H_z, \quad E_z = -\eta_3 Z H_x \qquad (y = 0+) \\ E_x &= -\eta_2 Z H_z, \quad E_z = \eta_4 Z H_x \qquad (y = 0-) \end{aligned} \tag{15}$$

corresponding to the surface impedances

$$\bar{\bar{\eta}}_+ = \eta_1 \hat{x}\hat{x} + \eta_3 \hat{z}\hat{z}, \quad \bar{\bar{\eta}}_- = \eta_2 \hat{x}\hat{x} + \eta_4 \hat{z}\hat{z} \tag{16}$$

on the upper and lower faces respectively.

If the incident field is independent of z and either E- or H- polarized, the problem for each polarization involves only one pair of impedances and is indistinguishable from that for an isotropic surface. We can now use the boundary conditions for either E_z and H_z or E_y and H_y. Thus, for E polarization the problem is a scalar one for the component E_z satisfying

$$\frac{\partial E_z}{\partial y} + \frac{ik}{\eta_3} E_z = 0 \quad (y = 0+), \qquad \frac{\partial E_z}{\partial y} - \frac{ik}{\eta_4} E_z = 0 \quad (y = 0-) \tag{17}$$

and also for the component H_y satisfying

$$\frac{\partial H_y}{\partial y} + \frac{ik}{\eta_3} H_y = 0 \quad (y = 0+), \qquad \frac{\partial H_y}{\partial y} - \frac{ik}{\eta_4} H_y = 0 \quad (y = 0-) \tag{18}$$

but because the latter were obtained from (15) by a process of tangential differentiation, the solution of the boundary problem (18) may require special care to ensure that the radiation and/or edge conditions are fulfilled. Similarly for H polarization we can use either H_z or E_y satisfying

$$\left(\frac{\partial}{\partial y} + ik\eta_1\right) \begin{matrix} H_z \\ E_y \end{matrix} = 0 \quad (y = 0+), \qquad \left(\frac{\partial}{\partial y} - ik\eta_2\right) \begin{matrix} H_z \\ E_y \end{matrix} = 0 \quad (y = 0-). \tag{19}$$

For a more general incident field including a plane wave at skew incidence, our options are more limited. If the impedances are isotropic, the components E_y and H_y satisfying (19) and (18) with $\eta_3 = \eta_1$, $\eta_4 = \eta_2$ provide a scalarization of the problem, but for arbitrary anisotropic impedances no scalar boundary conditions have been found. The best that we can then do is to express the four conditions (15) in terms of two field components, for example

$$\begin{aligned}
&\left\{\frac{\partial}{\partial y} + \frac{ik}{\eta_3}\left(1 + \frac{1}{k^2}\frac{\partial^2}{\partial z^2}\right)\right\} E_z + \frac{i}{k}\frac{\partial^2}{\partial x \partial z} ZH_z = 0 \\
&\qquad\qquad\qquad\qquad\qquad\qquad\qquad\qquad (y = 0+) \\
&\left\{\frac{\partial}{\partial y} + ik\eta_1\left(1 + \frac{1}{k^2}\frac{\partial^2}{\partial z^2}\right)\right\} ZH_z - \frac{i}{k}\frac{\partial^2}{\partial x \partial z} E_z = 0 \\
&\left\{\frac{\partial}{\partial y} - \frac{ik}{\eta_4}\left(1 + \frac{1}{k^2}\frac{\partial^2}{\partial z^2}\right)\right\} E_z + \frac{i}{k}\frac{\partial^2}{\partial x \partial z} ZH_z = 0 \\
&\qquad\qquad\qquad\qquad\qquad\qquad\qquad\qquad (y = 0-) \\
&\left\{\frac{\partial}{\partial y} - ik\eta_2\left(1 + \frac{1}{k^2}\frac{\partial^2}{\partial z^2}\right)\right\} ZH_z - \frac{i}{k}\frac{\partial^2}{\partial x \partial z} E_z = 0
\end{aligned} \tag{20}$$

or

$$\begin{aligned}
&\frac{\partial}{\partial x}\left(\frac{\partial}{\partial y} + ik\eta_1\right)E_y - \eta_1 \frac{\partial}{\partial z}\left(\frac{\partial}{\partial y} + \frac{ik}{\eta_1}\right)ZH_y = 0 \\
&\qquad\qquad\qquad\qquad\qquad\qquad\qquad (y = 0+) \\
&\frac{\partial}{\partial x}\left(\frac{\partial}{\partial y} + \frac{ik}{\eta_3}\right)ZH_y + \frac{1}{\eta_3} \frac{\partial}{\partial z}\left(\frac{\partial}{\partial y} + ik\eta_3\right)E_y = 0 \\
&\frac{\partial}{\partial x}\left(\frac{\partial}{\partial y} - ik\eta_2\right)E_y + \eta_2 \frac{\partial}{\partial z}\left(\frac{\partial}{\partial y} - \frac{ik}{\eta_2}\right)ZH_y = 0 \\
&\qquad\qquad\qquad\qquad\qquad\qquad\qquad (y = 0-) \\
&\frac{\partial}{\partial x}\left(\frac{\partial}{\partial y} - \frac{ik}{\eta_4}\right)ZH_y - \frac{1}{\eta_4} \frac{\partial}{\partial z}\left(\frac{\partial}{\partial y} - ik\eta_4\right)E_y = 0\,.
\end{aligned} \tag{21}$$

Some factors that might motivate one choice or the other have been discussed by Senior [28].

Nevertheless, there are some special anisotropic impedance values for which the boundary conditions simplify. If $\eta_1 = \eta_2 = 0$ it follows rather trivially from (15) that

$$\begin{aligned}
&E_x = 0, \quad \frac{\partial H_x}{\partial y} + ik\eta_3\left(1 + \frac{1}{k^2}\frac{\partial^2}{\partial x^2}\right)H_x = 0 \qquad (y = 0+) \\
&E_x = 0, \quad \frac{\partial H_x}{\partial y} - ik\eta_4\left(1 + \frac{1}{k^2}\frac{\partial^2}{\partial x^2}\right)H_x = 0 \qquad (y = 0-)
\end{aligned} \tag{22}$$

and the components E_x and H_x now provide a scalarization for any η_3 and η_4. Similarly, if $\eta_3 = \eta_4 = \infty$

$$\begin{aligned}
&H_x = 0, \quad \frac{\partial E_x}{\partial y} + \frac{ik}{\eta_1}\left(1 + \frac{1}{k^2}\frac{\partial^2}{\partial x^2}\right)E_x = 0 \qquad (y = 0+) \\
&H_x = 0, \quad \frac{\partial E_x}{\partial y} - \frac{ik}{\eta_2}\left(1 + \frac{1}{k^2}\frac{\partial^2}{\partial x^2}\right)E_x = 0 \qquad (y = 0-)
\end{aligned} \tag{23}$$

for any η_1 and η_2. On the other hand, if $\eta_1 = \eta_2 = \infty$

$$\begin{aligned}
&H_z = 0, \quad \frac{\partial E_z}{\partial y} + \frac{ik}{\eta_3}\left(1 + \frac{1}{k^2}\frac{\partial^2}{\partial z^2}\right)E_z = 0 \qquad (y = 0+) \\
&H_z = 0, \quad \frac{\partial E_z}{\partial y} - \frac{ik}{\eta_4}\left(1 + \frac{1}{k^2}\frac{\partial^2}{\partial z^2}\right)E_z = 0 \qquad (y = 0-)
\end{aligned} \tag{24}$$

for any η_3 and η_4, and if $\eta_3 = \eta_4 = 0$

$$E_z = 0, \quad \frac{\partial H_z}{\partial y} + ik\eta_1\left(1 + \frac{1}{k^2}\frac{\partial^2}{\partial z^2}\right)H_z = 0 \qquad (y = 0+)$$

$$E_z = 0, \quad \frac{\partial H_z}{\partial y} - ik\eta_2\left(1 + \frac{1}{k^2}\frac{\partial^2}{\partial z^2}\right)H_z = 0 \qquad (y = 0-) \tag{25}$$

for any η_1 and η_2, and in both these cases the components E_z and H_z provide the desired scalarization. These same components are also effective if $\eta_3 = 1/\eta_2$ and $\eta_4 = 1/\eta_1$, or if $\eta_4 = -\eta_3$ and $\eta_2 = -\eta_1$, and in the latter case the boundary conditions (15) are identical on the two sides of the half plane.

As evident from (9), an impedance sheet supports both electric and magnetic currents, and in the particular case of a half plane (or, indeed, any infinitesimally thin sheet) confined to the plane $y = 0$, the boundary conditions (15) imply

$$E_x(+) + E_x(-) = \frac{\eta_1 - \eta_2}{\eta_1 + \eta_2} J_z^* + \frac{2\eta_1\eta_2}{\eta_1 + \eta_2} ZJ_x$$

$$E_z(+) + E_z(-) = -\frac{\eta_3 - \eta_4}{\eta_3 + \eta_4} J_x^* + \frac{2\eta_3\eta_4}{\eta_3 + \eta_4} ZJ_z$$

$$H_x(+) + H_x(-) = \frac{\eta_3 - \eta_4}{\eta_3 + \eta_4} J_z + \frac{2}{\eta_3 + \eta_4} YJ_x^* \tag{26}$$

$$H_z(+) + H_z(-) = -\frac{\eta_1 - \eta_2}{\eta_1 + \eta_2} J_x + \frac{2}{\eta_1 + \eta_2} YJ_z^*$$

where

$$\underline{J} = \left[H_z\right]_-^+ \hat{x} - \left[H_x\right]_-^+ \hat{z}$$

$$\underline{J}^* = -\left[E_z\right]_-^+ \hat{x} + \left[E_x\right]_-^+ \hat{z} \tag{27}$$

are the total electric and magnetic currents respectively borne by the sheet. The conditions (26) are entirely equivalent to (15) and show the coupling between the electric and magnetic currents that can occur with an anisotropic sheet.

The acoustically resistive and conductive membranes also have their electromagnetic analogs supporting only an electric or a magnetic current. An electrically resistive sheet is simply an electric current sheet whose strength is proportional to the tangential electric field at its surface, and in recent years the concept of such a sheet has found many useful applications. As first noted by Levi-Civita (see [1], p. 19), its electromagnetic properties are completely specified by its resistivity, and in the general case when this is anisotropic, the conditions at the surface of the sheet are

$$\begin{aligned} [\hat{n}\wedge\underline{E}]_-^+ &= 0 \\ Z\bar{\bar{R}} \cdot [\hat{n}\wedge\underline{H}]_-^+ &= -\hat{n}\wedge(\hat{n}\wedge\underline{E}) \end{aligned} \tag{28}$$

(cf. the conditions (6)) where $\hat{n}$ is now the outward normal to the upper (positive) side and $Z\bar{\bar{R}}$ is the resistivity in ohms per square. If the sheet is confined to (say) the plane $y = 0$, the fact that the tangential components of the scattered magnetic field are asymmetrical across the sheet enables us to replace the second of the conditions (28) by the equivalent boundary condition

$$2Z\bar{\bar{R}} \cdot \hat{n}\wedge(\underline{H} - \underline{H}^i) = -\hat{n}\wedge(\hat{n}\wedge\underline{E}). \tag{29}$$

However, the first condition still remains and even a planar resistive sheet is necessarily a transition (rather than a boundary value) problem unless the incident field is such that $\hat{n}\wedge\underline{E} = 0$.

The dual problem is that of a (magnetically) conductive sheet supporting only a magnetic current, and by analogy with (28) the conditions at its surface are

$$\begin{aligned} [\hat{n}\wedge H]_-^+ &= 0 \\ Y\bar{\bar{R}}^* \cdot [\hat{n}\wedge\underline{E}]_-^+ &= \hat{n}\wedge(\hat{n}\wedge\underline{H}) \end{aligned} \tag{30}$$

where $Y\bar{\bar{R}}^*$ is an anisotropic conductivity in ohms per square. By writing the second of the conditions (28) and (30) in terms of the sum fields on both sides of the sheet we can also define a combined resistive-conductive sheet at which the conditions are

$$
\begin{aligned}
2Z\bar{\bar{R}} \cdot [\hat{n}_\wedge \underline{H}]_-^+ &= -\hat{n}_\wedge(\hat{n}_\wedge\{E(+) + \underline{E}(-)\}) \\
2Y\bar{\bar{R}}^* \cdot [\hat{n}_\wedge \underline{E}]_-^+ &= \hat{n}_\wedge(\hat{n}_\wedge\{\underline{H}(+) + \underline{H}(-)\})
\end{aligned}
\tag{31}
$$

and for a particular relationship between the resistivity and conductivity tensors the combined sheet is equivalent to an impedance boundary condition one. This is most easily seen in the case of a planar sheet confined to (say) the surface $y = 0$. Manipulation of (31) then gives

$$
\begin{aligned}
E_x(\pm) &= Z\left(R_1 \pm \frac{1}{4R_3^*}\right)H_z(+) - Z\left(R_1 \mp \frac{1}{4R_3^*}\right)H_z(-) \\
E_z(\pm) &= -Z\left(R_3 \pm \frac{1}{4R_1^*}\right)H_x(+) + Z\left(R_3 \mp \frac{1}{4R_1^*}\right)H_x(-)
\end{aligned}
\tag{32}
$$

where

$$
\bar{\bar{R}} = R_1\hat{x}\hat{x} + R_3\hat{z}\hat{z}, \quad \bar{\bar{R}}^* = R_1^*\hat{x}\hat{x} + R_3^*\hat{z}\hat{z}.
$$

Provided

$$
2R_1^* = \frac{1}{2R_3}, \quad 2R_3^* = \frac{1}{2R_1}, \tag{33}
$$

(32) are identical to the impedance boundary conditions (15) with

$$
\eta_1 = \eta_2 = 2R_1, \quad \eta_3 = \eta_4 = 2R_3, \tag{34}
$$

but only for this resistivity-conductivity relationship is the combination sheet equivalent to an impedance sheet [27]. For other combination sheets and for a resistive or conductive sheet in isolation, the problem is fundamentally a transition one.

III. METHODS OF SOLUTION

There are two basic methods of solution of our half plane problems and one or other is applicable in all cases for which solutions have been obtained so far. Each has a number of variations which are of interest more for the physical insight that they afford than for any increase in capability, and there are in addition methods which are effective in rather special cases and which may then provide a rather elementary approach. An example of the latter is the method proposed by Tan and Cheng [29] for a class of half plane problems. The method assumes a solution consisting of a linear combination of Fresnel integrals of the form appearing in the solution for a metallic half plane, and then chooses the coefficients to satisfy the boundary conditions. In most instances the boundary conditions on the two sides of the half plane must be identical, and though Tan and Cheng refer to one of the geometries as an impedance half plane, we remark that the impedances must be such that $\bar{\bar{\eta}}_- = -\bar{\bar{\eta}}_+$.

The two basic methods have very different starting points and, superficially at least, are quite distinct. Nevertheless, they do have many similarities and with the developments that have taken place in recent months their capabilities are now comparable as regards the type of problem considered here. It is convenient to describe the methods in terms of their application to the simple problem of a plane acoustic wave

$$V^i = e^{-ik\rho\,\cos(\phi - \phi_o)} \tag{35}$$

incident normally on an impedance half plane $y = 0$, $x \gtrless 0$. The boundary conditions are as shown in (3) or (4) with (possibly) $\eta = \eta_3$ on the upper face and $\eta = \eta_4$ on the lower. We shall then examine how the methods apply to the other problems involving impedance half planes in acoustics and electromagnetics, and conclude with a few comments about the transition problems of resistive and conductive half planes.

The first method is that of Maliuzhinets [16] and is the more general of the two inasmuch as it is applicable to wedge-shaped regions as well. It is equivalent to the approach which Senior [25] developed based on Peters' [19] analysis of a problem in hydrodynamics, but is more elementary and straightforward, and we shall confine attention to it. The method assumes a representation of the total field in cylindrical polar coordinates in the

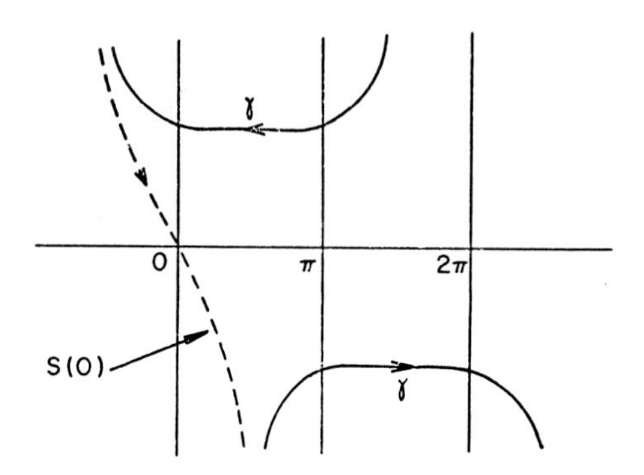

form of a Sommerfeld integral, viz.

$$V = \frac{1}{2\pi i} \int_{\gamma} e^{ik\rho \cos \alpha} s(\alpha - \phi)d\alpha \tag{36}$$

where γ is the Sommerfeld contour as shown in Fig. 1 and $s(\alpha - \pi) - (\alpha - 2\pi + \phi_o)^{-1}$ is analytic in the strip $0 \leq Re.\ \alpha \leq 2\pi$. On applying the boundary conditions (4) and eliminating the derivatives using integration by parts, we find

$$\begin{aligned} &\int_{\gamma} e^{ik\rho \cos \alpha} (1 - \eta_3 \sin \alpha)\, s(\alpha)d\alpha = 0 \\ &\int_{\gamma} e^{ik\rho \cos \alpha} (1 + \eta_4 \sin \alpha)\, s(\alpha - 2\pi)d\alpha = 0 \end{aligned} \tag{37}$$

and for these to be satisfied it is both necessary and sufficient [15] that the integrands be even about $\alpha = \pi$. Hence

$$\begin{aligned} (1 - \eta_3 \sin \alpha)\, s(\alpha) &= (1 + \eta_3 \sin \alpha)\, s(2\pi - \alpha) \\ (1 + \eta_4 \sin \alpha)\, s(\alpha - 2\pi) &= (1 - \eta_4 \sin \alpha)\, s(-\alpha) \end{aligned} \tag{38}$$

and if we now write

$$s(\alpha) = \sigma(\alpha) \frac{\Psi(\alpha)}{\Psi(\pi - \phi_o)} \tag{39}$$

where

$$\sigma(\pm\alpha) = \sigma(\pm 2\pi \mp \alpha), \tag{40}$$

we can choose

$$\sigma(\alpha) = \frac{1}{2} \sin \frac{\phi_o}{2} \left(\sin \frac{\alpha}{2} - \cos \frac{\phi_o}{2} \right)^{-1} \tag{41}$$

provided $\Psi(\alpha)$ is analytic and free of zeros in the strip.

From (38) by a process of elimination,

$$(1 - \eta_3 \sin \alpha)(1 - \eta_4 \sin \alpha)\Psi(\alpha + 4\pi)$$
$$= (1 + \eta_3 \sin \alpha)(1 + \eta_4 \sin \alpha)\Psi(\alpha) \tag{42}$$

This is a first order difference equation of interval 4π, but if $\eta_4 = \eta_3$ the equation can be reduced to

$$(1 - \eta_3 \sin \alpha)\Psi(\alpha + 2\pi) = (1 + \eta_3 \sin \alpha)\Psi(\alpha) \tag{43}$$

whose interval is 2π. The solution of (42) having the required properties is

$$\Psi(\alpha) = \psi_\pi(\alpha + \pi + \chi_3)\psi_\pi(\alpha + \pi - \chi_3)\psi_\pi(\alpha - \pi + \chi_4)\psi_\pi(\alpha - \pi - \chi_4) \tag{44}$$

where

$$\eta_{3,4} = \sec \chi_{3,4} \tag{45}$$

and $\psi_\pi(\alpha)$ is the special meromorphic function

$$\psi_\pi(\alpha) = exp\left\{-\frac{1}{8\pi} \int_o^\alpha (\pi \sin v - 2\sqrt{2}\, \pi \sin \frac{v}{2} + 2v) \frac{dv}{\cos v}\right\}. \tag{46}$$

$\psi_\pi(\alpha)$ has been computed by Bucci [4] for a range of complex arguments, and many of its properties are listed by Maliuzhinets [16] and Bowman [3].

If the contour γ is closed using two steepest descents paths through $\alpha = 0$ and 2π, V can be expressed as the sum of the residues at the included poles. The corresponding terms are the incident and reflected waves of geometrical optics existing in the appropriate regions, together with any surface wave contributions that may occur. The remaining (diffracted) field is then

$$V^d = -\frac{1}{2\pi i} \int_{S(0)} e^{ik\rho \cos \alpha} \{s(\alpha - \phi) - s(2\pi + \alpha - \phi)\} d\alpha \tag{47}$$

where S(0) is a path from $\alpha = -\frac{\pi}{2} + i\infty$ to $\alpha = \frac{\pi}{2} - i\infty$ passing through the origin, and on inserting the expression for $s(\alpha)$, we have

$$V^d = \frac{1}{2\pi i}\int_{S(0)} e^{ik\rho\cos\alpha}\frac{\sin\frac{\phi_o}{2}}{\cos(\alpha-\phi)+\cos\phi_o}\left\{\left(\sin\frac{\alpha-\phi}{2}+\cos\frac{\phi_o}{2}\right)\right.$$

$$\left.\cdot\frac{\Psi(\alpha-\phi)}{\Psi(\pi-\phi_o)}+\left(\sin\frac{\alpha-\phi}{2}-\cos\frac{\phi_o}{2}\right)\frac{\Psi(2\pi-\alpha-\phi)}{\Psi(\pi-\phi_o)}\right\}d\alpha \tag{48}$$

From the relations satisfied by the function $\psi_\pi(\alpha)$, it can be shown that

$$\Psi(2\pi+\alpha) = \left\{\frac{\cos\frac{1}{2}\chi_3 - \cos\frac{1}{2}\left(\alpha+\frac{\pi}{2}\right)}{\cos\frac{1}{2}\chi_3 - \cos\frac{1}{2}\left(\alpha-\frac{\pi}{2}\right)}\right\}$$

$$\cdot\left\{\frac{\cos\frac{1}{2}\chi_4 - \cos\frac{1}{2}\left(\alpha+\frac{\pi}{2}\right)}{\cos\frac{1}{2}\chi_4 + \cos\frac{1}{2}\left(\alpha-\frac{\pi}{2}\right)}\right\}\Psi(\alpha) \tag{49}$$

and in particular, when $\eta_4 = \eta_3$ ($= \eta$, say)

$$\Psi(2\pi+\alpha) = \frac{1+\eta\sin\alpha}{1-\eta\sin\alpha}\Psi(\alpha) \tag{50}$$

in which case

$$V^d = \frac{1}{\pi i}\int_{S(0)} e^{ik\rho\cos\alpha}\frac{\Psi(2\pi+\alpha-\phi)}{\Psi(\pi-\phi_o)}\frac{1}{1+\eta\sin(\alpha-\phi)}$$

$$\cdot\frac{\sin\frac{\alpha-\phi}{2}\sin\frac{\phi_o}{2}}{\cos(\alpha-\phi)+\cos\phi_o}\left(1-2\eta\cos\frac{\alpha-\phi}{2}\cos\frac{\phi_o}{2}\right)d\alpha. \tag{51}$$

Away from the geometrical optics boundaries and at large distances ($k\rho \gg 1$) V^d has the character of an edge wave. By a steepest descents evaluation of the integral in (48),

$$V^d \sim \sqrt{\frac{2}{\pi k\rho}}\, e^{i(k\rho-\pi/4)}P(\phi,\phi_o)$$

with [2]

$$P(\phi,\phi_o) = \frac{i}{2}\,\frac{\sin\frac{\phi_o}{2}}{\cos\phi+\cos\phi_o}\left\{\left(\sin\frac{\phi}{2}-\cos\frac{\phi_o}{2}\right)\frac{\Psi(-\phi)}{\Psi(\pi-\phi_o)}\right.$$
$$\left.+\left(\sin\frac{\phi}{2}+\cos\frac{\phi_o}{2}\right)\frac{\Psi(2\pi-\phi)}{\Psi(\pi-\phi_o)}\right\} \tag{52}$$

In certain cases the right-hand side can be expressed in terms of trigonometric functions alone. Thus, for a hard half plane ($\eta_4 = \eta_3 = \infty$, implying $\chi_4 = \chi_3 = \pi/2$),

$$\Psi(\alpha) = \frac{1}{2}\cos\frac{\alpha}{2}\left\{\psi_\pi(\pi/2)\right\}^2$$

giving

$$\frac{\Psi(-\phi)}{\Psi(\pi-\phi_o)} = -\frac{\Psi(2\pi-\phi)}{\Psi(\pi-\phi_o)} = \frac{\cos\frac{\phi}{2}}{\sin\frac{\phi_o}{2}}.$$

Hence

$$P^{(h)}(\phi,\phi_o) = -i\,\frac{\cos\frac{\phi}{2}\cos\frac{\phi_o}{2}}{\cos\phi+\cos\phi_o} \tag{53}$$

in accordance with the known result [3]. Similarly, for a soft half plane ($\eta_4 = \eta_3 = 0$, implying $\chi_4 = \chi_3 = i\infty$), $\Psi(\alpha)$ is independent of α and therefore

$$P^{(s)}(\phi,\phi_o) = i\,\frac{\sin\frac{\phi}{2}\sin\frac{\phi_o}{2}}{\cos\phi+\cos\phi_o} \tag{54}$$

([3], p. 315). If the half plane is soft on top ($\chi_3 = i\infty$) but hard on the bottom ($\chi_4 = \pi/2$), then

$$\frac{\Psi(\alpha)}{\Psi(\pi-\phi_o)} = \frac{\cos\frac{1}{4}(\pi-\alpha)}{\cos\frac{\phi_o}{4}}$$

so that

$$\frac{\Psi(-\phi)}{\Psi(\pi-\phi_o)} = \frac{\cos\frac{1}{4}(\pi+\phi)}{\cos\frac{\phi_o}{4}}, \quad \frac{\Psi(2\pi-\phi)}{\Psi(\pi-\phi_o)} = \frac{\cos\frac{1}{4}(\pi-\phi)}{\cos\frac{\phi_o}{4}}$$

giving

$$P^{(s,h)}(\phi,\phi_o) = i\sqrt{2}\,\frac{\sin\frac{\phi}{4}\sin\frac{\phi_o}{4}}{\cos\phi+\cos\phi_o}\left(1+\cos\frac{\phi}{2}+\cos\frac{\phi_o}{2}\right). \qquad (55)$$

Conversely, if the half plane is hard on top ($\chi_3 = \pi/2$) and soft on the bottom ($\chi_4 = i\infty$),

$$P^{(h,s)}(\phi,\phi_o) = i\sqrt{2}\,\frac{\cos\frac{\phi}{4}\cos\frac{\phi_o}{4}}{\cos\phi+\cos\phi_o}\left(1-\cos\frac{\phi}{2}-\cos\frac{\phi_o}{2}\right) \qquad (56)$$

and we observe that

$$P^{(h,s)}(\phi,\phi_o) = P^{(s,h)}(2\pi-\phi,2\pi-\phi_o)$$

as expected. The idea of backing a soft (absorbing) surface with a hard surface to provide rigid support is important in the design of acoustic baffles, and (55) and (56) are useful for assessing their performance. The expression (55) for $P^{(s,h)}(\phi,\phi_o)$ is equivalent to, but a good deal simpler than, the one obtained by Rawlins [21] using his extension of the standard Wiener-Hopf technique.

In Cartesian coordinates the half plane is a two-part boundary value problem, and if Green's theorem or some other approach is used to represent the scattered field as a surface integral, application of the boundary conditions leads to a system of Wiener-Hopf integral equations for the "currents" induced in the half plane. In principle at least these equations can be solved using a Fourier or bilateral Laplace transform supplemented by function-theoretic techniques, and we shall refer to this as the Wiener-Hopf method.

To illustrate this second basic method, we again consider the scalar plane wave (35) incident on an impedance half plane but with the same impedance η at both faces. The boundary conditions are now (3), and by applying Green's theorem to the region exterior to a surface surrounding the half plane and then collapsing the surface to the half plane itself, we have

$$V = V^i - \frac{i}{4}\int_o^\infty \left\{f_1(x') + f_2^*(x')\frac{\partial}{\partial y}\right\} H_o^{(1)}\left(k\sqrt{(x-x')^2 + y^2}\right)dx' \quad (57)$$

where

$$f_1(x) = \left[\frac{\partial V}{\partial y}\right]_-^+, \quad f_2^*(x) = [V]_-^+ \quad (58)$$

and the notation has been chosen to correspond to that of Senior [28]. On adding the limits as $y \to \pm 0$ and using the first of the eqs. (3),

$$-\frac{\eta}{k} f_1(x) = 2iV^i + \frac{1}{2}\int_o^\infty f_1(x') H_o^{(1)}(k|x-x'|)dx' \quad (59)$$

valid for $x \gtrless 0$, and this is an inhomogeneous Wiener-Hopf integral equation for $f_1(x)$. Similarly, by differentiating (57) with respect to y and then adding the limits,

$$\frac{k}{\eta} f_2^*(x) = 2i\frac{\partial V^i}{\partial y} + \frac{1}{2}\lim_{y\to 0}\int_o^\infty f_2^*(x')\frac{\partial^2}{\partial y^2} H_o^{(1)}\left(k\sqrt{(x-x')^2 + y^2}\right)dx' \quad (60)$$

valid for $x \gtrless 0$, and this is a Wiener-Hopf equation for $f_2^*(x)$.

The procedure for solving equations of this type is fully documented in the literature ([3], pp. 41-49). Taking for example (59), the equation is extended to apply for all x by taking $f_1(x) = 0$, $x < 0$, with $\Omega(x)$ equal to the right hand side of (59) for $x < 0$ and 0 otherwise. If a Fourier transform is defined as

$$g(\xi) = \int_{-\infty}^\infty e^{-i\xi x} g(x)dx$$

with inverse

$$g(x) = \frac{1}{2\pi}\int_{-\infty}^\infty e^{i\xi x} g(\xi)d\xi,$$

it then follows that $f_1(\xi)$ is analytic in a lower half plane and $\Omega(\xi)$ in an upper half plane. If, moreover, k is assumed to have a positive imaginary part (later allowed to vanish), there exists a strip of analyticity permitting the transformation of the extended integral equation. The result is

$$\eta\Omega(\xi) = 2ikV^i(\xi) + \left(\eta + \frac{k}{k^2 - \xi^2}\right)f_1(\xi). \tag{61}$$

The key step is now the factorization (or "splitting") of the coefficient of $f_1(\xi)$ into lower and upper functions having overlapping half planes of analyticity. If

$$\eta + \frac{k}{k^2 - \xi^2} = \frac{K_-(\xi)}{K_+(\xi)} \tag{62}$$

implying

$$K_-(\xi)K_+(-\xi) = 1$$

[23], $K_-(\xi)$ and $K_+(\xi)$ can be determined using Cauchy's theory, and from the known behavior of $f_1(\xi)$ at infinity, Liouville's theorem then enables us to find $f_1(\xi)$ and $\Omega(\xi)$.

The analysis for (60) is very similar and for future reference we remark that if (61) and its analog are written as a pair of simultaneous equations, the matrix $G(\xi)$ that must be factored is a diagonal one having

$$G_{11}(\xi) = \eta + \frac{k}{k^2 - \xi^2}, \quad G_{22}(\xi) = \frac{k^2 - \xi^2}{k} \quad \eta + \frac{k}{k^2 - \xi^2}.$$

The functions $f_1(x)$ and $f_2^*(x)$ can be obtained from $f_1(\xi)$ by Fourier inversion, and when their expressions are inserted into (57) we have

$$V = V^i - \frac{1}{2\pi i}\int_C \frac{K_+(-k\cos\phi_o)}{K_-(k\cos\alpha)} \frac{1 - 2\eta\frac{y}{|y|}\cos\frac{\alpha}{2}\cos\frac{\phi_o}{2}}{\cos\alpha + \cos\phi_o}$$

$$\cdot\ e^{ik(x\cos\alpha + |y|\sin\alpha}\, d\alpha$$

where $\xi = k\cos\alpha$ and the path C in the α plane runs from $i\infty$ to the origin and thence to π and $\pi - i\infty$, passing above the pole at

$\alpha = \pi - \phi_o$. On introducing the polar coordinates (ρ,ϕ), the integrand is seen to have a saddle point at $\alpha = \phi$ $(y \geq 0)$ or $\alpha = 2\pi - \phi$ $(y < 0)$, and when the path is deformed into a steepest descents path through this point, the contribution of the pole at $\alpha = \pi - \phi_o$ combines with V^i to yield the geometrical optics terms. The resulting diffracted field is

$$V^d = -\frac{1}{2\pi i}\int_{S(0)} e^{ik\rho\cos\alpha}\,\frac{K_+(-k\cos\phi_o)}{K_-(k\cos\{\alpha-\phi\})}\cdot\frac{1-2\eta\cos\frac{\alpha-\phi}{2}\cos\frac{\phi_o}{2}}{\cos(\alpha-\phi)+\cos\phi_o}\,d\alpha \qquad (63)$$

and when this is compared with (51) it is obvious that the split functions must be closely related to Maliuzhinets' function $\Psi(\alpha)$ for $\eta_4 = \eta_3 = \eta$. From the expressions for $K_+(\xi)$ and $K_-(\xi)$ given by Senior [23], it can be shown that

$$K_+(k\cos\alpha) = \cos\frac{\alpha}{2}\, exp(2p/\pi)\{(\sqrt{2}+1)^2(\eta/2)^{1/2}\Psi(\alpha)\}^{-1}$$
$$K_-(k\cos\alpha) = \frac{1+\eta\sin\alpha}{\sin\frac{\alpha}{2}}\, exp(2p/\pi)\{(\sqrt{2}+1)^2(2\eta)^{1/2}\Psi(\alpha)\}^{-1} \qquad (64)$$

where $p = 0.9159656...$ is Catalan's constant, and with this identification (51) and (63) are identical. A further consequence of (64) is that $K_-(\xi)$ is also the solution of a difference equation, and by treating the (cut) ξ plane as one sheet of a two-sheeted Riemann surface, we have [25]

$$K_-(\xi e^{2i\pi}) = \frac{k+\eta\sqrt{k^2-\xi^2}}{k-\eta\sqrt{k^2-\xi^2}}\,K_-(\xi) \qquad (65)$$

c.f. (50), which is a difference equation with (angular) interval 2π.

A Wiener-Hopf problem such as this can be formulated in several different ways but a requirement which is common to them all is the factorization of a function of the transform variable ξ.

An approach due to Jones [9] and embraced by Noble [18] is to apply the transform to the partial differential equation before imposing the boundary conditions. The complex variable equation in the transform plane is thereby obtained directly without the necessity of formulating an integral equation for (say) the currents induced in the half plane. In still another approach the rather cumbersome derivation of the integral equation is circumvented by using separation of variables to develop dual integral equations for a quantity related to the far field amplitude. This procedure was expounded by Karp [10] and exploited by Clemmow [6,7], and in the case of the present problem the representation employed is

$$V - V^i = \int_C \{Q_1 \ (\cos\alpha) \mp \sin\alpha \ Q_2 \ (\cos\alpha)\} \ e^{ik\rho\cos(\alpha\mp\phi)} d\alpha$$

for $y \gtrless 0$, where C is the path defined earlier and $Q_1(\xi)$, $Q_2(\xi)$ are angular spectra proportional to the Fourier transforms of $f_1(x)$, $f_2^*(x)$ respectively. If C is now deformed into a steepest descents path through $\alpha = \phi$ $(y \geq 0)$ or $2\pi - \phi$ $(y < 0)$, a simple change of variable produces a result strikingly similar to the representation (47) inherent in Maliuzhinets' method.

In this discussion of the Wiener-Hopf method we have assumed that both sides of the half plane have the same impedance. If, instead, the impedances on the upper and lower faces are η_3 and η_4 respectively with $\eta_4 \neq \eta_3$, application of the boundary conditions to the representation (57) gives

$$-\frac{1}{k}\frac{2\eta_3\eta_4}{\eta_3+\eta_4} f_1(x) + i\frac{\eta_3-\eta_4}{\eta_3+\eta_4} f_2^*(x)$$

$$= 2iV^i + \frac{1}{2}\int_o^\infty f_1(x')H_o^{(1)}(k|x-x'|)dx'$$

$$-i\frac{\eta_3-\eta_4}{\eta_3+\eta_4} f_1(x) + \frac{2k}{\eta_3+\eta_4} f_2^*(x) \tag{66}$$

$$= 2i\frac{\partial V^i}{\partial y} + \frac{1}{2}\lim_{y\to 0}\int_o^\infty f_2^*(x')\frac{\partial^2}{\partial y^2} H_o^{(1)}\left(k\sqrt{(x-x')^2+y^2}\right)dx'$$

which now constitute a pair of coupled Wiener-Hopf integral equations for $f_1(x)$ and $f_2^*(x)$. In the plane of the transform variable ξ, the matrix $G(\xi)$ multiplying the column vector $\{f_1(\xi),\ ik\ f_2^*(\xi)\}$ is

$$G(\xi) = \begin{pmatrix} \dfrac{2\eta_3\eta_4}{\eta_3+\eta_4} + \dfrac{k}{\Gamma}, & -\dfrac{\eta_3-\eta_4}{\eta_3+\eta_4} \\ \dfrac{\eta_3-\eta_4}{\eta_3+\eta_4}, & \dfrac{2}{\eta_3+\eta_4} + \dfrac{\Gamma}{k} \end{pmatrix} \qquad (67)$$

where $\Gamma = \sqrt{k^2 - \xi^2}$, and though in principle it is possible to factor this into matrices having overlapping planes of analyticity, no direct method for doing so exists. The matrix can be simplified somewhat by adding γ times the first equation to the second, with $\gamma = 1/\eta_3$ and $-1/\eta_4$, in which case

$$G(\xi) = \begin{pmatrix} 1 + \dfrac{k}{\eta_3\Gamma}, & \dfrac{1}{\eta_3} + \dfrac{\Gamma}{k} \\ -1 - \dfrac{k}{\eta_4\Gamma}, & \dfrac{1}{\eta_4} + \dfrac{\Gamma}{k} \end{pmatrix} \qquad (68)$$

but even so it is only recently that a method of factorization has been found.

A clue as to how we might proceed can be obtained from Maluizhinets' solution for this same diffraction problem. It will be recalled that $s(\alpha)$ and the split functions are closely related if $\eta_4 = \eta_3$ and $s(\alpha)$ then satisfies a difference equation of interval 2π. When $\eta_4 \neq \eta_3$, however, the interval is 4π (see 42), suggesting that we should now invoke a relationship between the values of $G(\xi)$ on adjacent sheets of a Riemann surface. In the particular case of an impedance half plane soft on top ($\eta_3 = 0$) and hard on the bottom ($\eta_4 = \infty$) for which (68) reduces to

$$G(\xi) = \begin{pmatrix} \dfrac{k}{\Gamma}, & 1 \\ -1, & \dfrac{\Gamma}{k} \end{pmatrix},$$

Rawlins [21] achieved a factorization using an intuitive argument based on the change of sign of the radical on crossing the branch cut emanating from $\xi = -k$. More recently still, Hurd [8] has developed a general theory for matrices $G(\xi)$ whose only singularities are branch points such that

$$G(\xi)\left[G^{(-)}(\xi)\right]^{-1} = \begin{pmatrix} 0, & \sigma_1 \\ \sigma_2, & 0 \end{pmatrix} \qquad (69)$$

for some σ_1, σ_2, where $G^{(-)}(\xi)$ differs from $G(\xi)$ in having $-\Gamma$ in place of Γ. The conditions are satisfied by (68) (though not by (67)), and Hurd has presented the explicit factorization for any η_3 and η_4. With this extension, the Wiener-Hopf and Maliuzhinets methods now have the same capabilities as regards the acoustic problem of an impedance half plane.

In addition to this problem there are a number of others involving impedance half planes, and it is of interest to see how the basic methods apply to them. If the acoustic plane wave is incident obliquely, with

$$V^i = e^{ik\rho \sin\beta \cos(\phi - \phi_o) ikz \cos\beta} \qquad (70)$$

in place of (35), the total field must have this same z dependence, and since the boundary conditions (4) can be written as

$$\frac{1}{\rho}\frac{\partial V}{\partial \phi} \pm \frac{ik \sin\beta}{\eta'_{3,4}} V = 0$$

with

$$\eta'_{3,4} = \eta_{3,4} \sin\beta,$$

both methods work as before. Indeed, the analyses for $V e^{-ikz \sin\beta}$ differ from those for V in the normal incidence case $\beta = \pi/2$ only in having k replaced by $k \sin\beta$ and $\eta_{3,4}$ by $\eta'_{3,4}$.

For an electromagnetic wave at normal incidence on an anisotropic half plane with the impedances (16) it is obviously sufficient to consider E- and H-polarizations separately. The anisotropy then produces no difficulty at all and for each polarization the problem is a scalar one for (say) the z component of the field. Thus, for an E-polarized plane wave with E_z^i given by (35) the total field component E_z satisfies the boundary condition (17) and is identical to the solution V. The analogous result for H-polarization follows on replacing η_3 by $1/\eta_1$ and η_4 by $1/\eta_2$. For oblique incidence however, the problem is not quite so simple and it is desirable to examine the two basic methods separately.

In order to build on the success of Maliuzhinets' method in the above cases, it would seem natural to use the boundary conditions (20) with $\partial/\partial z$ replaced by $ik \cos\beta$, but since these couple E_z and H_z regardless of the anisotropy, we are now forced to represent E_z and H_z simultaneously in the form

$$E_z = \frac{1}{2\pi i}\int_\gamma e^{ik\rho \sin\beta \cos\alpha}\, s_1(\alpha-\phi)\, d\alpha$$
$$ZH_z = \frac{1}{2\pi i}\int_\gamma e^{ik\rho \sin\beta \cos\alpha}\, s_2(\alpha-\phi)\, d\alpha. \tag{71}$$

Application of the boundary conditions then leads to the four functional equations

$$(\sin\beta - \eta_3 \sin\alpha)s_1(\alpha) - \cos\alpha \cos\beta\, s_2(\alpha)$$
$$= (\sin\beta + \eta_3 \sin\alpha)s_1(2\pi-\alpha) - \cos\alpha\cos\beta\, s_2(2\pi-\alpha)$$

$$(\sin\beta - 1/\eta_1 \sin\alpha)s_2(\alpha) + \cos\alpha\cos\beta\, s_1(\alpha)$$
$$= (\sin\beta + 1/\eta_1 \sin\alpha)s_2(2\pi-\alpha) + \cos\alpha\cos\beta\, s_1(2\pi-\alpha)$$

$$(\sin\beta + \eta_4 \sin\alpha)s_1(\alpha-2\pi) + \cos\alpha\cos\beta\, s_2(\alpha-2\pi) \tag{72}$$
$$= (\sin\beta - \eta_4 \sin\alpha)s_1(-\alpha) + \cos\alpha\cos\beta\, s_2(-\alpha)$$

$$(\sin\beta + 1/\eta_2 \sin\alpha)s_2(\alpha-2\pi) - \cos\alpha\cos\beta\, s_1(\alpha-2\pi)$$
$$= (\sin\beta - 1/\eta_2 \sin\alpha)s_2(-\alpha) - \cos\alpha\cos\beta\, s_1(-\alpha)$$

relating s_1 and s_2, and from these either function can be eliminated. If we again introduce a (common) singular function $\sigma(\alpha)$ as shown in (39), we obtain after must tedious manipulation

$$F(\eta_1,1/\eta_3)F(\eta_2,1/\eta_4)\Psi(\alpha+4\pi) - \{F(\eta_1,1/\eta_3)F(-\eta_1,-1/\eta_3) + \Lambda$$
$$+ F(\eta_2,1/\eta_4)F(-\eta_2,-1/\eta_4)\}\Psi(\alpha) + F(-\eta_1,-1/\eta_3)F(-\eta_2,-1/\eta_4)\Psi(\alpha-4\pi) = 0 \qquad (73)$$

where $\Psi(\alpha) = \Psi_1(\alpha)$ or $\Psi_2(\alpha)$ with

$$F(a,b) = \cos^2\alpha\,\cos^2\beta + (\sin\alpha + a\sin\beta)(\sin\alpha + b\sin\beta)$$

$$\Lambda = \sin^2\beta\,(\eta_1 + \eta_2 - 1/\eta_3 - 1/\eta_4)^2\,\sin^2\alpha - \left(\frac{\eta_1}{\eta_3} - \frac{\eta_2}{\eta_4}\right)^2 \sin^2\beta\ .$$

This is a second order difference equation of interval 4π, but if $\eta_4 = \eta_3$ and $\eta_2 = \eta_1$ implying $\bar{\bar{\eta}}_+ = \bar{\bar{\eta}}_-$, the equation can be reduced to

$$F(\eta_1,1/\eta_3)\Psi(\alpha+2\pi) + \sqrt{2}(\eta_1 - 1/\eta_3)\sin\beta\,\Psi(\alpha)$$
$$- F(-\eta_1,-1/\eta_3)\Psi(\alpha-2\pi) = 0 \qquad (74)$$

whose interval is only 2π. The decrease in interval is analogous to that achievable in the acoustic problem, but unfortunately the equation is still a second order one for which there is no known method of solution. The only obvious circumstances in which (73) reduces to a first order equation (and is therefore soluable) is when $\Lambda = 0$, requiring

$$\eta_1 = -\eta_2,\quad \eta_3 = -\eta_4$$

or

$$\eta_1 = 1/\eta_4,\quad \eta_2 = 1/\eta_3$$

which are the cases noted in Section II. Nevertheless, if the impedances are isotropic ($\eta_3 = \eta_1$, $\eta_4 = \eta_2$) it is known that the components E_y and H_y provide a simple scalarization of the problem, implying that the linear combinations

$$\sin\beta\cos\beta\,s_1(\alpha) + \cos\alpha\,s_2(\alpha),\quad \sin\beta\cos\beta\,s_2(\alpha) - \cos\alpha\,s_1(\alpha)$$

satisfy first order difference equations of the form (42), but this fact is hardly evident from the functional equations (71).

In the special case of isotropic impedances the problem is greatly simplified by working with the components E_y and H_y from the outset. These satisfy the (decoupled) scalar boundary conditions (19) and (18) respectively and if

$$
\begin{aligned}
E_y &= \frac{1}{2\pi i} \int_\gamma e^{ik\rho \sin\beta \cos\alpha} s_1(\alpha - \phi) d\alpha \\
ZH_y &= \frac{1}{2\pi i} \int_\gamma e^{ik\rho \sin\beta \cos\alpha} s_2(\alpha - \phi) d\alpha
\end{aligned}
\tag{75}
$$

it follows immediately that $s_2(\alpha)$ satisfies the difference equation (42) whilst $s_1(\alpha)$ satisfies the same equation with $1/\eta_3$ replaced by η_3 $(= \eta_1)$ and $1/\eta_4$ by η_4 $(= \eta_2)$. When the other field components are expressed in terms of $s_1(\alpha)$ and $s_2(\alpha)$, the integrands contain a factor $\cos^2\alpha + \cot^2\beta$ in the denominator, and to ensure that the radiation condition is fulfilled it is now necessary to modify the solution (39) that was appropriate in the acoustic problem. The modification consists of the addition of source-free solutions whose strengths are chosen to suppress the inhomogeneous plane waves contributed by the poles at $\cos\alpha = \pm i \cot\beta$. The details can be found in Bucci and Franceschetti [5] where the solution is derived for an isotropic impedance half plane with impedances differing on the two sides.

Unfortunately the representations (75) are no more successful than (71) in solving the anisotropic problem, and when the boundary conditions (21) are imposed, the resulting difference equations are again second order ones with coefficients even more complicated than those in (73). It would therefore appear that an intrinsic coupling of the field components via the (vector) boundary conditions manifests itself in second order difference equations, and since there is no procedure for obtaining their solutions, Maliuzhinets' method no longer works.

We now consider the application of the Wiener-Hopf method to this same problem of a plane electromagnetic wave at oblique incidence. If Green's theorem or a Hertz vector representation is used to express the total field in terms of the electric and magnetic currents $\underline{J}$ and $\underline{J}^*$ induced in the half plane, application of the boundary conditions (26) leads to four coupled Wiener-Hopf integral equations for the four (tangential) components of $\underline{J}$ and $\underline{J}^*$. Each equation actually involves only three unknowns but we can increase the symmetry of the system by combining the equations in pairs in a manner similar to that used in deriving (68) from (67). When this is done, the coefficient of the column vector $\{J_x(\xi), J_x(\xi), YJ_x^*(\xi), YJ_z^*(\xi)\}$ in the ξ plane is the 4 by 4 matrix

$$G(\xi) = \begin{pmatrix} 1 + \dfrac{k^2 - \xi^2}{\eta_2 k\Gamma}, & \dfrac{\xi \cos\beta}{\eta_2 \Gamma}, & \dfrac{\xi \cos\beta}{\Gamma}, & \dfrac{1}{\eta_2} + \dfrac{k \sin^2\beta}{\Gamma} \\ 1 + \dfrac{k^2 - \xi^2}{\eta_1 k\Gamma}, & \dfrac{\xi \cos\beta}{\eta_1 \Gamma}, & \dfrac{-\xi \cos\beta}{\Gamma}, & -\left(\dfrac{1}{\eta_1} + \dfrac{k \sin^2\beta}{\Gamma}\right) \\ -\dfrac{\xi \cos\beta}{\eta_3 \Gamma}, & -\left(1 + \dfrac{k \sin^2\beta}{\eta_3 \Gamma}\right), & \dfrac{1}{\eta_3} + \dfrac{k^2 - \xi^2}{k\Gamma}, & \dfrac{\xi \cos\beta}{\Gamma} \\ -\dfrac{\xi \cos\beta}{\eta_4 \Gamma}, & -\left(1 + \dfrac{k \sin^2\beta}{\eta_4 \Gamma}\right), & \dfrac{1}{\eta_4} + \dfrac{k^2 - \xi^2}{k\Gamma}, & \dfrac{\xi \cos\beta}{\Gamma} \end{pmatrix} \tag{76}$$

where $\Gamma = \sqrt{k^2 \sin\beta - \xi}$. For normal incidence ($\beta = \pi/2$), $G(\xi)$ separates into two 2 by 2 matrices of the form shown in (68).

In the special case when the impedances are isotropic ($\eta_1 = \eta_3$, $\eta_2 = \eta_4$) it proves convenient to combine the equations in pairs using multiplying factors ξ and $\pm k \cos\beta$. The procedure is directly analogous to that used by Senior [24] and gives

$$G(\xi)=\begin{pmatrix} k\cos\beta\left(1+\frac{k}{\eta_4\Gamma}\right), & \xi\left(1+\frac{k}{\eta_4\Gamma}\right), & -\xi\left(\frac{1}{\eta_4}+\frac{\Gamma}{k}\right), & k\cos\beta\left(\frac{1}{\eta_4}+\frac{\Gamma}{k}\right) \\ \xi\left(1+\frac{\Gamma}{\eta_4 k}\right), & -k\cos\beta\left(1+\frac{\Gamma}{\eta_4 k}\right), & k\cos\beta\left(\frac{1}{\eta_4}+\frac{k}{\Gamma}\right), & \xi\left(\frac{1}{\eta_4}+\frac{k}{\Gamma}\right) \\ k\cos\beta\left(1+\frac{k}{\eta_3\Gamma}\right), & \xi\left(1+\frac{k}{\eta_3\Gamma}\right), & \xi\left(\frac{1}{\eta_3}+\frac{\Gamma}{k}\right), & -k\cos\beta\left(\frac{1}{\eta_3}+\frac{\Gamma}{k}\right) \\ \xi\left(1+\frac{\Gamma}{\eta_3 k}\right), & -k\cos\beta\left(1+\frac{\Gamma}{\eta_3 k}\right), & -k\cos\beta\left(\frac{1}{\eta_3}+\frac{k}{\Gamma}\right), & -\xi\left(\frac{1}{\eta_3}+\frac{k}{\Gamma}\right) \end{pmatrix} \tag{77}$$

and if we now introduce new unknowns $g_{1,2}(\xi)$ and $g^*_{1,2}(\xi)$ such that

$$g_1(\xi) = k\cos\beta\, J_x(\xi) + \xi J_z(\xi)$$

$$g_2(\xi) = \xi J_x(\xi) - k\cos\beta\, J_z(\xi)$$

with corresponding expressions for $g_1^*(\xi)$ and $g_2^*(\xi)$, the matrix coefficient of the column vector $\{g_1(\xi),\ g_2(\xi),\ Yg_1^*(\xi),\ Yg_2^*(\xi)\}$ again separates into two 2 by 2 matrices. We remark that g and g^* are the functions that would have occurred naturally if we had expressed the boundary conditions in terms of the field components E_y and H_y (see (19) and (18)), and have the same half planes of analyticity as the original current transforms. For impedances which are the same on both sides of the half plane, i.e., $\eta_4 = \eta_3$, addition and subtraction of the equations making up each pair leads to the separated Wiener-Hopf equations obtained by Senior [24], whereas for $\eta_4 \neq \eta_3$ the 2 by 2 matrices can be factored using Hurd's [8] extension of the Wiener-Hopf techniques.

If the impedances are anisotropic but the same on both sides, i.e., $\eta_2 = \eta_1$, $\eta_4 = \eta_3$, $\eta_1 \neq \eta_3$, (76) can be simplified by addition and subtraction of the equations in pairs. Once again $G(\xi)$ separates into two 2 by 2 matrices, and the coefficient of the column vector

$$\left\{J_x(\xi),\ \frac{\eta_3}{\eta_1} J_z(\xi)\right\}$$

is then

$$G(\xi) = \begin{pmatrix} k \cos\beta\left(1+\dfrac{k}{\eta_1\Gamma}\right), & \xi\left(1+\dfrac{k}{\eta_3\Gamma}\right) \\ \xi\left(1+\dfrac{\Gamma}{\eta_1 k}\right), & -k\cos\beta\left(1+\dfrac{\Gamma}{\eta\ k}\right) \end{pmatrix} \tag{78}$$

This cannot be factorized directly, nor does it satisfy the condition (69) required for the application of Hurd's method except in the trivial case of normal incidence ($\beta = \pi/2$). Although it is not impossible that some further simplification could be made such that the resulting matrix would satisfy (69), this does not seem very likely. We are therefore at an impasse and the situation is even more hopeless for a general anisotropic impedance. Like Maliuzhinets' method, the Wiener-Hopf approach is ineffective when the impedance is anisotropic, and the recent extensions of this approach have not overcome the limitation.

Our final remarks concern the transition problem of a resistive or conductive half plane. From a Wiener-Hopf analysis for an impedance half plane we can deduce the corresponding resistive or conductive sheet solution by merely suppressing the contribution of the magnetic or electric current respectively. Since the only solutions which are relevant are those for impedances which are the same on both faces, Hurd's extension is no longer required and if the resistivity (or conductivity) is isotropic, the problem for even a plane wave at oblique incidence is easily obtained in terms of the field components E_y and H_y [28]. In contrast, Maliuzhinets' method is ill-suited to this type of problem, and to illustrate the difficulty that occurs, consider the problem of the acoustic wave (35) incident on a resistive half plane for which the boundary conditions are (6). If we again represent the total field in the form (36), the continuity of V across the sheet requires

$$s(\alpha) - s(\alpha - 2\pi) = s(2\pi - \alpha) - s(-\alpha),$$

whilst the jump condition gives

$$\left(\frac{2}{\eta} - \sin\alpha\right)s(\alpha) + \sin\alpha\ s(\alpha - 2\pi) = \left(\frac{2}{\eta} + \sin\alpha\right)s(2\pi - \alpha) - \sin\alpha\ s(-\alpha)$$

Each equation now involves four functions, and though they can still be manipulated to produce a first order difference equation of interval 2π, it is no longer evident how the singular function $\sigma(\alpha)$ (see (39)) can be determined. Of course, for anisotropic sheets at oblique incidence, both methods fail for the reasons stated earlier.

IV. CONCLUSIONS

These problems of imperfect half planes are of interest to the mathematician as well as to the engineer, and present a variety of challenges to both. To the radar or acoustic engineer, the results are of practical importance in the design and application of absorbers, and are also needed if we are to employ effectively and economically the computer programs which are now in use. To the mathematician, the class of problems created by the various boundary conditions provides an admirable vehicle for exploring the similarities and differences of two distinct methods of analysis, and when two methods are available, each may serve to expand the capability of the other. We have already noted how the work of Hurd [8] has extended the scope of the Wiener-Hopf method by providing a factorization for a matrix of the form shown in (69). If a similar method can be developed for the more general matrix (78), this would not only yield the solution for an anisotropic sheet, but could also result in the solution of the corresponding difference equation (74) in Maliuzhinets' method. We remark that this difference equation is not substantially different from that which arises when a wedge with imperfect surfaces is illuminated by a plane electromagnetic wave at oblique incidence.

V. REFERENCES

1. Bateman, H., "Electrical and Optical Wave Motion," Cambridge University Press, Cambridge, England (1915).
2. Bowman, J.J., High Frequency Backscattering from an Absorbing Infinite Strip with Arbitrary Face Impedances, *Can J. Phys. 45*, 2409-2430 (1967).
3. Bowman, J.J., T.B.A. Senior and P.L.E. Uslenghi, "Electromagnetic and Acoustic Scattering by Simple Shapes," North Holland Pub. Co. (1969).
4. Bucci, O.M., On a Function Occurring in the Theory of Scattering from an Impedance Half Plane, Laboratorio di Onde Elettromagnetiche, Instituto di Elettrotecnica, Napoli, Italy (1974).
5. Bucci, O.M. and G. Franceschetti, Electromagnetic Scattering by a Half Plane with Two Face Impedances, *Radio Sci. 11(1)*, 49-59 (1976).
6. Clemmow, P.C., A Method for the Exact Solution of a Class of Two-Dimensional Diffraction Problems, *Proc. Roy. Soc. (London) A205*, 286-308 (1951).
7. Clemmow, P.C., "The Plane Wave Spectrum Representation of Electromagnetic Fields," Pergamon Press, London (1966).
8. Hurd, R.A., The Wiener-Hopf-Hilbert Method for Diffraction Problems, *Can J. Phys. 54(7)*, 775-780 (1976).
9. Jones, D.S., A Simplifying Technique in the Solution of a Class of Diffraction Problems, *Q. J. Mech. 3*, 189-196 (1952).
10. Karp, S.N., Wiener-Hopf Techniques and Mixed Boundary Value Problems, *Commun. Pure Appl. Math. 3*, 411-426 (1950).
11. Knott, E.F., V.V. Liepa and T.B.A. Senior, Non-Specular Radar Cross Section Study, *The University of Michigan Radiation Laboratory Report No. 011062-1-F* (1973).
12. Knott, E.F. and T.B.A. Senior, Non-Specular Radar Cross Section Study, *The University of Michigan Radiation Laboratory Report No. 011764-1-T* (1974).
13. Knott, E.F. and T.B.A. Senior, Non-Specular Radar Cross Section Study, *The University of Michigan Radiation Laboratory Report No. 011764-1-F* (1974).
14. Liepa, V.V., E.F. Knott and T.B.A. Senior, Scattering from Two-Dimensional Bodies with Absorber Sheets, *The University of Michigan Radiation Laboratory Report No. 011764-2-T* (1974).
15. Maliuzhinets, G.D., Inversion Formula for the Sommerfield Integral, *Sov. Phys.-Dokl. 3(4)*, 752-755 (1959).
16. Maliuzhinets, G.D., Excitation, Reflection and Emission of Surface Waves from a Wedge with Given Face Impedances, *Sov. Phys.-Dokl. 3(4)*, 752-755 (1959).
17. Morse, P.M. and K.U. Ingard, "Encyclopedia of Physics, Acoustics I," Springer-Verlag, Berlin (1961).
18. Noble, B., "Methods Based on the Weiner-Hopf Technique," Pergamon Press, London (1958).

19. Peters, A.S., Water Waves over Sloping Beaches and the Solution of a Mixed Boundary Value Problem for $\Delta^2\phi - k'\phi = 0$ in a Sector, *Comm. Pure Appl. Math. 5*, 87-108 (1952).
20. Rawlins, A.D., Acoustic Diffraction by an Absorbing Half Plane in a Moving Fluid, *Proc. Roy. Soc. (Edinburgh) A72*, 337-357 (1974).
21. Rawlins, A.D., The Solution of a Mixed Boundary Value Problem in the Theory of Diffraction by a Semi-Infinite Plane, *Proc. Roy. Soc. (London) A346*, 469-484 (1975).
22. Rawlins, A.D., Diffraction of Sound by a Rigid Screen with a Soft or Perfectly Absorbing Edge, *J. Sound & Vib. 45(1)*, 53-67 (1976).
23. Senior, T.B.A., Diffraction by a Semi-Infinite Metallic Sheet, *Proc. Roy. Soc. (London) A213*, 436-458 (1952).
24. Senior, T.B.A., Diffraction by an Imperfectly Conducting Half Plane at Oblique Incidence, *Appl. Sci. Res. 8(B)*, 35-61 (1959).
25. Senior, T.B.A., Diffraction by an Imperfectly Conducting Wedge, *Comm. Pure Appl. Math. 12(2)*, 337-372 (1959).
26. Senior, T.B.A., Some Extensions of Babinet's Principle, *J. Acoust. Soc. Am. 58(2)*, 501-503 (1975).
27. Senior, T.B.A., Half-Plane Edge Diffraction, *Radio Sci. 10(6)*, 645-650 (1975).
28. Senior, T.B.A., Diffraction Tensors for Imperfectly Conducting Edges, *Radio Sci. 10(10)*, 911-919 (1975).
29. Tan, H.S. and D.K. Cheng, Solutions to a Class of Half-Plane Diffraction Problems, *Radio Sci. 5(8,9)*, 1191-1196 (1970).

MULTIPLE SCATTERING OF WAVES BY PERIODIC AND BY RANDOM DISTRIBUTIONS[1]

Victor Twersky
Mathematics Department
University of Illinois at Chicago Circle

I. SUMMARY AND INTRODUCTION

We discuss several recent developments for scattering of waves by periodic and random distributions, as well as applications to biomedical diagnostics. The deterministic cases include the line of equally spaced identical scatterers, and the doubly periodic planar array. For the statistical cases, we consider uncorrelated and pair-correlated distributions of discrete obstacles.

The solution for the periodic line was obtained by specializing the multiple scattering representation for arbitrary three dimensional configurations. Aside from its intrinsic interest (e.g., to the conical-cylindrical modes of long chain molecules) the periodic line is an essential component of the planar array, which we decomposed into a set of equally spaced lines and analyzed by earlier procedures for the grating of parallel cylinders. Similarly, the periodic plane (of interest to the multimode systems of regular mono-molecular layers and of analogous surfaces) is a component of the plane-bounded three-dimensional lattice, analyzed earlier as a finite number of equally spaced planar scatterers. Marked departures from single-scattering correspond to multi-mode coupling for near-grazing modes at spacings moderate or large compared to wavelength, and to scatterer multipole coupling for spacings small compared to wavelength. At low frequencies,

[1]Work supported in part by National Science Foundation Grants GP-33368X and MPS 75-07391.

ISBN 0-12-709650-7

the leading packing effects for an array of spheres can be represented as single scattering by an equivalent ellipsoid corresponding to elongation along the array and contraction at right angles; the pseudo-Brewster effects are anisotropic for the rectangular lattice. For very low frequencies, the explicit closed forms include up to octupole-octupole coupling (of interest to dense packing effects, and to hypochromisim and related oscillator-coupling phenomena in molecular biology).

For random distributions, earlier work on slab regions of large tenuous absorbing scatterers was extended to include incoherent interface effects; for particular ranges of the parameters, the back-space incoherent intensity may be much larger than the coherent reflection. Explicit results for spheroids have applications to geophysical propagation studies, to biomedical diagnostics (e.g., clinical determinations of the ratio of oxygenated to regular hemoglobin within red blood cells), and to environmental monitoring (remote determination of increasing chlorophyll content of bodies of water resulting from nitrogenous wastes). Low frequency results for the coherent phase in dense distributions of aligned anisotropic ellipsoids were used to analyze form and intrinsic birefringence (e.g., for polarization studies on biological tissues and cells). Corresponding explicit intensity results for dense distributions of parallel slabs, cylinders, and spheres were obtained on evaluating the pair-correlation integrals in the attenuation coefficients by using statistical mechanics theorems and the scaled-particle equations of state. In particular, the theory for cylindrical scatterers was applied to account for the transparency of the normal cornea (whose collagen fibers are the scatterers) and for its opacity when swollen.

II. THE SINGLE OBSTACLE

For a plane wave $\phi e^{-i\omega t}$,

$$\phi = e^{i\underline{k} \cdot \underline{r}}, \quad \underline{k} = \hat{k}k = \hat{k}2\pi/\lambda, \quad \underline{r} = \hat{r}r, \tag{1}$$

incident on an obstacle with center (the center of the smallest circumscribing sphere, a sphere of radius a) at the phase origin, we write the solution of the reduced wave equation as $\psi = \phi + u$. The scattered wave u is the radiative function

$$u(\underline{r}) = c\int [h_0(k|\underline{r}-\underline{r}'|)\nabla u(\underline{r}') - u\nabla h_0] \cdot d\underline{\mathcal{S}}(r') \equiv \{h_0, u\};$$

$$h_0(x) = h_0^{(1)}(x),\ H_0^{(1)}(x),\ e^{i|x|};\ c = \frac{k}{i4\pi},\ \frac{1}{i4},\ \frac{1}{i2k} \qquad (2)$$

with $\underline{\mathcal{S}} = \mathcal{S}\hat{n}$ as the obstacle's surface, $\hat{n}$ as the outward normal, and $\nabla = \nabla_{r'}$. The three forms of h_0 and c correspond to 3-, 2-, 1-dimensional problems (i.e., to bounded obstacles [1], cylinders [2], and slabs respectively). In one-dimension, the brace operation represents the sum of the values at the entrance ($z' = -a$) and exit ($z' = a$) faces. Asymptotically, for $r \sim \infty$, we have

$$u \sim h(kr)g(\hat{r},\hat{k}),\ g(\hat{r},\hat{k}) = \{e^{-i\underline{k}_r \cdot \underline{r}'}, u\} \equiv g\{\underline{k}_r, \underline{k}\},\ \underline{k}_r = k\hat{r} \qquad (3)$$

with h as the asymptotic form of h_0, and $g(\hat{r},\hat{k})$ as the conventional scattering amplitude. For penetrable scatterers with interior (v) field ψ' specified by propagation parameter K', and with surface ($\mathcal{S}$) conditions $\psi = \psi'$, $\partial_n\psi = B'\partial_n\psi'$, we may rewrite g as the volume integral

$$g(\hat{r},\hat{k}) = -c\int\left[\left(B'K'^2 - k^2\right)e^{-i\underline{k}_r\cdot\underline{r}'}\psi' - (B'-1)\nabla e^{-i\underline{k}_r\cdot\underline{r}'} \cdot \nabla\psi'\right]dv(\underline{r}')$$

$$\equiv [\![e^{-i\underline{k}_r \cdot \underline{r}'}, \psi']\!] \equiv g[\![\underline{k}_r, \underline{k}]\!] \qquad (4)$$

with v as the volume of the obstacle.

The normalization for g is such that for lossless scatterers, the energy theorem reduces to $-Re g(\hat{k},\hat{k}) = M|g(\hat{r},\hat{k})|^2$ with M as the mean value over all directions $\hat{r}$; we have $M_3 = (1/4\pi)\int d\Omega(\theta,\alpha)$, $M_2 = (1/2\pi)\int d\theta$, and M_1 (for $\hat{k} = \hat{z}$ perpendicular to the slab) is one-half the forward and back scattered values. For lossy obstacles,

$$-\sigma_0 Re g(\hat{k},\hat{k}) = \sigma_A + \sigma_S,\ \sigma_S = \sigma_0 M|g(\hat{r},\hat{k})|^2;\ \sigma_0 = \frac{4\pi}{k^2},\ \frac{4}{k},\ 2 \qquad (5)$$

with σ_A and σ_S as the absorption and scattering cross-sections respectively, and $\sigma_0 = 1/ick$.

Using the complex spectral representation for h_0 in (2), we have (at least for $r > a$ for all $\hat{r}$, and for $r > (\hat{r} \cdot \underline{r}')_{max}$ for given $\hat{r}$),

$$u(\underline{r}) = \int_c e^{i\underline{k}_c \cdot \underline{r}} g(\hat{r}_c, \hat{k}), \quad \underline{k}_c = k\hat{r}_c. \tag{6}$$

In three-dimensions, $\hat{r}_c = \hat{r}(\theta_c, \phi_c)$ and $\int_c = (1/2\pi)\iint d\Omega(\theta_c, \phi_c)$ with contours [1] as for $h_0^{(1)}$; in two, $\hat{r}_c = \hat{r}(\theta_c)$ and $\int_c = (1/\pi)\int d\theta_c$ with contour [2] as for $H_0^{(1)}$; in one, $\underline{k}_c = \pm k\hat{z}$ and $\int_c$ selects the sign corresponding to $z = \pm|z|$.

III. CONFIGURATION OF N-OBSTACLES

For a fixed configuration of N obstacles with centers at $\underline{r}_s$ for $s = 1, 2, \ldots, N$, we write the solution at $\underline{r} = \underline{r}_a$ as [1,2]

$$\Psi^a = \phi^a + \sum_s U_s^a, \quad U_s^a = U_s(\underline{r}_a - \underline{r}_s; \underline{r}_1, \underline{r}_2, \ldots, \underline{r}_N), \tag{7}$$

where $U_s^a \sim G_s h(k|\underline{r}_a - \underline{r}_s|)$, and G_s is the corresponding multiple scattering amplitude. Symbolically, we have

$$U_s^a = u_s^a \cdot \left[\phi^s + \sum_t^1 U_t^s\right], \quad \sum_t^1 = \sum_{t \neq s}, \tag{8}$$

and the iterated form

$$\Psi^a = \phi^a + \sum u_s^a \phi^s + \sum\sum^1 u_s^a \cdot u_t^s \phi^t + \ldots \tag{9}$$

shows the structure for successive scatterings.

Equivalently, with reference to scatterer t,

$$\Psi = \Psi_t = \Phi_t + U_t, \quad \Phi_t = \phi + \sum_s^1 U_s, \tag{10}$$

where Φ_t may be regarded as the net excitation for scatterer t. We have $U_t = \{h_0, U_t\}_t$, and

$$U_t(\underline{r} - \underline{r}_t) = \int_c e^{i\underline{k}_c \cdot (\underline{r} - \underline{r}_t)} G_t(\hat{r}_c). \tag{11}$$

The functions Ψ_t, Φ_t, U_t satisfy the same relations at the surface (s_t) and interior (v_t) of scatterer t as ψ, ϕ, u for the isolated obstacle [1-4] and an arbitrary direction of incidence $\hat{k}_1$. Thus $\{\psi,\Psi_t\}_t = 0$; consequently $G_t(-\hat{k}_1) = \{\phi,U_t\}_t = \{\Phi_t,u\}_t$, and we obtain [1-3]

$$G_t(\hat{r}) = g_t(\hat{r},\hat{k})e^{i\underline{k} \cdot \underline{r}_t} + \sum_s^{1} \int g_t(\hat{r},\hat{r}_c)G_s(\hat{r}_c)e^{i\underline{k}_c \cdot \underline{R}_{ts}},$$

$$\underline{R}_{ts} = \underline{r}_t - \underline{r}_s. \tag{12}$$

This system of functional equations $G[g]$ determines G if g and the configuration are known.

For the analogous vector electromagnetic problems and an incident field $\hat{e}\phi$, with $\hat{e}$ as the electric polarization, we use dyadic representations [3]

$$\tilde{\phi} = (\tilde{I} - \hat{k}\hat{k})\phi, \quad \tilde{\phi} \cdot \hat{e} = \hat{e}\phi, \quad \hat{e} \cdot \hat{k} = 0;$$

$$\tilde{u} \sim \tilde{g}h, \quad \tilde{U}_t \sim \tilde{G}_t h, \quad \tilde{\Psi}_t = \tilde{\Phi}_t + \tilde{U}_t, \tag{13}$$

etc., such that $\underline{E} = \tilde{\Psi} \cdot \hat{e}$ is the required electric field. In particular, the analog of (12) involves $\tilde{G}_t$, $\tilde{g}_t$ and $\tilde{g}_t \cdot \tilde{G}_s$.

IV. REGULAR DISTRIBUTIONS

The functional equation $G[g]$ of (12) has been applied in detail to two obstacles [1-3], gratings of parallel cylinders [5], the periodic line of bounded identical obstacles [6], and to the doubly periodic plane [7]. As illustrations, we include several key forms and results for the rectangular planar lattice

$$\underline{r}_s = \underline{b}_s = \hat{x}s_1b_1 + \hat{y}s_2b_2; \quad s_i = 0,\pm1,\pm2, \ldots; \quad i = 1,2 \tag{14}$$

and an incident wave

$$\phi_0 = e^{i\underline{k}_0 \cdot \underline{r}}, \ \hat{k}_0 = \hat{x}\xi_0 + \hat{y}\eta_0 + \hat{z}\zeta_0, \ \zeta_0 = (1 - \xi_0^2 - \eta_0^2)^{\frac{1}{2}} = |\zeta_0| > 0. \tag{15}$$

We express the solution in terms of plane wave modes ϕ_ν (propagating and evanescent) specified by

$$\hat{k}_\nu = \hat{x}\xi(\nu_1) + \hat{y}\eta(\nu_2) + \hat{z}\zeta(\nu_1,\nu_2); \quad \nu_i = 0, \pm 1, \pm 2, \ldots;$$

$$\begin{Bmatrix}\xi\\ \eta\end{Bmatrix} = \begin{Bmatrix}\xi_0\\ \eta_0\end{Bmatrix} + \lambda\begin{Bmatrix}\nu_1/b_1\\ \nu_2/b_2\end{Bmatrix};$$

$$\zeta = (1 - \xi^2 - \eta^2)^{1/2} = \begin{Bmatrix}|\zeta|\\ i|\zeta|\end{Bmatrix} \text{for } \xi^2 + \eta^2 \lessgtr 1, \tag{16}$$

where $\nu(\nu_1,\nu_2)$ represents a double-infinite discrete set of modes. The values $\zeta = |\zeta|$, $i|\zeta|$ correspond to propagating, evanescent modes; the limit $\zeta \to 0$ (which occurs only for special values of k, b_i, ξ_0, η_0) corresponds to grazing modes.

We obtain

$$\Psi = \phi_0 + C\textstyle\sum_\nu \phi_\nu G_{\nu 0}/\zeta_\nu, \quad G_{\nu 0} = G(\hat{k}_\nu, \hat{k}_0),$$

$$C = 2\pi/k^2 b_1 b_2, \quad \zeta_\nu = \zeta(\nu_1,\nu_2) \tag{17}$$

with the multiple scattered amplitude determined in terms of $g_{\nu\alpha} = g(\hat{k}_\nu, \hat{k}_\alpha)$ by

$$G_{\nu 0} = g_{\nu 0} + CSg_{\nu\alpha}G_\alpha/\zeta_\alpha, \quad S = \lim_{\varepsilon\to 0}\Big(\sum_{\nu_1}\sum_{\nu_2} - \int d\nu_1 \int d\nu_2\Big)e^{ik\varepsilon\zeta(\nu_1,\nu_2)} \tag{18}$$

where $\int d\nu_1 \int d\nu_2$ represents integration over the corresponding continuous set of modes, and the exponential convergence factor approaches zero with increasing $|\nu_1|$ or $|\nu_2|$.

If there are no modes near grazing, then except for small kb_i we may use

$$G_{\nu 0} \simeq g''_{\nu 0}, \quad g''_{\nu\alpha} = g_{\nu\alpha} + CS_p g_{\nu\mu} g''_{\mu\nu}/\zeta_\mu, \quad S_p = \textstyle\sum_p - \int_p, \tag{19}$$

where S_p, the propagating part of the operator in (18), corresponds to $\zeta(\nu_1,\nu_2) = |\zeta|$. The amplitude g'' differs from g in that its radiative losses are restricted to the propagating modes of the array. For kb_i large, $\sum_p \simeq \int_p$, and (19) reduces to the single-scattered value $G_{\nu 0} \simeq g_{\nu 0}$. More generally, we require forms for G which include the coupling effects of the near-grazing modes. The phenomena for near-grazing evanescent modes ($\zeta = \zeta_N = i|\zeta_N| \simeq 0$) are analogous to the Wood anomalies for the grating [5], and the parameters ($\lambda_R, \hat{k}_R$, etc.) that specify the grazing modes are the analogs of those considered by Rayleigh for the grating.

We specify the grazing modes $\zeta_R = 0$ in terms of a lattice vector $\underline{v} = \hat{x}\nu_1/b_1 + \hat{y}\nu_2/b_2$ in the space reciprocal to $\underline{b}_s$, by

$$\lambda_R = \frac{1}{v}[(\hat{k}_0 \cdot \hat{v})^2 + (\hat{k}_0 \cdot \hat{z})^2]^{\frac{1}{2}} - \hat{k} \cdot \hat{v},$$

$$\hat{k}_R = \hat{x}\left(\xi_0 + \lambda_R \frac{\nu_1}{b_1}\right) + \hat{y}\left(\eta_0 + \lambda_R \frac{\nu_2}{b_2}\right) \tag{20}$$

where $v = [(\nu_1/b_1)^2 + (\nu_2/b_2)^2]^{\frac{1}{2}} = n/d$ with d as the separation of the direct lattice lines perpendicular to $\hat{v}$, i.e., $\underline{v} \cdot \underline{b}_s = s_1\nu_1 + s_2\nu_2 = (s_1\mu_1 + s_2\mu_2)n$ is constant for variable s_i and fixed $\nu_i = n\mu_i$, with n as the common integer factor. In particular if there is only one mode (N) near grazing, we replace (19) by

$$G_{\nu 0} \simeq g''_{\nu 0} + \frac{Cg''_{\nu N}g''_{N0}/\zeta_N}{1 - Cg''_{NN}/\zeta_N}. \tag{21}$$

For the near-grazing mode itself ($\nu = N$), we have

$$G_{N0} \simeq g''_{N0}/(1 - Cg''_{NN}/\zeta_N), \tag{22}$$

which when substituted into (17) shows that the field Ψ has no singularities in the limit of a grazing mode $\zeta_N \to \zeta_R = 0$. For the remaining modes, $G_{\nu 0}$ of (21) shows the same resonances for $\zeta_N = i|\zeta_N| \simeq 0$ as discussed for the grating [5]. In particular $|G_{\nu 0}|$ may show a maximum for parameters such that

$$1 - CImg''_{NN}/|\zeta_N| = 0, \quad G_{\nu 0} \simeq g''_{\nu 0} - g''_{\nu N}g''_N/Reg''_{NN} \tag{23}$$

where the coupling term may be orders of magnitude larger than $g''_{\nu 0} \simeq g_{\nu 0}$. See Ref. 7 for more complete forms and for detailed discussion of degeneracies.

To discuss multipole coupling effects, we expand g in spherical harmonics in terms of the isolated scattering coefficients a_n^m, and similarly,

$$G(\hat{r},\hat{k}) = \sum_{nm} A_n^m(\hat{k})Y_n^m(\hat{r}); \ \sum_{nm} = \sum_{n=0}^{\infty}\sum_{m=-n}^{n} . \tag{24}$$

For spherically symmetric obstacles, we obtain from (18),

$$A_n^m = (-1)^m a_n \left[Y_n^{-m}(\hat{k}) + \sum_{\sigma\tau} A_\sigma^\tau \sum_\ell d_\ell \binom{-m}{n};\binom{\tau}{\sigma} H_\ell^{\tau-m} \right] \tag{25}$$

where $n + \sigma + \ell$ is even, ℓ changes by steps of 2 from $|n - \sigma|$ (or from $|\tau - m|$ if it is the larger) to $n + \sigma$, and the known numbers [8,9] d_ℓ arise from the expansion

$$Y_n^{-m} Y_\sigma^\tau = \sum_\ell d_\ell \left({-m \atop n}; {\tau \atop \sigma} \right) Y_\ell^{\tau-m} \tag{26}$$

The functions H are the lattice sums

$$H_n^m = \sum_s{}^1 e^{i\underline{k}\cdot\underline{b}_s} i^n h_n^{(1)}(kb_s) Y_n^m(-\hat{b}_s), \ \underline{b}_s = \hat{x}s_1 b_1 + \hat{y}s_2 b_2 \tag{27}$$

where $\sum_s^1$ excludes the central element $s_1 = s_2 = 0$. Rapidly converging forms [7] of H follow from the earlier developments of the simpler sums that arose for the grating [5] and for the periodic line [6].

As an illustration we apply (24) to normal incidence on a lattice with spacings b_i small compared to λ, for the boundary condition $\partial_n \Psi = 0$ on spherical obstacles of radius a. Keeping only monopole and dipole terms we write the transmission and reflection coefficients as

$$T = 1 + CG(\hat{z},\hat{z}) = 1 + C(A_0 + A_1), \ R = CG(-\hat{z},\hat{z}) = C(A_0 - A_1),$$
$$C = 2\pi/k^2 b_1 b_2, \tag{28}$$

and obtain

$$T \simeq 1 - \frac{iQ}{1+iQ} + \frac{i3Q/2}{1 - i3Q/2 - v}, \quad R \simeq \frac{-iQ}{1+iQ} - \frac{i3Q/2}{1 - i3Q/2 - v},$$

$$Q = \frac{2\pi ka^3}{3b_1 b_2}, \quad v = \frac{1}{2}\left(\frac{a}{b_1}\right)^3 L_2^0(b_2/b_1) = \frac{1}{2}R^3 L_2^0(\rho), \tag{29}$$

where we introduced $R = a/b_1 < 1/2$, and $\rho = b_2/b_1 \geq 1$. The low frequency lattice function in (29) is given by

$$L_2^0(\rho) = 2\zeta(3) + \frac{4\zeta(2)}{\rho^2} - 4\sum_{\nu=1}^{\infty}\sum_{t=1}^{\infty}(2\pi\nu)^2[K_0(\nu t2\pi\rho) - K_2(\nu t2\pi\rho)], \tag{30}$$

where $\zeta(n)$ is the Riemann ζ-function, and K_n is the modified Hankel function; the double sum is very rapidly converging [7]. The forms in (29) satisfy $|R \pm T| = 1$ as required by the scattering theorems [10].

In the low frequency limit $k \to 0$, the ratios of the multiple-scattered to single-scattered coefficients involved in (28) satisfy

$$A_0/a_0 \to 1, \; A_1/a_1 \to 1/D, \; k \to 0, \tag{31}$$

where the monopole term shows no frequency independent packing effect but the dipole does. Corresponding to (29), which includes only dipole-dipole coupling, we write $D = D_1 = 1 - v$, or equivalently

$$D_1 = 1 - R^3 d_1(\rho), \; R = a/b_1 < \frac{1}{2}, \; \rho = b_2/b_1 \geq 1 \tag{32}$$

where we used $d_1 = L_2^0/2$; we have, e.g., $d_1(1) \simeq 4.52$, $d_1(2) \simeq 2.02$, and $d_1(3) \simeq 1.57$. If we include up to octupole-octupole coupling effects for $k \to 0$, we obtain the form [7],

$$D_2 \simeq 1 - R^3 d_1 - R^{10} d_2 - R^{17} d_3, \; d_n = d_n(\rho) \tag{33}$$

where the $d_n(\rho)$ in terms of the appropriate lattice sums [7] specify the anisotropic packing effects.

Were we to plot A_1/a_1 based either on (32) or (33) as a function of R with ρ as the parameter, the highest curve $\rho = 1$ would correspond to the square lattice cell, and the lowest to the most elongated rectangular cell. For a given ρ, the curve based on (32) is lower than that based on (33); such pairs of curves differ little to $R \simeq 1/4$, and the differences that arise with R increasing to 1/2 are not large. The biggest differences occur at $R = 1/2$, and although more complete forms than either D_1 or D_2 are required at this physical limit, we note for comparison, that then $D_1^{-1}(1) \simeq 2.3$, $D_2^{-1}(1) \simeq 2.44$, and $D_1^{-1}(3) \simeq 1.24$, $D_2^{-1}(3) \simeq 1.25$; thus the coupling effects and the influence of the octupoles are more pronounced for the square cell. The corresponding normalized reflection coefficient of the lattice is

$$-R/iQ \to 1 + 3A_1/2a_1 \to 1 + 3/2D, \; k \to 0, \tag{34}$$

where $-iQ \simeq Ca_0 \simeq CA_0$. See Ref. 7 for additional applications.

V. STATISTICAL ENSEMBLE OF CONFIGURATIONS

The average of Ψ of (7) over a statistically homogeneous ensemble [11-15] of configurations of N identical and aligned obstacles whose centers are uniformly distributed in V is given by

$$\langle\Psi(\underline{r})\rangle = \phi + \rho\int_V \langle U_s(\underline{r}-\underline{r}_s)\rangle_s d\underline{r}_s, \; \rho = N/V, \; \langle U_s\rangle_s = \langle\Psi_s\rangle_s - \langle\Phi_s\rangle_s;$$

$$\langle U_s(k)\rangle_s = \int_C \langle G_s(\hat{r}_C)\rangle_s e^{i\underline{k}_C \cdot (\underline{r}-\underline{r}_s)}. \tag{35}$$

If $\underline{r}$ is within $v_s = v$, we use $\langle\Psi_s(K')\rangle_s - \langle\Phi_s(k)\rangle_s$, and if not, the radiative form $\langle U_s(k)\rangle_s$ with $\langle G_s\rangle_s$ as the average of (12) over all variables but $\underline{r}_s$ (now a dummy). We write the pair distribution function as $\rho f(\underline{R})$ such that

$$f(\underline{R}) = 0 \text{ if } R < b(\hat{R}), \; f(\underline{R}) \sim 1 \text{ if } R \sim \infty, \tag{36}$$

where $b(\hat{R})$, the minimum separation of centers, is a function of $\hat{R} = \hat{R}_{ts}$, and $\underline{R}_{ts} = \underline{b}(\hat{R})$ specifies the exclusion surface (S inclosing volume v) around one center. Then, from (12) we have

$$\langle G_t(\hat{r})\rangle_t = g_t(\hat{r},\hat{k})e^{i\underline{k}\cdot\underline{r}_t}$$

$$+ \rho\int_{V-v} d\underline{r}_s\, f(\underline{R}_{ts})\int_c g_t(\hat{r},\hat{r}_c)\langle G_s(\hat{r}_c)\rangle_{st}\, e^{ik_c\cdot\underline{R}_{ts}} \qquad (37)$$

where $\langle G_s\rangle_{st}$, the average over all variables but $\underline{r}_s$ and $\underline{r}_t$, may be expressed in terms of g_s, $\langle G_t\rangle_{st}$ and an integral over $d\underline{r}_m$ of $\langle G_m\rangle_{stm}$ and the three particle distribution function $\Gamma(R_{ts},R_{sm})$. We truncate the hierarchy system by means of $\langle G_s\rangle_{st} \simeq \langle G_s\rangle_s$; this approximation, analogous to $\langle\Phi_s\rangle_{st} \simeq \langle\Phi_s\rangle_s$, as used by Lax [12], excludes various scattering processes for fixed sets of obstacles. Working with the symbolic form [15] $U_s(\underline{r}_a - \underline{r}_s) = U_s^a = u_s^a \cdot \Phi_s$, we restore the processes for two fixed obstacles by (48) of Ref. 15:

$$\langle\Phi_s\rangle_{st} \simeq \left[1 - u_t^s \cdot u_s^t\right]^{-1} \cdot \left[\langle\Phi_s\rangle_s + u_t^s \cdot \langle\Phi_t\rangle_t\right]$$

$$= \langle\Phi_s\rangle_s + u_t^s \cdot \langle\Phi_t\rangle_t + \ldots,$$

which we introduce within $\int d\underline{r}_s$ of (37) as

$$\langle G_s(\hat{r})\rangle_{st} \simeq \langle G_s(\hat{r})\rangle_s + \int_c g(\hat{r},\hat{r}_c)\langle G_t(\hat{r}_c)\rangle_t e^{i\underline{k}_c\cdot\underline{R}_{st}} + \ldots, \qquad (38)$$

etc. To restore scattering processes for three fixed obstacles would require double integrals $\int d\underline{r}_s \int d\underline{r}_m$ involving $f(\underline{R}_{ts})f(\underline{R}_{sm})$ and $\Gamma(\underline{R}_{ts},\underline{R}_{sm})$. See Keller [13] and Ref. 15 for procedures not based on the hierarchy integrals.

For a slab region distribution $z = 0$ to d, except for boundary layers (with thickness of the order of a scatterer's projection on $\hat{z}$), the average internal field has the form [14]

$$\langle\Psi\rangle = \Psi_I = \Psi_1 + \Psi_2 = \sum\Psi_i, \quad \Psi_i = A_i e^{i\underline{K}_i\cdot\underline{r}}$$

$$K_i/k = \eta_i, \quad \underline{K}_i\cdot\hat{t} = \underline{k}\cdot\hat{t} \qquad (39)$$

such that $\hat{t}\cdot\hat{z} = 0$. The bulk propagation vectors $\underline{K}_i$ and bulk indices η_i associated with the average wave (the coherent field) specify the effects within an unbounded distribution. The average transmitted and reflected fields have the forms [14]

$$\Psi_T = Te^{i\underline{k}\cdot\underline{r}} = T\phi, \quad \Psi_R = Re^{i\underline{k}'\cdot\underline{r}} = R\phi', \quad \underline{k}' = \underline{k} - 2\underline{k}\cdot\hat{z}\hat{z} \tag{40}$$

with ϕ' as the image of ϕ in $z = 0$. For distributions of identical scatterers, the fields $<>_s$ corresponding to (39) may be written

$$<\Psi_s(\underline{r}_s + \underline{r}')>_s = \sum\psi^i(\underline{r}')\Psi_i(\underline{r}_s), \quad <\Phi_s>_s = \sum\phi^i\Psi_i, \quad <U_s>_s = \sum u^i\Psi_i \tag{41}$$

with $\underline{r}'$ as the local vector from the fixed center $\underline{r}_s$. The fields ψ^i, ϕ^i, U^i satisfy the same conditions at $\mathcal{S}_s$ and $\mathcal{V}_s$ as for an isolated obstacle and may be interpreted as the fields for an equivalent scatterer [4]. We write the multiple scattering amplitude as

$$<G_s(\hat{r})>_s = <G(\underline{r}_s;\hat{r})>_s = \sum g^i\Psi_i = \sum g(\underline{k}_r|\underline{K}_i)\Psi_i(\underline{r}_s), \tag{42}$$

$$g(\underline{k}_r|\underline{K}_i) = \left\{e^{-i\underline{k}_r\cdot\underline{r}'}, u^i\right\} = [\![e^{-i\underline{k}_r\cdot\underline{r}'}, \psi^i]\!], \tag{43}$$

where we use $g(\underline{k}_r|\underline{K}_i)$ or g^i to distinguish the present amplitude from the earlier two-space function [14] $g(\underline{k}_r, \underline{K})$.

Comparison of the forms (39) and (40) with results for an equivalent uniform slab leads to approximations for the coefficients A_i, R, T in terms of associated impedances Z_i. For $i = 1$, with $K_1 = K$, $Z_1 = Z$, we have the forms as in (14) and (17) of Ref. 14 (p. 726),

$$\begin{Bmatrix} Z \\ 1 \end{Bmatrix} = -\frac{\rho}{2c\underline{k}\cdot\hat{z}}\left[\frac{g(\underline{k}|\underline{K})}{(\underline{K}-\underline{k})\cdot\hat{z}} \pm \frac{g(\underline{k}'|\underline{K})}{(\underline{K}+\underline{k})\cdot\hat{z}}\right]; \tag{44}$$

for $i = 2$, we replace Z by $-Z_2$ and $\underline{K}$ by $\underline{K}_2$.

Substituting (39) and (42) in (35), we obtain extinction (of ϕ) and cancellation (of ϕ') relations (which reproduce (44) if boundary layers are negligible), as well as the relation [4]

$$K_i^2 - k^2 = -\frac{\rho}{c}[\![e^{-i\underline{K}_i\cdot\underline{r}'}, \psi^i]\!] \equiv -\frac{\rho}{c} g[\![\underline{K}_i|\underline{K}_i]\!] \tag{45}$$

where the argument on $g[\![\underline{K}|\underline{K}]\!]$ indicates explicit restriction to the volume integral from (4). Special cases and approximations for this representation were considered earlier, particularly by

Lax [12]. In terms of the external surface integral form (2) we have [4]

$$K_i^2 - k^2 = \frac{-(\rho/c)g\{\underline{K_i}|\underline{K_i}\}}{1-\rho\int e^{-i\underline{K_i}\cdot\underline{r}'}(\psi^i - \phi^i)dv}, \quad g\{\underline{K_i}|\underline{K_i}\} \equiv \{e^{-i\underline{K_i}\cdot\underline{r}'}, u^i\} \tag{46}$$

The g forms in (45) and (46) with argument $\underline{K}|\underline{K}$ are not equivalent, and the equations for $K_i^2 - k^2$ suggest different interpretations and different expansion procedures. See Ref. 4 for additional results and for representations for the corresponding bulk parameters B and $B\eta^2$.

Substituting (42) into (37) with $\langle G_s\rangle_{st}$ replaced by $\langle G_s\rangle_{s'}$, and introducing

$$F(\underline{k_r},\underline{k_c}|\underline{K_i}) = g(\hat{r},\hat{r}_c)g(\underline{k_c}|\underline{K_i}), \quad U_i = \int_c F(\underline{k_r},\underline{k_c}|\underline{K_i})e^{i\underline{k_c}\cdot\underline{R}},$$

$$F\{\underline{k_r},\underline{K_i}|\underline{K_i}\} = \{e^{-i\underline{K_i}\cdot\underline{R}}, U_i\}_S \tag{47}$$

we obtain [4]

$$g(\underline{k}|\underline{K}) = \frac{-\rho F\{\underline{k_r},\underline{K}|\underline{K}\}}{c(K^2-k^2)} + \int_c F(\underline{k_r},\underline{k_c}|\underline{K})M(\underline{k_c},\underline{K}),$$

$$M(\underline{k_c},K) = \rho\int_{V-v}[f(\underline{R})-1]\, e^{i(\underline{k_c}-\underline{K})\cdot\underline{R}}d\underline{R}, \tag{48}$$

where, because $f(\underline{R}) \sim 1$ as $R \sim \infty$, we let V represent all space.

To indicate relations with existing forms, for uncorrelated distributions ($M = 0$), for sparse concentrations ($\rho \simeq 0$), if we approximate $F\{\,\}$ by $g(\hat{r},\hat{k})g(\underline{k}|\underline{K})$, then from (48) for $\hat{r} = \hat{k}$,

$$K^2 - k^2 \simeq -\rho g(\hat{k},\hat{k})/c = -\rho ik\sigma_0 g(\hat{k},\hat{k}) \equiv 2k(K_R - k), \tag{49}$$

where K_R is essentially Rayleigh's result [16] generalized to arbitrary scatterers. From (49) and (5), the coherent attenuation coefficient is

$$2ImK_R = -\rho\sigma_0 Reg(\hat{k},\hat{k}) = \rho(\sigma_A + \sigma_S) \tag{50}$$

and the corresponding phase is specified by

$$ReK_R/k = Re\eta_R = 1 + (\rho\sigma_0/2k)Img(\hat{k},\hat{k}). \quad (51)$$

Explicit approximations follow from known results for $g(\hat{k},\hat{k})$.

For small symmetrical scatterers (with radius, or half-width, a small compared to λ) we may represent K_R as series in powers of ka. In particular, if the scatterers are characterized by $\varepsilon' = \eta'^2$, such that $\eta' \simeq 1$ (tenuous scatterers) with $Re\eta' >> Im\eta'$, then (51) reduces to

$$Re\eta_R \simeq 1 + \omega(Re\eta' - 1), \ \omega = \rho v; \ v_3 = \frac{4}{3}\pi a^3, \ v_2 = \pi a^2, \ v_1 = 2a \quad (52)$$

with ω as the fraction of space filled by scattering material. Similarly, to lowest order in k, the absorption effects are also shape-independent,

$$\rho\sigma_A \simeq 2k\omega Im\eta'. \quad (53)$$

The corresponding attenuation effects resulting from scattering are specified by

$$\rho\sigma_S \simeq 2\omega|\delta|^2 L, \quad \delta = k(\eta' - 1)$$

$$L_3 = k^2 v_3/3\pi, \quad L_{2\ell} = kv_2/2, \quad L_{2t} = kv_2/4, \quad L_1 = v_1 \quad (54)$$

where the L's have the dimension of a length; the two values of L_2 correspond to $\hat{E}$ lengthwise or transverse to the cylinder's axis.

On the other hand, for large tenuous scatterers (negligible back scattering), from (4) we construct

$$g[\![\underline{k},\underline{k}]\!] \simeq -c2k^2(\eta' - 1)\int e^{-ikz'}\psi' dv$$

$$\psi' \simeq e^{ikz_1 + iK'(z' - z_1)}, \ \hat{k} = \hat{z}, \quad (55)$$

with $z_1(x,y)$ as the ray entrance point on the scatterer's surface. Thus

$$g(\hat{k},\hat{k}) \simeq g_0(1 + iL\delta), \ g_0 = -c2k\delta v, \ L = \frac{1}{v}\int(z - z_1)dv \quad (56)$$

where we integrate from $z = z_1(x,y)$ to the exit point $z_2(x,y)$ and then over x,y. We have $K_R - k \simeq \omega\delta(1 + iL\delta)$, and obtain (52) and (53), as well as the form in (54) in terms of the present length parameter L.

To modify (50) and include correlations, we write

$$M(\underline{k}_r,\underline{k}) = \rho\int [f(\underline{R}) - 1]e^{ik(\hat{r}-\hat{k})\cdot\underline{R}} d\underline{R} \equiv W(\hat{r},\hat{k}) - 1 . \tag{57}$$

For spherically symmetric distributions, $f(\underline{R}) = f(R)$ is the radial distribution function, and $W[f(R)]$ is a standard form in x-ray diffraction by liquids [17]. Using (49) and (57) in (48), for $Img >> Reg$ we obtain

$$ImK \simeq \rho\sigma_A + \rho\sigma_0 M[|g(\hat{r},\hat{k})|^2 W(\hat{r},\hat{k})] \tag{58}$$

which includes the leading effects of pair-correlation.

From (58), for small scatterers with average center spacing small compared to λ, to lowest order in k we obtain [18]

$$2ImK \simeq \rho(\sigma_A + \sigma_S W), \quad W \equiv 1 + \rho\int [f(\underline{R}) - 1]d\underline{R} \tag{59}$$

where W is proportional to the variance ($\mathcal{V}$) of the number (n) of particles in a central region (V_C) of the distribution. We have

$$\mathcal{V} = \langle n^2\rangle - \langle n\rangle^2 = \langle n\rangle W \tag{60}$$

where $\langle n\rangle$ is the average number of particles in V_C such that $\langle n\rangle/V_C = N/V = \rho$. From statistical thermodynamics [17],

$$\frac{\mathcal{V}}{\langle n\rangle} = \rho k_B T_A \zeta_T = k_B T_A \left(\frac{\partial\rho}{\partial p}\right)_T = W, \tag{61}$$

where k_B is Boltzmann's constant, T_A is the absolute temperature, ζ_T is the isothermal compressibility, and p is the pressure. The last equality relates W to a derivative of the equation of state (E) of the fluid. For impenetrable inflexible particles of volume v such that $W = W(w)$ with $w = \rho v$, we write the equation of state as

$$p/k_B T_A = E(\rho,v) \tag{62}$$

and differentiate with respect to ρ to construct

$$\left(\frac{\partial E}{\partial \rho}\right)^{-1} = W(w), \quad w = \rho v. \tag{63}$$

If we use the scaled particle approximations for rigid symmetrical particles [19],

$$E_3 = \frac{\rho(1 + \rho v + \rho^2 v^2)}{(1 - \rho v)^3}, \quad E_2 = \frac{\rho}{(1 - \rho v)^2}, \quad E_1 = \frac{\rho}{1 - \rho v}, \tag{64}$$

(where $v = v_3$ is a sphere, v_2 a cylinder, and v_1 a slab) then we obtain simple rational functions of w for the correlation-packing factor W:

$$W_3 = \frac{(1 - w)^4}{(1 + 2w)^2}, \quad W_2 = \frac{(1 - w)^3}{1 + w}, \quad W_1 = (1 - w)^2, \tag{65}$$

such that W_3 decreases faster than W_2 which in turn decreases faster than W_1 with increasing W. See Ref. 18 for detailed discussion of W and of the related function

$$S_m = wW_m = \frac{w(1 - w)^{m+1}}{[1 + (m-1)W]^{m-1}}; \quad m = 3,2,1. \tag{66}$$

The maximum of S_3 is at $W_\wedge \simeq 0.128$, of S_2 at $W_\wedge \simeq 0.215$, and of S_1 at $W_\wedge = 1/3$. We take the radius of v as half the minimum separation (b) of scatterer centers, such that $b/2 \gtrsim a$. (In subsequent applications we assume that $\frac{1}{2}b - a$ is a transparent shell having the same index of refraction as the imbedding medium.)

The forms (65) and (66) apply for essentially rigid symmetrical particles, and take into account the space-occupying property of the particles as well as the increase in local order with increasing w resulting from the well-defined shape of the neighbors. For uncorrelated flexible particles that can pack to fill all space, the symmetrical function $S_s = \omega(1 - \omega)$ that arose in the development [14] for large tenuous scatterers appears to include the essential occupancy effects. Using an approximation of (44) for negligible coherent reflection, i.e.,

$$K - k \simeq -\frac{\rho}{2kc} g(\underline{k}|\underline{K}), \tag{67}$$

we interpreted [20] g as a two-space isolated scattering amplitude $g[\![\underline{k},\underline{K}]\!]$ corresponding to e^{iKz} incident, and replaced the internal field ψ' of (55) by $e^{iKz_1 + iK'(z' - z_1)} \equiv \psi_1$. Thus

$$g[\![\underline{k},\underline{K}]\!] \simeq g_0[1 + iL(K' - K)], \; K - k \simeq \omega(K' - k)[1 + iL(K' - K)] \tag{68}$$

from which

$$K - k \simeq \omega\delta[1 + iL\delta(1 - \omega)] \equiv K_s - k, \; 2ImK_s = \rho\sigma_A + \rho\sigma_s(1 - \omega),$$

$$\delta = k(\eta' - 1), \; \omega = \rho v \tag{69}$$

where $\rho\sigma_s(1 - \omega) = \rho\sigma_s W_s = 2|\delta|^2 LS_s$. We obtain the same result from (45) in the form $K - k \simeq -(\rho/2kc)g[\![\underline{K}|\underline{K}]\!]$, if we use $g[\![\underline{K}|\underline{K}]\!] \simeq -c2k^2(\eta' - 1)\int e^{iKz'}\psi^i dv$ with $\psi^i \simeq \psi_1$.

Alternatively, if in (48) for $\hat{r} = \hat{k}$ we introduce $F\{\underline{k}_r, \underline{K}|\underline{K}\} \simeq g(\hat{r},\hat{K})g(\underline{k}_K|\underline{K})$, then

$$K^2 - k^2 \simeq -\frac{\rho}{c} g(\hat{K},\hat{K}) \left[1 - \frac{1}{g(\underline{k}_K|\underline{K})} \int_C g(\hat{K},\hat{r}_c) g(\underline{k}_c|\underline{K}) M(\underline{k}_c,\underline{K})\right]^{-1},$$

$$\underline{k}_K = k\hat{K}. \tag{70}$$

Essentially this form in terms of one g-function with $k\hat{K}$ replaced by $\underline{K}$ is given by Lax [12]. The leading terms

$$K^2 - k^2 \simeq -\frac{\rho}{c}\left[g(\hat{K},\hat{K}) + \int_C g(\hat{K},\hat{r}_c) g(\hat{r}_c,\hat{K}) M(\underline{k}_c,\underline{K})\right] \tag{71}$$

reduce to (58) under the same restrictions.

The general relation (48) enables us to derive more complete results by analytical procedures. Thus the low frequency limit ($k \to 0$) for incidence along an axis of aligned anisotropic triaxial ellipsoids (in a similar ellipsoidal distribution) characterized by principal indices η'_i and depolarization integrals q_i is

$$\eta_i^2 = 1 + \frac{\omega(\eta_i'^2 - 1)}{1 + (1 - \omega)q_i(\eta_i'^2 - 1)} \tag{72}$$

See Ref. 21 for an elementary construction of this form and for detailed applications to birefringence studies. For $\eta_i' = \eta'$ and $q_i = 1/3$, eq. (72) reduces to the Maxwell-Clausius-Mossotti-Lorenz-Lorentz form.

For symmetrical scatterers, we analyze (48) by multipole decompositions, and relate η^2 directly to the known isolated scattering coefficients a_n of the standard [1-3] Fourier or Legendre expansions for g. For the scalar problems we obtain

$$\eta^2 - 1 = -C\sum_n A_n, \quad A_n = a_n \eta^{2n}(1 + \sum_\nu h_{n\nu} A_\nu), \quad h_{n\nu} = \frac{H_{n\nu}}{\eta^{n+\nu}}, \tag{73}$$

where the A_n are determined by a_n and $H_{n\nu}$ (analogs of lattice sums). For circular cylinders

$$C = i\rho 4\pi/k^2, \quad 2H_{n\nu} = C\eta^{\nu-n} \sum_{m=0}^{n-1} \eta^{2m} + H_{\nu-n} + H_{\nu+n}$$

$$H_n = H_{-n} = 2\pi\rho\int_0^\infty [f(R) - 1] J_n(KR) H_n^{(1)}(kR) R dR \tag{74}$$

In general the dependence of H_n on K is secondary, and we reduce (73) by iterative procedures starting with $J_n(kR)$ in (74). Similarly for spheres

$$C = i\rho 4\pi/k^3, \quad H_{n\nu} = \sum_{|n-\nu|}^{|n+\nu|} d_m \begin{pmatrix} 0 & 0 \\ n; & \nu \end{pmatrix} \left[CN_m^{\nu n} + H_n \right],$$

$$N_m^{\nu n} = \frac{\eta^{\nu+n} - \eta^m}{\eta^2 - 1} = \eta^m + \eta^{m+2} + \ldots + \eta^{\nu+n-2},$$

$$H_n = 4\pi\rho\int_0^\infty [f(R) - 1] j_n(KR) h_n^{(1)}(kR) R^2 dR, \tag{75}$$

with the coefficients d_m as discussed for (24).

For the electromagnetic problem of spheres characterized by electric (b_n) and magnetic (c_n) isolated scatterer coefficients [3], we obtain

$$\eta^2 - 1 = -C\sum (B_n + C_n), \quad C = i\rho 4\pi/k^3,$$

$$B_n = b_n \eta^{2n}\left[1 + \sum\left(B_\nu h^1_{n\nu} + C_\nu h^2_{n\nu}\right)\right],$$

$$C_n = c_n \eta^{2n}\left[1 + \sum\left(C_\nu h^1_{n\nu} + B_\nu h^2_{n\nu}\right)\right], \quad h^i_{n\nu}\,\eta^{n+\nu} = H^i_{n\nu} \tag{76}$$

where the coefficients B_n and C_n are determined by the pair of algebraic systems. The functions $H^i_{n\nu}$ are of the same form as $H_{n\nu}$ of (75) but involve the coefficients $d^i_m\begin{pmatrix}-1 & 1\\ n & ; \nu\end{pmatrix}$ that arise in the corresponding expansions of the scalar products of spherical Hanson functions $\underline{C}^{\pm 1}_n$ and $\underline{B}^{\pm 1}_n$, i.e., d^1_m corresponds to $\underline{C} \cdot \underline{C}$, $\underline{B} \cdot \underline{B}$ and d^2_m to $\underline{B} \cdot \underline{C}$, $\underline{C} \cdot \underline{B}$.

The results for $K = k\eta$ with $\eta^2 = \varepsilon$ determine the coherent field in an unbounded distribution. For a slab region distribution of thickness d, for values of the relative parameters such that coherent interface effects are negligible, we may approximate the internal average wave (39) by

$$\langle\Psi\rangle \simeq e^{iKz} \tag{77}$$

and the transmitted wave of (40) by

$$\Psi_T = T\phi \simeq e^{i(K-k)d}\phi \tag{78}$$

We write the corresponding transmitted coherent intensity as

$$C = |\langle\Psi\rangle|^2 = |T|^2 \simeq e^{-2dmKd} \equiv e^{-(\gamma+\beta)d} \tag{79}$$

where γ corresponds to absorption and β to scattering losses, e.g., for (59) as well as (69) we have

$$\gamma = \rho\sigma_A, \quad \beta = \rho\sigma_S W = \beta_R W \tag{80}$$

with $\beta_R = \rho\sigma_S$ as Rayleigh's approximation.

The average total transmitted intensity

$$T = \langle|\Psi|^2\rangle = C + I, \quad I = \langle|\Psi - \langle\Psi\rangle|^2\rangle \tag{81}$$

is the sum of the coherent (C) and incoherent (I) contributions. In an insertion loss geometry with negligible coherent and incoherent interface effects [20], we may approximate T by C provided the incoherent scattering $I(\alpha)$ into the detector (a cone of half-angle α) is negligible. If $I(\alpha)$ is not negligible, we use [14, 18,20]

$$T = C + I(\alpha) = e^{-(\gamma+\beta)d} + q(\alpha)[e^{-\gamma d} - e^{-(\gamma+\beta)d}]. \qquad (82)$$

For large tenuous scatterers, $q(\alpha)$ is the received fraction of the forward half-space scattering of one obstacle on a line between transmitter and detector, i.e.,

$$q(\alpha) = \frac{\sigma(\alpha)}{\sigma(\pi/2)} = \frac{\sigma(\alpha)}{\sigma}, \quad \sigma(\alpha) = \frac{1}{k^2}\int_\alpha |\underline{g}(\hat{r},\hat{z})|^2 d\Omega(\hat{r}) \qquad (83)$$

with $\int_\alpha d\Omega$ representing integration over a cone of half-angle α around $\hat{z}$. If the net back-space scattering is negligible, then $\sigma \simeq \sigma_s$. We sketch an elementary derivation of the form in (82) to indicate its physical aspects as well as its limitations, and then consider more complete forms for T and corresponding forms for the reflected flux R.

For a plane wave normally incident on a half-space distribution, we write the total intensity at a distance z from the entry face ($z = 0$) for negligible interface effects and negligible back-half-space scattering, as the sum of two terms

$$T(z) = C(z) + I(z); \; T(0) = C(0) = 1, \; I(0) = 0, \qquad (84)$$

where $C(z) \simeq |\langle\Psi\rangle|^2$ and $I(z) \simeq \langle|\Psi|^2\rangle - |\langle\Psi\rangle|^2$ are the coherent and incoherent components. The coherent intensity (the main beam) is attenuated by absorption and by scattering, i.e.,

$$\frac{dC}{dz} = -(\gamma+\beta)C, \; C(0) = 1, \qquad (85)$$

from which $C(z) = e^{-(\gamma+\beta)z}$ (and γ and β as in (80) are known). The incoherent scattering is similarly attenuated but it is enriched by scattering gains

$$\frac{dI}{dz} = -(\gamma+\beta)I + \beta T = -\gamma I + \beta C; \quad I(0) = 0, \; C = e^{-(\gamma+\beta)z} \qquad (86)$$

The final form for dI/dz stresses that the incoherent intensity is attenuated solely by absorption (I's scattering losses being replenished by corresponding scattering gains) and that C is the source of I. Thus $\left(\frac{dI}{dz} + \gamma I\right)e^{\gamma z} = \beta e^{-\beta z}$, and consequently

$$I(z) = e^{-\gamma z} - e^{-(\gamma + \beta)z} \tag{87}$$

is the incoherent intensity within the distribution at a point receiving all forward half-space radiation ($\sigma \simeq \sigma_S$) from 0 to z. (To the present approximation, the total intensity $T(z) = e^{-\gamma z}$ shows only absorption losses, the scattering losses being balanced by corresponding gains.) If we truncate the semi-infinite distribution at $z = d$, and represent an external receiver by a cone of half-angle α, then (82) follows from

$$T = C(d) + q(\alpha)I(d) = C + I(\alpha), \tag{88}$$

where $q = \sigma(\alpha)/\sigma \simeq \sigma(\alpha)/\sigma_S$.

We may obtain more complete results for T, and corresponding forms for the back-space intensity R, by applying the symbolic integral equation approximations [15]

$$\langle\Psi^a\rangle = \phi^a + \rho\int u_s^a \cdot \langle\Psi^s\rangle d\underline{r}_s\,, \tag{89}$$

$$\langle|\Psi^a|^2\rangle = |\langle\Psi^a\rangle|^2 + \rho\int |v_s^a|^2 \cdot \langle|\Psi^s|^2\rangle d\underline{r}_s\,,$$

$$v_s^a = u_s^a + \rho\int u_t^a \cdot v_s^t\, d\underline{r}_t\,. \tag{90}$$

The equation for $\langle|\Psi|^2\rangle$ has source term $|\langle\Psi\rangle|^2$; its scattering kernel $|v|^2$ corresponds to a radiative function v satisfying the integral equation for $\langle\Psi\rangle$ with source u instead of ϕ. In particular, we considered situations where the coherent interface effects were negligible so that (77) sufficed, and applied (90) in terms of

$$|\langle\Psi\rangle|^2 \simeq |e^{iKz}|^2 = e^{-Lz}, \quad L = 2Imk = \gamma + \beta, \tag{91}$$

to determine the incoherent interface effects for large tenuous scatterers [20].

In terms of $\sigma(\alpha)$ of (83) for scattering into a cone in the forward direction, such that $\sigma(\pi) = \sigma_S$ is the scattering cross section, and $\sigma(\pi/2) = \sigma$ the forward half-space scattering, we write $\sigma'(\alpha) = \sigma(\pi) - \sigma(\pi - \alpha)$ for the back scattering into a cone α, and $\sigma'(\pi/2) = \sigma'$ for the net scattering into the back half-space. From these we construct the ratio $q = \sigma(\alpha)/\sigma$ as in (83), as well as $p = \sigma'(\alpha)/\sigma'$, and $q' = \sigma'(\alpha)/\sigma$. The leading term of the transmitted intensity for $\sigma' << \sigma$ is the form (82),

$$T_0 = T_0[q] = e^{-Ld} + q(\alpha)\left[e^{-Md} - e^{-Ld}\right], \quad M \simeq \gamma + \frac{\gamma 2\rho\sigma'(1-w)}{2\gamma + \beta}, \tag{92}$$

where M reduces to γ as in (82) if σ' is neglected. For a semi-infinite distribution, the net back-space scattering equals the incoherent interface coefficient P; the back scattering into a cone α is the fraction

$$R_\infty[p] = \frac{\sigma'(\alpha)}{\sigma'}P = p(\alpha)P \simeq \frac{q'(\alpha)\beta}{2\gamma + \beta}, \quad q' = \frac{\sigma'(\alpha)}{\sigma}, \quad P \simeq \frac{\sigma'\beta}{\sigma(2\gamma + \beta)}. \tag{93}$$

For a slab region of arbitrary thickness, the analog of $T_0[q]$ for R is

$$R_0 = R_\infty[p]\left\{1 - e^{-Md}T_0[q/p]\right\} \tag{94}$$

where $T_0[q/p]$ is (92) with q replaced by q/p. More generally, including multiple incoherent interface effects, we have to all orders of P,

$$R = R_0\left[1 - P^2 e^{-2Md}\right]^{-1}, \quad T = T_0 - Pe^{-Md}R \tag{95}$$

See Ref. 20 for corresponding expressions for off-axis receiving cones.

VI. BIO-MEDICAL APPLICATIONS

In the following we illustrate the applicability of several scattering relations for bio-medical purposes.

A. Transparency of the Normal Cornea and its Opacity when Swollen

As discussed in detail by Maurice [22], discounting transparent surface layers, the central region of the cornea consists of about 200 parallel layers, each made up of parallel collagen fibers (index η_c) in an adherent ground substance (imbedding medium, η_m). On the average, there are 40 fibers along a layer thickness with axial separation under normal conditions of 2 diameters. The angle between fiber axes in successive layers is in general large, so the overall effect is that of a cross-banded shell of fiber-reinforced laminations. When the cornea is placed in an aqueous solution, the cross-banding restricts swelling to the direction perpendicular to the layers.

For optical transmission studies, the central region of the normal cornea (thickness d_1) may be represented [22] as a set of layers of parallel cylinders (radius a) with the orientation of the axis in successive layers taken as uniformly random. At $\lambda = 5 \times 10^{-4}$mm, Maurice determined average values for the key parameters,

$$\eta_c \simeq 1.47,\ \eta_m \simeq 1.345,\ a \simeq 1.55 \times 10^{-5}\text{mm},$$
$$w_1 = \pi\rho_1 a^2 \simeq 0.23,\ d_1 \simeq 0.46\text{mm}, \tag{96}$$

and compared transmission data with essentially

$$T_R = e^{-\beta_R d},\quad \beta_R = \frac{\rho}{2}(\sigma_{s\ell} + \sigma_{st}),$$

$$\beta_R \simeq \frac{\rho}{8}k^3\pi^2a^4\left(\eta_r^2 - 1\right)^2\left[1 + \frac{2}{\left(\eta_r^2 + 1\right)^2}\right],\quad \eta_r = \frac{\eta_c}{\eta_m},\quad k = \frac{2\pi\eta_m}{\lambda}, \tag{97}$$

where $\frac{1}{2}(\sigma_{s\ell} + \sigma_{st})$, the mean scattering cross section for $\hat{E}$ lengthwise and transverse to a cylinder's axis, is appropriate both for unpolarized and for polarized light (because of the randomness of the fiber axes in successive layers). The values in (96) give

$$\beta_R d_1 \simeq 2.58,\ T_R \simeq 0.08 \tag{98}$$

as compared to the measured value $T_n \simeq 0.9$ for the normal cornea, and Maurice [22] therefore concluded that transparency depends on regularity; the opacity of the cornea caused by swelling arises from a resulting loss of regularity.

We regard [18] the regularity within a layer of parallel collagen fibers as essentially that of a two-dimensional dense gas of impenetrable disks with diameter b (the closest seperation of axes) larger than the fiber diameter $2a$. The shell of thickness $\frac{1}{2}b - a$ is assumed to be tough adherent ground substance (relative index $\eta_r = 1$), but although transparent for the scattering problem, the coat determines the size of the statistical mechanical particles whose correlations influence the net transmission. The value of b, an average in the same sense as a of (96), is obtained by fitting to transmission data for the normal cornea. In terms of W_2 of (65) we model transmission through the central region of the cornea in the course of a swelling process by

$$T(w) \simeq e^{-\beta_R d W(w)} = T_R^{\,W},$$

$$W(w) = \frac{(1-w)^3}{1+w}, \quad w = \frac{w_1 d_1}{d} = \frac{w_1}{t} = w_t, \tag{99}$$

where $w_1 = \rho_1 \pi b^2/4 = \omega_1 b^2/4a^2 \simeq 0.23(b/2a)^2$ applies when the cornea has normal thickness d_1, and $w_1 d_1/d = w_1/t$ when swollen to $t = d/d_1$ times normal.

The value w_1 (the normal volume fraction of the coated collagen fibers) is determined by fitting T of (99) to experimental data for T_n. The choice $w_1 \simeq 0.6$ is appropriate for Maurice's transmission data [22]; thus from (98) and (99),

$$W(0.6) \simeq 0.04, \quad T(0.6) = T_1 \simeq 0.90, \tag{100}$$

which corresponds to outer-coat radius $(b/2)$ about 60 percent larger than fiber radius, i.e., from $w_1 \simeq 0.6 \simeq 0.23(b/2a)^2$, we have $b/2a \simeq 1.62$. This value for $b/2a$, which provides an average gap of about $3a/4$ between coated fibers as compared to $2a$ between bare fibers, is also consistent with Maurice's data for diffusion of large molecules within the cornea.

When swollen to 1.5 times normal thickness, $w = w_{1.5} = 0.4$, and (99) gives

$$W(0.4) \simeq 0.154, \quad T_{1.5} \simeq 0.67. \tag{101}$$

The values $T_1 \simeq 0.90$ and $T_{1.5} \simeq 0.67$ are also in accord with data of Cox et al [23] that a 50 percent increase of corneal thickness decreased transmission from 0.88 ± 0.02 to 0.67 ± 0.03. Similarly when swollen to twice normal thickness, $w = w_2 = 0.3$, and

$$W(0.3) \simeq 0.264, \quad T_2 \simeq 0.51. \tag{102}$$

The choice of the value of w_1 is not too critical. For a 5 percent increase

$$T(0.63) = T_1 \simeq 0.92, \quad T_{1.5} \simeq 0.70, \quad T_2 \simeq 0.53; \tag{103}$$

for a 5 percent decrease

$$T(0.57) = T_1 \simeq 0.88, \quad T_{1.5} \simeq 0.64, \quad T_2 \simeq 0.48, \tag{104}$$

where the present T_1 and $T_{1.5}$ are also in accord with measurements [23]. See Ref. 18 for additional discussion and illustrations.

We based the above on the scaled particle approximation for the equation of state for rigid disks to obtain a simple functional dependence of the packing correlation function $W(w)$ on volume fraction. However, we may also use alternative equations of state or else seek to apply the statistical thermodynamical relations (61) with appropriate models for the compressibility or pressure.

B. Scattering and Absorption by Blood

We consider [24] optical studies on blood seeking the oxygenated fraction of hemoglobin within the red blood cells (RBC). For such purposes, whole blood is essentially a suspension of biconcave discoid RBC in a physiological salt solution about 0.9 percent NaCL. On dilution with water to about 0.6 percent, the cells accumulate water and swell to spheres, and at about 0.3 percent practically all the cells lose their hemoglobin to the imbedding liquid and their residual structure disintegrates (osmotic hemolysis). For concreteness, rough ranges for the blood parameters at normal physiological values are: volume fraction (hematocrit) $\omega = \rho v = 0.44 \pm 0.05$ and concentration $\rho = (5 \pm 1) \times 10^6/(\mathrm{mm})^3$; the average values give $v = 88\mu^3$. The normal RBC has maximum diameter 8.5μ, greatest thickness 2.4μ, and least thickness 1μ. For some purposes, we compare data for RBC's with results for spheriods with major axis 4 times the length of the minor (symmetry) axis; if we take the minor axis as 1.1μ, then the spheroid's volume and surface are $v = 8.91\mu^3$ and $s = 138\mu^2$ as compared with RBC ranges $v = 87 \pm 7\mu^3$ and $s = 160 \pm 10\mu^2$.

The oxygen saturation of blood, the oxygenated fraction (f) of its hemoglobin, is a principal parameter of interest in clinical medicine. If a sample of blood is hemolyzed, f for the resulting hemoglobin solution can be determined by optical absorption measurements based on the Lambert-Beer relation

$$T_a = e^{-\gamma d}, \quad \gamma = \Gamma\omega, \quad \Gamma(\lambda) = f\Gamma_o(\lambda) + (1-f)\Gamma_r(\lambda), \tag{105}$$

where Γ is proportional to the sum of the weighted molecular-absorption coefficients of oxygenated and regular hemoglobin, and ω is the volume fraction of RBC before hemolysis. Thus, f can be isolated without prior knowledge of ω and d from simultaneous measurements of $-\ln T_a = \gamma d$ at two wavelengths λ_1 and λ_2, i.e., f is linear in the ratio $M = \gamma(\lambda_2)/\gamma(\lambda_1)$. It is convenient to choose λ_1 so that $\Gamma_o(\lambda_1) = \Gamma_r(\lambda_1)$ (an isobestic point, e.g., $\lambda_1 = 805$ nm), and λ_2 so that Γ_o and Γ_r differ greatly (e.g., $\lambda_r > 5\lambda_o$ at $\lambda_2 = 660$ nm); then [25]

$$f = \frac{\Gamma_r(\lambda_2) - \Gamma_r(\lambda_1)M}{\Gamma_r(\lambda\ ) - \Gamma_o(\lambda_2)} \tag{106}$$

in which the ratios of the Γ's are known to high accuracy, and M is obtained by measurements.

Ideally, f would be determined from measurements on whole blood within the body. However, for nonhemolyzed blood (RBC suspensions), the Lambert-Beer relation is not valid experimentally

[26]. The transmitted flux T does not in general decay linearly with ω and d, and the corresponding optical density at certain wavelengths has been reported [25,26] to be as high as 20 times that for hemolyzed blood. To compare data with theory, the scattering effects of the large blood cells must be taken into account.

For a transmission measurement (transmission oximetry), if interface scattering is negligible, we use $T \simeq T_0$ of (82) and (92) in the form

$$T = C + I(q) = e^{-\gamma d}\left[e^{-\beta d} + q(1 - e^{-\beta d})\right] \equiv T_a \cdot T_s;$$

$$\gamma = \Gamma\omega, \quad \beta = B\omega(1 - \omega); \quad 0 \lesssim q \lesssim 1. \tag{107}$$

If $q \simeq 1$, then practically all the scattered flux is detected, and $T \simeq T_a$. We also have $T \simeq T_a$ if $\omega \simeq 1$ is realizable; on the other hand, near the extreme $\omega \simeq 0$, the form $T \simeq 1 - \omega d[\Gamma + B(1 - q)]$ involves scattering as well as absorption.

We separate the absorption and scattering effects by

$$\tau d = -lnT = \Gamma\omega d - lnT_s, \tag{108}$$

which, as shown by Anderson and Sekelj [27], can be applied to data to determine Γ and f essentially as for the hemolyzed case, if B and q are practically constant for the λ's in question. If the parameters depend on λ we use [24] $B = C_1|\eta'(\lambda) - 1|^2/\lambda^2 = C_1 N(\lambda)/\lambda^2$; for large α, we have $q = 1 - C_2\lambda^2$ [and for α small, $q = C_3\lambda^{-2}exp(-C_4\lambda^{-2})$]. The C_n are fixed for the purpose at hand. In particular, for small Bd (or near the extremes of q), we may use

$$T = e^{-\tau d}, \quad \tau \simeq \gamma + \beta(1 - q) \tag{109}$$

where $B(1 - q) = C_5 N(\lambda)$ for large α. Thus in terms of $M_\tau = \tau(\lambda_2)/\tau(\lambda_1)$ we obtain a more general linear form in M_τ analogous to (106),

$$f = \frac{\Gamma_r(\lambda_2) + CN_2 - [\Gamma_r(\lambda_2) + CN_1]M_\tau}{\Gamma_r(\lambda_2) - \Gamma_o(\lambda_2)},$$

$$C \sim \frac{(1-\omega)s}{8\nu \sin^2(\alpha/2)}, \qquad N_n = |\eta'(\lambda_n) - 1|^2, \tag{110}$$

which includes scattering effects of the blood cells for large α. More complete expressions may be based on (108) with $B(\lambda)$ and $q(\lambda)$ as discussed in Ref. 24.

Anderson and Sekelj [27] stressed that the different functional dependence of γ and β on ω facilitated separating the absorption effects from the scattering, and reported accord between T of (107) and data for several Γ's ranging from $\Gamma > B$ to $\Gamma << B$, and ω ranging from 0.05 to about 0.6. Results for $\Gamma << B$ over an extensive range of ω from 0.08 to 0.85, were reported earlier by Loewinger et al [26] (who modified T_s of Ref. 14 by including two absorption coefficients).

For $\Gamma << B$ a pronounced maximum was observed [26,27] for data of τ vs ω near $\omega = 1/2$ (close to the normal value in the bloodstream). We see from (107) that if $\Gamma = 0$, then $\tau d = -\ln T_s$ has a maximum at $\omega = \omega_\wedge = 1/2$ corresponding to $\beta = B/4$. If $\Gamma \neq 0$, a maximum occurs at

$$\omega_\wedge = \frac{1}{2} + \frac{\Gamma}{2B(1-q)} \left[1 + q(e^{\beta d} - 1)\right], \quad \beta = \beta(\omega_\wedge), \tag{111}$$

provided $\omega_\wedge < 1$. Thus for q not too close to unity, for thin samples such that $Bd/4$ is not large, τ will show a maximum if Γ/B is small. From (111), or its approximation for small $Bd/4$,

$$\omega_\wedge \simeq \frac{1}{2} + \frac{\Gamma}{2B(1-q)} \left(1 + \frac{qBd}{4}\right) \tag{112}$$

we see that the maximum is shifted to $\omega > 1/2$ with Γ and increasing d in accord with the data of Loewinger et al [26].

For reflection oximetry [28] with $d \simeq \infty$, we use R_∞ of (93) in the form [24]

$$R_\infty = \frac{q' B(1-\omega)}{2\Gamma + B(1-\omega)} \tag{113}$$

with $q' \simeq C_6 \lambda^2$. In terms of $M_R = R_\infty(\lambda_1)/R_\infty(\lambda_2)$, we proceed as for (110) to obtain

$$f = \frac{\Gamma_r(\lambda_2) + C'N_2/\lambda_2^2 - \left[\Gamma_R(\lambda_1) + C'N_1/\lambda_1^2\right](N_2/N_1)M_R}{\Gamma_r(\lambda_2) - \Gamma_o(\lambda_2)}$$

$$C' \simeq 4\pi^2 \langle L \rangle (1 - \omega) \tag{114}$$

with $\langle L \rangle$ as the average over orientations of the length parameter L of (56). Polanyi and Hehir [28] obtain a linear relation for f in terms of M_R based on an approximation of R_∞ for $\Gamma >> B$.

For arbitrary d, the practical analog of T of (107) is R_0 of (94) in the form

$$\begin{aligned} R &= R_\infty \left\{ 1 - e^{-2\gamma d}\left[e^{-\beta d} + \bar{q}\left(1 - e^{-\beta d}\right)\right]\right\} \\ &= R_\infty \left\{ 1 - T_a{}^2 T_s[\bar{q}] \right\}, \quad \bar{q} = \bar{q}(\alpha) \gtrsim 1 \end{aligned} \tag{115}$$

provided $R > 0$. The condition $q \gtrsim 1$ means that the fraction of the forward-space flux scattered into an axial cone α is greater than the fraction of the back-space flux backscattered into a cone α (i.e., $\bar{q}(\alpha) = q(\alpha)/p(\alpha) \gtrsim 1$).

If $d \simeq 0$, then

$$R \simeq R_\infty[2\gamma - \beta(\bar{q} - 1)]d \simeq q'\beta[1 - \bar{q}\beta/(2\gamma + \beta)]d \tag{116}$$

provided that $2\gamma > \beta(\bar{q} - 1)$. Thus, R increases from zero linearity in d and in ω. At the other extreme, if $d \to \infty$, then R increases to R_∞ in accord with data [29] at fixed ω. Eq. (113) shows that $R_\infty \to 0$ as $\omega \to 0$; if $B/2\Gamma$ is large, the major part of the decrease is at large ω. In seeking a maximum of R of (115) for ranges of the parameters such that $R_\infty(\omega)$ is slowly varying compared to the function in braces, we obtain the form (111) with γ replaced by 2γ and q by $\bar{q}$; for this case, because $\bar{q} > 1$, the maximum corresponds to $\omega_\wedge \lesssim 1/2$. If the term in Γ/B is very small, the maximum occurs at $\omega_\wedge \simeq 1/2$ in accord with the date of Enson et al [30]; with increasing Γ or d, the maximum is shifted to smaller values of ω in accord with the data of Kramer et al [29], and of Anderson and Sekelj [27]. For ranges of the parameters such that the variation of $R_\infty(\omega)$ in R is significant, a maximum may occur for $\omega > 1/2$.

Note that in using (115) together with (107) in an experimental arrangement, the value of q isolated from T data may differ from that isolated from R data because of beam divergence (which we neglect) and because the actual scatterers may lack inversion symmetry (which we assumed). See Ref. 24 for additional biomedical applications.

VII. REFERENCES

1. Twersky, V., *J. Math. Phys. 3*, 83-91 (1962).
2. Twersky, V., "Electromagnetic Waves," R.E. Langer (editor), University of Wisconsin Press, 361-389 (1962); J.E. Burke, D. Censor and V. Twersky, *J. Acoust. Soc. Am. 37*, 5-13 (1965).
3. Twersky, V., *J. Math. Phys. 8*, 589-610 (1967).
4. Twersky, V., *J. d'Analyse Mathematique 30*, 498-511 (1976). The generalization to anisotropic random media, and the detailed development of (47), (48) and (70)-(75), is given in V. Twersky, *J. Math. Phys. 18*, 2468-2486 (1977). The analogous results for the electromagnetic case have also appeared in *J. Math. Phys. 19*, 215-230 (1978).
5. Twersky, V., *IRE Trans. AP-4*, 330-345 (1956); *Arch. Ration. Mech. & Anal. 8*, 323-332 (1961); *IRE Trans. AP-10*, 737-765 (1962); *J. Opt. Soc. Am. 52*, 145-171 (1962); J.E. Burke and V. Twersky, *IEEE Trans. AP-14*, 465-480 (1966).
6. Twersky, V., *J. Acoust. Soc. Am. 53*, 96-112 (1973).
7. Twersky, V., *J. Math. Phys. 16*, 633-643, 644-657, 658-666 (1975).
8. Friedman, B. and J. Russek, *Q. Appl. Math. 12*, 13 (1954).
9. Stein, S., *Q. Appl. Math. 19*, 15 (1961).
10. Twersky, V., *J. Appl. Phys. 27*, 1118-1122 (1956).
11. Foldy, L.L., *Phys. Rev. 67 (2)*, 107-119 (1945).
12. Lax, M., *Rev. Mod. Phys. 23*, 287-310 (1951); *Phys. Rev. 88(2)*, 621-629 (1952).
13. Keller, J.B., *Proc. Symposium Appl. Math. 13*, 227-246 (1962); *16*, 145-170 (1964).
14. Twersky, V., *J. Math. Phys. 3*, 700-715, 724-734 (1962).
15. Twersky, V., *Proc. Sympsoium Appl. Math. 16*, 84-116 (1964).
16. Lord Rayleigh, *Philos. Mag. 47*, 375-383 (1899).
17. Green, H.S., "The Molecular Theory of Fluids," Interscience, New York, 62ff (1952).
18. Twersky, V., *J. Opt. Soc. Am. 65*, 524-530 (1975).
19. Reiss, H., H.L. Frisch and J.L. Lebowitz, *J. Chem. Phys. 31*, 369 (1959); E. Helfand, H.L. Firsch and J.L. Lebowitz, *J. Chem. Phys. 34*, 1037 (1961).
20. Twersky, V., *J. Opt. Soc. Am. 60*, 908-914 (1970).

21. Twersky, V., *J. Opt. Soc. Am. 65*, 239-245 (1975).
22. Maurice, D.M., *J. Physiol. (London) 136*, 263 (1957); "The Eye," H. Davson (editor), ch. 6, Academic Press, New York, (1968).
23. Hart, R.W. and R.A. Farrell, *J. Opt. Soc. Am. 59*, 766 (1969); J.L. Cox, R.A. Farrell, R.W. Hart and M.E. Langhorn, *J. Physiol. 210*, 601 (1970).
24. Twersky, V., *J. Opt. Soc. Am. 60*, 1084-1093 (1970).
25. Kramer, K., J.O. Elam, G.A. Saxton and W.N. Elam, *Am. J. Physiol. 165*, 229 (1951).
26. Loewinger, E., A. Gordon, A. Weinreb and H. Gross, *J. Appl. Physiol. 19*, 1179 (1964).
27. Anderson, N.M. and P.S. Sekelj, *Phys. Med. & Biol. 12*, 173, 185 (1967).
28. Polanyi, M.L. and R.M. Hehir, *Rev. Sci. 33*, 1050 (1962).
29. Ziljstra, W.G., "A Manual of Reflection Oximetry," Van Gorcum, Assen, Netherlands (1958); K. Kramer, K. Graf, W. Overbeck and H. Zaun, *Pflug. Arch. Physiol. 262*, 285 (1956).
30. Enson, Y., W.A. Briscoe, M.L. Polanyi and A. Cowin, *J. Appl. Physiol. 17*, 552 (1962).

THEORIES OF SCATTERING FROM WIRE GRID AND MESH STRUCTURES

James R. Wait
Cooperative Institute for Research in Environmental Sciences
University of Colorado

An analytical method for plane wave scattering from crossed wire grids is described. The two periodic grids of thin wires can be located in planes that are parallel to the surface of a flat homogeneous ground or half-space. By applying an impedance boundary condition at the individual wires, a doubly infinite set of linear algebraic equations are obtained for the coefficients of the current harmonics. Both perturbation and matrix inversion techniques can be used to solve these equations. However, when the inter-grid space is allowed to vanish corresponding to a bonded mesh, the perturbation method fails and the matrix inversion procedure becomes unwieldy unless circuital relations at the wire junctions are invoked. Calculations are presented for a number of practically important situations. Of particular significance is that the bondings of the wire junctions may not be desirable from both the standpoint of interference protection and antenna ground system design. Such conclusions are found to be consistent with the results of Kontorovich and his Soviet colleagues who employ an entirely different method.

I. INTRODUCTION

Scattering from periodic structures is a subject that has a long and illustrious history dating back to Lamb and Rayleigh in the last century. The subject can often be treated as a special case of the general theory of the "multiple scattering of waves". This is the topic of a review paper by Professor Victor Twersky

ISBN 0-12-709650-7

included in this conference proceedings. Therefore, I will bow to the "master" and deal specifically with structures characterized by wires thin in the sense that only axial currents play a significant role. But I will also consider some of the implications of locating such periodic wire structures in the vicinity of the earth. This aspect of the subject, of course, is important in dealing with the design of grounding systems for antennas and related shielding problems.

An excellent survey of the theory of single parallel wire grids was published by Tove Larsen [2] in 1962. She made a very exhaustive study of grid structures consisting of thin parallel wires and pointed out some of the pitfalls of using equivalent circuit representations when the grid is located in proximity to an interface. This question had a bearing on the desing parameters for matching dielectric lenses and in radial-wire ground systems for monopole antennas.

Even in the case of single parallel wire grids, the proper treatment for general incidence is not trivial [1,3,4]. Most references to "oblique incidence" consider only cases where the electric or magnetic vector is parallel to the grid wires. Theorems are then sometimes quoted to indicate that extension to general incidence is merely a matter of replacing the radial wave number u by $u \cos\theta$ where θ is the angle subtended by the incident wave normal and the axial direction of the grid wires. As it turns out, this procedure is not valid for imperfectly conducting grid wires nor is it applicable when the grid is parallel to the interface of a contrasting half-space such as the ground. The coupling between the TE (transverse electric) and the TM (transverse magnetic) modes is an important ingredient of such problems and this certainly modifies the physical nature of the interaction with the environment.

When the grid wires are buried in the earth, the role of insulation must be considered explicitly. This seemingly complex problem can be handled in a fairly straightforward manner, particularly when the diameter of the insulation is small compared with the grid spacing [5]. Of special interest here is to examine how the performance of the grid as a shielding or grounding device is influenced by its distance to the air/earth interface [4]. We find, for example, that the reflection properties are very sensitive to this distance when the grid is elevated above the ground. But the dependence on the depth of the buried grid is weak. In spite of certain thin wire idealizations in the theoretical formulation, there is no practical difficulty in considering the limiting case when the grid is located in the interface itself.

We now come to a really vital aspect of the subject of shielding. If we use two parallel grids whose wires are perpendicular to each other, what can be said about the overall reflection and transmission properties as a function of the geometry? The first and most straightforward approach is to extend our earlier analytical approach but now allow the wire grids to have a suitable doubly periodic Floquet form [6,7]. Not surprisingly, this leads to a doubly infinite set of linear algebraic equations to solve for the unknown coefficients for the grid currents. If the grids are not too closely spaced, we find that a simple perturbation procedure can be used to determine the currents [7]. However, when the grids approach one another, the interaction becomes very strong and the resulting matrix inversion cannot be avoided [8]. The question of the convergence of this process and the method of truncation of the matrix has been considered [8,9]. In particular, the limit of vanishing separation (i.e. bonded mesh) requires some special consideration. Here it is found that invoking of the Kirchhoff current law and the continuity of charge at the wire junctions will speed the convergence. However, there are still some open questions here on how one should deal with grid wires barely touching or in the case of imperfect bonding caused by corrosion of the contracting wires. Some of these complications need further examination and study.

Scattering from meshes and wire grids in free space have also been treated by the method of "averaged" boundary conditions that have been developed by Kontorovich and his Soviet colleagues [10-13]. Their method can be used to predict the reflection and transmission properties of mesh structures when the interwire spacing is small compared with a wavelength. In agreement with our work, they find that an *unbonded* wire mesh will have *much better* shielding properties than a *bonded* mesh. The difference is particularly striking at grazing angles of incidence. Their technique also lends itself to wire mesh structures where the bonding is imperfect even though the crossed wires may be touching. In fact, intentional impedance loading at the wire intersections can also be handled by this quasi circuit theory approach. Its validity has not really been established, although limited experimental confirmation has been provided. In contradiction to Otteni [6], we have verified that the method of averaged boundary conditions is consistent with the rigorous analytical approach even when the interwire spacing is of the order of one-quarter wavelength.

Another topic that has been dealt with is the physical analogy between a square wire mesh and a thin plasma sheet insofar as a guided surface wave is concerned [14]. This is an example of how one can use wire mesh structures to simulate a thin ionized layer with or without collisions. The anisotropic unbonded wire mesh also has a limited correspondence with the magnetized plasma sheet.

Finally, we should mention that the present work provided insight to the mechanism that allows energy to leak (intentionally or nonintentionally) through a loosely braided shield of coaxial cable [15-23]. The interaction of the currents on the counterwound helices in such a model is treated in a similar fashion to the planar wire mesh when the propagation direction is at 45° to the axes of the crossed wires. But even here bonding may play a role if the excitation of the leaky coaxial cable is not symmetrical and/or there is close coupling with the environment.

In the following text we describe our analytical method and present some typical numerical results for the wire mesh or crossed wire grid structure located over a homogeneous conducting half-space. We then present a brief exposition of the Soviet work mentioned and describe some of the important results that seem highly relevant to the topic of this conference proceedings. Finally we add a brief analytical discussion of the impedance loaded wire grid.

II. GENERAL SCATTERING FORMULATIONS

Instead of dealing with the subject chronologically, it seems to be more convenient to begin with the configuration shown in Fig. 1. Many of the wire grid and mesh structures considered in the past are special cases of this geometry. As indicated, two parallel wire grids are located over a homogeneous conducting ground. The various parameters and the (x,y,z) coordinate system are clearly identified in the figure.

The individual wires are all the same radius c that is small compared with all other physical dimensions of the problem. As indicated below, we will apply an impedance boundary condition to these wires.

The wires in the grids are mutually perpendicular to each other and the grids are separated by a distance h between the wire centers. The top grid is located at a height d above the air/ground interface.

Spacing between the x directed wires is b while the spacing between the y directed wires is a.

The electric vector of the incident plane wave is specified by

$$\vec{E}^{inc} = \vec{E}_o \, e^{ik(x \cos\phi + y \sin\phi)S} \times e^{ikzC} \tag{1}$$

where

$$S = \sin\theta$$

and

$$C = \cos\theta$$

in terms of the angle of incidence θ. The free space wave number is k and ϕ is the angle subtended by the plane of incidence and the x directed grid wires.

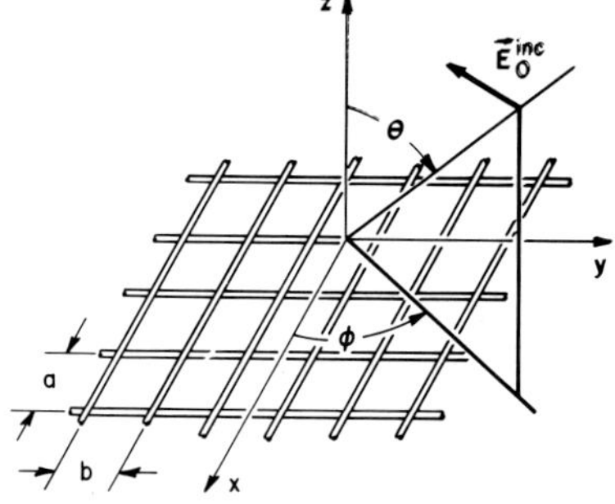

Fig. 1. Perpendicular wire grids parallel to a conducting half space.

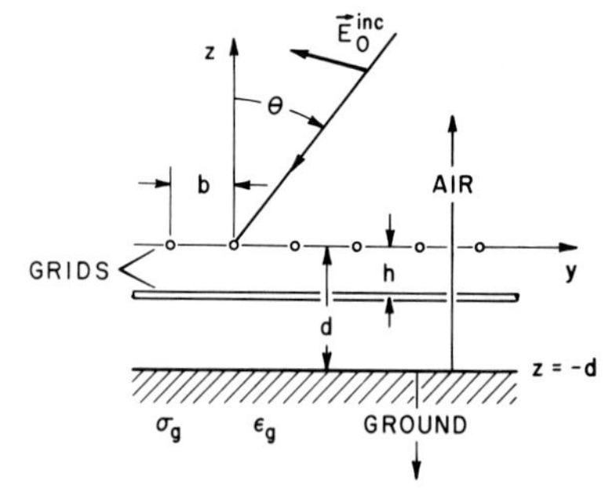

The excited grid currents are assumed to have th forms:

$$I_{xq} = \sum_{m=-\infty}^{+\infty} A_m e^{i2\pi mx/a} \times e^{ikS(x \cos\phi + qb \sin\phi)} \tag{2}$$

where

$$q = 0, \pm 1, \pm 2, \ldots$$

and

$$I_{yn} = \sum_{q=-\infty}^{+\infty} B_q e^{i2\pi qy/b} \times e^{ikS(y \sin\phi + ma \cos\phi)} \tag{3}$$

where

$$m = 0, \pm 1, \pm 2, \ldots$$

The fields due to these wire currents are obtained from:

$$\vec{E}^{w} = k^2\vec{\Pi} + \text{grad div}\vec{\Pi} - i\mu_o\omega \text{ curl } \vec{\Pi}^* \tag{4}$$

$$\vec{H}^{w} = k^2\vec{\Pi}^* + \text{grad div}\vec{\Pi}^* + i\varepsilon_o\omega \text{ curl}\vec{\Pi} \tag{5}$$

where $\vec{\Pi}$ and $\vec{\Pi}^*$ are the electric and magnetic Hertz vectors, respectively. We choose

$$\vec{\Pi} = (\Pi_x, \Pi_y, 0) \tag{6}$$

and

$$\vec{\Pi}^* = (\Pi_x^*, \Pi_y^*, 0) \tag{7}$$

where the half-space coupling is accounted for by $\vec{\Pi}^*$. But we also note that:

$$\Pi_{x,y} = \Pi\Big|_{\substack{\text{grid}\\ \text{currents}}} + \Pi\Big|_{\substack{\text{Grd.}\\ \text{Refl.}}} \tag{8}$$

In the case of a perfectly conducting ground or half-space, the magnetic Hertz vector would be zero.

The representations for the Hertz potentials have the form:

$$\begin{aligned}\Pi_x = -\frac{i\mu_o\omega}{2k^2b}\sum_m\sum_q A_m[&\exp(-\Gamma_{mn}|z|)\\ &+ R_{mq}\exp(-\Gamma_{mn}(z + 2d))]\Gamma_{mn}^{-1} \qquad \text{for } z > -d\\ &\times \exp[i\frac{2\pi m}{a} + kS\cos\phi)x\\ &+ i(\frac{2\pi q}{b} + kS\sin\phi)y]\end{aligned} \tag{9}$$

where

$$\Gamma_{mn} = i(k^2 - (\frac{2\pi m}{a} + kS\cos\phi)^2 - (\frac{2\pi q}{b} + kS\sin\phi)^2)^{\frac{1}{2}} \quad (10)$$

where the summations are over all integer values of m and q including zero. Similar forms for Π_x^*, Π_y and Π_y^* can be constructed and they need not be explicitly given. The coefficients $R_{m,q}$, etc. can be determined by the usual boundary conditions at the air-earth interface for each mode m,q. This process is carried out in an explicit fashion elsewhere [8a]. The next step is to determine the current amplitudes A_m and B_q from the impedance boundary conditions at the wires. These are:

$$E_x^{\text{total}} = I_{xq} Z_a \quad \begin{cases} y = qb \\ z = c \end{cases} \quad (11)$$

$$E_y^{\text{total}} = I_{ym} Z_b \quad \begin{cases} x = ma \\ z = -h-c \end{cases} \quad (12)$$

These couple the modes (spatial harmonics) m,q. The resulting equations can be solved by i) perturbation for large h and/or ii) truncation with matrix inversion [7,9]. To illustrate the nature of the problem, we write down the explicit form of the coupled equations for the square mesh ($a = b$) for bonded wires ($h \to 0$) for free space ($d \to \infty$):

$$\delta_{mo} E_{ox}\, e^{ikc\cos\phi} - A_m \left\{ \frac{i\mu_o\omega}{2k^2 a} [k^2 - (\frac{2\pi m}{a} + kS\cos\phi)^2] S_m \right\}$$

$$+ \frac{i\mu_o\omega}{2k^2 a} (\frac{2\pi m}{a} + kS\cos\phi) \sum_q B_q (\frac{2\pi q}{a} + kS\sin\phi) \frac{e^{-\Gamma_{mq}c}}{\Gamma_{mq}} = 0 \quad (13)$$

and

$$\delta_{qo} E_{oy}\, e^{ikc\cos\theta} - B_q \left\{ \frac{i\mu_o\omega}{2k^2 a} \left[k^2 - \left(\frac{2\pi q}{a} + kS\sin\phi\right)^2\right] S'_q \right\}$$

$$+ \frac{i\mu_o\omega}{2k^2 a}\left(\frac{2\pi q}{a} + kS\sin\phi\right) \sum_m A_m \left(\frac{2\pi m}{a} + kS\cos\phi\right) \frac{e^{-\Gamma_{mq} c}}{\Gamma_{mq}} = 0 \qquad (14)$$

where

$$S_m = \sum_q \frac{e^{-\Gamma_{mq} c}}{\Gamma_{mq}}, \quad S'_q = \sum_m \frac{e^{-\Gamma_{mq} c}}{\Gamma_{mq}}$$

and where

$$\delta_{mn} = \begin{cases} 1 (m = n) \\ 0 (m \neq n) \end{cases}$$

As indicated, this is a doubly infinite set. When the grids are located over a half-space and/or when they are separated, similar but more complicated forms arise [7,8a,8b]. Also, as indicated in these references, more convergent forms for the summations such as S_m and S'_q above can be constructed.

III. SOME CALCULATED RESULTS FOR MESH IN FREE SPACE

In Fig. 2 we show the essential geometry for describing the result of the scattered fields where the perpendicular ($\perp$) wire grids are in free space. The discussion is simplified if we restrict attention to the case where a and b are both less than the free-space wavelength. There are no grating lobes present, although we should emphasize that the theory is not restricted to this case.

The first thing we would like to show is the transmission coefficient $T_{\theta\theta}$ for the case where the electric vector is in the plane of incidence [9]. Such results are shown in Fig. 3 where the matrix of the coefficients is solved by direct matrix inversion after truncation. The two cases shown are depicted as

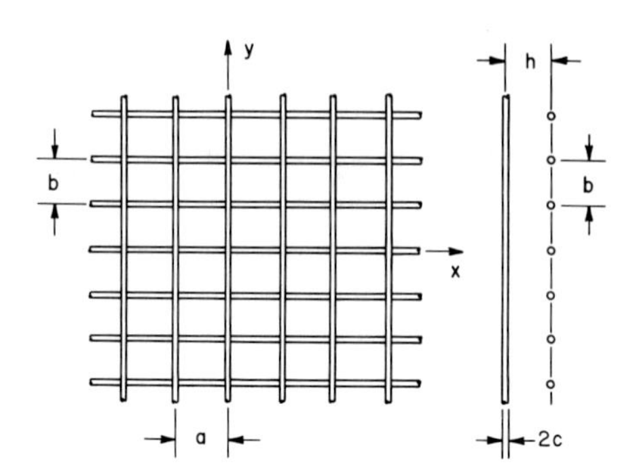

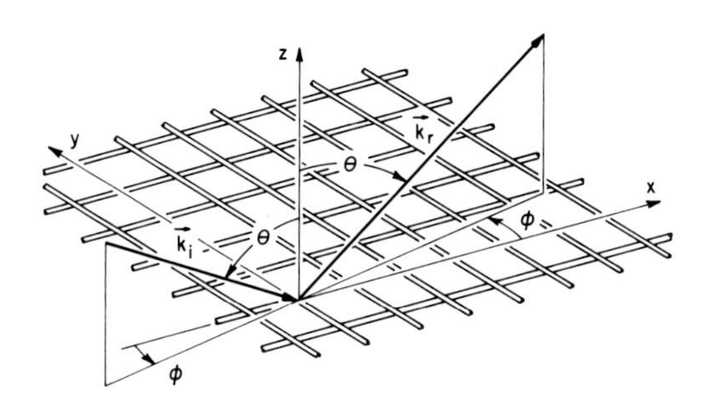

Fig. 2. The incident and reflected wave number vector $\vec{k}_i$ and $\vec{k}_r$ are shown. The corresponding reflecting coefficient matrix is

$$\begin{vmatrix} R_{\theta\theta} & R_{\theta\phi} \\ R_{\phi\theta} & R_{\phi\phi} \end{vmatrix}$$

where the first subscript denotes the polarization of the incident E vector and second subscript refers to the reflected E vector. The transmission coefficient matrix

$$\begin{vmatrix} T_{\theta\theta} & T_{\theta\phi} \\ T_{\phi\theta} & T_{\phi\phi} \end{vmatrix}$$

is defined in the same fashion.

bonded where the grid separation is set equal to zero and unbonded where the grid separation is set equal to 3 wire radii. This graph shows very dramatically the very poor convergence of the matrix inversion solution in the case of bonded wire junctions. It is probable this fact was not fully appreciated by Otteni [6] in his work. The experimental results of Kontorovich et al [10] are consistent with our calculations as indicated in Fig. 3. It is possible that their data for "unbonded" wire junctions are influenced partly by intermittent wire contact since the calculated result for unbonded wires yields a lower transmission coefficient.

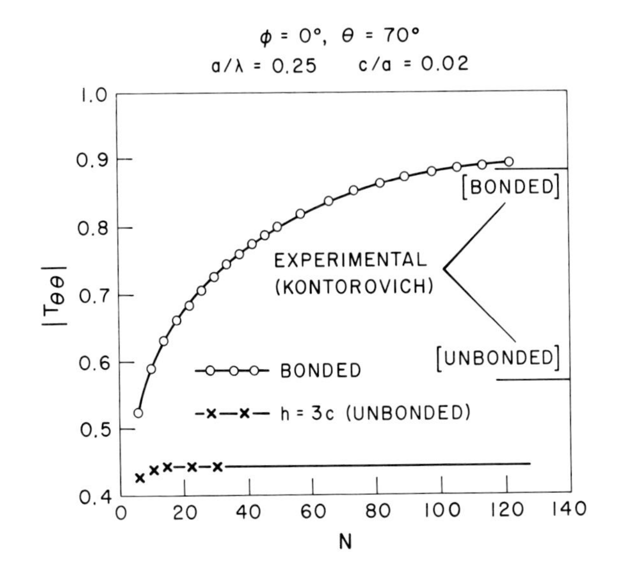

Fig. 3. Transmission coefficients versus number of unknowns N after truncation for direct matrix inversion for square mesh.

The matrix inve sion process can be checked by confirming that Kirchhoff's law and charge (density) continuity at wire junctions are satisfied. These can be written

$$I_{xo}(0^{+}) - I_{xo}(0^{-}) + I_{yo}(0^{+}) - I_{yo}(0^{-}) = 0 \qquad (15)$$

and

$$\left.\frac{\partial I_{xo}}{\partial x}\right|_{x=0^{-}} = \left.\frac{\partial I_{xo}}{\partial x}\right|_{x=0^{+}} = \left.\frac{\partial I_{yo}}{\partial y}\right|_{y=0^{-}} = \left.\frac{\partial I_{yo}}{\partial y}\right|_{y=0^{+}} \qquad (16)$$

Such equations have been used to improve the convergence of the original coupled set [9]. We stress, however, that such circuital relations are not mandatory. They are a consequence and not a basic ingredient of the boundary value treatment.

To illustrate the significant effects that take place at a bonded wire junction, we show in Fig. 4 current I_x on the x directed grid wire over one cell in the case of grazing incidence and where the E vector of the incident wave is perpendicular to the mesh. As indicated, there is a significant discontinuity or jump of the current at the junction. There is, of course, a

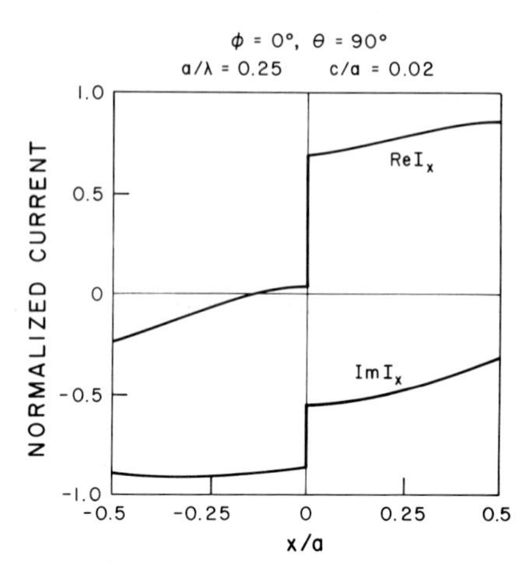

Fig. 4. The current jump or discontinuity at a junction for a bonded wire mesh for a "vertically polarized" wave at grazing incidence along the x directed wires.

corresponding jump of current in the y directed wires in such a manner that Kirchhoff's current law is satisfied.

In view of the practical importance of designing meshes to reduce the transmission of electromagnetic waves (i.e. interference protection) it is useful to exhibit some concrete calculated results for both bonded and unbonded meshes. These data are based on the above boundary value formulation [7,8a,8b,9]. Examples of such calculations are shown in Fig. 5 where the elements of the transmission matrix are shown plotted as a function of the angle ϕ subtended by the x directed wires and the plane of incidence. Both bonded (h = 0) and unbonded (h = 3c) cases are shown. The angle of incidence is 70° (measured from the normal to plane of the mesh) and the wire spacing is chosen such that $a = b = \lambda/4$. The important observation from these results is that $T_{\theta\theta}$ and $T_{\phi\phi}$ are virtually independent of azimuthal angle ϕ in the case of the bonded mesh. Also in this case the cross-polarized components are almost zero (i.e. $T_{\theta\theta} \simeq T_{\theta\phi} \simeq 0$ for $h = 0$). On the other hand, the unbonded mesh exhibits a significant variation with ϕ. Also, in this case, there are significant cross polarized components (i.e. $T_{\theta\phi}$ and $T_{\phi\theta}$ except when $\phi = 0°$ and 45°. In these two special cases the bonded and unbonded meshes behave almost in identical fashion.

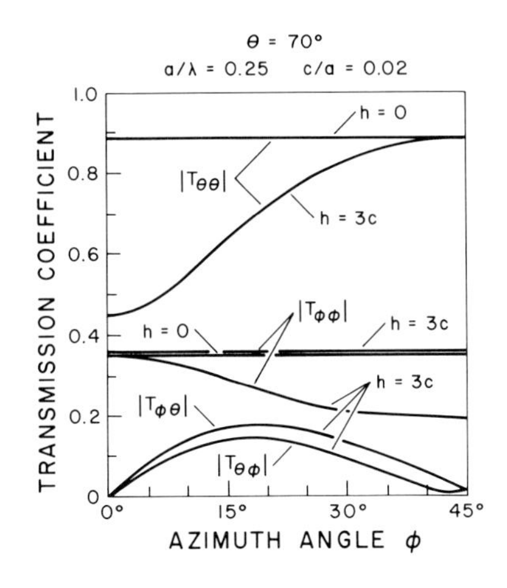

Fig. 5. The elements of the transmission coefficient matrix for a square mesh located in free space.

An important observation from the curves in Fig. 5 and from many other calculations of this type, is that the unbonded grid has a lower overall transmission than for the bonded grid of the same dimensions. We emphasize, however, this conclusion may not be applicable in all cases to wire meshes of limited extent. Also, the cross-polarized components are always larger for the unbonded mesh.

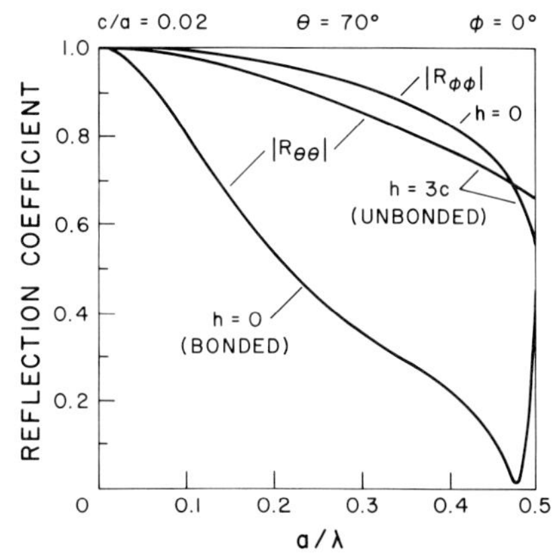

Fig. 6. The reflection coefficients as a function of the wire spacing for a square mesh located in free space.

Some similar calculations have been carried out for the reflection coefficient matrix. Examples are illustrated in Fig.6 where $R_{\theta\theta}$ and $R_{\phi\phi}$ (i.e. vertically and horizontally polarized reflection coefficients) are plotted as a function of a/λ for the square mesh. In this particular case, because $\phi = 0^\circ$, both $R_{\theta\phi}$ and $R_{\phi\theta}$ are identically zero. These curves show that $R_{\theta\theta}$ is markedly dependent on the condition of the bonding where $R_{\phi\phi}$ for $\phi = 0^\circ$, is virtually independent of such bonding. Also, $R_{\theta\theta}$ is generally much smaller for the bonded than for the unbonded mesh. This is consistent with our earlier remarks for the transmission properties. In fact, energy conservation can be easily checked [7,9] since, for all the calculations described here the grid wires themselves are assumed to be perfectly conducting (i.e. incident power = transmitted + reflected power). Specifically we require that

$$|R_{\theta\theta}|^2 + |R_{\theta\phi}|^2 + |T_{\theta\phi}|^2 + |T_{\theta\theta}|^2 = 1$$

and

$$|R_{\phi\phi}|^2 + |R_{\phi\theta}|^2 + |T_{\phi\phi}|^2 + |T_{\phi\theta}|^2 = 1$$

IV. INFLUENCE OF A CONDUCTING HALF-SPACE

The effect of locating the mesh over the ground surface is illustrated in Fig. 7 for the parameters indicated where the vertically polarized reflection coefficient is plotted as a function of the angle of incidence. The case for the homogeneous ground alone (i.e. absence of mesh) is shown by the solid curve. It exhibits the characteristic Brewster-angle minimum at an angle of approximately 18° from grazing. For comparison, the mesh located in free space is also shown for the two cases of $\phi = 0^\circ$ and 45° by the dashed curves. The mesh located over the ground at a height $d = 0.1a$, where $a = \lambda/20$ is then indicated by the broken curve. Strictly speaking, the calculations in Fig. 7 apply to an unbonded mesh since the grid separation $h = 3c$. However, the case for $\phi = 45^\circ$ is also applicable within graphical accuracy to a bonded square mesh for all angles of ϕ.

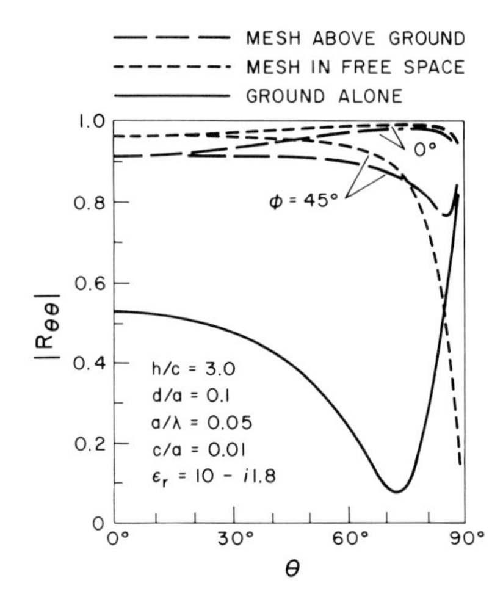

Fig. 7. Vertically polarized reflection coefficient for the three cases of: i) ground surface without mesh, ii) mesh in free space, and iii) mesh above ground surface.

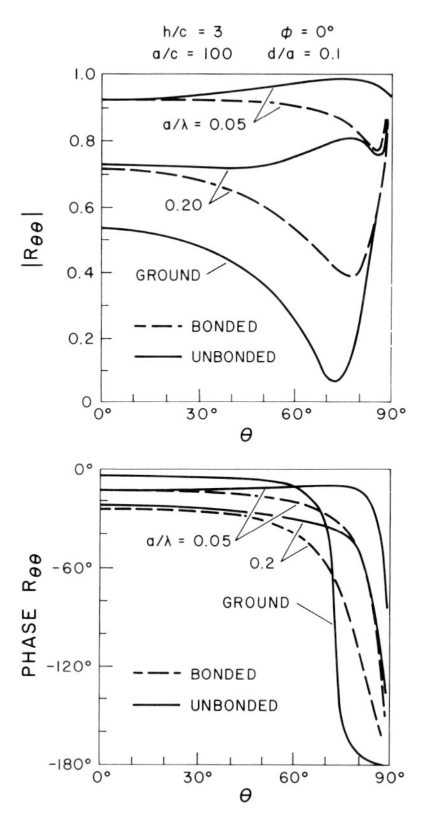

Fig. 8. The amplitude and phase of the reflection coefficient for a square mesh of interwire spacing a, located over a homogenous ground.

Additional calculations of the vertically polarized reflection coefficient for a square mesh are exhibited in Fig. 8 for various values of interwire spacing a. Again, the reflection coefficient for the ground of complex dielectric constant $\varepsilon_r = 10 - i1.8$ in the absence of the mesh is shown for comparison. Not surprisingly the smaller wire spacings are associated with higher reflection and smaller phase shifts. The calculations [8a] shown in Fig. 8 are actually carried out for a grid separation $h = 3c$ but, for $\phi = 45°$, the results would be indistinguishable from a bonded square mesh at all angles of ϕ. Thus we have labelled such curves in Fig. 9 as applying to bonded square mesh for $\phi = 0°$.

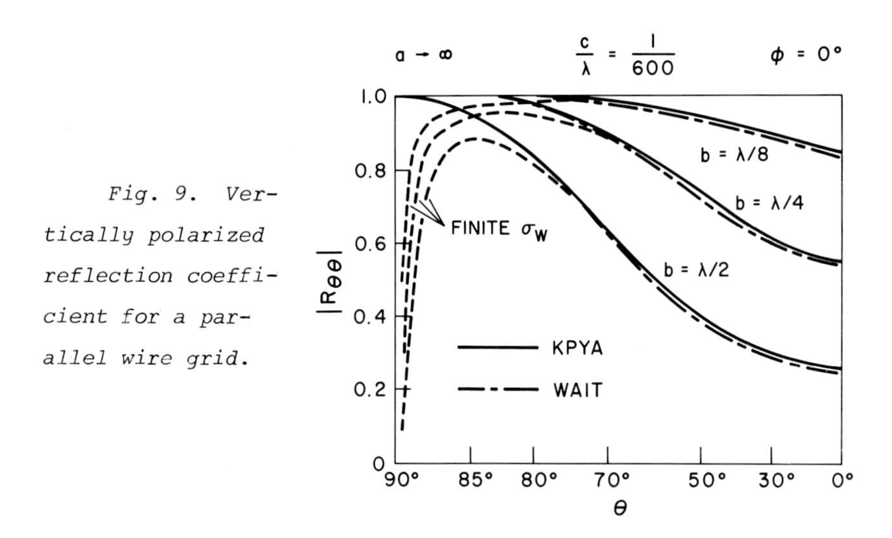

Fig. 9. Vertically polarized reflection coefficient for a parallel wire grid.

V. THE AVERAGE BOUNDARY CONDITION METHOD

Many of the problems discussed above have been considered by Soviet investigators. Following the relatively early analysis of Kontorovich [10,12], they use the method of averaged boundary conditions. We give here a brief exposition of the method and then summarize some of the results of the calculations for grid and mesh structures.

In Kontorovich's method [12] for a square mesh, we first need to define an average over a square cell of dimension as follows:

$$\overline{\phi}(x,y,z) = \frac{1}{a} \int_{x-a/2}^{x+a/2} \phi(x,y,z)\,dx \tag{17}$$

Thus, if ϕ is continuous

$$\frac{\partial \overline{\phi}}{\partial x} = \overline{\frac{\partial \phi}{\partial x}} \tag{18}$$

Now

$$\vec{E} = (k^2 + \text{grad div})\vec{\Pi} \tag{19}$$

where

$$\vec{\Pi} = \frac{1}{4\pi i \varepsilon_o \omega} \int_V \vec{J}\psi\,dv \tag{20}$$

and

$$\psi = \frac{e^{-ikR}}{R}\,(e^{i\omega t}) \tag{21}$$

is the appropriate Green's function for free space. We integrate the expression for $\vec{\Pi}$ by parts twice and change the order of differentiation and integration. Then

$$\vec{E} = -\frac{i\mu_o\omega}{4\pi}\left\{ \int_V [\vec{J} + \frac{1}{k^2}\ \text{grad div}\ \vec{J}]\psi\,dv \right.$$

$$- \frac{1}{k^2}\left[\sum_i \int_{S_i} [(\text{div}\ \vec{J})_1 - (\text{div}\ \vec{J})_2]\psi\vec{n}\ ds \right. \tag{22}$$

$$\left.\left. - \sum_i \text{grad} \int_{S_i} \left(J_n^{(1)} - J_n^{(2)}\right) \psi\ ds\right]\right\}$$

The sums are over all surfaces S_i at which there are discontinuities in current density and their first derivatives. Kontorovich [12] applies this to a planar mesh with a surface current density $\vec{j}(x,y)$. It is necessary to restrict the length of the element a to be much less than the free space wavelength λ. Furthermore, we need to assume a quadratic polynomial distribution for the currents on the wires within one cell:

$$I_x = a_{\kappa,\ell} + b_{\kappa,\ell}x' + c_{\kappa,\ell}(x')^2 \tag{23}$$

and

$$I_y = \alpha_{\kappa,\ell} + \beta_{\kappa,\ell}y' + \gamma_{\kappa,\ell}(y')^2 \tag{24}$$

where the integers κ,ℓ denote nodal or junction points at the corners of the cells. The coefficients $a_{\kappa,\ell}$, etc. are yet to be determined. We now invoke the following conditions:

- Sum of currents at nodal points are finite (Kirchhoff's law)
- Charge density has no discontinuities even at nodal points.

Thus we find that

$$a_{-1,0} + b_{-1,0}a + c_{-1,0}a^2 = -\alpha_{0,1} - \beta_{0,-1}a - \gamma_{0,1}a^2 + \alpha_{0,0} + a_{0,0} \tag{25}$$

$$b_{-1,0} + 2c_{-1,0}a = b_{0,0} \tag{26}$$

$$\beta_{0,1} + 2\gamma_{0,1}a = \beta_{0,0} \tag{27}$$

We may also connect an impedance Z between the conductors at a nodal point. Thus approximately

$$Z(a_{-1,0} + b_{-1,0}a + c_{-1,0}a^2 - a_{0,0}) \simeq -\frac{1}{\nu}(b_{0,0} - \beta_{0,0}) \tag{28}$$

where ν is a constant, i.e. potential difference, $\propto$ difference in charge density at the junction. We note that

$$Z \to 0 \text{ [bonded junction]}$$

$$Z \to \infty \text{ [Unbonded junction]}$$

Kontorovich's [12] final results for the "smoothed" fields in the plane of the mesh are:

$$\begin{aligned} \bar{\bar{E}}_x &\cong \frac{i\mu_o \omega a}{2\pi}\left(\ell n\,\frac{a}{c} - 1.85\right) \\ &\times \left\{ \bar{\bar{j}}_x + \frac{1}{k^2(2+\chi)}\,\frac{\partial}{\partial x}\left[\operatorname{div}\vec{\bar{\bar{j}}} + \chi\,\frac{\partial \bar{\bar{j}}_x}{\partial x}\right]\right\} \end{aligned} \tag{29}$$

and

$$\begin{aligned} \bar{\bar{E}}_y &= \frac{i\mu_o \omega a}{2\pi}\quad \ell n\,\frac{a}{c} - 1.85 \qquad \text{at } z = 0 \\ &\times \left\{ \bar{\bar{j}}_y + \frac{1}{k^2(2+\chi)}\,\frac{\partial}{\partial y}\left[\operatorname{div}\vec{\bar{\bar{j}}} + \chi\,\frac{\partial \bar{\bar{j}}_y}{\partial y}\right]\right\} \end{aligned} \tag{30}$$

where

$$\chi = Zi\omega C \qquad \text{and } C = \frac{2\pi\varepsilon_o a}{\ell n\frac{a}{c} - 1}$$

These field quantitites are double averaged as indicated by the double bar. The results were generalized to a rectangular grid ($a \neq b$) by Astrakhan [11], to an orthogonal grid by Kontorovich and Zhukov [24] and to a two-dimensional slot array by Zolotukhina [25].

Equations (29) and (30) are the conditions that can be used to describe the effective role of the square mesh with a prescribed impedance contact at the wire junctions. It is important to emphasize that this has been derived on the basis that the inter-wire spacing a is small compared with a wavelength. This, of course, is not a restriction in the direct boundary value method described earlier. Also, the averaged field method is not applicable when the mesh is located on or over a conducting half space.

Kontorovich, Petrunkin, Yesepkina and Astrakhan [10] in their development of the average field method, made a direct comparison with the rigorous solution by Wait [26] for a parallel wire grid at oblique incidence. The geometry here corresponds to that shown in Fig. 2 with the spacing a set equal to infinity (i.e. absence of y directed wires). The magnitude of the reflection coefficient $R_{\theta\theta}$ is shown plotted as a function of θ for interwire spacing b equal to $\lambda/8$, $\lambda/4$, $\lambda/2$. The average field result [10] is indicated by KPYA and seen to be very close to the rigorous solution. In this figure, we also show the typical modification of the results at near grazing angles for $\phi = 0°$ when the conductivity of the wires are taken into account [10,26]. The relatively close agreement between the average field solution and the rigorous method is somewhat fortuitous here. It is partly a consequence of the parallel wire grid assumption where the current varies smoothly along the wires. Some similar comparison calculations due to Astrakhan [13] are shown in Fig. 10. Here we again see a surprisingly close correspondence even when the interwire spacing in the grid is as large as one-half wavelength N. However, recent calculations [27] for the square mesh using both methods are not quite so optimistic. Nevertheless, provided $a/\lambda \geq 0.25$, it appears that the error in the reflection coefficient in the average field method is not more than 10% in magnitude and 10° in phase shifts.

VI. DISCUSSION OF WIRE MESH COMPUTATIONS USING THE AVERAGE FIELD METHOD

To illustrate the influence of various contact impedances at the wire junctions, results for $|R_{\theta\theta}|$ are shown in Figs. 11 and 12 based on Kontorovich's et al [10] calculations. Here the limit of the bonded mesh corresponds to $\chi = 0$ where the unbonded mesh is for $\chi = \infty$. Not surprisingly, when $\phi = 45°$, the

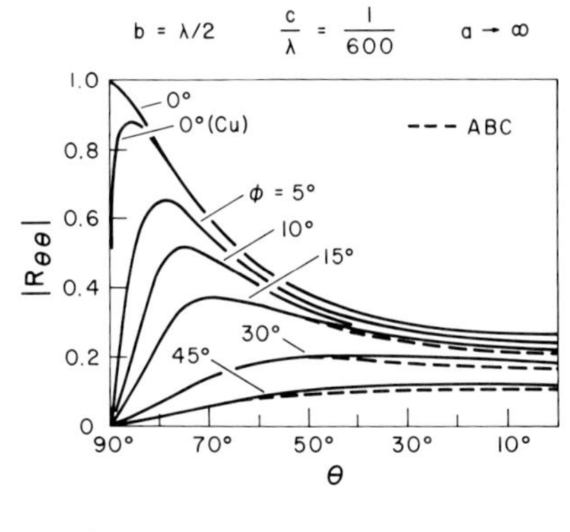

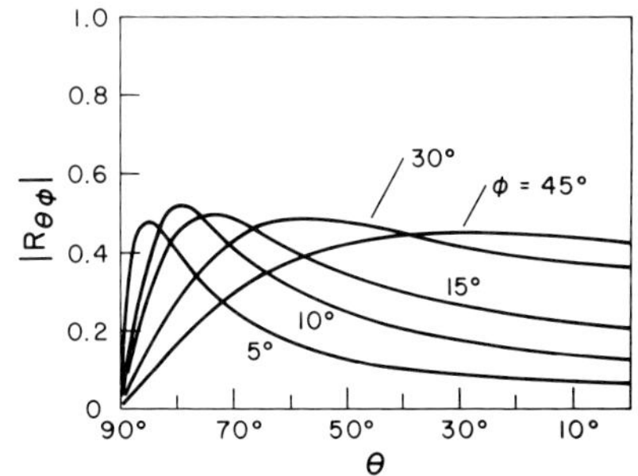

Fig. 10. Reflection and conversion coefficient for vertically polarized incidence on a wire grid at various angles ϕ. Comparisons of Wait's formulation with the average boundary condition results are shown. The curve for ϕ = 0° (Cu) designates that this is the only case where metal wires of finite conductivity such as copper would differ from the perfect conductivity assumption.

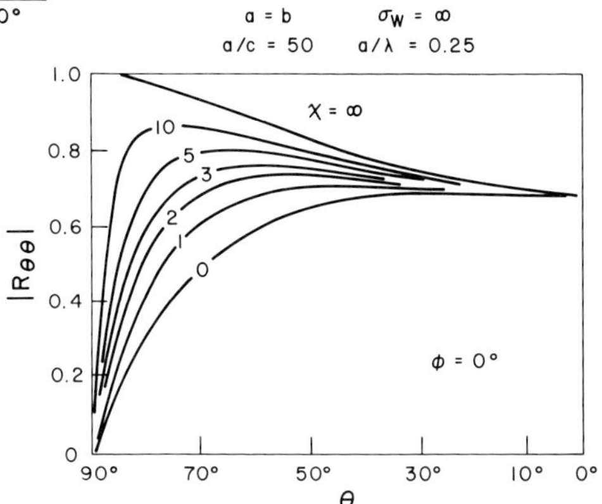

Fig. 11. Influence of the wire contact impedance on the reflection coefficient $R_{\theta\theta}$ for a square mesh for ϕ = 0 and 15°.

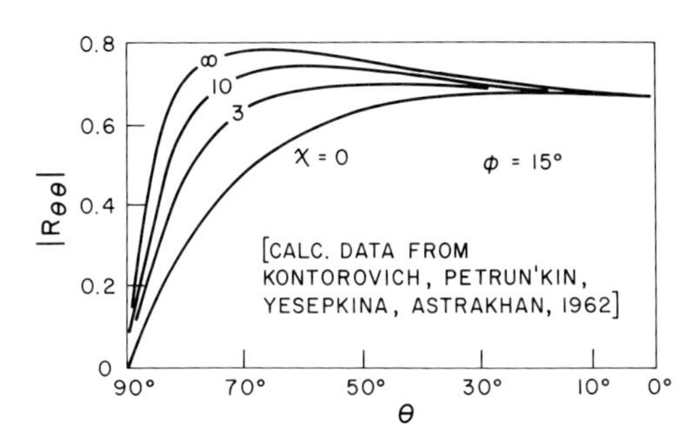

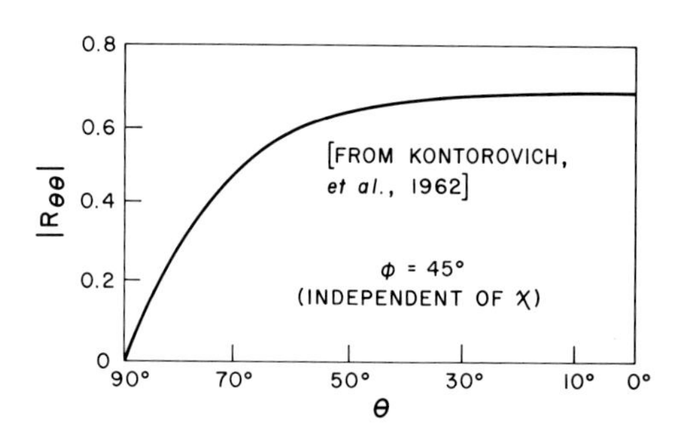

Fig. 12. Influence of the wire contact impedance on the reflection coefficient $R_{\theta\theta}$ for a square mesh for ϕ = 30 and 45°.

reflection coefficient is independent of the impedance contact. This is a special characteristics of a square mesh made up of identical wires.

Further square mesh calculations [13] of this type are shown in Fig. 13 for $\left|R_{\theta\theta}\right|$ and $\left|R_{\theta\phi}\right|$ and in Fig. 14 for $\left|R_{\phi\phi}\right|$ and $\left|R_{\phi\theta}\right|$. Of particular interest here is that $R_{\phi\phi}$ is independent of χ when $\phi = 0°$. It appears that these calculations are indeed appropriate for $a/\lambda = 0.25$ rather than $a/\lambda = 0.5$ as labelled by Astrakhan [13]. This fact was confirmed by a direct numerical calculation [27] for the two cases $\chi = 0$ and $\chi = \infty$.

A convenient presentation of Astrakhan's results is given in Figs. 15 and 16 to show the importance of the parameter $g = (2a/\lambda)\ell n(a/2\pi c)$. This format is valid in an approximate sense provided a/λ is somewhat less than $\lambda/4$. Also, the mesh wires are perfectly conducting although this would only make a difference in the case of the unbonded mesh for $R_{\theta\theta}$ and $\phi = 0°$ and near

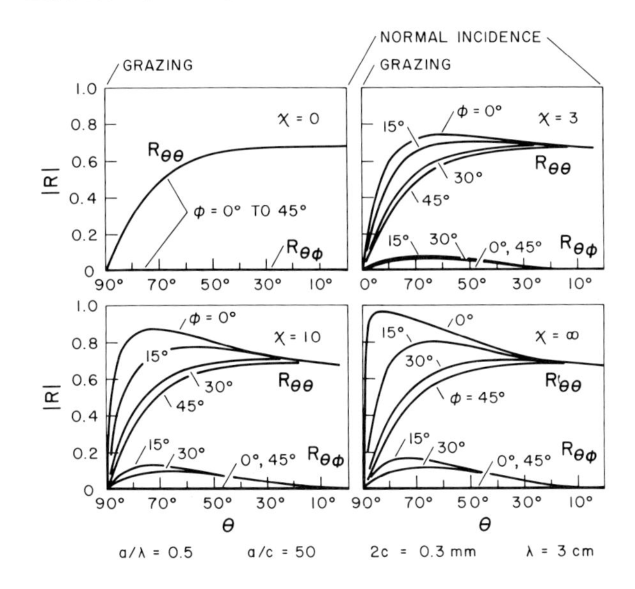

Fig. 13. *Reflection and conversion coefficients* $R_{\theta\theta}$ *and* $R_{\phi\phi}$ *for a square mesh for various impedance contacts at the wires.*

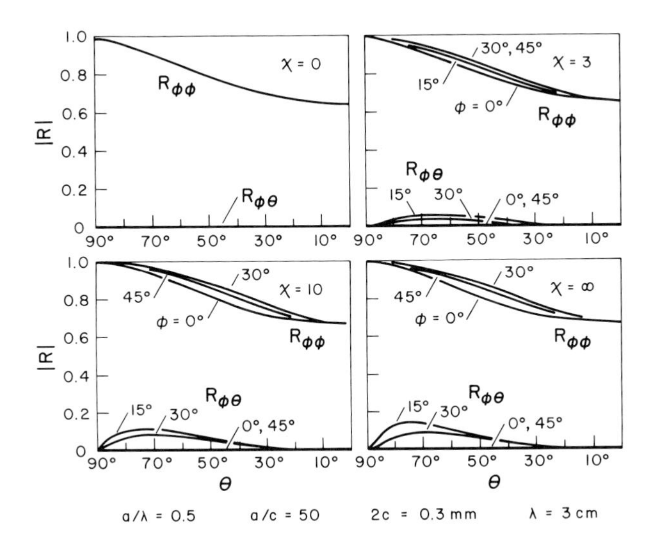

Fig. 14. *Reflection and conversion coefficients* $R_{\phi\phi}$ *and* $R_{\phi\theta}$ *for a square mesh for various impedance contacts at the wires.*

grazing incidence. The two cases shown in Fig. 16 again illustrate the important properties of square meshes. Namely, $R_{\phi\phi}$ is independent of the impedance contact if $\phi = 0°$ which $R_{\theta\theta}$ is independent of this parameter if $\phi = 45°$.

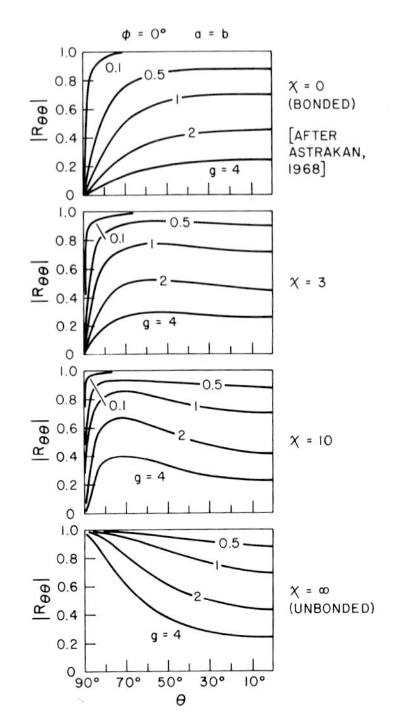

Fig. 15. $|R_{\theta\theta}|$ *as a function of the angle of incidence showing influence of parameter*

$$g = \frac{2a}{\lambda} \ln \frac{a}{2\pi c}$$

for a square mesh with perfectly conducting wires and with

$$a \leq \lambda/4.$$

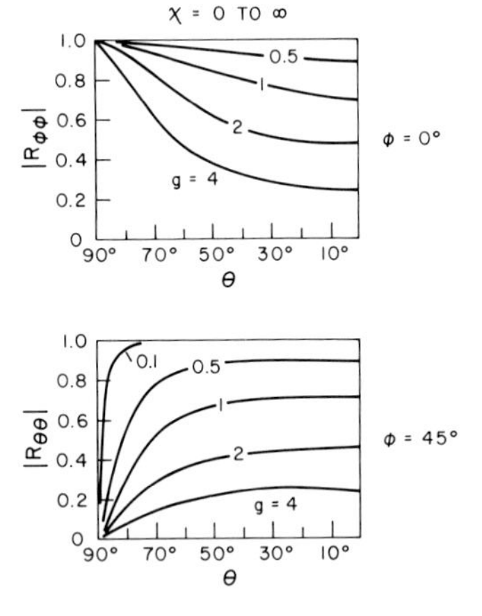

Fig. 16. $|R_{\phi\phi}|$ *and* $|R_{\theta\theta}|$ *as a function of the angle of incidence for two special cases. Note* $g = \frac{2a}{\lambda} \ln \frac{a}{2\pi c}$.

VII. SOME SPECIAL CONFIGURATIONS CONSIDERED BY THE SOVIETS

There have been several Soviet studies of grids in the vicinity of conducting and dielectric interfaces. For example, Vinichenko et al [28] considers the geometry indicated in Fig. 17. The reflection of an E polarized plane wave from this configuration was analyzed by Wait [29] over twenty years ago. The situation is an example where the evanenscent field of the grid

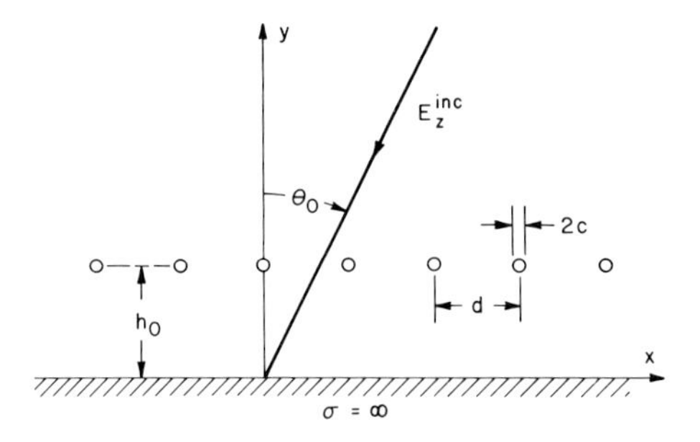

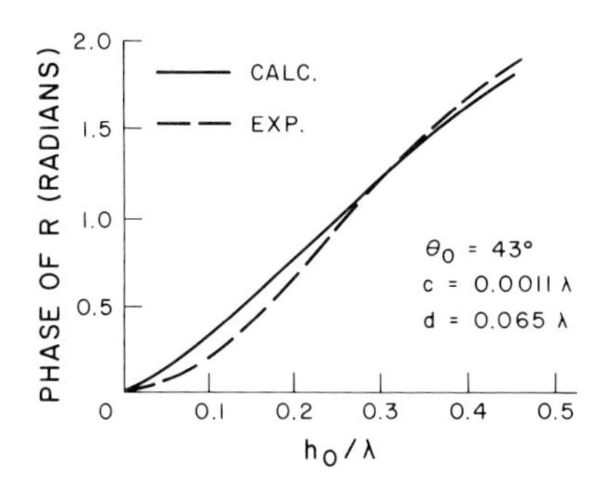

Fig. 17. The geometry considered by Wait (1954) and some recent calculated and experimental data compared.

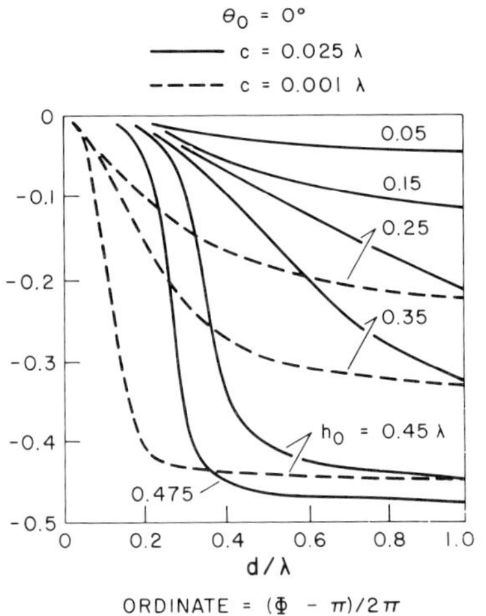

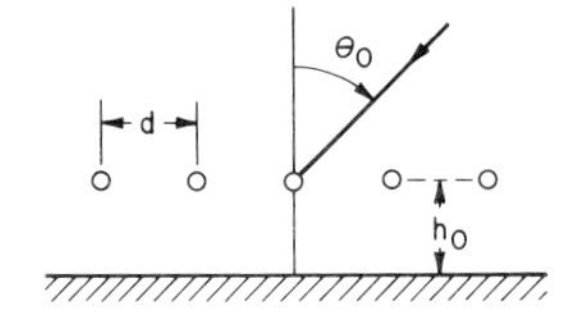

Fig. 18. The phase shift Φ for normal incidence as a function of various parameters for the configuration indicated.

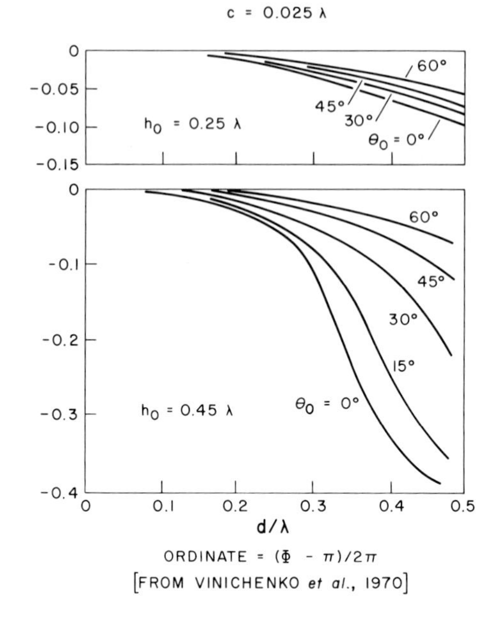

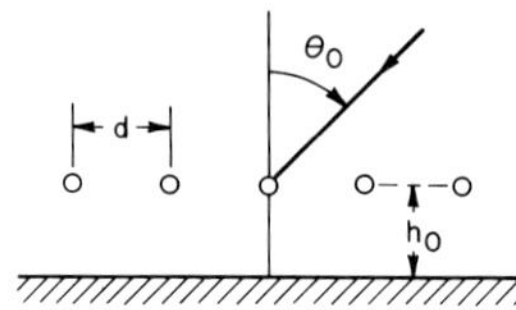

Fig. 19. The phase shift Φ for oblique incidence as a function of various parameters.

may interact with the adjacent ground plane. Earlier equivalent circuit representations of such problems neglected this effect. The experimental data [28] and the calculated results [29] are also shown in Fig. 17 for the conditions indicated. Here the phase of the resultant reflection coefficient is shown plotted as a function of the distance h_o from the grid to the conducting plane. The agreement is quite reasonable. The small departures are probably due to the assumption of vanishingly thin wires in the grid.

Further calculations [28] for the same configuration considered above are shown in Figs. 18 and 19 where Φ is the phase of the reflection coefficient. These curves indicate that the phase can be controlled over a wide range.

A related configuration shown in Fig. 20 was considered by Kvadadze et al [30]. Here the wire grid is located parallel to a dielectric slab of finite thickness. Using an extension of Wait's formulation [3] for a single interface, they computed the resultant transmission coefficient for a normally incident plane wave. Some of these results are indicated in Fig. 21 for various values of the complex dielectric constant $\varepsilon/\varepsilon_o$ of the slab. The ordinate here is equal to $|E_z^{trans}/E_z^{inc}|^2$ and the abscissa is the interwire spacing d. The dashed curve corresponds to the power transmittance with no slab present (i.e. $s \to 0$). Some experimental data [30] for the configuration are shown in Fig. 22 for two values of d. Again the agreement with the calculations seems quite reasonable.

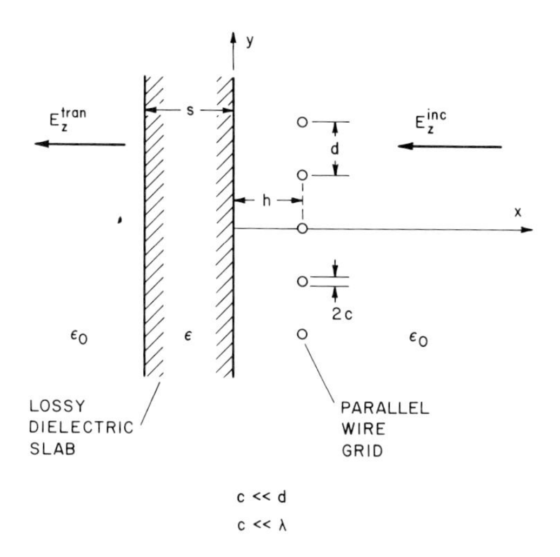

Fig. 20. Transmission of a plane wave through a wire grid/ dielectric slab combination.

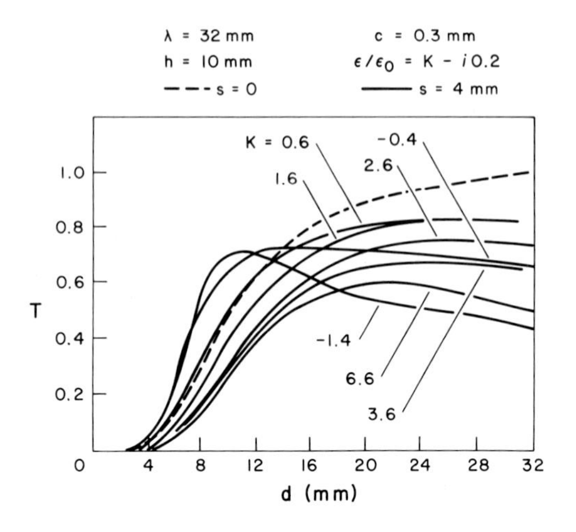

Fig. 21. Some calculated results of the power transmittance due to Kvavadze et al (1970) for the configuration indicated in Fig. 20.

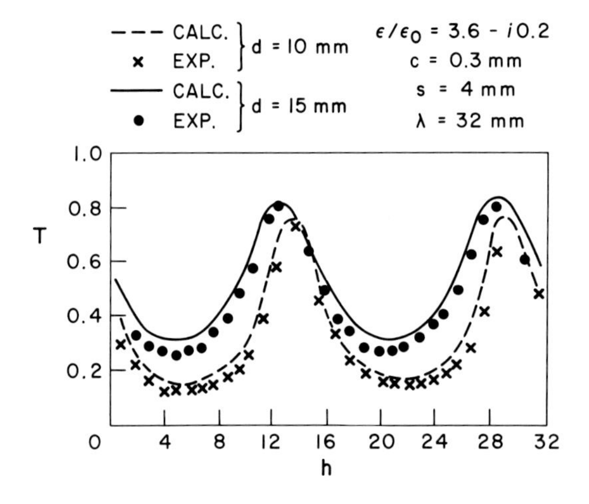

Fig. 22. Comparison between calculated and experimental data for configuration indicated in Fig. 20.

VIII. PERIODICALLY LOADED WIRE GRID

A useful configuration to tie together some of the concepts raised in the previous discussions is a wire grid that is periodically loaded with impedance elements. The configuration is indicated in Fig. 23. The interwire spacing is b while the elements are spaced at a distance a. We outline the pertinent analysis for this case.

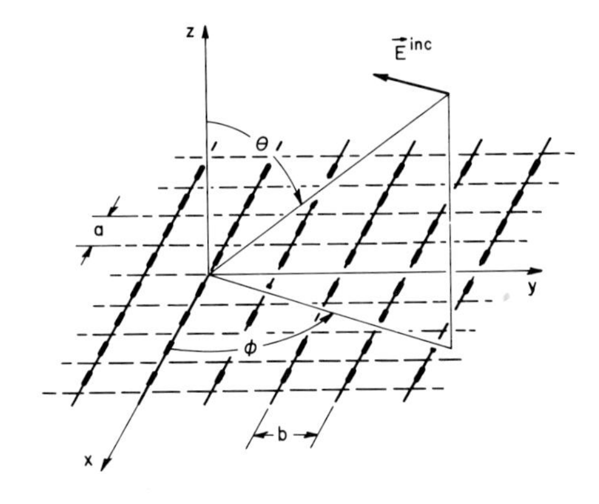

Fig. 23. Periodically loaded wire grid.

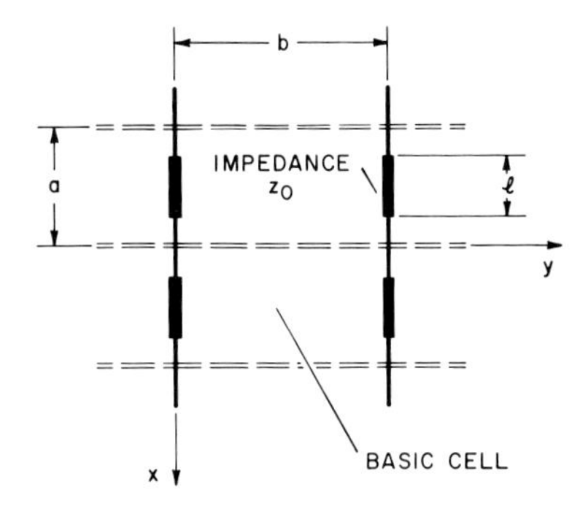

Fig. 24. The idealization of the impedance loading by finite length sections of resistive wire.

In Fig. 24 we indicated a section of the loaded grid where we imagine that the impedance elements are finite sections of wires with a series impedance per unit length different from the intervening portions of the wires. For example, we can say that the effective series impedance $Z(x)$ for the wires of one cell is given by

$$Z(x) = Z_W \quad \text{for } 0 < x < \frac{a-\ell}{2}, \; = \frac{Z_o}{\ell} \quad \text{for } \frac{a-\ell}{2} < x < \frac{a+\ell}{2}, \qquad (31)$$

$$= Z_W \quad \text{for } \frac{a+\ell}{2} < x < a$$

where Z_W is the series impedance of the (unloaded) cylindrical wires.

Then

$$Z(x) = \sum_{n=-\infty}^{+\infty} Z_n \exp(i2\pi nx/a) \tag{32}$$

where

$$Z_n = \left(\frac{Z_o}{a} - \frac{Z_w \ell}{a}\right) \frac{\sin\frac{\pi n\ell}{a}}{\frac{\pi n\ell}{a}} e^{-i\pi n} + Z_w \tag{33}$$

$$\simeq \frac{Z_o}{a} e^{-i\pi n} + Z_w \quad (\text{for } \ell/a <<< 1)$$

We now consider the incident plane wave

$$E_x^{inc} = E_{ox} e^{ik \sin\theta \cos\phi\, x} e^{ik \sin\theta \cos\phi\, y} \tag{34}$$

$$\times e^{ik \cos\theta\, z} \quad (e^{i\omega t})$$

Then the excited grid currents given by

$$I_p = \sum_m A_m e^{i2\pi mx/a} e^{ik \sin\theta \cos\phi\, x} \tag{35}$$

$$\times e^{ik \sin\theta \sin\phi \cdot pb}$$

for the path wire. Then, the average current density over the grid is

$$j_x^{av} = \frac{1}{ab} \int_o^a \int_o^b j_x(x,y)\,dx\,dy = \frac{A_o}{b} e^{ik \sin\theta \cos\phi\, x} \tag{36}$$

$$\times e^{ik \sin\theta \sin\phi \cdot y}$$

The Hertz vector due to the grid currents has only an x component that is given by

$$\Pi_x = \frac{i\mu_o \omega}{2bk^2} \sum_m \sum_q A_m e^{i2\pi mx/a} \frac{e^{-\Gamma_{mq}|z|}}{\Gamma_{mq}}$$

$$\times\ e^{i2\pi qy/b}\ e^{ik\ \sin\theta\ \sin\phi\ y}\ e^{ik\ \sin\theta\ \cos\phi\ x} \tag{37}$$

$$\Gamma_{mq} = \left[\left(\frac{2\pi m}{a} + k\ \sin\theta\ \cos\phi\right)^2 + \left(\frac{2\pi q}{b} + k\ \sin\theta\ \sin\phi\right)^2 - k^2\right]^{\frac{1}{2}}$$

when, as usual, the summations extend over all integers from $-\infty$ to $+\infty$ including zero.

The boundary condition at grid wires are now applied in the usual manner. Thus,

$$E_x^{\text{total}} = Z(x) \cdot I_p(x) \tag{38}$$

for all x at $y = pb$ and $z = c$. We then obtain

$$A_m \hat{Z}_m + \sum_n{}' \ Z_n A_{m-n} = \delta_{mo} E_{ox} \tag{39}$$

where

$$\hat{Z}_m = \frac{i\mu_o\omega}{2k^2b}\left[k^2 - \left(\frac{2\pi m}{a} + k\ \sin\theta\ \cos\phi\right)^2\right] \tag{40}$$

$$\times \left[\frac{1}{\Gamma_{mo}} + \frac{b}{\pi}\left(\ell n\frac{b}{2\pi c} + \Delta_m\right)\right] + Z_o$$

and where

$$\Delta_m = \sum_q{}' \ \frac{2\pi}{b\Gamma_{mq}} - \frac{1}{|q|} \tag{41}$$

is a quasi-static correction (the prime indicates omission of the $q = 0$ term).

We now can define an effective transfer impedance at $z = 0$ by

$$E_x^{av} = Z_g\ j_x^{av} \tag{42}$$

Thus we find that

$$\frac{Z_g}{b} = \frac{E_{ox}}{A_o} - \frac{\eta_o}{2b}\left(\frac{1 - \sin^2\theta\ \cos^2\phi}{\cos\theta}\right) \tag{43}$$

A limiting case is when we neglect coupling, then

$$A_m = 0 \quad \text{for} \quad m \neq 0$$
$$= E_{ox}/\hat{Z}_o \quad \text{for } m = 0 \tag{44}$$

where

$$Z_o = \frac{\eta_o}{2b} \frac{1 - \sin^2\theta \cos^2\phi}{\cos\theta} \tag{45}$$

$$+ \frac{i\mu_o\omega}{2\pi}(1 - \sin^2\theta \cos^2\phi)(\ln \frac{b}{2\pi c} + \Delta_o) + Z_o \tag{46}$$

where

$$Z_o = \frac{z_o}{a} + Z_w$$

$$\Delta_o = \sum_{q=1}^{\infty} \frac{2\pi}{b\Gamma_{oq}} - \frac{1}{q} \tag{47}$$

and where

$$\Gamma_{oq} = \left[\left(\frac{2\pi q}{b} + k \sin\theta \sin\phi\right)^2 - k^2\right]^{\frac{1}{2}}$$

If Δ_o is neglected, we have

$$E_x^{av} \simeq \frac{i\mu_o\omega b}{2\pi} \ln \frac{b}{2\pi c}\left(1 + \frac{1}{k^2} \frac{\partial^2}{\partial x^2}\right) j_x^{av} + Z_o b\, j_x^{av} \tag{48}$$

This is consistent with results obtained by Kontorovich [12] and Zolotukhina [25] using the average field method. To illustrate the effect of coupling we assume here that

$$Z(x) = Z_o\left(1 + M \cos \frac{2\pi x}{a}\right) \tag{49}$$

Then, without difficulty, we find that

$$A_m P_m + A_{m+1} + A_{m-1} = \delta_{mo}\, 2E_{ox}/MZ_o \tag{50}$$

where

$$P_m = \frac{i\mu_o\omega}{k^2 b}\left[k^2 - \left(\frac{2\pi m}{a} + k\sin\theta\cos\phi\right)^2\right]$$
$$\times\left[\frac{1}{\Gamma_{mo}} + \frac{b}{\pi}\left(\ell n\frac{b}{2\pi c} + \Delta_m\right)\right]\frac{1}{MZ_o} + \frac{2}{M} \tag{51}$$

The solution of (50) is conveniently written

$$A_o = \frac{2E_{ox}}{MZ_o}\,\frac{1}{P_o + \chi_1 + \chi_{-1}} \rightarrow \frac{E_{ox}}{\hat{Z}_o} \quad \text{if } M \ll 1 \text{ and } \frac{b}{\lambda} \ll 1 \tag{52}$$

where

$$\chi_{\pm} = \frac{A_{\pm 1}}{A_o} = -\cfrac{1}{P_{\pm 1} - \cfrac{1}{P_{\pm 2} - \cfrac{1}{P_{\pm 3}}}} \tag{53}$$

is a rapidly convergent continued fraction.

IX CONCLUDING REMARKS

Much work still needs to be done if we are to understand fully how wire mesh structures interact with electromagnetic fields. As we have seen, the practice of bonding the wire junctions [31] may be questioned from the standpoint of both interference protection and antenna ground systems. Of course, the analytical model we employ does not consider edge effects. These would be important in many cases when dealing with meshes and grids of finite lateral extent.

X. ACKNOWLEDGMENT

I would like to thank Dr. David A. Hill for his contributions to this paper.

XI. REFERENCES

1. Twersky, V., *IRE Trans. AP-10,* 737-765 (1962).
2. Larsen, T. *IRE Trans. Microwave Theory Tech. MIT-10,* 191-201 (1962).
3. Wait, J.R., *Appl. Sci. Res. 7,* 355-360 (1958).
4. Wait, J.R., *IRE Trans. Antennas Propag. AP-10,* 538-542 (1962).
5. Wait, J.R., *Can. J. Phys. 50,* 2149-2157 (1972) and *52,* 2302-2303 (1974).
6. Otteni, G.A., *IEEE Trans. Antennas & Propag. AP-21,* 843-851 (1973).
7. Hill, D.A. and J.R. Wait, *Can. J. Phys. 52,* 227-237 (1974).

8a. Wait, J.R. and D.A. Hill, Electromagnetic Scattering by Two Perpendicular Grids over a Conducting Half Space, *Radio Sci. 11,* 725-730 (Aug./Sept. 1976).

8b. Hill, D.A. and J.R. Wait, Electromagnetic Scattering from an Unbonded Rectangular Wire Mesh Located near the Air-Ground Interface, *AFWL Interaction Note*, in press.

9. Hill, D.A. and J.R. Wait, *Can. J. Phys. 54,* 353-361 (1976).
10. Kontorovich, M.I., V.Y. Petrunkin, N.A. Yesepkina and M.I. Astrakhan, *Radio Eng. & Electron. Phys. 2,* 222-231 (1962).
11. Astrakhan, M.I. *Radio Eng. & Electron. Phys. 8,* 1239-1241 (1964).
12. Kontorovich, M.I., *Radio Eng. & Electron. Phys. 8,* 1242-1245 (1964).
13. Astrakhan, M.I., *Radio Eng. 23,* 76-83 (1968).
14. Hill, D.A. and J.R. Wait, Electromagnetic Surface Wave Propagation over a bonded wire mesh, *AFWL Interaction Note*, in press.
15. Casey, K.F., Effects of Braid Resistance and weatherproofing Jackets on Coaxial Cable Shielding, *AFWL Interaction Note 192* (August 31, 1974).
16. Salt, H., The Surface Transfer Impedance of Coaxial Cables, *IEEE/EMC Symp. Record* (1968).
17. Vance, E.F., *IEEE Trans. Electromagn. Compat. EMC-17,* 71-77 (1975).
18. Whitmer, R.M., *IEEE Trans. Electromagn. Compat. EMC-15,* 180-187 (1973).
19. Wait, J.R. and D.A. Hill, *IEEE Trans. Microwave Theory & Tech. MTT-23,* 401-405 (1975).
20. Wait, J.R., Electromagnetic Theory of the Loosely Braided Coaxial Cable, Part I, *AFWL Interaction Note 266* (Feb. 23, 1976).
21. Wait, J.R., Electromagnetic Field Analysis for a Coaxial Cable with Periodic Slots, *AFWL Interaction Note 265* (Feb. 23, 1976).
22. Mahmoud, S.F. and J.R. Wait, *IEEE Trans. Commun. COM-24,* 82-87 (1976).
23. Wait, J.R. and D.A. Hill, Influence of Spatial Dispersion of the Shield Transfer Impedance of a Braided Coaxial Cable,

IEEE Trans. Microwave Theory & Tech. in press (1976).

24. Kontorovich, M.I. and A.D. Zhukov, *Radio Eng. & Electron. Phys. 18,* 1793-1799 (1973).
25. Zolotukina, N.M., *Radio Eng. & Electron. Phys. 20,* 112-114 (1975).
26. Wait, J.R., *Appl. Sci. Res. B4,* 393-400 (1954).
27. Hill, D.A. and J.R. Wait, Analytical and Numerical Procedures for Electromagnetic Scattering Problems Involving Wire Mesh Structures, presented at the National Conference on Electromagnetic Scattering, Chicago, Illinois, (June 15-18, 1976).
28. Vinichenko, Yu. P., I.N. Zakaryev and A.A. Lemanskiy, *Radio Eng. & Electron. Phys. 15,* 2196-2203 (1970).
29. Wait, J.R., *Can. J. Phys. 32,* 571-579 (1954).
30. Kuadadze, D.K., V.P. Kopaleyshvili and R.S. Popovidi, *Radio Eng. & Electron. Phys. 15,* 1184-1188 (1970).
31. White, D.R.J., "A Handbook on Electromagnetic Shielding Materials and Performance," p. 2.19-2.27. Don White Consultants, Inc., Germantown, Maryland (1975).

ELECTROMAGNETIC SCATTERING

ELECTROMAGNETIC INVERSE PROBLEM

Vaughan H. Weston
Division of Mathematical Sciences
Purdue University

The various problems and techniques associated with the inverse problem of Maxwell's equations, the wave equation, or other related equations are reviewed. Special emphasis is placed upon the determination of the properties of non-homogeneous media in multidimensional space, from various observed data, as opposed to the determination of the shapes of obstacles.

There are various asymptotic techniques for the determination of the refractive index $n(x)$ associated with the wave equation

$$\Delta u = \frac{n^2}{c^2} u_{tt}$$

in three dimensional space. As originally formulated in geophysics, the time of arrival of pulses at various locations, generated by sources at different locations is used to formulate the problem in terms of a non-linear functional equation. In practice the problem is linearized by treating $n(x)$ as a linear combination of an unknown perturbed and a prescribed unperturbed part. The non-uniqueness of the resulting equation is used by Backus and Gilbert to develop an optimum strategy when the measured data is limited.

The scattering problem for a non-homogeneous medium at a fixed frequency is formulated in terms of a non-linear problem. This non-linear formulation leads to some previous results when the appropriate asymptotic approximation is made. When the Born approximation is made for almost transparent media, previous

ISBN 0-12-709650-7

results of Wolf are obtained, in which the inverse problem is connected with holography.

The Gelfand-Levitan type of equation has been formulated directly in the time domain for the propagation of waves in an anisotropic half-space, or a lossy slab, where in each case the coefficients of the equation are functions of one spatial variable only. Recently, Faddeev has outlined a generalization of the Gelfand-Levitan treatment for the three dimensional Schrödinger equation.

Not all exact treatments of the inverse problem have led to linear equations to be solved. Half space ($z \geq 0$) problems with boundary conditions, data and a point source specified at the interface $z = 0$, for the telegraph equation or the wave equation with absorption have led to non-linear integral-differential equations. To determine the coefficients of the equations which are functions of z only, contraction mapping techniques are employed.

I. INTRODUCTION

A review of various recent techniques associated with inverse problems of non-homogeneous media are presented. Although the main concern here is the vector inverse problem associated with Electromagnetic Scattering, the inverse problem for the scalar wave equation and similar equations are presented, since in many instances the techniques can be immediately generalized to the vector problem. In other cases, the vector Electromagnetic inverse scattering problem can be reduced to a scalar problem. An example of this is the case of a stratified half-space, with a dipole situated on the boundary plane and oriented perpendicular to this plane. In this connection the techniques of Blagovenskii and Buzdin discussed in the next section could be applied.

Because of the amount of different ideas that will be discussed here, the review of various methods will be somewhat terse, and to obtain a complete understanding the reader should go to the references cited. However it is hoped enough detail is given in order to obtain an idea of the approach used.

In the next section, a review of some exact techniques associated with problems formulated in the time domain directly (as opposed to the frequency domain) are given. The majority of these problems either are one-dimensional in the spatial variable to begin with, or are such that they can be reduced to such, and many of them led to some generalizations or modifications of the Gelfand-Levitan type equations. Others are non-linear and require appropriate mathematical techniques.

In section III, the inverse problem for the time dependent wave equation associated with the non-homogeneous index of refraction $n(\underline{x})$ is formulated in terms of the arrival times, at various locations on a surface, of signals, generated by point sources at different locations on the surface. Although this is a long-standing prolem (in geophysics), some recent practical techniques have been developed, for the case where there are a finite number of measurements (measurement set is incomplete), and there are errors in the measurements. The concepts used here can be generalized to Electromagnetic scattering problems, and in particular, as will be mentioned, to fixed frequency problems given in the next section.

In section IV, the scattering problem for a non-homogeneous medium at a fixed frequency is formulated. It is pointed out that this is a non-linear problem. The non-linear formulation leads to some previous results when the appropriate asymptotic approximation is made. When the Born approximation is used for almost transparent media, previous results of Wolf are obtained, in which the inverse problem is connected to Holography. When the scatterer is a smooth convex obstacle, high frequency asymptotic results lead to the results of Bojarski. It is pointed out that the general non-linear problem, and its corresponding vector representation, needs to be explored further.

In section V, a brief review of a recent generalization to

three dimensions of the Gelfand-Levitan equation associated with inverse scattering potential, is given. Apart from some mathematical problems involving conditions on the potential that have to be resolved, the results when extended to two dimensions can be applied to Electromagnetic scattering by a cold plasma cylinder with polarization parallel to the axis.

II. EXACT NON-STATIONARY INVERSE PROBLEMS

There has been considerable development on the exact non-stationary inverse problem, that is inverse problems based directly upon quantities measured in the time domain, as opposed to the frequency domain or the spectral properties.

The Gelfand-Levitan equation for one-dimensional potential scattering has been generalized in many cases. Kay [1,2] obtained the Gelfand-Levitan time dependent equation for potential scattering by taking the Fourier transform of the Gelfand-Levitan equations expressed in the frequency domain. Time dependent approaches have been used by Sondhi and Gopinath [3], and by Balanis [4]. In the latter paper, the case of electromagnetic scattering of transverse waves incident upon a stratified cold plasma is treated, where in particular the equation considered is

$$\frac{\partial^2 E}{\partial z^2} - \frac{1}{c^2}\frac{\partial^2 E}{\partial t^2} - V(z)E = 0$$

where $V(z)$, the potential-like term, is related to plasma density. In this case, if $R(ct)$ the reflected wave, the Gelfand-Levitan equation to be solved is

$$R(z + y) + K(z,y) + \int_{-z}^{z} K(z,z')R(z' + y)dz' = 0, \quad -z < y < z$$

with $V(z)$ obtained from the relation

$$V(z) = \frac{2dK(z,z)}{dz} .$$

The inverse problem for the following more generalized system

$$u_{xx} - u_{tt} + A(x)u_x + B(x)u_t = 0$$

where the coefficients A, B are C^1 functions of x vanishing outside $0 < x < \ell$, has been treated by Weston [5], [6]. A generalization of the Gelfand-Levitan type equations was developed, involving the reflected coefficients $R_\pm$, and transmitted wave coefficients $T_\pm$ for waves incident upon the slab in positive x-direction (+ sign) and the negative x-direction (- sign). For an incident short pulse of the form of a delta function $u^i = \delta(x - t)$, the reflected wave for $x \leq 0$, will be given by $R_+(x + t)$ and the transmitted wave for $x \geq \ell$ would be given as the sum of a delta function component with amplitude $e^{-\gamma_-}$, and the step function $e^{-\gamma_-} T_+(x - t)$. Similar results hold for incident short pulse in the opposite direction.

The generalized Gelfand-Levitan equations developed are given by

$$g(x,t) + \int_x^t T_+(s - t)g(x,s)ds = -G(x)^{-1}T_+(x - t) + R_+(x + t)$$

$$- \int_{-x}^x R_+(y + t)f(x,y)dy \quad for \quad x \leq t \leq 2\ell - x,$$

$$f(x,t) + \int_t^x T_-(x - t)f(x,s)ds = T_-(x - t) - G(\ell)G(x)^{-1}R_-(x + t)$$

$$- G(\ell) \int_x^{2\ell-x} R_-(y + t)g(x,y)dy \quad for \quad -x \leq t \leq x.$$

The equations contain two unknown functions $g(x,t)$ and $f(x,t)$ which have to be found. The parameter $G(\ell)$ is given explicitly

$$G(\ell) = \exp[\gamma_- - \gamma_+]$$

and the unknown factor $G(x)$ in the equation is eliminated by use of the auxilliary equation

$$G^{-1}(x) = 1 + \int_{x}^{2x} R_{+}(s)ds - \int_{-x}^{x} f(x,y) \; [1 + \int_{0}^{x+y} R_{+}(s)ds]dy$$

If this is substituted into the above two integral equations, one obtains a coupled system of integral equations involving only the unknown functions $g(x,t)$ and $f(x,t)$.

The approach to the resulting integral equations is the same as the approach for the Gelfand-Levitan equations. Hold x *fixed* and solve the equations for variable t. Hence find $g(x,t)$ and $f(x,t)$ for a fixed x. Then change the variable x. Once $g(x,t)$ and $f(x,t)$ are found, one obtains the coefficients of the wave equation in the following manner.

(1) The function $G^{-1}(x)$ is determined from the auxilliary equation, once $f(x,t)$ is known. From this $B(x)$ may be determined by the relation

$$B(x) = - \frac{d}{dx}[\ell n \; G(x)]$$

(2) The coefficient $A(x)$ is then determined from the relation

$$\frac{2df(x,x)}{dx} + \frac{1}{2}B'(x) - \frac{1}{4}B^2(x) = -\frac{1}{2}A'(x) - \frac{1}{4}A^2(x)$$

The above results can be used to obtain the electrical properties of a stratified slab $0 \leq z \leq L$, from measurements associated with short electromagnetic pulses incident normally on the slab, provided that ϵ, σ are sufficiently smooth. In this case the equation for the time dependent electrical field is given by

$$\frac{\partial^2 E}{\partial z^2} - \epsilon\mu_0 \frac{\partial^2 E}{\partial t^2} - \sigma\mu_0 \frac{\partial E}{\partial t} = 0$$

where the permittivity $\epsilon(z)$ is a constant ϵ_0 for $z \leq 0$ or $z \geq L$, and $\sigma(z)$ vanishes there. With the change of variable

$$x = \int_0^z [\mu_0 \epsilon(s)]^{\frac{1}{2}} ds$$

the above equation reduces to that of the previous with

$$A(x) = -\frac{d}{dz}(\mu_0 \epsilon)^{-\frac{1}{2}}$$

$$B(x) = -\sigma/\epsilon$$

and the length ℓ given by

$$\ell = \int_0^L [\mu_0 \epsilon(s)]^{\frac{1}{2}} ds.$$

Once the generalized form of the Gelfand Levitan equation is solved for $A(x)$ and $B(x)$, $\epsilon(z)$ and $\sigma(z)$ are recovered, by first finding z as a function of x,

$$z = (\mu_0 \epsilon_0)^{-\frac{1}{2}} \int_0^x \exp[-\int_0^\tau A(s)ds]d\tau$$

then

$$[\epsilon(z)/\epsilon_0]^{\frac{1}{2}} = \exp \int_0^{x(z)} A(s)ds,$$

$$\sigma(z) = -\epsilon(z)B(x(z)).$$

For the case where the electrical properties are not sufficiently smooth, the techniques outlined above have to be considerably modified. Recent analysis of Kreuger [7] can be used to take into account discontinuities.

A set of half-space inverse problems have been treated by Blagovescenskii. These problems involve a second order partial differential equation or systems of equations in the variables (x_1, x_2, x_3, t) with the coefficients functions of x_1 only. Here the half space is given by $x_1 > 0$. In particular, Blagovenscenskii [8] treats seismic problem involving the set of three Lamb equations

$$(\lambda + \mu)\frac{\partial}{\partial x_i}\text{ div } u + \mu\Delta u_i + \frac{\partial\mu}{\partial x_1}\left(\frac{\partial u_1}{\partial x_i} + \frac{\partial u_i}{\partial x_1}\right) + \delta_{i1}\frac{\partial\lambda}{\partial x_1}\text{ div } u = \rho\frac{\partial^2 u_1}{\partial t^2}$$

where the unknows λ, μ, ρ are to be found as functions of x_1. The hyperbolic equation

$$c^2\frac{\partial^2}{\partial x_1^2} + 2b\frac{\partial^2 u}{\partial x_1 \partial x_2} + a\frac{\partial^2 u}{\partial x_2^2} = \frac{\partial^2 u}{\partial t^2}$$

with a, b, c to be found as functions of x_1 only, and similar equations are treated by Blagovescendkii [9]. In these papers with the initial value conditions imposed that all quantities vanish for $t < 0$, the boundary conditions and data are given on the interface $x_1 = 0$, where in particular a delta function source is imposed.

The main approach is to use the concept of moments with respect to a transverse variable, i.e., the nth moment of the quantity u is given by

$$U^{(n)}_{(x_1,t)} = \iint u(x_1, x_2, x_3, t)x_2^n dx_2 dx_3.$$

With this, the equations are reduced to a set of equations involving the variables x_1 and t only. Then by making an appropriate transformation from the variable x_1 to a new variable y, the equations for the moments are reduced to the form

$$U_{tt} = U_{yy} - K(y)U_y + F(y,t)$$

where the zero'th order moment satisfies the homogeneous system with $F = 0$, and with boundary conditions of the form at $y = 0$ (corresponding to $x_1 = 0$) $U = f(t)$, $\frac{\partial U}{\partial y} = \delta(t)$ where $f(t)$ is a given function. The equations for the higher moments are given by the non-homogeneous equation with the non-homogeneous term $F(y,t)$ expressed in terms of the higher order moments. Thus with this technique the system is reducible to a set of one dimensional problems, for which usual techniques can be employed. Various techniques are described for the one dimensional non-stationary problem associated with the first moment equation [10].

The moment approach is used [11] to reduce the inverse problem for the equation

$$u_{tt} = u_{xx} + u_{zz} + q(x,z)u$$

for the half-space $z > 0$, with $q(x,z)$ in the form

$$q(x,z) = \sum_{n=0}^{N} \alpha^n x^n q_n(z)$$

where α is some small parameter.

Buzdin [12] used the same concept of moments for the half-space ($x_1 > 0$) inverse problem associated with the wave equation with absorption,

$$u_{tt} + \sigma(x_1)u_t = c^2(x_1)\Delta u$$

where it is assumed that the unknown coefficient $\sigma(x_1) > 0$. Again the initial conditions are imposed implying that u vanishes for $t < 0$. As before a delta function source is prescribed

$$\frac{\partial u}{\partial x_1} = \delta(t)\delta(\underline{x}), \quad x_1 = 0$$

and the data to solve the inverse problem is given by the value of u at $x_1 = 0$

$$u = F(t, x_2, x_3).$$

Even though the moment technique simplifies the equation to a system of equation in the two variables x_1, t only, the inverse problem is not solvable in a simple linear fashion. Using the property of the characteristics, the inverse problem is reduced to a system of non-linear integral-differential equations of the general form

$$\varphi = K\varphi$$

where φ is a vector valued function of seven components. It is shown that for a sufficiently small domain of x_1, contraction mapping can be used to solve the system.

Romanov [13] also uses contraction mapping to solve an inverse problem when posed in form of a non-linear equation. In particular he treats the half-space ($x_1 > 0$) problem for the Telegraph equation

$$u_{tt} = \Delta u + a(x_1)u + 4\pi\, \delta(\underline{x},t)$$

associated with a delta-function source, initial conditions $u = 0$ for $t < 0$, and boundary condition $\frac{\partial u}{\partial x_1} = 0$ at $x_1 = 0$.

The inverse problem in this case is to determine the coefficient $a(x_1)$ which is a function of x_1 only, given the value of $u(0, x_2, x_3; t)$ at a fixed point on the interface. With an even extension of u, the problem is extended to the whole space. Then u is split into the two parts

$$u = 2\,\frac{\delta(t - |x|)}{|x|} + v(x,t)$$

and Kirchhoff's formula is used to obtain an integral equation for v and

$$w = \frac{(t^2 - |x|^2)}{t}\,\frac{\partial v}{\partial t}$$

If the value of v and v_t is known at a particular point on the interface, or more precisely if $f(t)$ is known where

$$v + w = f(t)$$

then an equation is developed in the form

$$a\left(\frac{1}{2}\sqrt{t^2 - |x|^2}\right) = f(t) + K(v, w_t).$$

This coupled with the equations for v, w yield a system of non-linear integral equations which is solved by contraction mapping.

III. TIME OF ARRIVAL TECHNIQUES

We will review techniques devised originally for geophysical applications, for the time dependent wave equation in three dimensional space. In particular we consider the initial-value problem

$$n^2(\underline{x})n_{tt} = \Delta u + \delta(\underline{x} - \underline{x}^0, t)$$

$$u\big|_{t<0} = 0$$

associated with a point source. The measured quantity of interest here is the arrival time $\tau(\underline{x}, \underline{x}^0)$ at the point $\underline{x}$ produced by the source at $\underline{x}^0$, given by

$$\tau(\underline{x}; \underline{x}^0) = \int_{\Gamma(\underline{x},x^0)} n(\underline{x})d|\underline{x}| \tag{1}$$

where the integral is the arc-length $ds = n(\underline{x})d|\underline{x}|$ taken over the geodesic from $\underline{x}_0$ to $\underline{x}$. The statement of the inverse problem follows.

"Let D be a compact simply-connected domain with boundary S. The problem is to determine $n(x)$ in D, if $\tau(\underline{x}, \underline{x}_0)$ is known for values of $\underline{x}$ and $\underline{x}_0$ on S."

In many applications the problem is simplified by restricting that $n(x)$ be a function of one variable only. For instance the domain D could be the half-space $z \geq 0$ and n is a function of z only. Then one would want to determine $n(z)$ from values of τ for $\underline{x}$ and $\underline{x}_0$ on the plane $z = 0$. It has been shown that the solution is unique for a class of functions $n(z)$ monotone in z. However, as pointed out by Gerver and Markusevic [14], there is non-uniqueness when $n(z)$ is not monotone. Some results for the non-one dimensional problem is given by Belonosova and Alekseev [15]. Originally the problem was formulated for a spherical domain with n as a function of r only. It was observed by Herglotz [16] and Wiechert [17] that if $rn(r)$ is a monotone increasing in function of r, the inversion amounts to solving a form of Abel's integral equation. For the case where the points $\underline{x}$ and $\underline{x}_0$ lie on a great circle of radius unity, and angular distance θ apart, the integral equation has the form

$$\tau = 2\int_{r_p}^{1} \frac{n^2(r)rdr}{[r^2n^2(r)-p^2]^{\frac{1}{2}}} \tag{2}$$

where r_p is the minimum radius reached by the ray, $r_p = r_p(\theta)$, and p is the ray parameter given by $p = r_p n(r_p)$. An elementary solution is given by Jeffreys [18].

When the geometry is more complicated or the measured data is limited, a practical approach in solving Eq (1) which is essentially non-linear in nature, is to linearize it. One treats $n(x)$ as perturbation of a known quantity $n_0(\underline{x})$. The linearized form is given by

$$\delta\tau = \int_{\Gamma^0(x,\, x^0)} \delta n \; d|x| \tag{3}$$

where δn is the Frechet derivative of n, and the integral is taken over ray path (or geodesic) Γ^0 corresponding to the unperturbed value $n_0(\underline{x})$. $\delta\tau$ is the difference in time travel between the unperturbed and perturbed signals.

If $n_0(\underline{x})$ is a function of one variable z, and D is the space $0 \le z \le H$, it is shown in Romanov [13] that the linearized problem has a unique solution for δn if certain monotonicity restrictions are placed on $n_0(z)$. For the case where δn is a function of three variables (x, y, z), one needs δT for all pairs of points $\underline{x}$, $\underline{x}^0$ lying lines which lie in the surface $z = 0$, and pass through one fixed point. Extensions to spherical geometry and many other cases of the linearized problem Eq (3) are given in Romanov.

The more practical problem consists of solving the linearized Eq (3) for δn when there is insufficient data. An approach has been developed by Bachus and Gilbert in a series of papers [19-22] for the spherical earth case. To briefly illustrate their approach we will consider the linearized version of Eq (2) given by

$$\delta\tau = 2\int_{r_p}^{1} \delta n \frac{n_0(r)r}{[r^2 n_0^2(r)-p^2]^{\frac{1}{2}}} dr.$$

Instead of solving for δn we will solve for the relative change

$$m(r) = \frac{\delta n}{n_0} .$$

The equation is then placed in the form

$$\delta\tau = \int_0^1 K(r)m(r)dr$$

where

$$K(r) = \begin{cases} 2r(r^2 n_0^2(r)-p^2)^{-\frac{1}{2}}, & r_p < r \leq 1 \\ 0 & , 0 \leq r < r_p \end{cases}$$

In order to get a square-integral data kernel, one sets

$$G(r) = \int_0^r K(r)dr$$

and thus one obtains

$$\delta\tau = m(1)G(1) - \int_0^r m'(r)G(r)dr$$

The problem consists of finding $n(r)$ or $m(r)$ when a finite set of data is given, i.e., the arrival time δT_i is given for a set of n different locations specified by the angular distance θ_i. For each case, the rays corresponding to the unperturbed model $n_0(r)$ will travel different routes characterized by the parameter p_i. Thus the kernel $G(r)$ will be different in each case, and with the set of n observations one has the following set of linear functional equations

$$\delta\tau_i = m(1)G_i(1) - \int_0^r m'(r)G_i(r)dr, \quad i = 1, 2, \ldots n.$$

Since $\delta\tau_i$ represents the difference in time travel between observed data and the calculated value associated with the unperturbed model of $n_0(r)$, $\delta\tau_i$ will contain some error due to error in the observed data. If the observed data has error of zero mean and variance σ_i^2, then the above equations can be written in the following form of inequalities

$$\gamma_i - \sigma_i \leq m(1)G_i(1) - \int_0^1 m'(r)G_i(r)dr \leq \gamma_i + \sigma_i.$$

The solution to this system is non-unique. Backus and Gilbert make use of this non-uniqueness to obtain an optimum choice of solution. They look for the solution that is closest in the least squares sense to the original or starting model. Thus the problem now is to find $m'(r)$ which lies in the subspace of $L_2(0,1)$ spanned by the vectors $G_i(r)$, and which minimizes

$$\frac{1}{2} \int_0^1 [m'(r)]^2 dr + \frac{1}{2}[m(1)]^2,$$

subject to the inequalities above. With the solution in the form

$$m'(r) = \sum \lambda_j \; G_j(r)$$

the problem is reduced to an algebraic system. Once the λ_j's are found, $m'(r)$ and hence $n(r)$ can be found.

The process is repeated. One takes as the new unperturbed index $n_0(r) + \delta n(r)$ and uses this to calculate the theoretical travel times and the data kernels $G_i(r)$. With these new values, a new choice of $\delta n(r)$ is computed as indicated above. Hence a sequence of δn are computed. The process terminates when the calculated travel times agree with the observed travel times to within a standard error.

The mathematical aspect of the above technique and some

extensions employing the simplex method of linear programming, together with the various possible strategies for optimization, has recently been considered by Sabatier [23,24].

IV. FIXED FREQUENCY

In contrast to the previous section we will consider the multidimensional inverse problem for a fixed frequency or monochromatic wave. In addition to referring to some previous efforts on the subject we will indicate some future areas of activity and the possible methods that could be carried out.

We will consider scalar wave equation for fixed frequency ω or wave number k

$$\Delta u + k^2 n^2(\underline{x})u = 0$$

where the index of refraction $n(x) = 1$ outside some bounded region representing the scatterer. The above equation is transformed into the integral equation

$$u(\underline{x}) = u^i(\underline{x}) + \frac{1}{4\pi} \int k^2[n^2(\underline{y}) - 1] \frac{e^{ikR}}{R} u(\underline{y})d\underline{y}$$

where $R = |\underline{x} - \underline{y}|$ and $u^i(\underline{x})$ represents the incident wave

$$u^i(\underline{x}) = \exp i(\underline{k}^i \cdot \underline{x})$$

with $\underline{k}^i$ the incident propagation vector of length k. In the direction indicated by the vector $\underline{k}^s$, the far scattered field is expressed in terms of the complex scattering amplitude $f(\underline{k}^i, \underline{k}^s)$ by the relation

$$u^s \sim \frac{e^{ik|x|}}{|x|} f(\underline{k}^i, \underline{k}^s)$$

with

$$f(\underline{k}^i, \underline{k}^s) = \frac{k^2}{4\pi} \int [n^2(y) - 1]e^{-i\underline{k}^s \cdot \underline{y}} u(\underline{y}, \underline{k}^i)d\underline{y}$$

Note that f may be obtained from measurements of the scattered field over a large sphere enclosing the scatterer or equivalently on two parallel planes enclosing the scattering. The latter result is indicated by E. Wolf [25,26] who in addition relates the scattering problem for transparent objects to holography. More comments of Wolf's results are given below.

If $f(\underline{k}^i, \underline{k}^s)$ is known for all directions of k^s for a fixed $\underline{k}^i$, then one can obtain

$$\tilde{f}(\underline{x}_1, \underline{k}^i) = \int e^{i\underline{k}^s \cdot \underline{x}_1} f(\underline{k}^i, \underline{k}^s) w(\underline{k}^i, \underline{k}^s) d\Omega_s$$

where $d\Omega_s$ is the element of solid angle in the scattered direction, $w(\underline{k}^i, \underline{k}^s)$ is a prescribed weight function; and $\underline{x}_1$ is an arbitrary point. It can be shown that the above reduces to

$$\tilde{f}(\underline{x}_1, \underline{k}^i) = \frac{k^2}{4\pi} \int [n^2(\underline{y}) - 1] e^{i\underline{k}^i \cdot \underline{y}} T[\underline{x}_1, \underline{y}_1, \underline{k}^i] d\underline{y} \tag{4}$$

where $T[\underline{x}_1, \underline{x}, \underline{k}^i]$ is the solution to

$$T[\underline{x}_1, \underline{x}, \underline{k}^i] - \frac{k^2}{4\pi} \int \frac{e^{ikR}}{R} [n^2(\underline{y}) - 1] T[\underline{x}_1, \underline{y}, \underline{k}_i] d\underline{y}$$

$$= \int e^{i\underline{k}^s \cdot (\underline{x}_1 - \underline{x})} w(\underline{k}^i, \underline{k}^s) d\Omega_s .$$

When the weight factor w is taken to be the identity, then the right-hand side of the above equation reduces to

$$\frac{4\pi \sin(k|\underline{x}_1 - \underline{x}|)}{k|\underline{x}_1 - \underline{x}|}$$

and T is independent of the vector $\underline{k}^i$.

For a fixed direction of incidence, Eq. (4) represents a nonlinear equation for $n(x)$ and is somewhat similar to the arrival time equations in the preceding section.

If in addition, the scattered field is known for all

directions of incidence, then one can form from observed data

$$F(\underline{x}_1, \underline{x}_2) = \int e^{-i\underline{k}^i \cdot \underline{x}_2} f(\underline{x}_1, \underline{k}^i) d\Omega_i$$

where x_2 is arbitrary and $d\Omega_i$ is the element of solid angle corresponding to the direction of incidence. When $\underline{x}_1 = \underline{x}_2$ the above has the form

$$F(\underline{x}_1, \underline{x}_1) = \int [n^2(\underline{y}) - 1]\, R(\underline{x}_1, \underline{y}) d\underline{y} \tag{5}$$

where

$$R(\underline{x}_1, \underline{y}) = \frac{k^2}{4\pi} \int e^{i\underline{k}^i \cdot (\underline{y} - \underline{x}_1)} T[\underline{x}_1, \underline{y}, \underline{k}^i] d\Omega_i$$

Since $R(\underline{x}_1, \underline{y})$ depends implicitly on $n(x)$, Eq. (5) is a nonlinear equation for $n(x)$ with the left-hand side a known function. The properties of this equation remain to be studied in detail. However, some preliminary results are given below.

If the weighting factor $w(\underline{k}^i, \underline{k}^s)$ is chosen as follows

$$w(\underline{k}^i, \underline{k}^s) = \frac{1}{(\pi\sqrt{2})^3} (k^2 - \underline{k}^i \cdot \underline{k}^s)^{\frac{1}{2}}$$

and if the Born (from the physicists standpoint) or Neumann approximation to T is employed, i.e.,

$$T[\underline{x}_1, \underline{x}, \underline{k}^i] \sim \int e^{i\underline{k}^s \cdot (\underline{x}_1 - \underline{x})} w(\underline{k}^i, \underline{k}^s) d\Omega_s$$

then it can be shown that

$$R(\underline{x}_1, \underline{x}) = \frac{k}{\pi^2 |\underline{x} - \underline{x}_1|} \left[\frac{\sin(2k|\underline{x} - \underline{x}_1|)}{2k|\underline{x} - \underline{x}_1|} - \cos(2k|\underline{x} - \underline{x}_1|)\right]$$

To point out the significance of this result note that $R(\underline{x}_1, \underline{x})$ given above, forms a delta sequence for a suitable class of test functions as $k \to \infty$, i.e.,

$$\lim_{k\to\infty} \int \psi(\underline{x})R(\underline{x}_1, \underline{x})d\underline{x} = \psi(\underline{x}_1)$$

Thus when the Born approximation is valid, $F(\underline{x}_1, \underline{x})$ yields a smooth approximation to $(n^2(\underline{x})-1)$. This result for the Born approximation was obtained by Wolf [26] who worked directly with the Born approximation initially and used Fourier Transforms. His result was expressed in terms of the Fourier Transform of $[n^2(\underline{x}) - 1]$.

However the Born approximation only holds for $k^2[n^2(\underline{x}) - 1]$ sufficiently small, hence one cannot take the limiting process as $k \to \infty$. An interesting question remains as to the behavior of $R(\underline{x}_1, \underline{x})$ as $k \to \infty$ for $[n^2(\underline{x}) - 1]$ small, and hence the behavior of the right-hand side of Eq (5). An attempt to answer this using the Rytov approximation was given in [27]. However, no precise estimates were given.

Another problem that arises when the scattered field is given for a few (N) directions of incidence, then one has a set of N non-linear equations of the form

$$f(\underline{x}_1, \underline{k}_n^i) = \frac{k^2}{4\pi} \int [n^2(\underline{y}) - 1]e^{i\underline{k}_n^i \cdot \underline{y}} T[\underline{x}_1, \underline{y}, \underline{k}_n^i]d\underline{y}$$

when $n = 1, 2, \ldots$ there the techniques of Backus and Gilbert may be employed, i.e.: perform a linear perturbation of the right-hand side. In this connection, it is interesting to note that the linearization about free space, $n(x) \equiv 1$, yields the Born approximation, i.e.:

$$\delta\tilde{f}(\underline{x}_1, \underline{k}_n^i)$$

$$= \frac{k^2}{4\pi} \int\int \delta(n^2(y))w(\underline{k}^i, \underline{k}^s)e^{i\underline{k}_n^i \cdot \underline{y} + i\underline{k}^s \cdot (\underline{x}_1 - \underline{y})} d\underline{y}d\Omega_s$$

In addition to investigating the scalar problem the

associated vector problem can be treated in the same way.

One interesting fact remains to be noted. Inverse scattering for fixed frequency from obstacles can be treated in the same manner as above. In this case the basic integral equation for the field quantities involves strictly the surface of the obstacle (whether Dirichlet or Neumann condition are involved). One can proceed as before and formulate the quantity $F(\underline{x}_1, \underline{x}_1)$ from the far scattered field. Without going into detail it can be formally shown, that if we take the weight function

$$w(\underline{k}^i, \underline{k}^s) = \frac{k^2}{(\pi\sqrt{2})^3} (k^2 - \underline{k}^i \cdot \underline{k}^s)^{-\frac{1}{2}}$$

and assume the obstacle is strictly a smooth convex shape, then

$$\lim_{k\to\infty} \text{Real } F(\underline{x}_1, \underline{x}_1) = \chi(\underline{x}_1)$$

where $\chi(\underline{x}_1)$ is the characteristic function of the body, defined by

$$\chi(\underline{x}_1) = \begin{cases} 1, & \text{for } \underline{x}_1 \text{ inside} \\ 0, & \text{for } \underline{x}_1 \text{ outside.} \end{cases}$$

This result is recognized as that obtained previously by Bojarski [27] using direct methods for the special case of scattering from convex smooth bodies. Thus the general formulation given here includes two special asymptotic results on the opposite extremities of the problem, namely the transparent body results of Wolf and the convex obstacle results of Bojarski.

V. MULTIDIMENSIONAL GELFAND-LEVITAN EQUATION

Recently Faddeev [28] has sketched a formal treatment of the Gelfand-Levitan Equation generalized to the multi-dimensional case (with emphasis on $n = 3$) for the inverse problem associated with the Schrodinger Operator

$$H = -\Delta + v(\underline{x})$$

The potential $v(\underline{x})$ is required to satisfy certain smooth conditions, vanish sufficiently rapidly as $|\underline{x}| \to \infty$, and most important, satisfy an implicit condition which Faddeev calls "Condition C". This will be stated below.

Before proceeding to the actual form of the generalized Gelfand-Levitan equation, some preliminary material is needed.

From Ikebe [29], the operator H given above has a continuous spectrum characterized by the vector $\underline{k}$, where $\underline{k}/|\underline{k}|$ takes on values on the unit sphere. Associated with the continuous spectrum are the "generalized eigenfunctions", $u^{\pm}$ which satisfy the stationary equation

$$\Delta u^{\pm} + k^2 u^{\pm} = v(x)u^{\pm}$$

with behavior as $|x| \to \infty$ given by

$$u^{\pm} = e^{i(k,x)} + \frac{e^{\pm i|k||x|}}{|x|} f^{\pm}(\underline{k},\underline{n}) + O\left(\frac{1}{|x|}\right)$$

where $\underline{n}$ is a unit vector in the scattered direction. In addition to the continuous spectrum, H may have discrete spectrum which is real and negative. In the development of the inverse problem, Faddeev treats the case where the potential is such that H has a single, simple eigenvalue $-\kappa^2$, with real eigenfunction u,

$$Hu = -\kappa^2 u.$$

The "generalized eigenfunctions" $u^{\pm}$ are related by

$$u^{+}(\underline{x},\underline{k}) = u^{-}(\underline{x},\underline{k}) - 2\pi i \int f(\underline{k},\underline{\ell})\delta(k^2 - \ell^2)u^{-}(\underline{x},\underline{\ell})d\ell$$

in terms of the scattering amplitude $f(\underline{k},\underline{\ell})$ which is defined by

$$f(\underline{k},\underline{\ell}) = \frac{1}{(2\pi)^3} \int e^{-i(\underline{\ell},\underline{x})} v(\underline{x})u^{+}(\underline{x},\underline{k})d\underline{x}.$$

For $|k| = |\ell|$, the scattering amplitude is related to $f^{+}(\underline{k},\underline{n})$ by

$$f^{+}(\underline{k},\underline{n}) = -2\pi^2 f(\underline{k},\underline{\ell}), \quad k^2 \quad \ell^2.$$

The "generalized eigenfunctions" $u^{\pm}$ give rise to tranformation operators $U^{\pm}$ with the property that

$$HU^{\pm} = U^{\pm}H_0$$

Where H_0 is the unperturbed operator H with $v = 0$.
The scattering amplitude can be used to define a scattering operator S such that $U^{\pm}$ are related by

$$U^{-} = U^{+}S.$$

The inverse problem consists of determining $v(x)$ from knowledge of the scattering amplitude $f(\underline{k},\underline{\ell})$ for $|k| = |\ell|$, the eigenvalues $-\kappa_1^2$ and some residue expressions associated with the eigenvalues. In order to obtain a Volterra integral equation for this, a Volterra transformation is required. Since $U^{\pm}$ do not have this property other transformation operators are required:

Faddeev [30,31] obtained a class of Volterra transformations U_γ characterized by the Volterra direction $\underline{\gamma}$, where $\underline{\gamma}$ is a unit vector. The form of the operator U_γ is given by

$$U_\gamma\psi(\underline{x}) = \psi(\underline{x}) + \int_{(y-x,\gamma)>0} A_\gamma(\underline{x},\underline{y})\psi(\underline{y})d\underline{y}$$

The existence of such an operator depends upon the potential $v(\underline{x})$ satisfying condition C, namely that the equation

$$h(\underline{x}) - \int G_\gamma(\underline{x}-\underline{y},\underline{k})v(\underline{y})h(\underline{y})d\underline{y} = 0$$

does not have a non-trivial bounded solution for all real $\underline{k}$.
Here G_γ is theGreen's function given by

$$G_\gamma = \frac{1}{(2\pi)^3}\int \frac{e^{i(\underline{\ell},\underline{x})}}{k^2 - \ell^2 + i0(\underline{k}-\underline{\ell},\gamma)}\,d\underline{\ell}$$

where $(x+i0\alpha)^{-1}$ is the distribution $(x + i0)^{-1}$ for $\alpha > 0$, and $(x - i0)^{-1}$ for $\alpha < 0$. The transformation operator U_γ is related to $U^{\pm}$ as follows

$$U_{\gamma} = U^{\pm}N_{\gamma}^{\pm}$$

where

$$N^{\pm}H_0 = H_0 N^{\pm}$$

The Gelfand-Levitan equation follows from the completeness relation for $u^{\pm}$ and the eigenfunctions u. Expressed in terms of operators $U^{\pm}$ and hence Volterra transformation operators U_{γ}, this is given by

$$U_{\gamma}W_{\gamma}U_{\gamma}^{*} = I$$

where the weight operator W_{γ} has the form

$$W = (N^{\pm})^{-1}(N^{\pm *})^{-1} \quad \chi \boxtimes \chi$$

when H has the single simple eigenvalue $-\kappa^2$ with eigenfunction u, and χ is given by

$$u = U_{\gamma}\chi.$$

Rewriting the expression in the form

$$U_{\gamma}W_{\gamma} = (U_{\gamma}^{*})^{-1}.$$

The explicit form of the Gelfand-Levitan equation has the form

$$A_{\gamma}(\underline{x},\underline{y}) + \Omega_{\gamma}(\underline{x},\underline{y}) + \int_{(z-x,\gamma)>0} A_{\gamma}(x,y)\Omega_{\gamma}(z,y)\,dz = 0$$

for $(x,y) < (y,x)$

where

$$\Omega_{\gamma}(\underline{x},\underline{y}) = W_{\gamma}(\underline{x},\underline{y}) - \delta(\underline{x}-\underline{y})$$

This is an equation for $A_{\gamma}(\underline{x},\underline{y})$ as a function of the variable y, with x and γ taking on the role of parameters.

The kernel of the equation is constructed from knowledge of

the scattering amplitude $f(\underline{k},\underline{\ell})$ for $|\underline{k}| = |\underline{\ell}|$ and the eigenvalues. To obtain it, one first has to solve the following integral equation

$$h(\underline{k},\underline{\ell}) = f(\underline{k},\underline{\ell}) + 2\pi i \int h_\gamma(\underline{k},\underline{m}) f(\underline{m},\underline{\ell})\, \Theta\,[(\underline{m} - \underline{k},\, \gamma)]\, \delta(m^2 - k^2)\, d\underline{m}$$

for the function $h(\underline{k},\underline{\ell})$ for $|k| = |\ell|$ and for arbitrary γ. In the above, $\Theta(x)$ stands for the Heaviside step function. Then the regularized fredholm determinant $\Delta_\gamma(\underline{k})$ is found in terms of $h(\underline{k},\underline{\ell})$ and the eigenvalues. The appropriate expressions will not be given here for this quantity. Faddeev gives it for the case mentioned where there is one simple eigenvalue. For the case where there are no eigenvalues the kernel of the weight operator is given by

$$W_\gamma(\underline{x},\underline{y}) = \frac{1}{(2\pi)^3} \int e^{-i(\underline{k},\underline{x})}\, \tilde{W}_\gamma(\underline{k},\underline{\ell})\, e^{i(\underline{\ell},\underline{y})}\, d\underline{k}\, d\underline{\ell}$$

where

$$\tilde{W}(\underline{k},\underline{\ell}) = \frac{1}{\Delta_\gamma(-\underline{k})} \int Q^{\pm}_{-\gamma}(\underline{k},\underline{m})\, \overline{Q^{\pm}_{-\gamma}(\underline{\ell},\underline{m})}\, d\underline{m}\, \frac{1}{\Delta_\gamma(\underline{\ell})}$$

where

$$Q^{\pm}(\underline{k},\underline{\ell}) = \delta(\underline{k} - \underline{\ell}) \pm 2\pi i\, \delta(k^2 - \ell^2)\, \Theta\,[(\underline{k} - \underline{\ell},\gamma)]\, h_\gamma(\underline{k},\underline{\ell})$$

When the operator H has eigenvalues, the weight kernel $W_\gamma(\underline{x},\underline{y})$ contains additional terms corresponding to these eigenvalues.

Once the Gelfand-Levitan Equation has been solved for $A_\gamma(x,y)$, then the potential is recovered from the relationship

$$V(\underline{x})\,\delta(\underline{x} - \underline{y}) = -2\delta\,((\underline{x},\gamma) - (\underline{y},\gamma))\, \frac{\partial}{\partial(\underline{x},\gamma)} \left[A_\gamma(\underline{x},\underline{y}) \,\Big|\, (x,\gamma) = (y,\gamma) \right]$$

In concluding the outline of the multidimensional Gelfand-Levitan approach, we should of course mention that as indicated by Faddeev, there are a number of mathematical problems that have to be investigated for the rigorous development of the theory. Among them, an explicit condition on potential $v(\underline{x})$ is needed so that it satisfies "condition C". An indirect approach to this was given by Perla Menzala [32] who from the nonstationary viewpoint, showed that if $v(\underline{x})$ is non-negative and sufficiently small, then the scattering operator uniquely determines the scatterer.

The multidimensional Gelfand-Levitan equation could be applied to Electromagnetic Scattering from a plasma cylinder. In this case the equation for the electric field E is

$$\Delta E + k^2 E = k_p^2(\underline{x})E$$

where $k_p = \omega_p/c$ and $\omega_p(\underline{x})$ is the plasma frequency. The term $k_p^2(\underline{x})$ corresponds to the potential term $v(\underline{x})$.

VI. CONCLUSIONS

Briefly the main areas of future research in inverse scattering should be the following (among others).

(1) The non-linear inverse scattering problem associated with fixed frequency, for non-homogeneous media (and obstacles) as outlined in section 4, and the generalization to the associated vector problem should be studied.

(2) The application of the multidimensional Gelfand-Levitan equation to Electromagnetic scattering should be pursued. Also it would be worth while to formulate it in the time domain.

VII. REFERENCES

1. Kay, I., *Commun. Pure, Appl. Math. 13,* 371-393 (1960).
2. Kay, I., The Inverse Scattering Problem for Transmission Lines, presented at Workshop on the Mathematics of Profile Inversion, Ames Research Center (July 1971).
3. Sondhi, M.M. and B. Gopinath, *J. Acoust. Soc. Am. 49,* 1867- (1971).
4. Balanis, G.N., *J. Math. Phys.13,* 1001-1005 (1972).
5. Weston, V.H., *J. Math. Phys. 13,*1952-1956 (1972).
6. Weston, V.H., *J. Math. Phys. 15,* 209-213 (1974).
7. Kreuger, R.J., An Inverse Problem for a Dissipative Hyperbolic Equation with Discontinuous Coefficients, to be published *Q. Appl. Math. 34,* 129-148 (1976).
8. Blagovescenskii, A.S., "Topics in Mathematics & Physics Vol. 1," (English translation) p. 55-67 (1967).
9. Blagovescenskii, A.S., *Proceedings of Steklov Institute of Math. 115,* 42-62 (1971).

10. Blagovescenskii, A.S., *Proceedings Steklov Institute of Math. 115,* 30-41 (1971).
11. Blagovescenskii, A.S., *Proceedings Steklov Institute of Math. 115,* 63-76 (1971).
12. Buzdin, A.A., *Proceedings Steklov Institute of Math. 115,* 77-85 (1971).
13. Romanov, V.G., "Integral Geometry and Inverse Problems for Hyperbolic Systems," Springer-Verlag (1974).
14. Gerver, M.L. and V.M. Markusevic, *Dokl. Akad. Nauk. SSSR 163,* 1337-1380 (1965).
15. Belonosova, A.V. and A.S. Alekseev, in the collection: "Some Methods and Algorithms for Interpreting Geophysical Data," p. 137-154. Moscow, Nauka (1967).
16. Herglotz, G., *Z. für Math. & Phys. 52, 3,* 275-299 (1905).
17. Wiechert, E. and K. Zoeppritz, *Nachr. Akad. Wiss. Goettingen 4,* 415-549 (1907).
18. Jeffrys, H., "The Earth," 4th Ed. Cambridge Univ. Press, London & New York (1962).
19. Backus, G.E. and J.E. Gilbert, *Geophys. J.R. Astron. Soc. 16,* 169-205 (1968).
20. Backus, G.E. and J.E. Gilbert, *Bull. Seismog. Soc. Am. 59,* 1407-1414 (1969).
21. Backus, G.E. and J.E. Gilbert, *Philos. Trans.,* 123-192 (1970).
22. Gilbert, J.F., *Geophys. J.R. Astron. Soc. 23,* 125-128 (1971).
23. Sabatier, P.C., Positivity Constraints in Linear Inverse Problems, preprint (Oct. 1975).
24. Sabatier, P.C., Positivity Constraints in Linear Inverse Problems, II, preprint (Nov. 1975).
25. Wolf, E., Determination of the Amplitude and the Phase of Scattered Fields by Holography, *J. Opt. Soc. Am.*
26. Wolf, E., *Opt. Comm. 1, 4,* 153-156 (L969).
27. Bojarski, N.N., Electromagnetic Inverse Scattering Theory, Sect. 2.3, p. 15-22. Syracuse Univ. Res. Corp., Special Projects Lab. Report SPL-TR68-70 (Dec. 1968). AD 711 644.
28. Ikebe, T., *Arch. Ration. Mech. & Anal. 5,* 1-34 (1960).
29. Faddeev, L.D., Itogi Nauk. *& Tekh. Sovremennye Problemy Mathematiki 3,* 98-180 (1974).
30. Faddeev, L.D., *Sov. Phys.-Dokl. (*English Trans.) *10,* 1033-1035 (1966).
31. Faddeev, L.D. *Sov. Phys.-Dokl. (*English Trans.)*11,* 209-211, (1966).
32. Perla, M.G., *J. Differ. Equations 20,* 233-247 (1976).
33. A nonlinear approach to inverse scattering has recently been developed by Weston, in a paper to appear in *J. Math. Phys.* (1979).

MOMENT-METHOD TECHNIQUES IN ELECTROMAGNETICS FROM AN APPLICATIONS VIEWPOINT[1]

E. K. Miller and A. J. Poggio
Lawrence Livermore Laboratory

The various issues of concern in the practical application of Moment-Method techniques in electromagnetics are discussed. Special attention is devoted to the steps involved in the development and use of a computer model. A survey of the state-of-the-art in connection with problem types most often encountered is also given. A recommendation is made for a methodological framework to be developed for EM problem solving to more effectively exploit analytical, computational and experimental procedures.

I. INTRODUCTION AND SUMMARY

There has been a literal explosion in the use of computer-oriented techniques in electromagnetics over the past decade. Of course, computations have always had a place in obtaining results for the classical problems of EM, for example, the Mie series for the sphere. Such problems presented relatively simple computational requirements, however, and could be evaluated using desk calculators, an approach which lasted into the 1960's. The increasing power and availability of the digital computer have not only replaced such burdensome methods, but have vastly expanded the repertoire of EM problem solving.

Of particular importance in this regard has been the role of

[1]This work was performed under the auspices of the U.S. Energy Research and Development Administration under contract number W-7405-ENG-48.

ISBN 0-12-709650-7

integral-equation formulations and their numerical treatment via the method of moments (MoM). While the first published numerical results of MoM EM solutions occurred only as recently as the mid 60's, they have already come to occupy a central place in electromagnetics. Because the use of MoM has become so prevalent, and it has contributed so significantly to our problem solving capability, it has become increasingly difficult to maintain a proper perspective concerning just what such numerical techniques can be realistically expected to do. This is particularly the case when MoM techniques are considered from an applications viewpoint, where the idealization, approximation and limitations inherent in the original formulation may not be appreciated or understood by the user.

In this paper, we intend to discuss MoM techniques in electromagnetics from the viewpoint of the user and his application. We will begin by presenting a brief analytical background to establish the basic ingredients of MoM techniques. Next to be considered is a state-of-the-art review of these techniques in the context of computer requirements and limitations. This section is intended to scope the computational aspects of the MoM, and the implications that present and projected computer technology may have for future MoM capabilities.

The presentation will continue by dissecting the MoM into its component parts, and evaluating the impact of each on the end application. These parts, which will be discussed in turn, have been identified as: 1) formulation; 2) manipulation and approximation; 3) numerical implementation; 4) computation/application; and 5) validation. Each of these steps will be examined and summarized to show how it relates to the capabilities and limitations of numerical techniques based upon the MoM.

The particular procedure known as computer modeling will then be considered in some detail. We will focus on the decisions faced by the modeler in the course of using a computer code, and how these decisions may be affected by the formulation, the approximations and assumptions employed, the numerical treatment used and the application. This will be done by following the steps involved in developing a model for a series of increasingly more complicated problems, beginning with a simple straight wire, to show the uncertainties, inadequacies and limitations to which the development of a model is subject.

The point of this discussion will be to emphasize the degree to which modeling is an art, and to identify areas where further theoretical effort may be productive. Where appropriate, numerical examples will be included. The section on modeling will conclude with a summarization of generic EM problems amenable to MoM treatment and an assessment of their current status, giving consideration to application requirements.

The synergistic roles of analysis, computation and experimentation will next be considered. It should have become apparent by this point in the presentation that MoM techniques, though powerful, cannot alone solve all EM problems. Instead, MoM is but one of several tools available and should be considered in this light. Still more powerful techniques might be derived by combining various individual methods into hybrid approaches. One example is provided by a recent development in which a MoM technique was combined with some results from the geometrical theory of diffraction to permit the modeling of a wire antenna located on a two-dimensional wedge. The general area of hybrid methods appears to be one where innovative ideas like this could be especially productive. Other possibilities and examples will be suggested.

A concluding section of the paper will summarize the main points considered and provide some recommendations for future work. To reiterate the main ideas of the presentation, they are: 1) EM problem solving has become increasingly computer oriented; 2) the impetus for this has resulted primarily from order-of-magnitude advances in computer technology; 3) future computer developments alone are unlikely to make it possible to sustain or match recent advances in EM computational techniques; 4) formulation, approximation and validation continue to be indispensable ingredients of numerical methods; 5) computer modeling is as much art as science; 6) careful interpretation is needed in order to relate results obtained from idealized models to real-world problems; 7) hybrid techniques which combine analysis, computation and/or experimentation appear to be increasingly promising.

One additional point should be noted. Although the general topic of the conference is electromagnetic scattering, the MoM itself is of course largely independent of the excitation employed. That is, it applies equally well to radiation as to scattering problems. This is due to the fact that the formulation and numerical treatment need not be made specific to either. Because of this commonality of radiation and scattering in the MoM context, our presentation will include both areas. This will allow us to discuss the general aspects of MoM applications without unnecessary restrictions.

II. ANALYTICAL BACKGROUND

It is not our intent to devote this section to an involved derivation of the typical integral equations which arise in exterior boundary value problems. Rather, we will outline the procedure, establish the nomenclature and point out some of the analytical issues which must be resolved when using numerical techniques for the solution of integral equations.

A. The Classical Problem

The classical exterior problem with which we are concerned is shown in Fig. 1. An obstruction, imbedded in a linear, isotropic, homogeneous medium, is excited by an electromagnetic field. The desired result is generally the scattered field or an induced surface quantity such as current or charge density. Often times, the desired result is a secondary parameter derivable from these field-related quantities, for example, terminal voltage, current or input impedance for an antenna.

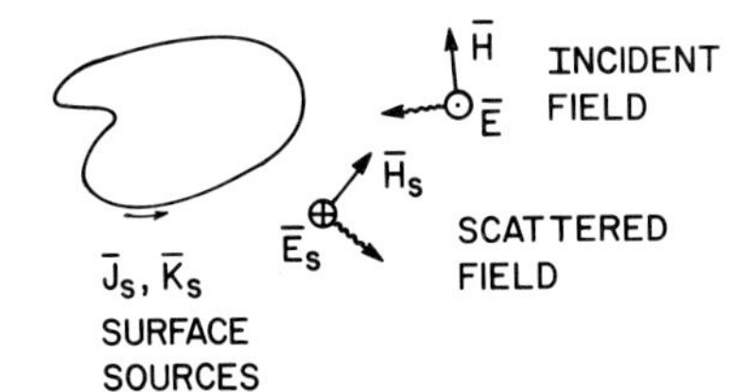

Fig. 1. The exterior scattering problem can be developed in terms of induced sources.

Several approaches are available for the solution of the stated problem. Here, we will deal only with the approach which reduces to the solution of unknowns over the surface containing the obstacle. These solutions, over a two-dimensional domain, can then be used to construct any other desired quantities.

The generality of this approach is at once both its forte and its weakness. The surface current density, for instance, is a fundamental or basic ingredient for evaluation of other field parameters (e.g., charge density, near and far fields, etc.). As a result, it is an extremely useful quantity. But, its determination is not at all simple and the effort involved in finding it may not be commensurate with the use to which it is put. For example, the detail and care with which one must carry out computations for the current density when evaluating input impedance may far outweigh that which would be necessary in a variational approach. This, however, is not the case when the surface current density is the quality of interest. Here, the surface approach is the most viable. Concisely, it is the ultimate breadth of applicability of this as yet unspecified tool that is the characteristic that is so valuable to the user. We will now present the rudimentary theory behind this technique, which is now fairly well known.

B. Integral Representations

A derivation of some integral representations for electromagnetic fields in terms of sources within and fields on a closed surface S can be found in several texts [14],[3],[12]. We will not dwell on these derivations here but merely write two particular integral representations in terms of electric and magnetic field. They are

E-field integral representation -

$$\bar{E}(\bar{x}) = T\bar{E}^{inc}(\bar{x}) - \frac{T}{4\pi}\int_V \{j\omega\mu J\phi + K \times \nabla'\phi - \frac{\rho}{\varepsilon}\nabla'\phi\}dv$$
$$- \frac{T}{4\pi}\int_S (j\omega\mu\bar{J}_S\phi + K_S \times \nabla'\phi - \frac{\rho_S}{\varepsilon}\nabla'\phi)ds' \tag{1}$$

H-field integral representation -

$$\bar{H}(\bar{x}) = T\bar{H}^{inc}(\bar{x}) + \frac{T}{4\pi}\int_V \{-j\omega\varepsilon K\phi + \bar{J} \times \nabla'\phi + \frac{m}{\mu}\nabla'\phi\}dv'$$
$$+ \frac{T}{4\pi}\int_S (-j\omega\varepsilon\bar{K}_S\phi + \bar{J}_S \times \nabla'\phi + \frac{m_S}{\mu}\nabla'\phi)ds' \tag{2}$$

with

$$\phi = \frac{e^{-jk|\bar{x} - \bar{x}'|}}{|\bar{x} - \bar{x}'|}$$

where $\bar{x}$ = field point; $\bar{x}'$ = source point; V = volume of interest bounded by S and the surface at infinity; $T = 1$ for $\bar{x}$ in V but not on S and 2 for $\bar{x}$ on S; $⨍_S$ = principal value integral over S; J, K, ρ, m = densities of electric and magnetic current and charge, respectively.

The integral representations we have used as illustrations are by no means exhaustive of those available. Rather, they are only some widely used examples which are meant to convey to the reader the types of terms with which one must deal in practice.

These integral representations are extremely useful in that they allow evaluation of a field at a point in a volume to be realized by merely integrating over sources in the volume and over certain field components (or equivalent sources) on the surfaces bounding the volume. We use the notation of an incident field to represent the integral over the bounding surface which contains the sources of the exciting field.

C. Integral Equations Derived from the Representations

One of the strong points of such integral representations is their suitability for use in generating integral equations. By introducing the boundary conditions imposed on the fields at the surface by the characteristics of the material on the two sides of the boundary, one can develop an expression relating equivalent surface sources to the exciting field on the surface. Rather than belabor this point, we refer the reader to [12] and here consider only one of the interesting cases, that of a perfectly conducting obstacle. For simplicity, we will drop the driven sources within V.

For a perfect electric conductor, $\hat{n} \times \bar{E}(\bar{x}) = 0$, that is, the tangential electric field on the surface must vanish. Hence two distinct integral equations for the surface current density can be written, viz.,

Electric Field Integral Equation (EFIE)

$$0 = \hat{n} \times \bar{E}^{inc} - \frac{1}{4\pi} \hat{n} \times \int_S (j\omega\mu \bar{J}_S \phi - \frac{\rho_S}{\varepsilon} \nabla' \phi) ds' \qquad \bar{r} \in S \tag{3}$$

Magnetic Field Integral Equation (MFIE)

$$\bar{J}_S = 2\hat{n} \times \bar{H}^{inc} + \frac{1}{2\pi} \hat{n} \times \int_S \bar{J}_S \times \nabla' \phi ds' \qquad \bar{r} \in S \tag{4}$$

For simplicity in our future discussions, we write these in the form

$$\hat{n} \times \bar{E}^{inc} = \bar{\bar{M}} \cdot \bar{J}_S \qquad \bar{r} \in S \tag{5}$$

$$\hat{n} \times \bar{H}^{inc} = (\frac{1}{2} \bar{\bar{I}} - \bar{\bar{L}}) \cdot J_S \qquad \bar{r} \in S \tag{6}$$

The integral equations which have been developed are only two out of a large number that can be written. For example, the vector differential operators have been written in the integrand but we could have written the equations with the operators outside the integral. This would have led to the potential form of the equations. Alternatively for a wire, the electric field equation could have been written in terms of a magnetic vector potential by solving the inhomogeneous vector differential to give rise to Hallen's integral equation.

D. The Method of Moments and the Numerical Solution of Integral Equations

Solutions to the integral equations which arise in electromagnetics can rarely be achieved through analytical means. The number of closed-form solutions is exceedingly small so that we must generally resort to numerical means. Several techniques have been developed, but we will focus on a single broad class which is referred to as the Method of Moments (MoM). References for this method abound in the literature, so we will only review the nomenclature in this section and apply the method to the solution of the operator equation (see [4] for example),

$$L\{A(\bar{x}')\} = B(\bar{x})$$

The essence of the Method of Moments is:

1) Expand the unknown function using a set of basis functions $\{\bar{f}_n;\ n = 1, 2, \ldots, N\}$ in the domain of the operator

$$\bar{A}(\bar{x}) \simeq \sum_{n=1}^{N} A_n \bar{f}_n(\bar{x}) \tag{7}$$

2) Define a suitable inner product of tangential vector functions over the surface S as

$$<\bar{P} \cdot \bar{Q}> = \int_S \bar{P} \cdot \bar{Q} da \tag{8}$$

3) Define a set of weight functions in the range of the operator L, i.e.,

$$\{\bar{w}_m(\bar{x});\ m = 1, 2, \ldots, M\} \tag{9}$$

4) Take the inner product of the weight functions with the operator equation in which the expansion has been substituted

$$\sum_{n=1}^{N} A_n \langle w_m \cdot L(\bar{f}_n(x))\rangle = \langle w_m \cdot B\rangle \qquad m = 1, 2, \ldots, M \tag{10}$$

The last equation defines M relationships for the N unknown values a_n. In essence, it equates the projections of the operator equation on the space of weight functions. Since it is a linear system of equations, it is ideally suited for numerical computation. In fact, the application of the procedure to (5) with

$$\bar{J}_s = \sum_{n=1}^{N} A_n \bar{J}_n$$

leads to

$$\sum_{n=1}^{N} A_n \langle \bar{w}_m \cdot \bar{\bar{M}} \cdot \bar{j}_n\rangle = \langle \bar{w}_m \cdot \hat{n} \times \bar{E}^{inc}\rangle \qquad m = 1, 2, \ldots, M \tag{11}$$

which can be written as

$$ZI = V \tag{12}$$

with

$$[I]_n = A_n \quad , \quad [Z]_{mn} = \langle \bar{w}_m \cdot \bar{\bar{M}} \cdot \bar{j}_n\rangle \quad , \quad [V]_m = \langle \bar{w}_m \cdot \hat{n} \times \bar{E}^{inc}\rangle$$

During the remainder of this paper, reference to numerical solution of an integral equation is a reference to (11), the embodiment of MoM.

The issues involved in the choice of bases and weight functions are numerous and complex. One one hand one wishes to choose bases which can closely resemble the unknown function while, on the other hand, one wants functions which allow $L(f_n)$ to be readily evaluated. Examples of typical bases and weights are provided in [2],[12],[8].

A great deal has been written about the analytical foundations of the method of moments. We will now look at this technique from the viewpoint of the user and will try to solidify the framework into which this tool must fit to be of use in an applications-oriented sense.

III. STATE-OF-ART REVIEW

A. The Impact of Computer Technology upon EM Problem Solving

We have made the observation above that the use of computers in electromagnetics problem solving has increased dramatically during the past decade or so. This has been instrumental in substantially advancing our problem solving capability, as Fig. 2a conceptually demonstrates. There we have depicted problems representative of the pre-computer (∿1954) era to indicate the solution capability then prevalent. This should be compared with Fig. 2b, where problems are shown to typify present (∿1978) solution capabilities. The increased capability which these figures imply has come about largely because of advances in both computer speed and storage that have occurred in the interim, as manifested by Fig. 3. Relative to the technology of 1954, computer storage is now $\sim 10^3$ larger and computer speed is $\sim 10^4$ faster. Note, however, that the growth curves of both these attributes of computer performance give evidence that the order-of-magnitude increases enjoyed in the early stages of computer development may not continue.

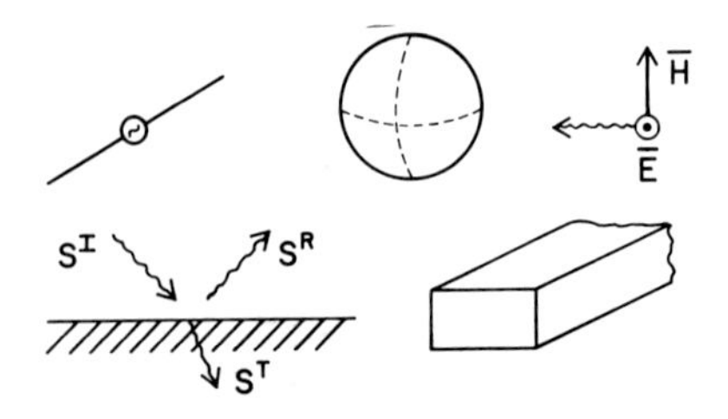

Electromagnetic analysis capability has advanced substantially in recent years:

Fig. 2a. Then (1954).

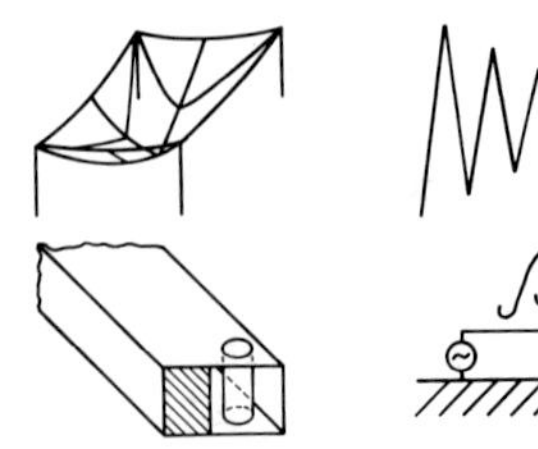

Fig. 2b. Now (1978).

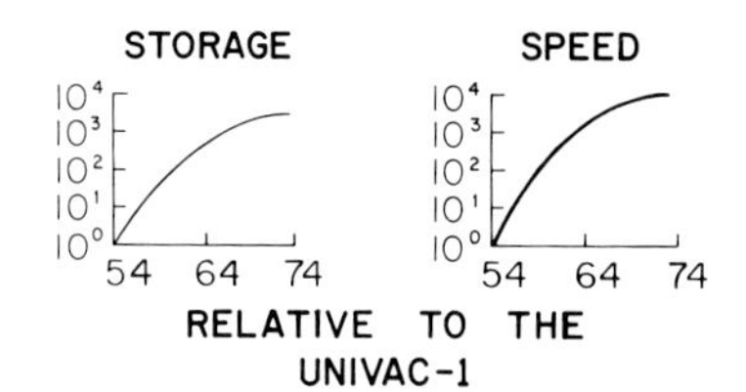

Fig. 3. Increased analysis capability due to computer development.

The latter observation has important implications because the numerical methods used for EM applications can require a substantial computer capability. If we consider an integral equation technique for example, the actual problem is replaced by an N-port numerical equivalent (see Fig. 4 or (11) with $M = N$) using a representation which requires N^2 mutual and self admittances where N is a function of object geometry and size in wavelengths. This can lead to a need for both extensive computer storage and long computer running times. As shown by Fig. 5, these requirements grow rapidly with increasing object size as measured in terms of length in wavelengths for a wire, or area in square wavelengths for a surface object. Even if continued order-of-magnitude increases in computer size and speed could be anticipated, the curves of Fig. 5 demonstrate that the maximum object size that might be routinely treated would not increase in proportion. Further, in view of the observation that such computer advances are unlikely to be realized, we conclude that the past growth rate in EM problem solving capability which could be credited to improved computer technology is itself unlikely to continue.

Fig. 4. Integral equation methods require substantial computer capability. Representation via an N*-port model requires* N^2 *mutual self admittances.*

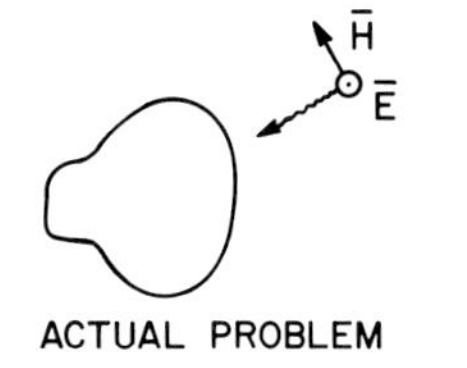

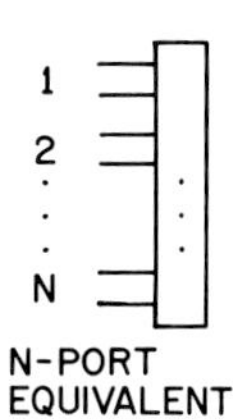

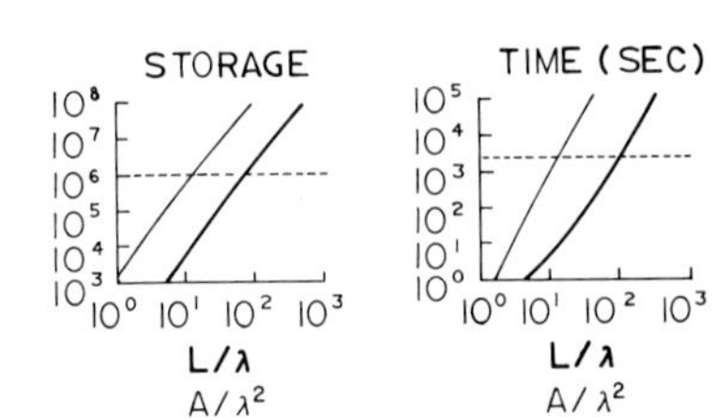

Fig. 5. Further computer advances are unlikely to substantially increase problem size. Wire (at right); surface (at left).

B. Some Practical Constraints

The discussion above has briefly addressed the general impact of computer technology upon EM problem solving and identified some practical limitations which should be kept in mind concerning present capabilities and future expectations. Let us now examine more specifically what these present capabilities are. Our attention will be directed to integral equation techniques for the most part since they have formed the basis for the majority of EM computer-based methods.

A cursory examination of the literature in the area of EM computer modeling will reveal that many more results are available for wires than for surface objects. Although there may be several reasons why this is so, the most important is probably that the number of numerical ports required to model a wire is proportional to its length, and for a surface is proporational to its area, as depicted in Fig. 6. This means that for given wire and surface objects, the storage, which is proportional to N^2, will grow as the square and the fourth power of the frequency, respectively. Similarly, the computer time, which is proportional to N^2 to N^3, increases as the square to the cube of the frequency for wires and from the fourth to the sixth power of the frequency for surfaces. The practicality of doing wide-bandwidth calculations for surfaces is obviously much less than is the case for wires.

Looked at another way, if the surface object has an area

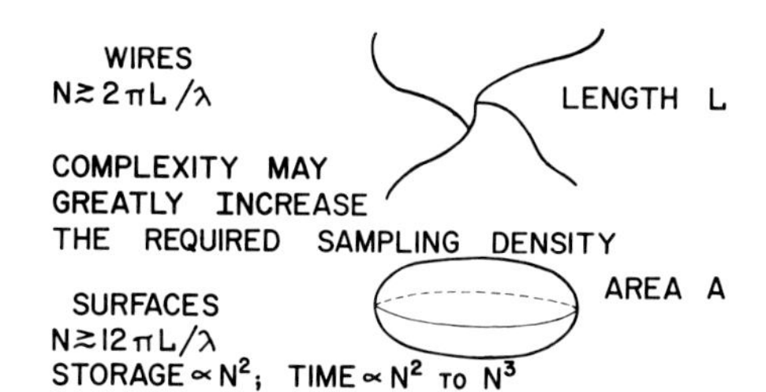

Fig. 6. Integral equation methods are generally better suited for wires than surfaces.

$\sim 6\ L^2$, i.e., is a cube of length L on a side, while the wire object is straight and of length L, then

$$\frac{N_s}{N_w} \sim 2 \times 36\ L/\lambda$$

where the subscripts s and w refer to the surface and wire objects respectively.[2] When account is taken of the storage and time dependencies upon N, the computational advantage of treating wires over surfaces is obvious. It should be mentioned, however, that surface objects having rotational symmetry are much more tractable to handle, because N_s is then reduced to a more nearly linear rather than area dependency. Symmetry can be an important factor in many problems in reducing the computational effort required of a numerical solution.

Object size is not the only factor which determines N. Also important is the degree of object configurational complexity, which can increase N over that indicated above because of the need to adequately resolve the object's geometry. As a matter of fact, even describing the object's geometry to the computer, let alone obtaining a numerical solution for its EM behavior, can be a demanding task. The result is that in many cases, configurational complexity can be more influential than size alone. Whichever is the dominant factor in a given problem, we are led to conclude that in general, size and complexity limit the applicability of direct computer modeling, as graphically depicted in Fig. 7.

---

[2] Assuming $\sim$6 samples/wavelength are used.

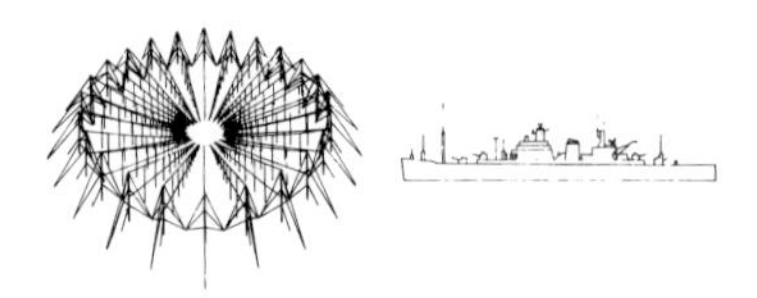

L $\geq$ *10λ at lowest frequency.* A $\geq$ *10λ^2 at 10 MHz.*

Fig. 7. Structural size and complexity limit the applicability of direct computer modeling.

C. The Set of Simpler Problems

For purposes of discussion, it is useful to identify a set of "simpler" problems, which on the basis of their relative simplicity and computational expense, are those for which routine computer modeling can be expected to be most effective. In this connection, we will define, somewhat arbitrarily, the set of "simpler" problems to be those which require no more than 10 minutes' time, and which can be solved in core, using the CDC-7600 computer. On the basis of the computer time and storage curves previously shown, we conclude that this set includes wires up to $\sim 50\lambda$ in length and surfaces up to $\sim 5\lambda^2$ in area. These values can be significantly decreased by configurational complexity, and increased by symmetry.

Though this definition may seem to be rather restrictive, the solution capability which this set of problems embraces is not insignificant, however. The kinds of wire problems which can be solved with these constraints include lossy media, interface effects (i.e., the ground problem), ground screens, layered grounds, the effects of impedance loads and both time- and frequency-domain formulations. For surface objects, generally the same kinds of problems can be handled, but the preponderance of applications have been for free space.

Computer modeling of these simpler problems provides, of course, access to a wide range of physically significant results, many of which are difficult to measure. Although these may be obvious to someone acquainted with MoM techniques, they may bear repeating here. They include input and transfer impedance (or admittance), current and charge distributions, near fields, and far scattered and radiated fields, as a function of frequency and time. Such results are available for any problem for which a valid computer model can be developed and numerically solved.

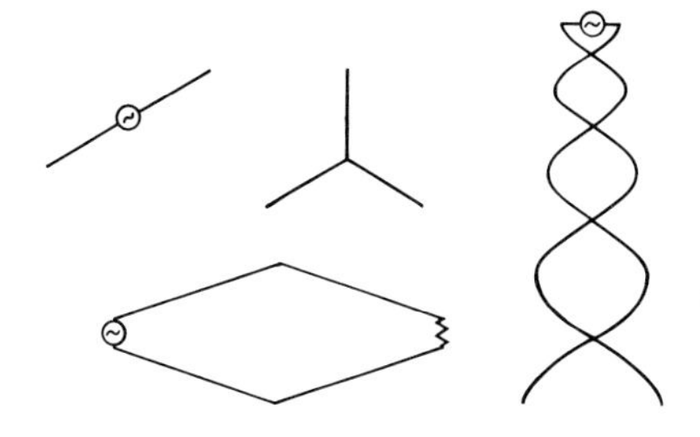

Fig. 8. For some problems the computer model is readily apparent.

The development of a model, then, is a key element in computer modeling. For some problems, such as those represented in Fig. 8, a suitable computer model is readily apparent, since the object geometry is naturally suited to approximation via a piecewise linear collection of wires. Unfortunately, however, few of even these "simpler" problems lead to unambiguous models. For example, it is fairly common to employ wire models to perform calculations for aircraft. At sufficiently low frequencies the aircraft cross-section will be small compared with the wavelength, and the aircraft may therefore be approximated as a "wire" object which can be modeled using a wire integral equation. But even when this approximation is selected, the most appropriate wire model is not obvious, as demonstrated in Fig. 9. The model detail which is required depends upon the application, size-to-wavelength ratio, kind of information sought, the accuracy desired, etc.

We might expect, of course, that model ambiguity would increase with both the configurational and environmental complexity of the problem being treated. Consequently, the resemblance between the real problem and its computer model may become more

Fig. 9. Even few "simpler" problems lead to unambiguous computer models. Model detail required depends on application, size-wavelength ratio, etc.

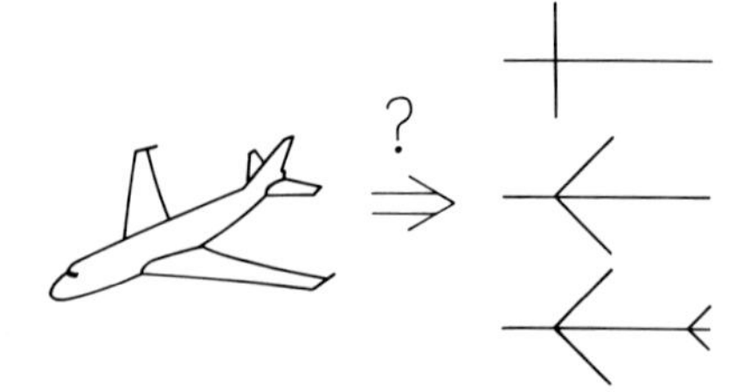

remote, as illustrated in Fig. 10 by the original problem and the computer model used for it [5]. In spite of the idealization employed in this case, which included: 1) eliminating the source antenna and including its effect by the incident field it produces as measured at the test-antenna midpoint; 2) ignoring the support towers of the test antenna; 3) approximating the test antenna by the simplified straight-wire model shown; and 4) accounting for the test-antenna's interaction with the ground by using the reflection-coefficient approximation, useful results can be obtained as demonstrated by Fig. 11. More elaborate antenna configurations such as that shown in Fig. 12 can be treated in a reasonably straightforward manner [7]. By using wire-grid approximations to non-wire objects, even more complex problems can be treated, as the wire-grid models shown in Fig. 13 attest. Although such wire-grid representations must be employed with care, they can be quite useful for various applications and can meet the qualifications previously given for the simpler problem set.

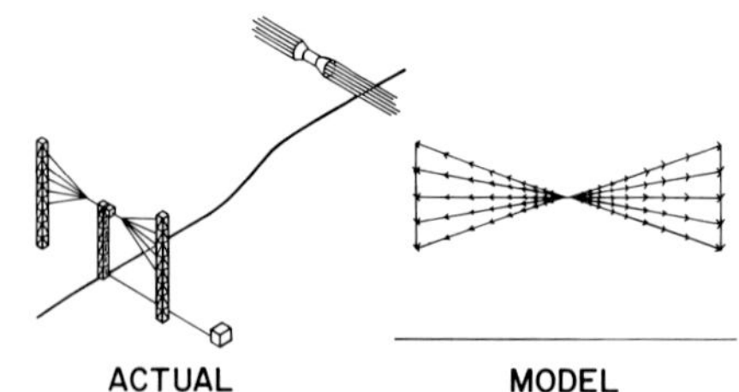

Fig. 10. Simple model may suffice for complex problems.

The computer modeling of these simpler problems can be effective in several ways. First of all, such computations can

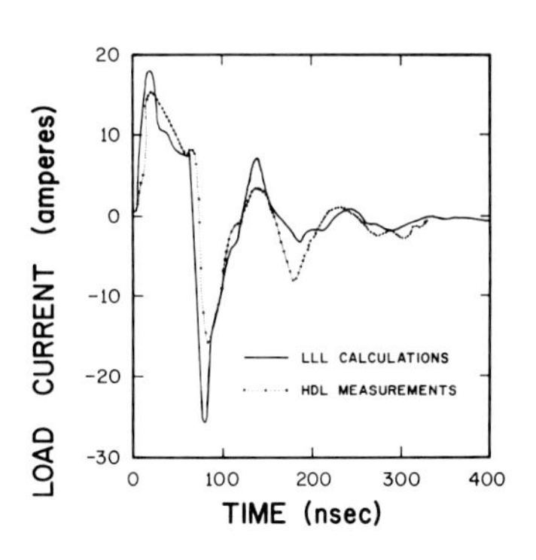

Fig. 11. In spite of idealization, useful results can be obtained.

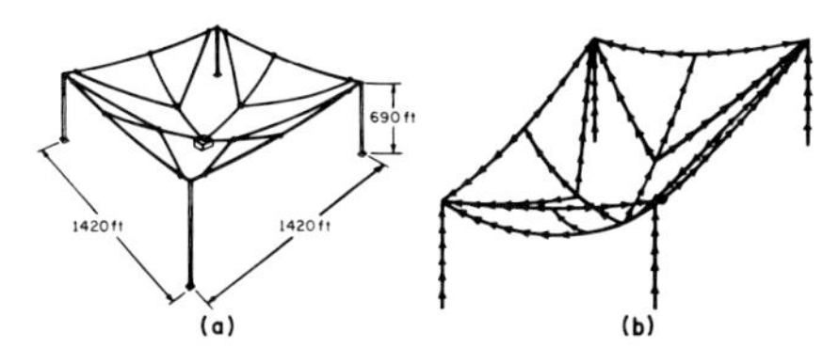

Fig. 12. A piecewise linear model of as complex an antenna as this is easy to develop and analyze. The Sectionalized Loran Transmitter is shown here.

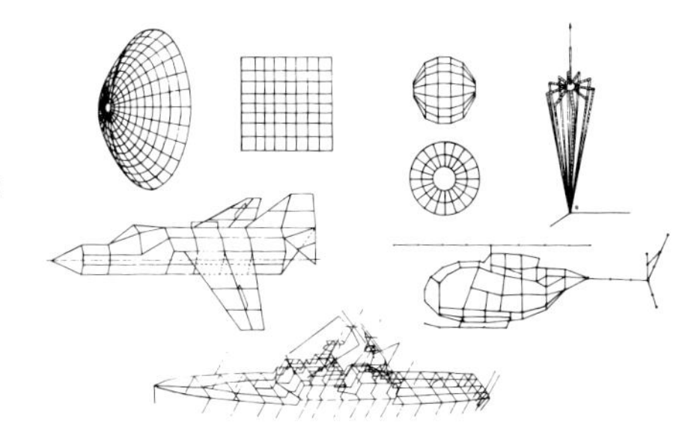

Fig. 13. More complex models can also be developed and analyzed.

provide more complete information than is readily available from measurement. This can result in increased insight and understanding, and consequently a greater appreciation for the physics of the problem. A second advantage of computer modeling is the capability it provides for performing parametric, optimization and sensitivity studies. Finally, as we have tried to show above, a reasonable degree of realism, complexity and scope is made possible.

D. More Complex Problems

Problems which fall outside the simpler set as previously defined may also be suitable for computer modeling. The difficulty in their treatment stems from the fact that their configurational complexity and/or size may increase N beyond what can be routinely accomodated in terms of computer cost and complexity of input data generation. Because the integral equation treatment will in general provide all N^2 admittances associated with the numerical model, but at the expense of a computer time which eventually grows as N^3; the computational effort per admittance values also grows as N.[3] Thus, while the integral-equation approach yields a complete solution, in the sense that all admittances are found, it does so only at a disproportionately increased expense.

But the actual information needs of these more complex problems may not markedly exceed those of the simpler problems. Consider the situation illustrated in Fig. 14, for example.

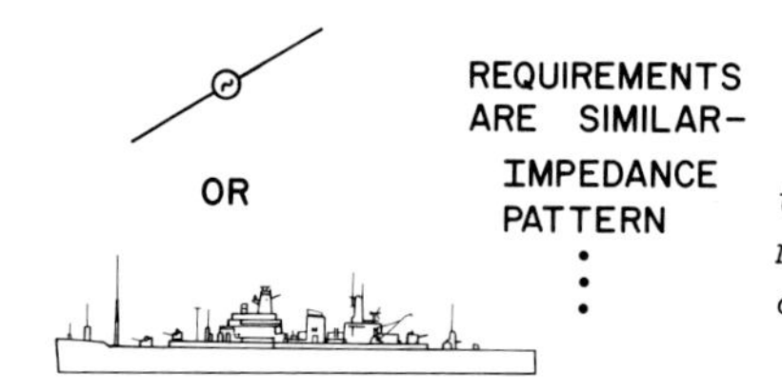

Fig. 14. Specific information needs do not increase markedly with problem size and complexity.

Whether dealing with a single dipole, or the antenna suite on a large ship, the information sought is similar in nature, and would generally include input impedance, radiation pattern, efficiency, etc. Although there are of course several antennas located in a complicated environment in the latter case (resulting in the need for mutual coupling and other characteristics of several antennas compared with the single dipole), the

[3] The computer time required for iteration solutions increases in principle less rapidly with N than N^3, but has not been used with wide success.

the information needs in quantitative terms for the two problems are not nearly so different as the number of admittances required for their complete computer models. The problem is that a decreasing proportion of these admittances represents useful results as problem size and/or complexity increases. This means that an increasing proportion of the computed information is relatively extraneous. As shown by Fig. 15, where we plot the relationship between units of information obtained (as represented by the admittance) as a function of computer effort involved, this extraneous information is increasingly wasteful in terms of the effort required in its generation.

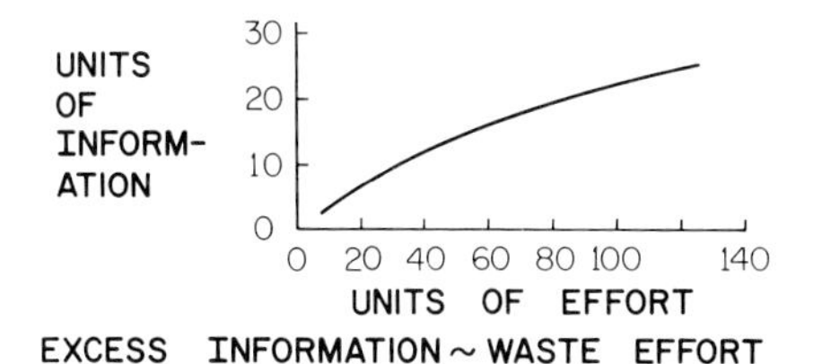

Fig. 15. Required solution effort thus sensitive function of information quantity.

When using direct integral equation techniques, then, we conclude that one reason large, complex problems can become impractical to treat is that the integral equation approach leads to a complete solution whether or not one is sought. This complete solution involves a quantity of information which can impose an unacceptable burden on the processing capability of the computer. The fact that not all the information generated is needed indicates from a practical viewpoint that the procedure is inefficient in providing the information actually desired. It is therefore relevant to ask what role direct integral-equation techniques might play in complex problem solving, and whether modifications, extensions, etc., might be developed to circumvent the disadvantages outlined above.

First of all, we should recognize that certain applications can be such as to justify expensive, inefficient computer calculations, where integral-equation techniques could play an important role. These might include the numerical evaluation of canonical approximations, performance of sensitivity studies and the assessment of approximate methods. Such applications would essentially represent one-of-a-kind problems, or those for which extensive computations are not routinely performed; but where instead the computer model is used to validate other results or to establish the most important parameters.

Second, it should be noted that variations of the direct approach might lead to more practicable techniques. These might include modifications of the direct approach such as sparse matrix techniques; introduction of asymptotic procedures based upon physical optics, geometrical theory of diffraction, etc.; hybrid analytical combinations; or even hybrid analytical-experimental combinations. These are areas which are either unexplored or just beginning to be considered.

Finally, we must think about the possibility of developing more innovative techniques which will circumvent the problems discussed above. What form these might take, or whether any significantly different approaches are indeed feasible, is now unknown. Stated simply, however, the problem is to minimize the solution effort required, commensurate with the accuracy and detail required of the result.

IV. COMPUTER MODELING: A MULTI-STEP PROCESS

These are several individual steps which can be identified in the development and application of a computer model. For purposes of discussion, it is convenient to consider them roughly in the order in which they are first encountered in the overall numerical treatment, as: 1) formulation, 2) approximation, 3) numerical implementation, 4) computation, and 5) validation. As a general observation, we can mention that the overall goal of the computer model is that it provide results of accuracy and detail commensurate with the problem requirements with a maximum of flexibility and minimum of formulational effort and computational cost. The implications of this goal will become more evident in the following discussion.

A. Formulation

With the increasing emphasis which is being placed on numerical analysis and computation, there is a danger that inadequate attention is being directed to problem formulation. It should be obvious that the formulation is the cornerstone of the entire process, and that the most powerful computer is incapable of solving the simplest problem if that problem cannot be stated in theoretical terms.

A very large proportion of the computer models being used today are based upon integral equations which are generally used for linear, homogeneous media. An alternative is provided by the differential form of Maxwell's equations, which permits the treatment of nonlinear, inhomogeneous media.

Whatever are the specifics of the formulation, it ultimately involves describing the relevant physics of the problem as integral or differential field relations plus the associated boundary conditions. Nominally, we might expect increased analytical effort at this stage to eventually result in decreased computational effort later on. For example, a solution of the external electromagnetic problem for a spherical object can be compactly expressed in analytical terms using spherical Bessel functions, Legendre polynomials and circular harmonics. These classical functions will require numerical evaluation in general, but their computation is relatively straightforward and efficient. As an alternative, an integral equation technique might be used, which replaces the Bessel function and Legendre polynomial part of the analytical solution with an entirely numerical computation. This approach can, however, become computationally intractable for large spheres, whereas the analytical solution remains numerically manageable. As a further simplification in the analytical treatment, we might employ a differential equation formulation for the external problem. But this would result in a substantial increase in computation time and consequently a further decrease in the sphere size which could be handled.

There is thus seen to be a tradeoff to be made between analytical and computational effort. But there are other factors to be considered as well. One of these is the problem specificity of a given formulation. The analytical solution for the sphere problem as described above is applicable only to spheres located in linear, homogeneous media. The integral equation treatment can be applied to non-spherical axisymmetric objects, and with decreased computational efficiency, can also be used for non-symmetric three-dimensional objects. Inhomogeneous, non-linear media and non-symmetric object shapes can be handled by the differential equation approach.

An additional difficulty of analytic solutions, like that available for the sphere, is that they are of course obtainable for only a limited set of separable geometries. And even for some of these special geometries, considerably more computational work than mere function evaluation is required to obtain numerical results. As one example of this, the various modes associated with a prolate spheroid are coupled in general and may require the solution of a matrix of high order [17]. Another example is provided by the bi-cone antenna [18].

We thus conclude that classical formulational techniques, dealing primarily with separation-of-variables boundary-value solutions, cannot alone be very effective in developing treatments for general problems. But purely integral-equation or differential equation formulations may not be the most suitable ones for any given problem either. There are aspects of all these approaches (and other approaches, too, that have not been mentioned) that may in combination provide a better formulation than any one. In the formulation step of the overall computer modeling procedure, then, we should attempt to identify what approach or combination of approaches may be best for a particular problem set. Several examples are given in [19].

Ideally, the formulation, numerical procedures, approximations, etc., should be consistent with the overall goal of the computer model stated above. Realistically, the spectrum of choices available to us is not continuous, so that not all the various aspects of the computer model can be optimized, if indeed any can. Thus, we must try to establish those factors that seem most important in a given applicational context, and consider them first. In most computer modeling work done to date, the greatest share of the effort has probably been devoted to numerical analysis and computational questions, with the formulation remaining relatively simple and unsophisticated. As we attempt to solve more demanding problems, we are finding that increased attention is likely to be needed in the formulation area, and innovative ideas in particular will be necessary if computer limitations are to be overcome. One example of an innovative approach is provided by the combination of geometrical theory of diffraction and an integral equation technique to handle wire antenna on finite ground planes [16]. Other approaches along this line will be necessary to rectify some of the problems outlined below.

B. Approximation and Errors

Because of the fact that the few practical problems can be formulated, let alone solved with any degree of exactness, it is mandatory to resort to various kinds of approximations in the course of computer modeling. These approximations arise in several ways. First of all, the formulation may itself only approximate the actual problem of interest. Second, the numerical implementation must inevitably provide the framework for only an approximate solution to the formulation employed. Third, the actual computation is necessarily limited in accuracy by the characteristics of the machine which is used. Finally, the computer model may only provide an incomplete characterization of the problem being treated.

It is convenient to briefly discuss each of these approximations in turn.

1. *Formulation Approximations*

Representative of approximations in this category are:

a. *The thin-wire approximation*. This involves neglecting both circumferentially directed current and the circumferential variation of the longitudinal current on a wire, and the use of a reduced kernel in the integral equation to obtain a line rather than an area integration. It is one of the most widely used approximations, and appears in many different forms.

b. *Surface-impedance approximation*. This is used for imperfectly conducting objects to avoid the need for treating both the tangential electric and magnetic fields as unknowns by establishing their ratio via a surface impedance.

c. *Reflection-coefficient approximation*. The analysis of antennas located near the earth's surface requires that the fields reflected from the ground be accounted for. This approximation involves using the Fresnel plane-wave reflection coefficients to obtain these fields, rather than the rigorous, but much more computationally expensive, Sommerfeld integrals.

2. *Numerical Approximations*

Here we encounter all those approximations associated with the numerical reduction of the analytical representation. These include:

a. *Basis function*. The unknown field components must be given some mathematical representation in order to attempt their solution. Since their actual behavior is of course unknown, the basis or expansion function employed can obviously only approximate them.

b. Weight function. The field equations can be satisfied in only a sampled sense. The sampling is established by the weight or test functions employed, and provides only an approximate matching of the original analytical description.

c. Integration and differentiation. In either integral or differential form, the field equations require integration or differentiation of the unknown field quantities. This is accomplished approximately in the context of the basis and weight function employed in the numerical reduction, and the ultimate accuracy achievable will be intimately dependent upon them.

3. Computation Approximations

In this category fall those approximations (or errors) which occur in the actual computation process itself. The most important of these is probably matrix factorization (or inversion) round-off error. Due to the large number of multiply and divide operations which occur in this process, round-off errors can accumulate to the point where insufficient numerical accuracy is maintained. This is especially critical for ill-conditioned systems, where the linear independency for the equations being solved is marginal.

4. Model Approximations

Perhaps the greatest uncertainty attendant to computer applications is due to discrepancies between the actual problem and the computer model which is used for it. It is here where computer modeling is closer to art than science. Clearly, most real world problems are sufficiently complex that it is not practical, for both analytical and computational reasons, to employ a model that is complete in every detail. Instead, the modeler must make judicious choices based upon intuition, experience or prior knowledge in developing a model. This exercise is subject to several constraints, including the time available to prepare the problem for the computer and the expense of doing the actual calculations; the kind and amount of information sought and how it is to be used; and the accuracy and confidence bounds appropriate to that particular problem.

Clearly, approximation pervades the entire modeling process, from formulation through validation.

C. Numerical Implementation

In this step, the analytic description developed in the formulation is reduced to a basis suitable for computer (or numerical) solution. As such, it involves the application of numerical analysis techniques such as finite difference approximations for differentiation, numerical quadrature for integral evaluation and interpolation and extrapolation.

The analytical formulation for the numerical treatment is the ubiquitous method of moments, which appears in various forms in almost all the numerical methods used in electromagnetics, whether or not explicitly acknowledged. In essence, the moment method transforms the original analytic description to a numerical one involving a linear system of equations, which can then be solved using standard matrix techniques [4].

Aside from variations in this process which originate from the original formulation, the major differences in the subsequent treatment are due to the various combinations of basis and weight functions which may be employed. The choices concerning which of the latter to use in turn depend upon several factors, including the problem dimensionality, formulation characteristics and the degree of generality desired in the resulting computer algorithm. Generally speaking, increased efficiency and scope of application are obtained at the expense of decreased ease of use and modification.

This particular step in computer modeling generates perhaps more debate concerning the "best" approach than does any other, including the formulation. This derives from the fact that there is a wider range of apparently equally plausible possibilities in the numerical implementation than there are in the other steps. Much of the debate, however, stems more from appearance than from substance regarding apparent differences. For example, in reducing the integral equation for a straight wire, interchanging the basis and weight functions produces results identical to those obtained from their original arrangement. A similar observation can be made about a comparison of the finite element method and a moment method solution based upon Galerkin's technique (use of the same basis and weight functions). It is unfortunately too easy at this point for the physics of the original problem to be obscured by manipulation associated with the numerical treatment.

D. Computation

It is at the computation step where most users first encounter a given code, and to which most theoreticians and code developers tend to devote less attention by comparison. Consequently, the users are often the first to discover errors in the code, and to become more familiar with its limitations.

There are several attributes associated with the code and computation process that deserve mention for the insight they provide concerning computer modeling. We have tried to conceptually illustrate in Fig. 16 the relative advantage of those we feel are more important as determined by the degree of code complexity or development. The judgments implied there are highly subjective and non-specific, and are included primarily to generate an awareness of some of the factors which may influence code usage. In general terms we observe that:

1. Efficiency is likely to be higher for a more complex code, because, for example, it will contain higher order basis and weight functions that permit fewer unknowns to be used to obtain a given numerical accuracy.
2. Scope of applicability, i.e., the variety of problems for

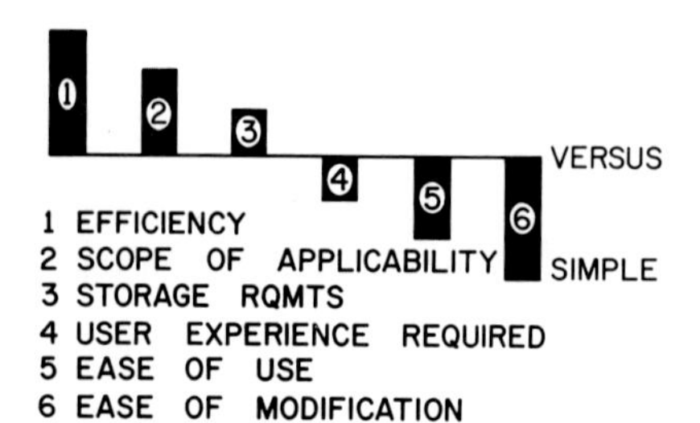

Fig. 16. What kind of code is best depends upon many factors.

which the code can be used, will likewise be expected to be greater for the more developed code, as it should thus include more features which can be called upon.

3. Storage requirements of more developed codes might be concluded to be greater simply because they contain more features, but might be somewhat offset by reduced storage due to the possibility of using fewer unknowns.
4. User experience necessary to successfully operate a code will usually be less for those which are less complex, because fewer choices need be made and thus fewer mistakes are likely.
5. Ease of use should be greater for the simpler codes, again because they impose fewer demands upon the user.
6. Ease of modification depends highly upon the code structure and interconnections, which generally tends to favor the simpler code when the user wishes to make changes.

Satisfactory code usage may be concluded to depend upon many factors besides the problem for which the code is being used. The importance of adequate documentation of the code in this respect cannot be over stressed, particularly as to explicitly stating the code's limitations. Equally important is documentation of user experience, since very often one user can learn from the experience of another. Of special use in this regard, is the telling of which has not worked, as well as what has, since negative information can help others avoid the same problems, as well as to indicate areas where further development effort may be appropriate.

We might conclude by observing that many codes have been described as "user oriented". Unfortunately, no uniform standards as to documentation, comment cards, sample problems, etc., are routinely adhered to. Very often as a result, new codes are developed to do the work of those already available, because the effort anticipated by the prospective user to develop his own code is less than that he perceives to be necessary in order to employ an already existing one.

E. Validation

Of the five steps we have associated with computer modeling using the moment method, validation may ultimately absorb as much effort in the course of using a given code as the other four together. This is due to a number of factors. Perhaps most important is the difficulty of obtaining independent analytical, numerical or experimental results of the requisite accuracy to provide a validation benchmark. Then too, when such results may be available, they very often introduce their own element of uncertainty because of error or approximations. Furthermore, each of the other steps as previously outlined is relatively well defined and closed ended, i.e., involves a rather clear sequence of events, the conclusion of which marks the completion of that particular step. Validation, on the other hand, tends to be open ended, since new problems continue to be encountered to which the code has not been applied, and a multitude of coding errors can occur to invalidate even a routine calculation in subtle ways.

For purposes of discussion, it is useful to associate two essentially independent errors with the computer modeling process. These are:

1. The physical modeling error, ε_p, which results from approximating the actual object with a (usually) simpler configuration for analytical and/or computational purposes.

2. The numerical model error, ε_N, which results from obtaining only an approximate numerical solution for the configuration actually selected.

It is usually necessary to conduct experiments to evaluate ε_p, since the actual structure is rarely amenable to rigorous numerical treatment. On the other hand, ε_N may often be estimated by studying the numerical convergence of a given code, or by comparing results from two or more independent codes applied to the same configuration.

In the course of attempting to validate a computer model then, we must keep in mind both experimental and computational methods. Experimental validation is generally the most convincing, but measurement has its own difficulties which quite often frustrate its effectiveness in this role. While comparison of experimental and computed data in an absolute sense is probably preferable, much can be done on a relative basis by, for example, comparing the difference between calcuations for two numerical models with that obtained for measurements for two corresponding experimental models. Also on a relative basis, two experimental models might be measured. The first should correspond as closely as possible to the numerical model to provide a check on the numerical validity of the calculation. The second should resemble the real problem as closely as possible to provide a check on the physical validity of the numerical model, which for all but the simplest problems will incorporate some degree of approximation, or which due to the formulation may completely alter the problem description. As an example of the former, we might use circular cylinders and flat plates to model an aircraft; while in the latter instance, we might instead use a wire-grid model.

The idea of numerical validation may appear somewhat self-contradictory, since it seems that numerical results of themselves cannot be used to infer their own validity. Actually there are several ways in which the apparent validity of numerical results can be assessed. These include:

1. Reciprocity and energy conservation checks, which are, however, necessary, but not sufficient conditions for solution validity.
2. Boundary condition checks to see the degree to which the required boundary conditions are satisfied.
3. Examination for "non-physical" behavior, such as oscillations in the current distribution which are related to the inter-sample spacing and not the wavelength.
4. Numerical convergence checks as mentioned, which show the solution behavior as the number of unknowns is increased. This exercise must be used with care, however, as the results may eventually diverge.
5. Independent numerical results.

There are also a number of programming checks which can be performed in troubleshooting a code. These may depend upon characteristics of the code itself, such as program conventions, data input and running requirements, as well as the specifics of the computer and compiler being used. In this latter category we might find dimension and mode statements, word size, declaration hierarchy, complex function definitions, library subroutines and variable initialization. One of the most frequently encountered troubleshooting problems occurs when code modifications have been made, and the subsequent output appears suspect. It would perhaps be useful to include as a part of the code itself, a few test problems and results that could be used as an internal consistency check on the modified version.

Finally, we should mention that successful use of a code depends upon realistic expectations and informed awareness about its capabilities and limitations. These include knowledge of its mathematical foundation, guidelines about suitable computer models and an understanding of the physical approximations which are involved. In wire modeling, for example, we can identify the following general aspects to be kept in mind:

1. Mathematical - Based upon length (L)/diameter (D) $\gtrsim$ 10;

$D \leq \lambda/2\pi$; negligible circumferential current; no circumferential variation of longitudinal current; and use of the thin-wire kernel.

2. Physical - Model consists of a piece-wise connection of linear segments; use of an equivalent cylinder for non-circular sections.
3. Computer - Recommended model characteristics are the segment length $\Delta \leq \lambda/6$; $\Delta \geq 3D$; segment length may be dependent upon object geometry as well as size; avoidance of abrupt segment length changes is desirable; include only necessary details in model.

V. SOME SPECIFIC ISSUES IN MODELING AND COMPUTATION

In keeping with our concentration on application-oriented issues, we will now consider a succession of problems, progressing from the simple to the difficult, raising at appropriate points the questions of concern to the analyst. It is not our intent to delve deeply into the details of each problem, but rather to state the concerns which accompany them.

An important portion of this section is devoted to the philosophy of modeling. For each real problem there will be two models developed. The first is the physical model, which represents a reduction of reality to a tractable level for analysis. The second is the numerical model, which is actually used in computation and which encompasses the details of the mathematical and computational approximations.

A. A Progression of Problems

1. Straight Wire - Scattering

The straight wire represents perhaps the simplest of problems attacked using MoM and, indeed, numerous computer programs exist which are capable of solving straight wire problems. Furthermore, the physical model for the conducting wire is straightforward, as it is identical to reality. The numerical model using MoM requires the definition of several parameters including length, radius and frequency. The questions of importance are whether the number of segments N is sufficient to adequately represent the current distribution and whether N is so large as to invalidate the thin wire kernel. Furthermore, the importance of the end cap region must be assessed and a decision made as to whether to attempt to model the region or to ignore it.

2. Straight Wire - Antenna

The only difference between the scattering and antenna cases is the definition of the source. For scattering, the source is generally easily defined as a continuous exciting field distribution over the structure. For antennas the source definition may be more difficult, since discretization may preclude a δ-gap source. In this case, the real situation may be a dipole with a narrow gap driven by a line, the physical model may be a conducting wire driven by a δ-gap and the numerical model a wire driven by a tangential field source over a single segment of the antenna.

One might ask to what degree the three models (real, physical and numerical) might differ with regard to the ultimate results they provide. Experience and perusal of the literature indicates that there is generally little measurable difference.

3. Curved Wire

A curved wire presents somewhat more difficulty than the straight one. The problem is associated with the contour of integration of the thin-wire integral equation. The physical model is not affected by the curvature, except perhaps in that the azimuthal invariance and symmetry requirements of the thin wire reduced kernel are violated. On the other hand, the numerical model has most often been constructed using a piecewise linear approximation to the curved center line, since there are advantages to be realized performing the integrations over straight lines.

These include the ease of performing the numerical integrations as well as the fact that closed form expressions for fields are available for certain current behaviors on straight elements [8]. Nonetheless, integral equations formulations have been pursued and results obtained for the actual curved contour [6]. However, later computations have shown that the differences between piece-wise straight and curved lines are quite small.

4. *Loaded Straight Wire*

The loaded straight wire is conceptually the same as the antenna case. Here, the voltage drop across the load may be represented by a dependent source $(-I(z_L)Z(z_L))$. Hence, the same questions must be asked as were asked in the antenna case. Specifically, one must be concerned with the spatial extent of the load in the various models and the effects of these variations on the final results.

5. *Thick Conducting Wire*

When the radius of a conducting wire is increased, one may eventually reach a point where the assumptions which form a part of the thin-wire approximation are violated. One must ask whether the radius is large enough to allow azimuthal variation of axial current or existence of a circumferential current. The specified thick conducting cylinder should have a physical model with close resemblance. But this requires close consideration of the end caps as well as the effects of non-azimuthally symmetric field quantities. The numerical model based on thin wire theory is generally not sufficiently advanced to allow for end caps and definitely cannot cope with azimuthal asymmetries. The main question is again how much these variations affect the ultimate results.

6. *Bent Conducting Wire*

The bent conducting wire introduces to thin wire theory one of its first difficulties - minor though it may be. Reality here would be physically modeled by a bent conducting wire, again very similar in detail. But a numerical model based on thin wire theory would have some (possibly strong) approximations. Thin wire theory assumes azimuthal symmetry, yet at the bend this is not possible. The main question is then whether the effect of the bend region extends a great distance along the structure or is localized. Also, if the effect is indeed local, we can ask to what extent our results (e.g., current density) may be valid near the bend. Clearly we see that our numerical model has several associated issues.

Experience has shown that bend region effects, especially for included angles which are not too acute, are quite local and that computed results for current density at distance of order five radii agree very well with experiment. However, we can infer a restriction in that we cannot rely on quantities closer to the bend than this since the assumed azimuthally independent field distributions and those that occur in reality may be significantly different.

7. *Wire Structures with Multiple Wire Junctions and Radii Discontinuities*

The multiple wire junction represents a difficulty in modeling. The physical model of reality is again very similar, but the numerical model for MoM in the thin wire sense is not so straightforward. The junction introduces difficulties regarding azimithal symmetry and Kirchhoff's current law.

We know that Kirchhoff's current law (sum of currents is zero) must be satisfied at the junction. But many codes require information regarding the charge density at the junction as well and, in the context of thin wire theory, this is not available. Several schemes are in use and each has shown some success. Naturally, the results computed using this model can only be relied upon at sufficient distances from the junction, again several wire radii.

The questions that face the user in this case are probably the most pressing that we have encountered in this paper. Junction effects, due to the constraints and approximations in the thin wire model, have been the source of much discussion and uncertainty, as their adequate treatment is often questionable.

Thin wire theory also suffers from a deficiency in the case of wires with discontinuous radii changes. Here again a great deal of discussion has occurred and only recently has some experimental data been available for model validation. The subtleties of discontinuous radii are beyond the scope of this paper, but the potential existence of difficulties must be clear to the user.

8. Wire Above Ground

When asked to analyze a wire structure over imperfectly conducting ground, the analyst must immediately begin to study alternatives. Reality for this case is easily defined, but the physical model raises several possibilities. For instance, the ground may be conductive enough to be represented by a perfect conductor. In this case the numerical model is straightforward. On the other hand, one might choose to make the physical model conform more closely to reality. Then the numerical model can take on several forms. The effects of the imperfectly conducting ground can be represented quite precisely through use of the Sommerfeld Theory, so that when coupled with MoM the resulting model is reasonably accurate. Due to the complexity and cost of such a model, it may be advantageous to apply another model for the ground. One alternative uses Fresnel Reflection Coefficients to account for the ground interaction and, though not as precise as the Sommerfeld approach, is reasonably accurate for structures not too close to the ground.

For a wire above ground we see the branches of the decision tree which the analyst must negotiate. And, we have not even mentioned the possibilities of the structure touching or penetrating the ground, which introduce further complexities, decisions and uncertainties.

9. Smooth Closed Body

The evaluation of electromagnetic field variables for scattering by a smooth conducting body can be undertaken in several ways. We can physically model reality using a wire grid model and apply MoM to that structure. Or, we can use integral equations for the closed surfaces in our physical model (as discussed in Section II) and proceed with MoM solution using the solid surface methods for our numerical model. Of course, the analyst must be aware of the difficulties which might arise in using various formulations such as the internal resonance difficulties with the MFIE.

Wire gridding has been widely used in the recent past because the numerical model is easily generated and the status of wire analysis methods is well beyond that of surface analysis. However, the uncertainties of this approach and the difficulty of interpretation of the output (how is the wire current converted into surface current density?) have left many uneasy. For instance, certain loop currents have been observed in the grids which are not possible for the solid surface.

The solid surface approach has not been as widely used. Geometry definition is more difficult and there are two different equations which can be applied - each with its own difficulties. The analyst must decide whether he would prefer to work with EFIE, a Fredholm integral equation of the first kind with a highly singular kernel, or with the MFIE, a Fredholm integral equation of the second kind with a better-behaved kernel but with an internal resonance difficulty. The user must weigh these issues before proceeding with analysis.

10. Smooth Open Body

Our previous discussions were based on closed bodies. When the body is open as in a shell, the user is restricted from using

MFIE, as it is invalid for bodies with vanishing enclosed volume. The user then is faced with using the EFIE in either wire-gridding or solid surface form. Other approaches are also available but are too specialized to be useful at the present time.

11. Body With Edges

The presence of edges on a conducting body presents yet another difficulty for the user. The integral equations may not be strictly valid at the edge but are valid slightly removed. Not only is the surface tangent function discontinuous, but certain components of the surface current density may be unbounded. In order to assure reasonably accurate results, the user must be aware of these items and should be prepared to make adjustments. Reality and the physical model may be the same, but the numerical model might have more dense surface discretization near edges (without edge match points for collocation) or might even round the edges (as they often are found to be in the real world).

12. Bodies With Apertures

This problem is very similar to that of the shell, although in this case the body generally more completely encloses a volume, thereby forming a cavity and may indeed not be an infinitely thin sheet. The user, in developing the model, might replace the aperture-perforated body with a finite-thickness wall with a model which involves: 1) an infinite zero-thickness screen with a hole; 2) an infinite finite-thickness screen with a hole; and 3) the actual body. The mathematical model for the latter might be 1) surface integral equations such as EFIE over the conducting surface or 2) coupled integral equations for the various distinct regions in the problem. No precise prescription can be made regarding solution of these problems - only the user can decide.

13. A Body With Appendages

Both solid body and wire problems have been mentioned. Here, one is faced with a solid body to which a wire is appended.

Reality and the physical model are remarkably similar, as indeed each portion of the total structure can be well modeled. The numerical model presents some interesting problems. The structure must be partitioned into two portions, i.e., the wire and body with the EFIE used for wire modeling and MFIE for surface modeling. Naturally, interactions between the two portions must be handled in terms of the appropriate field quantities, e.g., the magnetic field due to the wire current must be used to provide the coupling to the MFIE model of the body. Furthermore, a difficulty may be encountered at and near the point of contact of the wire and body. The flow of current from the wire onto the surface should be carefully modeled lest the local errors that might result invalidate the entire computation.

Obviously, the combination of the two problems into one can lead to difficulty unless care is taken in formulating the overall problem. Coupling and interface problems in the model must be adequately modeled. The user would be well advised to exercise extreme care when undertaking such combined methods.

14. Dielectric Bodies

The treatment of dielectric bodies using integral equation techniques is somewhat more involved than that for conductors. Although the physical model for the real homogeneous dielectric obstruction is straightforward, the construction of the numerical model is not so simple. The analyst can proceed in two distinct ways [12], viz., the polarization current approach and the coupled integral equation approach. But, the former is a three-dimensional volumetric integral equation while the latter is a surface integral equation. On the other hand, the former can cope with inhomogeneous materials while the latter must have the surface enclosing a homogeneous region. Additional questions which arise include those relating to the sampling density requirements for the dielectric body and whether the sampling densities are determined by the material with the largest dielectric constant.

15. Summarizing the Specifics of Modeling

We have only scratched the surface regarding the problem that might be posed. Of course, we do not expect very many realistic problems to fall precisely into one of the mentioned categories, but they can certainly fall into some combination of categories. But this is not the point we wanted to make. Rather, we wished to make it eminently clear to the reader that the user is faced with many decisions in the modeling process. In so doing, we can instill some appreciation of being faced with a problem to be solved with a fixed set of resources. These resources may include funds, time or manpower as well as available state-of-the-art technology such as computer programs and available results. No fixed rules can be established for arriving at the solution, for only the modeler can weigh the alternatives. However, the implication is that the modeler be aware of these subtleties in the modeling process.

VI. THE SYNERGISM OF THEORETICAL ANALYSIS, NUMERICAL ANALYSIS AND EXPERIMENTATION

There is no clear demarcation separating the realms of theory (an area based upon equations and very simple number manipulation), numerical analysis (an area requiring extensive number manipulation) and experimentation. In approaching a realistic problem, we must remember that experiment and theory first went hand in hand in expanding the theory, which was then applied to perhaps more complicated problems within its range of applicability. We have learned from theory how to view problems and how to approximate. From experiment, we learned validation of theory and approximation and even how to extend the theory. Now we have numerical approaches which can be viewed as adjuncts to both - requiring on one hand the formulation capability of theory, while on the other hand the validative capability of experiment. In fact, in some cases experimental results provide an integral portion of a model. The totality of the three realms represents an extensive resource which can be applied to problems.

We are well aware that the problem type, the information needed and the resources available are determining factors for the choice of approach to a problem. Certainly, there are areas where experimentation is difficult and costly and where numerical analysis would be more cost effective. The same can be said of all three realms. The synergistic realtionship is not meant to imply replaceability, but rather an intimate relationship.

An effective problem solution can be achieved by matching the problem type, information needed and resources available to the arsenal of techniques provided by theory, numerical analysis and experimentation. Of course, the analyst must be aware of all of these - a very demanding requirement. It is here that an *Electromagnetic Methodology* will come to the forefront as a "highly organized yet adaptive practice of precise and orderly thought and procedure for EM problem management." It will be a procedure for systematic application of tools and techniques to define and solve EM problems [11].

VII. CONCLUSIONS

Several points have been brought out in this paper, not the least of which include:

1) EM problem solving has become more computer oriented.
2) Order of magnitude advances in computer technology have enhanced this trend.
3) Future advances in computer technology are unlikely to sustain recent achievement rate increases in EM computational capabilities.
4) Formulation, approximation and validation are indispensable ingredients for numerical techniques.
5) Computer modeling in EM involves "art" as well as science.

6) The reduction of the realistic problem to a physical model and then to a numerical model leads to a need for careful interpretation of results.
7) Hybrid methods combining theoretical analysis, numerical analysis and experimentation appear promising.
8) A realistic expectation on the part of the user regarding both capabilities and limitations of the techniques is necessary.

In the overall context of problem solving, the MoM has achieved a central position. But it does not represent a tool which can solve all problems, as its computer resource requirements can become burdensome. As a result, the future will do well to bring together tools and problems in an appropriate framework (a methodology) so that cost-effective problem solving can be achieved.

VIII. REFERENCES

1. Diaz, M. Fig. 2-47 from Chapter 2 of Computer Techniques for Electromagnetics and Antennas, short course notes, University of Illinois, September 28-October 1 (1970).
2. Fenlon, F.H., Calculation of the Acoustic Radiation Field at the Surface of a Finite Cylinder by the Method of Weighted Residuals, *Proc. IEEE 57*, 291 (1969).
3. Harrington, R.F., "Time Harmonic Electromagnetic Fields," McGraw-Hill, New York (1961).
4. Harrington, R.F., "Field Computation by Moment Methods," MacMillan Co., New York (1968).
5. Landt, J.A., F.J. Deadrick, E.K. Miller and R. Kirchofer, Computer Analysis of the Far Doublet Antenna, Lawrence Livermore Laboratory Report UCRL-74846 (1973).
6. Mei, K.K., On the Integral Equations of Thin Wire Antennas, *IEEE Trans. Antennas & Propag. AP-13*, 374 (1965).
7. Miller, E.K., F.J. Deadrick and W.O. Henry, Computer Evaluation of Large, Low-Frequency Antennas, *IEEE Trans. Antennas & Propag. AP-21(3)*, 386-389 (1973).
8. Miller, E.K. and F.J. Deadrick, Some Computational Aspects of Thin-Wire Modeling, in "Numerical and Asymptotic Techniques in Electromagnetics," (R. Mittra, ed.), Springer-Verlag, New York (1973).
9. Miller, E.K. and B.J. Maxum, Mathematical Modeling of Aircraft Antennas and Supporting Structures, Final Report, ECOM Contract ADDB07-68-C-0456, Report No. ECOM-0456-1 (1970).

10. Miller, E.K. and J.B. Morton, The RCS of a Metal Plate with a Resonant Slot, *IEEE Trans. Antennas & Propag. AP-18,* 290 (1970).
11. Miller, E.K. and A.J. Poggio, A Case for an EM Methodology, IEEE International Symposium on Electromagnetic Compatibility, Washington, D.C. (July 1976).
12. Poggio, A.J. and E.K. Miller, Integral Equation Solutions of Three-Dimensional Scattering Problems, in "Computer Techniques for Electromagnetics," (R. Mittra, ed.), Pergamon Press, New York (1973).
13. Richmond, J.H., A Wire-Grid Model for Scattering by Conducting Bodies, *IEEE Trans. Antennas & Propag. AP-14,* 782 (1966).
14. Stratton, J.A., "Electromagnetic Theory," McGraw-Hill, New York (1941).
15. Thiele, G.A., and M. Diaz, Radiation of a Monopole Antenna on the Base of a Conical Structure, "Proc. Conf. on Environmental Effects on Antenna Performance," (J.R. Wait, ed.), Environmental Science Services Administration, Boulder, Colorado, 20 (1969).
16. Thiele, G.A. and T.H. Newhouse, A Hybrid Technique for Combining Moment Methods with the Geometrical Theory of Diffraction, *IEEE Trans. Antennas & Propag. AP-23*, 62.
17. Wait, J.R., Electromagnetic Radiation from Spheroidal Structures, in "Antenna Theory, Part 1, (R.E. Collin and F.J. Zucker, eds.), McGraw-Hill, New York (1969).
18. Wait, J.R., Electromagnetic Radiation from Conical Structures, in "Antenna Theory, Part 1, (R.E. Collin and F.J. Zucker, eds.), McGraw-Hill, New York (1969).
19. Bevensee, R.M., J.N. Brittingham, F.J. Deadrick, T.H. Lehman, E.K. Miller and A.J. Poggio, *IEEE Trans. Antennas and Propag. AP-26*, 156 (1978).

FINITE METHODS IN ELECTROMAGNETIC SCATTERING[1]

Kenneth K. Mei, Michael E. Morgan and Shu-Kong Chang
Department of Electrical Engineering and Computer Sciences
University of California, Berkeley

Recent developments in the unimoment method [1] have brought finite difference [2] and finite element methods [3] into the computational techniques of electromagnetic scattering. In this paper the various finite methods and their potentials in the scattering computations are examined. A section on programming technique is included for those uninitiated and the applications of the finite methods in two-dimensional and three-dimensional scattering problems, together with some of the associated computational subject matters are presented.

I. INTRODUCTION

Recent developments in the unimoment method [1] have brought finite difference [2] and finite element methods [3] into the computational techniques of electromagnetic scattering. Finite difference had been a dominant computational method of solving partial differential equations, until it was overshadowed by the finite element methods in the mid-sixties. Recent development in sparse matrix algorithms has also changed the manner the finite methods are implemented. Originally, the relaxation methods were the main tools of solving finite difference equations, and now

---

[1]Research sponsored by the U.S. Army Research Office Grant DAAG29-77-G-0021, and the U.S. Army Mobility Equipment Development Research and Development Center Contract DAAK02-75-C-002.

ISBN 0-12-709650-7

almost all finite equations are inverted instead of iteratively solved. The finite element method, which was first developed by the structure engineers, is now applied in almost all engineering disciplines. It is now well understood. Mathematical analysis [4] and sophisticated techniques [5],[6] have been developed for it. In antenna and scattering computations the finite techniques are still like new-found friends, and the acquaintence with them is yet to be developed. The purpose of this paper is to examine those methods from the standpoint of one who is interested in electromagnetic scattering computations. A section on programming technique is included for those uninitiated, and the applications of the finite methods in two-dimensional and three-dimensional scattering problems, together with some of the associated subject matter, are presented.

II. FINITE METHODS

In this paper the term "finite methods" is meant to include finite difference, weighting function, Galerkin's and finite element (or Ritz finite element) methods. Literally any numerical method that approximates continuum mathematics by discrete mathematics may be termed a finite method; however, that is not usually what we mean. For instance, the moment method of solving integral equations is not classified as a finite method. A finite method is traditionally used to solve differential or partial differential equations directly.

During the last fifteen years, computation in electromagnetic scattering problems has been actively pursued almost entirely in terms of integral equations. Needless to say, the drift to integral equations is very natural - and for good reasons. In integral equations, the computations are limited to the scatterer or antenna; while in the finite methods they are generally spread over the entire space. In integral equations the radiation conditions are automatically satisfied, while in the finite methods they require special numerical treatments which are often unsatisfactory. Recent advances in communications and remote sensing have urgently demanded the results of scattering by dielectric and lossy inhomogeneous bodies. The only practical approach to such problems appears to be direct solution of partial differential equations rather than solution by integral equations, the formulation of which in an inhomogeneous medium is already a difficult task. Recent developments in the application of finite methods [1] have essentially cleared the way for general applications of the finite methods to electromagnetic scattering.

For the sake of brevity, we shall discuss the methods first in one dimension. Let L be a differential operator; more specifically for illustration purposes we set $L = \frac{d^2}{dx^2}$. The objective of the finite methods is to solve the equation

$$L\phi(x) = f(x) \tag{1}$$

with boundary conditions $\phi(0) = a$ and $\phi(1) = b$.

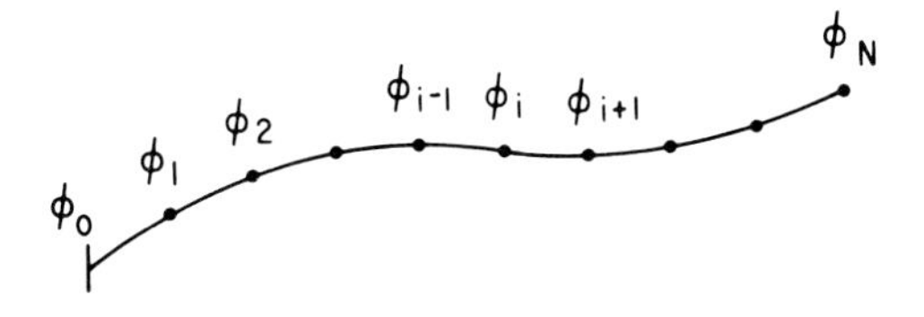

Fig. 1. Discretization of an independent variable.

In most numerical methods, the independent variable is discretized as shown in Fig. 1. The formulations of the finite methods are illustrated in the following:

$$L = \frac{d^2}{dx^2}$$

Finite Difference

$$\frac{\phi_{i+1} - 2\phi_i + \phi_{i-1}}{(\Delta x)^2} = f_i \qquad i = 1,\ldots,N-2 \tag{2}$$

Weighting Function Method I (WFM I)

$$\int_0^1 \frac{d^2\phi(x)}{dx^2} W_i(x)dx = \int_0^1 f(x)W_i(x)dx \qquad i = 1,2,\ldots,M \tag{3}$$

Weighting Function Method II (WFM II)

$$\phi^1(x)W_i(x)\Big]_0^1 - \int_0^1 \frac{d\phi}{dx}\frac{dW_i}{dx}\,dx = \int_0^1 f(x)W_i(x)dx \tag{4}$$

Galerkin's Method

$$\phi(x) = \sum_{i=1}^{N} a_i\zeta_i(x) \text{ and set } W_i(x) \text{ in (4) to } \zeta_i(x)$$

(Ritz) Finite Element

$$\text{Minimize } I = \int_0^1 \left[\left(\frac{d\phi}{dx}\right)^2 + 2f\phi\right] dx \tag{5}$$

In most textbooks weighting function is not distinguished, whether integration by parts is used or not. For convenience of the following discussion, we shall call it weighting function method II when integration by parts is used, otherwise weighting function method I.

We shall name the set of functions which are used to approximate the solution $\phi(x)$ the trial functions, thus

$$\phi(x) = \sum_{i=1}^{N} a_i\zeta_i(x) \tag{6}$$

where ζ_i are trial functions. $W_i(x)$ are a set of weighting functions.

The following are a few definitions of commonly used terms:

Patch function - functions which are zero outside an element and its close neighbors, such as those shown in Fig. 2.

Element - a line segment (one dimension), area (two dimensions) or volume (three dimensions) as a result of discretization.

Interpolate - a set of trial functions is termed interpolative if it is unity at one node and zero at any other node. An interpolative trial function has the property that $\phi(x) \simeq \sum_{i=1}^{N} a_i \zeta_i(x)$ where a_i represents $\phi(x_i)$.

Conformity - a set of trial functions is termed conforming if each of its members is continuous.

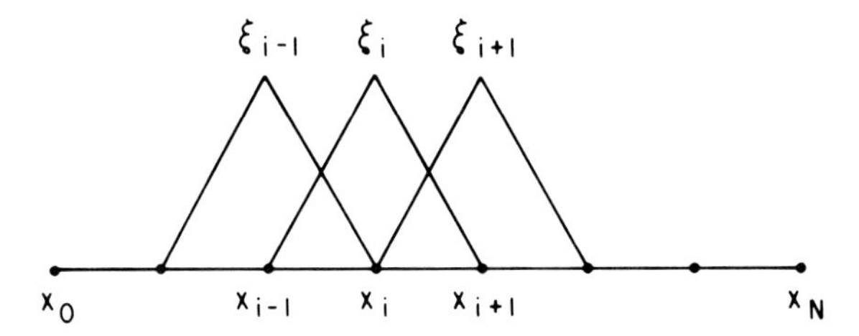

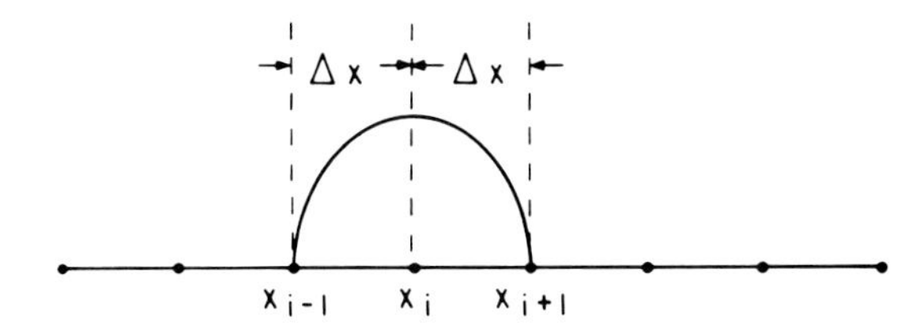

$$1 - \frac{(x - x_i)^2}{\Delta x^2}$$

Fig. 2. Patch functions. (a) Linear. (b) Quadratic.

Only those methods that use patch functions as trial functions and weighting functions will be considered as finite methods.

It is well known that the finite difference equations are derived from Taylor's series. It is not difficult to relate the finite difference equation to the W.F. method. Consider a quadratic patch function

$$Q_i(x) = A_i + B_i(x - x_i) + C_i(x - x_i)^2 \tag{7}$$

passing through $\phi(x_{i-1})$, $\phi(x_i)$, $\phi(x_{i+1})$. It is immediately evident $A_i = \phi(x_i)$; $B_i = \dfrac{\phi(x_{i+1})-\phi(x_{i-1})}{2\Delta x}$; $C_i = \dfrac{\phi(x_{i+1})-2\phi(x_i)+\phi(x_{i-1})}{2(\Delta x)^2}$.

In (3) let $\phi(x) = \sum_{i=1}^{N} Q_i(x)$ and the weighting functions $W_i(x)$ be $\delta(x - x_i)$; we find that (3) reduces to the finite difference equation (2). On the other hand, in (5) let $\phi(x) = \sum_{i=1}^{M} a_i\xi_i(x)$, then

$$I = \int_0^1 \left\{\left[\sum_{i=1}^{M} a_i\xi_i'(x)\right]^2 + 2f \sum_{i=1}^{M} a_i\xi_i(x)\right\}dx,$$

$$\frac{\partial I}{\partial a_j} = \int_0^1 \left\{2\left[\sum_{i=1}^{M} a_i\xi_i'(x)\right]\xi_j'(x) + 2f\xi_j(x)\right\}dx = 0\ , \quad j = 1,2,\ldots,M \tag{8}$$

which is equivalent to (4) if the trial functions $\xi_i(x)$ and the weighting functions $W_i(x)$ are the same set of functions, and $W_i(0) = W_M(1) = 0$.

II. NUMERICAL RESTRICTIONS

While W.F. methods I and II seem to be only a simple mathematical step away from each other, their numerical constraints are quite different. In the simple example of (3) and (4), it is evident that in W.F. method I the trial functions must be at least quadratic and the weighting function must be integrable; while in W.F. method II, the trial functions can be linear and the weighting function must belong to C', ($W'(x)$ integrable). The consequence is that one can use linear functions as trial functions in W.F. method II, but not in method I.

Another important observation is that the union of the set of trial functions must cover the entire region, while the weighting functions are not so required. Thus, in the W.F. methods, the weighting functions may be sampling functions, e.g., pulses or patch functions which do not overlap, such as shown in Fig. 3, as opposed to those patch functions of Fig. 2 which are used as trial functions.

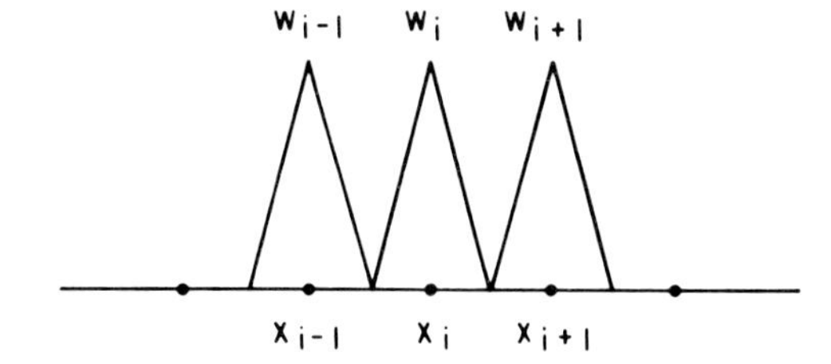

Fig. 3. Patch functions which can only be used as weighting functions.

One important property of sampling is that discrepancies in the trial functions may be suppressed by the weighting functions. An obvious discrepancy in the W.F. method I using piecewise quadratic trial functions is that the trial functions are not conforming, as illustrated in Fig. 4. If we use the left of the quadratic for $x_i - \frac{\Delta x}{2} \leq x \leq x_i + \frac{\Delta x}{2}$ and the right quadratic for $x_{i+1} - \frac{\Delta x}{2} \leq x \leq x_{i+1} + \frac{\Delta x}{2}$, the discontinuity is obvious. Yet it causes no ill effect because the weighting functions are zero at the discontinuities.

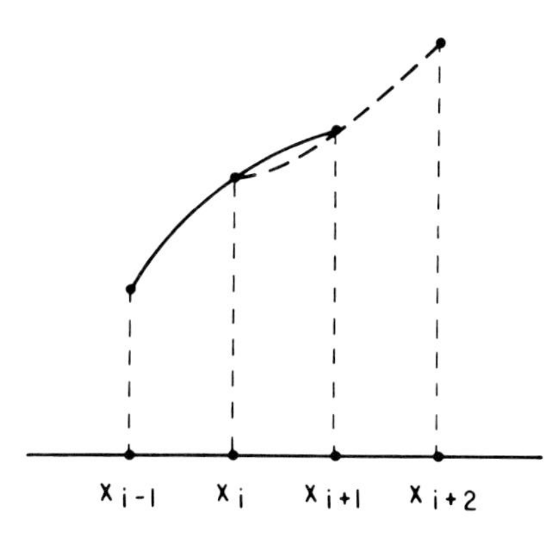

Fig. 4. Nonconforming trial functions.

The case for discontinuity becomes very serious for the finite element method, since the trial functions are now chosen to be the weighting functions and the trial functions must cover the entire region. The derivatives at the discontinuities result in delta functions, and the squares of delta functions are not integrable. That is why elements must be conforming in the finite element method. However, being nonconforming is still not fatal in the finite element method. Frequently the irregularities are ignored and the computations still may yield good results, although just as frequently they fail. Nonconforming elements are often referred to as "variational crime" in finite element literatures; and there exists a test, known as the "patch test," which may be used to predetermine whether a specific set of nonconforming elements would converge to the correct result. In one dimension, conforming elements are easily obtained even for high order polynomial approximations. Any piecewise polynomial function which is non-zero at only one node is a conforming element. In two or higher dimensions conforming requires extra effort, which may result in increasing the number of nodes or the band of the matrix or both.

III. ONE-DIMENSION FINITE METHODS

Consider the W.F. method II using linear functions for both $\xi_i(x)$ and $W_i(x)$, such as those shown in Fig. 2a. The integral in (4) from x_{i-1} becomes

$$\int_{x_{i-1}}^{x_i} \frac{\phi_i - \phi_{i-1}}{\Delta x} \cdot \frac{1}{\Delta x}\, dx - \int_{x_{i-1}}^{x_{i+1}} \frac{\phi_{i+1} - \phi_i}{\Delta x} \cdot \frac{1}{\Delta x}\, dx$$

$$= - \frac{\phi_{i+1} - 2\phi_i + \phi_{i-1}}{\Delta x} \tag{9}$$

which is almost identical to the finite difference equation (2), except that the right hand side of (2) is collocation and that of (4) is weighted by a linear patch function. Indeed the base of the weighting function may be reduced without any effect on the equation.

Now, consider a quadratic trial function used in W.F. method II, i.e.,

$$\xi_i(x) = \phi_i + \frac{\phi_{i+1} - \phi_i}{2\Delta x} (x - x_i) + \frac{\phi_{i+1} - 2\phi_i + \phi_{i-1}}{2\Delta x^2} (x - x_i)^2 \tag{10}$$

then,

$$\frac{d\xi_i(x)}{dx} = \frac{\phi_{i+1} - \phi_i}{\Delta x} + \frac{\phi_{i+1} - 2\phi_i + \phi_{i-1}}{\Delta x^2}(x - x_i) \quad . \tag{11}$$

If the weighting function is even with respect to x_i, $\frac{dW_i}{dx}$ will be odd function w.r.t. x_i; thus the first term in (11) will have no contribution in the integral of the product $\frac{d\xi_i(x)}{dx}\frac{dW_i(x)}{dx}$, and we shall get an equation again identical to the finite difference equation.

It seems that all finite methods result in the same equations. The fact is that, in this particular example, the second order polynomial operated by the differential operator results in a constant. As a result, a sampling weighting function set is as good as any distributed weighting function set. Indeed, the quadratic functions are the best approximation functions one can get to solve this particular problem, because the differential equation is satisfied not only at the nodes but also between nodes, and any attempt to use higher order approximation usually makes the approximation worse rather than better.

IV. TWO-DIMENSION PROBLEMS

The above is the story for one-dimensional equations. In two dimensions, many of the above discussions are still valid, but there are other complications.

A. Conformity

Conformity of elements in one dimension is never a problem, but in two dimensions it is a major problem when high order approximations are to be considered. Consider a typical finite element discretization in Fig. 5. The elements are triangular and the trial functions are pryamid-like linear patch functions. A typical trial function consists of six elements and seven nodes. Referring to Fig. 5, a linear trial function centering at the

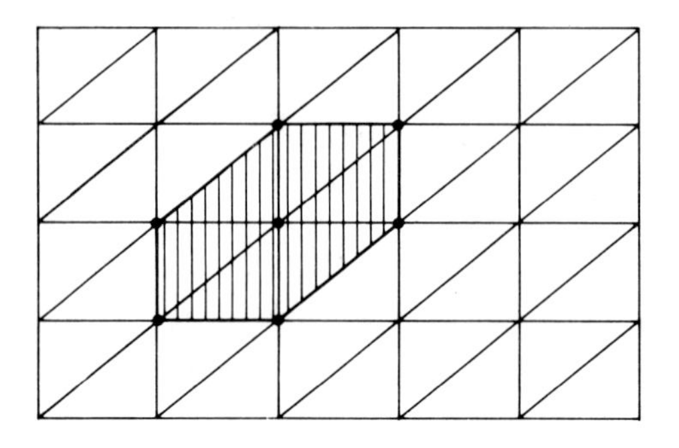

Fig. 5. Typical discretization in a two-dimension finite element method.

point i consists of six planes which have values one at the point i and zero at the other nodes, such as shown in Fig. 6.

Fig. 6. A typical linear trial function for two-dimensional finite element method.

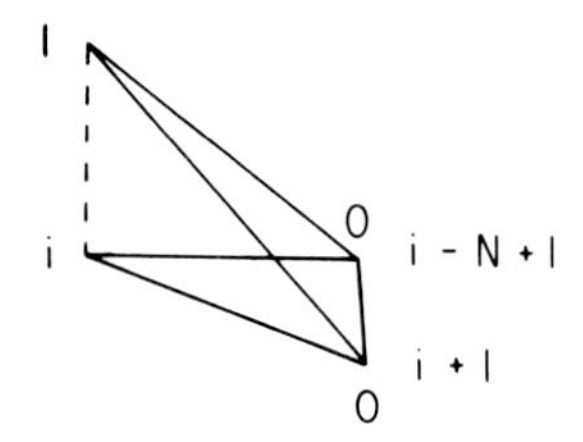

It is evident that the linear elements are conforming; i.e., any linear combination of such trial functions shall have no discontinuity.

The second order approximation in the finite element methods uses a piecewise quadratic function for $\phi(x,y)$ within each element:

$$\phi(x,y) = a_0 + a_1x + a_2y + a_3xy + a_4x^2 + a_5y^2 . \tag{12}$$

The quadratic function of (x,y) has six degrees of freedom. The three nodes of a triangular element are not sufficient to determine the six coefficients of the quadratic. Extra nodes need to be added, such as shown in Fig. 7, in which a node is designated on each side of an element. These trial functions are now conforming. Quadratics which are unity at one node and zero at the rest of the nodes of a triangular element can be found either by inverting a 6×6 matrix or by using an area coordinate system (for details see [5] and [7]). The second order approximation in the finite element method gives better accuracy, but it also increases the number of nodes and increases the bandwidth of the matrix.

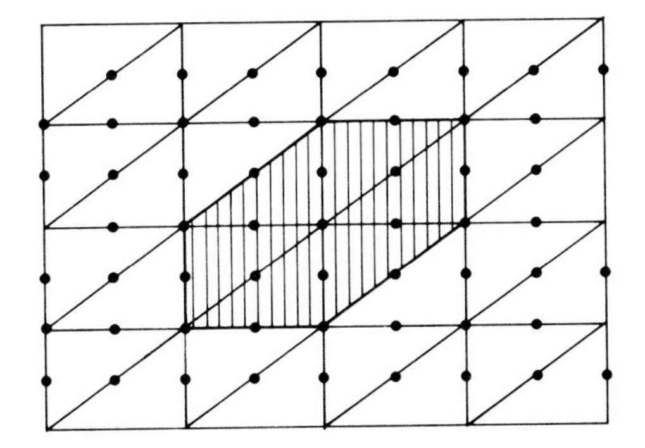

Fig. 7. Discretization and nodes for quadratic elements in a two-dimension finite element method.

In finite difference, assuming square mesh, the usual five point equation for the Laplacian operator gives (referring to Fig. 5)

$$\phi_{i+N} + \phi_{i+1} - 4\phi_i + \phi_{i-1} + \phi_{i-N} = 0 \quad . \tag{13}$$

It is a result of quadratic approximation of $\phi(x,y)$ near ϕ_i, but it is not a complete quadratic. Indeed, a five point equation cannot represent a complete quadratic which has six coefficients. The missing term in (13) is a_3xy, which can be omitted if the mesh is right angle.

The trial functions resulting in (13) are not conforming, but the discrepancy is again suppressed by the weighting functions. Again, for solving Laplace's equation the quadratic equation of (13) satisfies the differential equation between nodes along the mesh, and it should be a better approximation than the nine point finite difference equations (fourth order polynomial) which satisfy the differential equation only at the nodes. Many engineers have been surprised to find that their nine point finite difference equations give worse results than the five point finite difference equations when dealing with Laplace equations.

B. Consistency

Because of the sampling property of the finite difference equations, there are frequent inconsistencies, particularly when dealing with inhomogeneous media. In the case of a Helmholtz equation for a continuously inhomogeneous medium, the finite difference equation is

$$\phi_{i+N} + \phi_{i+1} + (k_i^2 h^2 - 4\phi_i) + \phi_{i-1} + \phi_{i-N} = 0 \qquad (14)$$

where the field points ϕ_{i+N}, ϕ_{i+1}, ϕ_i, ϕ_{i-1}, ϕ_{i-N} are assumed to be in a homogeneous medium ε_i. Letting i be replaced by $i + 1$, we find ϕ_i, ϕ_{i+1} are now assumed to be in a homogeneous medium ε_{i+1}. This type of inconsistency, while tolerable, is not very satisfying theoretically. The problem becomes more serious when a node is situated at a material discontinuity, such as shown in Fig. 8. There are several ways of handling material discontinuities in the finite difference method, but in general they are not satisfying.

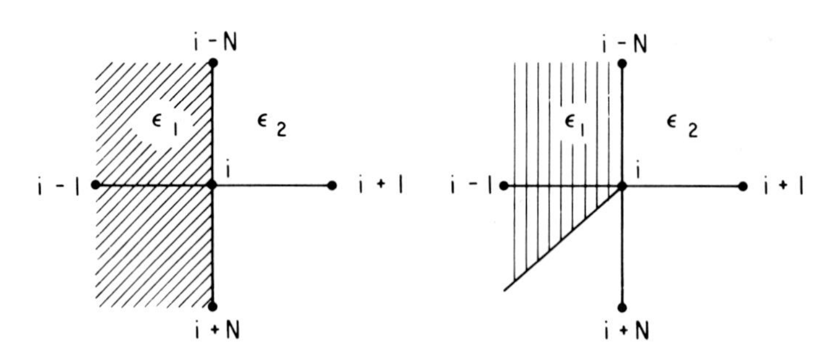

Fig. 8. Nodes on material discontinuities.

In the finite element method, the fields are clearly identified in each patch; hence there is no ambiguity as in the case of the finite difference. Furthermore, the quadratic function actually consists of six piecewise quadratic functions, while in the finite difference method the requirement of twice differentiability on the trial functions necessitates a single polynomial to cover a group of connected nodes.

C. Flexibility

Because the polynomials in the finite elements are complete, the method is essentially independent of the shape of the triangular element. In the finite difference method, the polynomial is incomplete; hence it is restricted to rectangular meshes. It can be made to be independent of mesh geometry, however, by using a six point difference mesh such as shown in Fig. 9. Indeed, for a six point finite difference, since it is now independent of mesh geometry, the mesh is no longer needed. The only requirement is that the nodes are relatively uniformly distributed. A survey of recent developments in variable grid finite difference methods can be found in [8].

The usual applications of the finite difference method use rectangular mesh, with the boundary nodes specially treated by extrapolation or interpolation [9]. The merit of such a scheme is the simplified programming, and the disadvantages are reduced accuracy and the difficulties involved in finding the normal derivatives from the solution. In the finite element method, using triangular elements, it is necessary to store information about the locations of the nodes, and also the information about the elements, i.e., the nodes which form a particular element. Finite difference method using variable mesh should require the same information, i.e., the information about the neighboring nodes of a control node must be available. Consequently, finite difference method using variable mesh is about as flexible as the finite element method, and the degree of complexity in programming for both methods is about the same.

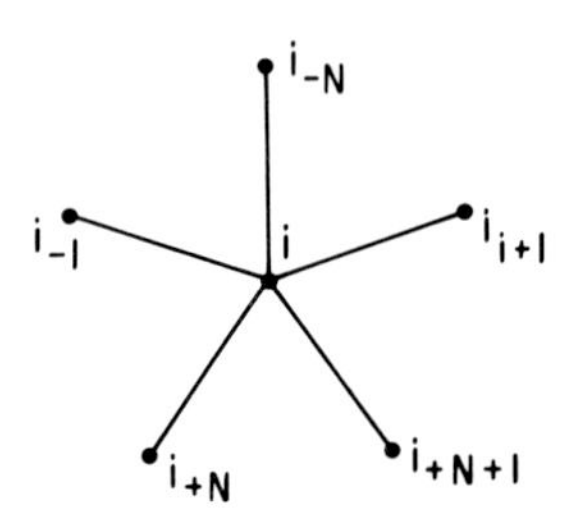

Fig. 9. A six point finite difference mesh.

The conclusion is that finite difference is about as good as linear approximation in finite element. The higher order approximation in finite element should be more accurate than the corresponding approximations in finite difference, yet the finite elements method would require more nodes. The finite difference using rectangular grid is simple to program but less flexible; and by using star meshes it can be made more flexible, but its complexity is raised to that of the linear finite element method. The method seldom used is the W.F. method II, with which one may use piecewise polynomial trial functions like in the finite element method, yet with a much relaxed conformity requirement. It will be a worthwhile project to investigate whether the relaxed conformity requirement on W.F. method II can reap some of the benefits of higher order approximations without increasing the nodes or bandwidth of the resulting matrix.

V. PROGRAMMING TECHNIQUE

To implement the finite methods one need first to describe the geometry and construct the mesh, followed by generating the matrix and solving the matrix. These basic steps are described in the following.

A. Mesh Construction

In most practical problems, the node numbers are large and the mesh topology is complicated. Therefore, it is almost an impossible task to instruct the computer of an exact mesh geometry.

The best strategy is to construct a rule which the program can follow to generate the mesh automatically. There are many ways of automatic mesh generation. In the following we shall describe one of them, which we have used very effectively:

1. Construct a Regular Mesh Inside the Circular Boundary

This can be done in many ways, a few of which are shown in Fig. 10. These meshes can be generated automatically. Fig. 10a is used if the center part of the circle is included. Fig. 10b is used when the center part of the circle is not included in the finite method.

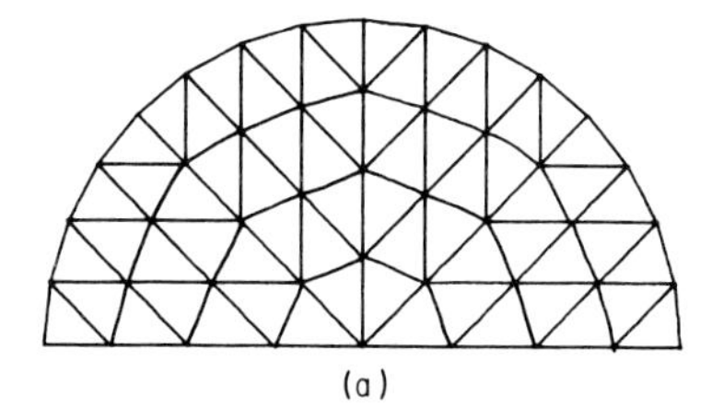
(a)

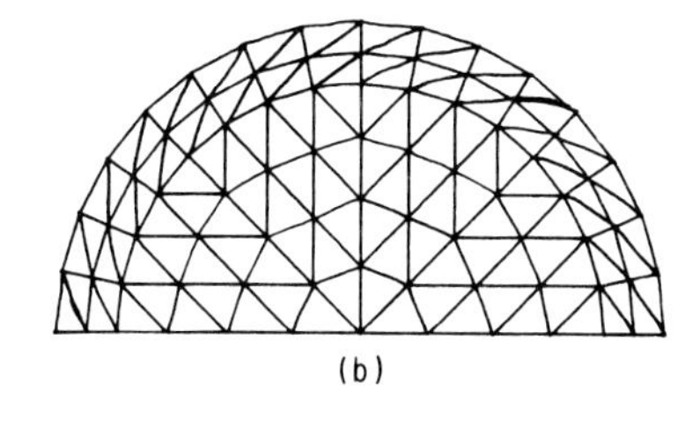
(b)

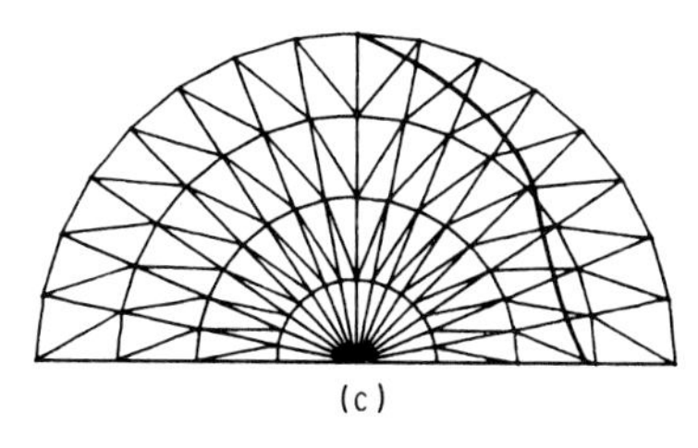
(c)

Fig. 10. Regular meshes in a circular region.

2. Find the Intersection of the Scatterer Contour with the Mesh Line

If the mathematical formula of the scatterer contour is known, it is relatively easy to find the intersections of the contour with the mesh lines. For the mesh geometry of Fig. 10b, it is a particularly easy task, because we only need to find the intersections between the contour and the radial lines. One of the two nearest nodes to an intersection point along the radial line is now shifted to the intersection point. The rest of the nodal points along the particular radial lines are now shifted so as to made the perturbation equally distributed. It is noticed that there are two possible orientations of the diagonal hypotenuse of the triangular elements between two adjacent radial lines. We shall call those elements of Fig. 11a left handed (LH) elements, and those of Fig. 11b right handed (RH) elements. The choice of the type of element may be made based on the rule:

$$R_{i+1} \geq R_i \rightarrow \text{LH element} \qquad R_{i+1} < R_i \rightarrow \text{RH element} .$$

Using the above rules, the contour of a distorted raindrop is discretized in Fig. 12.

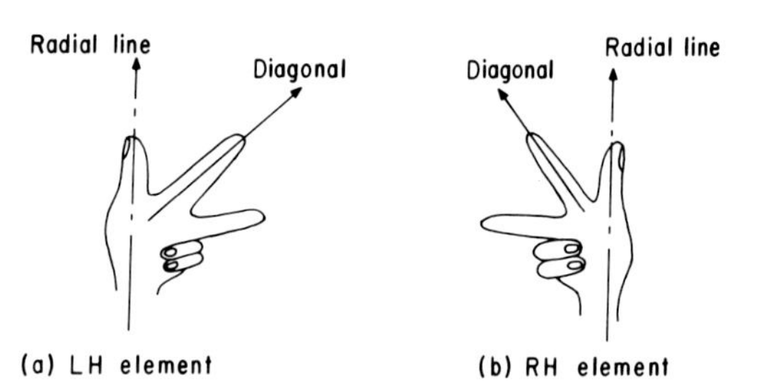

Fig. 11. (a) Left handed and (b) right handed elements.

B. Computation of Matrix Elements

The key to the programming of any finite method with variable elements is the name list of nodes, which form a particular element. Referring to Fig. 13, we generate an array ELEM(I,J), where I is the numbering of the element and the dimension for J is three. This array is used to store the list of the nodes which form the particular element I. For example, ELEM(12,1) = 3; ELEM(12,2) = 6, ELEM(12,3) = 7. Of course, the coordinate locations of each node should be stored in another array.

With the information of the elements and nodes available, we

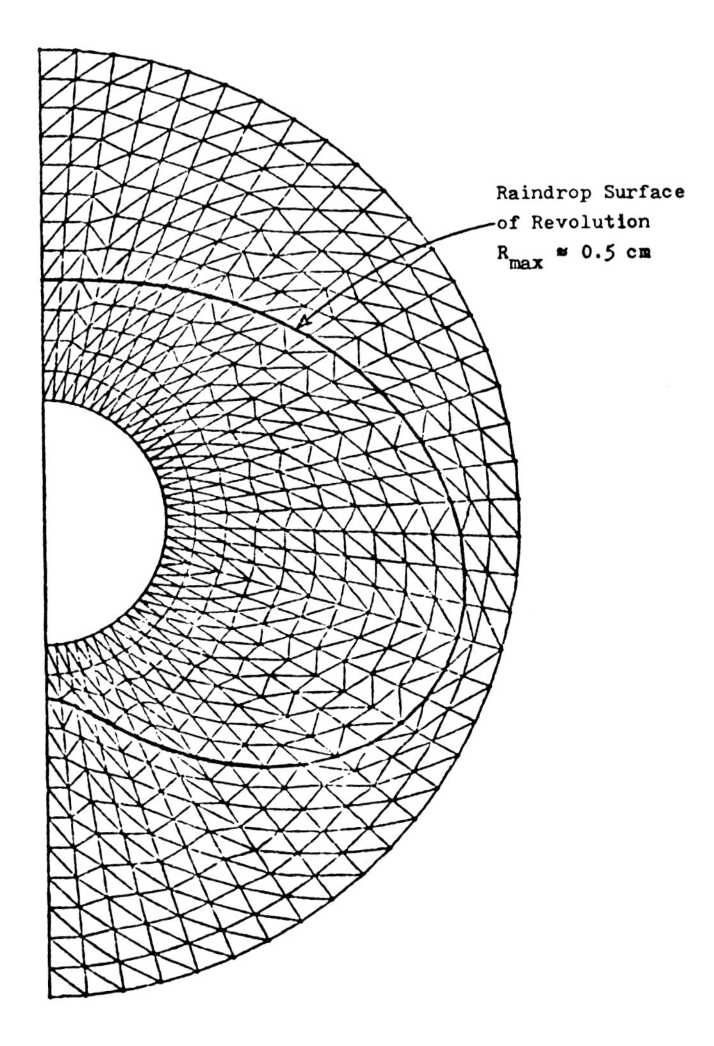

Fig. 12. Discretization of a distorted raindrop.

now consider the generation of the matrix elements. In the following, we shall consider the generation of the matrix element of the two-dimension Helmholtz equation using the weighting function method II. Thus we compute the integral:

$$\int_S (\nabla U \cdot \nabla W_i + k^2 U W_i) ds \tag{15}$$

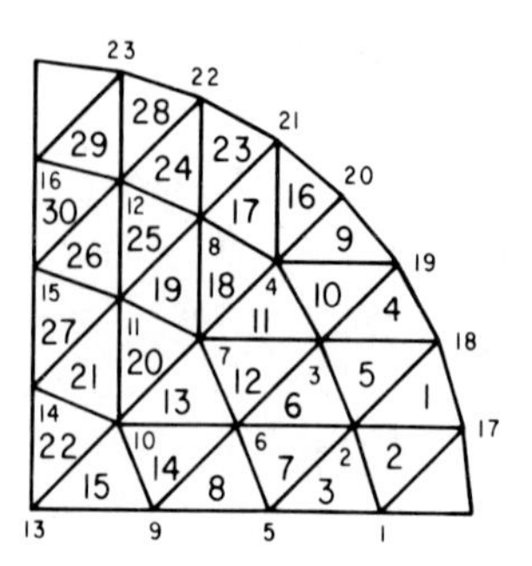

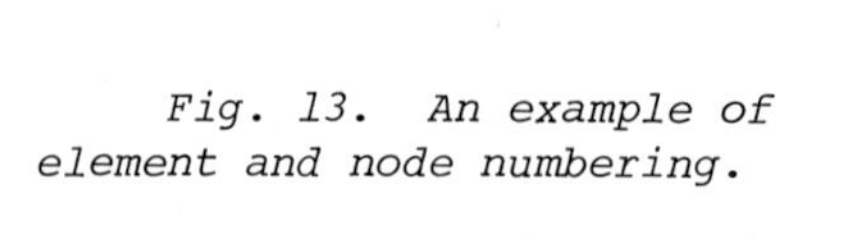

Fig. 13. An example of element and node numbering.

Both U and W will be piecewise linear functions; therefore the result will be identical to the finite element method.

A typical linear element h, with nodes $(h)_1$, $(h)_2$, $(h)_3$ should have the field within the element described by a linear function

$$U_{(h)} = a_{(h)1} + a_{(h)2}x + a_{(h)3}y \tag{16}$$

The coefficients $a_{(h)i}$ are found by matrix inversion

$$\bar{a}_{(h)} = N_{(h)}^{-1} \bar{U}_{(h)} \tag{17}$$

The vector $\bar{U}_{(h)}$ consists of the field at the nodal points; hence, $U_{(h)i}$, $(i = 1,2,3)$, and

$$N_{(h)} = \begin{bmatrix} 1 & x_{(h)1} & y_{(h)1} \\ 1 & x_{(h)2} & y_{(h)2} \\ 1 & x_{(h)3} & y_{(h)3} \end{bmatrix}$$

Let us consider W_i to be a linear pyramid function which is unity at node i and zero at the surrounding nodes, such as shown in Fig. 14. Throughout the entire process of weighting integrals, each element will be weighed by three different weighting functions, each of which may be represented by a linear function which is unity at one node of the element and zero at the other two. The function which is unity at node $(h)i$ is actually part of the

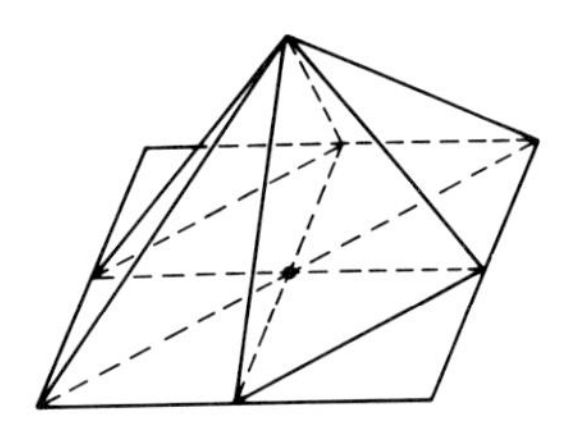

Fig. 14. A linear weighting function.

weighting function $(h)i$, and the coefficients generated by it are associated with the $(h)i^{th}$ equation or the $(h)i^{th}$ row of the matrix.

To each $U_{(h)}$ in an element h, there are three associated weighting functions: $W_{(h)j}$ $(j = 1,2,3)$. Let $W^j_{(h)}$ be part of $W_{(h)j}$, which is associated with the element h and which has value at the node $(h)j$ and zero at the other two nodes of element h.

$$W^j_{(h)} = b^j_{(h)1} + b^j_{(h)2}x + b^j_{(h)3}y \tag{18}$$

and $\bar{b}^j_{(h)} = N^{-1}_{(h)}\ \bar{n}_j$ where $\bar{n}_1 = \begin{bmatrix}1\\0\\0\end{bmatrix}$, $\bar{n}_2 = \begin{bmatrix}0\\1\\0\end{bmatrix}$, $\bar{n}_3 = \begin{bmatrix}0\\0\\1\end{bmatrix}$. The gradient of $U_{(h)}$ gives:

$$\nabla U_{(h)} = \begin{bmatrix}0 & 0 & 0\\0 & 1 & 0\\0 & 0 & 1\end{bmatrix}\begin{bmatrix}a_{(h)1}\\a_{(h)2}\\a_{(h)3}\end{bmatrix} = \begin{bmatrix}0 & 0 & 0\\0 & 1 & 0\\0 & 0 & 1\end{bmatrix} N^{-1}_{(h)}\ \bar{U}_{(h)} \tag{19}$$

and the gradient of $W^j_{(h)}$ is:

$$\nabla W^j_{(h)} = \begin{bmatrix}0 & 0 & 0\\0 & 1 & 0\\0 & 0 & 1\end{bmatrix} N^{-1}_{(h)}\ \bar{n}_j \tag{20}$$

The integral $\int_{(h)} \nabla U_{(h)} \cdot \nabla W^j_{(h)}\ ds$ is

$$\int_{(h)} U_{(h)}{}^T \begin{bmatrix}0 & 0 & 0\\0 & 1 & 0\\0 & 0 & 1\end{bmatrix} N^{-1}_{(h)}{}^T \begin{bmatrix}0 & 0 & 0\\0 & 1 & 0\\0 & 0 & 1\end{bmatrix} N^{-1}_{(h)}\ \bar{n}_j\ ds$$

$$= \bar{U}_{(h)}{}^T \cdot \bar{P}^j_{(h)}\ A_{(h)} \tag{21}$$

where $A_{(h)}$ is the area of the element h, and $\bar{p}^j = \left[p^j_{(h)1}, p^j_{(h)2}, p^j_{(h)3}\right]^T$. $A_{(h)} p^j_{(h)i}$ is recognized as the contribution to the matrix element $G[(h)j, (h)i]$ $(i = 1,2,3)$. This value should be accumulated in the array G in the location $G[(h)j, (h)i]$.

Similarly the second integral of (15) can be carried out by

$$k^2 \int_{(h)} U_{(h)} W^j_{(h)} ds = k^2 \int_{(h)} \bar{U}^T_{(h)} N^{-1\,T}_{(h)} \begin{bmatrix} 1 \\ x \\ y \end{bmatrix} [1\ x\ y]\, N^{-1}_{(h)} \bar{n}_j ds$$

$$= k^2 \int_{(h)} \bar{U}^T_{(h)} N^{-1\,T}_{(h)} \begin{bmatrix} 1 & x & y \\ x & x^2 & xy \\ y & xy & x^2 \end{bmatrix} N^{-1}_{(h)} \bar{n}_j ds \tag{22}$$

$$= k^2 \int \bar{U}^T_{(h)} \cdot \bar{q}^j_{(h)}\, ds \tag{23}$$

The ith element of the vector $\bar{q}^j_{(h)}$ is thus the contribution to the $(h)j^{\text{th}}$ row and $(h)i^{\text{th}}$ column of the matrix. The integrals

$$I_{ij} = \int_{(h)} x^i y^j ds \tag{24}$$

may be found by using the local coordinates: $u = x - x_c$; $v = y - y_c$; $x_c = (x_{(h)1} + x_{(h)2} + x_{(h)3})/3$; $y_c = (y_{(h)1} + y_{(h)2} + y_{(h)3})/3$.

The following table gives the most useful results [10]:

$i + j$	I_{ij}
0	A (area of triangle)
1	0
2	$\frac{A}{12}(u_1^i v_1^j + u_2^i v_2^j + u_3^i v_3^j)$
3,4	$\frac{A}{30}(u_1^i v_1^j + u_2^i v_2^j + u_3^i v_3^j)$
5	$\frac{2A}{105}(u_1^i v_1^j + u_2^i v_2^j + u_3^i v_3^j)$

where $u_i = x_{(h)i} - x_c$; $v_i = y_{(h)i} - y_c$; and $A = |\det N_{(h)}|/2$. If $k^2 = k_0^2\varepsilon$ where $\varepsilon(x,y)$ is a function of position, we may represent $\varepsilon(x,y)$ by piecewise polynomial functions, and the integrals can be found accordingly.

C. Solving the Linear Equations

The dimension of the unknown vector $\bar{u}$ is usually very large. Fortunately, the matrix $G(I,J)$ is usually sparse and banded. Referring to Fig. 13, we notice that the nodes (1,2,3,4) are only connected to the interior nodes (5,6,7,8) which are only connected to the interior (1,2,3,4) and (9,10,11,12), etc. Representing these nodal values of the field by vectors: $\bar{v}_1 = (u_1,u_2,u_3,u_4)$; $\bar{v}_2 = (u_5,u_6,u_7,u_8)$; $\bar{v}_3 = (u_9,u_{10},u_{11},u_{12})$; $\bar{v}_4 = (u_{13},u_{14},u_{15},u_{16})$, we can represent the banded matrix by the vector equations

$$B_i\bar{v}_{i-1} + A_i\bar{v}_i + C_i\bar{v}_{i+1} + \bar{d}_i = 0 \tag{25}$$

The matrix equation is thus (referring to Fig. 13)

$$G\bar{U} = \begin{bmatrix} A_1 & C_1 & 0 & 0 \\ B_2 & A_2 & C_2 & 0 \\ 0 & B_3 & A_3 & C_3 \\ 0 & 0 & B_4 & A_4 \end{bmatrix} \begin{bmatrix} \bar{v}_1 \\ \bar{v}_2 \\ \bar{v}_3 \\ \bar{v}_4 \end{bmatrix} = \begin{bmatrix} \bar{d}_1 \\ \bar{d}_2 \\ \bar{d}_3 \\ \bar{d}_4 \end{bmatrix} \tag{26}$$

where all A_i's are 4×4 sparse matrices, C_i's are lower triangular sparse matrices (frequently diagonal) and B_i's are upper triangular sparse matrices (frequently diagonal); $B_1 = C_4 = 0$ and the vector $\bar{d}_i$ represent the contributions from the boundary values.

The solution can be obtained by the following block-by-block elimination method. Assume

$$\bar{v}_{i-1} = R_i\bar{v}_i + \bar{s}_i \qquad i = 1,2,\ldots \tag{27}$$

Equation (25) can be rewritten as

$$\bar{v}_i = -(A_i + B_iR_i)^{-1}C_i\bar{v}_{i+1} - (A_i + B_iR_i)^{-1}(B_i\bar{s}_i + \bar{d}_i) \tag{28}$$

Comparing (28) with (27) with $i = i+1$, we get

$$R_{i+1} = -(A_i + B_i R_i)^{-1} R_i \tag{29}$$

and

$$\bar{s}_{i+1} = -(A_i + B_i R_i)^{-1} (B_i \bar{s}_i + \bar{d}_i) \tag{30}$$

The above two recurrence formulas can be initiated with $i = 1$ in (25), which results in

$$\bar{v}_1 = -A_1^{-1} C_1 \bar{v}_2 - A_1^{-1} \bar{d}_1 \tag{31}$$

Comparing (31) with (27) for $i = 2$, we get

$$\left.\begin{aligned} R_2 &= -A_1^{-1} C_1 \\ \bar{s}_2 &= -A_1^{-1} \bar{d}_1 \end{aligned}\right\} \tag{32}$$

Comparing (32) with (29) and (30), it is clear that

$$\left.\begin{aligned} R_1 &= 0 \\ s_1 &= 0 \end{aligned}\right\} \tag{33}$$

The end condition at $i = N$ is

$$B_N \bar{v}_{N-1} + A_N \bar{v}_N + \bar{d}_N = 0 \tag{34}$$

Substituting (27), with $i = N$, into (34) we get

$$\bar{v}_N = -(A_N + B_N R_N)^{-1} (B_N \bar{s}_N + \bar{d}_N) = \bar{s}_{N+1} \tag{35}$$

To summarize, we have the following procedure:

1) Use the initial conditions (33)
2) To generate R_i, $\bar{s}_i$ $(i = 1,2,\ldots,N+1)$ from the recurrence formulas (29) and (30).
3) Use the end condition (25)
4) To calculate $\bar{v}_i$ $(i = N, N-1, N-2,\ldots, 1)$ from (27).

VI. APPLICATIONS

A. Two Dimensional Scattering

Equation (29) clearly indicates that all the R_i matrices can be computed independently of the boundary values. Therefore, we may obtain a set of the solutions of the interior problem based on a set of specified boundary values from the same R_i's. The vectors $\bar{s}_i$'s are dependent on the boundary values, but their computations are not as time consuming as those for R_i's. From the interior solutions the normal derivatives for each Dirichlet problem can be found.

Let the exterior solution be expanded in a finite series of cylindrical harmonics, assuming symmetry with respect to the x-axis

$$u(\bar{r}) = \sum_{n=0}^{N} a_n \cos n\phi \, H_n^{(2)}(kr) \qquad r \geq a \tag{36}$$

and the interior solution be expanded into $N + 1$ linearly independent solutions described in the previous section

$$u(\bar{r}) = \sum_{n=0}^{N} b_n u_n(\bar{r}) \qquad r \leq a \tag{37}$$

where a is the radius of the circumscribing circle, and the normal derivatives $\frac{\partial U_n}{\partial r}(a)$ are found via the solutions $u_n(\bar{r})$. To find the coefficients a_n and b_n, we equate (36) and (37), and their normal derivatives at $r = a$, to get

$$\int_0^{\pi/2} \left\{ U^{inc}_{(a,\phi)} + \sum_{n=0}^{N} a_n \cos n\phi \, H_n^{(2)}(ka) \right\} \cos m\phi \, d\phi$$
$$= \sum_{n=0}^{N} b_n \int_0^{\pi/2} U_n(a,\phi) \cos m\phi \, d\phi \qquad (m = 0,1,2,\ldots,N) \tag{38}$$

$$\int_0^{\pi/2} \left\{ \frac{\partial U^{inc}}{\partial r}(a,\phi) + \sum_{n=0}^{N} a_n \cos n\phi \frac{\partial H_n^{(2)}}{\partial r}(ka) \right\} \cos m\phi \, d\phi =$$

$$= \sum_{n=0}^{N} b_n \int_0^{\pi/2} \frac{\Delta U_n(a,\phi)}{\Delta r} \cos m\phi \; d\phi \qquad (m = 0,1,\ldots,N) \tag{39}$$

The above should result in a $2(N+1) \times 2(N+1)$ matrix from which the coefficients can be found. A word of advice here regarding the normal derivatives $\frac{\partial U^{inc}_{(r)}}{\partial r}$ and $\frac{\partial H_n^{(2)}}{\partial r}(kr)$: In general, the exact values of the normal derivatives of the incident field and the Hankel's functions are known on $r = a$; however, it is advisable to use their approximate values $\frac{\Delta U^{inc}_{(a)}}{\Delta r}$ and $\frac{\Delta H_n^{(2)}}{\Delta r}(ka)$ instead. The reason is that the right side of (39) contains approximate derivatives obtained from the finite difference of the nodal values. They are not equal to the exact derivatives. The equality of the approximate and the exact derivatives can be enforced through the alteration of the coefficients from their proper values. Indeed, (39) is more meaningful if both sides have the same order of approximation.

A few solutions of the two-dimensional scattering are shown in Figs. 15 and 16. The details of the two-dimensional solution can be found in [3].

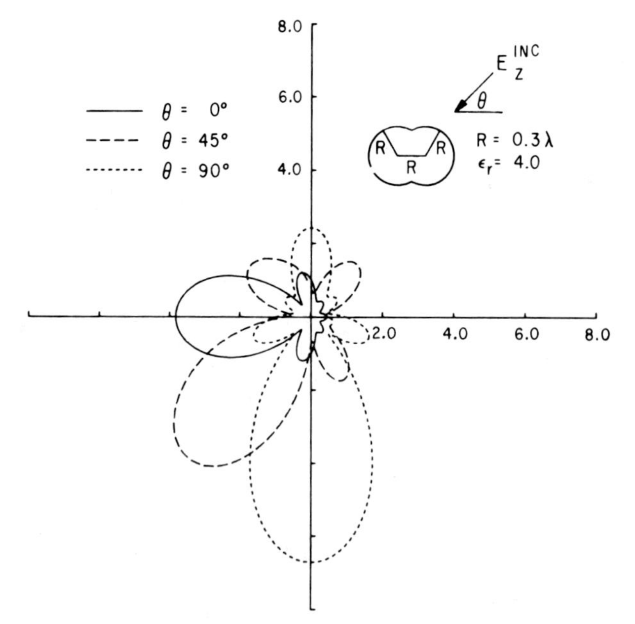

Fig. 15. Scattering by two overlapping dielectric cylinders: E_Z^{inc}.

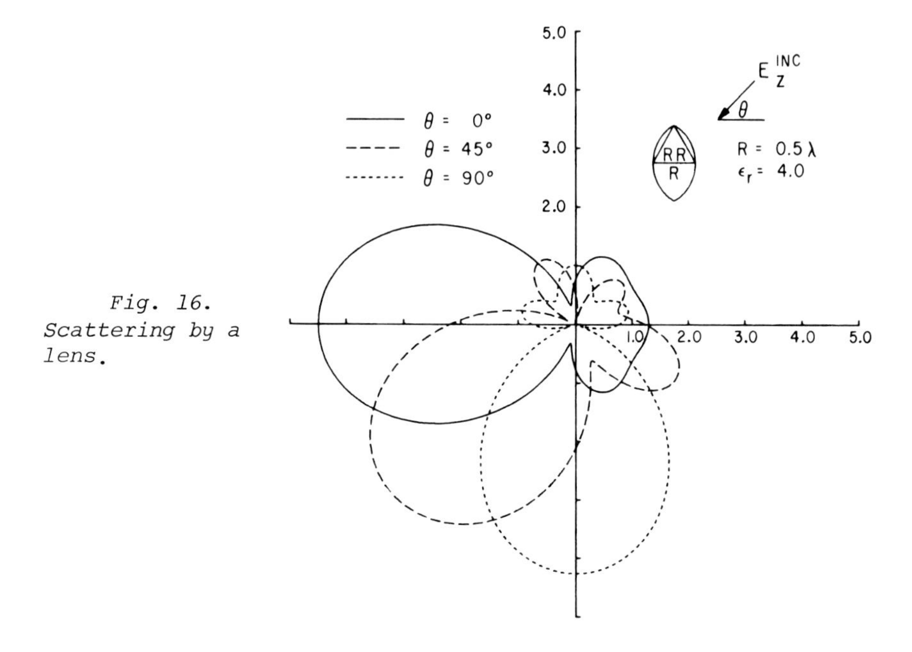

Fig. 16. Scattering by a lens.

B. Three Dimensional Scattering

1. Formulation

A truly arbitrary three dimensional scattering problem of reasonable size is not going to be solved economically in the foreseeable future using any presently available technique. The three dimensional scatterers that can be investigated reasonably economically are axially symmetrical. An axially symmetrical scatterer with non-axially symmetric incident fields can be decomposed into aximuthal modes

$$\bar{E}(\rho, z, \phi) = \sum_{m=-\infty}^{\infty} \bar{e}_m(\rho, z) e^{jm\phi} \tag{40}$$

$$\eta_0 \bar{H}(\rho, z, \phi) = \sum_{m=-\infty}^{\infty} \bar{h}_m(\rho, z) e^{jm\phi} \tag{41}$$

where $\eta_0 = 120\pi$ ohms. We shall use the cylindrical coordinates for the interior problem instead of spherical coordinates so that the integrals resulting from the finite methods will be more manageable. The above azimuthal components can be solved independently of one another.

Let $R = k\rho$ and $Z = kz$, and $\psi_{m,1}$ and $\psi_{m,2}$ be the two scalar potentials:

$$\frac{\psi_{m,1}}{R} = \hat{\phi} \cdot \bar{e}_m \tag{42}$$

$$\frac{\psi_{m,2}}{R} = \hat{\phi} \cdot \bar{h}_m \tag{43}$$

The rest of the field components can be derived from the potentials by the following formulas [9]:

$$\hat{\phi} \times \bar{e}_m = jf_m(m\hat{\phi} \times \nabla\psi_{m,1} - R\mu_r \nabla\psi_{m,2}) \tag{44}$$

$$\hat{\phi} \times \bar{h}_m = jf_m(m\hat{\phi} \times \nabla\psi_{m,2} + R\varepsilon_r \nabla\psi_{m,1}) \tag{45}$$

where

$$f_m = [\mu_r(R,Z)\varepsilon_r(R,Z)R^2 - m^2]^{-1} \tag{46}$$

The differential equations of $\psi_{m,1}$ and $\psi_{m,2}$ are

$$\nabla \cdot [f_m(R\varepsilon_r \nabla\psi_{m,1} + m\hat{\phi} \times \nabla\psi_{m,2})] + \varepsilon_r\psi_{m,1}/R = 0 \tag{47}$$

$$\nabla \cdot [f_m(R\varepsilon_r \nabla\psi_{m,2} - m\hat{\phi} \times \nabla\psi_{m,1})] + \mu_r\psi_{m,2}/R = 0 \tag{48}$$

These differential equations can be recast into the minimization of the functional,

$$F = \int_S L(R,Z,\psi_1,\psi_2,\nabla\psi_1,\nabla\psi_2)\,dR\,dZ \tag{49}$$

where

$$L = f_m[\nabla\psi_{m,1} \cdot (R\varepsilon_r\nabla\psi_{m,1} + m\hat{\phi} \times \nabla\psi_{m,2}) + \nabla\psi_{m,2}(R\mu_r\nabla\psi_{m,2} - m\hat{\phi} \times \nabla\psi_{m,1})] - (\varepsilon_r\psi_{m,1}^2 + \mu_r\psi_{m,2}^2)/R \tag{50}$$

It is noticed that $f_m(R,Z)$ is singular at the surfaces where $R\sqrt{\mu_r\varepsilon_r} = |m|$. This can happen when μ_r and ε_r are both real. The nodal fields do not, however, behave arratically at these surfaces and, in fact, except at points which correspond to material interfaces, the nodal field components are uniformly holomorphic.

2. Singular Integrals

In the application of finite methods to (47) and (48) or (50), singular integrals of the type

$$Q_{r,s} = \int_{\Omega} \frac{R^2 Z^s}{\varepsilon_r \mu_r R^2 - m^2} \, dR dZ \tag{51}$$

are often encountered. The integration of (51) may be effected as follows:

Consider the integral

$$I = \int_{\Omega} g(R) Z^s dR dZ \tag{52}$$

Using the two dimensional Stoke's theorem

$$\int_{\Omega} \nabla \times \bar{v} \cdot \overline{ds} = \oint_{\partial\Omega} \bar{v} \cdot \overline{d\ell} \tag{53}$$

and letting

$$\bar{v}(R,Z) = - \frac{g(R) Z^{s+1}}{s+1} \hat{R} \tag{54}$$

then $\nabla \times \bar{v} = -g(R) Z^s \hat{\phi}$ and $\overline{ds} = -\hat{\phi} dR dZ$ and (53) becomes

$$\int_{\Omega} g(R) Z^s dR dZ = \frac{1}{s+1} \oint_{\partial\Omega} g(R) Z^{s+1} \, dR \tag{55}$$

Applying (55) to (51) we get

$$Q_{r,s} = \frac{1}{s+1} \oint_{\partial\Omega} \frac{R^r Z^{s+1}}{\varepsilon_r \mu_r R^2 - m^2} \, dc \tag{56}$$

For a typical element the clockwise contour of (56) is shown in Fig. 17. Consider a typical integration along a line segment, such as from the ith node to the jth node. Along this path the linear functional dependence of $Z(R) = \alpha R + \beta$ where

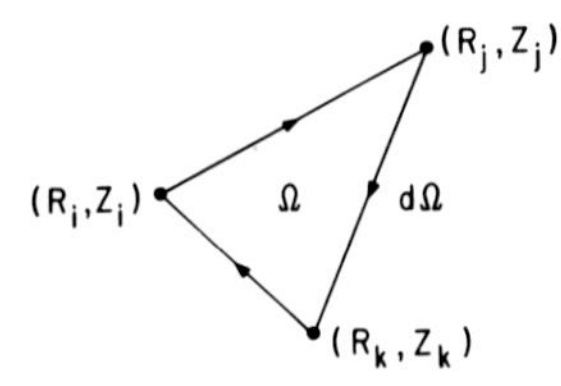

Fig. 17. A contour integral around an element.

$$\alpha = \frac{(Z_j - Z_i)}{(R_j - R_i)} \text{ and } \beta = Z_i - \alpha R_i \tag{57}$$

is substituted into the integrand of (56) and the term $Z^{s+1} = (\alpha R + \beta)^{s+1}$ is expanded into binomial series in powers of R. The resultant integrals will be of the form

$$I_Q(n,m) = \int_{R_i}^{R_j} \frac{R^n}{k^2R^2 - m^2}\, dR \tag{58}$$

where $n = 0,1,\ldots,r + s + 1$; for $n = 0,1$ we have

$$I_Q(0,m) = \begin{cases} -(k^2R^2)^{-1} \Big|_{R_i}^{R_j}, & m = 0 \\ \frac{1}{2km}\{\ln(kR - m) - \ln(kR + m)\} \Big|_{R_i}^{R_j}, & m \neq 0 \end{cases} \tag{59}$$

$$I_a(1,m) = \frac{1}{2k^2}[\ln(kR - m) + \ln(kR + m)] \Big|_{R_i}^{R_j}, \text{ all } m \tag{60}$$

The remainder of the integrals may be generated using the recurrence formula

$$I_Q(n,m) = I_Q(n,0) + \frac{m^2}{k^2} I_Q(n - 2,m) \tag{61}$$

An important result to be noted is the fact that, even for the case of lossless media, where $k = \sqrt{\varepsilon_r\mu_r}$ is real, the quadrature in I_Q will produce complex values when the integration passes through the simple pole singularity in the integrand. The pole locations for the case of real and complex k are illustrated in Fig. 18, and hence the integration path deformation necessary for real k, where $\delta \to 0$. In pathing through the pole, with increasing R, the appropriate natural log function in (59) and (60) will pick up an additional $j\pi$ to add to the integral, which is contributed by the residue of the integral. The Global System matrix will thus include complex elements even in lossless media cases.

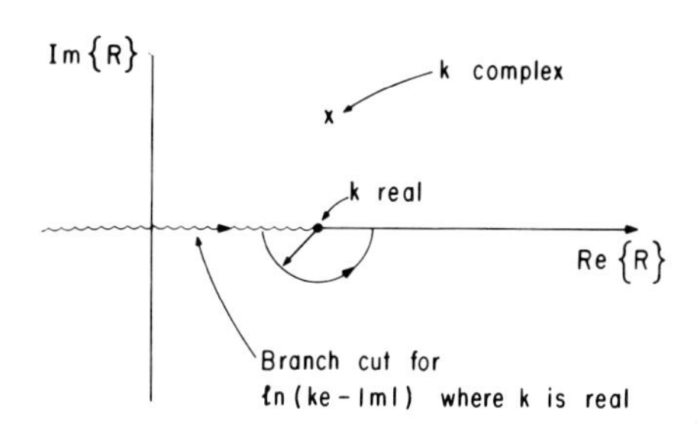

Fig. 18. The pole locations of real and complex k.

VII. RESULTS

Using the above described formulas and computational techniques, we are able to compute the scattering of plane waves by dielectric bodies of revolution of various shapes [10]. To verify the computed results, we first compute the scattering by a dielectric sphere, the exact result of which is known. In the computations, we deliberately offset the center of the scatterer from the origin, so that computationally speaking the geometry is no longer regular. The comparison of the results is shown in Fig. 19 for the scattering amplitude and in Fig. 20 for the phase.

As computational techniques become more sophisticated, verification of the computed results soon becomes a problem, because classical solution of a few special geometries will not be sufficient to confirm the computations of complex problems. Experimental verification may eventually be the only way to confirm the validity of general programs. We have both computed and measured the bistatic scattering of a finite dielectric cylinder with a spherical void in the center, such as shown in Fig. 21

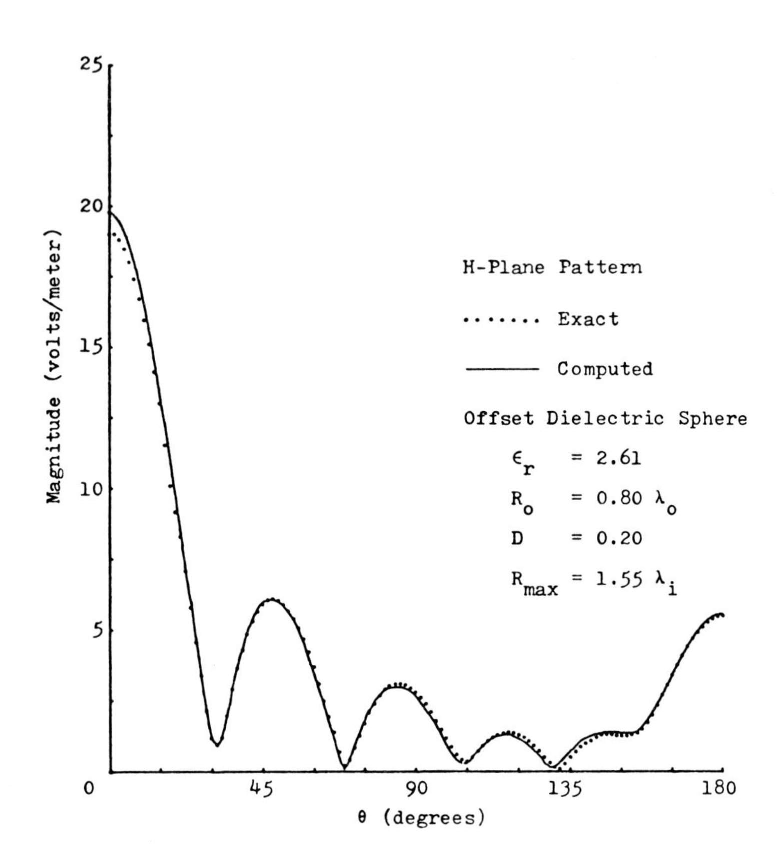

Fig. 19. The comparison of amplitudes of scattered fields by numerical and analytical solutions of a dielectric sphere.

The computed and measured results are shown in Figs. 22 and 23. Their agreement is truly remarkable. These results should definitively confirm the applicability of the unimoment method in electromagnetic scattering problems.

VIII. FURTHER DEVELOPMENT

At the present, using linear finite element methods, it is possible to solve axially symmetrical dielectric scattering problems which have a maximum dimension of about $4\lambda_0$, using the

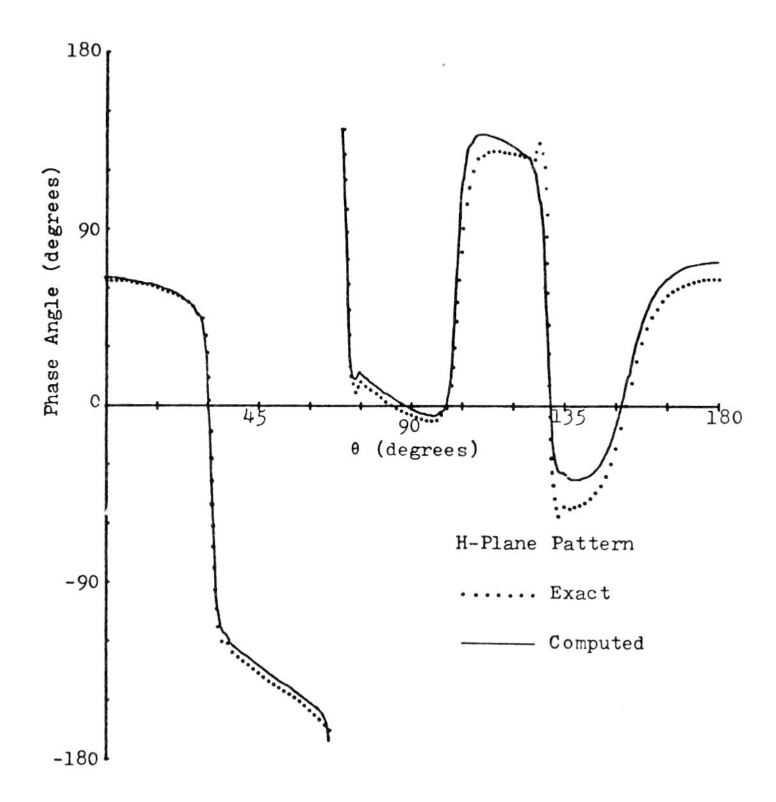

Fig. 20. The comparison of phases of scattered fields by numerical and analytical solutions of a dielectric sphere. H-Plane pattern: ····· Exact; ——— Computed.

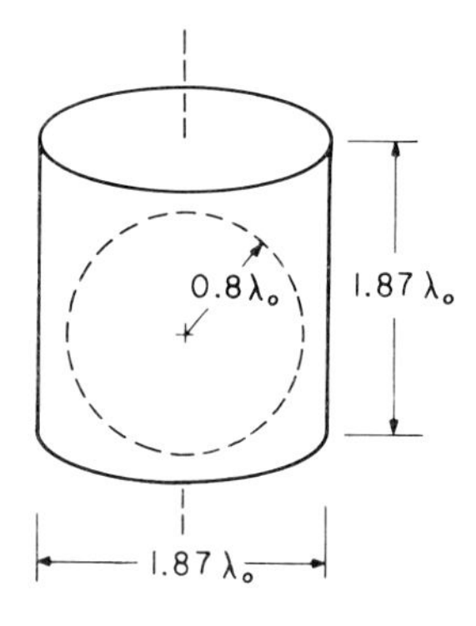

Fig. 21. A dielectric finite cylinder with a spherical void.

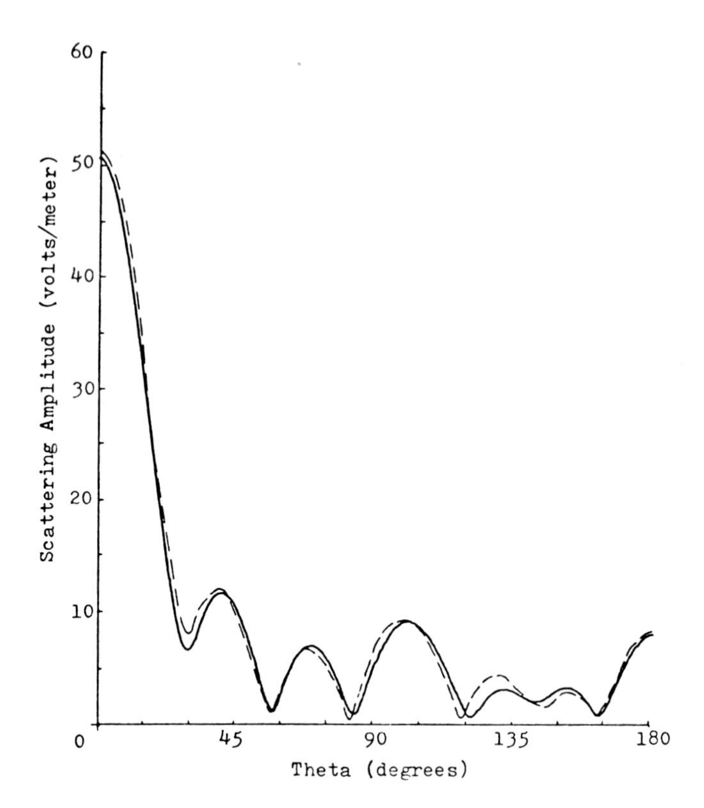

Fig. 22. Computed and measured amplitudes of scattered fields of a hollow dielectric cylinder illuminated by a 90° incident plane wave. H-*Plane Pattern: --- Measured; —— Computed.*

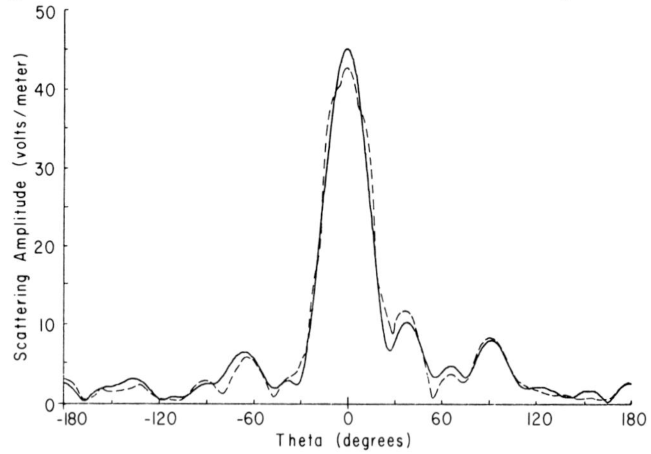

Fig. 23. Same as Fig. 22, with a 45° incident plane wave. TE incidence a = *45°.* H-*Plane Pattern: --- Measured; —— Computed.*

CDC 7600 computer. Actually, it is difficult to be exact regarding the limitation of the method because it should depend on the dielectric constant and geometry of the scatterer. At this stage, however, it is quite evident that the immediate limitation of the method is the capacity of the computer in handling finite element equations in a large closed region. Fortunately, finite element techniques have been extensively investigated by structure engineers and improvements of these techniques have been reported at each of the related conferences. The finite element technique described in this paper is quite elementary. Yet, with this elementary approach, we have already extended the art of scattering computation quite significantly. It is conceivable that more sophisticated finite methods could enhance the dimensions of solvable interior problems to the point that the size of the matrix, resulting from enforcing the continuity conditions on the separable surface, becomes the limiting factor of the computation. Therefore, the most needed effort is to gather the power of the finite element method and to popularize its use among the electromagnetic community.

The application of the technique is very broad indeed. Current problems, such as scattering by advanced composit, absorption by biological material, scattering by buried obstacles, just to name a few, can all be analyzed with the finite methods. It is also conceivable that the unimoment method can be combined with the method of moment to solve a variety of problems, which consist of wires and a body of revolution.

In conclusion, we would like to emphasize that the application of finite methods in antenna and scattering problems is not an alternative to the method of moment; rather it is its supplement. It is obvious that the finite method is most uneconomical where the method of moment is most economical, such as in thin wire problems, whereas the finite method is most attractive where the method of moment is most clumsy, such as in scattering by inhomogeneous material bodies.

IX. REFERENCES

1. Mei, K.K., Unimoment Method of Solving Antenna and Scattering Problems, *IEEE Trans. Antennas & Propag. AP-22*, 760-766 (1974).
2. Stovall, R.K. and K.K. Mei, Application of a Unimoment Technique to a Biconical Antenna with Inhomogeneous Dielectric Loading, *IEEE Trans. Antennas & Propag. AP-23(3)* (1975).
3. Chang, S.K. and K.K. Mei, Application of the Unimoment Method to Electromagnetic Scattering of Dielectric Cylinders, *IEEE Trans. Antennas & Propag. AP-24(1)*, 35-42 (1976).
4. Strang, G. and G.J. Fix, "An Analysis of the Finite Element Method," Prentice Hall, Englewood Cliffs, N.J. (1973).
5. Zienkiewicz, O.C. and Y.K. Cheung, "The Finite Element Method in Structural and Continuum Mechanics," McGraw Hill, London (1967).
6. Wachspress, E.L., "A Rational Finite Element Basis," Academic Press, New York (1975).
7. Norrie, D.H. and G. DeVries, "The Finite Element Method," Academic Press, New York (1973).
8. Jensen, P.S., A Survey of Some Recent Work in Variable Grid Finite Difference Methods for Partial Differential Equations, Lockheed Missiles and Space Co., Palo Alto, Calif., Report LMSC 6-78-70-24, 12 pp. (1970).
9. Morgan, M., S.K. Chang and K.K. Mei, Coupled Azimuthal Potentials for Electromagnetic Field Problems in Inhomogeneous Axially-Symmetric Media, *IEEE Trans. Antennas & Propag. AP-25(3)*, 413-417 (1977).
10. Morgan, M., Numerical Computation of Electromagnetic Scattering by Inhomogeneous Dielectric Bodies of Revolution, Ph.D. Thesis, Department of Electrical Engineering and Computer Sciences, University of California, Berkeley, Calif. (1976).

THE NUMERICAL SOLUTION OF TRANSIENT ELECTROMAGNETIC SCATTERING PROBLEMS

C. Leonard Bennett
Sperry Research Center

I. INTRODUCTION

The increasing interest in the solution of transient electromagnetic scattering problems has been prompted by a number of reasons. Over the last ten years there has been an increasing interest in wideband radar design for higher resolution and target classification problems. There has also been much interest in reducing the radar cross section of targets such as aircraft over a wide frequency band by placing radar absorbing materials in the vicinity of target scattering centers. The electromagnetic pulse (EMP) problem is an area which is essentially a transient problem. For target identification problems the transient response or impulse response is particularly useful, since it contains all the electromagnetic information about the target and, more importantly, it is closely related to the target geometry. Nonlinear scattering problems have become of interest in recent years and are more easily addressable by obtaining the transient solution directly.

With the advent of subnanosecond technology, it became clear that both electromagnetic scattering measurements, by virtue of hardware advancements, and electromagnetic scattering computations, by virtue of digital computer capability advances, have become possible directly in the time domain. The measurements or calculations are carried out by illuminating the target with a smoothed impulse waveform. The resulting smoothed impulse

ISBN 0-12-709650-7

response contains all the information about the electromagnetic scattering properties of the target over the frequency band that is defined by the spectrum of the incident smoothed impulse. The utility of using the impulse response as a characteristic signature was noted in 1965 [1]. This representation is useful for a number of reasons. First, all the electromagnetic information about the scattering properties of a target is contained in the impulse response. Second, the radar cross section or, equivalently, the frequency response can be obtained from the impulse response by a Fourier transform. Thirdly, the response of the target due to any radar waveform can be obtained from the impulse response by a simple convolution procedure. For the purpose of target classification, the impulse response is a particularly useful characterization, since it is closely related to the actual target geometry. Finally, the impulse response can provide a better understanding of electromagnetic transient phenomena.

This paper presents a review of the current status of transient electromagnetic scattering technology. It reviews both time domain scattering range technology and transient electromagnetic scattering computational techniques and applications.

II. TIME DOMAIN SCATTERING RANGE TECHNOLOGY

The conventional way for making measurements of the scattering characteristics or radar cross section of targets is to illuminate the target with a single frequency radar waveform [2]. This provides narrow antenna beamwidths as well as high signal to noise ratios. In these systems the extraneous returns from other objects are usually eliminated by placing the scattering range inside an expensive anechoic chamber. Furthermore, if broadband or transient responses are desired, then the conventional measurements must be repeated at all frequencies contained in the spectrum of the transient excitation. And, these measurements must maintain both amplitude and phase information accurately at all the measurement points.

The cost of an anechoic chamber, the difficulty in making coherent measurements over a broad frequency band and the advance of subnanosecond technology led in 1967 to the development of a scattering range that can measure the transient response directly in the time domain [3]. Subsequent to that time, the technique has been refined and automated [4-10]. This section will describe the basic operating principle of a typical time domain scattering range [7].

All of the time domain scattering ranges have been built on a ground plane to provide the isolation from equipment and cabling that is necessary for its operation. This ground plane acts as an electromagnetic mirror, and thus limits application to the measurement of targets with mirror symmetry.

The functional block diagram of the scattering range is shown in Fig. 1. The system signal source is a high-voltage switch which generates a 300 V step function with a risetime less than 100 psec. The signal is radiated from a vertical monopole antenna protruding through a circular ground plane 20 feet in diameter. This wave is then scattered by a target and received on a flush mounted coaxial horn antenna, which essentially smoothes and differentiates the signal, and thus provides the smoothed impulse response of the target. The received waveform is sampled by a sampling oscilloscope that has been triggered by the initial pulse and whose sampling gate deflection is under the control of a microprocessor. Unprocessed data are displayed on the oscilloscope CRT while the sampled-and-held waveform is passed through a low-pass filter, digitized, read into the microprocessor and stored on magnetic tape automatically. This system has been designed to correct for long-term timing drift and/or amplifier drift. In addition, the waveforms are stored in such a way that they are ready for the subsequent operations of averaging (to remove short-term noise) and baseline processing. The effects of a time varying baseline are subtracted from measured waveforms to improve system accuracy. Finally, the microprocessor is connected to a general purpose computer for more sophisticated and/or more rapid signal processing.

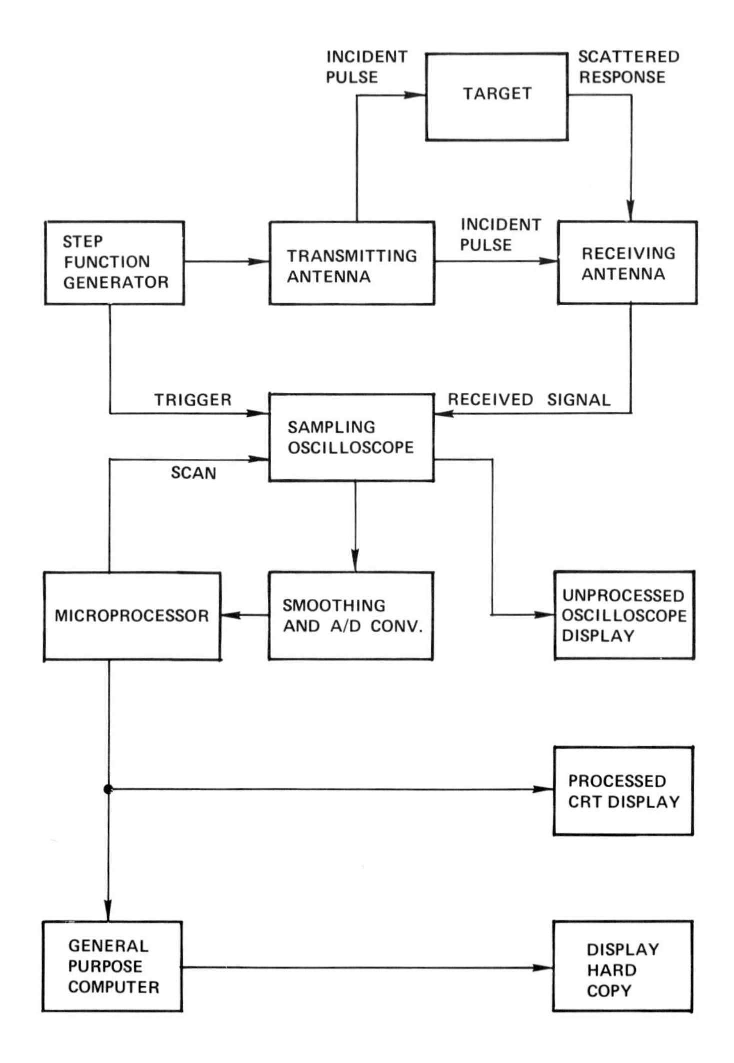

Fig. 1. Functional block diagram of time domain scattering range.

The salient characteristics of the range are the speed and simplicity with which transient responses or, equivalently, broadband frequency-domain data can be obtained. These advantages

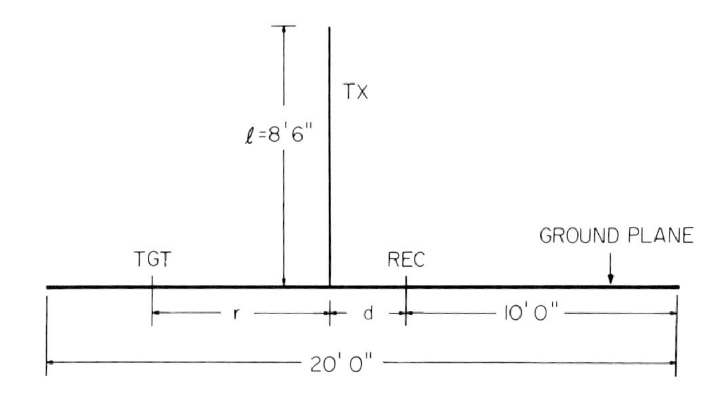

Fig. 2. Geometrical configurations of time domain scattering range (d = *18"*, r = *36").*

accrue because the time-domain scattering range yields an "uncontaminated" interval of time between the arrival of the direct wave and the arrival of unwanted reflections. The sketch in Fig. 2 shows the relative location of the elements on the ground plane, and the photographs in Fig. 3 show the range response as it appears at the oscilloscope (no data processing has been used at this point). The transmitted signal travels outward from the base of the wire antenna and is received at R at time $t_o = d/c$ (where c is the speed of light). This time is marked by the pulse at the left end of the trace in Fig. 3. The outgoing wave reaches the target at $t = r/c$, is reflected, and arrives at the receiver at $t_1 = (2r + d)/c = t_o + (2r/c)$. The targets are usually located anywhere from 2 to 5 feet from the transmitting antenna; therefore, the target returns lie in the region marked by the doublets at t_1 and t_2 in the lower photograph. The erratic response at the right edge of the trace which occurs after t_3 marks the arrival of the pulse reflected by the table edge and the effects of the pulse radiated from the tip of the transmitting antenna. The entire region between the direct transmission and the table edge response forms a convenient time "window" to view the target response and allows one to "gate out" (in time) unwanted reflections. Thus, undistorted transient target responses can be viewed without

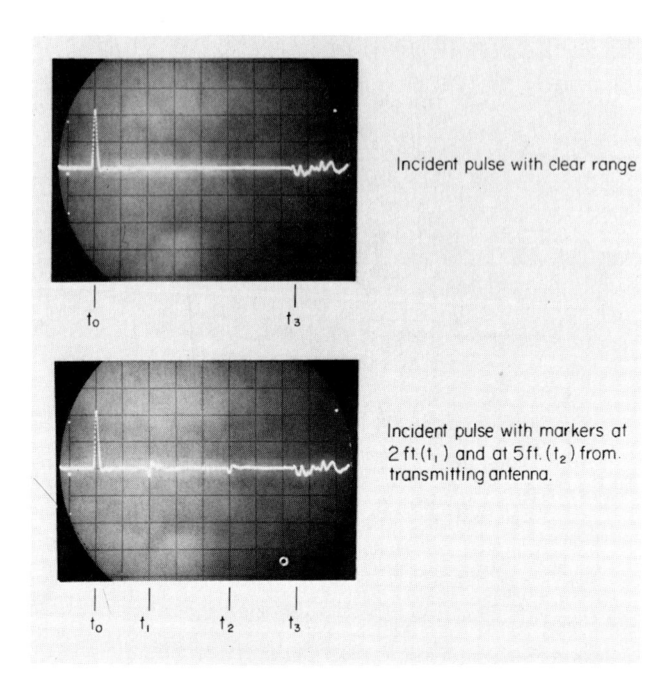

Fig. 3. Response of time domain scattering range showing incident pulse and time window (horizontal scale: 2 ns/div.; vertical scale: 200 mV/div.).

resorting to anechoic chambers. In addition, a single time-domain measurement obviates the requirement for measurement of the amplitude and phase response at many frequencies.

The accuracy of the measurement system can be estimated by measuring the standard deviation of the measured voltage. For the case where the peak of the incident pulse is approximately 500 mV and the 10 mV scale on the sampling oscilloscope is used, the standard deviation of the measured response has been measured as 0.06 mV or approximately 0.6 percent of the peak value of the measured return. In addition, the measured responses can be further processed on a general purpose computer by means of a convolution procedure to obtain the response due to a Gaussian shaped incident pulse (or any other incident waveform whose spectrum is contained within that of the measured incident pulse) rather than the approximate smoothed impulse used in the actual measurements. An added benefit of this process is the reduction of high frequency noise in the response where no signal is present anyway. The results of these measurements can be used to obtain scattering data directly, to obtain a better understanding of the transient phenomenon and to check the results obtained by new calculation techniques. Some measured results will be displayed along with calculated responses later in this paper.

III. TRANSIENT ELECTROMAGNETIC SCATTERING CALCULATIONS

Before the advent of high-speed digital computers, most of the electromagnetic scattering problems were solved in the frequency domain. The classical boundary value solution technique was applied to the problems where the target surface coincided with the coordinate surfaces of a separable coordinate system. This technique was limited to only a few target classes. Approximate techniques such as geometric optics, physical optics and the Rayleigh approximation were applied to other problems to obtain estimates of the scattered response. However, with the development of high-speed digital computers, frequency domain integral equation techniques were developed [11-12] which allowed solution of the scattering problem for simple bodies of arbitrary shape from near zero frequency well into the resonance region. These integral equation solutions have provided many interesting and useful results; however, certain difficulties exist near the resonances of the internal scattering problem [13]. Shortly thereafter, it was observed that an integral equation and its solution could also be obtained directly in the time domain [14-17]. The direct time domain solution had advantages where the solution of transient electromagnetic scattering problems were desired in addition to removing the internal resonant problem observed in the frequency domain solutions.

The choice of the frequency domain integral equation or the

time domain integral equation approach must depend on the specifics of the problem that needs to be solved [18-20]. It also depends upon the cost of computer time and the size of computer memory. For the surface scattering problem, if the total time response is desired and only a moderate number of incident angles are required, the time domain yields less running time. On the other hand, if only the response at a single frequency is needed, then the frequency domain approach is faster. The computer memory requirements for a given surface geometry are less for the time domain case. Since these time and memory results are algorithm and machine dependent, the development of better algorithms in the future could change the balance.

The remainder of this paper will be restricted to a discussion of the direct time approach, often called the space-time integral equation approach.

IV. SPACE TIME INTEGRAL EQUATION

The derivation of the integral equation for the current density on the surface of a perfectly conducting scatterer is described in this section. In Fig. 4 the geometry of the problem is described. There is an electromagnetic wave incident on a target which sets up currents on the surface of the target in such a way that the boundary condition on the surface is satisfied. These currents in turn produce a scattered field. Once the expression for the currents has been obtained, the problem is solved, since the field at any point in space can be computed from them.

The technique for obtaining an expression for the surface currents [4] is to start with the expression for the total field at an arbitrary point in space, due to the incident field and the surface currents. This arbitrary point in space is then moved to a point on the surface of the scatterer using a limiting procedure. Next the boundary conditions on the tangential components

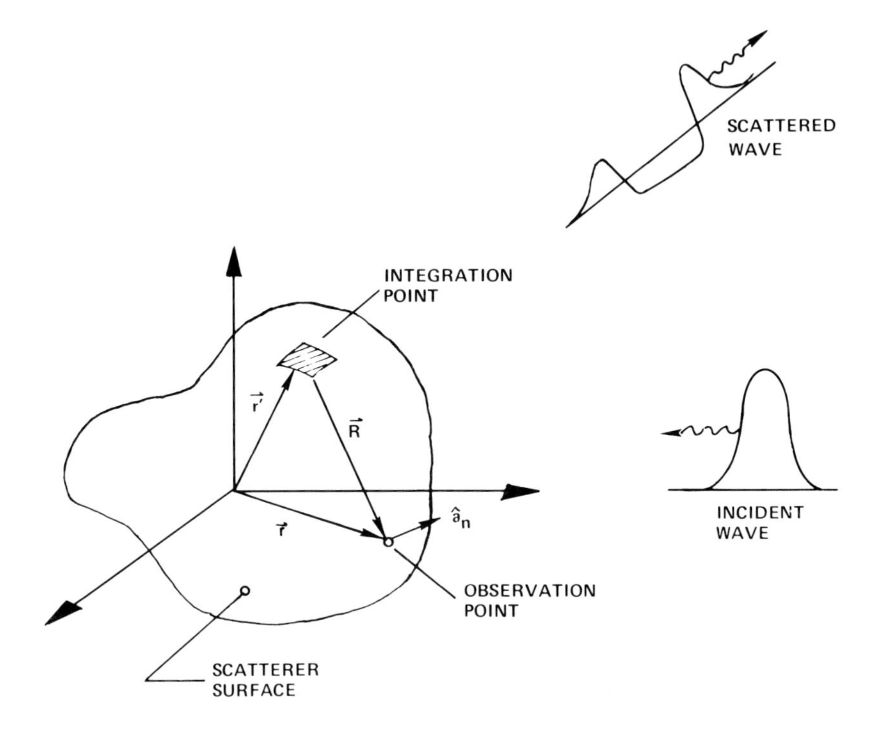

Fig. 4. Geometry of problem.

of the fields at the surface are applied and a space-time integral equation results which can be solved numerically by a stepping on in time procedure. Once the numerical representation of the surface current has been found, the field at any point in space, in particular the far field, can be calculated.

For the case of perfectly conducting scatterers being discussed here, the *H*-field boundary condition or the *E*-field boundary condition is sufficient to derive the integral equation. If the *H*-field boundary condition is used, the *H*-field or Maue (1948) integral equation [21] results. This is a second kind equation and no space derivatives on the surface are required. From a numerical solution standpoint, this equation is well suited for closed conducting bodies, but gives problems on open thin surfaces or wires where the thickness goes to zero.

If the *E*-field boundary condition is used, the *E*-field

integral equation results. This is a first kind equation and space derivatives on the surface are required. From a numerical standpoint it is well suited for targets with zero body thickness and can handle thin wires and open thin surfaces.

V. SOLID CONDUCTING BODIES

The solution of scattering from closed conducting surfaces is accomplished by numerical solution of the H-field integral equation given as

$$\vec{J}(\vec{r},t) = 2\hat{a}_n \times \vec{H}^i(\vec{r},t) + \frac{1}{2\pi}\int_S \hat{a}_n \times \left\{\left[\frac{1}{R^2} + \frac{1}{Rc}\frac{\partial}{\partial \tau}\right]\vec{J}(\vec{r}\,',\tau) \times \hat{a}_R\right\}_{\tau = t - R/c} dS' \quad (1)$$

where $\vec{r}$ = position vector to the observation point; $\vec{r}\,'$ = position vector to the integration point; $R = |\vec{r} - \vec{r}\,'|$; $\hat{a}_R = \frac{\vec{r} - \vec{r}\,'}{R}$; t = time in seconds; c = the speed of light; $\hat{a}_n$ = unit normal at $\vec{r}$; S = surface of the target; $\vec{H}^i$ = incident magnetic field; $\vec{J}$ = surface current density.

The first term in the right-hand side of (1) may be considered the source term and represents the direct influence of the incident field on the current at the observation point $(\vec{r},t)$. Moreover, this term, when applied to the illuminated side of the scatterer, yields the familiar physical-optics approximation for the surface current. The integral term on the right-hand side of (1) represents the influence of currents at other surface points on the current at $(\vec{r},t)$. The crucial observation here is that the influence of other currents on the current at $(\vec{r},t)$ is delayed by R/c, which makes the "marching on in time" numerical solution of (1) feasible.

Once the current density has been obtained, the field at any point in space can be computed. In most applications of interest, however, the quantity of interest is the far scattered field given by

$$\vec{H}^s(\vec{r},t) = \frac{1}{4\pi rc}\int_S \left\{\frac{\partial \vec{J}(\vec{r}\,',\tau)}{\partial \tau}\right\} \times \hat{a}_r \, dS' \qquad \tau = t - R/c \quad (2)$$

where $\vec{H}^s$ = far scattered $\vec{H}$ field; r = distance of observer from target; $\hat{a}_r$ = unit vector pointing from target to observer.

The technique for solution of the integral is to first write (1) in terms of its tangential components

$$J_u(\vec{r},t) = J_u^i(\vec{r},t) + I_u(\vec{r},t) \qquad J_v(\vec{r},t) = J_v^i(\vec{r},t) + I_v(\vec{r},t)$$

where $J_u^i = (2\hat{a}_n \times \vec{H}^i)_u$; $J_v^i = (2\hat{a}_n \times \vec{H}^i)_v$; I_u = u component of the second term on the right-hand side of (1); I_v = v component of the second term on the right-hand side of (1). Next the target surface is divided into patches and the current is computed at the center, $\vec{r}_i$, of each patch

$$J_u(\vec{r}_i,t) = J_u^i(\vec{r}_i,t) + \rho_{ui} J_u(\vec{r}_i,t) + \sum_{k \neq i} f_{uik} \, \Delta S_k \tag{3}$$

where $\rho_{ui} = \frac{1}{2}\sqrt{\frac{\Delta S_i}{4\pi}}\,(k_{ui} - k_{vi})$ = contribution due to patch at $\vec{r}_i$; k_{ui}, k_{vi} = the principle curvatures at $\vec{r}_i$; ΔS_i = the area of the patch at $\vec{r}_i$; f_{uik} = the contribution to $J_u(\vec{r}_i,t)$ due to currents on other patches at earlier points in time.

Rearrangement of terms in (3) yields the recurrence relation in time for the surface current

$$J_u(\vec{r}_i,t) = \frac{J_u^i(\vec{r}_i,t) + \sum_{k \neq i} f_{uik} \, \Delta S_k}{1 - \rho_{ui}} \tag{4}$$

The ideal excitation would be an electromagnetic impulse; however, it is not amenable to numerical solution. What has been used is a smoothed impulse or a regularization of an impulse. The Gaussian form

$$e(t) = \frac{a_n}{\sqrt{\pi}} \, exp(-a_n^2 t^2) \tag{5}$$

has been chosen because it has rapidly decaying frequency domain tails and because it is "time limited" and well suited for numerical solution. The resulting response computed using this excitation is the smoothed impulse response of the target.

For solution, the target geometry must be represented numerically. This is carried out by dividing the surface into patches which give a good geometry representation, are approximately curvilinear square and have approximately equal areas. The minimum spacing between patch centers should be less than 1/8 the width of the incident smoothed impulse.

The time variable must also be divided into increments for numerical solution. In the work to date, time has been divided into equal increments. The basic time increment should be less than 1/8 the width of the incident smoothed impulse. Moreover, the basic time increment should be chosen to be less than the minimum spacing between patch centers for stability reasons.

Equation (4) is solved on a digital computer for the current density by simply marching on in time. Once the current density has been obtained, the far scattered field is computed by performing the integration in (2) numerically. Procedures have been developed, implemented and tested for computing the smoothed impulse response of multiple cylindrical scatterers, smooth convex bodies and bodies with edges.

The case of twin cylinders with radius a and center-to-center spacing of $2a$ is considered [22] in Fig. 5. The TE smoothed impulse response of twin cylinders with end-on incidence is shown in Fig. 5a. In this space snapshot and those that follow, all the space dimensions are to scale except the distance to the scatterer. The basic space unit a (the radius of the cylinder in this case), and both the space variation of the scattered field and the size of the scattering body are drawn to the same scale. The large semicircle is the distance reference of the scattered field and represents the locus of points in space that the peak of the incident pulse would have reached if it had been reflected from the origin. The specular return in the backscatter direction exhibits itself as an initial positive pulse. The initial part of the return is identical with that obtained from a single circular cylinder. This is expected, since the incident wave does not reach the

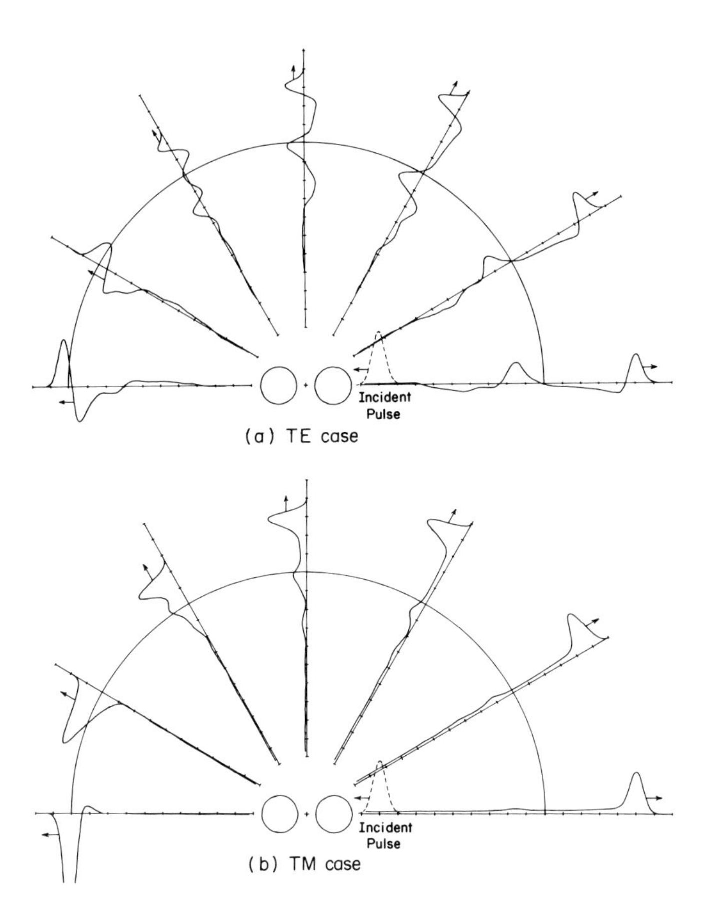

Fig. 5. Smoothed impulse response of twin cylinders with radius a *(end-on incidence).*

second cylinder until later in time. Although the second cylinder is "hidden" from the view of the observer in the backscatter direction, the second positive pulse that this observer sees can be attributed to a "reflection" of the incident wave from the front side of the back cylinder. The third positive pulse can be attributed to a wave traveling around the rear of the second cylinder.

The TM smoothed impulse response of twin cylinders with end-on incidence is shown in Fig. 5b. The initial part of the return in the backscatter direction is exactly the same as that obtained for a single cylinder. As expected, there is very little return in the backscatter direction that can be attributed to the shadowed cylinder.

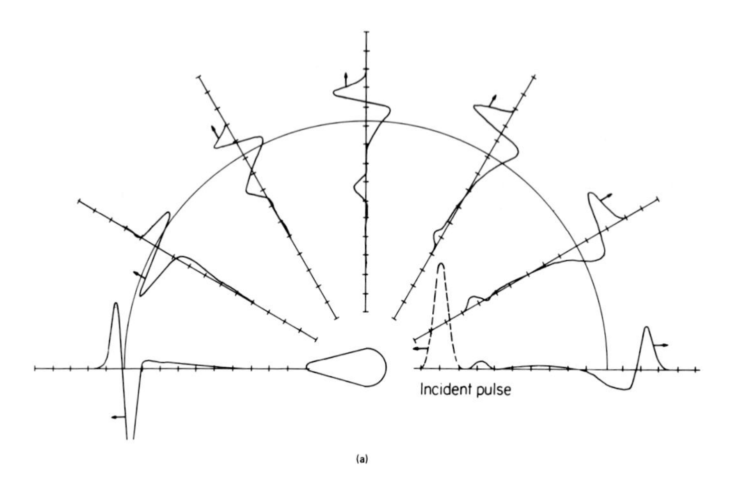

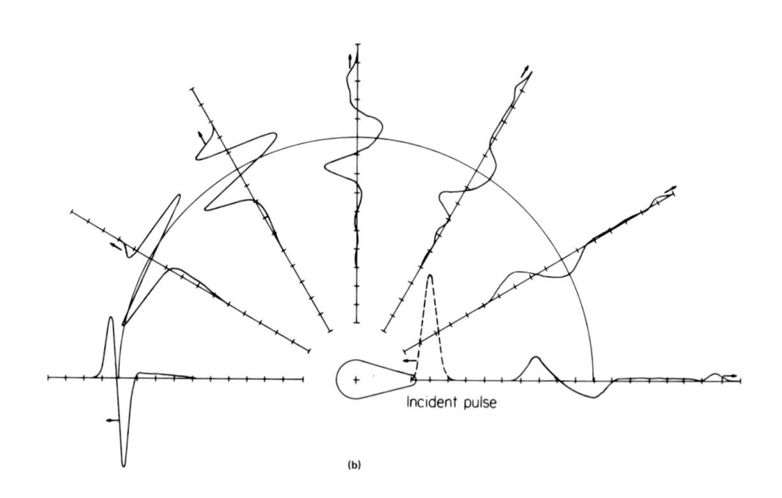

Fig. 6. E-*plane smoothed impulse response of blunt-nose cone sphere.*

In Fig. 6a the smoothed impulse response of a blunt nose cone sphere with backside incidence is shown. The initial portion of the response in the near backscatter directions is the return from the large sphere cap and is the same as was obtained for the sphere. The response then becomes small until the return from the wave traveling around the rear appears. The smoothed impulse response of the blunt nose cone sphere with nose on incidence is displayed in Fig. 6b. The small return from the spherical cap on the nose of the cone can be easily distinguished. Next, the sides of the cone appear to produce small, nearly constant return. Following this is a negative pulse that can be attributed to the join between the cone and the sphere cap. Finally, the wave traveling around the rear of the sphere exhibits itself as a positive pulse.

VI. THIN WIRES AND OPEN THIN SURFACES

The previous section described the integral equation, the solution technique used and some results that have been obtained for transient scattering from closed surfaces. This section will treat the case of thin wires and open thin surfaces. As mentioned earlier, the E field integral equation is more appropriate from a numerical standpoint for solution of this type of problem. The general space-time integral equation for a conducting body is given by

$$0 = \varepsilon \frac{\partial \vec{E}^{i}_{\tan}(\vec{r},t)}{\partial t} + \left\{ \left[\nabla(\nabla\cdot) - \frac{1}{c^2}\frac{\partial^2}{\partial t^2} \right] \left[\frac{1}{4\pi} \int_S \left\{ \frac{\vec{J}(\vec{r}\,',\tau)}{R} \right\}_{\tau = t - R/c} dS' \right] \right\}_{\tan} \tag{6}$$

where ε = permittivity of space; ∇ = del operator; tan = subscript denoting the component of the vector which is tangent to the surface at $\vec{r}$. This is an integral equation of the first kind, since the unknown, $\vec{J}$, appears only inside the integral on the right-hand side.

For the case of thin wires, (6) can be reduced by using the approximation that the current has no azimuthal component or variation. Using this technique, time domain solutions have been obtained for straight wires [16,17,23,24], for wires with loading [25], for bent wires and wire structures [26-30] and for wires with nonlinear loads [31-33].

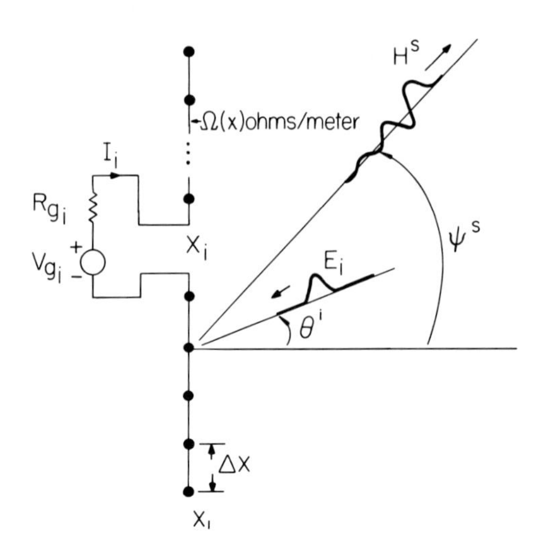

Fig. 7. Geometry of thin wire antenna as a scatterer, receiver or radiator.

For the case where a thin, straight wire is located on the x-axis as shown in Fig. 7, (6) reduces to

$$\left[\frac{\partial^2}{\partial x^2} - \frac{1}{\alpha}\sqrt{\frac{\varepsilon}{\mu}}\left(\Omega(x) + \frac{R_g(x)}{\Delta x}\right)\frac{\partial}{\partial t} - \frac{\partial^2}{\partial t^2}\right] I(x,t) \tag{7}$$

$$= -\frac{1}{\alpha}\left[\sqrt{\frac{\varepsilon}{\mu}}\left(\frac{\partial E_x^i(x,t)}{\partial t} + \frac{1}{\Delta x}\frac{\partial V_g(x,t)}{\partial t}\right)\right] + \frac{1}{4\pi}\,\Box^2\int_{\substack{\text{Non}\\ \text{Self}\\ \text{Segments}}} \left\{\frac{I(x', \tau)}{R}\right\} dx' \qquad \tau = t - R$$

where I = wire current; V_g = generator voltage; R_g = generator resistance; Ω = distributed resistance along the wire; μ = permeability of space; Δx = length of the segment that the wire is divided into; $\alpha = \frac{1}{2\pi}\, \ell n\left[(\Delta x + \sqrt{\Delta x^2 + a^2})\right]/a$; a = the wire radius; $\Box^2 = \left(\frac{\partial^2}{\partial x^2} - \frac{\partial^2}{\partial t^2}\right)$ = wave operator; t = time in units of light meters.

Note that the left-hand side of this equation contains the one-dimensional wave operator with a loss term, and the right-hand side contains known source terms or currents at earlier points in time. Equation (6) is solved numerically by dividing the wire into segments and computing the currents at the end of each segment on a digital computer by marching on in time. This solution [25] is carried out in a manner similar to that used by Courant, Friedrich and Levy [34] in their numerical solution of the one-dimensional wave equation.

Fig. 8. Comparison of step response of wire scattering computer by time domain techniques and frequency domain techniques.

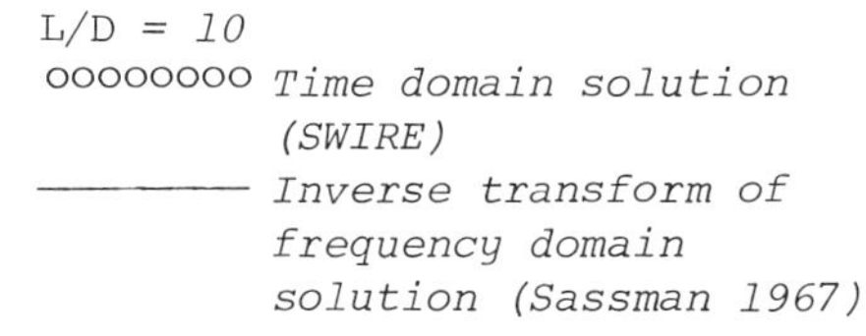

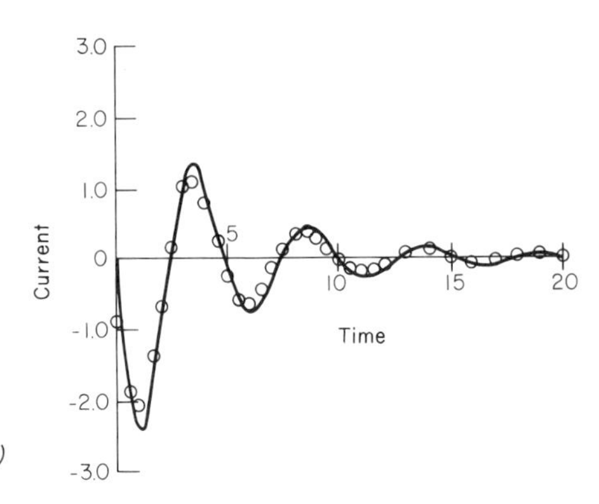

Fig. 8 displays the smoothed step response of a wire scatterer with length to diameter ratio of 10 and no loading. The comparison between the time-domain solution [24,25,35] and the alternate frequency domain solution [36] shows good agreement.

In performing direct time-domain measurements, it is desirable to be able to radiate a time-limited pulse of electromagnetic energy. In the scattering range described earlier, this was accomplished by using a monopole on a ground plane which was long enough so that the reflections from the end would be out of the time window of interest. However, by use of loading along a finite length monopole, a similar result may be achieved. Fig. 9 displays the far radiated field of a monopole on a ground plane that is driven by a smoothed step voltage from a 50Ω coaxial line for the case of three amounts of distributed loading on the wire. It can be seen that a loading of 500Ω/meter will yield a far field waveform that approximates a smoothed doublet and exhibits a "time limited" character.

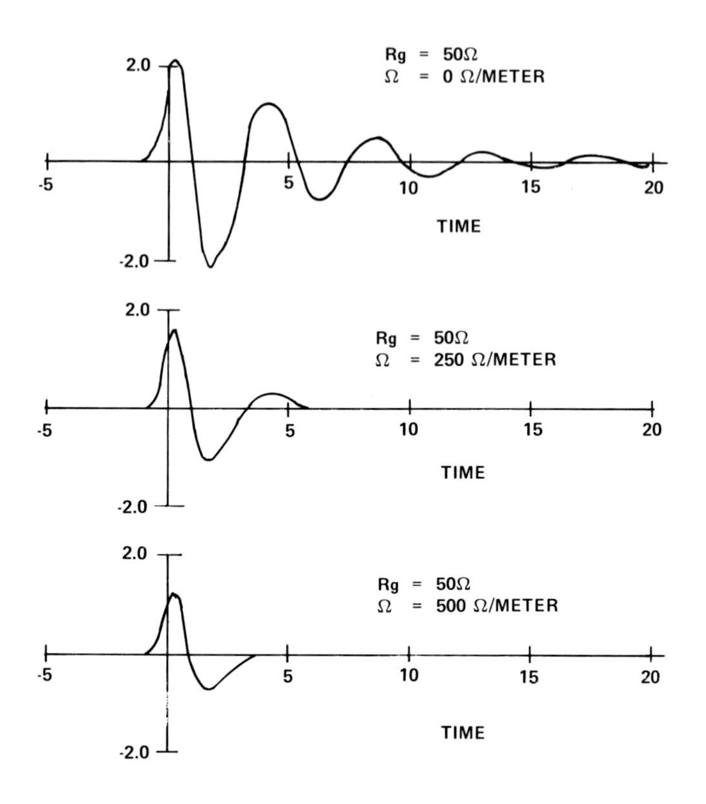

Fig. 9. Far field of monopole (L/D = *100) with smoothed step excitation and distributed loading.*

VII. FLAT PLATES

For the case of flat plates, (6) can be simplified by noting that the surface current flows in only one plane. The geometry of a rectangular flat plate is displayed in Fig. 10, in which the plate is located in the $y = 0$ plane. For this geometry, the components of the surface current are obtained from (6) as

$$\Box^2 J_X + \frac{\partial^2}{\partial_X \partial_Z} J_Z = \frac{1}{\gamma}\left[-\varepsilon \frac{\partial E_X^i}{\partial t} - (\Box^2 + \boxtimes) \frac{1}{4\pi} \int_{\substack{\text{Non} \\ \text{Self}}} \left\{\frac{J_X}{R}\right\} dS'\right]_{\tau = t - R} \tag{8}$$

$$\square^2 J_Z + \frac{\partial^2}{\partial_Z \partial_X} J_X = \frac{1}{\gamma} \left[-\varepsilon \frac{\partial E_Z^i}{\partial t} - (\square^2 + \boxtimes) \frac{1}{4\pi} \int_{\substack{\text{Non}\\ \text{Self}}} \left\{ \frac{J_Z}{R} \right\} dS' \right]_{\tau = t - R}$$

where $\boxtimes$ = the cross derivative terms in the $\nabla(\nabla \cdot)$ operator; $\gamma = 2 \Big/ \left(\sqrt{\frac{\Delta s}{\pi} - \left(\frac{\delta}{2}\right)^2} - \frac{\delta}{2} \right)$; Δs = patch area; δ = plate thickness.

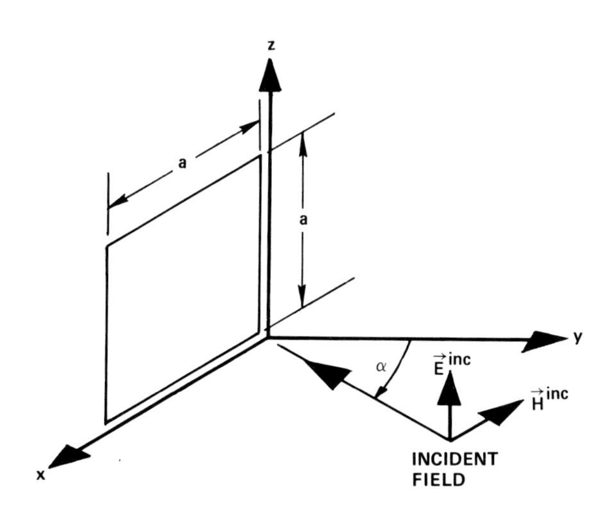

Fig. 10. Geometry of scattering from a rectangular flat plate.

The left-hand side of (8) contains a wave operator on the major component and a mixed space derivative operator operating on the other current component. It is this mixed derivative operator that provides the coupling between the two components of surface current in (8). On the right-hand side are terms due to the incident field and due to currents at other points on the surface but at earlier points in time. For numerical solution, the plate is divided into curvilinear square patches, and the current is computed at the center of all interior patches by marching on in time using the numerical representation of (8). The current on the edge and corner patches is obtained by applying the boundary conditions. For the cases studied to date, the component of the current perpendicular to the edge is set equal to zero on the edge patches. The component of the current parallel to the edge is obtained by noting that it must vary as one over the square root of the distance from the edge. Finally, both components of the current at the corner patches are set equal to zero.

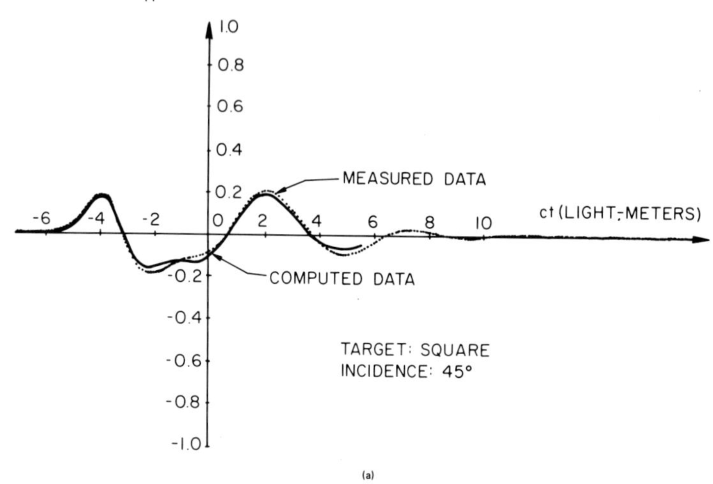

(a)

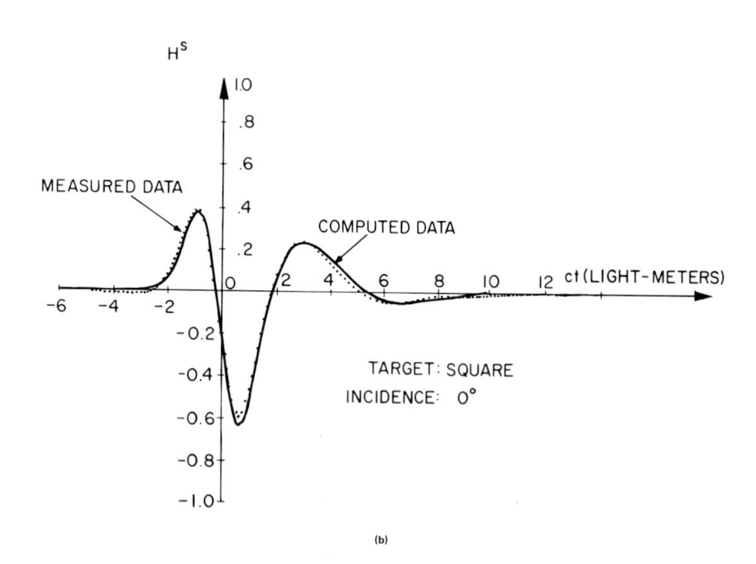

(b)

Fig. 11. Smoothed impulse response of a square flat plate.

As an example of the results which have been obtained using

this technique, the smoothed impulse response for a square flat plate is shown in Fig. 11. The calculated results for normal incidence are compared with results measured on the time domain scattering range and shown to be in good agreement in Fig. 11a. Note that the initial response approximates a smoothed doublet as expected. This is followed by a second positive pulse, which is due to a wave traveling across the plate face and returning to the observer. The calculated and measured results for a 45° angle of incidence also agree well, as can be seen in Fig. 11b. This technique has also been extended to the case of non-planar surfaces [8,37,38]. A sample result is displayed in Fig. 12 for a rectangular parabolic cylinder section. Again, the good agreement between calculations and measurements is apparent.

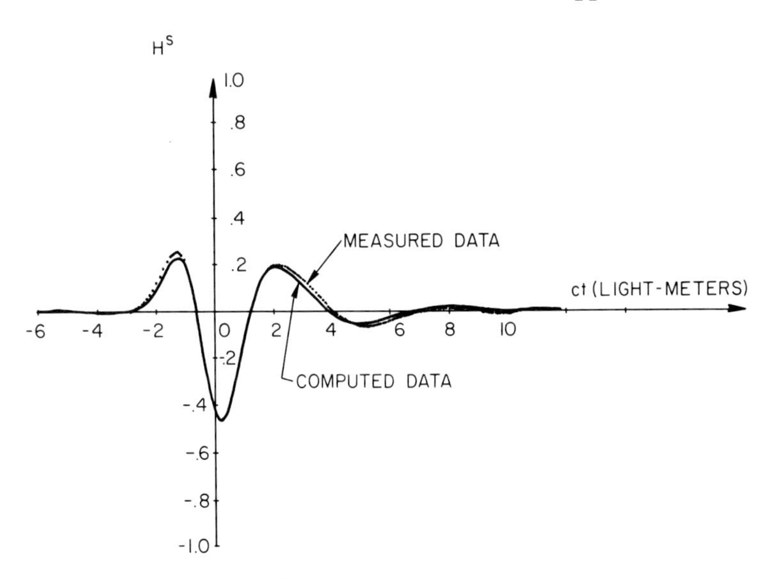

Fig. 12. Smoothed impulse response of rectangular parabolic cylinder section. Target: Parabolic Cylinder. Incidence: 0°.

VIII. BODIES WITH WIRE ANTENNAS AND STRUTS

The problem of determining the scattering properties of targets with wire antennas or struts is important in aircraft and satellite surveillance. This problem has been addressed both in the frequency domain [39,40] and in the time domain [24,41]. This section will describe the time domain solution which yields the smoothed impulse response of the target.

Since this problem contains both closed surfaces and thin wires, a hybrid integral equation approach is applied which uses the E field integral equation over the wire portions and the H field integral equation over the surface portions. The integral equations are:

For the wire current

$$\Box^2 I(\ell,t) = -\frac{1}{\alpha}\left[\sqrt{\frac{\varepsilon}{\mu}}\frac{\partial E^i_\ell(\ell,t)}{\partial t} + \sqrt{\frac{\varepsilon}{\mu}}\frac{\partial E_{s\ell}(\ell,t)}{\partial t} + \frac{1}{4\pi}\Box^2\int_{\substack{\text{Non}\\ \text{Self}\\ \text{Wire}}} \frac{I(\ell',\tau)}{R}\,d\ell'\right]_{\tau = t-R} \tag{9}$$

where ℓ = dimension along wire length; $E^i_\ell = \ell^{th}$ component of the incident E field; $E_{s\ell} = \ell^{th}$ component of the $\vec{E}$ field due to the surface currents; t = time in light meters;

And for the surface current

$$\vec{J}(\vec{r},t) = 2\hat{a}_n \times \vec{H}^i(\vec{r},t) + 2\hat{a}_n \times \vec{H}_w(\vec{r},t) + \frac{1}{2\pi}\int_S \hat{a}_n \times \left\{\left[\frac{1}{R^2} + \frac{1}{R}\frac{\partial}{\partial\tau}\right]\vec{J}(\vec{r}',\tau) \times \hat{a}_R\right\}dS'_{\tau = t-R} \tag{10}$$

where $\vec{H}_w = \vec{H}$ field due to the wire currents; t = time in light meters.

The numerical solution of the coupled space time integral equations (9) and (10) is carried out by first dividing the surface into patches and the wires into segments. Next, the current at the center of each patch on the surface and at the end of each segment on the wire is computed by marching on simultaneously in time the numerical representation of the coupled equations (9) and (10) in a manner similar to that used for wires alone and surfaces alone. The current at both the wire free end and at the wire surface end are obtained by applying the boundary conditions for those two points.

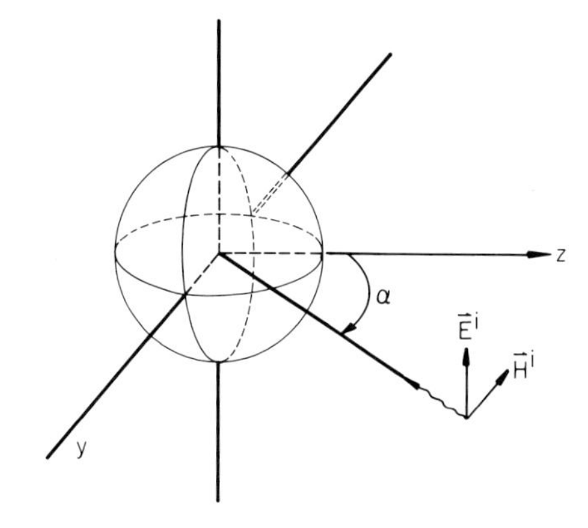

Fig. 13. SSS satellite model (length-to-diameter ratio of each antenna is 32).

As an example of an application of this technique, a model of the small scientific satellite (SSS) shown in Fig. 13 is considered. This target consists of a sphere located at the origin with four antennas attached, each a sphere diameter long, coincident with the x and y axis. Both the E plane and H plane smoothed impulse responses of this target are displayed in Fig. 14 for the case of an incident wave traveling in the negative z direction.

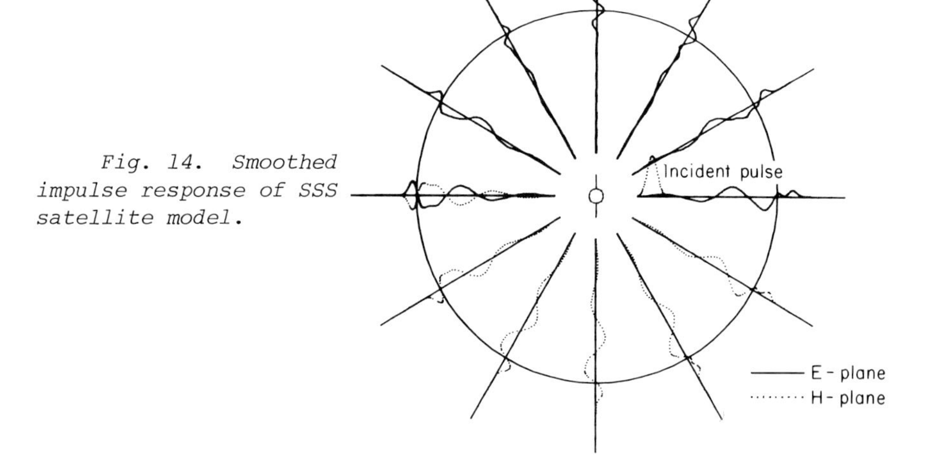

Fig. 14. Smoothed impulse response of SSS satellite model.

It is interesting to relate the response in the backscatter direction to the target geometry. The specular return from the nose of the sphere appears first and is of the same value as was obtained for the case of scattering by a sphere alone. Next, there is a second peak that can be attributed mainly to the specular return from the wire antennas. This is then followed by a damped oscillation with a spacing between zero crossings that is approximately 2.5 sphere diameters. This is less than the distance between the opposite free tips of the wire (3 sphere diameters), but more than the length of an image representation of each antenna (2 sphere diameters). Moreover, the decay of these oscillations is much more rapid than would occur for the wires alone in free space with the same length-to-diameter ratio. This demonstrates a significant interaction between the antennas and the sphere surface.

IX. BODIES WITH FINS ATTACHED

The problem of determining the scattering properties of targets with fins attached is important in missile detection and classification. This problem has been addressed directly in the time domain [42] by using a hybrid integral equation approach. This formulation used an $\vec{E}$ field integral equation over the fin portions and an $\vec{H}$ field integral equation over the surface portions. The integral equations are:

For the surface current

$$\vec{J}_S(\vec{r},t) = 2\hat{a}_n \times \vec{H}^i(\vec{r},t) + \frac{1}{2\pi}\int_S \hat{a}_n \times \left\{\left[\frac{1}{R^2}+\frac{1}{R}\frac{\partial}{\partial\tau}\right]\vec{J}_S(\vec{r}',\tau) \times \hat{a}_R\right\}_{\tau=t-R} dS'$$

$$+ \frac{1}{2\pi}\int_F \hat{a}_n \times \left\{\left[\frac{1}{R^2}+\frac{1}{R}\frac{\partial}{\partial\tau}\right]\vec{J}_F(\vec{r}',\tau) \times \hat{a}_R\right\}_{\tau=t-R} dS' \tag{11}$$

where $\vec{J}_S$ = surface current; $\vec{J}_F$ = fin current.

And for the fin current

$$\square^2 \vec{J}_F(\vec{r},t) + \boxtimes \vec{J}_F(\vec{r},t) = \frac{1}{\gamma}\left\{-\varepsilon \frac{\partial \vec{E}^i}{\partial t} - \square^2\left[\vec{A}_{F_{NS}}(\vec{r},t) + \vec{A}_S(\vec{r},t)\right] - \boxtimes\left[\vec{A}_{F_{NS}}(\vec{r},t) + \vec{A}_S(\vec{r},t)\right]\right\}_{tan} \tag{12}$$

where $\vec{A}_S = \frac{1}{4\pi}\int\!\!\int_S \left\{\frac{\vec{J}_S(\vec{r}\,',\tau)}{R}\right\}_{\tau = t - R} dS'$; $\vec{A}_{F_{NS}} = \frac{1}{4\pi}\int_{\substack{\text{Non-}\\ \text{Self}\\ \text{Fin}}} \left\{\frac{\vec{J}_F(\vec{r}\,',\tau)}{R}\right\}_{\tau = t - R} dS'$;

$\gamma = \dfrac{2}{\left[\sqrt{\dfrac{\Delta S}{\pi} - \left(\dfrac{\delta}{2}\right)^2} - \dfrac{\delta}{2}\right]}$; ΔS = fin patch area; δ = fin thickness.

The numerical solution of the coupled space-time integral equation given in (11) and (12) is carried out by marching on in time using a digital computer after first dividing the surface and fin into patches. The surface current is computed at the center of each surface patch using a numerical representation of (11) in a manner similar to that used for the solution by surfaces alone. The fin current is computed at the center of each interior fin patch using a numerical representation of (12) in a manner similar to that used for a flat plate. The currents on the free space edge patches and on the surface join patches of the fin are established by application of the appropriate boundary conditions.

As an example of an application of this technique, the finned cylinder target shown in Fig. 15 is considered. This target consists of a right cylinder of radius a with length-to-diameter ratio of two-to-one with its right axis coinciding with the z-axis and centered at the origin. Four fins (one cylinder diameter in a side) are arranged symmetrically about the cylinder body coinciding with the $x = 0$ and $y = 0$ planes. A vertically polarized smoothed impulse whose basewidth is equal to the cylinder length is used to illuminate this target. The smoothed impulse response that has been calculated for this target is compared with the measured result in Fig. 16 for an axial angle of incidence. The agreement between the calculation and the measurement

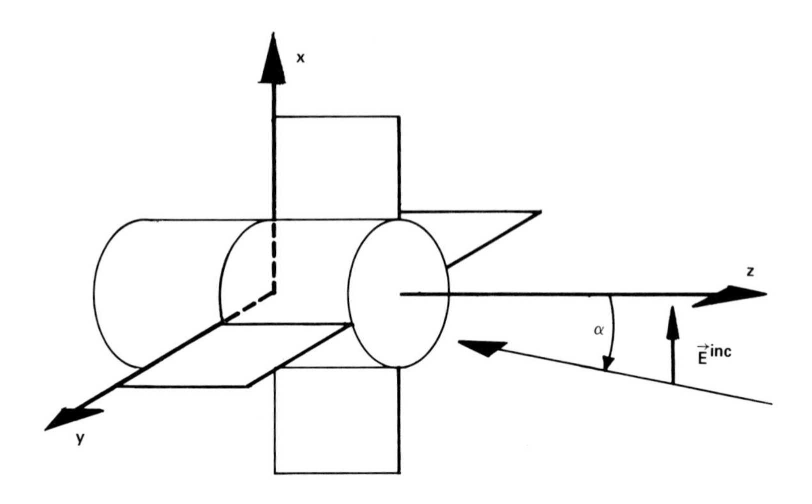

Fig. 15. Geometry of cylinder with fin attached.

is good. The initial portion of the response is due to the return from the face of the cylinder and the leading vertical fin edges. This is followed by the return from the far edge of the vertical fins, a creeping wave return and oscillation due to the interaction between the fins and the cylinder body.

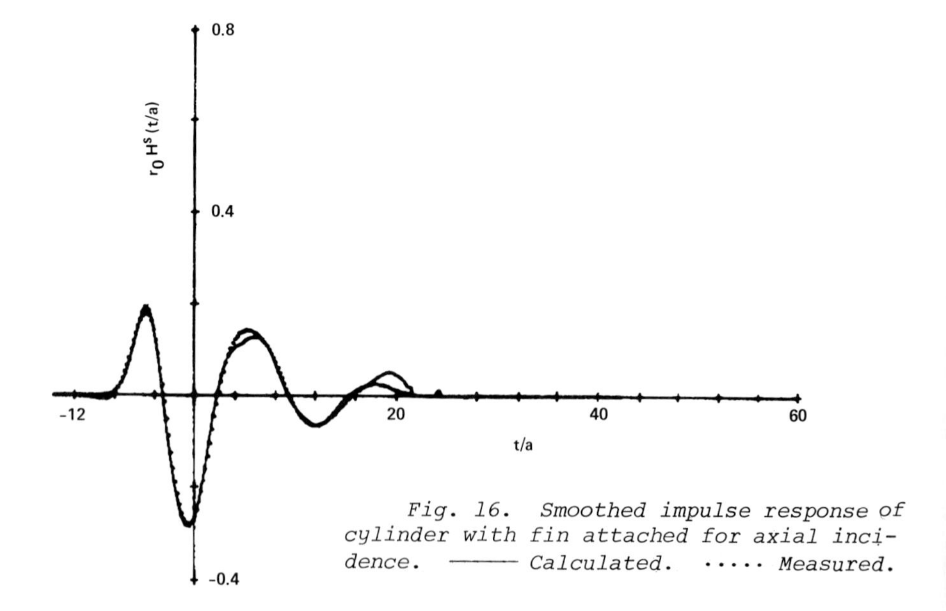

Fig. 16. Smoothed impulse response of cylinder with fin attached for axial incidence. ——— Calculated. ····· Measured.

X. IMPULSE RESPONSE AUGMENTATION TECHNIQUE

What has been described up to now in this paper is the space-time integral equation, its numerical solution and results obtained using a smoothed impulse excitation. This numerical solution of the integral equation yields the smoothed impulse response of target sizes up to several pulsewidths. If transformed into frequency domain terms, this technique yields results from zero frequency up well into the resonance region of the target. The impulse response augmentation technique, first suggested in 1968 [15] and later developed in 1973 [7,43], couples the calculated smoothed impulse response with knowledge about the singular portion of the impulse response, such as location, type and size of singularity, and knowledge about the asymptotic variation in frequency to obtain an estimate of the impulse response or, equivalently, an estimate of the frequency response over the entire spectrum. The philosophy motivating this approach is to make use of all information available about the scattering problem in order to obtain an estimate of the "total" response. This may be viewed as a hybrid approach to the EM scattering problem, which combines results obtained by numerical techniques with results obtained by asymptotic techniques. Using this technique, results have been obtained for smooth convex bodies (sphere, sphere-capped cylinders and prolate spheroid), for bodies with edges (right circular cylinders) and for open thin surfaces (square flat plates) [7,8,42-45]. To date, the technique has dealt only with the far field, although it could also be applied to the near field or the surface currents.

As an example of the results obtained with this technique, the case of a sphere is considered. Fig. 17a displays the far field smoothed impulse response of a sphere with radius a in the backscatter direction. The width of the incident pulse is twice the sphere diameter. Fig. 17b displays a comparison between the frequency response obtained by transforming the calculated

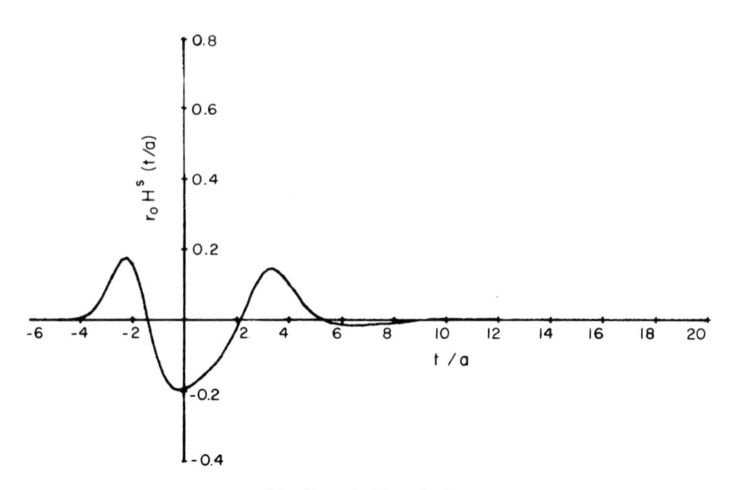

(a) Smoothed Impulse Response

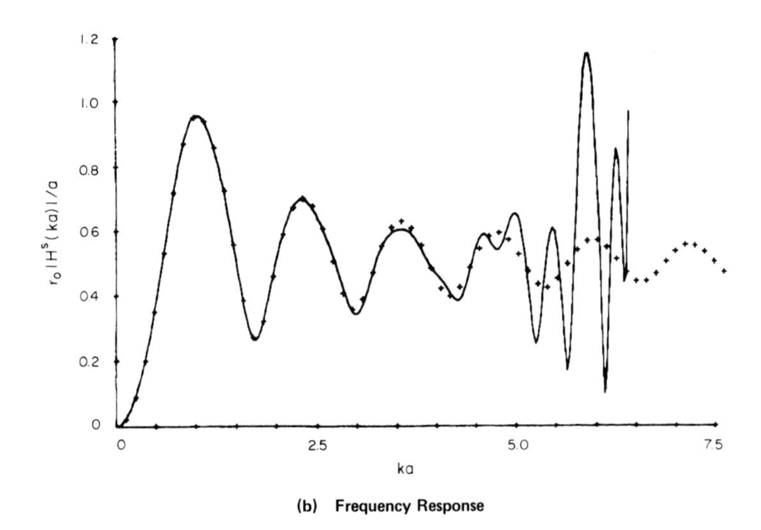

(b) Frequency Response

——— *Impulse response augmentation technique.* +++++ *Theoretical.*

Fig. 17. Response of sphere with radius a.

smoothed impulse response with the familiar Mie series solution. The two results agree well up to a ka of 4.5. Fig. 18 compares the impulse response and frequency response obtained by use of the impulse augmentation technique with the theoretical results

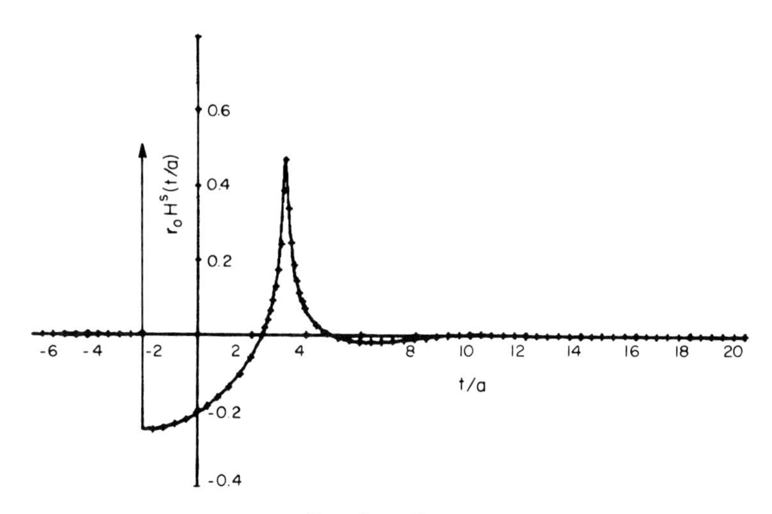

a. Impulse Response.

——— *Impulse response augmentation technique.* +++++ *Theoretical.*

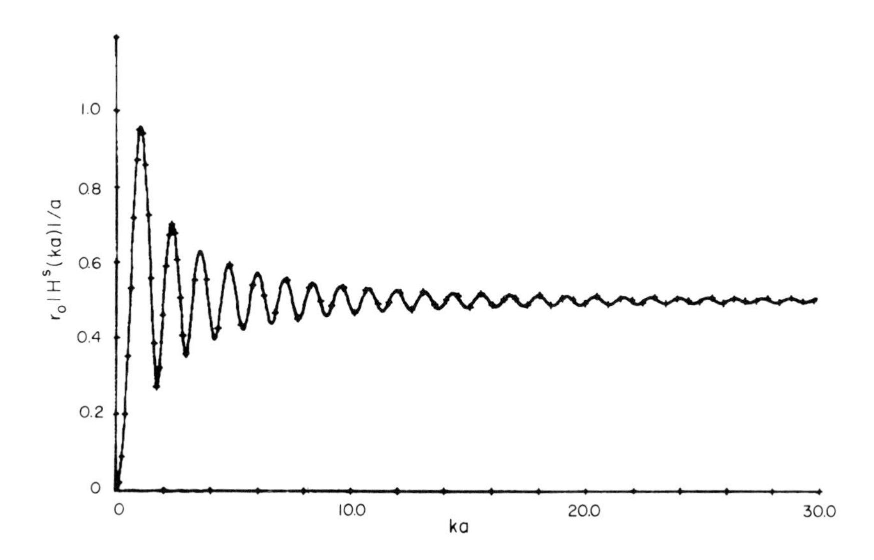

b. Frequency Response.

——— *Impulse response augmentation technique.* +++++ *Theoretical.*

Fig. 18. Response of sphere with radius a.

and shows very good agreement. It is interesting to relate the features of the impulse response in Fig. 18a to the geometry of the target. The initial return is an impulse of amplitude 0.5 that is due to the specular reflection from the nose of the sphere. This is followed by a negative step of amplitude 0.25 that is due to the return from the vicinity of the specular point and would be predicted by a physical optics approximation. It is interesting to note that the slope of the waveform following the negative step is a function of the difference in the principle curvatures at the specular point [42,44]. Finally, the second positive pulse at $t/a \simeq 3.1$ is due to the creeping wave that travels around the rear of the sphere and returns to the observer.

XI. INVERSE SCATTERING

The previous discussion in this paper has been directed towards the direct scattering problem; i.e., given the incident waveform and the target geometry, find the scattered field. This section addresses the inverse scattering problem in which it is desired to find the target geometry when only the incident waveform and the scattered waveform are given. The solution of this problem has direct application to the problem of classification and identification of radar targets. This discussion addresses the time domain solution of this problem.

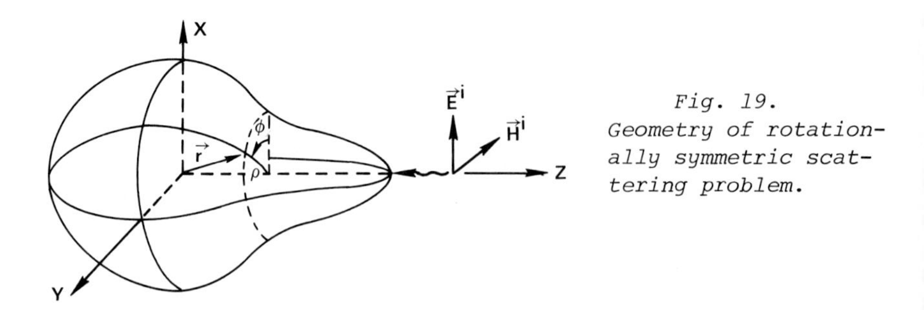

Fig. 19. Geometry of rotationally symmetric scattering problem.

The specific case considered is that of a rotationally symmetric scattering problem which consists of a rotationally symmetric target illuminated by a wave from the axial direction, as shown in Fig. 19. This target geometry can be represented simply by its contour function ρ, as defined in Fig. 19. The space-time integral equation for the solution of this inverse problem is given by

$$\rho(z,t) = \left[2r_o \, H_R^s(\vec{r},t) - \frac{1}{2\pi} \int_S \left\{ \vec{J}_{CR}(\vec{r}\,',\tau) \times \hat{a}_r \right\} dS' \right]^{\frac{1}{2}} \qquad (13)$$

$$\tau = t - R$$

where ρ = contour function of the target; r_o = distance of observer from the target; H_R^S = far scattered field ramp response of target; H_R^i = incident ramp H field; $\vec{J}_R$ = total current; $\vec{J}_{CR} = \vec{J}_R - 2\hat{a}_n \times \vec{H}_R^i$.

This expression represents the exact relationship between the scatterer response, H_R^S, and the target geometry, ρ. In particular, the target contour function, ρ, is given in terms of the scattered far field H_R^S and in terms of "correction currents" $\vec{J}_{CR}$, that appear at earlier computed surface points at earlier points in time. This space-time integral equation has been solved by iteration of the marching on in time approach [7,46-48].

An example of the application of this technique is shown in Fig. 20 for the case of a sphere cap flat end cylinder when viewed from the sphere cap end. Note that the first iteration gives a good estimate of the spherical end cap and sides of the cylinder; but it took 13 iterations to give a reasonable estimate of the far end of the target. Although there is some rounding of the far edges, this technique can provide a powerful tool in obtaining information concerning the geometry of a target in the shadow region and, in this case, shows good agreement with the actual contour in Fig. 20e.

XII. SUMMARY

The advances in the state-of-the-art software and hardware technology have brought to the engineer much more powerful tools

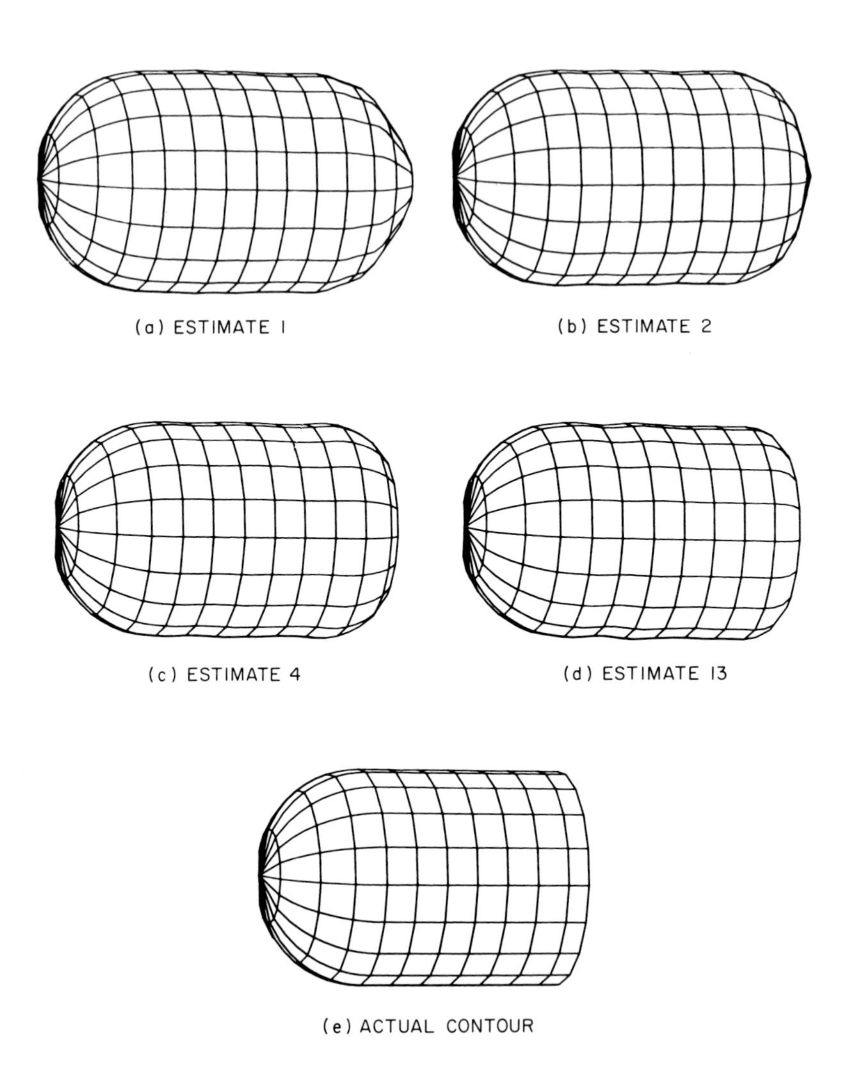

Fig. 20. Contours of sphere capped flat end cylinder obtained by time domain solution of inverse scattering problems.

for the solution of electromagnetic scattering problems, both from a measurement standpoint and also from a calculation standpoint. This paper described time domain scattering measurement techniques and presented a review of the time-domain solution for both the direct scattering problem and the inverse scattering problem. The numerous examples presented demonstrate the power and versatility of a technique that has been made possible by the advent of the large, high-speed digital computer.

XIII. REFERENCES

1. Kennaugh, E.M. and D.L. Moffatt, Transient and Impulse Response Approximations, *Proc. IEEE 53,* 893-901 (1965).
2. Blacksmith, P., Jr., R.E. Hiatt and R.B. Mack, Introduction to Radar Cross-Section Measurements, *Proc. IEEE 53,* 901-920 (1965).
3. DeLorenzo, J.D., A Range for Measuring the Impulse Response of Scattering Objects, *1967 NEREM Record 9,* 80-81 (1967).
4. Bennett, C.L. and J.D. DeLorenzo, Short Pulse Response of Radar Targets, 1969 International Antenna and Propagation, Austin, Texas, 9-11 December, 1969.
5. Smith, R.S., J.D. DeLorenzo and C.L. Bennett, Wideband Surface Current Analysis, Sperry Research Center, Sudbury, Mass., Final Report on Contract No. F30602-69-C-0357, July 1970.
6. Nicolson, A.M., C.L. Bennett, D. Lamensdorf and L. Susman, Applications of Time Domain Metrology to the Automation of Broadband Microwave Measurement, SCRC-RR-21, August 1971, *IEEE Trans. PGMTT, MTT-20(1),* 3-9 (1972).
7. Bennett, C.L., A.M. Auckenthaler, R.S. Smith and J.D. DeLorenzo, Space-Time Integral Equation Approach to the Large Body Scattering Problem, Final Report on Contract No. F30602-71-C-0162, RADC-CR-73-70, AD 763 794, May 1973.
8. Bennett, C.L., K.S. Menger and R. Hieronymus, Space-Time Integral Equation Approach for Targets with Edges, Final Report on Contract No. F30602-73-C-0124, SCRC-CR-74-3, March 1974.
9. Bennett, C.L., H. Cronson and A.M. Nicolson, Recent Advances in Time Domain Measurement Techniques, 1974 URSI Meeting, Atlanta, Ga., 11-13 June, 1974.
10. Deadrick, F.J., E.K. Miller and H.G. Hudson, The LLL Transient Electromagnetics Measurements Facility, UCRL-51933, October 1975.
11. Andreasen, M.G., Scattering from Parallel Metallic Cylinders with Arbitrary Cross-Section, *IEEE Trans. Antennas & Propag. AP-12,* 746-754 (1964).
12. Andreasen, M.G., Scattering from Bodies of Revolution, *IEEE Trans. Antennas & Propag. AP-13,* 303 (1965).
13. Mitzner, K.M., Numerical Solution of the Exterior Scattering Problem at the Eigenfrequencies of the Interior Problem, 1968 Fall URSI Meeting, Boston, Mass., 10-12 September, 1968.
14. Bennett, C.L., A Technique for Computing Approximate Electromagnetic Impulse Response of Conducting Bodies, Ph.D.

Thesis, School of Electrical Engineering, Purdue University, August 1968 (University Microfilms Order No. 69-7420).
15. Bennett, C.L. and W.L. Weeks, Electromagnetic Pulse Response of Cylindrical Scatterers, 1968 International Antenna and Propagation Symposium, Boston, Mass., September 1968.
16. Sayre, E.P. and R.F. Harrington, Transient Response of Straight Wire Scatterers and Antennas, 1968 International Antenna and Propagation Symposium, Boston, Mass, September 1968.
17. Sayre, E.P., Transient Response of Wire Antennas and Scatterers, Ph.D. Thesis, Syracuse University, Syracuse, New York, June 1969.
18. Bennett, C.L. and E.K. Miller, Some Computational Aspects of Transient Electromagnetics, 1972 Spring URSI Meeting, Washington, D.C., 13-15 April 1972.
19. E.K. Miller, Some Computational Aspects of Transient Electromagnetics, Lawrence Livermore Lab. Rep. UCRL-51276, September 1972.
20. Bennett, C.L. and R.M. Hieronymus, Time Domain vs. Frequency Domain Solution of EM Scattering Problems, 1975 IEEE APS/URSI Meeting, Urbana, Illinois.
21. Maue, A.W., Zur Formulierung eines allgemeinen Beugungs: problems durch eine Integralgleichung, Zeitschrift fur Physik, Bd 126, S. 601-618 (1949).
22. Bennett, C.L. and W.L. Weeks, Transient Scattering from Conducting Cylinders, *IEEE Trans. Antennas & Propag. AP-18 (5)*, (1970).
23. Bennett, C.L. and J. Martine, A Space-Time Integral Equation Solution for Currents on Wire Structures with Arbitrary Excitations, 1970 URSI Fall Meeting, Columbus, Ohio, 15-17 September 1970.
24. Bennett, C.L., et al, Integral Equation Approach to Wideband Inverse Scattering, Final Report on Contract No. F30602-69-C-0332; TR SCRC-CR-70-16, June 1970, Vol. 1, AD 879 849, Vol. 2 AD 876 627.
25. Bennett, C.L. and A.M. Auckenthaler, Transient and Time Domain Solutions for Antennas and Scatterers, 1971 IEEE International Convention, New York, 22-25 March 1971.
26. Miller, E.K., A.J. Poggio and D.J. Burke, An Integro-Differential Equation Technique for the Time-Domain Analysis of Thin-Wire Structures, Part I - The Numerical Method, *J. Comput. Phys. 12*, 24 (1973).
27. Poggio, A.J., E.K. Miller and D.J. Burke, An Integro-Differential Equation Technique for the Time-Domain Analysis of Thin Wire Structures, Part II - Numerical Results, *J. Comput. Phys. 12*, 210-233 (1973).
28. Liu, T.K., Time Domain Analysis of Linear Antennas and Scatterers, Ph.D. Thesis, University of California, Berkeley, Calif., June 1972.
29. Liu, T.K. and K.K. Mei, A Time-Domain Integral Equation

Solution for Linear Antennas and Scatterers, *Radio Sci. 8*, 797-804 (1973).
30. Solman, F.J., Time-Domain Response of Wire Antenna Arrays, Ph.D. Thesis, Purdue University, Lafayette, Indiana, December 1973.
31. Schuman, H., Time-Domain Scattering from a Nonlinearly Loaded Wire, *IEEE Trans. Antennas & Propag. AP-22*, 611-613 (1974).
32. Tesche, F.M. and T.K. Liu, Transient Response of Antennas with Non-Linear Loads, *Electron. Lett. 11*, 18-19, 9 January 1975.
33. Liu, T.K. and F.M. Tesche, Analysis of Antennas and Scatterers with Non-Linear Loads, *IEEE Trans. Antennas & Propag. AP-24(2)*, 131-138 (1976).
34. Courant, R., K. Friedricks and H. Levy, Uber die partiellen Differenzengleichungen der mathematischen Physik, *Mathematische Annalen 100*, 32 (1928).
35. Auckenthaler, A.M. and C.L. Bennett, Computer Solution of Transient and Time Domain Thin-Wire Antenna Problems, *IEEE Trans. Microwave Theory & Tech. MTT-19(11)*, 892-893 (1971).
36. Sassman, R.W., The Current Induced on a Finite, Perfectly Conducting, Solid Cylinder in Free Space by an Electromagnetic Pulse, EMP Interaction Notes, Note XI, 25 July 1967.
37. Menger, K.S., C.L. Bennett, D. Peterson and C. Maloy, The Space-Time Integral Differential Equation Solution of Scattering by Open, Thin Surfaces, 1974 URSI Meeting, Atlanta, Ga. 11-13 June,1974.
38. Bennett, C.L. and R. Hieronymus, Numerical Solution of Space-Time Integral Equation for Scattering by Open, Thin Surfaces, 1975 Annual URSI Meeting, Boulder, Co.
39. Albertsen, N.C., J.E. Hansen and N. Jensen, Numerical and Experimental Investigations of the Influence of Spacecraft Structures on Antenna Radiation Patterns, Report on Contract No. ESTEC 1340/71/AA, The Technical University of Denmark, Lyngby, February 1972.
40. Albertsen, N.C., J.E. Hansen and N.E. Jensen, Numerical Analysis of Radiation from Wire Antennas on Conducting Bodies, 1972 Spring URSI Meeting, Washington, D.C., 13-15 April, 1972.
41. Bennett, C.L., A.M. Auckenthaler and J.D. DeLorenzo, Transient Scattering by Three-Dimensional Conducting Surfaces with Wires, 1971 International Antenna and Propagation Symposium, Los Angeles, Calif., 22-24 September,1971.
42. Bennett, C.L. And R.M. Hieronymus, Integral Equation Solution, Final Report on Contract No. F30602-75-C-0040, Sperry Research Center, Sudbury, Mass., December 1975.
43. Bennett, C.L., The Impulse Response Augmentation Technique, 1973 International IEEE/G-AP Symposium, Boulder, Co., August 1973.
44. Bennett, C.L., Effects of Polarization on Electromagnetic

Impulse Response, 1974 Annual URSI Meeting, Boulder, Co., 14-17 October 1974.

45. Bennett, C.L., Application of the Impulse Response Augmentation Technique to Scattering by a Sphere-Capped Cylinder, 1975 IEEE/URSI Meeting, Urbana, Illinois, 3-5 June, 1975.
46. Hieronymus, R.M., C.L. Bennett and J.D. DeLorenzo, Time-Domain Approach to Inverse Scattering, 1974 URSI Meeting, Atlanta, Ga., 11-13 June, 1974.
47. Chaudhuri, S.K. and W.M. Boerner, Application of Polarizational Aspects of Space-Time Integral Solution to Electromagnetic Inverse Scattering, 1975 IEEE AP-S International Symposium, Urbana, Illinois, 2-4 June 1975.
48. Chaudhuri, S.K. and W.M. Boerner, Utilization of Polarization-Depolarization Characteristics in Profile Inversion of a Perfectly Conducting Prolate Spheroid, Paper presented at the USNC/URSI-IEEE Meeting, Boulder, Co., 20-23 October, 1975.

COMPUTATIONAL METHODS FOR TRANSMISSION OF WAVES THROUGH APERTURES

Roger F. Harrington and Joseph R. Mautz
Department of Electrical and Computer Engineering
Syracuse University

The general problem of two regions isolated except for coupling through an aperture is considered. The operator equation is derived and solved via the method of moments. The aperture characteristics are expressed in terms of two aperture admittance matrices, one for each region. The admittance matrix for one region is independent of the other region, and hence can be used for any problem involving that region and aperture. The solution can be represented by two generalized n-port networks connected in parallel with current sources. The current sources are related to the tangential magnetic field which exists over the aperture region when the aperture is closed by an electric conductor. The problem of an aperture in a conducting plane excited by a plane wave is considered in detail. Explicit formulas and sample computations are given for a rectangular aperture in a conducting plane. The problem of transmission of electromagnetic waves from a waveguide into half space through an aperture is also considered in detail. Explicit formulas and sample computations are given for a rectangular waveguide feeding a rectangular aperture.

I. INTRODUCTION

This paper considers the general problem of two regions electromagnetically coupled through an aperture. There are many specific applications of this theory, such as apertures in a

ISBN 0-12-709650-7

conducting plane, waveguide-fed apertures, cavity-backed apertures, waveguide to waveguide coupling, waveguide to cavity coupling, and cavity-to-cavity coupling. The literature on these problems is extensive. Formulation of specific problems can be found in many books, of which [1-4] are typical. These books give many additional references to the literature.

The equivalence principle and the method of moments are used in this paper to obtain a general formulation for aperture problems. It is shown that the problem separates into two parts, namely, the regions on each side of the aperture. The only coupling is through the aperture, whose characteristics can be expressed by aperture admittance matrices, one for each region. These admittance matrices depend only on the region being considered, being independent of the other region. The aperture coupling is then expressible as the sum of the two independent aperture admittance matrices, with source terms related to the incident magnetic field. This result can be interpreted in terms of generalized networks as two n-port networks connected in parallel with current sources. The resultant solution is equivalent to an n-term variational solution.

Since the problem is divided into two mutually exclusive parts, one can separately solve a few canonical problems, such as apertures in conducting screens, in waveguides, and in cavities, and then combine them in various permutations mentioned in the first paragraph. Computer programs can be developed for treating broad classes of canonical problems, such as apertures of arbitrary shape in conducting planes, in rectangular waveguides, and in rectangular cavities. Such programs can then serve as broad and versatile tools for designing electromagnetic systems with aperture coupling.

We start with a general formulation of the problem, and then specialize it to two cases (1) electromagnetic transmission through a rectangular aperture in a perfectly conducting plane, and (2) transmission from a rectangular waveguide into half space through a rectangular aperture. Versatile computer programs have been written for these two cases, and are available in research reports [5,6]. These particular problems are only representative of the many cases for which solutions can be obtained and general purpose computer programs written.

II. GENERAL FORMULATION

Figure 1 represents the general problem of aperture coupling between two regions, called region *a* and region *b*. In region *a* there are impressed sources $\mathbf{J}^i$, $\mathbf{M}^i$ and region *b* is assumed source free. The more general case of sources in both region *a* and region *b* can be treated as the superposition of two problems, one with sources in region *a* only, plus one with sources in region *b* only. Each region of Fig. 1 is shown to be bounded by an electric conductor, although other types of electromagnetic isolation may be used. Region *a* is shown closed and region *b* is shown open to infinity, although each region may be open or closed. The equivalence principle [2, Sec. 3-5] is used to divide the problem into two equivalent problems, as shown in Fig. 2. In region *a*, the field is produced by the sources $\mathbf{J}^i$, $\mathbf{M}^i$, plus the equivalent magnetic current.

$$\mathbf{M} = \mathbf{n} \times \mathbf{E} \tag{1}$$

over the aperture region, with the aperture covered by an electric conductor. In region *b*, the field is produced by the equivalent magnetic current $-\mathbf{M}$ over the aperture region, with the aperture covered by an electric conductor. The fact that the equivalent current in region *a* is $+\mathbf{M}$ and that in region *b* is $-\mathbf{M}$ ensures that the tangential component of electric field is continuous across the aperture. The remaining boundary condition to be applied is continuity of the tangential component of magnetic field across the aperture.

The tangential component of magnetic field in region *a* over the aperture, denoted $\mathbf{H}_t^a$, is the sum of that due to the

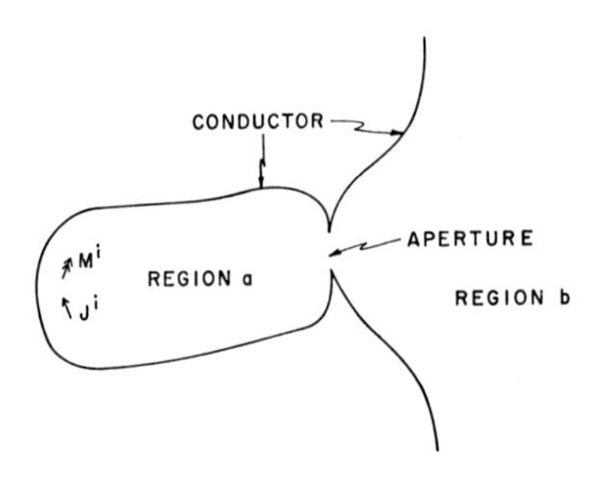

Fig. 1. The general problem of two regions coupled by an aperture.

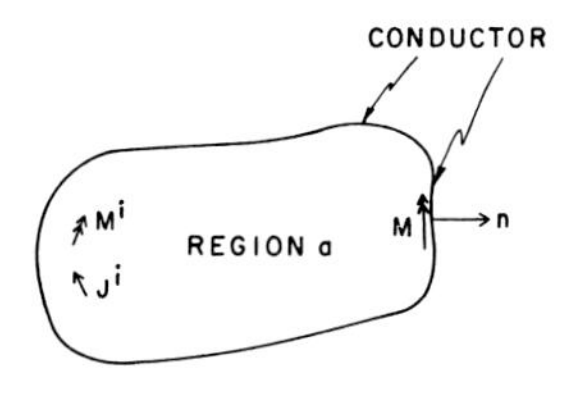

(a) EQUIVALENCE FOR REGION a.

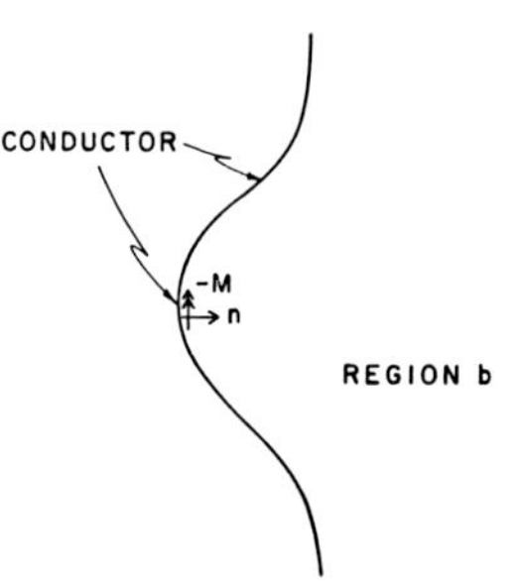

(b) EQUIVALENCE FOR REGION b.

Fig. 2. The original problem divided into two equivalent problems.

impressed sources, denoted $\underset{\sim}{H}_t^i$, plus that due to the equivalent source $\underset{\sim}{M}$, denoted $\underset{\sim}{H}_t^a(M)$, that is

$$\underset{\sim}{H}_t^a = \underset{\sim}{H}_t^i + \underset{\sim}{H}_t^a(\underset{\sim}{M}) \tag{2}$$

Note that $\underset{\sim}{H}_t^i$ and $\underset{\sim}{H}_t^a(M)$ are both computed with a conductor covering the aperture. A similar equation holds for region b, except that the equivalent sources $-\underset{\sim}{M}$ are the only sources. Hence, the tangential component of magnetic field in region b over the aperture is

$$\underset{\sim}{H}_t^b = \underset{\sim}{H}_t^b(-\underset{\sim}{M}) = -\underset{\sim}{H}_t^b(\underset{\sim}{M}) \tag{3}$$

where $\underset{\sim}{H}_t^b(\underset{\sim}{M})$ is computed with a conductor covering the aperture. The last equality in (3) is a consequence of the linearity of the $\underset{\sim}{H}_t^b$ operator. The true solution is obtained when $\underset{\sim}{H}_t^a$ of (2) equals $\underset{\sim}{H}_t^b$ of (3), or

$$\underset{\sim}{H}_t^a(\underset{\sim}{M}) + \underset{\sim}{H}_t^b(\underset{\sim}{M}) = -\underset{\sim}{H}_t^i \tag{4}$$

This is the basic operator equation for determining the equivalent magnetic current $\underset{\sim}{M}$.

If (4) were satisfied exactly, we would have the true solution. We use the method of moments [7] to obtain an approximate solution. Define a set of expansion functions $\{\underset{\sim}{M}_n,\ n=1,2,\ldots,N\}$, and let

$$\underset{\sim}{M} = \sum_n V_n \underset{\sim}{M}_n \tag{5}$$

where the coefficients V_n are to be determined. Substitute (5) into (4) and use linearity of the H_t operators to obtain

$$\sum_n V_n \underset{\sim}{H}_t^a(\underset{\sim}{M}_n) + \sum_n V_n \underset{\sim}{H}_t^b(\underset{\sim}{M}_n) = -\underset{\sim}{H}_t^i \tag{6}$$

Next, define a symmetric product

$$\langle A,B\rangle = \iint_{\text{apert.}} \underset{\sim}{A} \cdot \underset{\sim}{B}\, ds \tag{7}$$

and a set of testing functions $\{\underset{\sim}{W}_n,\ n=1,2,\ldots,N\}$, which may or may not be equal to the expansion functions. We take the symmetric product of (6) with each testing function $\underset{\sim}{W}_m$, and use the linearity of the symmetric product to obtain the set of equations

$$\sum_n V_n \langle W_m, H_t^a(M_n)\rangle + \sum_n V_n \langle W_m, H_t^b(M_n)\rangle = -\left[W_m, H_t^i\right] \tag{8}$$

$m=1,2,\ldots,N$. Solution of this set of linear equations determines the coefficients V_n and the magnetic current $\underset{\sim}{M}$ according to (5). Once $\underset{\sim}{M}$ is known, the fields and field-related parameters may be computed by standard methods.

The above solution can be put into matrix notation as follows:

Define an admittance matrix for region a as

$$[Y^a] = [\langle -W_m, H_t^a(M_n)\rangle]_{N\times N} \tag{9}$$

and an admittance matrix for region b as

$$[Y^b] = [\langle -W_m, H_t^b(M_n)\rangle]_{N\times N} \tag{10}$$

The minus signs are placed in (9) and (10) on the basis of power considerations. Define a source vector

$$\vec{I}^i = [\langle W_m, H_t^i\rangle]_{N\times 1} \tag{11}$$

and a coefficient vector

$$\vec{V} = [V_n]_{N\times 1} \tag{12}$$

Now the matrix equation equivalent to equations (8) is

$$[Y^a + Y^b]\vec{V} = \vec{I}^i \tag{13}$$

This can be interpreted in terms of generalized networks as two networks $[Y^a]$ and $[Y^b]$ in parallel with the current source $\vec{I}^i$ as shown in Fig. 3. The resultant voltage vector

$$\vec{V} = [Y^a + Y^b]^{-1}\vec{I}^i \tag{14}$$

is then the vector of coefficients which give $\underset{\sim}{M}$ according to (5).

It is important to note that computation of $[Y^a]$ involves *only* region a, and computation of $[Y^b]$ involves *only* region b. Hence, we have divided the problem into two parts, each of which may be formulated independently. Once $[Y]$ is computed for one region, it may be combined with $[Y]$ for any other region, making it useful for a wide range of problems. For example, the same aperture admittance matrix for radiation into half-space would be useful for plane-wave excitation of the aperture, waveguide excitation, and cavity excitation.

III. MEASUREMENT

A linear measurement is defined as a number which depends linearly on the source. Examples of linear measurements are components of the field at a point, voltage along a given contour, and current crossing a given surface. Measurements made in region b will depend linearly only on the equivalent current $-\underset{\sim}{M}$. Measurements made in region a will depend linearly on the impressed sources $\underset{\sim}{J}^i$, $\underset{\sim}{M}^i$, as well as on the equivalent current $\underset{\sim}{M}$. We now illustrate these concepts with a particular example.

Consider the measurement (computation) of a component H_m of magnetic field at a point $\underset{\sim}{r}_m$ in region b. It is known that this component can be obtained by placing a magnetic dipole $\underset{\sim}{K\ell}_m$ at $\underset{\sim}{r}_m$, and applying the reciprocity theorem to its field and to the original field [2, Sec. 3-8]. The original field in region b is given by the solution to Fig. 2b. The problem involving the magnetic dipole, called the adjoint problem, is shown in Fig. 4. Application of the reciprocity theorem to these two cases yields

$$H_m K\ell_m = - \iint_{\text{apert.}} \underset{\sim}{M} \cdot \underset{\sim}{H}^m ds \tag{15}$$

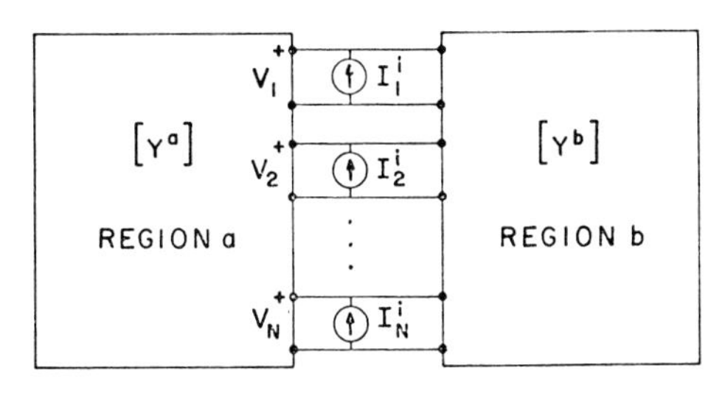

Fig. 3. The generalized network interpretation of equation (13).

Fig. 4. The adjoint problem for determining H_m at $\underset{\sim}{r}_m$.

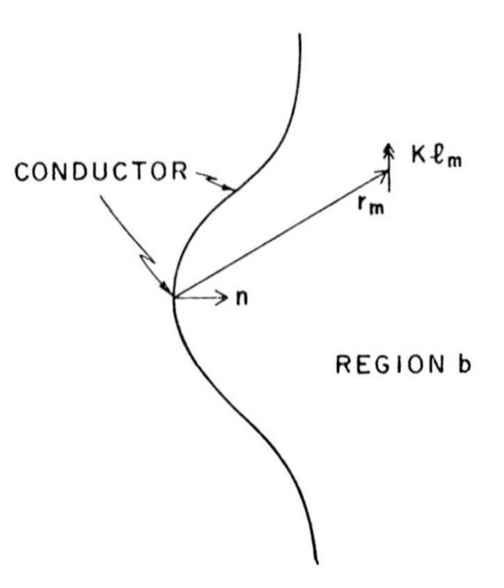

Here $\underset{\sim}{H}^m$ is the magnetic field from $K\underset{\sim}{\ell}_m$ in the presence of a complete conductor, and H_m is the component in the direction of $K\underset{\sim}{\ell}_m$ of the magnetic field at $\underset{\sim}{r}_m$ due to $-\underset{\sim}{M}$ in the presence of a complete conductor. To evaluate (15), substitute for $\underset{\sim}{M}$ from (5) and obtain

$$H_m K\ell_m = \sum_n V_n <-M_n, H^m> \tag{16}$$

This can be written in matrix form as

$$H_m K\ell_m = \tilde{I}^m \vec{V} \tag{17}$$

where $\tilde{I}^m$ is the transpose of a measurement vector

$$\vec{I}^m = [<-M_n, H^m>]_{N\times 1} \tag{18}$$

Note that the elements of $\vec{I}^m$ are similar in form to those of $\vec{I}^i$ given by (11), except that $-\underset{\sim}{M}_n$ replaces $\underset{\sim}{W}_m$. The minus sign difference reflects the fact that the equivalent source in region *b* is $-\underset{\sim}{M}$, in contrast to that in region *a* which is $+\underset{\sim}{M}$. Now substitute (14) into (17) to obtain

$$H_m K\ell_m = \tilde{I}^m [Y^a + Y^b]^{-1} \vec{I}^i \tag{19}$$

If the magnetic dipole is of unit moment, then (19) gives H_m at $\underset{\sim}{r}_m$ directly.

Every linear measurement in region b will be of the form (19). For example, if a component of $\underset{\sim}{E}$ at $\underset{\sim}{r}_m$ were desired, we would place an electric dipole at $\underset{\sim}{r}_m$ and apply reciprocity. In general, a linear measurement involves applying reciprocity to the original problem and to an adjoint problem. A determination of the sources of the adjoint problem is a part of the formulation of the problem.

If a linear measurement is made in region a, it will involve a contribution from the impressed sources $\underset{\sim}{J}^i$, $\underset{\sim}{M}^i$ added to that from the equivalent sources $\underset{\sim}{M}$. For example, instead of (19) we would have

$$H_m K\ell_m = H_m^i K\ell_m + \tilde{I}^m[Y^a + Y^b]^{-1}\vec{I}^i \tag{20}$$

where H_m^i is the magnetic field from $\underset{\sim}{J}^i$, $\underset{\sim}{M}^i$ in the presence of a complete conductor. Also, in region a we would define the measurement vector to be

$$\vec{I}^m = [\langle M_n, H^m\rangle]_{N\times 1} \tag{21}$$

instead of (18), because the equivalent sources are $+\underset{\sim}{M}$ in region a in contrast to $-\underset{\sim}{M}$ in region b. Note that it is the difference field $\underset{\sim}{H}-\underset{\sim}{H}^i$ in region a (due to $\underset{\sim}{M}$) that is directly analogous to the transmitted field $\underset{\sim}{H}$ in region b (due to $-\underset{\sim}{M}$).

A quadratic measurement is one which depends quadratically on the sources. Examples of quadratic measurements are components of the Poynting vector at a point, power crossing a given surface, and energy within a given region. A particular quadratic measurement of considerable interest is the power transmitted through the aperture, which we now consider.

The complex power P_t transmitted through the aperture is basically

$$P_t = \iint_{\text{apert.}} \underset{\sim}{E} \times \underset{\sim}{H}^* \cdot \underset{\sim}{n}\, ds \tag{22}$$

where the asterisk denotes complex conjugate. Substituting from (1), we have

$$P_t = \iint_{\text{apert.}} \underset{\sim}{M} \cdot \underset{\sim}{H}^*\, ds \tag{23}$$

This involves only the tangential component of $\underset{\sim}{H}$, which in region b we denoted by $\underset{\sim}{H}_t^b(-\underset{\sim}{M})$. For $\underset{\sim}{M}$ we use the linear combination (5) and obtain

$$\underset{\sim}{H}_t^b(-\underset{\sim}{M}) = -\sum_n V_n \underset{\sim}{H}_t^b(\underset{\sim}{M}_n) \tag{24}$$

Substituting this for $\underset{\sim}{H}$ and (5) for $\underset{\sim}{M}$ into (23), we obtain

$$P_t = -\sum_m \sum_n V_m V_n^* \iint_{\text{apert.}} \underset{\sim}{M}_m \cdot \underset{\sim}{H}_t^{b*}(\underset{\sim}{M}_n)\, ds \tag{25}$$

If $\underset{\sim}{M}_m$ are real, the conjugate operations can be taken outside the integrals. Moreover, if $\underset{\sim}{M}_m = \underset{\sim}{W}_m$ (Galerkin's method), then the negative of the integrals in (25) are Y_{mn}^{b*} as defined by (10), and

$$P_t = \sum_m \sum_n V_m V_n^* Y_{mn}^{b*} \tag{26}$$

This can be written in matrix form as

$$P_t = \tilde{V}[Y^b]^* \vec{V}^* \tag{27}$$

Note that this is the usual formula for power into network $[Y^b]$ of Fig. 3.

IV. APERTURES

Consider a conducting plane covering the $z = 0$ plane except for an aperture, as shown in Fig. 5. The two regions $z > 0$ and $z < 0$ are identical half spaces, and hence their admittance

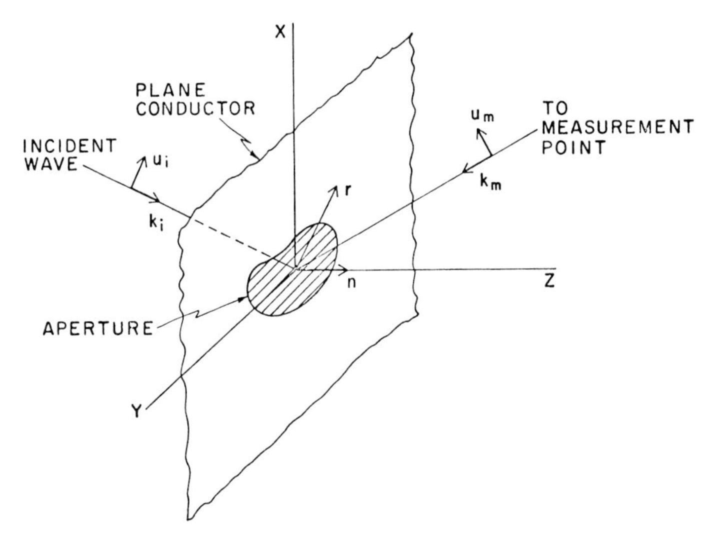

Fig. 5. Aperture in a plane conductor.

matrices are the same. Therefore, we let

$$[Y^a + Y^b] = 2[Y^{hs}] \tag{28}$$

where $[Y^{hs}]$ denotes the aperture admittance for the aperture opening into half space, say $z > 0$. When the aperture is covered by a conductor, the $z = 0$ plane is a complete conducting plane, and image theory applies. The magnetic current expansion functions are on the surface of the $z = 0$ plane. Their images are equal to them and are also on the $z = 0$ plane [2, Sec. 3-6]. The result is that $[Y^{hs}]$ is twice the admittance matrix obtained using expansion functions $\underset{\sim}{M}_n$ radiating into free space everywhere. This problem is dual to that for the impedance matrix of a plane conductor, a problem considered recently in the literature [8].

The original excitation of the aperture is by the impressed sources $\underset{\sim}{J}^i$, $\underset{\sim}{M}^i$ in the region $z < 0$. The impressed field $\underset{\sim}{H}_t^i$ used in the operator equation (4) is the tangential magnetic field due to $\underset{\sim}{J}^i$, $\underset{\sim}{M}^i$ with the aperture covered by a conductor (Fig. 2a). In this case the $z = 0$ plane is a complete conductor, and image theory again applies. The result is that the tangential component of $\underset{\sim}{H}$ over the $z = 0$ plane when it is covered by a conductor is just twice what it is for the same sources in free space. Hence,

$$\underset{\sim}{H}_t^i = 2\underset{\sim}{H}_t^{io} \tag{29}$$

where $\underset{\sim}{H}_t^{io}$ is the tangential component of the magnetic field over the aperture due to the sources $\underset{\sim}{J}^i$, $\underset{\sim}{M}^i$ in free space. The components of the excitation vector $\vec{I}^i$ defined by (11) are now

$$I_m^i = 2 \iint_{\text{apert.}} \underset{\sim}{W}_m \cdot \underset{\sim}{H}_t^{io} \, ds \tag{30}$$

where $\underset{\sim}{W}_m$ is the mth testing function.

A case of special interest is that of plane wave excitation. A unit plane wave is given by

$$\underset{\sim}{H}^{io} = \underset{\sim}{u}_i e^{-j\underset{\sim}{k}_i \cdot \underset{\sim}{r}} \tag{31}$$

where $\underset{\sim}{u}_i$ is a unit vector specifying the direction of $\underset{\sim}{H}^{io}$, $\underset{\sim}{k}_i$ is the propagation vector of magnitude $2\pi/\lambda$ and pointing in the direction of propagation, and $\underset{\sim}{r}$ is the radius vector to an arbitrary field point. These vectors are shown in Fig. 5. The components (30) of the plane wave excitation vector are then

$$P_m^i = 2 \iint_{\text{apert.}} \underset{\sim}{W}_m \cdot \underset{\sim}{u}_i \, e^{-j\underset{\sim}{k}_i \cdot \underset{\sim}{r}} \, ds \tag{32}$$

The symbol $\vec{P}^i$ has been used for this particular vector to distinguish it from the more general excitation vector (30).

Similar simplifications apply to the adjoint (measurement) problem. For the evaluation of a component of magnetic field at a point $\underset{\sim}{r}_m$, a magnetic dipole $K\underset{\sim}{\ell}_m$ is placed at the measurement point $\underset{\sim}{r}_m$. This radiates in the presence of a complete conductor over the $z = 0$ plane, and hence, analogous to (29), we have

$$\underset{\sim}{H}_t^m = 2\underset{\sim}{H}_t^{mo} \tag{33}$$

Here $\underset{\sim}{H}_t^m$ denotes the tangential component of $\underset{\sim}{H}$ over the aperture from $K\ell_m$ when the $z = 0$ plane is covered by a conductor, and $\underset{\sim}{H}_t^{mo}$ denotes that from $K\ell_m$ when radiating in free space. The components of the measurement vector $\vec{I}^m$ defined by (18) are now

$$I_n^m = -2 \iint_{\text{apert.}} \underset{\sim}{M}_n \cdot \underset{\sim}{H}_t^{mo}\, ds \tag{34}$$

where $\underset{\sim}{M}_n$ is the nth expansion function.

A case of special interest is that of far-field measurement. This is obtained by a procedure dual to that used for radiation and scattering from conducting wires [9]. To obtain a component of $\underset{\sim}{H}$ on the radiation sphere, we take a source $K\ell_m$ perpendicular to $\underset{\sim}{r}_m$ and let $r_m \to \infty$. At the same time we adjust $K\ell_m$ so that it produces a unit plane wave in the vicinity of the origin. The required dipole moment is given by

$$\frac{1}{K\ell_m} = \frac{-j\omega\varepsilon}{4\pi r_m}\, e^{-jkr_m} \tag{35}$$

and the plane-wave field it produces in the vicinity of the origin is

$$\underset{\sim}{H}^{mo} = \underset{\sim}{u}_m\, e^{-j\underset{\sim}{k}_m \cdot \underset{\sim}{r}} \tag{36}$$

Here $\underset{\sim}{u}_m$ is a unit vector in the direction of $\underset{\sim}{H}^{mo}$, $\underset{\sim}{k}_m$ is the propagation vector, and $\underset{\sim}{r}$ is the radius vector to an arbitrary field point. Again these vectors are shown in Fig. 5. The components (34) of the far-field measurement vector are then

$$P_n^m = -2 \iint_{\text{apert.}} \underset{\sim}{M}_n \cdot \underset{\sim}{u}_m\, e^{-j\underset{\sim}{k}_m \cdot \underset{\sim}{r}}\, ds \tag{37}$$

The symbol $\vec{P}^m$ is used for this particular measurement vector to distinguish it from the more general measurement vector (34). The far-zone magnetic field is now given by (19) with $K\ell_m$ given by (35), $\vec{I}^m = \vec{P}^m$, $\vec{I}^i = \vec{P}^i$, and the aperture admittance given by

(28). Hence

$$H_m = \frac{-j\omega\varepsilon}{8\pi r_m} e^{-jkr_m} \tilde{P}^m [Y^{hs}]^{-1} \vec{P}^i \tag{38}$$

The usual two radiation components H_θ and H_ϕ are obtained by orienting $K\ell_m$ in the θ and ϕ directions, respectively.

A parameter sometimes used to express the transmission characteristics of an aperture is the transmission cross section τ. It is defined as that area for which the incident plane wave contains sufficient power to produce the radiation field H_m by omnidirectional radiation over half space. For unit incident magnetic field, this is

$$\tau = 2\pi r_m^2 \, |H_m|^2 \tag{39}$$

Substituting from (38), we obtain

$$\tau = \frac{\omega^2 \varepsilon^2}{32\pi} \, |\tilde{P}^m [Y^{hs}]^{-1} \vec{P}^i|^2 \tag{40}$$

Note that τ depends upon the polarization and direction of the incident wave (via $\vec{P}^i$), and upon the polarization measured and direction to the measurement point (via $\tilde{P}^m$).

Another parameter used to express the transmission characteristics of an aperture is the transmission coefficient T, defined as

$$T = \frac{P_{\text{trans.}}}{P_{\text{inc.}}} \tag{41}$$

where $P_{\text{trans.}}$ is the time-average power transmitted by the aperture, and $P_{\text{inc.}}$ is the free space power incident on the aperture. For unit plane wave excitation, the incident power is

$$P_{\text{inc.}} = \eta S \cos \theta_{\text{inc.}} \tag{42}$$

where $\eta = \sqrt{\mu/\varepsilon}$ is the intrinsic impedance of free space, S is the aperture area, and $\theta_{\text{inc.}}$ is the angle between k_i and n. The transmitted power is

$$P_{\text{trans.}} = Re(P_t) \tag{43}$$

where $Re(P_t)$ denotes the real part of P_t, given by (27) with $[Y^b] = [Y^{hs}]$. Hence

$$T = \frac{1}{\eta\, S \cos\theta_{\text{inc.}}} Re(\tilde{V}[Y^{hs}]\vec{V}^{*}) \tag{44}$$

Note that T depends on both the direction of incidence and on the polarization of the incident wave.

Finally, because of symmetry about the $z = 0$ plane, the difference field $\underset{\sim}{H} - \underset{\sim}{H}^i$ which exists in the region $z < 0$ is simply related to the transmitted field which exists in the region $z > 0$. The difference field in the region $z < 0$ is produced by an equivalent current $\underset{\sim}{M}$ on a plane conductor over the $z = 0$ plane. By image theory, it is also the field produced in the region $z < 0$ by the source $2\underset{\sim}{M}$ in free space. The transmitted field in the region $z > 0$ is produced by the source $-2\underset{\sim}{M}$ in free space. Hence, the difference field in the region $z < 0$ and the negative of the transmitted field in the region $z > 0$ are both produced by the same magnetic current $2\underset{\sim}{M}$ radiating in free space.

Previous studies of apertures in plane conductors include those for small apertures [10,11] and those for circular apertures [12,13]. Some numerical results for apertures of arbitrary shape have been obtained using Babinet's principle plus a wire grid approximation to the complementary conducting plane [14,15]. The reader may consult these papers for other references.

V. RECTANGULAR APERTURE IN A PLANE CONDUCTOR

Fig. 6 shows the problem to be considered and defines the coordinates and parameters to be used. The infinitely conducting plate covers the entire $z = 0$ plane except for the aperture, which is rectangular in shape with side lengths $L_x\Delta x$ and $L_y\Delta y$

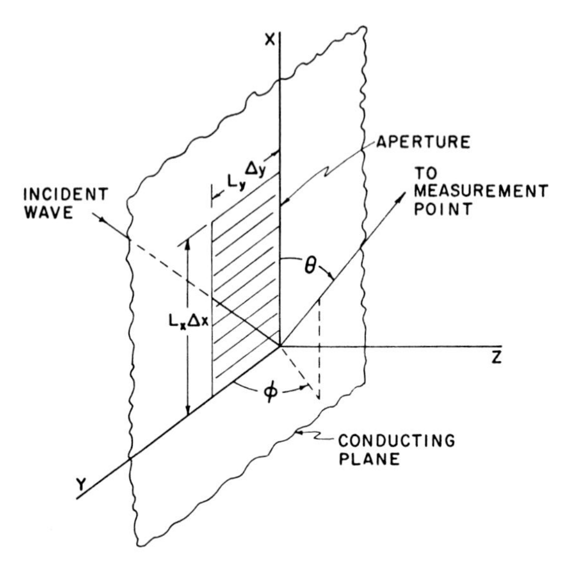

Fig. 6. Rectangular aperture in a conducting plane.

in the x and y directions, respectively. The excitation of the aperture is a uniform plane wave incident from the region $z < 0$. The field to be computed is the far zone magnetic field in the region $z > 0$, at the angles θ,ϕ.

The solution is expressed in terms of the equivalent magnetic current $\mathbf{M} = \hat{\mathbf{z}} \times \mathbf{E}$, where z is the unit z-directed vector and $\mathbf{E}$ is the electric field in the aperture. To compute $\mathbf{M}$, we use a linear expansion in terms of basis functions $\mathbf{M}_i$ and evaluate the coefficients by the method of moments. This involves determining a generalized admittance matrix considered in Subsection A below, and an excitation vector. To determine the field produced by $\mathbf{M}$, we need a measurement vector. The excitation and measurement vectors for the present problem are of the same form, and are considered in Subsection B below.

A. Admittance Matrix

According to (10) and (28), the elements of the admittance matrix $[Y]$ are given by

$$Y_{ij} = (Y^a + Y^b)_{ij} = -4\langle W_i, H(M_j)\rangle \tag{45}$$

where $H(M_j)$ is the magnetic field produced by M_j radiating in free space. The magnetic field $H(M_j)$ can be expressed in terms of an electric vector potential F and magnetic scalar potential ϕ as [16]

$$H(M_j) = -j\omega F_j - \nabla\phi_j \tag{46}$$

where

$$F_j = \frac{\varepsilon}{4\pi} \iint_{\text{apert.}} M_j \frac{e^{-jk|r-r'|}}{|r-r'|} ds \tag{47}$$

$$\phi_j = \frac{1}{4\pi\mu} \iint_{\text{apert.}} \rho_j \frac{e^{-jk|r-r'|}}{|r-r'|} ds \tag{48}$$

$$\rho_j = \frac{\nabla \cdot M_j}{-j\omega} \tag{49}$$

where r and r' are respectively the vectors to the field and source points in the aperture, ω is the angular frequency, ε is the capacitivity of free space, μ is the permeability of free space, and $k = \omega\sqrt{\mu\varepsilon}$ is the propagation constant in free space. Substituting (7) and (46) into (45), we obtain

$$Y_{ij} = 4 \iint_{\text{apert.}} W_i \cdot (j\omega F_j + \nabla\phi_j) ds. \tag{50}$$

Because of the identity

$$0 = \iint_{\text{apert.}} \nabla \cdot (\phi_j W_i) ds = \iint_{\text{apert.}} W_i \cdot \nabla\phi_j ds + \iint_{\text{apert.}} \phi_j \nabla \cdot W_i ds \tag{51}$$

(50) becomes

$$Y_{ij} = 4j\omega \iint_{\text{apert.}} (\mathbf{F}_j \cdot \mathbf{W}_i + \phi_j \rho_i) ds \tag{52}$$

where

$$\rho_i = \frac{\nabla \cdot \mathbf{W}_i}{-j\omega} \tag{53}$$

We choose the set of testing functions $\mathbf{W}_i$ equal to the set of expansion functions $\mathbf{M}_i$. The rectangular aperture $0 \le x \le L_x \Delta x$, $0 \le y \le L_y \Delta y$ where L_x and L_y are integers is divided into rectangular subareas of length Δx in x and Δy in y. The set $\mathbf{M}_i$ of expansion functions is split into a set $\mathbf{M}_i^x$ of x directed magnetic currents and a set $\mathbf{M}_i^y$ of y directed magnetic currents defined by

$$\mathbf{M}^x_{p+(q-1)(L_x-1)} = \hat{\mathbf{x}}\, T^x_p(x)\, P^y_q(y), \quad \begin{matrix} p = 1,2,\ldots L_x - 1 \\ q = 1,2,\ldots L_y \end{matrix} \tag{54}$$

$$\mathbf{M}^y_{p+(q-1)L_x} = \hat{\mathbf{y}}\, T^y_q(y)\, P^x_p(x) \quad \begin{matrix} p = 1,2,\ldots L_x \\ q = 1,2,\ldots L_y - 1 \end{matrix} \tag{55}$$

where $T^x_p(x)$ are triangle functions defined by

$$T^x_p(x) = \begin{cases} \dfrac{x - (p-1)\Delta x}{\Delta x} & (p-1)\,\Delta x \le x \le p\Delta x \\ \dfrac{(p+1)\Delta x - x}{\Delta x} & p\Delta x \le x \le (p+1)\Delta x \\ 0 & |x - p\,x| \ge \Delta x \end{cases} \tag{56}$$

and $T^y_q(y)$ are similarly defined. $P^x_p(x)$ are pulse functions defined by

$$P^x_p(x) = \begin{cases} 1 & (p-1)\,x \le x < p\,x \\ 0 & \text{all other } x \end{cases} \tag{57}$$

and $P^y_q(y)$ are similarly defined. The magnetic charge sheets associated with $\mathbf{M}^x$ and $\mathbf{M}^y$ are obtained from (49), the equation of continuity. Introduction of the two types of expansion functions $\underset{\sim}{M}_j^x$ and $\underset{\sim}{M}_j^y$ and the two types of testing functions $\underset{\sim}{M}_i^x$ and $\underset{\sim}{M}_i^y$ into (52) gives rise to four Y submatrices defined by

$$Y_{ij}^{uv} = 4j\omega \iint_{\text{apert.}} (\underset{\sim}{F}_j^v \cdot \underset{\sim}{M}_i^u + \phi_j^v \rho_i^u) ds \tag{58}$$

where u is either x or y and v is either x or y. In (58), $\underset{\sim}{F}_j^v$ and ϕ_j^v are the electric vector and magnetic scalar potentials due to $\underset{\sim}{M}_j^v$.

The integrations over the "field" magnetic current and charge $\underset{\sim}{M}_i^u$ and ρ_i^u explicit in (58) are approximated by sampling the integrands at the center of each pulse function. The exponential functions in (47) and (48) are approximated by the following four term approximation

$$e^{-jkr} \approx e^{-jkr_o}[1 - jk(r-r_o) - \frac{k^2}{2}(r-r_o)^2 + \frac{jk^3}{6}(r-r_o)^3] \tag{59}$$

where

$$r = \sqrt{x^2 + y^2} \tag{60}$$

$$r_o = \sqrt{(s\Delta x)^2 \quad (t\Delta y)^2}. \tag{61}$$

where s is the difference between the subscripts p of the "source" and "field" pulses. Similarly, t is defined such that $t\Delta y$ is the difference between the y coordinates of the centers of the "source" and "field" pulses $P_q^y(y)$. The resultant integrals can then be evaluated in terms of known integrals. The details can be found in [5].

B. Plane Wave Excitation and Measurement Vectors

The plane wave excitation vector $\vec{P}^i$ of (32) and the plane wave measurement vector $\vec{P}^m$ of (37) are of the same form except for a minus sign. We therefore need to evaluate only one of them, say the measurement vector $\vec{P}^m$. We specialize it to four principal plane patterns as

$$\left(P^{mx}_{p+(q-1)(L_x-1)}\right)_{\theta y} = 2\Delta x\Delta y \sin\theta \left(\frac{\sin\frac{k\Delta x\cos\theta}{2}}{\frac{k\Delta x\cos\theta}{2}}\right)^2 e^{jpk\Delta x\cos\theta}, \tag{62}$$

$$p = 1,2,\ldots L_x - 1$$

$$q = 1,2,\ldots L_y$$

$$(P^{my}_i)_{\theta y} = 0,\ i = 1,2,\ldots L_x(L_y - 1) \tag{63}$$

$$(P^{mx}_i)_{yy} = o,\ i = 1,2\ldots.(L_x - 1)L_y \tag{64}$$

$$\left(P^{my}_{p+(q-1)L_x}\right)_{yy} = -2\Delta x\Delta y \left(\frac{\sin\frac{k\Delta x\cos\theta}{2}}{\frac{k\Delta x\cos\theta}{2}}\right) e^{jk(p-\frac{1}{2})\Delta x\cos\theta}, \tag{65}$$

$$p = 1,2,\ldots L_x$$

$$q = 1,2,\ldots L_y - 1$$

$$(P^{mx}_i)_{\phi x} = 0,\ i = 1,2,\ldots\ (L_x - 1)L_y \tag{66}$$

$$\left(P^{my}_{p+(q-1)L_x}\right)_{\phi x} = 2\Delta x\Delta y \sin\phi \left(\frac{\sin\frac{k\Delta y\cos\phi}{2}}{\frac{k\Delta y\cos\phi}{2}}\right)^2 e^{jkq\Delta y\cos\phi}, \tag{67}$$

$$p = 1;2,\ldots L_x$$

$$q = 1,2\ldots.L_y - 1$$

$$\left(P^{mx}_{p+(q-1)(L_x-1)}\right)_{xx} = -2\Delta x\Delta y \left(\frac{\sin\frac{k\Delta y\cos\phi}{2}}{\frac{k\Delta y\cos\phi}{2}}\right) e^{jk(q-\frac{1}{2})\Delta y\cos\phi}, \tag{68}$$

$$p = 1,2,\ldots L_x - 1$$

$$q = 1,2,\ldots L_y$$

$$(P_i^{my})_{xx} = 0, \; i = 1,2,\ldots L_x(L_y - 1) \tag{69}$$

The superscripts x and y and the subscripts $p+(q-1)(L_x-1)$ and $p+(q-1)L_x$ on the left-hand sides of (62) - (69) are the same as those in (54) and (55). The subscript θy denotes a θ polarized measurement in the $y = 0$ plane, yy denotes a y polarized measurement in the $y = 0$ plane, ϕx denotes a ϕ polarized measurement in the $x = 0$ plane and xx denotes an x polarized measurement in the $x = 0$ plane. The coordinate system and angles θ and ϕ appearing on the right-hand sides of (62) to (69) are defined in Fig. 6.

VI. REPRESENTATIVE COMPUTATIONS

A versatile computer program has been developed using the formulas obtained from Section V. This program is described and listed in [5]. Some representative computations obtained with this program are given in this section.

The first example is for a narrow slot, of width $\lambda/20$ and of variable length L. The far-zone quantity plotted was the transmission cross section, defined by (39). We use the notation:

$$\tau_{\theta y} = 2\pi r^2 |H_\theta|^2 \quad \text{in the } y = 0 \text{ plane,}$$
$$\tau_{xx} = 2\pi r^2 |H_x|^2 \quad \text{in the } x = 0 \text{ plane.} \tag{70}$$

For the case being considered, the orthogonal components of $\underline{H}$ in these two planes were zero. Fig. 7 shows plots of $\tau_{\theta y}$ and τ_{xx} for x-directed slots of width $\lambda/20$ and length (a) $L = \lambda/4$, (b) $L = \lambda/2$, (c) $L = 3\lambda/4$, and $dL= \lambda$. In all cases, the excitation was due to a plane wave normally incident on the conducting plane with the magnetic field in the x direction. Note the large transmission cross section for $L = \lambda/2$, case (b), due to the slot being near resonance. The plots of τ are of the same form as scattering cross

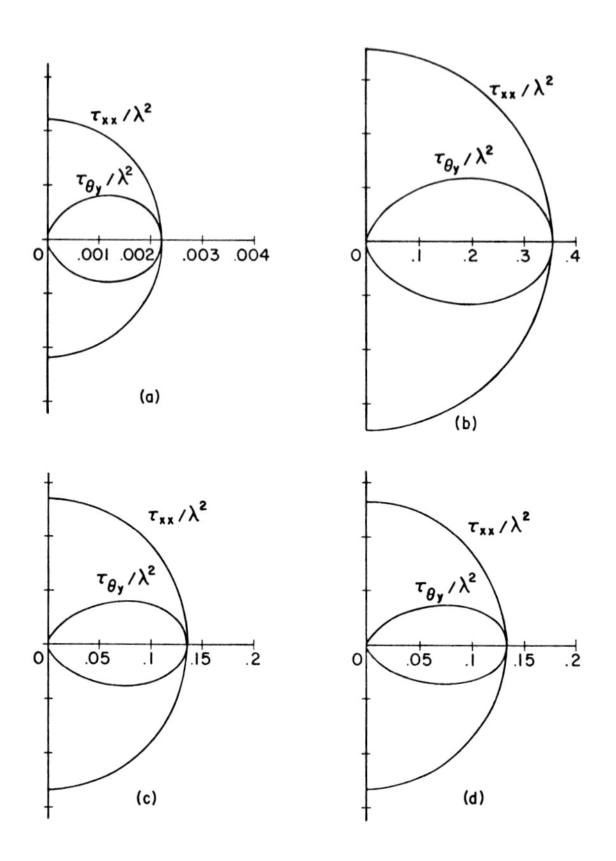

Fig. 7. Transmission cross section for slots of length L in the x direction and width λ/20 in the y direction. (a) L = λ/4, (b) L = λ/2, (c) L = 3λ/4, (d) L = λ. Excitation is by a plane wave normally incident on the conducting plane with magnetic field in the x direction.

section from the complementary conducting strips, as known from Babinet's principle.

Fig. 8. shows plots of the equivalent magnetic current in the aperture region for the same slots. Since $\underset{\sim}{M} = \hat{\underset{\sim}{z}} \times \underset{\sim}{E}$, they are also plots of the tangential component of $\underset{\sim}{E}$ in the slots. Again note the large value of $\underset{\sim}{M}$ for the case of $L = \lambda/2$, which is near resonance. Note also that for short slots ($L \leq 3\lambda/4$), the $\underset{\sim}{M}$ is almost equiphasal and closely approximated by a half sine wave.

Next, computations were made to test the rate of convergence of the solution as the number of subsections was increased. A slot of width $\lambda/10$ and length 2λ was chosen for the study. Again the excitation is a plane-wave normally incident on the conducting plane with the magnetic field in the x direction. Figure 9 shows plots of $\tau_{\theta y}$ and τ_{xx} for the cases (a) 39, (b) 19, (c) 9, and (d) 4 triangular expansion functions respectively, Note that the patterns (a) and (b) are essentially the same, and pattern (c) is only slightly different. They differ appreciably from (d), which results from only 4 expansion functions. The difference in the solutions as the number of expansion functions is decreased is better illustrated by plots of $\underset{\sim}{M}$, as shown in Fig. 10. These are for the same cases as the corresponding cases of Fig. 9. It can be seen clearly how the computed equivalent current in the slot region changes as the number of subsections is reduced. As a rule of thumb, for near-field quantities (such as $\underset{\sim}{M}$) one should use subareas of length $\lambda/10$ or less and for far-field quantities (such as τ) length $\lambda/5$ or less.

Other computations were made for wider apertures and by waves not normally incident on the conducting plane. Examples of these can be found in [5].

VII. WAVEGUIDE-FED APERTURES

Consider now a uniform waveguide feeding an aperture in a conducting plane, as shown in Fig. 11. In general, the aperture may be of different size and shape than the waveguide cross section. The half-space region $z > 0$ is the same as in the previous problem, Fig. 5, and the analysis of Section IV applies. An analysis of the waveguide region is given here.

Let the excitation of the waveguide be a source which produces a single mode, of unit amplitude, incident on the

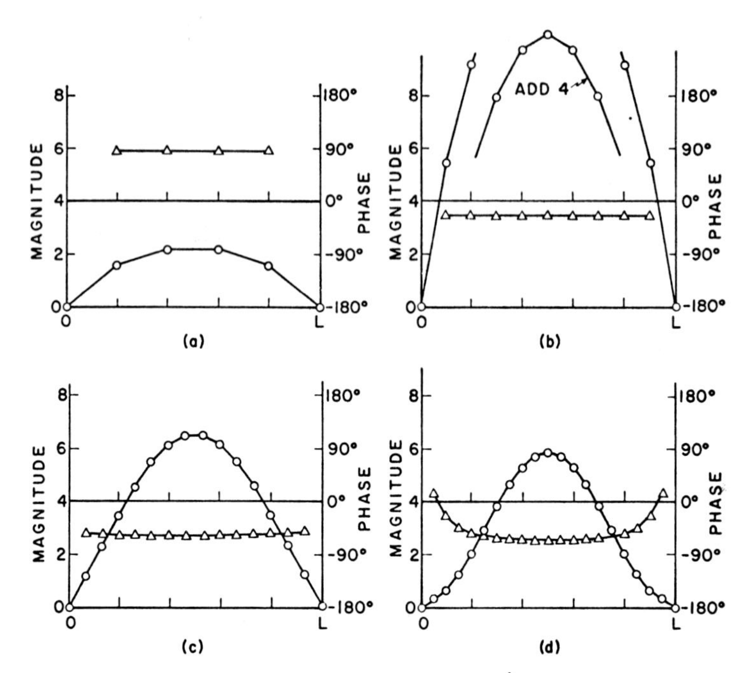

Fig. 8. Magnitude and phase of $|M/E^i|$, where M is the x-directed magnetic current and E^i is the incident electric field, for the same slots as for Fig. 7. (a) $L = \lambda/4$, (b) $L = \lambda/2$, (c) $L = 3\lambda/4$, (d) $L = \lambda$. Circles denote magnitude, triangles denote phase.

aperture. This mode (usually the dominant mode) is denoted by the index o. The field transverse to the z-direction can then be expressed in modal form as [2, Sec. 8-1].

$$
\begin{aligned}
E_t &= e^{-\gamma_o z} \underset{\sim}{e}_o + \sum_i \Gamma_i e^{\gamma_i z} \underset{\sim}{e}_i \\
H_t &= Y_o e^{-\gamma_o z} \underset{\sim}{u}_z \times \underset{\sim}{e}_o - \sum_i \Gamma_i Y_i e^{\gamma_i z} \underset{\sim}{u}_z \times \underset{\sim}{e}_i
\end{aligned}
\tag{71}
$$

It is assumed that all modes, *TE* and *TM*, are included in the summation. The γ_i are modal propagation constants.

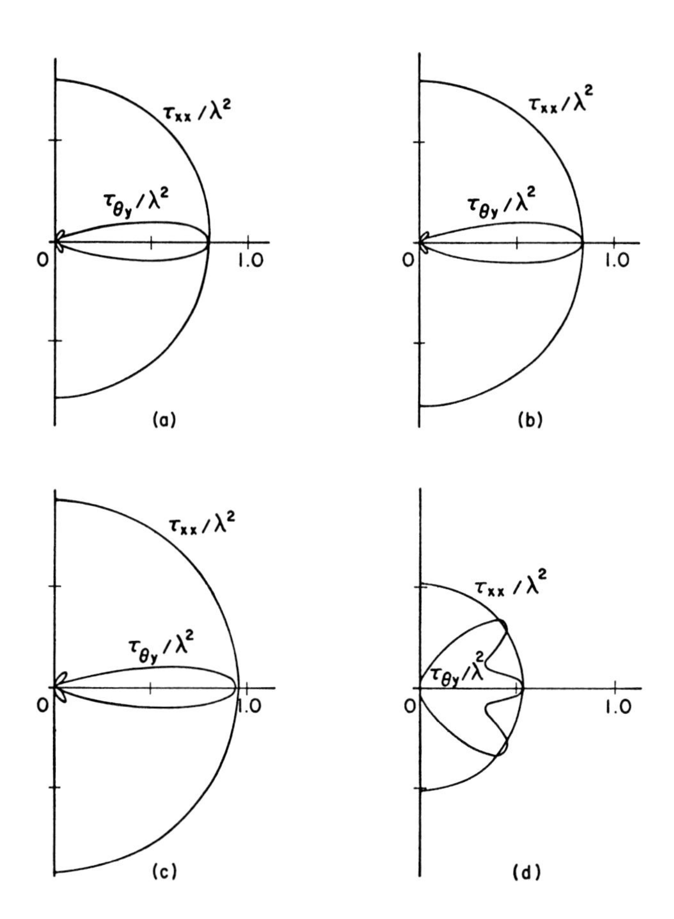

Fig. 9. Transmission cross section when the number of expansion functions is (a) 39, (b) 19, (c) 9, and (d) 4. Computations are for a slot of length 2λ in the x direction and width λ/10 in the y direction. Excitation is by a plane wave normally incident on the conducting plane with magnetic field in the x direction.

$$\gamma_i = \begin{cases} j\beta_i = jk\sqrt{1 - (\lambda/\lambda_i)^2} & \lambda < \lambda_i \\ \alpha_i = k_i\sqrt{1 - (\lambda_i/\lambda)^2} & \lambda > \lambda_i \end{cases} \tag{72}$$

where λ_i is the i-th mode cut-off wavelength, and $k_i = 2\pi/\lambda_i$ is the i-th mode cut-off wavenumber. The Y_i are the modal characteristic admittances

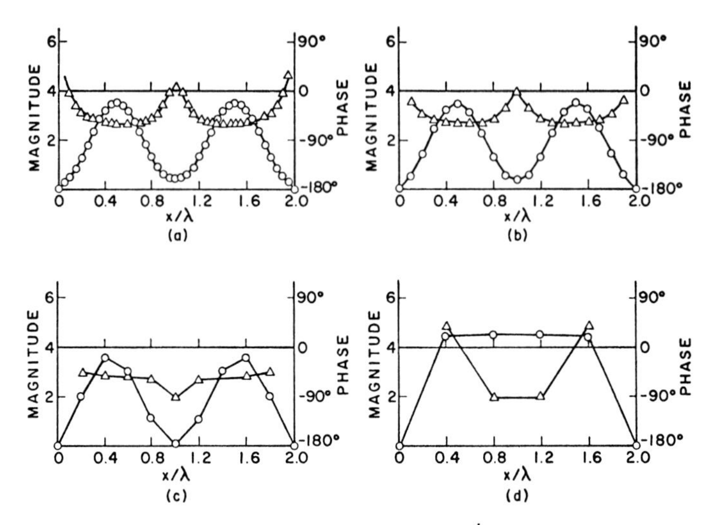

Fig. 10. Magnitude and phase M/E^i, where M is the x-directed magnetic current and E^i is the incident electric field, when the number of expansion functions is (a) 39, (b) 19, (c) 9, and (d) 4. Circles denote magnitude, triangles denote phase. Computations are for the same slot as for Fig. 9.

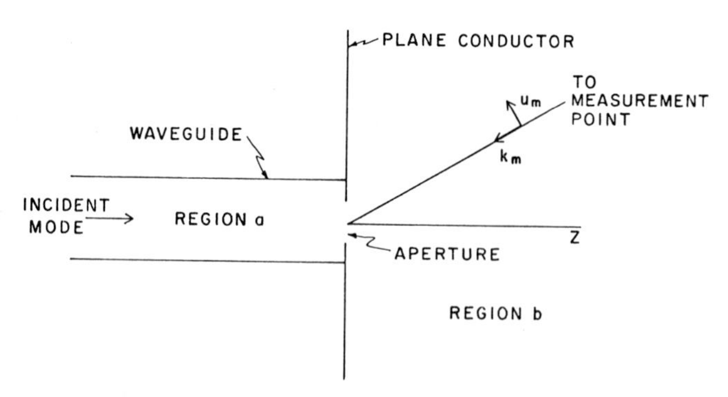

Fig. 11. Waveguide-fed aperture in a conducting plane.

$$Y_i = \begin{cases} \gamma_i / j\omega\mu \ , & TE \text{ modes} \\ j\omega\varepsilon / \gamma_i \ , & TM \text{ modes} \end{cases} \tag{73}$$

Γ_o is the reflection coefficient for the o-th mode, and Γ_i is the complex amplitude of the $-z$ traveling component of the i-th mode. The $\mathbf{e}_i$ are normalized modal vectors, so that the modal orthogonality relationahips are

$$\iint_{\text{guide}} \mathbf{e}_i \cdot \mathbf{e}_j ds = \begin{cases} 0 & i \neq j \\ 1 & i = j \end{cases} \tag{74}$$

where the integration is over the waveguide cross section.

To evaluate the aperture admittance (9) in the waveguide region, we consider a single expansion function $\mathbf{M}_n$ on the $z = 0$ plane in the waveguide region. The tangential field produced by $\mathbf{M}_n$ will be of the form (71), except that there is no incident wave. Hence, this field is

$$\mathbf{E}_t^a(\mathbf{M}_n) = \sum_i A_{ni} \, e^{\gamma_i z} \mathbf{e}_i \tag{75}$$

$$\mathbf{H}_t^a(\mathbf{M}_n) = -\sum_i A_{ni} Y_i \, e^{\gamma_i z} \mathbf{u}_z \times \mathbf{e}_i$$

where the A_{ni} are modal amplitudes. At $z = 0$ we have

$$\mathbf{M}_n = \mathbf{u}_z \times \mathbf{E}_t^a \Big|_{z=0} = \sum_i A_{ni} \mathbf{u}_z \times \mathbf{e}_i \tag{76}$$

Multiply each side of this equation scalarly by $\mathbf{u}_z \times \mathbf{e}_j$ and integrate over the waveguide cross section, obtaining

$$\iint_{\text{guide}} \mathbf{M}_n \cdot \mathbf{u}_z \times \mathbf{e}_j ds = \sum_i A_{ni} \iint_{\text{guide}} (\mathbf{u}_z \times \mathbf{e}_i) \cdot (\mathbf{u}_z \times \mathbf{e}_j) ds \tag{77}$$

By orthogonality (74), all terms of the summation are zero except the $i = j$ term. Hence,

$$A_{ni} = \iint_{\text{apert.}} \underset{\sim}{M}_n \cdot \underset{\sim}{u}_z \times \underset{\sim}{e}_i ds \tag{78}$$

We have replaced the integral over the waveguide cross section by one over the aperture, since $\underset{\sim}{M}_n$ exists only in the aperture region. The elements of the aperture admittance matrix (9) are now given by

$$Y_{mn}^{wg} = - \iint_{\text{apert.}} \underset{\sim}{W}_m \cdot \underset{\sim}{H}_t^a(\underset{\sim}{M}_n) ds \tag{79}$$

where the superscript *wg* denotes waveguide. The $\underset{\sim}{H}_t^a$ of (79) is given by the second equation of (75) evaluated at $z = 0$, hence

$$Y_{mn}^{wg} = \sum_i A_{ni} Y_i \iint_{\text{apert.}} \underset{\sim}{W}_m \cdot \underset{\sim}{u}_z \times \underset{\sim}{e}_i ds \tag{80}$$

Now define the constants

$$B_{mi} = \iint_{\text{apert.}} \underset{\sim}{W}_m \cdot \underset{\sim}{u}_z \times \underset{\sim}{e}_i ds \tag{81}$$

which are similar in form to the A_{ni} of (78). The elements (80) then are given by

$$Y_{mn}^{wg} = \sum_i A_{ni} B_{mi} Y_i \tag{82}$$

Hence, all elements of the waveguide admittance matrix $[Y^{wg}]$ are linear combinations of the modal characteristic admittances Y_i. For Galerkin's method, $\underset{\sim}{W}_n = \underset{\sim}{M}_n$ and the A_{ni} and B_{ni} are equal.

We next evaluate the equivalent magnetic current $\underset{\sim}{M}$, given by (5). The incident field is given by the first term on the right-hand side of (71). When the aperture is covered by a conductor, the waveguide is terminated by a conducting plane. According to image theory, the tangential magnetic field at $z = 0$ is then just twice that of the incident wave, or

$$\underset{\sim}{H}_t^i = 2Y_o \underset{\sim}{u}_z \times \underset{\sim}{e}_o \tag{83}$$

This is the $\underset{\sim}{H}_t^i$ used in (11) to evaluate the excitation vector $\vec{I}^i$.

Hence, the components of the excitation vector are

$$I_m^i = 2Y_o \iint_{\text{apert.}} \underset{\sim}{W}_m \cdot \underset{\sim}{u}_z \times \underset{\sim}{e}_o \, ds = 2Y_o B_{mo} \tag{84}$$

The total aperture admittance matrix is

$$[Y^a + Y^b] = [Y^{wg} + Y^{hs}] \tag{85}$$

where $[Y^{wg}]$ is the waveguide aperture admittance and $[Y^{hs}]$ is the half-space aperture admittance. The coefficient matrix $\vec{V}$ is given by (14) with the admittance matrix given by (85), or

$$\vec{V} = [Y^{wg} + Y^{hs}]^{-1} \vec{I}^i \tag{86}$$

Finally, the equivalent magnetic current $\underset{\sim}{M}$ is given by (5) where the coefficients V_n are the components of $\vec{V}$.

Once $\underset{\sim}{M}$ is found, the modal amplitudes Γ_i in (71) can be evaluated from (1) and the orthogonality properties of the modes. From (1) and (71), we have

$$\underset{\sim}{M} = \underset{\sim}{u}_z \times \underset{\sim}{E}_t \Big|_{z=0} = \underset{\sim}{u}_z \times \underset{\sim}{e}_o + \sum_i \Gamma_i \underset{\sim}{u}_z \times \underset{\sim}{e}_i \tag{87}$$

Now multiply each side scalarly by $\underset{\sim}{u}_z \times \underset{\sim}{e}_j$ and integrate over the waveguide cross section. By the orthogonality relationships (74), all terms of the summation vanish except the term $i = j$. The result is

$$\iint_{\text{guide}} \underset{\sim}{M} \cdot \underset{\sim}{u}_z \times \underset{\sim}{e}_i \, ds = \begin{cases} 1 + \Gamma_o & i = 0 \\ \Gamma_i & i \neq 0 \end{cases} \tag{88}$$

Here the integration over the guide can be changed to that over the aperture because $\underset{\sim}{M} = 0$ except in the aperture. Substituting for $\underset{\sim}{M}$ from (5) into (88), and using the definition (78), we have

$$\sum_n V_n A_{no} = 1 + \Gamma_o$$

$$\sum_n V_n A_{ni} = \Gamma_i \quad i \neq 0 \tag{89}$$

Finally, by defining modal measurement vectors as

$$\vec{A}_i = [A_{ni}]_{N\times 1} \tag{90}$$

and using (86), we can write (89) as

$$1 + \Gamma_o = \tilde{A}_o [Y^{wg} + Y^{hs}]^{-1} \vec{I}^i \tag{91}$$

and, for $i \neq 0$,

$$\Gamma_i = \tilde{A}_i [Y^{wg} + Y^{hs}]^{-1} \vec{I}^i \tag{92}$$

The parameter of most interest is Γ_o, the reflection coefficient of the incident mode. This is often expressed in terms of an admittance

$$Y_{ap} = \frac{1 - \Gamma_o}{1 + \Gamma_o} Y_o \tag{93}$$

which is the equivalent aperture admittance seen by the incident mode.

The region $z > 0$ for waveguide-fed apertures is the same half-space region as existed in the previous problem of an aperture in a conducting plane. Hence, evaluation of the fields in terms of $\underset{\sim}{M}$ in this region is done the same way as in Section IV. For example, the $\underset{\sim}{u}_m$ component of the far-zone magnetic field at a point $\underset{\sim}{r}_m$ is given by

$$H_m = \frac{-j\omega\varepsilon}{4\pi r_m} e^{-jkr_m} \tilde{P}^m [Y^{wg} + Y^{hs}]^{-1} \vec{I}^i \tag{94}$$

which is (38) with the term $[2Y^{hs}]^{-1}\vec{P}^i$ replaced by (86). The excitation vector $\vec{I}^i$ has elements given by (84), and the far-field measurement vector $\vec{P}^m$ has elements given by (37). The power gain pattern is the ratio of the radiation intensity in a given direction to the radiation intensity which would exist if the total power P_t were radiated uniformly over half space, or

$$G = \frac{2\pi r_m^2 \eta |H_m|^2}{Re(P_t)} \tag{95}$$

Substituting for H_m from (94), we have

$$G = \frac{\omega^2 \varepsilon^2 \eta}{8\pi Re(P_t)} |\tilde{P}^m [Y^{wg} + Y^{hs}]^{-1} \vec{I}^i|^2 \tag{96}$$

where P_t is given by (27). Note that this gain is a function of the $\underset{\sim}{H}$ component measured, as well as direction to the field point.

VIII. RECTANGULAR WAVEGUIDE FEEDING A RECTANGULAR APERTURE IN A PLANE CONDUCTOR

Figure 12 shows the problem to be considered and defines the coordinates and parameters to be used. The perfectly conducting plate covers the entire $z = 0$ plane except for the aperture which is rectangular in shape with side lengths $L_x \Delta x$ and $L_y \Delta y$ in the x and y directions respectively. L_x and L_y are positive integers and $Lx \geq 2$. The aperture is fed by a rectangular waveguide. The excitation of the waveguide is a source which produces one mode, of unit amplitude, which travels toward the aperture.

The general method of solution is to cover the aperture with a perfect electric conductor, to place magnetic current sheets $+\underset{\sim}{M}$ and $-\underset{\sim}{M}$ respectively on the left-hand and right-hand sides of this conductor, to obtain an integral equation for $\underset{\sim}{M}$ by equating the tangential magnetic fields on both sides of this conductor, and to solve this integral equation using the method of moments. The testing functions are the same as the expansion functions for $\underset{\sim}{M}$ and are denoted by $\underset{\sim}{M}_i$. Each $\underset{\sim}{M}_i$ is a triangle in the direction of current flow and a pulse in the direction perpendicular to current flow.

Expression (78) for A_{ij} requires a knowledge of the

expansion functions $\underset{\sim}{M}_i$ and waveguide modes $\underset{\sim}{e}_j$.

The set $\underset{\sim}{M}_i$ of expansion functions is split into a set $\underset{\sim}{M}_i^x$ of x directed magnetic currents and a set $\underset{\sim}{M}_i^y$ of y directed magnetic currents defined by

$$\underset{\sim}{M}^x_{p+(q-1)(L_x-1)} = \underset{\sim}{u}_x\, T_p^x(x-x_1)\, P_q^y(y-y_1) \quad \begin{cases} p = 1,2,\ldots L_x - 1 \\ q = 1,2,\ldots L_y \end{cases} \tag{97}$$

$$\underset{\sim}{M}^y_{p+(q-1)L_x} = \underset{\sim}{u}_y\, T_q^y(y-y_1) P_p^x(x-x_1) \quad \begin{cases} p = 1,2,\ldots L_x \\ q = 1,2,\ldots L_y - 1 \end{cases} \tag{98}$$

where $T_p^x(x)$ are triangle functions defined by

$$T_p^x(x) = \begin{cases} \dfrac{x - (p-1)\Delta x}{\Delta x} & (p-1)\Delta x \le x \le p\Delta x \\[2ex] \dfrac{(p+1)\Delta x - x}{\Delta x} & p\Delta x \le x \le (p+1)\Delta x \\[2ex] 0 & |x - p\Delta x| \ge \Delta x \end{cases} \tag{99}$$

with $T_q^y(y)$ obtained by replacing x by y and p by q, and $P_p^x(x)$ are pulse functions defined by

$$P_p^x(x) = \begin{cases} 1 & (p-1)\Delta x \le x < p\Delta x \\ 0 & \text{all other } x \end{cases} \tag{100}$$

with $P_q^y(y)$ obtained by replacing x by y and p by q.

The set $\underset{\sim}{e}_j$ of modes for the rectangular waveguide is split into a set $\underset{\sim}{e}_j^{TE}$ of TE modes given by [2]

$$\underset{\sim}{e}^{TE}_{m+n(L_m+1)} = \sqrt{\frac{ab\,\varepsilon_m \varepsilon_n}{(mb)^2 + (na)^2}}\left[\underset{\sim}{u}_x \frac{n}{b}\cos\frac{m\pi x}{a}\sin\frac{n\pi y}{b} - \underset{\sim}{u}_y \frac{m}{a}\sin\frac{m\pi x}{a}\cos\frac{n\pi y}{b}\right] \tag{101}$$

$$m = 0,\ 1,2,\ldots L_m \qquad m + n \neq 0$$

$$n = 0,\ 1,2,\ldots L_n$$

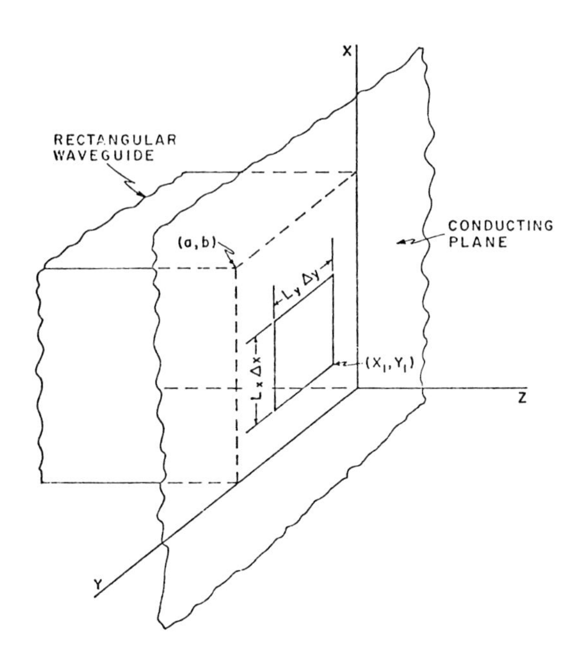

Fig. 12. A rectangular waveguide radiating through a rectangular aperture into half space bounded by an electric conductor.

$$\varepsilon_m = \begin{cases} 1 & m = 0 \\ 2 & m = 1,2,\ldots \end{cases}$$

and a set $\underset{\sim}{e}_j^{TM}$ of *TM* modes given by

$$\underset{\sim}{e}^{TM}_{m+(n-1)L_m} = 2\sqrt{\frac{ab}{(mb)^2 + (na)^2}}\,[\underset{\sim}{u}_x \frac{m}{a} \cos\frac{m\pi x}{a} \sin\frac{n\pi y}{b} + \underset{\sim}{u}_y \frac{n}{b} \sin\frac{m\pi x}{a} \cos\frac{n\pi y}{b}] \tag{102}$$

$$m = 1,2,3,\ldots L_m$$

$$n = 1,2,3,\ldots L_n$$

With the separation of $\underset{\sim}{M}_i$ into $\underset{\sim}{M}_i^x$ and $\underset{\sim}{M}_i^y$ and $\underset{\sim}{e}_j$ into $\underset{\sim}{e}_j^{TE}$ and

$\underset{\sim}{e}_j^{TM}$, A_{ij} of (78) expands to

$$A_{ij}^{xTE} = \iint_{\text{apert.}} \underset{\sim}{M}_i^x \cdot \underset{\sim}{u}_z \times \underset{\sim}{e}_j^{TE} ds \tag{103}$$

$$A_{ij}^{yTE} = \iint_{\text{apert.}} \underset{\sim}{M}_i^y \cdot \underset{\sim}{u}_z \times \underset{\sim}{e}_j^{TE} ds \tag{104}$$

$$A_{ij}^{xTM} = \iint_{\text{apert.}} \underset{\sim}{M}_i^x \cdot \underset{\sim}{u}_z \times \underset{\sim}{e}_j^{TM} ds \tag{105}$$

$$A_{ij}^{yTM} = \iint_{\text{apert.}} \underset{\sim}{M}_i^y \cdot \underset{\sim}{u}_z \times \underset{\sim}{e}_j^{TM} ds \tag{106}$$

Here, $\underset{\sim}{M}_i^x$, $\underset{\sim}{M}_i^y$, $\underset{\sim}{e}_j^{TE}$, and $\underset{\sim}{e}_j^{TM}$ are given by (97), (98), (101), and (102), respectively.

A tedious but straightforward evaluation of the integrals appearing in (103) to (106) leads to

$$\sqrt{\frac{2\pi}{\Delta x \Delta y}}\, A_{ij}^{xTE} = \sqrt{\frac{4\pi \Delta x \Delta y \varepsilon_n}{ab}} \left(\frac{m\pi}{k_j a}\right) \left(\frac{\sin \frac{m\pi\Delta x}{2a}}{\frac{m\pi\Delta x}{2a}}\right)^2 \left(\frac{\sin \frac{n\pi\Delta y}{2b}}{\frac{n\pi\Delta y}{2b}}\right) \sin \frac{m\pi(x_1 + p\Delta x)}{a} \cos \frac{n\pi(y_1 + (q-1/2)\Delta y)}{b} \tag{107}$$

$$\sqrt{\frac{2\pi}{\Delta x \Delta y}}\, A_{ij}^{yTE} = \sqrt{\frac{4\pi \Delta x \Delta y \varepsilon_m}{ab}} \left(\frac{n\pi}{k_j b}\right) \left(\frac{\sin \frac{n\pi\Delta y}{2b}}{\frac{n\pi\Delta y}{2b}}\right)^2 \left(\frac{\sin \frac{m\pi\Delta x}{2a}}{\frac{m\pi\Delta x}{2a}}\right) \cos \frac{m\pi(x_1 + (p-1/2)\Delta x)}{a} \sin \frac{n\pi(y_1 + q\Delta y)}{b} \tag{108}$$

$$\sqrt{\frac{2\pi}{\Delta x \Delta y}}\, A_{ij}^{xTM} = -\sqrt{\frac{8\pi \Delta x \Delta y}{ab}} \left(\frac{n\pi}{k_j b}\right) \left(\frac{\sin \frac{m\pi\Delta x}{2a}}{\frac{m\pi\Delta x}{2a}}\right)^2 \left(\frac{\sin \frac{n\pi\Delta y}{2b}}{\frac{n\pi\Delta y}{2b}}\right) \sin \frac{m\pi(x_1 + p\Delta x)}{a} \cos \frac{n\pi(y_1 + (q-1/2)\Delta y)}{b} \tag{109}$$

$$\sqrt{\frac{2\pi}{\Delta x \Delta y}}\, A_{ij}^{yTM} = \sqrt{\frac{8\pi \Delta x \Delta y}{ab}}\, \left(\frac{m\pi}{k_j a}\right) \left(\frac{\sin \frac{n\pi \Delta y}{2b}}{\frac{n\pi\Delta y}{2b}}\right)^2 \left(\frac{\sin \frac{m\pi\Delta x}{2a}}{\frac{m\pi\Delta x}{2a}}\right)$$

$$\cos \frac{m\pi (x_1 + (p-1/2)\Delta x)}{a} \sin \frac{n\pi (y_1 + q\Delta y)}{b} \tag{110}$$

where

$$k_j = \sqrt{\left(\frac{m\pi}{a}\right)^2 + \left(\frac{n\pi}{b}\right)^2} \tag{111}$$

In (107) and (109),

$$i = p + (q-1)(L_x - 1) \qquad \begin{array}{l} p = 1,2,\ldots L_x - 1 \\ q = 1,2,\ldots L_y \end{array} \tag{112}$$

whereas, in (108) and (110),

$$i = p + (q-1)L_x \qquad \begin{array}{l} p = 1,2,\ldots L_x \\ q = 1,2,\ldots L_y - 1 \end{array} \tag{113}$$

In (107) and (108)

$$j = m + n(L_m + 1) \qquad \begin{array}{l} m = 0,1,2,\ldots L_m \\ n = 0,1,2,\ldots L_n \end{array} \qquad m+n \neq 0 \tag{114}$$

whereas in (109) and (110),

$$j = m + (n-1)L_m \qquad \begin{array}{l} m = 1,2,\ldots L_m \\ n = 1,2,\ldots L_n \end{array} \tag{115}$$

If m or n is zero in (107) or (108), the resulting $(\frac{\sin 0}{0})$ is to be replaced by unity.

The characteristic admittances Y_j of the rectangular wave-guide with relative dielectric constant ε_r and relative permeability unity are classified as either *TE* admittances Y_j^{TE} or *TM* admittances Y_j^{TM} given by [2]

$$-j\eta Y_j^{TE} = \begin{cases} -\sqrt{\left(\frac{k_j}{k}\right)^2 - \varepsilon_r} & k\sqrt{\varepsilon_r} < k_j \\ -j\sqrt{\varepsilon_r - \left(\frac{k_j}{k}\right)^2} & k\sqrt{\varepsilon_r} > k_j \end{cases} \tag{116}$$

$$-j\eta Y_j^{TM} = \begin{cases} \dfrac{\varepsilon_r}{\sqrt{\left(\frac{k_j}{k}\right)^2 - \varepsilon_r}} & k\sqrt{\varepsilon_r} < k_j \\ \dfrac{j\varepsilon_r}{\sqrt{\varepsilon_r - \left(\frac{k_j}{k}\right)^2}} & k\sqrt{\varepsilon_r} > k_j \end{cases} \tag{117}$$

In (116) and (117), η is the characteristic impedance of free space, k is the free space wavenumber, and k_j is the cutoff wavenumber given by (111). Strictly speaking, we should have defined separate cutoff wavenumbers, say k_j^{TE} and k_j^{TM}, for *TE* and *TM* modes because the relationship between j, m, and n in (111) is given by (114) for *TE* modes and by (115) for *TM* modes.

IX. REPRESENTATIVE COMPUTATIONS

A computer program using the formulas derived in the preceding sections has been written. It is described and listed in reference [6]. In this section we give some examples of the computations that can be made using the general program.

Figure 13 shows the equivalent magnetic current for a rectangular waveguide of dimensions λ by $\lambda/2$ radiating into half space through a narrow centered rectangular slot of dimensions λ by $\lambda/10$. Figure 13(a) shows the x-component of equivalent magnetic current, which is also equal to the y-component of tangential $\underline{E}$ field in the slot. No y-component of magnetic current was

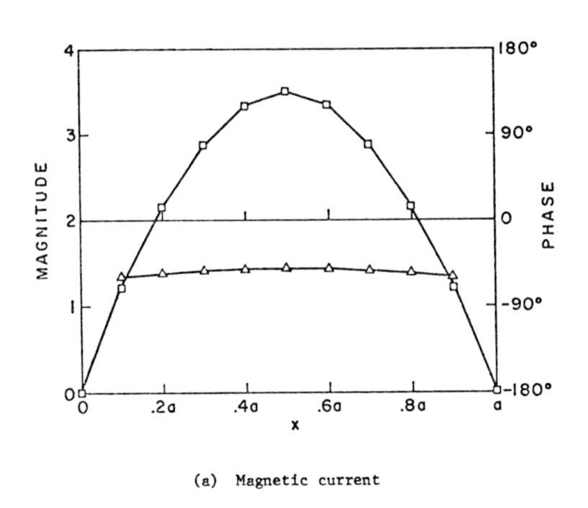

(a) Magnetic current

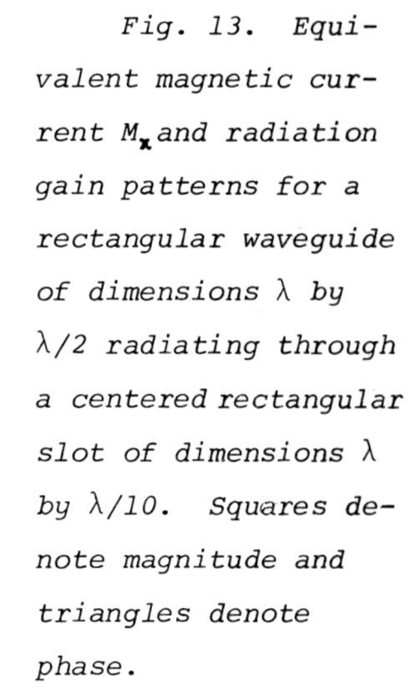

Fig. 13. Equivalent magnetic current M_x and radiation gain patterns for a rectangular waveguide of dimensions λ by λ/2 radiating through a centered rectangular slot of dimensions λ by λ/10. Squares denote magnitude and triangles denote phase.

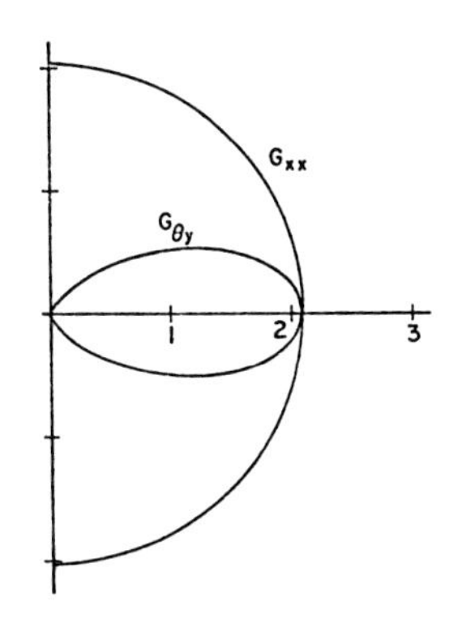

(b) Gain patterns

obtained because only one pulse in y was used. $\underset{\sim}{M}$ is normalized with respect to

$$\sqrt{\frac{1}{ab} \iint_{wg} |\underset{\sim}{E}^i|^2 \, dx \, dy} \tag{118}$$

where the integral is over the waveguide cross section. In other words, the normalization factor is the root-mean-square value of the incident $\underset{\sim}{E}$ field. The phase of $\underset{\sim}{M}$ is with respect to that of the incident electric field at the aperture. All computations are for dominant TE_{10} mode excitation. Fig. 13(b) shows the radiation gain patterns in the two planes $x = 0$ and $y = 0$. The notation $G_{\theta y}$ denotes the gain pattern due to H_θ in the $y = 0$ plane. The notation G_{xx} denotes the gain pattern due to H_x in the $x = 0$ plane. The horizontal axis in Fig. 13(b) is the z axis.

Other computations were made for open-ended waveguides radiating into half space bounded by a plane conductor. Plots of the equivalent magnetic current and radiation patterns for some cases are given in [6]. Finally, Fig. 14 shows a plot of the equivalent aperture admittance seen by the dominant mode for an open-ended square waveguide radiating into half space. This aperture admittance is defined by

$$Y_{ap} = \frac{1 - \Gamma_o}{1 + \Gamma_o} Y_o$$

where Γ_o is the reflection coefficient and Y_o is the characteristic wave admittance, both for the dominant mode. Our computations are compared to some previously obtained by Cohen, Crowley, and Levis [17]. Figure 14 shows the results for a square waveguide of side length a. Measured results were given in [17] for a square waveguide, and these are also shown in Fig. 14.

X. DISCUSSION

A general formulation for aperture problems using the method of moments has been given. The solution is expressed in terms of aperture admittance matrices, one for each region. While the exposition has been written in terms of a single aperture, the formulation applies equally well to multiple apertures. The extension of the theory to several regions is also relatively straightforward. An example of this more general case is many waveguides radiating into half space, a problem treated by Galindo and Wu [3]. Another example is a coaxial line feeding a cavity, which in turn radiates into half space through an aperture. In this case the generalized network equivalent would be a network representing the cavity, connected through some ports to a network representing the coaxial line, and connected through other ports to a network representing half space.

The division of an aperture admittance into two parts, one for each region, has been previously noted when the aperture field is assumed [2,4,18]. That is equivalent to using a one-term moment solution, in which case the aperture admittances are scalars; This paper shows that a similar property is valid in terms of general moment solutions, where the fields are represented by an N-term expansion with unknown coefficients. The aperture admittances then become N by N matrices. If one region is changed, only the aperture admittance matrix pertaining to that region need be changed. This changes the total aperture admittance matrix and the tangential electric field in the aperture, but not the aperture admittance matrix pertaining to the unchanged region.

The computer program for rectangular apertures in a plane conductor given in [5] is written explicitly for rectangular apertures, but the formulas are valid for any aperture composed of rectangular subsections. Other apertures, such as L-shaped, T-shaped, square O-shaped, etc., could be treated by appropriately changing the computer program. Apertures of arbitrary shape could be treated by approximating them by rectangular subsections. As with all moment solutions, the size of the apertures which can be treated depends upon the size of the matrix which can be computed and inverted. The examples indicate that the rectangular subsections should have side lengths not greater than 0.2 wavelengths for reasonable accuracy.

The computer program for rectangular waveguides radiating into half space through rectangular apertures is given in [6]. When the aperture dimensions are small compared to the waveguide dimensions, many waveguide modes may be required to accurately obtain the waveguide admittance matrix. In this case it would be advantageous to have an analytic approximation to the sum of higher-order mode contributions. However, we have not

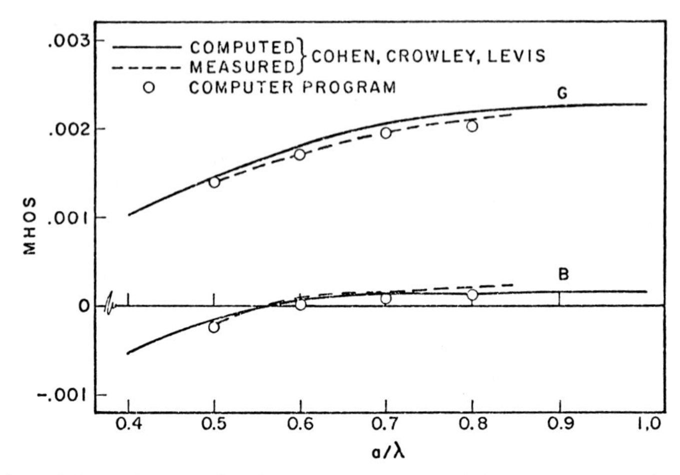

Fig. 14. The equivalent aperture admittance seen by the dominant mode for an open-ended square waveguide of width a radiating into half space. Our computed results are compared to those calculated and measured by Cohen, Crowley, and Levis [17].

investigated this possibility. Our program gives accurate results even for relatively small apertures, although large numbers of higher-order modes may be required. The program is written so that the excitation may be any mode desired, not necessarily the dominant mode. The aperture can be located anywhere within the waveguide cross section. The principal limitation to the use of the program is one set by the cross-sectional size (in square wavelengths) of the waveguide and aperture. As with most moment solutions, when the size of the aperture becomes too large then too many expansion functions are required for a solution. As a rule of thumb, for reasonable accuracy we need at least 5 expansion functions per wavelength for each component of current, or 50 expansion functions per square wavelength. Hence, even on large computers, one is limited to apertures of the order of a few square wavelengths in size.

Although we make no assumption as to the distribution of

tangential electric field (or magnetic current) in the aperture, for narrow slots of length one-half wavelength or less the distribution is very nearly a half sine wave. This distribution is often taken as an assumption for variational solutions to aperture problems [2]. For open-ended waveguides, the tangential electric field in the aperture is often assumed to be that of the dominant waveguide mode [2,4,18]. Waveguide modes as expansion functions have been used in a moment solution to calculate the aperture admittance of open-ended waveguides radiating into half-space [19]. Aperture penetration is reviewed in [20].

XI. REFERENCES

1. Collin, R.E., "Field Theory of Guided Waves," McGraw-Hill Book Company, New York (1960).
2. Harrington, R.F., "Time-Harmonic Electromagnetic Fields," McGraw-Hill Book Company, New York (1961).
3. Amitay, N., V. Galindo and C.P. Wu, "Theory and Analysis of Phased Array Antennas," Wiley-Interscience, New York (1972).
4. Croswell, W.F., Antennas and Wave Propagation, "Electronics Engineers' Handbook," (D.G.Fink and A.A. McKenzie, Eds.) McGraw-Hill Book Company, New York (1975).
5. Mautz, J.R. and R.F. Harrington, Electromagnetic Transmission through a Rectangular Aperture in a Perfectly Conducting Plane, Scientific Report No. 10, Contract No. F19628-73-C-0047 with Air Force Cambridge Research Laboratories, Hanscom A.F.B., Mass. (February 1976).
6. Mautz, J.R. and R.F. Harrington, Transmission from a Rectangular Waveguide into Half Space through a Rectangular Aperture, Scientific Report No. 12, Contract No. F19628-73-C-0047 with Air Force Cambridge Research Laboratories, Hanscom A.F.B., Mass. Also avail. as TR-76-5, ECE Dept. Syracuse University, (May, 1976).
7. Harrington, R.F., "Field Computation by Moment Methods," Macmillan Company, New York (1968).
8. Wang, N.N., J.H. Richmond and M.C. Gilreath, *IEEE Trans. Antennas & Propag. AP-23,* 376-382 (1975).
9. Harrington, R.F., *Proc. IEEE 55,* 136-149 (1967).
10. Bethe, H.A. *Phys. Rev. 66,* 163-182 (1944).
11. Bouwkamp, C.J., *Rep. Prog. Phys. 17,* 35-100 (1954).
12. Bekefi, G., *J. Appl. Phys. 24,* 1123-1130 (1953).
13. Bouwkamp, C.J., *IEEE Trans. Antennas & Propag. AP-18,* 152-176 (1970).
14. Adams, A.T., C.B. Varnado and D.E. Warren, Aperture Coupling by Matrix Methods, "1973 IEEE EMC Symposium Record," New

York, p. 226-240 (June, 1973).

15. Lin, J-L., W.L. Curtis and M.C. Vincent, *IEEE Trans. Antennas & Propag. AP-22*, 467-471 (1974).
16. Papas, C.H., "Theory of Electromagnetic Wave Propagation," McGraw-Hill Book Company, New York, p. 23 (1965).
17. Cohen, M., T. Crowley and K. Levis, The Aperture Admittance of a Rectangular Waveguide Radiating into Half-Space, Antenna Lab. Rep. ac 21114 S.R. No. 22, Ohio State Univ. (1953).
18. Bailey, M.C., *IEEE Trans. Antennas & Propag. AP-18*, 596-603 (1970).
19. Carson, C.T. and G.K. Cambrell, *Institution of Radio and Electronics Engineers (Australia) Proceedings 34*, 22-26, (1973).
20. Butler, C.M., Y. Rahmat-Samii and R. Mittra, *IEEE Trans. Antennas and Progag. AP-26*, 82-93 (1978).

THE ROLE OF SCATTERING THEORY IN ELECTROMAGNETIC INTERFERENCE PROBLEMS

Carl E. Baum
Air Force Weapons Laboratory

This paper considers one of the most complicated aspects of scattering theory, that of electromagnetic interference. The inherent difficulty of this problem results from the fact that electronic systems are not designed with this phenomenon as the dominant consideration. This results in a very complicated analysis problem. While one cannot solve the relevant scattering problems with any degree of precision for complicated systems, one can still devise ways of making the problem more tractable. Topological concepts are useful for subdividing the system into smaller parts which can be separately analyzed. Such concepts also help one in designing aspects of the system to minimize the interference problem. Consideration of some of the general properties of the electromagnetic scattering from complicated objects also gives ways to think about the scattering, at least from the point of view of organizing experimental results.

I. INTRODUCTION

Many electromagnetic problems involve some fundamental elements of simplicity. Such include most design problems. For example, an antenna has a certain geometry of conductors, insulators, loading impedances, etc., which are placed as they are because of certain desirable effects on the electromagnetic response. Viewed another way, many electromagnetic problems concern an object which was made a certain way <u>because a concept (mathematical) existed which said, "If you build it this way the following</u>

ISBN 0-12-709650-7

performance parameters will result."

The electromagnetic interference problem is quite different. Consider some military (or civilian) piece of hardware (system) designed to perform some mission. Assume this system has some electronic equipment which may or may not include antennas. This equipment may be designed to operate at certain frequencies, through certain antennas, cables, etc., at certain signal levels.

Suppose now that some other electromagnetic sources are present. Such are in general not part of the design considerations of the equipment of concern. They may operate at different frequencies or be transient in nature. They may also be physically remote and perhaps not even be known to exist or to possibly exist in the future. Such electromagnetic interference phenomena presently go by many names, including the nuclear electromagnetic pulse (EMP), lightning, electronic warfare (EW), electromagnetic compatibility (EMC), etc., each emphasizing different types of sources and receivers. As time goes on one expects even more such problem areas to appear.

This brings us to a definition. Electromagnetic interference is: 1) the propagation into an object (system) of an electromagnetic signal which 2) was not desired (by the system operator) to be there, 3) usually by an unidentified path (unidentified before the fact and perhaps even after the fact), and 4) brings about some change in the performance of the object (system), specifically a change undesired by the system operator, this change being either temporary or permanent.

A similar problem is that of unintended transmission of an electromagnetic signal from an object (system). This has some similarities to the interference problem except that the propagation is in the reverse direction.

The electromagnetic interference problem is then one of unintentional response, i.e. unintentional on the part of the designer. Therefore it is one of undesigned response. This implies in general that the electromagnetic interference problem is one of complexity, becuase very little consideration was given to making the problem simple. Stated another way, the electromagnetic interference problem has not been one of synthesis, but rather one of analysis involving very great detail.

The processes by which electromagnetic energy propagates onto and into the system (scatterer) can be defined as the electromagnetic interaction problem. Presently on complicated real systems this interaction problem is only solved with high confidence experimentally. We would like to use electromagnetic scattering theory (analysis) to solve such problems, or failing this to at least partly solve such problems and/or to help understand and organize relevant experimental data on the electromagnetic response of complicated systems.

In the electromagnetic interaction problem we are then faced with analyzing the complex system given to us. It is not in general designed to conform to the surfaces of coordinate systems for which the solutions of Maxwell's equations separate, or to have any other such convenience. Hence one must look for general properties of electromagnetic scatterers to simplify the analysis of the problem. Assumptions concerning the electromagnetic nature of the scatterer should be limited to approximately valid general properties such as linearity, passivity (or more generally stability), reciprocity, etc., although some interesting cases would negate the use of some such assumptions (including corona and arc formation as examples).

In attacking the interaction problem, then, let us look for ways to divide the overall problem into a set of smaller ones. Specifically, this division should be according to electromagnetic properties of general scatterers or general classes of scatterers so that the division is a natural one, i.e., one in which 1) the electromagnetic descriptions of the pieces are in some sense simpler, and 2) the electromagnetic descriptions of the pieces can fit together in a manner which gives an efficient description of the electromagnetic behavior as a whole.

There are various concepts useful for accomplishing this division. They are all associated with various mathematical properties of Maxwell's equations in various forms. These include:

1. General electromagnetic theorems (group theoretic or symmetry properties of the equations, such as reciprocity, conservation of energy, duality, etc.).
2. Alternate and approximate forms of Maxwell's equations, including:
   a. Integral equations.
   b. Transmission line representations.
   c. Circuit (lumped element) representations.
3. Topology of the scatterer.
   a. Approximately conducting surfaces and associated enclosed volumes.
   b. Paths of transmission line conductors (tubes) and associated junctions.
   c. Branches and nodes for lumped elements.
4. Diagonalizations.
   a. Integral operators.
   b. Matrices.
5. Analytic properties of the response as a function of the complex frequency, s.
   a. Low frequencies (power series or more generally, asymptotic series).
   b. Intermediate frequencies (singularity expansion method involving resonances and other singularity terms).
   c. High frequencies (asymptotic expansions involving ray concepts).

Ths paper emphasizes some of the concepts appropriate to item 3 above, i.e., topology. However item 2 is closely related to this and so also receives some attention. Items 4 and 5 are only treated briefly since they are more familiar to the readers, and some review and summary expositions have appeared [6,11], and there are other papers in this book treating these subjects [7,8].

II. SOME ASPECTS OF THE RESPONSE OF GENERAL SCATTERING OBJECTS

The starting point for a general consideration of interaction with complex scatterers is of course Maxwell's equations:

$$\begin{aligned} \nabla \times \vec{E}(\vec{r},t) &= -\frac{\partial}{\partial t}\vec{B}(\vec{r},t) - \vec{J}_m(\vec{r},t) \\ \nabla \times \vec{H}(\vec{r},t) &= \frac{\partial}{\partial t}\vec{D}(\vec{r},t) + \vec{J}(\vec{r},t) \\ \nabla \cdot \vec{J}(\vec{r},t) &= -\frac{\partial}{\partial t}\nabla \cdot \vec{D}(\vec{r},t) = -\frac{\partial}{\partial t}\rho(\vec{r},t) \\ \nabla \cdot \vec{J}_m(\vec{r},t) &= -\frac{\partial}{\partial t}\nabla \cdot \vec{B}(\vec{r},t) = -\frac{\partial}{\partial t}\rho_m(\vec{r},t) \end{aligned} \qquad (2.1)$$

with magnetic currents and charges included for generality. The constitutive relations (including Ohm's law) are

$$\begin{aligned} \tilde{\vec{D}}(\vec{r},s) &= \tilde{\vec{\vec{\varepsilon}}}(\vec{r},s) \cdot \tilde{\vec{E}}(\vec{r},s), \quad \tilde{\vec{J}}(\vec{r},s) = \tilde{\vec{\vec{\sigma}}}(\vec{r},s) \cdot \tilde{\vec{E}}(\vec{r},s) \\ \tilde{\vec{B}}(\vec{r},s) &= \tilde{\vec{\vec{\mu}}}(\vec{r},s) \cdot \tilde{\vec{H}}(\vec{r},s), \quad \tilde{\vec{J}}_m(\vec{r},s) = \tilde{\vec{\vec{\sigma}}}_m(\vec{r},s) \cdot \tilde{\vec{H}}(\vec{r},s) \end{aligned} \qquad (2.2)$$

with the current densities in these equations not including sources. Here the Laplace transform with respect to time (two-sided) is indicated by a tilde above a quantity and is defined by

$$\tilde{F}(s) \equiv \int_{-\infty}^{\infty} F(t)e^{-st}dt, \quad F(t) = \frac{1}{2\pi j}\int_{\Omega_0-j\infty}^{\Omega_0+j\infty} \tilde{F}(s)e^{st}ds \qquad (2.3)$$

where $F(t)$ may be a scalar, vector, dyadic, or any rank tensor function of time. The Laplace transform variable is

$$s = \Omega + j\omega \qquad (2.4)$$

with $e^{j\omega t}$ time convention obtained by setting $\Omega = 0$. The inversion contour C_0 from $\Omega_0 - j\infty$ to $\Omega_0 + j\infty$ lies in a strip of convergence of the transform integral in the s plane.

More general forms of constitutive relations than in (2.2) are possible. However our present considerations will use this type of time invariant, linear, and localized medium description.

Consider a general integral equation for a scatterer or antenna as

$$\langle \tilde{\overleftrightarrow{\Gamma}}(\vec{r},\vec{r}\,';s);\tilde{\vec{J}}_s(\vec{r}\,',s)\rangle = \tilde{\vec{I}}(\vec{r},s), \quad \vec{r} \in S \tag{2.5}$$

Here $\tilde{\vec{I}}$ is related to the incident field in some form and $\tilde{\overleftrightarrow{\Gamma}}$ is the corresponding kernel related to the Green's function [3]. The domain of integration indicated by $\langle \; , \; \rangle$ is taken as some surface S of interest and $\tilde{\vec{J}}_s$ is the corresponding surface current density.

This form of surface integral equation is particularly appropriate in the context of the topological concept for complicated scatterers discussed in this paper. It applies to the exterior response for approximately closed highly conducting surfaces, as well as to similarly restricted interior surfaces (cavity walls). This allows one to separately consider the response characteristics of the separate volumes (say through electrically small and physically small apertures and other conductor penetrations), and finally combine the separate solutions to give the total solution for the complicated scatterer.

III. TRANSMISSION-LINE REPRESENTATIONS

It is a common feature of electrical engineering education to introduce topological concepts in the context of circuit theory. As illustrated in Fig. 3.1, this consists of a set of nodes with branches connecting some of the node pairs, but not generally all of the node pairs. This graph, as it is sometimes referred to, is not necessarily a planar graph, i.e. the nodes can be points in a three-dimensional Euclidean space with the branches as lines between these points (non-intersecting other than at nodes). In our diagram (Fig. 3.1) the nodes are large dots and the branches are lines connecting these dots.

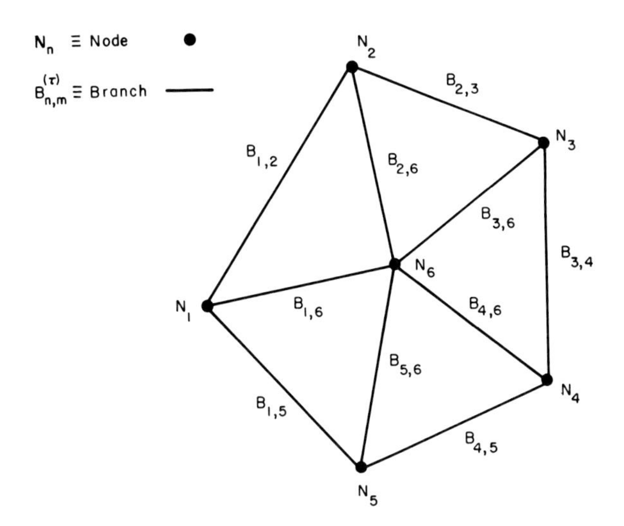

Fig. 3.1. Circuit topology.

Corresponding to the circuit topology there is a certain way of writing equations in terms of quantities defined by this topology. These are normally referred to as Kirchhoff's laws. One defines voltages $\tilde{V}_n$ at each node N_n and currents $\tilde{I}^{(\tau)}_{n,m}$ leaving N_n to N_m along the branch $B^{(\tau)}_{n,m}$ ($\tau = 1,2,...,$ labeling which branch between the nodes) with the condition

$$\sum_{m,\tau} \tilde{I}^{(\tau)}_{n,m} \equiv 0, \quad \tilde{I}_{n,n} \equiv \tilde{I}^{(1)}_{n,n} \equiv 0 \tag{3.1}$$

The voltages and currents are related by

$$\tilde{V}_n - \tilde{V}_m = \tilde{Z}^{(\tau)}_{n,m}\tilde{I}^{(\tau)}_{n,m} - \tilde{V}^{(\tau)}_{s_{n,m}} \tag{3.2}$$

where $\tilde{V}_{s_{n,m}}$ is a voltage source (increasing from n to m) and $\tilde{Z}_{n,m}$ is some impedance (assumed linear); in a degenerate case the branch current might be specified by a current source. Typically the superscript τ may be suppressed for any cases that its maximum value for any node pair is 1. If for a node pair N_n and N_m there is more than one branch $B^{(\tau)}_{n,m}$ (i.e., $\tau_{max} > 1$), then by summing parallel admittances one can always reduce this to a single branch $B_{n,m}$.

Corresponding to the circuit topology, there are matrices relating the nodes and branches [2]. These topological matrices are in turn related to the mesh and node forms of the matrix equations for the circuit network under consideration, giving appropriate impedance and admittance forms of network matrix representations.

In a similar manner one can define topological concepts for transmission-line networks as in Fig. 3.2. The basic quantities

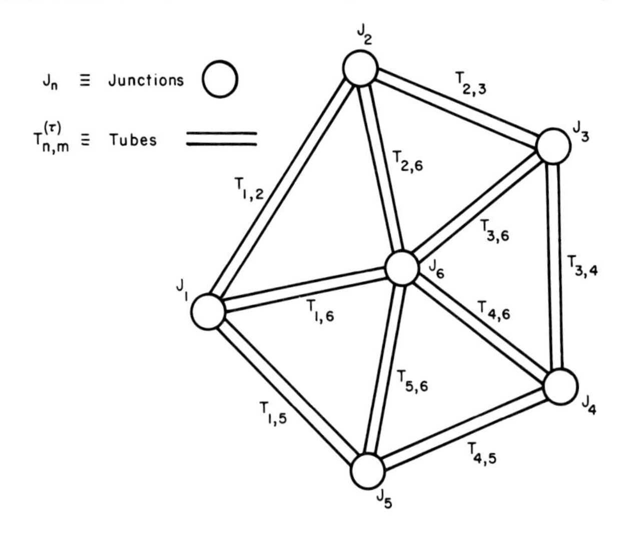

Fig. 3.2. Transmission-line topology.

here are junctions J_n and tubes $T^{(\tau)}_{n,m}$, analogous to nodes and branches for circuits. However the mathematical treatment is quite different than that for the circuit case. One can note that circuit concepts are used in formulating each junction and each incremental length of tube. The transmission-line topology then deals with large aggregates of circuits, both in discrete (junctions) and continuous (tubes) forms. Note that in general (unlike the circuit topology) parallel tubes $T^{(\tau)}_{n,m}$ ($\tau_{max} > 1$) between junctions J_n and J_m cannot be reduced to a single tube which has a single set of per-unit-length equations.

A tube represents an N-wire transmission line consisting of

N conductors with a ground return. By definition there is no net current on the tube, at least in the mathematical idealization. The tube is described by per-unit-length parameters with the general equations

$$\begin{aligned}\frac{\partial}{\partial x}(\tilde{V}_n(x,s)) &= -(\tilde{Z}'_{n,m}(x,s)) \cdot (\tilde{I}_n(x,s)) + (\tilde{V}'_{s_n}(x,s)) \\ \frac{\partial}{\partial x}(\tilde{I}_n(x,s)) &= -(\tilde{Y}'_{n,m}(x,s)) \cdot (\tilde{V}_n(x,s)) + (\tilde{I}'_{s_n}(x,s))\end{aligned} \tag{3.3}$$

$(\tilde{Z}'_{n,m}(x,s)) \equiv$ impedance per unit length matrix $(N \times N)$

$(\tilde{Y}'_{n,m}(x,s)) \equiv$ admittance per unit length matrix $(N \times N)$

$(\tilde{V}'_{s_n}(x,s)) \equiv$ voltage source per unit length vector (N)

$(\tilde{I}'_{s_n}(x,s)) \equiv$ current source per unit length vector (N)

$(\tilde{V}_n(x,s)) \equiv$ voltage vector (N)

$(\tilde{I}_n(x,s)) \equiv$ current vector (N)

It simplifies matters somewhat if the impedance and admittance per unit length matrices are independent of x. The equivalent circuit (per unit length consistent with (3.3)) is given in Fig. 3.3.

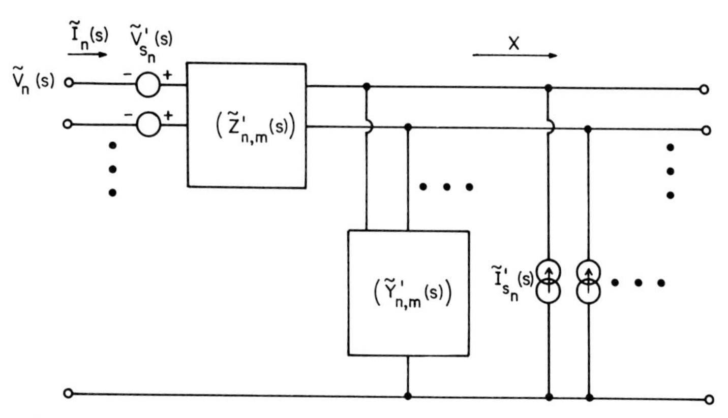

Fig. 3.3. Tube representation (per unit length).

Tubes terminate (both ends) in junctions. One or more tubes may terminate in any given junction. A junction with M ports is illustrated by the equivalent circuit in Fig. 3.4. Here M is

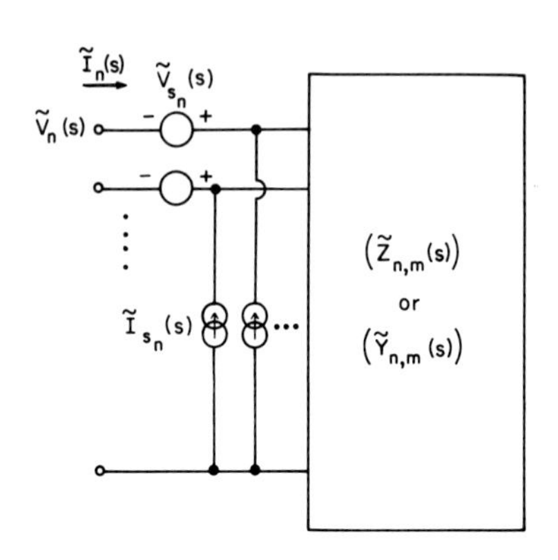

Fig. 3.4. Junction representations.

taken as the sum of the wires on each of the tubes terminating in this junction. This junction is characterized by the general equations

$$\begin{aligned}(\tilde{V}_n(s)) &= (\tilde{Z}_{n,m}(s)) \cdot [(\tilde{I}_n(s)) + (\tilde{I}_{s_n}(s))] - (\tilde{V}_{s_n}(s)) \\ (\tilde{I}_n(s)) &= (\tilde{Y}_{n,m}(s)) \cdot [(\tilde{V}_n(s)) + (\tilde{V}_{s_n}(s))] - (\tilde{I}_{s_n}(s))\end{aligned} \tag{3.4}$$

$(\tilde{Z}_{n,m}(s)) \equiv$ impedance matrix $(M \times M)$

$(\tilde{Y}_{n,m}(s)) \equiv (\tilde{Z}_{n,m}(s))^{-1} \equiv$ admittance matrix $(M \times M)$

$(\tilde{V}_{s_n}(s)) \equiv$ voltage source vector (M)

$(\tilde{I}_{s_n}(s)) \equiv$ current source vector (M)

$(\tilde{V}_n(s)) \equiv$ voltage vector (M)

$(\tilde{I}_n(s)) \equiv$ current vector (M)

This form in Fig. 3.4 allows for both voltage and current sources, which is a useful form from a physical point of view because of the different kinds of low-frequency (quasi static) coupling to local fields. However, this can be readily changed to an open circuit (Thevenin) form with only voltage sources by noting

$$(\tilde{V}_n(s))_{o.c.} = (\tilde{Z}_{n,m}(s)) \cdot (\tilde{I}_{s_n}(s)) - (\tilde{V}_{s_n}(s)) \tag{3.5}$$

and a short circuit (Norton) form with only current sources by noting

$$(\tilde{I}_n(s))_{s.c.} = (\tilde{Y}_{n,m}(s)) \cdot (\tilde{V}_{s_n}(s)) - (\tilde{I}_{s_n}(s)) \tag{3.6}$$

so that the number of independent sources (as far as terminal properties are concerned) can be reduced to a single M-vector.

By partitioning the M ports at the junction according to the tubes terminating at the junction one can find a set of scattering matrices at the junction. Together with the sources and tube propagation characteristics the entire cable network can be formulated according to the topological matrices describing the interconnection of junctions and tubes and the various waves on the tubes. This leads to the BLT equation, an equation written conveniently in terms of four tensors or supermatrices (matrices of matrices) and the corresponding two tensors or supervectors (vectors of vectors) [13,14]. This is a fairly extensive and detailed topic which will not be delved into here. F. Tesche, T. Liu, and this author have extensive results on this question which will be reported in the future.

IV. REPRESENTATIONS OF THREE-DIMENSIONALLY EXTENDED SYSTEMS

In principle any three dimensional region of space can be split up into smaller volumes. Each of these smaller volumes can be formulated for its electromagnetic response in terms of the fields on the boundaries because of the electromagnetic uniqueness theorem [1]. However, for use in general electromagnetic engineering design and analysis of complex systems certain assumptions can be made to simplify the application of such concepts.

For complex systems such as aircraft, ships, missiles, etc., one can often identify conducting surfaces in the form of metal panels. These surfaces can also often be considered as closed surfaces, except for some imperfections or holes, referred to as penetrations.

A. Scatterer Topology

Let us then define a scatterer topology for three-dimensional objects as indicated in Fig. 4.1. Analogous to nodes for circuit

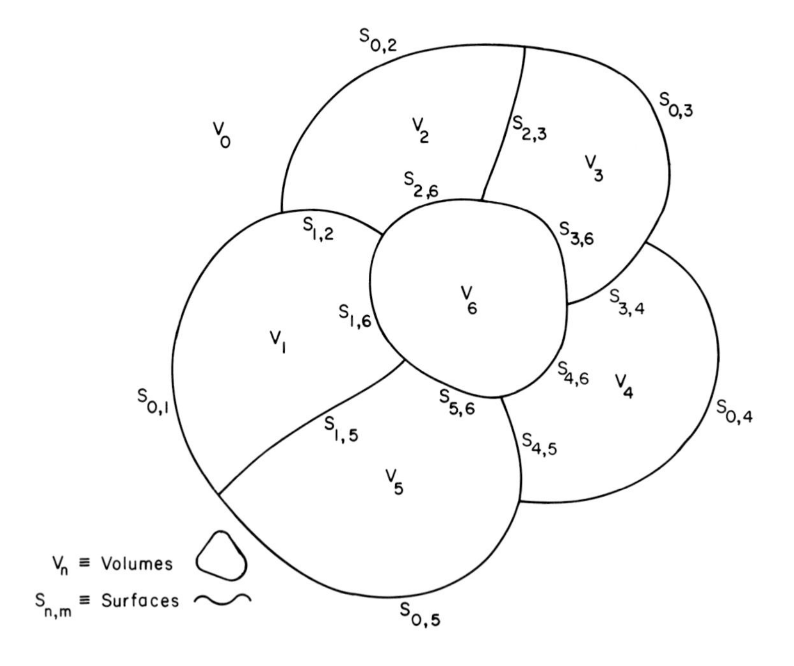

Fig. 4.1. Scatterer topology.

topology and junctions for transmission-line topology we have volumes for scatterer topology; analogous to branches for circuit topology and tubes for transmission-line topology we have surfaces for scatterer topology. Volumes and surfaces form our scatterer topology, and this should not be surprising noting how common theorems of electromagnetic quantities relate volume integrals of one quantity to surface integrals over a closed surface (surrounding the original volume) of a related quantity. Such concepts include charge conservation, energy conservation, and reciprocity to name a few; numerous types of relations between volume and surface quantities can be found [9].

Our basic concept is then to define a set of topological quantities as

$$V_n \equiv \text{volumes}, \quad n = 0,1,2,\ldots,N_V$$

$$S_{n,m} \equiv \text{surfaces representing boundaries between } V_n \text{ and } V_m$$

$$n,m = 0,1,2,\ldots,N_V \tag{4.1}$$

Here V_0 represents the space exterior to our scatterer for completeness and

$$\bigcup_{m=1}^{N_V} S_{0,m} = \bigcup_{n=1}^{N_V} S_{n,0} \equiv \text{outer scatterer boundary} \tag{4.2}$$

Note that for some n,m pairs $S_{n,m}$ is empty, i.e. contains no points since V_n and V_m have no common boundary in such cases. In the union (symbol $\bigcup$) (4.2) one should include any dividing lines (curves) separating pieces of the outer scatterer boundary.

Having divided the scatterer into a set of volumes one can form a mathematical description of the electromagnetic characteristics of any given volume. According to the electromagnetic uniqueness theorem [1] the response in a passive volume is controlled by electromagnetic parameters on the closed surrounding surface. Assuming the volume is linear, it can be represented by an $N_p \times N_p$ matrix in s domain corresponding to a generalized N_p-port network. This matrix might be an impedance or admittance matrix; such matrices are symmetric for the common case that reciprocity applies. Of course other forms of matrices are possible since the N_p port network has $2N_p$ variables at the N_p ports, and the matrix merely represents the linear relationship of one set of N_p of these variables to the remaining set of N_p variables; there are many ways to form the two ordered sets of N_p variables.

In some cases a particular port n is represented by a voltage $\tilde{V}_n$ and a current $\tilde{I}_n$. Another port (an electrically small aperture) may be represented by a pair of equivalent electric dipole moments $\vec{p}_{a_{in}}$ and $\vec{p}_{a_{out}}$ or the corresponding short circuit surface charge densities $\rho_{s_{in}}$ and $\rho_{s_{out}}$ which produce these equivalent electric dipole moments through an appropriate aperture electric polarizability $\tilde{P}_a(s)\vec{1}_s\vec{1}_s$ where $\vec{1}_s$ is the unit vector normal to the surface at the port (short circuited). Similarly one can consider a pair of equivalent magnetic dipole moments $\vec{m}_{a_{in}}$ and $\vec{m}_{a_{out}}$ which are parallel to the surface at the port (shorted). Each of these magnetic dipole moments has two components and the associated polarizability $\tilde{\overleftrightarrow{M}}_a$ has in general four non-zero components, but two "independent" ones (the corresponding eigenvalues in general). Hence an aperture can be thought of as three ports into the volume of interest, although some of these ports may have negligible importance in some cases.

Properly assigning, then, the number of ports n_p (and their description) to each penetration one can sum over the n_p to find N_p, leading to the desired matrix description of the volume. Having the matrix descriptions of all the volumes one can combine them in a way to obtain a description of the entire scatterer. Note in combining the volume representations that voltage quantities keep the same sign in the simplest conventions in going from one volume to the next while current quantities change sign (usually being into the volume). This combination may take various forms. It is interesting to note that one form is basically the BLT equation [13,14] as used for transmission line networks; here the tubes shrink to zero length and the junction matrix descriptions correspond to the volume matrix descriptions.

B. Hierarchical Scatterer Topology

Having defined the concept of scatterer topology let us extend the concept to hierarchical scatterer topology as indicated

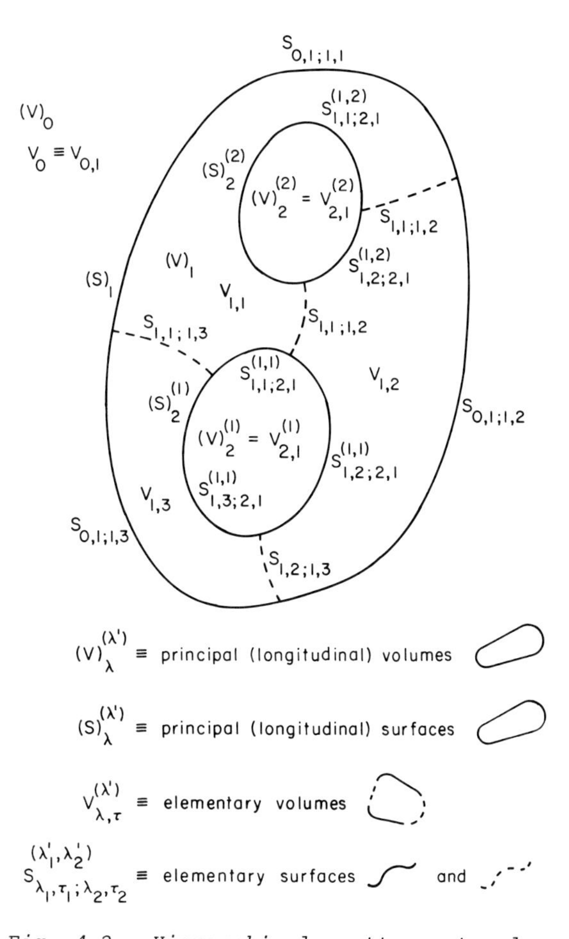

Fig. 4.2. Hierarchical scatterer topology.

in Fig. 4.2. The hierarchical topology is an ordering of the scatterer topology according to the number of volumes one must pass through in going from the outside to the inside of the scatterer [10,12]. In this regard first define the principal volumes related to the layered nature of the hierarchical topology as

$$(V)_\lambda \equiv \text{principle volume or longitudinal volume}$$

$$\lambda = 0,1,2,\ldots,\lambda_{max}$$

$$\lambda \equiv \text{longitudinal index} \equiv \text{shielding layer number}$$

$$\lambda_{max} \equiv \text{shielding order of system} \tag{4.3}$$

$$(S)_\lambda \equiv \text{principle surface or longitudinal surface}$$
$$= \text{surface between } (V)_{\lambda-1} \text{ and } (V)_\lambda$$
$$\lambda = 1,2,\ldots,\lambda_{max}$$

Here $(S)_0$ has not been included but it could be used to represent some surface at infinity completely surrounding the scatterer. Note the use of ( ) to indicate the principal volumes and surfaces, distinguishing them from the simple volumes and surfaces defined in (4.1). Since the $(S)_\lambda$ are closed surfaces we have the fact that they contain (symbol $\supset$) certain volumes (used in a generalized sense, since in a set-theoretic sense a surface contains other surfaces) which can be written as

$$(S)_\lambda \supset (V)_{\lambda+n} \quad , \quad n = 1,2,\ldots,\lambda_{max} - \lambda \tag{4.4}$$

or indicating all the volume within $(S)_\lambda$

$$(S)_\lambda \supset \left[\bigcup_{n=1}^{\lambda_{max}-\lambda} (V)_{\lambda+n} \right] \tag{4.5}$$

Relating the outer boundary $(S)_1$ of the scatterer to previous concepts in (4.2) we have

$$(S)_1 = \bigcup_{m=1}^{N_V} S_{0,m} = \bigcup_{n=1}^{N_V} S_{n,0} \tag{4.6}$$

In the case that $(V)_\lambda$ for some λ is not continuous, i.e., comes in two or more disjoint pieces (no common boundary) let us add a superscript or set of superscripts to distinguish which part is intended, and similarly for the respective bounding surfaces as

$$\begin{aligned} (V)_\lambda^{(\lambda')} &\equiv \lambda'\text{th part of the } \lambda\text{th principal volume} \\ (S)_\lambda^{(\lambda')} &\equiv \lambda'\text{th part of the } \lambda\text{th principal surface} \end{aligned} \tag{4.7}$$

This situation is illustrated in Fig. 4.2 for two parts for $\lambda = 2$.

Having considered the longitudinal (or outside to inside) decomposition of the scatterer, next consider the transverse decomposition. Associated with each layer or longitudinal volume $(V)_\lambda$ there can be further subdivisions associated with surfaces (conducting surfaces) linking the principal surfaces $(S)_\lambda^{(\lambda')}$ and $(S)_{\lambda+1}^{(\lambda_1')}$. This produces a set of elementary volumes $V_{\lambda,\tau}^{(\lambda')}$ with elementary surfaces for these smaller volumes as

$$\begin{gathered} S_{\lambda_1,\tau_1;\lambda_2,\tau_2}^{(\lambda_1',\lambda_2')} \equiv \text{boundary between } V_{\lambda_1,\tau_1}^{(\lambda_1')} \text{ and } V_{\lambda_2,\tau_2}^{(\lambda_2')} \\ \tau \equiv \text{transverse index} = 1,2,3,\ldots \end{gathered} \tag{4.8}$$

where the use of τ can be suppressed if in the layer λ of concern there is only one transverse volume, the index 1 being understood. Similarly $\lambda' = 1$ only may sometimes by suppressed.

The principle volumes can be written in terms of the elementary volumes as

$$\begin{gathered} (V)_\lambda^{(\lambda')} = \bigcup_\tau V_{\lambda,\tau}^{(\lambda')} \\ (V)_\lambda = \bigcup_{\lambda'} (V)_\lambda^{(\lambda')} = \bigcup_{\tau,\lambda'} V_{\lambda,\tau}^{(\lambda')} \end{gathered} \tag{4.9}$$

The principal surfaces can be written as

$$\begin{gathered} (S)_\lambda^{(\lambda')} = \bigcup_{\tau_1,\tau_2} S_{\lambda_1,\tau_1;\lambda_2,\tau_2}^{(\lambda_1',\lambda_2')} \\ \lambda_1 \equiv \lambda - 1, \quad \lambda_2 \equiv \lambda, \quad \lambda_2' \equiv \lambda', \quad \tau_2 \equiv \tau \\ (S)_\lambda = \bigcup_{\lambda'} (S)_\lambda^{(\lambda')} \end{gathered} \tag{4.10}$$

In forming the unions above care should be taken to include any dividing surfaces in the case of volumes and any dividing lines (curves) in the case of surfaces.

Fig. 4.3 indicates some of the kinds of penetration of electromagnetic energy into a scatterer with a hierarchical topology. One of the primary reasons for considering a hierarchical topology is that it is a shielding topology; λ_{max} is the shielding

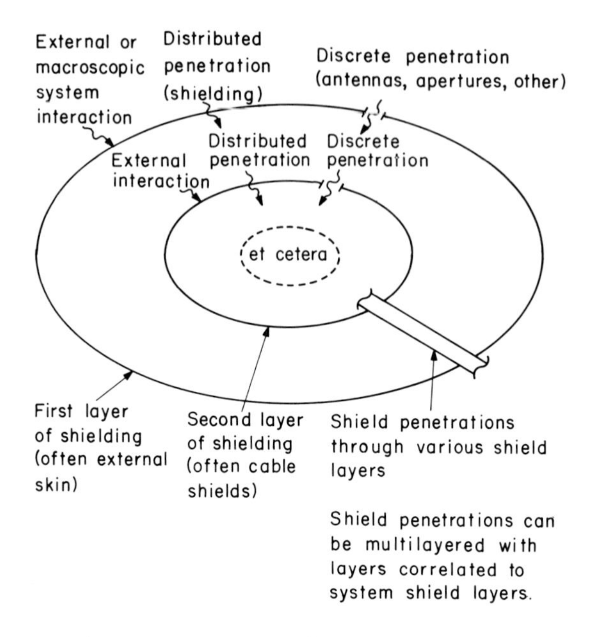

Fig. 4.3. Topological shielding.

order of the scatterer (system). A familiar example to some is a double-wall screen room for instrumentation; here λ_{max} would be 2. Note that for RF purposes the diffusion of electromagnetic energy through continuous metal walls can be neglected; generally the various holes and conductor penetrations are more important to consider. It is this property that allows us to treat each volume in terms of a set of discrete ports. In any event the electromagnetic energy is constrained to pass through some number λ_{max} of separate shielding surfaces (principal surfaces) in reaching the innermost volume; this energy can be attenuated on passing through each such surface for greater shielding of the innermost principal volume(s) $(V)_{\lambda_{max}}^{(\lambda'_{max})}$.

C. Interaction Sequence Diagrams

The concept of a hierarchical topology leads to an interaction sequence diagram for a scatterer [10,12]. The interaction sequence diagram is a diagram of the electromagnetic signal flow from the outside to the inside of the scatterer, i.e., through the various shielding layers or principal volumes and surfaces. This diagram can be a simple approximation of the most important paths for signals getting inside, or it may be as complicated as one wishes, say representing the coupling between every pair of ports in every elementary volume. For well-shielded systems (meaning the signal going through each shield layer is small compared to the driving signal on the outside of the layer) some of these port-to-port interactions may be neglected; scatterers that meet this requirement are those for which a hierarchical topology is meaningful.

An interaction sequence diagram can be considered as being orthogonal to a topological diagram as in Fig. 4.3. Figs. 4.4 through 4.7 give some examples of different shielding orders λ_{max} of scatterers (systems). Fig. 4.4 shows a case of a completely

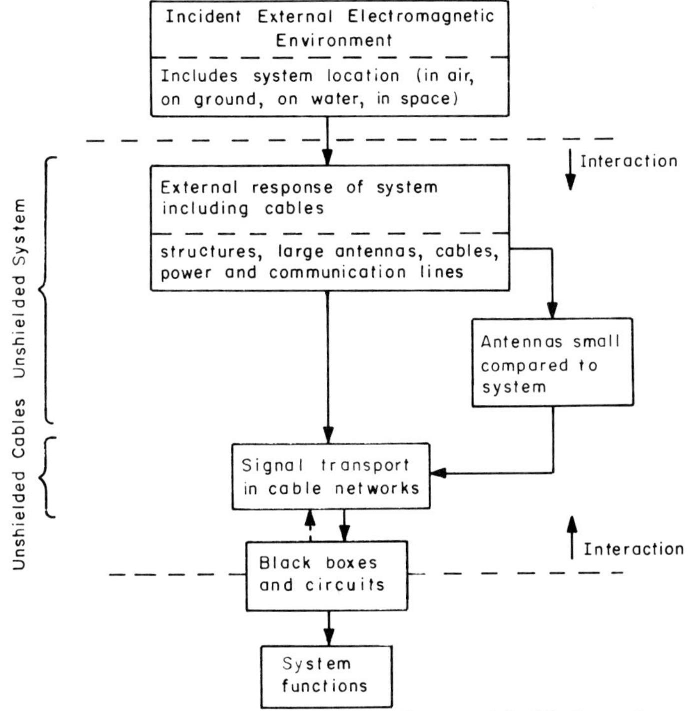

Fig. 4.4. Interaction sequence for unshielded system with unshielded cables: shielding order = 0.

unshielded system ($\lambda_{max} = 0$); this example is characterized by the lack of any system level shielding (overall closed metal envelope) and the lack of any cable shields (braid, solid, conduit, etc.). In such a case the signal cable wires ("interior") are directly exposed to the incident electromagnetic environment and there is no effective electromagnetic division of interior from exterior. If shields are added to the cables, the situation depicted in Fig. 4.5 is produced giving a case of $\lambda_{max} = 1$.

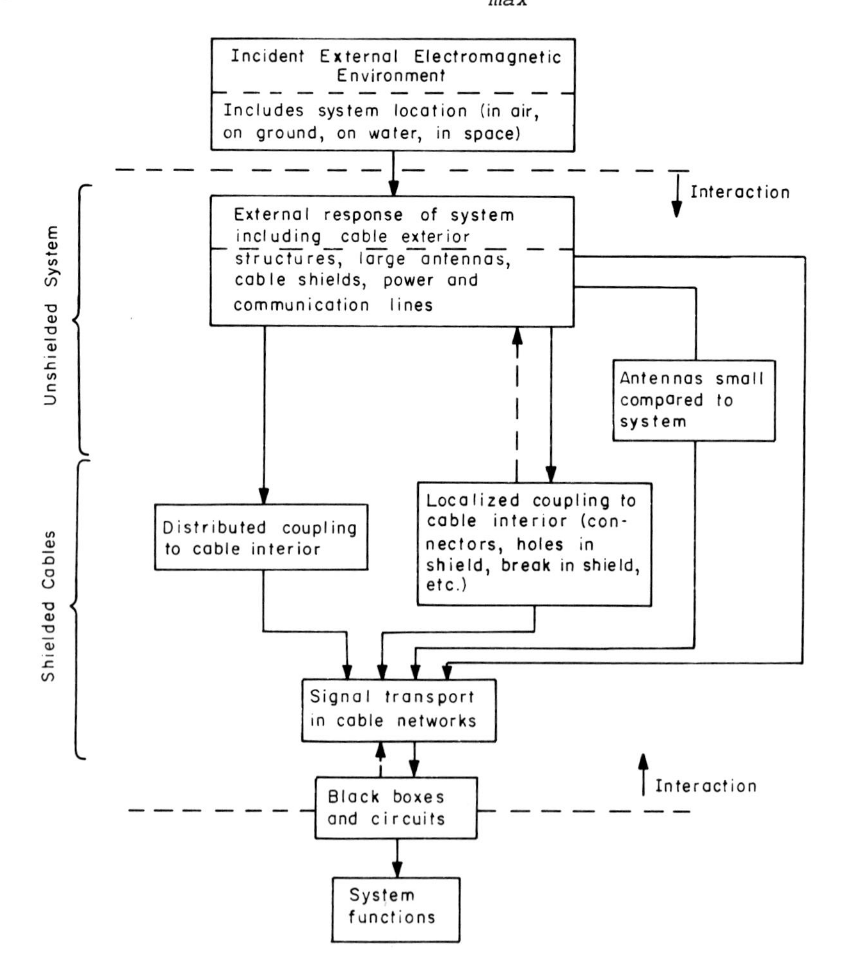

Fig. 4.5. Interaction sequence for unshielded system with shielded cables: shielding order = 1.

Alternately λ_{max} = 1 is achieved by providing the scatterer with an overall metal envelope instead of shielding the cables; this situation is illustrated in Fig. 4.6. If both an overall metal

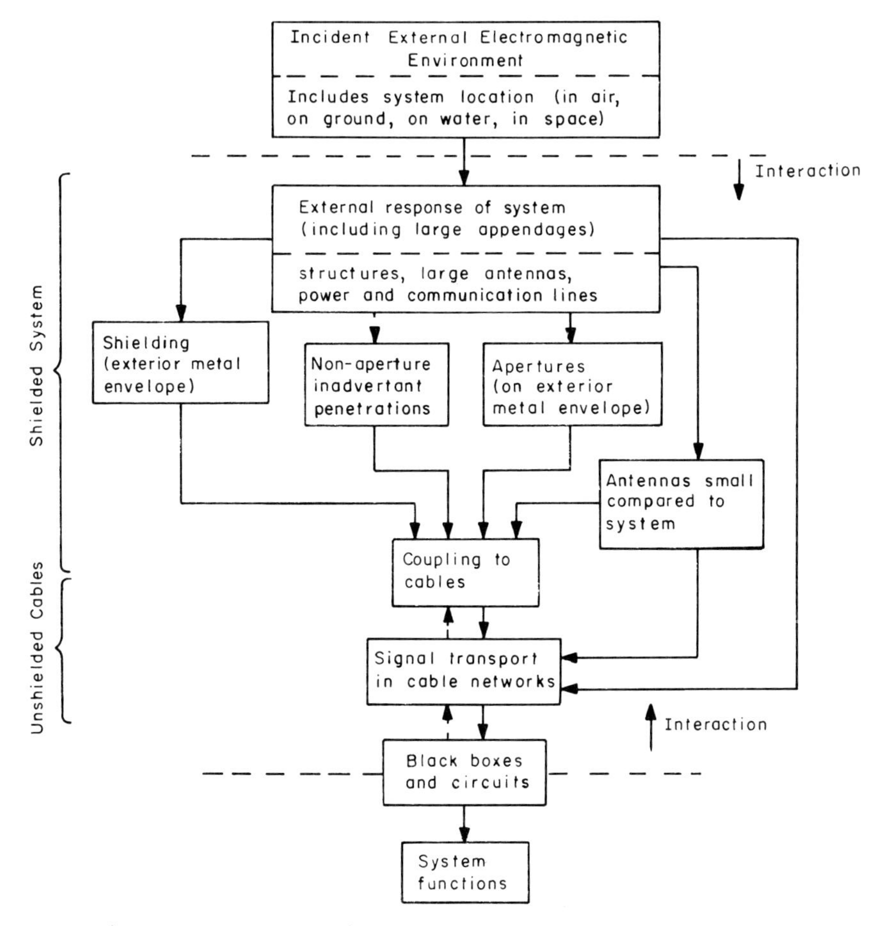

Fig. 4.6. Interaction sequence for shielded system with unshielded cables: shielding order = 1.

envelope and cable shields are provided one has a second order shielded system (λ_{max} = 2) as illustrated in Fig. 4.7. Note in the progression of increasing shielding order the number of steps to be considered in an interaction sequence diagram increases because of the increase in the number of principal volumes and surfaces through which the signals are transported.

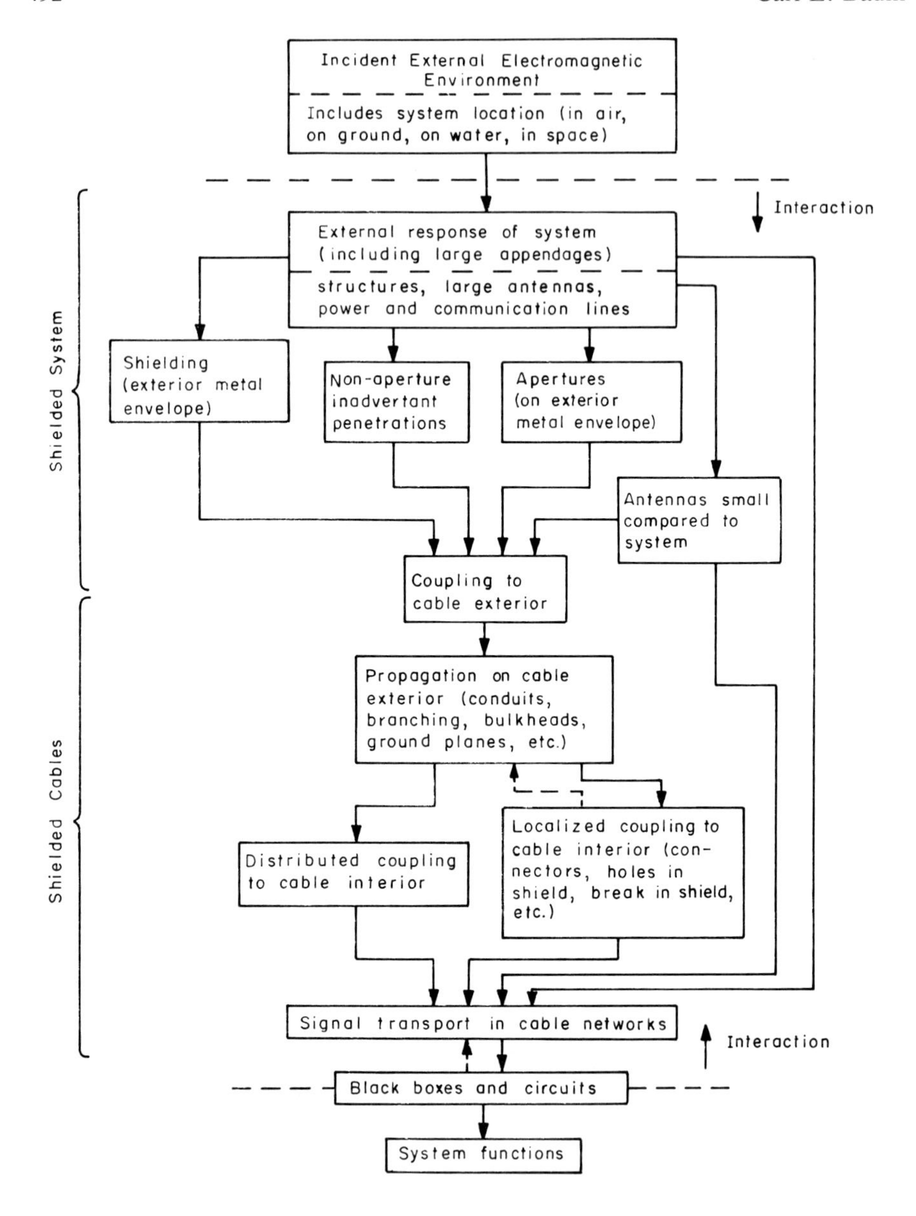

Fig. 4.7. Interaction sequence for shielded system with shielded cables: shielding order = 2.

In considering the interaction sequence diagrams in Figs. 4.4 through 4.7 one should note that ultimately it is some system function (such as lowering a landing gear of an aircraft) with which one is concerned. Such functions are controlled (or influenced) by some electronic equipment usually referred to as "black boxes". From an electromagnetic point of view it is the signals getting into these boxes that interest us. The incident electromagnetic environment outside the system (scatterer) is our given. Starting with the exterior system response and ending with the signals into the black boxes (circuit level) is what electromagnetic interaction is all about; it is this process that an interaction sequence diagram depicts.

D. Some Applications

Understanding some of the ways of dividing up the analysis of scatterer response according to the scatterer topology can lead to some concepts of scatterer (system) design for reduced electromagnetic interference. The hierarchical topology concept with its principal surfaces can be used as a design concept involving shielding surfaces. Such shielding (principal) surfaces $(S)_\lambda$ can be defined as part of the initial system design concept. Then all penetrations through such surfaces (apertures, antennas, signal wires, other conductors, etc.) can be identified and rigorously controlled. Selected shielding surfaces might be at the black box (electronic package) inputs (including limiting, filtering, various kinds of buffering, etc.), at the cables (shielding), perhaps additional shields around the cable raceways and electronic equipment areas (conduits and special bays), and/or an overall conducting envelope around the system (such as an aircraft or ship hull, etc.). A selected set of shielding surfaces can bring some order into the electromagnetic "hardening" of a system and provide a shielding management concept for both design and maintenance of the system in a configuration insensitive to various forms of electromagnetic interference (at least those forms not associated with the normal operational characteristics (frequencies, etc.) of the system).

Besides serving as part of a shielding concept, principal surfaces can serve as part of a "grounding" concept. Grounding is also a topological concept. Unfortunately this concept as practised is still at an elementary circuit stage. "Ground" is often viewed as a point (single point ground) to which all signals are referenced. While this is a topological concept, it is one valid strictly only at zero frequency (DC). As frequency is increased, the concept of an electrostatic potential (on which a signle point ground is based) loses practical meaning. For low frequencies (non-zero) there are potential differences in the circuit sense induced by time changing magnetic fluxes through loops from magnetic fields associated with time varying currents. As frequency is increased and dimensions of the system (scatterer) conductors become an appreciable fraction of a wavelength, the concept of a "single point ground" becomes meaningless; referenced to some other point on a connection to "ground" in the system such a ground may even be an open circuit at some frequencies. At such frequencies the electromagnetic pickup in the "ground" conductors is not "shorted out" at all.

Stated another way, at frequencies above where DC concepts apply more general grounding concepts must be applied if they are to be successful; ideally such high frequency concepts are also consistent with DC requirements. The foregoing discussion of scatterer (system) topology indicates that there are topological concepts of electromagnetic systems more general than circuit concepts. Since these are distributed concepts (volumes and surfaces) instead of circuit concepts (nodes and branches) these would seem a more general set of concepts around which to form a concept of grounding.

Let us then also consider principal surfaces as "grounding surfaces". If we are forming a ground reference in $(V)_{\lambda}^{(\lambda')}$, let it be to the boundaries of this principal volume, i.e., at least one of the principal surfaces $(S)_{\lambda}^{(\lambda_1')}$ and $(S)_{\lambda+1}^{(\lambda_2')}$. Furthermore, let the conducting path to such a surface lie entirely within

$(V)_{\lambda}^{(\lambda_1')}$, so that it is not exposed to signals in $(V)_{\lambda-1}^{(\lambda_0')}$, and does not introduce signals into $(V)_{\lambda+1}^{(\lambda_2')}$. One might have a "single point ground" concept as long as it is localized to apply separately in each principle volume subdivided as $(V)_{\lambda}^{(\lambda')}$. Alternately (perhaps depending on such things as how highly conducting the principal surface and connections to it are) one might have a "single conducting surface" ground concept of more general validity and utility; more rigorously this could be a "single closed conducting surface" ground.

V. DIVISION OF RESPONSE INTO DIFFERENT FREQUENCY REGIMES

In addition to the topological (geometric) decomposition of the system one can also simplify matters by decomposing the general form of the electromagnetic response into smaller pieces. While the general response follows from Maxwell's equations (2.1) or an integral equation (2.5) for the scatterer response, what is needed is some perhaps approximate concepts that more directly give the response (or portions of the response) in terms of the geometrical and other electromagnetic input parameters of the scatterer.

Such concepts for simplifying the form of the response do exist. Considering the response in the complex frequency plane there is a low frequency method (LFM) based on expansions near $s = 0$, a singularity expansion method (SEM) based on the singularities in the s plane and applying to intermediate frequencies (resonance region) and low frequencies, and a high frequency method (HFM) based on asymptotic expansions as $s \to \infty$ in the right half plane and on the $j\omega$ axis. There are also eigenmode (diagonalization) concepts for the integral equation (2.5) for the scatterer; these shed some insight on the LFM and SEM. These topics have been reviewed in a recent paper [11] and are only briefly summarized here.

As indicated in Fig. 5.1 we begin with some definition of

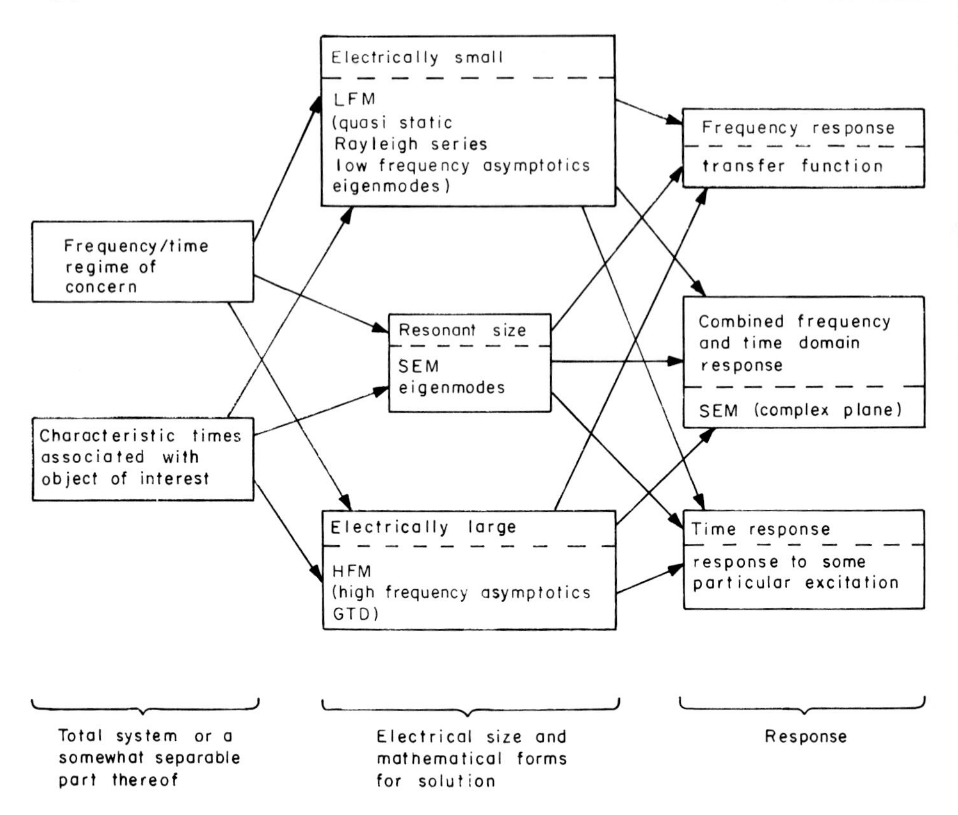

Fig. 5.1. Mathematical decomposition of electromagnetic interaction.

the frequency and/or time regimes of interest together with the characteristic times (transit times, etc.) associated with some part of the system of interest. Comparing these two aspects we consider the response of that portion of the system (a step in an interaction sequence diagram) from the viewpoint of LFM, SEM, and/or HFM points of view. From this, approximate frequency and/or time domain responses for portions of the system emerge.

The low frequency method (LFM) is based on asymptotic expansion about $s = 0$. For finite size conducting objects in free space and other situations equivalent to this, such as apertures or other perturbations on conducting surfaces, a power series in s can be formed. The leading terms in the scattering are described by electric and magnetic dipole moments related to the incident or short circuit fields (in the case of a highly conducting surface near a penetration) as

$$\begin{aligned}\tilde{\vec{p}}(\vec{r}\,',s) &= \tilde{\vec{\vec{P}}}(s) \cdot \tilde{\vec{D}}_{inc}(\vec{r}\,',s)\\ \tilde{\vec{m}}(\vec{r}\,',s) &= \tilde{\vec{\vec{M}}}(s) \cdot \tilde{\vec{H}}_{inc}(\vec{r}\,',s)\end{aligned} \tag{5.1}$$

where $\vec{r}\,'$ is some effective center of the scatterer or portion thereof of interest. Here the effects of the scatterer are contained in the polarizabilities $\tilde{\vec{\vec{P}}}(s)$ and $\tilde{\vec{\vec{M}}}(s)$. The associated scattered fields are given by

$$\tilde{\vec{E}}_{sc}(\vec{r},s) = -\frac{\gamma^2}{\varepsilon_0}\tilde{\vec{\vec{G}}}_0(\vec{r},\vec{r}\,';s) \cdot \tilde{\vec{p}}(\vec{r}\,',s) - \gamma Z_0 \nabla\tilde{G}_0(\vec{r},\vec{r}\,';s) \times \tilde{\vec{m}}(\vec{r}\,',s)$$

$$\tilde{\vec{H}}_{sc}(\vec{r},s) = \gamma c \nabla\tilde{G}_0(\vec{r},\vec{r}\,';s) \times \tilde{\vec{p}}(\vec{r}\,',s) - \gamma^2\tilde{\vec{\vec{G}}}_0(\vec{r},\vec{r}\,';s) \cdot \tilde{\vec{m}}(\vec{r}\,',s)$$

$$\tilde{G}_0(\vec{r},\vec{r}\,';s) = \gamma\,\frac{e^{-\zeta}}{4\pi\zeta}$$

$$\nabla\tilde{G}_0(\vec{r},\vec{r}\,';s) = \vec{1}_R\,\frac{\gamma^2}{4\pi}\left[-\zeta^2 - \zeta^{-1}\right]e^{-\zeta}$$

$$\begin{aligned}\tilde{\vec{\vec{G}}}_0(\vec{r},\vec{r}\,';s) &= \left[\vec{\vec{1}} - \gamma^{-2}\nabla\nabla\right]\tilde{G}_0(\vec{r},\vec{r}\,';s)\\ &= \frac{\gamma}{4\pi}\Bigg\{\left[-2\zeta^{-3} - 2\zeta^{-2}\right]e^{-\zeta}\vec{1}_R\vec{1}_R\\ &\quad + \left[\zeta^{-3} + \zeta^{-2} + \zeta^{-1}\right]e^{-\zeta}\left[\vec{\vec{1}} - \vec{1}_R\vec{1}_R\right]\Bigg\}\\ &\quad + \frac{1}{3\gamma^2}\,\delta\,(\vec{r} - \vec{r}\,')\vec{\vec{1}}\end{aligned} \tag{5.2}$$

$$R \equiv |\vec{r} - \vec{r}\,'|, \quad \zeta = \gamma R, \quad \gamma \equiv \frac{s}{c} = s\sqrt{\mu_0\varepsilon_0}, \quad Z_0 \equiv \sqrt{\frac{\mu_0}{\varepsilon_0}}$$

For intermediate frequencies the singularity expansion method (SEM) already has a considerable literature [6]. The important pole terms are derived from an integral equation (2.5) via

$$\langle \tilde{\vec{\vec{\Gamma}}}(\vec{r},\vec{r}\,';s_\alpha);\vec{j}_{s_\alpha}(\vec{r}\,')\rangle = \vec{0}, \quad \vec{r}\in S$$

$$\langle \vec{\mu}_\alpha(\vec{r});\tilde{\vec{\vec{\Gamma}}}(\vec{r},\vec{r}\,';s)\rangle = \vec{0}, \quad \vec{r}\in S \tag{5.3}$$

$$\tilde{\eta}_\alpha(s_\alpha) = \frac{\langle \mu_\alpha(\vec{r});\tilde{\vec{I}}(\vec{r},s_\alpha)\rangle}{\langle \vec{\mu}_\alpha(\vec{r});\frac{\partial}{\partial s}\tilde{\vec{\vec{\Gamma}}}(\vec{r},\vec{r}\,';s)\Big|_{s=s_\alpha};\vec{j}_{s_\alpha}(\vec{r}\,')\rangle}$$

These give the first order pole terms in the response as

$$\tilde{\vec{J}}_s(\vec{r},s) = \sum_\alpha \tilde{f}_p(s_\alpha)\tilde{\eta}_\alpha(s)\vec{j}_{s_\alpha}(\vec{r})(s - s_\alpha)^{-1} \tag{5.4}$$
$$+ \text{ other singularity terms}$$

where $\tilde{f}_p(s)$ is the waveform of the incident field (in frequency domain and polarization dependent). The coupling coefficient $\tilde{\eta}_\alpha(s)$ contains the dependence of the residues on the incident wave parameters while $\vec{j}_{s_\alpha}(\vec{r})$ gives the dependence on position over the scatterer. The natural frequency s_α is independent of both observation position and incident wave conditions.

The high frequency method (HFM) is based on asymptotic expansions as $s \to \infty$ and uses ray optical concepts [4, 5]. This is illustrated by the typical form of a "diffracted" ray as

$$\tilde{\vec{E}}_d(\vec{r},s) = \tilde{\vec{\vec{D}}}(\vec{1}_{inc},\vec{1}_d,\gamma,R,R_1,R_2)\cdot\tilde{\vec{E}}_{inc}(\vec{r}_d,s)$$

$\tilde{\vec{E}}_{inc}(\vec{r},s) \equiv$ incident wave propagating in direction $\vec{1}_{inc}$ when passing the diffraction point $\vec{r}_d$

$\tilde{\vec{E}}_d(\vec{r},s) \equiv$ diffracted wave propagating in direction $\vec{1}_d$ when passing the observation point $\vec{r}$ (5.5)

$$R \equiv |\vec{r} - \vec{r}_d|$$

$R_1, R_2 \equiv$ two radii of curvature of incident wave at $\vec{r}_d$

$\tilde{\vec{\vec{D}}} \equiv$ dyadic diffraction function

where the scattered field has the asymptotic behavior of the diffracted ray as

$$\tilde{\vec{E}}_{sc}(\vec{r},s) = \tilde{\vec{E}}_d(\vec{r},s) + o(\tilde{\vec{E}}_d(\vec{r},s)) \quad \text{as } s \to \infty \text{ with } Re[s] \geq 0 \tag{5.6}$$

Now the dyadic diffraction function can often be split up as

$$\tilde{\vec{\vec{D}}}(\vec{1}_{inc},\vec{1}_d,\gamma,R,R_1,R_2) = \sum_n \vec{\vec{d}}_n(\vec{1}_{inc},\vec{1}_d)\tilde{f}_n(\gamma,R,R_1,R_2)$$

$$\vec{\vec{d}}_n \equiv \text{dyadic diffraction coefficient} \tag{5.7}$$

$$\tilde{f}_n(\gamma,R,R_1,R_2) \equiv \text{diffraction propagation function}$$

Here the diffraction coefficients contain (by definition) only local scatterer information (from near $\vec{r}_d$); they may depend on more parameters than the incidence and diffraction angles, such as local radii of curvature (and perhaps also γ to keep things dimensionless). This is a fairly simplified point of view of high frequency approximations such as are used in GTD [4]. It is merely intended to illustrate that simplification of the form of the results by exhibiting terms with dependences on some but not all the parameters of the problem is possible for high frequency expansions as it is for low frequency and singularity expansions. There are many types of high frequency diffraction phenomena which need to be considered in detail.

VI. SUMMARY

The kinds of topics discussed in this paper are chosen for their capability for decomposing a very complex scattering problem into some set of smaller problems. Problem complexity here is a relative term; the individual problems may be simpler, but not in general trivial. Note after decomposing the problem and solving the parts one has to recombine the solutions for the parts into a solution for the whole.

Emphasis here has been on two basic concepts for such problem division. Provided the presence of approximate conducting surfaces allows it, scatterer topology based on these surfaces and their enclosed volumes is a powerful concept for a general dividing up of a scatterer. Matrix concepts can be used to recombine the solutions for the individual volumes. Similar topological concepts also apply to transmission-line networks which can be important for a part of the scatterer (system) response. Fig. 6.1 summarizes these various topological concepts. Conceivably other topologies relevant to electromagnetic problems can be constructed. For example, one may wish to construct hybrid topologies which combine circuit, transmission-line, and general scatterer topological concepts; this might be employed in attempting to describe parts of the scatterer using different representations.

	Topology	Basic Topological Quantity and Symbol	Interconnecting Topological Quantity and Symbol	Diagramatic form of Topology
1	Circuit	Node N_n	Branch $B_{n,m}^{(\tau)}$	(fig. 3.1)
2	Transmission Line	Junction J_n	Tube $T_{n,m}^{(\tau)}$	(fig. 3.2)
3	Scatterer	Volume V_n	Surface $S_{n,m}$	(fig. 4.1)
4	Hierarchical Scatterer	Principal Volume $(V)_{\lambda}^{(\lambda')}$	Principal Surface $(S)_{\lambda}^{(\lambda')}$ (closed but sometimes in more than one part)	(fig. 4.2)
		Elementary Volume $V_{\lambda,\tau}^{(\lambda')}$	Elementary Surface $S_{\lambda_1,\tau_1;\lambda_2,\tau_2}^{(\lambda_1',\lambda_2')}$ (usually open and, but sometimes closed)	

Fig. 6.1. Illustration of the correspondence amongst the various topological quantities.

Another kind of decomposition is that based on relative size of physical features of the geometry of interest compared to wavelength. This leads to low frequency expansions, singularity expansions, and high frequency expansions with some decomposition of the solutions with terms depending on some but not all parameters of the problem. Fig. 5.1 summarizes this kind of frequency decomposition.

These decomposition concepts appear to be in general mutually compatible. Perhaps additional decomposition concepts can also be developed which can be used in conjunction with those discussed here.

While the initial thrust in formulating scatterer decomposition has been one of analyzing the electromagnetic response of complicated scatterers, it also has use for scatterer (system) design (synthesis). Concepts such as shielding (of various orders λ_{max}) and grounding (with validity extended to high frequencies) can be specified and controlled in the system design, testing, and maintenance in field operation. The implications of the general electromagnetic concepts need to be reduced to engineering practice.

VII. REFERENCES

1. Stratton, J.A., "Electromagnetic Theory," McGraw-Hill, New York (1941).
2. Guillemin, E.A., "Introductory Circuit Theory," Wiley, New York (1953).
3. Tai, C-T., "Dyadic Green's Functions in Electromagnetic Theory," Intext, New York (1971).
4. Kouyoumjian, R.G., The Geometrical Theory of Diffraction and Its Application, in "Numerical and Asymptotic Techniques in Electromagnetics" (R. Mittra, Ed.), Springer Verlag, New York (1975).
5. Felson, L.B., Propagation and Diffraction of Transient Fields in Non-Dispersive and Dispersive Media, in "Transient Electromagnetic Fields" (L.B. Felsen, Ed.), Springer Verlag (1975).
6. Baum, C.E., The Singularity Expansion Method, in "Transient Electromagnetic Fields" (L.B. Felsen, Ed.), Springer Verlag (1975).
7. Baum, C.E., Toward an Engineering Theory of Electromagnetic Scattering: The Singularity and Eigenmode Expansion Methods, National Conference on Electromagnetic Scattering, University of Illinois at Chicago Circle (June 1976).
8. Dolph, C.L., and R.A. Scott, Recent Developments in the Use of Complex Singularities in Electromagnetic Theory and Elastic Wave Propagation, National Conference on Electromagnetic Scattering, University of Illinois at Chicago Circle (June 1976).

9. Baum, C.E., Electromagnetic Reciprocity and Energy Theorems for Free Space Including Sources Generalized to Numerous Theorems, to Combined Fields, and to Complex Frequency Domain, Mathematics Note 33 (December 1973).
10. Tesche, F.M., M.A. Morgan, B. Fishbine, and E.R. Parkinson, Internal Interaction Analysis: Topological Concepts and Needed Model Improvements, Interaction Note 248 (July 1975).
11. Baum, C.E., *Proc. IEEE 64,* 1598-1616 (1976).
12. Baum, C.E., How to Think about EMP Interaction, Proc. 1974 Spring FULMEN Meeting, 12-23 (April 1974).
13. Baum, C.E., Coupling into Coaxial Cables from Currents and Charges on the Exterior, USNC/URSI Meeting, Amherst, Massachusetts (October 1976).
14. Tesche, F.M., A General Multi-Conductor Transmission Line Model, USNC/URSI Meeting, Amherst, Massachusetts (October 1976).

RECENT DEVELOPMENTS IN THE USE OF COMPLEX SINGULARITIES IN ELECTROMAGNETIC THEORY AND ELASTIC WAVE PROPAGATION[1]

Charles L. Dolph
Department of Mathematics
The University of Michigan

and

Richard A. Scott
Department of Applied Mechanics and Engineering Science
The University of Michigan

INTRODUCTION

While complex singularities or eigenvalues as they are now called appear to have been first introduced by Lord Kelvin in 1884 as part of his solution of the scattering from a perfectly conducting sphere their serious consideration in physics probably dates from Gamow's paper on α-decay which appeared in 1928. However, as the discussion in Dolph [G8] shows, they are already implicitly present in Lord Rayleigh's 1881 solution of plane wave scattering from a cylinder in acoustics.

In electromagnetic theory they play an important role in the theory of anomalous propagation as they do in underwater and elastic wave transmission. However, it was not until the mid 1960's that a beginning of a rigorous set of results concerning them was established and then mainly for the cases of the reduced wave equation or the Schroedinger equation. Later some of the results were generalized to the electromagnetic case and to the n-body problem in quantum mechanics. These results are still far from

---

[1]Dedicated to the memory of James M. Wolf, Ph.d., a modern Renaissance man and magnificent friend.

ISBN 0-12-709650-7

complete if one takes either Sturm-Liouville Theory or that of self-adjoint operators as examples of almost complete theories. At any rate this will be subject of Part I where section A is devoted to the scalar and electromagnetic theory while in section B, Schroedinger's equation or, almost equivalently, the reduced wave equation with a potential, is treated.

In view of space limitations, detailed discussion, in Part I without proof, will be limited to the simplest cases of given method or technique and the interested reader will be referred to the references for details in more general cases.

In view of possible unfamiliarity with the material, the discussion in Part II is more detailed. Two major topics are treated, namely, the Cagniard-de Hoop method for inverting integral transforms, and the exploitation of poles lying on unphysical sheets.

While no claim for being *au courant* or complete is made for the bibliography, we do believe that it is sufficiently close to satisfying both of these criteria that an interested reader will have no difficulty in bridging any gap which may exist.

Part I

C.L. Dolph

Before discussing any topics it is necessary to alert the reader to the different conventions used for "complex singularities". In electrical engineering, Laplace transforms are often employed so that the complex singularities occur in the left-hand plane. Most physicists use complex Fourier transforms so that they occur in the lower half-plane, if the one dimensional Fourier transform is defined via the equation

$$f(k) = \frac{1}{\sqrt{2\pi}} \int_{-\infty}^{\infty} e^{ikx} f(x)dx.$$

Lax and Phillips and Van Winter, however, have them in the upper half-plane. This is the usual convention for the Hardy class H^2 of mathematics. Since this author has experienced the confusion which can arise when a lecturer used to one convention attempts to interpret in another, papers under discussion will employ the convention at hand with an occasional reminder of other conventions.

A.1. VARIATIONAL PRINCIPLES IN THE SCALAR CASE

The author first became interested in complex singularities because of the phenomena of anomalous radar propagation observed during World War II. There, as discussed by Kerr [A1.1], after separation of variables, one has an ordinary differential equation looking as if it belonged to the Sturm-Liouville type but with different boundary conditions so that the problem is non-self-adjoint. The mathematical theory of such problems, in contrast to solvable examples as found in Kerr, began with Phillip's report [A1.2] and was continued by his student Sims [A1.3]. For certain problems of this type involving a real potential but complex boundary conditions it is possible to set up a variational principle for determination of the eigenvalues that looks like that of Rayleigh-Ritz except that a *real* inner product and *real* symmetry are involved even though the quantities are complex. Such a principle is due to MacFarlane [A1.4]. While no complete justification has ever been given, Dolph, McLaughlin and Marx [A1.5] were able to give a min-max principle for the quadratic forms which resulted from any finite approximation. In it one has complex-valued real symmetric matrices which can be diagonalized as long as the minimum polynomial associated with them has simple roots. The case of the infinite dimensional problem remains open and it should be stressed that the min-max characterization of the "complex" eigenvalue is only for the imaginary part.

Real symmetry for complex-valued functions also provides the basis for the Schwinger variational principles for the wave equation and for the Schroedinger equation. As shown by Dolph and Ritt [A1.6] and Dolph [A1.7] for *real* $\kappa(\equiv|\underset{\sim}{k}|)$ the dissipative nature of the Green's function allows a saddle type characterization of the imaginary or real part of the stationary point. Unfortunately no method for determining the orientation of the saddle has yet been found, so error estimates are not available.

The interplay between real and Hermitian symmetry also occurs in the now classic papers by Weyl [A1.8], Müller [A1.9] and Ikebe [A1.10], the last author using it to prove the real symmetry of the resolvent Green's function for the two body problem. A systematic investigation of this relationship does not appear to have been made and it would be highly desirable for problems in mathematical physics.

A similar situation occurs in regard to the "complex eigenvalues" of exterior problems for the wave equation. Morawetz [A1.11] and the author [A1.12] found a variational principle for part of the "complex eigenvalues". In her normalization

$$u = \psi(\underset{\sim}{x},\mu)e^{-\mu t},$$

a solution of the wave equation

$$\frac{\partial^2 u}{\partial t^2} - \Delta u = 0, \quad t > 0 \quad \text{in } E \qquad \text{(A1.1)}$$

in the exterior of the obstacle is said to be outgoing if $\psi(\underset{\sim}{x},\mu)$ admits the representation in E

$$\psi(\underset{\sim}{x},\mu) = \int_{\partial E} \left\{ \frac{e^{\mu\rho}}{\rho} \frac{\partial\psi}{\partial n} (\underset{\sim}{x}',\mu) - \psi(\underset{\sim}{x}',\rho) \frac{\partial}{\partial n} (\frac{e^{\mu\rho}}{\rho}) \right\} d^2x'$$

for $\rho = |\underset{\sim}{x} - \underset{\sim}{x}'|$. This formula holds for $Re\ \mu \leq 0$ and hence, by analytic continuation for $Re\ \mu > 0$. If one now assumes that

$$v\ e^{\mu(r-t)} \qquad \text{(A1.2)}$$

is an outgoing solution for the wave equation for $t > 0$ in the exterior region E and if $v = 0$ on ∂E, the boundary of E, then

$$\Delta v = -2\mu \frac{\partial (rv)}{\partial r}, \text{ in } E.$$

Moreover, the *real* part of μ can be characterized as the stationary points of the expression

$$\frac{1}{4}\left\{2\int_{\partial E}(\underset{\sim}{x}\cdot\underset{\sim}{n})\left|\frac{\partial v}{\partial n}\right|^2 d^2x + \int_E |\nabla v|^2 d^3x\right\}\left[\int_E \left|\frac{1}{r}\frac{\partial}{\partial r}(rv)\right|^2 d^3x\right]^{-1}$$

[In the usual physicist normalization this would characterize the imaginary part of the "complex eigenvalues κ" in the lower half-plane.]

Transformations like equation (A1.2) have been used by Vainberg [A1.13] and Kleinman [A1.14] to obtain equations linear in κ rather than κ^2. Apart from the paper by Garabedian [A1.15] their use has been limited to the low-frequency range, based on a Green's function satisfying Laplace's equation as a parametrix.

A.2 RESULTS FROM THE LAX-PHILLIPS THEORY

Lax and Phillips [A2.1] were the first to give a reasonably general rigorous time-dependent interpretation of the complex-eigenvalues associated with the scalar wave equation subject to Dirichlet or Neumann boundary conditions. They were also the first to prove the first theorem on existence of complex eigenvalues as well as information concerning the asymptotic distribution of the imaginary part of the eigenvalues [N.B. the convention].

Here comments will be limited to that part of the theory which involves scattering from "non-confining" objects, the part which includes scattering from all convex objects. The theory fails if the object is "confining"--see Ralston for details [A2.2]

For classical interior problems the solution to the wave equation (A1.1) subject to Dirichlet, Neumann or Robin boundary conditions on some smooth bounded object is well known to be of the form

$$u(\underset{\sim}{x},t) = \sum_{1}^{\infty} (a_k e^{i\lambda_k t} + b_k e^{-i\lambda_k t})\, v_k(x)$$

where λ_k are the real eigenvalues of $-\Delta$ satisfying

$$\Delta v_k + \lambda_k^2 v_k = 0$$

in the interior, and the appropriate boundary conditions corresponding to the three classical problems.

In contrast, for exterior problems only continuous spectra are involved and the above solution for the scattered wave is replaced by one of the form

$$u(\underset{\sim}{x},t) = (2\pi)^{3/2} \int_{-\infty}^{\infty} w(\underset{\sim}{x},\underset{\sim}{k})[\alpha(\underset{\sim}{k})e^{i|\underset{\sim}{k}|t} + \beta(\underset{\sim}{k})e^{-i|\underset{\sim}{k}|t}]d^3k$$

In this expression $w(\underset{\sim}{x},\underset{\sim}{k})$ is an outgoing distorted plane wave of the form

$$e^{i\underset{\sim}{k}\cdot\underset{\sim}{x}} + v(\underset{\sim}{x},\underset{\sim}{k}\quad |\underset{\sim}{k}|\ )$$

where $v(\underset{\sim}{x}, \underset{\sim}{k}, |\underset{\sim}{k}|)$ is some solution of a Fredholm integral equation involving a single layer, a double layer, a complex combination or a volume distribution. (See Dolph [G.8]).

While some examples were known, Lax-Phillips [A2.3] were the first to show that asymptotically the solution of equation (A1.1) was of the form

$$u(\underset{\sim}{x},t) \sim \sum_{1}^{\infty} c_k\, e^{\mu_k t}\, w_k(\underset{\sim}{x})$$

where $$0 > Re\,\mu_1 > Re\,\mu_2 > \ldots$$

and where $\Delta w_k = \mu_k^2 w_k$, in the exterior of E,

and $w_k(\underset{\sim}{x})$ satisfied homogenous boundary condition of the Dirichlet or Neumann problem on ∂E. This they do by showing that the μ_k and w_k are eigenvalues and eigenfunctions of the infinitesimal generator of a certain semi-group. Moreover they show that the eigenvalues are related to the poles of the Heisenberg scattering matrix. (See [A2.1] page 9). In n-dimensions it is a linear operator on $L_2\,(S^2)$, the Hilbert space of square integrable functions on the unit sphere in E^3. It is of the form

$$S(z) = I + K(z) \tag{A2.1}$$

where $K(z)$ is an integral operator with kernel

$$\frac{iz}{2\pi}\; t(\underset{\sim}{\omega},-\theta;z)$$

$t(\underset{\sim}{\omega},\theta;z)$ being the transmission coefficient, which for $n = 3$ is determined by the solution of, for $Im z \leq 0$,

$$\Delta v + z^2 v = 0, \text{ in } E$$

$$v(\underset{\sim}{x},\underset{\sim}{\omega};z) = e^{iz\underset{\sim}{x}\cdot\underset{\sim}{\omega}} \quad \text{on } \partial E$$

via the asymptotic expansion

$$v(\underset{\sim}{x},\underset{\sim}{\omega};z) = \frac{e^{-iz\rho}}{\rho}\left[t(\underset{\sim}{\omega},\underset{\sim}{\theta};z) + O(\frac{1}{\rho})\right]$$

This scattering operator $S(z)$ has poles on the positive imaginary axis at z if and only if $S(\bar{z})$ has a zero there; that is if $S(\bar{z})$ has a non-trivial null space.

The scattering operator *uniquely* determines the scatterer.[1] (See Lax-Phillips [A2.1], Theorem 5.7, page 173). However as in section A.1 general results (apart from the fact that in odd dimensions the scattering operator is a meromorphic function in the compex plane) are limited to the imaginary part of the complex eigenvalues. The poles of $S(z)$ occur at the points $-i\lambda$ for which $i\bar{\lambda}$ is a zero of $S(z)$. Consequently it is convenient to set

---

[1] Significant progress in determining the approximate shape of a convex xcatterer from the scattering matrix has recently been made by A. Majda [A2.19] and A. Majda and M. Taylor [A2.20].

$\sigma = i\,z$ in order to discuss the zeros on the negative imaginary axis. For these zeros (hence for the poles on the positive imaginary axis) Lax and Phillips in [A2.3] proved that:

i) If an obstacle is contained in a star-shaped[2] obstacle whose scattering matrix has n purely imaginary zeros $\sigma_k^s \leq \sigma$, then the scattering matrix of the obstacle also has n purely imaginary zeros $-i\sigma_k$, with $\sigma_k \leq \sigma$.

ii) If $C(\sigma)$ denotes the number of purely imaginary zeros of the scattering matrix for a sphere of radius R which are less than or equal to σ in absolute value, then for either Dirichlet or Neumann boundary conditions

$$C(\sigma) \sim \frac{1}{2}\left(\frac{\sigma R}{\gamma_0}\right)^2, \quad \gamma_0 = 0.66274$$

iii) Moreover if the obstacle contains a sphere of radius R_1 and is contained in a sphere or radius R_2, then under the same boundary conditions

$$\liminf_{\sigma \to \infty} \frac{C(\sigma)}{\sigma^2} \geq \frac{1}{2}\left(\frac{R_1}{\gamma_0}\right)^2$$

while if the obstacle is star-shaped

$$\limsup_{\sigma \to \infty} \frac{C(\sigma)}{\sigma^2} \leq \frac{1}{2}\left(\frac{R_2}{\gamma_0}\right)^2$$

Many of the Lax-Phillips results have been extended to Robin boundary conditions by Beale [A2.4]. Schmidt [A2.5],[A2.6] extended the Lax-Phillips theory of scattering to the electromagnetic case. In the absence of current density, Schmidt treated Maxwell's equation subject to the boundary conditions

$$(\underset{\sim}{E} - \lambda \underset{\sim}{H}) \times \underset{\sim}{n} = 0$$

or

$$\underset{\sim}{H} \times \underset{\sim}{n} = 0$$

---

[2] For definition see Spivak [G19], p. 93.

He also proved in [A2.7] that, as in the scalar case, a scattering matrix exists which is an operator-valued meromorphic function in the complex plane, analytic in the lower half-plane (in the Lax-Phillips normalization) where z is a pole in the upper half-plane if and only if $\bar{z}$ is a zero of the scattering operator in the lower half plane. The scattering operator was shown to be of the form of equation (A2.1) except that for Maxwell's equation the integral operator is a 3 x 3 matrix valued function which is related to the transmission coefficient in this case. Although Saxon [S.8] introduced the scattering matrix earlier, it had not been used extensively before the work of Schmidt. His work provides a rigorous proof for the often stated result in the SEM theory that only poles occur for convex objects. Beale, also in [A2.4], was able to extend many of the results of Lax and Phillips to the electromagnetic case in Schmidt's framework.

The above results of Lax and Phillips and Beale constitute the most *general existence theorems* so far known for "complex eigenvalues" for the scalar wave equation and the equations of electromagnetic theory.

The theory of "complex eigenvalues" is clearly not nearly as complete nor as rich as that of eigenvalues for self-adjoint operators. For example one sees that even in the separated case for the problem of the sphere, no poles (complex eigenvalues) exist in the magnetic case if the angular momentum is zero. (See Beck and Nussenzweig [G.10] for details).

There is also an interesting relationship between the Lax-Phillips theory of scattering and the developing theory of non-self-adjoint operators. While as noted by Combes [G.6] in his Denver talk, Dolph [G.12] suggested that non-self adjoing operators would have a role in scattering theory, M. G. Krein [G.14] noted explicitly the relationship of the Lax-Phillips theory to that of the non-self-adjoint theory of B. Sz-Nagy and Foias [G.15]. Krein's students Adamjan and Arov [G.16] have further investigated this relationship.

A.3 THEOREMS ON NATURE AND LOCATION OF THE POLES

A few general results about the location of the poles of the scattering matrix have been established. For the equation, with q denoting a potential,

$$\frac{\partial^2 u}{\partial t^2} - c^2 \Delta u + qu = 0$$

Lax and Phillips [A3.1] showed that there exists constants a and b such that the complex poles z_p of the scattering matrix satisfy the inequality

$$Im\, z_p \geq a + b \ln |z_p| \;, \; b > 0 \tag{A3.1}$$

if any of the following holds:

i) If the obstacle is smooth, $c = 1$, $q = 0$ for $|\underset{\sim}{x}| > R$, the energy is positive definite and all rays go to infinity,

ii) $c = 1$ and $q = 0$ outside a convex body on which $u = 0$, and

iii) $c = 1$, q bounded and measurable, $q = 0$ for $|\underset{\sim}{x}| > R$, with an energy form that is not necessarily positive definite. The above are generalizations of earlier work by Ramm [A3.2, A3.3].

Some results concerning the absence of poles in given regions are also to be found in Lax-Phillips [A3.1]. Related results on the so-called "empty beer bottle problem" can also be found in Beale [A3.4]. LaVita [A3.5] has also investigated analytical perturbations of obstacles depending on a parameter ε and shown that the poles are algebraic functions of the parameter.

Up to now the scattering operator has provided the underlying structure. In many cases this can be replaced by suitable Fredholm operators and in part B the explicit relationship between the two will be exhibited in certain cases.

Apart from the partial results due to Tamarkin [G.17] for Fredholm integral operators depending analytically on a parameter z, and the result contained in Dunford and Schwartz [G.18], the first general results are due to Steinberg [A3.6], who gave the following necessary and sufficient conditions for the simplicity of the poles:

Let β be a Banach space and $B(\beta)$ the set of bounded operators on β and Z a subset of the complex plane which is open and connected. If $T(z)$, an abstract compact operator, maps Z into $B(\beta)$, it is said to be analytic in Z if for each z_o in Z

$$T(z) = \sum_{o}^{\infty} T_n (z-z_o)^n$$

Here T_n is in $B(\beta)$ and the series converges uniformly in the uniform topology in some neighborhood of z_o.

$T(z)$ is said to be meromorphic in Z if it is defined as analytic in Z except for a discrete set of points. Suppose at an exceptional point z_o, $T(z)$ has the representation

$$T(z) = \sum_{-N}^{\infty} T_n (z-z_o)^n$$

for $N \gtrless 0$, where the series converges in the uniform operator topology. Steinberg proved;

i) If $T(z)$ is an analytic family of compact operators for z, then either $[I-T(z)]$ is nowhere invertible or else $[I-T(z)]^{-1}$ is meromorphic in Z. (The theorem proved in Dunford and Schwartz [G.18] does not include the fact that the inverse is meromorphic, but Tamarkin did prove it for the subclass of Fredholm operators.)

ii) Let $T(z)$ be an analytic family of compact operators for z near z_o such that

$$I-T(z) = \sum_{o}^{\infty} A_n (z-z_o)^n$$

and let
$$P(z) = \frac{1}{2\pi i} \int_C [z'-T(z)]^{-1} dz'$$

where the integration is around a small circle $|z'-1| < \eta$, for small η. If $[I-T(z)]$ is invertible at some point but not invertible at z_o, then it will have a pole of order 1 at z_o whenever

$A_o^2 v = 0$ implies that $A_o v = 0$ for all v and $A_o v = 0$ implies that $P(z_o)A_1 \neq 0$ for all v.

Then $$[I-T(z)]^{-1} = \sum_{-1}^{\infty} B_n (z-z_o)^n$$

Steinberg was also able to use this result in [A3.7] to establish that bound states for a certain class of potentials in quantum mechanics were simple. As far as known to this writer no application to "complex eigenvalues" have yet been made. His paper also contains results on perturbation theory for operators $T(z,\varepsilon)$ analytic in z and jointly continuous in (z,ε) but will not be given explicitly here.

Subsequently Howland [A3.8] considered equivalent families of the form, in his notation,

$$A(z) = I + K(z)[= I-T(z)]$$

He defined z_o to be a weak zero of $A(z)$ if it was not invertible at z_o. If

$$K(z) = K_o(z_o) + (z-z_o)K_1 + \ldots,$$

then $A(z_o)^{-1}$ has a simple pole whenever the space β can be decomposed as the orthogonal sum of the range of $A(z_o)$ and K_1 times the null space of $A(z_o)$, a condition equivalent to stating that the algebraic multiplicity at z_o is equal to the geometric multiplicity at z_o. More important from the viewpoint of applications, he also showed that the inverse of $A(z)$ has a simple pole if and only if the inverse of

$$I + K_o + (z-z_o)K_1$$

has a simple pole. Other equivalent conditions involving more machinery can also be found in [A3.8].

SECTION B

Up until now the perturbation problems involved perturbations by obstacles. Here the perturbations of the operators will be caused by potentials. To our knowledge the methods to be described have not as yet been applied extensively in obstacle scattering or in elastic scattering but it is clear that they should be applicable in some cases[3].

B.1 SPECTRAL CONCENTRATION

Titchmarsh [Bl.1] discovered this phenomena of spectral concentration in connection with his work on singular Sturm-Liouville problems. It is associated not with a single operator but has invariant significance only for a family of operators usually taken in the form

$$H(\varepsilon) = H_o + \varepsilon H_1$$

Here ε is a small parameter and it is assumed that the eigenvalues λ_k^o and eigenfunctions ϕ_k^o of the unperturbed operator H_o are known. The usual perturbation theory leads one to expect that the perturbed eigenvalues and eigenfunctions admit the expansions

$$\lambda_\varepsilon = \lambda_k^o + \varepsilon\lambda_k^1 + \varepsilon^2\lambda_k^2 + \ldots, \qquad k = 1, 2, 3, \ldots$$

$$\phi_\varepsilon = \phi_k^o + \varepsilon\phi_k^1 + \varepsilon^2\phi_k^2 + \ldots$$

and many results when these formal series converge in an ordinary sense or at least asymptotically are known and can be found in the book by Kato [G.1]. However it may also happen that the unperturbed eigenvalue disappears upon perturbation and that the perturbed operator may have only a continuous spectrum. Kato has shown that if the series are truncated after N steps, then λ_ε and ϕ_ε will become pseudo-eigenvalues and eigenfunctions in a Hilbert space satisfying

$$||(H(\varepsilon) - \lambda_\varepsilon)\phi_\varepsilon|| = o(\varepsilon^N)$$

and that part of the spectrum of $H(\varepsilon)$ will be concentrated in

[3] L. Eyges and A. Nelson, Perturbation Theory of Scattering from Irregular Bodies, *Ann. Phys.* *100*, 37 (1976).

intervals of width $0(\varepsilon)$ centered around λ_k^0.

In his treatment Titchmarsh used an analytic continuation of the resolvent Green's function and in his examples he showed that below the eigenvalue of the unperturbed operator a complex pole would appear. The intuitive idea is as follows: For non-singular Sturm-Liouville operators which involve only denumerable discrete spectra Parseval's equation is

$$\int |f|^2 \, dx = \sum |\langle f, \phi_n^0 \rangle|^2$$

In the event that only a continuous spectrum occurred for the perturbed operator $H(\varepsilon)$, this would be replaced by

$$\int |f|^2 \, dx = \frac{1}{\pi} \int_{-\infty}^{\infty} |g(\lambda,\varepsilon)|^2 \, dm(\lambda,\varepsilon)$$

In his examples the right-hand side of the above could, in turn, be replaced by

$$\frac{1}{\pi} \int_{E(\varepsilon)} |g(\lambda,\varepsilon)|^2 \; dm(\lambda,\varepsilon) + F(f,\varepsilon) \qquad \text{(B1.1)}$$

Here the integration is over the set $E(\varepsilon)$ which converges to the unperturbed spectrum as $\varepsilon \to 0$ and which, in many cases, can be taken to consist of intervals of the form

$$\lambda_k^0 + \varepsilon\lambda_k^1 - \varepsilon^p, \quad \lambda_k^0 + \varepsilon\lambda_k^1 + \varepsilon^p, \; p \geq 1,$$

and where $F(f,\varepsilon)$ tends to zero with ε for a fixed f. His simplest example occurs in [B1.1] and is the differential equation

$$\frac{d^2y}{dx^2} + (\lambda + \varepsilon x)y = 0$$

subject to the boundary condition

$$y(0,\lambda)\cos\alpha + \frac{dy}{dx}(0,\lambda)\sin\alpha = 0$$

If $\cot\alpha > 0$ the unperturbed equation ($\varepsilon=0$) has a single negative eigenvalue at

$$\lambda = -\cot^2 \alpha$$

and purely continuous spectrum for positive λ.

For $\varepsilon \neq 0$ the perturbed equation can be solved in terms of Bessel functions of one-third order and it possesses a purely continuous spectrum $(-\infty,\infty)$. On the other hand the usual perturbation procedure produces the "pseudo-eigenvalue"

$$\lambda_\varepsilon = -\cot^2\alpha - \frac{1}{2}\varepsilon \tan \alpha + O(\varepsilon^2)$$

To interpret this, Titchmarsh noted that there were two Green's functions for both the unperturbed and the perturbed problem. The poles of that for the unperturbed problem give the eigenvalues but even for the unperturbed problem it should be recalled that while the Green's function is analytic in both the upper and lower half plane, the two functions need not be analytic continuations of each other. For the perturbed problem the Green's function for the upper half-plane could be continued to the lower half-plane where, although no longer a Green's function, it has a pole there near the eigenvalue of the unperturbed problem. The pseudo-eigenvalue above is a real number because the distance of it from the real axis is smaller than any power of ε and so cannot appear in any expansion in terms of powers of ε. In fact the complex pole is

$$\lambda_\varepsilon = -\cot^2\alpha - \frac{1}{2}\varepsilon \tan \alpha - [4 \cot^2_\alpha + O(\varepsilon)]\exp[\frac{\pi i}{6} - \frac{4 \cot 3\alpha}{3\varepsilon} - 1]$$

To see the effect of this pole, let

$$\lambda_\varepsilon = a - ib$$

so that the integral in equation (B1.1) becomes

$$\frac{a}{\pi} \int_I |g(\lambda,\varepsilon)|^2 \left[\frac{b}{(\lambda-a)^2 + b^2}\right] d\lambda$$

the integration interval I including the point a. For small b this integral has a sharp maximum at $\lambda = a$ which tends to zero as $b \to 0$, except in the neighborhood of a. The spectrum is thus seen concentrated in the neighborhood of a. Subsequently an abstract definition of spectral concentration was formulated by Conley and Rejto [G.3] and by Kato [G.1]. See also Riddell [B1.2].

McLeod in [G.3] asked the natural question as to whether all examples of spectral concentration were due to poles. While Howland [B1.3] gave some sufficient conditions for this to be the case recently in [B4.2] he has also given an example of spectral concentration where no pole was involved. Further brief comments on this subject will be found in section B.4 in connection with his treatment of the Livšic matrix. Those using the SEM method should be aware of this distinction, particularly if experimental data is involved.

B.2 LOCAL DISTORTION TECHNIQUES

In quantum mechanics analytic continuation is frequently employed. Here and in the next section only the two body problem will be discussed even though many of the techniques have been used for the n-body problem. As is proved, for example, in the book by Taylor [G.4] the two body problem is essentially equivalent to scattering by a potential of a single particle. Usually the scattering operator is defined in a time-dependent fashion in terms of the Møller wave operators but here Schroedinger's operator will be discussed only in the time-independent case. (For the time dependent treatment, the interested reader is referred to the review article by Kato and Kuroda (see [G.5]).) The scattering operators exist provided that, (assuming Hölder continuity except for a finite number of singularities) the potential $g(\underset{\sim}{x})$ behaves at infinity like, $O(r^{-1-\varepsilon})$, as was shown by Kato [B2.1]. It does not exist for Coulomb potentials and more singular ones, as was shown by Dollard [G.5].

For analytic continuation to be possible, even as in the spherically symmetric case, the potential must fall off exponentially. For a compact potential this implies a meromorphic continuation to an entire half-plane while for exponential decay it is limited to a strip half the width of the decay constant. it therefore is sufficient to limit consideration to the results contained in two papers by Ikebe [A1.10, B2.2]. In the first of these, really one of the classical papers in rigorous scattering theory and one which seriously influenced this author and his collaborators, the Schroedinger equation was written as

$$\Delta u + \kappa^2 u = q(\underset{\sim}{x})u \tag{B2.1}$$

(In operator form its self adjoint extension is $H = H_o + V$, where $H_o = -\Delta$ and V is the operator corresponding to the potential.) Here the energy $E = |\underset{\sim}{k}|^2 \equiv \kappa^2$. Subsequently κ may take on complex values. In arriving at equation (B2.1) it was assumed that

$$q = O(r^{-2-\varepsilon}).$$

The discussion was based on the Lippman-Schwinger integral equation

$$\phi(\underset{\sim}{x},\underset{\sim}{k}) = e^{i\underset{\sim}{k}\cdot\underset{\sim}{x}} - \frac{1}{4\pi}\int_{E_3} \frac{e^{i\kappa|\underset{\sim}{x}-\underset{\sim}{y}|}}{|\underset{\sim}{x} - \underset{\sim}{y}|} q(\underset{\sim}{y})\phi(\underset{\sim}{y},\underset{\sim}{k})d^3y$$

Defining a generalized Fourier transform in terms of the distorted plane wave solution of this equation by

$$\hat{f}_g(\underset{\sim}{k}) = \frac{1}{(2\pi)^{3/2}} \text{ l.i.m.} \int_{E_3} \overline{\phi(\underset{\sim}{x},\underset{\sim}{k})} f(\underset{\sim}{x})d^3x$$

and noting that the Fourier coefficients corresponding to bound states are given in terms of the eigenfunctions (and associated eigenvalues μ_n) by the formula

$$\hat{f}_n = \int_{E_3} \overline{\phi_n(\underset{\sim}{x})} f_n(\underset{\sim}{x}) d^3x$$

Ikebe was able to prove that, if M_3 denotes wave number space,

$$||f||^2 = \int_{M_3} |\hat{f}_g(\underset{\sim}{k})|^2 \, d^3k + \sum_1^\infty |\hat{f}_n|^2 \qquad \text{(B2.2)}$$

and

$$Hf = \frac{1}{(2\pi)^{3/2}} \text{ l.i.m.} \int_{M_3} |\underset{\sim}{k}|^2 \phi(\underset{\sim}{x},\underset{\sim}{k}) \hat{f}_g(\underset{\sim}{k}) d^3k + \underset{N \to \infty}{\text{l.i.m.}} \sum_1^N \mu_n \hat{f}_n \phi_n(\underset{\sim}{x})$$

The last two relations are valid for any f in the domain of H, $D(H) = D(H_o)$, if and only if

$$|\underset{\sim}{k}|^2 \hat{f}_g(\underset{\sim}{k}) \subset L_2(M_3) \text{ and } \sum_1^\infty \mu_n^2 |\hat{f}_n|^2 < \infty$$

Ikebe defined

$$U_+ f = \frac{1}{(2\pi)^{3/2}} \text{ l.i.m.} \int_{M_3} \phi(\underset{\sim}{x},\underset{\sim}{k}) \hat{f}_g(\underset{\sim}{k}) d^3k$$

$$U_{-f} = U_{+f}$$

and showed that $U_{\pm}$ were isometric operators with domains $L_2(E_3)$ and range $PL_2(E_3)$ where the projection is defined by the first term in eq. (B2.2). Moreover $U_{\pm}$ were shown to be identical to the Møller wave operators and that the scattering operator as defined by

$$S = U_+^* U_-$$

was unitary.

Bound states were discussed in terms of the homogeneous Fredholm equation

$$T(\kappa)\phi = -\frac{1}{4\pi}\int_{E_3}\frac{e^{i\kappa|\underset{\sim}{x}-\underset{\sim}{y}|}}{|\underset{\sim}{x}-\underset{\sim}{y}|}q(\underset{\sim}{y})\phi(\underset{\sim}{y})d^3y \qquad \text{(B2.3)}$$

by using the auxiliary Banach space of uniformly continuous functions.

In this space the operator $T(\kappa)$ is compact for $Im\kappa>0$ and it has a non-trivial solution if and only if $-\kappa$ is an eigenvalue of H. He also establishes that the resolvent operator

$$(H - \lambda I)^{-1}$$

is of the Carleman type for $Im\lambda \neq 0$ with a kernel which is *real* symmetric like the free-space Green's function.

In the second paper, under the hypothesis, (more than adequate for the purpose here) that

$$q(\underset{\sim}{x}) = O(r^{-3-\varepsilon}),$$

he showed that the Fourier transform of the scattering operator was of the form

$$(\widehat{Su})\,\underset{\sim}{k} = \hat{u}(\underset{\sim}{k}) - i\kappa\int_{\Omega}F(\kappa,\underset{\sim}{\omega},\underset{\sim}{\omega}')\hat{u}(\kappa,\underset{\sim}{\omega}')d^2\omega'$$

where F is related to the transmission coefficient as in the Lax-Phillips theory (note the normalization change), where asymptotically

$$\phi(\underset{\sim}{x},\underset{\sim}{k}) \sim e^{i\underset{\sim}{k}\cdot\underset{\sim}{x}} - \frac{2\pi}{|\underset{\sim}{x}|}e^{i\kappa r}F(|\underset{\sim}{k}|,-\underset{\sim k}{\omega},-\underset{\sim k}{\omega}) + O(r^{-1})$$

In the above, Ω is the set of unit vectors $\underset{\sim}{\omega}' \subset E_3$, and

$$\underset{\sim k}{\omega} = \underset{\sim}{k}/\kappa\,, \quad \underset{\sim x}{\omega} = \underset{\sim}{x}/|\underset{\sim}{x}|$$

His proof involved time-dependent methods and was reasonably difficult. Subsequently Schmidt [A2.11] gave a proof which was similar to that given by Lax and Phillips for obstacle scattering in their book [A2.1]. In any event the result is that the Fourier transform of the scattering operator consists of the identity operator plus an integral operator of the Hilbert-Schmidt type.

Dolph, McLeod and Thoe [B2.3] appear to have been the first to rigorously investigate the solutions of equation (B2.3) in the other half plane $Im\,\kappa < 0$. Since even the solutions of the unperturbed equation ($q=0$) grow exponentially large, the so-called exponential catastrophe, they used a larger Hilbert space with an inner product defined by

$$\langle \phi, \psi \rangle = \int_{E_3} \phi(\underset{\sim}{x})\,\bar{\psi}(\underset{\sim}{x})e^{-\alpha(\underset{\sim}{x})}d^3x$$

$$= \int \phi(\underset{\sim}{x})\,\bar{\psi}(\underset{\sim}{x})dm(\underset{\sim}{x})$$

and where $dm(\underset{\sim}{x}) = e^{-\alpha|\underset{\sim}{x}|}d^3x$, so that the Hilbert space is $L_2[dm(\underset{\sim}{x})]$. The integral equation corresponding to equation (B2.3) is now of the form

$$T(\kappa)\phi = -\frac{1}{4\pi}\int \frac{e^{i\kappa|\underset{\sim}{x}-\underset{\sim}{y}|}}{|\underset{\sim}{x}-\underset{\sim}{y}|}\,q(\underset{\sim}{y})e^{\alpha|\underset{\sim}{y}|}\,\phi(\underset{\sim}{y})\,dm(\underset{\sim}{y})$$

$$= -\frac{1}{4\pi}\int \frac{e^{i\kappa|\underset{\sim}{x}-\underset{\sim}{y}|}}{|\underset{\sim}{x}-\underset{\sim}{y}|}\,q(\underset{\sim}{y})\phi(\underset{\sim}{y})d^3y$$

and it was investigated under the hypothesis that

$$\sup_{r \geq r_o}\left| e^{\alpha|\underset{\sim}{x}|}\,q(\underset{\sim}{x})\right| < \infty, \text{ for some } \alpha, \text{ and } r_o > 0,$$

and that $$\int |q(\underset{\sim}{x})|^2 d^3x < \infty.$$

For $Im\kappa > -\frac{\alpha}{2}$, $T(\kappa)$ was shown to be an analytic family of operators on the Hilbert space just defined whose norm was such that

$$||T(\kappa)|| = O[(Im\kappa)^{-\frac{1}{2}}], \text{ as } |\kappa| \to \infty.$$

The inverse $[I-T(\kappa)]^{-1}$ was shown to be a meromorphic operator-valued function in the strip $Im\kappa > -\frac{\alpha}{2}$ [hence the entire lower half-plane in the event that the potential was compact] with a pole at κ_0 if and only if there existed a ϕ in $L_2(dm)$ such that

$$\phi = T(\kappa_0)\phi$$

In this same paper it was shown that the scattering operator admitted an analytic continuation to the same region and that it was also meromorphic there. It was also shown that the poles were symmetrically placed with respect to the negative imaginary axis. Examples treated by the integral equation method for the radially symmetric case were also formally discussed for the delta function potential, a box potential and an exponentially decaying potential.

Subsequently different proofs of these results were given by Steinberg [A3.7] and Howland [B2.4] and they were extended to the n-dimensional case by McLeod [A3.10] who also showed in [A3.11] that only a finite number of poles could occur in any strip parallel to the real κ axis. This he did by introducing a special coordinate system and integrating by parts in a way reminiscent of that used in the higher Born approximation (See Newton [G2]).

The identification of the poles of the S-matrix and the non-trivial solution of the homogenous Fredholm equation is due to Shenk and Thoe [B2.5]. For simplicity we will state their theorem only in the case that for $Im\kappa < 0, \kappa^2$ is not an eigenvalue. The result is then the following:

1) If $S(\bar{\kappa})^* h = 0$, with $h \neq 0$, and if, by definition,

$$\zeta h(\underset{\sim}{x}) = \int_\Omega h(\underset{\sim}{\omega})\phi_-(\underset{\sim}{x},\kappa,\underset{\sim}{\omega})d^2\omega$$

where the distorted plane wave ϕ_- is a solution of Schroedinger's equation with the asymptotic property

$$\phi_-(\underset{\sim}{x},\kappa,\underset{\sim}{\omega})-e^{i\kappa\underset{\sim}{x}\cdot\underset{\sim}{\omega}} = O[\exp(Im\kappa\ |\underset{\sim}{x}|\ )],$$

then $u(\underset{\sim}{x}) = \zeta h(\underset{\sim}{x})$ is said to be resonant state at κ, with the asymptotic behavior

$$u(\underset{\sim}{x}) \sim \left(\frac{2\pi}{i\kappa r}\right)^{3/2} h(\theta)e^{i\kappa r}$$

Conversely, if $u(\underset{\sim}{x})$ is a resonant state at κ and if $h(\theta)$ is defined by the above asymptotic relation, then

$$S(\bar{\kappa})^* h = 0$$

and

$$[I-T(\kappa)]\zeta(\underset{\sim}{x}) = \int_{\Omega} e^{i\kappa \underset{\sim}{x}\cdot\underset{\sim}{\omega}} \; S(\bar{\kappa})^* h(\underset{\sim}{\omega})\, d^2\omega$$

In [B2.6], Jensen extended the above to operators of the form

$$H = H_o + Q$$

where Q, the interaction, is of the form

$$Q = e^{-\mu r} \, V \, e^{-\mu r}, \quad \mu > 0,$$

where V is a compact self-adjoint operator mapping the Sobolev space H^1 of E_3 into H^{-1} of E_3. His results were obtained by the local distortion techniques developed by Babbitt and Balslev [B2.7]. They were able to continue certain matrix elements of the resolvent $(H-\lambda)^{-1}$ for $\lambda = E = \kappa^2$ on to a subset of the second Riemann sheet (the so-called unphysical sheet in the κ-plane).

Thomas [B2.8] used local distortion techniques for relatively compact perturbations of the form $H = H_o + V$ by working in momentum rather than configuration space. While the definition of a relative compact operator can be found in Kato [G1], the characterization of them due to Combes [B2.9] is sometimes more convenient and for convenience is repeated here:

Let H be a self-adjoint operator on a Hilbert space and V some linear operator whose domain contains $D(H)$. Then V is H-compact (or relatively) compact on D, if and only if for some z in the resolvent set of H the operator $V(H-z)^{-1}$ is compact.

[N.B. If the operator is relatively compact, V is H-ε bounded in the sense of Kato [G1]].

Thomas exhibits a dense set of vectors for which the resolvent matrix elements $\langle\psi,(H-z)^{-1}\psi\rangle$ are meromorphic as z crosses the positive real axis (the essential spectrum[4] of H minus the origin) and travels into the second sheet. His work depends on some of the results to be described in the next section and involves a contour distortion technique which is independent of whether the potential is short range, repulsive, spherically symmetric or dilation analytic (for definition see the next section). In terms of configuration space his results imply that if V is multiplication in L_2 of a function of compact support then the resolvent matrix elements are meromorphic on an, in general, infinite sheeted Riemann surface. His results also contain those obtained by Dolph et. al [B2.3].

B.3 GLOBAL DISTORTION TECHNIQUES

The idea of using global analytic continuation was probably introduced in 1962 in the paper by Bottino, Longoni and Regge [B3.1] and some of the ideas introduced there have led to Regge poles and Regge trajectories. In this paper the radial part of the Schroedinger equation

$$\frac{d^2\psi}{dr^2} + \kappa^2\psi - \frac{\ell(\ell+1)}{r^2}\psi(r) - q(r)\psi(r) = 0$$

was used. The two real parameters ℓ and κ were allowed to be complex. If $\lambda \equiv \ell + \frac{1}{2}$, the resulting poles were classified as follows:

i) λ real, κ complex. If $\kappa = i\eta$, η real and $\eta > 0$, one has bound states. If $\kappa = -i\eta$, they were called anti-bound or virtual states.

---

[4] See Kato ([G1], p. 243) for definition. In the sequel, the word *essential* will be omitted.

ii) If $Im\kappa < 0$, and not purely imaginary, they were called resonances which were shown to occur in pairs of conjugate poles for the S matrix $S(\lambda,\kappa)$; this corresponds to the situation discussed in the last section and is the only one to be discussed here. For completeness, however, the case where κ is real λ complex should be noted:

iii) $Im\lambda \neq 0$ and $\kappa \gtrless 0$, these poles have been called shadow states. Considered as a function of the two complex variables κ and λ, resonances and shadow states are particular intersections of the same singular surface of $S(\lambda,\kappa)$ since analytic functions of two complex variables are never singular on isolated points but always on analytic surfaces of dimension 2.

Without separation of variables the first global results are due to Lovelace [B3.2], who was able to discuss global continuation for a superposition of Yukawa potentials in momentum space. This was followed by the pioneering papers of Combes [B3.3] and Aguilar and Combes [B3.4] for the two body problem and the paper by Balslev and Combes [B3.5] for the n-body case. This work has also provided the necessary background for that of Thomas [B2.8] described in the last section. One needs adequate structure to be sure that use of matrix elements have an invariant significance. Up to now this additional structure has been provided by either the scattering operator or by a Fredholm integral equation. As Howland [B1.3] has noted explicitly, without sufficient structure one can always find a function f in a Hilbert space for which the matrix element

$$\langle (H-z)^{-1}f, f \rangle = \int_{-\infty}^{\infty} \frac{|f(z')|^2}{z'-z} dz'$$

has a pole at z if, as is usually the case,

$$Hf = zf$$

Global distortion techniques use the structure provided by

Nelson's theorem [B3.6] as the starting point. He defined, for any linear operator A in any Banach space, an analytic vector f to be one for which the series $\sum_{0}^{\infty}(|\theta|^n/n!)||A^n f||$ converges. He then proved that if U was a unitary representation of a strongly continuous Lie group on a Hilbert space, the set of analytic vectors for U would be dense in the Hilbert space.

In all the applications known to the author, the group has been chosen to be the linear group.

$$\{(\underset{\sim}{\tau},z): \underset{\sim}{\tau} \subset E^{3n-1}, z \subset [0,\infty])\}$$

with the group law

$$\{U(\underset{\sim}{\tau},z)\phi\}\underset{\sim}{x} = z^{(3n-1)/2} e^{-i<\underset{\sim}{x},z\underset{\sim}{\tau}>} \phi(z,\underset{\sim}{x})$$

in which the inner product $<,>$ is in configuration space. If $z = 0$, this leads to the dilation group, and if $z = 1$ to the so-called boost group. Since only the dilation group has been used in the two body case, considerations here will be limited to it. Use of both the boost and dilation group can be found in the papers by Combes and Thomas [B3.7] and Combes [B3.8], [G6].

Important here is the concept of relative $(-\Delta)$- compact operators which admit continuation with respect to the dilation parameter θ. Precisely, a dilation analytic vector ψ is one for which $V(\theta)\psi$ admits an analytic continuation into a strip $|Im\theta|<a$ for those analytic vectors associated with the generator

$$A = \frac{1}{2i}(\underset{\sim}{x} \cdot \nabla + \nabla \cdot \underset{\sim}{x}) \tag{B3.1}$$

of the dilation group.

In n-dimensional Euclidean space if $U(\theta)$ is the strongly continuous unitary representation in a Hilbert space of the dilation group on E^n, then

$$(U(\theta)f)(x) = e^{\frac{n\theta}{2}} f(e^{\theta}x), \theta \subset R, f \subset L^2(E^n)$$

Moreover, for V and H_o compact operators from the domain of H_o to $L_2(E^n)$, define $V(\theta) = U(\theta)V\,U(\theta)^{-1}$ for θ any real number. A dilation analytic perturbation is defined on those H_o -compact operators V such that $V(\theta)$ has a H_o-compact analytic continuation in an open connected domain of the complex plane such that

$$V(\bar{\theta})^* = V(\theta).$$

Then if
$$R(z) \equiv (H_o + V - z)^{-1}$$

and if ϕ is any vector from the dense set of analytic vectors defined by Nelson's theorem, the matrix element $<\phi, R(z)\phi>$ has a meromorphic continuation through the positive real axis from above into the half lines arg $z = -2a$, where $-a < Imz < a, a > 0$, denotes the domain to which $V(\theta)$ can always be extended analytically in θ. If $R_o(z) \equiv [H_o(z)-z]^{-1}$, then it easily is seen that

$$U(\theta)R_o(z)U(\theta)^{-1} = e^{2\theta}R_o(e^{2\theta}z)$$

and that
$$<\phi, R(z)\phi> = <\phi, R_o(z)\ [I + VR_o(z)]^{-1}\phi>.$$

Suppose now that ϕ is any vector in the dense set, and define

$$\Theta^{\varepsilon} = \{z|Imz > \varepsilon,\ \varepsilon = -\tfrac{1}{2}\arg z\}$$

where z is in the analytic domain of $V(\theta)$. Then for θ in this domain

$$H(\theta) = V(\theta)\ HV(\theta)^{-1} = e^{-2\theta}H_o + V(\theta)$$

and this in turn implies that

$$\{e^{-2\theta}H_o + V(\theta)-z\}^{-1} = e^{2\theta}R_o(e^{2\theta}z)\ [1+e^{2\theta}V(\theta)\ R_o(e^{2\theta}z)]^{-1}.$$

If
$$\psi_z(\theta) \equiv <\phi, R(z)\psi>$$

then for a fixed θ contained in the domain of anlyticity $Im\theta > \theta$, $\psi_z(\theta)$ is a meromorphic function of z in

$$\{z | e^{2\theta} z \subset [Re z > 0, \quad Im z > 0]\}.$$

Moreover, $\psi_z(\theta)$ defines an analytic continuation into a domain containing the real axis from above.

It can be shown that the poles of the meromorphic function $\psi_z(\theta)$ do not depend upon θ and can accumulate at most at zero. In fact, because of unitary equivalence the spectrum of $H(\theta)$, for $\theta = \alpha + \beta i$ depends only on β in $|\beta| < a$, where $\arg z < -2a$, and the spectrum $[0,\infty]$ of H_o in the two body case is rotated to the spectrum $e^{-2i\beta}R^+$, where R^+ denotes the positive real axis of $H(\theta)$. The real eigenvalues of $H(\theta)$ remain the same as those of $H = H(0)$ while non-real eigenvalues (poles) depend only on $Im\ \theta$. Complex eigenvalues which are isolated from the spectrum of $H(\theta)$ lie in the spectrum of $H(\theta')$, if $Im\theta'$ is sufficiently near $Im\ \theta$.

The over-all picture then consists of the following: For $Im\,\theta = 0$ one has an infinite spectrum beginning at zero (the only threshold point in the two body case) and a set of negative eigenvalues. As $Im\ \theta$ is turned up from 0, the eigenvalues stay fixed but the continuous spectrum swings into the lower half-plane and as it does so it can uncover some complex eigenvalues. In the n-body case, new thresholds (like zero in the two body case) can appear and isolated points embedded in the continuous spectrum may appear as isolated points of the spectrum of $H(\theta)$. In fact this results from the continuation of the matrix elements $\psi_z(\theta)$ to the second sheet.

Two natural questions arise from this interesting rotation and uncovering, and possibly recovering, of the complex eigenvalues. Can one concretely characterize the class of dilation analytic potentials, and why did one choose the dilation group to begin with? The first now has a concrete answer, at least when V is a multiplicative potential. For the radial symmetric case a characterization was given by Simon [B3.10] and recently Babbitt and Balslev in [B3.11] have obtained one for the general multiplicative case. Let S_a denote the set $\{(\rho e^{i\phi}), 0 < \rho < \infty,$

$|\phi| < a\}$. Then dilation analytic vectors are the restriction to R^+ of those functions f from S_a to $L_2(\Omega)$ with the property that

$$\sup_{-a+\varepsilon\leq\theta\leq a-\varepsilon} \int_0^\infty \rho^2 |f(\rho e^{i\theta})|^2 d\rho < \infty$$

This last characterization is closely related to the class of functions considered by Van Winter in [B3.12, B3.13, B3.14, B3.15] for the n-body problem and in fact Babbitt and Balslev are able to show that the class of interactions considered by her in the n-body case are less general than those for dilation analytic potentials. Further details will be omitted here.

The matter concerning the second question is more obscure. In a conversation with the author, Combes was unable to remember why, among the infinite number of possibilities covered by Nelson's theorem, he picked the dilation group. A possible reason could lie in the remarks of Lavine [B3.18]. Classically the quantity $\mathfrak{A} = \underset{\sim}{x} \cdot \underset{\sim}{p}$, $\underset{\sim}{p}$ being momentum, is conserved in scattering and if,

$$H = p^2 + V,$$

then the quantum analogue of the Poisson bracket for $\mathfrak{A}$ and H involves A, the generator of the dilation group, equation (B3.1). At any rate this choice has produced most interesting and novel results which have great significance. Hopefully something similar could be done in the Singular Expansion Method.

B.4. THE M.S. LIVŠIC METHOD IN PERTURBATION THEORY

Howland in [B4.1] and [B4.2] succeeded in unifying many of the results described above via the matrix first introduced by Livšic [B4.3]. Physical applications of Livšic's approach can be found in [B4.4, B4.5, B4.6] and reference should also be made to Brodskii [B4.6] and Dolph and Penzlin [G11]. The general theory is quite complicated and so the treatment here will be sketchy.

Before describing Howland's approach in detail the comment of Arlen Brown in a lecture given in 1962, namely, that Livšic's general theory could be motivated by a generalization of the Hilbert resolvent relation, should be noted. Explicitly if

$$R(z) = (A-z)^{-1}$$

then
$$R_z - R^*_{z_1} = R^*_{z_1}(A^* - A - \bar{z}_1 + z)R_z$$

$$= (z - \bar{z}_1)R^*_{z_1} R_z - iR^*_{z_1}(A - A^*)R_z$$

$$= (z - \bar{z}_1)R^*_{z_1} R_z - 2iR^*_{z_1} Im(A)R_z$$

For $A = A^*$, and $R^*_{z_1} = R_{\bar{z}_1}$ (the self-adjoint case), the above reduces to the usual Hilbert relation and one sees very clearly how non-self-adjointness enters, at least for bounded operators.

Howland considers $R(z,\varepsilon) = (H_\varepsilon - z)^{-1}$ in which H_ε is a family of self-adjoint operators on a Hilbert space, where $H_\varepsilon = H_o + \varepsilon V$. Now as is well known if λ_0 is an isolated eigenvalue of H_o of multiplicity m the above resolvent will be analytic in ε and the usual perturbation method will produce m eigenvalues λ_ε of H_ε, which are analytic functions of ε (See Kato [G1]). It may happen, as in the example of Titchmarsh discussed in section B.1, that H_ε may have only a continuous spectrum for $\varepsilon \neq 0$. However if λ_0 is an embedded eigenvalue in the continuous spectrum, $R(z,\varepsilon)$ may still be analytic for sufficiently small perturbations V. Up until this treatment in [B4.2], most authors including Howland have used the theory of spectral concentration to treat the case of the embedded eigenvalue. As in section B.1, spectral concentration is said to occur if there exists sets J_η of appropriate small measure whose strong limit

$$\lim_{\varepsilon \to 0} E_\varepsilon(J_\eta) = P$$

P being the orthogonal projection on the null space of $(H_o - z)$, and where E_ε is the spectral resolution of the identity occuring in

$$H_\varepsilon = \int_{-\infty}^{\infty} \lambda \, dE_\varepsilon(\lambda)$$

In the regular case, as in Titchmarsh's example, one can take J_η to be the *m*-eigenvalues of H_ε near λ_0. [Once again the reader should be cautioned that spectral concentration has a unitary invariant asymptotic property only for a *family* of operators.] In every case treated by Titchmarsh, spectral concentration could be associated with a second-sheeted pole of the matrix element $\langle R(z,\varepsilon)\phi,\phi\rangle$. Here sufficient structure was furnished by the fact that H is an analytic family in ε and the fact that $\lambda(\varepsilon)$ is a pole of $\langle R(z,\varepsilon)\phi,\phi\rangle$ which is analytic in ε with $\lambda(0) = \lambda_0$. Thus for ε in a suitable sector of the complex plane, $\{\lambda(\varepsilon)\}$ falls on the first sheet and is therefore a point eigenvalue of the non-self-adjoint operator H_ε so that $\{\lambda(\varepsilon)\}$ has a unitary invariant significance for H_ε.

In the theory of Livšic, the resolvent is replaced by a matrix. If M is a finite dimensional, closed subspace of the Hilbert space $\mathfrak{H}$, M being its orthogonal complement, and if P is the orthogonal projection onto M then the compression $P(H-z)^{-1}P$ of $R(z) = (H-z)^{-1}$ to M is meromorphic on the complement of the continuous spectrum of H and it is also an invertible operator for $Im z \neq 0$. To see this assume that

$$P\,R(z)\,Px = 0, \text{ for some } x \subset M$$

$$x = Pf, \quad f \subset \mathfrak{H}.$$

Then $\quad \langle Rx,x \rangle = \langle RPf, Pf \rangle = \langle PRPf,f \rangle = 0$

and hence $\quad 0 = \langle R(z)x,x \rangle = \eta \left| [(H-\lambda)^2 + \eta^2]^{-\frac{1}{2}} x \right|^2,$

where $z = \lambda + i\eta$, so that $x = 0$. The formula

$$[B(z)-z]^{-1} = PR(z)P$$

therefore defines an operator $B(z)$ on M which is meromorphic in z in the complement of the continuous spectrum of H and moreover has only *real* singularities. Howland defines $B(z)$ as the Livšic matrix which has the property that $B(\bar{z})^* = B(z)$. If M is contained in the domain of the operator H, then H may be written as the matrix

$$H = \begin{pmatrix} T & \Gamma \\ \Gamma^* & A \end{pmatrix}$$

on the Hilbert space $\mathfrak{H} = M \oplus M^{\perp}$. Here $T = T^*$ on $M^{\perp}$, $A = A^*$ on M, and Γ is a bounded map of M into $M^{\perp}$. This follows from a detailed calculation for $H = H_o + V$ where $H_o = T \oplus A$, and $V = \Gamma + \Gamma^*$. If $R_T(z) = [T-z]^{-1}$ and $R_A(z) = [A-z]^{-1}$ then $B(z) = A - \Gamma^* R_T(z)\Gamma$, for z not in the (essential) spectrum of H. [Moreover it follows that $B(z)$ is a dissipative operator in $Im\ z > o$ -see Lax-Phillips [A2.1] or Kato [G1] for definition.]

If $B(z) = B$ is identically constant then the Livšic matrix yields the Lax-Phillips theory. In particular it follows that the spectral points of B are the poles of the S matrix as defined in section A.2.

If $B(z)$ is not identically constant and if it admits an analytic continuation from $Im\ z > 0$ to some domain D intersecting the real axis and $W(z) = B^+(z) - z$, the + implying continuation from above, then $PR(z)P$ has the meromorphic continuation $W^{-1}(z)$ from $Im\ z > 0$ to D, with poles at the zeros of $\det W(z)$. These poles Howland defines as the resonances of H on M.

The theory of regular non-isolated eigenvalues can be based on the above. If H_ε is a family of operators defined for $|\varepsilon| < \delta$, which is self-adjoint when ε is real, the perturbation will be called a regular perturbation if and only if $B(z,\varepsilon)$ admits an analytic continuation from the upper half-plane $Im\ z > 0$ to a neighborhood of λ_0; the continuation being analytic in (z,ε) near $z = \lambda_0$ and $\varepsilon = 0$. Under these circumstances there are (counting multiplicities) exactly m zeros of the function

$$\det [B(z,\varepsilon) - z]$$

and they may be labeled in such a way that each zero has a Puiseux series in ε.

It is usually difficult to establish the analyticity of $B(z,\varepsilon)$, since $[T-z]^{-1}$ is not known explicitly. However in the now standard case that a factorization is possible; that is if $H = H_o + \varepsilon FG$, then, if one defines $Q(z) = G(T-z)^{-1}F$, and if it is bounded and continuous and admits a continuation to the second Riemann sheet, then

$$B(z,\varepsilon) = \lambda_0 \; I_m + \varepsilon PF \; [I + \varepsilon Q(z)]^{-1} GP$$

will be analytic in (z,ε).

The Livšic matrix can also be used for dilation-analytic vectors as described in section B.3. If λ_0 is an eigenvalue of $H_o = T + V_o$ of multiplicity m [in the n body case it is necessary that it is not a threshold], then for the non-self-adjoint operator

$$H_\varepsilon(\theta) = e^{-2\theta} \; T + V_o(\theta) + \varepsilon \; V_1(\theta)$$

it can be shown that the Livšic matrix $B(z,\varepsilon)$ is essentially the same as that for H_ε (θ). For θ near zero, the projections on M (θ) will be close to the projection on $M(0)$, so that the associated Livšic matrices will be regular also [See Simon [B3.10]].

Howland also gives an example showing that spectral concentration can occur for an analytic family with an embedded eigenvalue for which there are no resonances at $\varepsilon = 0$ and thus it might happen that knowledge of the poles would be inadequate for the SEM method, particularly if experimental data were involved.

This author continues to hope that Howland can be persuaded to write an expository paper which would make his great insight into the method of Livšic more accessible to other than the specialist in mathematical perturbation theory. Certainly should this come to pass, one conjectures that the Livšic method would in time be used to advantage in electromagnetic theory.

Part II

R.A. Scott

§2.1 TRANSFORMED SOLUTIONS

The writer is aware that many of the prospective readers of this section are not fully conversant with the field of elastodynamics. As an aid to such an audience, the review will focus on a concrete problem. The setting chosen is that of an isotropic, homogeneous elastic layer perfectly bonded to an underlying isotropic, homogeneous elastic half space, the system being excited by a time-dependent line source applied at the surface of the layer (Fig. 2.1).

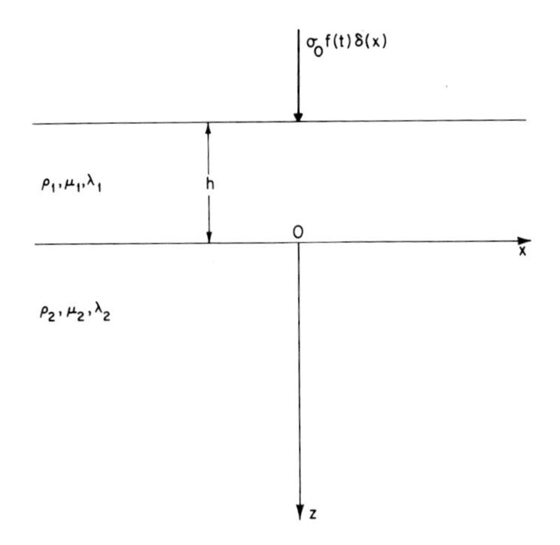

Fig. 2.1. Layer geometry.

For an elastic body, Newton's law of motion takes the form, in the absence of body forces, such as gravity,

$$\frac{\partial \sigma_{ij}}{\partial x_j} = \rho \frac{\partial^2 u_i}{\partial t^2}, \quad i = 1,2,3 \tag{2.1}$$

where x_j are inertial coordinates[1], σ_{ij} are the components of the stress tensor, u_i are the components of the displacement $\underset{\sim}{u}$ of a material point of the body, and ρ is the mass density (assumed constant). For isotropic solids, the constitutive law - Hooke's law - is

$$\sigma_{ij} = \lambda\ (\nabla\cdot\underset{\sim}{u})\delta_{ij} + \mu\left(\frac{\partial u_i}{\partial x_j} + \frac{\partial u_j}{\partial x_i}\right), \quad i,j = 1,2,3 \tag{2.2}$$

where δ denotes the Kronecker delta, and λ and μ are material parameters, constants for homogeneous bodies. Substitution of equation (2.1) into equation (2.2) leads to the vector displacement equation

$$(\lambda + 2\mu)\nabla(\nabla\cdot\underset{\sim}{u}) - \mu\ \nabla x(\nabla\ x\ \underset{\sim}{u}) = \rho\frac{\partial^2 \underset{\sim}{u}}{\partial t^2} \tag{2.3}$$

This vector equation may be solved in terms of potentials. Setting

$$\underset{\sim}{u} = \nabla\phi + \nabla x\underset{\sim}{A} \tag{2.4}$$

it is found that equation (2.3) is satisfied provided

$$\Delta\phi = \frac{1}{c_d^2}\frac{\partial^2\phi}{\partial t^2} \tag{2.5}$$

$$\Delta\underset{\sim}{A} = \frac{1}{c_s^2}\frac{\partial^2\underset{\sim}{A}}{\partial t^2} \tag{2.6}$$

where $c_d^2 = (\lambda + 2\mu)/\rho$, $c_s^2 = \mu/\rho$. It can be shown that the scalar potential ϕ describes longitudinal waves ($\equiv P$ waves), whereas the vector potential $\underset{\sim}{A}$ relates to transverse waves. Though both types of wave can exist independently in an infinite medium, they are in general coupled through interactions at an interface. The class of problems under consideration involve materials, geometries and interface conditions such that the ensuing notions are two dimensional, that is, all quantities are independent of

[1] Where convenient, x,y,z will be used in place of x_1, x_2, x_3 in in the sequel.

the coordinate y, a plane strain condition. Under these circumstances, and a specification of some gauge condition, the vector potential $\underset{\sim}{A}$ has only one component[2], $\underset{\sim}{A} = \psi \underset{\sim}{j}$ say, and equation (2.6) is replaced by

$$\Delta\psi = \frac{1}{c_s^2} \frac{\partial^2\psi}{\partial t^2} \tag{2.7}$$

After using the Laplace transform with respect to time and the Fourier transform with respect to x, equations (2.5) and (2.7) become,

$$\frac{d^2\tilde{\bar{\phi}}_m}{dz^2} + \eta^2_{dm}\,\tilde{\bar{\phi}}_m = 0, \quad m = 1,\, 2, \tag{2.8}$$

$$\frac{d^2\tilde{\bar{\psi}}_m}{dz^2} + \eta^2_{sm}\,\tilde{\bar{\psi}}_m = 0, \quad m = 1,\, 2, \tag{2.9}$$

where

$$\eta_{dm} = \left(\frac{s^2}{c_{dm}^2} + k^2\right)^{\frac{1}{2}}, \quad \eta_{sm} = \left(\frac{s^2}{c_{sm}^2} + k^2\right)^{\frac{1}{2}}, \quad m = 1,\, 2, \tag{2.10}$$

Here a quiescent initial state has been taken and the different potentials arise because of the two elastic media. In the above the bar denotes a Laplace transform, with parameter s and the tilde a Fourier transform. It should be noted from equations (2.10) that branch points occur. Specifying branches with positive real parts, and requiring boundedness as $z \to \infty$ the solutions to equations (2.8) and (2.9) are

$$\tilde{\bar{\phi}}_1 = A \exp(\eta_{d1} z) + B \exp(-\eta_{d1} z) \tag{2.11}$$

---

[2]This corresponds to a vertically polarized shear wave (≡SV wave). In three dimensions, a horizontally polarized shear wave (≡SH wave) also exists but it does not couple into P or SV waves.

$$\tilde{\bar{\psi}}_1 = D \exp(\eta_{s1} z) + E \exp(-\eta_{s1} z) \tag{2.12}$$

$$\tilde{\bar{\phi}}_2 = F \exp(-\eta_{d2} z) \tag{2.13}$$

$$\tilde{\bar{\psi}}_2 = G \exp(-\eta_{s2} z) \tag{2.14}$$

The boundary conditions for the problem are

$$\left.\begin{aligned} \sigma_{zz1} &= -\sigma_o f(t)\,\delta(x) \\ \sigma_{zx1} &= 0 \end{aligned}\right\} z = -h \tag{2.15}$$

where $\delta(x)$ denotes the Dirac delta function, and σ_o is a constant.

The interface conditions are

$$\left.\begin{aligned} \sigma_{zz1} &= \sigma_{zz2} \\ \sigma_{zx1} &= \sigma_{zx2} \\ u_{z1} &= u_{z2} \\ u_{x1} &= u_{x2} \end{aligned}\right\} z = 0 \tag{2.16}$$

Transforming equations (2.2), (2.4), (2.15) and (2.16), the six constants in equation (2.11) through (2.14) can be determined from the equations

$$A\mu_1 \left(\frac{s^2}{c_{s1}^2} + 2k^2\right)\exp(-\eta_{d1}h) + B\mu_1 \left(\frac{s^2}{c_{s1}^2} + 2k^2\right)\exp(\eta_{d1}h)$$

$$- 2Dik\mu_1\eta_{s1} \exp(-\eta_{s1}h) + 2Eik\mu_1\eta_{s1} \exp(\eta_{s1}h) = \frac{\sigma_o}{\sqrt{2\pi}} \bar{f}(s) \tag{2.17}$$

$$-2Aik\eta_{d1} \exp(-\eta_{d1}h) + 2Bik\ \eta_{d1} \exp(\eta_{d1}h) - D\left(\frac{s^2}{c_{s1}^2} + 2k^2\right) \times$$

$$\exp(-\eta_{s1}h) \ - \ E\left(\frac{s^2}{c_{s1}^2} + 2k^2\right)\exp(\eta_{s1}h) = 0 \qquad (2.18)$$

$$-ikA - ikB - \eta_{s1}D + \eta_{s1}E + ikF - \eta_{s2}G = 0 \qquad (2.19)$$

$$\eta_{d1}A - \eta_{d1}B - ikD - ikE + \eta_{d2}F + ikG = 0 \qquad (2.20)$$

$$- 2Aik\mu_1\eta_{d1} + 2Bik\mu_1\eta_{d1} - D\mu_1\left(\frac{s^2}{c_{s1}^2} + 2k^2\right) - E\mu_1\left(\frac{s^2}{c_{s1}^2} + 2k^2\right)$$

$$- 2F\mu_2 ik\eta_{d2} + G\mu_2\left(\frac{s^2}{c_{s2}^2} + 2k^2\right) = 0 \qquad (2.21)$$

$$A\mu_1\left(\frac{s^2}{c_{s1}^2} + 2k^2\right) + B\mu_1\left(\frac{s^2}{c_{s1}^2} + 2k^2\right) - 2Dik\mu_1\eta_{s1} + 2Eik\mu_1\eta_{s1}$$

$$- F\mu_2\left(\frac{s^2}{c_{s2}^2} + 2k^2\right) - 2Gik\mu_2\eta_{s2} = 0 \qquad (2.22)$$

Thus, in principle, the solutions in the transformed domain are determined. Appeal will be made to equations (2.17) through (2.22) in the various subcases to follow.

§2.2 CAGNIARD-DE HOOP TECHNIQUE

A powerful technique for inverting multi-integral transforms and which has received widespread use in elastodynamics is the Cagniard-de Hoop method. The original concepts can be found in the text by Cagniard [2.1]. For a more modern version, see Achenbach [2.2]. The technique will be reviewed here for the simpler geometry of a half-space, which can be simulated by taking A, D and h to be zero in equations (2.17) and (2.18). Then solving for E and B and substituting the results into equations (2.11) and (2.12) yields

$$\tilde{\bar{\psi}} = (\sigma_o \bar{f}(s)/\sqrt{2\pi\mu})\frac{2ik\eta_d}{M_-}e^{-\eta_s z},\quad \tilde{\bar{\phi}} = (\sigma_o \bar{f}(s)/\sqrt{2\pi\mu})\frac{\left(\frac{s^2}{c_s^2}+2k^2\right)}{M_-}e^{-\eta_d z}$$

where (2.23)

$$M_{\pm}(k,s) = \left(\frac{s^2}{c_s^2} + 2k^2\right)^2 \pm 4k^2\left(\frac{s^2}{c_d^2}+k^2\right)^{\frac{1}{2}}\left(\frac{s^2}{c_s^2}+k^2\right)^{\frac{1}{2}} \tag{2.24}$$

Inverting the Fourier transform, and changing to polar coordinates via $z = r\cos\theta$, $x = r\sin\theta$, with θ being measured from the z-axis, gives

$$\bar{\psi}(r,\theta,s) = \bar{f}(s)\int_{-\infty}^{\infty} k\frac{N(k,s)}{M_-(k,s)}\exp(-\eta_s r\cos\theta - ikr\sin\theta)dk \tag{2.25}$$

where
$$N(k,s) = \frac{\sigma_o i}{\pi\mu}\left(\frac{s^2}{c_s^2}+k^2\right) \tag{2.26}$$

Note that the subscript designating an elastic solid has been dropped. The first key step in the method is the variable change

$$k = \frac{s}{c_d}\zeta \tag{2.27}$$

Then equation (2.25) becomes

$$\frac{\bar{\psi}(r,\theta,s)}{\bar{f}(s)} \equiv \bar{I}(s) = \int_{-\infty}^{\infty}\zeta\frac{g(\zeta)}{R(\zeta)}\exp\left\{-s\left[\left(\zeta^2+\frac{c_d^2}{c_s^2}\right)^{\frac{1}{2}}\cos\theta + i\zeta\sin\theta\right]\frac{r}{c_d}\right\}d\zeta$$

where (2.28)

$$g(\zeta) = \frac{\sigma_o i}{\pi\mu}(1+\zeta^2) \tag{2.29}$$

$$R(\zeta) = \left(\frac{c_d^2}{c_s^2}+\zeta^2\right)^2 - 4\zeta^2(1+\zeta^2)^{\frac{1}{2}}\left(\frac{c_d^2}{c_s^2}+\zeta^2\right)^{\frac{1}{2}} \tag{2.30}$$

Note that in the above, s was taken to be real so that the contour remained unchanged. The source function $\bar{f}(s)$, and any terms involving s alone that may arise in other cases from the substitution, equation (2.27), can be handled by a convolution integral and attention in the following will be directed towards inversion of the expression $\bar{I}(s)$. Here an important feature should be noted, namely, s appears only in the exponential in the integral in equation (2.28) and merely as a multiplier of a function ζ. This structure, which stems from the fact that N and M_- in equations (2.24) and (2.26) are homogeneous functions of s and k, leads to an immediate interpretation of Laplace kernels later. One of the reasons for writing out equations (2.17) through (2.22) was to exhibit that the same is true for the more general case of the layered solid. Note that the pattern would be destroyed if, as occurs in certain inhomogeneous media, terms such as $(s^2 + k^2 + a^2)^{\frac{1}{2}}$, (a constant), were to appear. However in certain cases, the underlying concept can still be applied (see Karlsson and Hook [2.3]).

Before proceeding, the singularities of the integrand in equation (2.28) must be noted. There are branch points at $\pm i$ and $\pm i\ c_d/c_s$, and since the associated square roots occur in combinations other than products, the associated Riemann surface has four sheets. It is interesting to note that if discussion were confined to the denominator, $R(\zeta)$, a two-sheeted surface would suffice. As mentioned before, at this stage attention has been restricted to the sheet on which $Re\eta_d > 0$ and $Re\eta_s > 0$ $(\equiv (++)$ sheet). The associated branch cuts are shown on Fig. 2.2. There are also simple poles where $R(\zeta) = 0$. Though six in number, only two, denoted by $\pm ic_d/c_R$ lie on the (++) sheet. (These two also lie on the (--) sheet). These correspond to surface waves, called Rayleigh waves. They are nondispersive waves traveling parallel to the surface with constant phase velocity c_R and decaying exponentially into the interior.

The next step is the introduction of the new variable

$$T = \left(\zeta^2 + \frac{c_d^2}{c_s^2}\right)^{\frac{1}{2}} \cos\theta + i\zeta\sin\theta \qquad (2.31)$$

so that equation (2.28) may be written

$$\bar{I}(s) = \int_{-\infty}^{\infty} \left\{ \frac{\zeta g(\zeta)}{R(\zeta)} \frac{d\zeta}{dT} \right\} e^{-sT} dT \tag{2.32}$$

Now the central question in the Cagniard-de Hoop method is posed: can the original contour of integration - the real axis - be traded for a path(s) in the ζ-plane on which T has the properties of time, that is, it is real, monotonic increasing and has an infinite range? If so, then the integral in equation (2.32) is a Laplace-type integral and the inverse by inspection, is essentially

$$I(T) = \frac{\zeta(T) \quad g[\zeta(T)]}{R[\zeta(T)]} \frac{d\zeta}{dT}(T).$$

Happily, such paths do exist. THey are hyperbolas and are shown in Fig. 2.2. The upper one corresponds to negative values of T and so will not be further considered. The lower path intersects the imaginary axis at the point $-i\,\frac{c_d}{c_s}\sin\theta$, and in the case

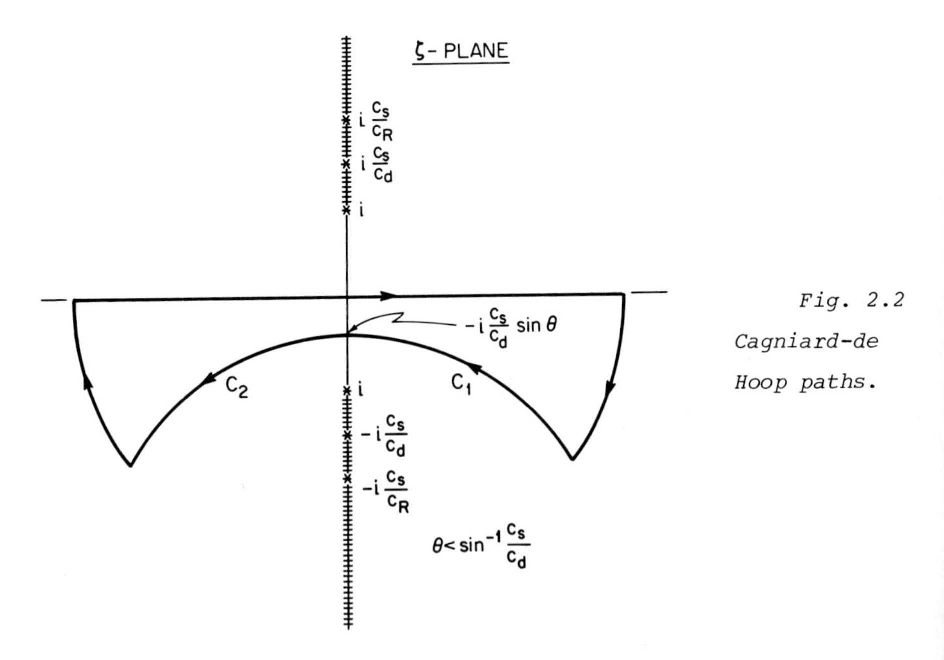

Fig. 2.2 Cagniard-de Hoop paths.

illustrated $\theta < \theta_c \equiv \sin^{-1} \frac{c_s}{c_d}$, an angle whose interpretation will shortly be given. The point of intersection corresponds to $T = r/c_s$. To every positive value of T greater than this, there are two image points in the ζ-plane. As $T \to \infty$, these points sweep out the portions of the hyperbola labeled C_1 and C_2 in Fig. 2.2. An interesting feature is that the path is one of steepest descents, since it has been specified by requiring $ImT = 0$.

The procedure is now as follows. A closed contour is constructed, consisting of the real axis, the Cagniard-de Hoop path $C_1 + C_2$ and circular arcs. The contributions from the latter vanish as their radii become infinite. On noting that on the (++) sheet $g(s^*) = g^*(s)$, where * denotes complex conjugate, C_1 and C_2 can be combined into a single integral. Then using Cauchy's integral theorem,

$$\bar{I}(s) = 2Re\int_{r/c_s}^{\infty} \left\{\frac{\zeta(T)g[\zeta(T)]}{R[\zeta(T)]} \frac{d\zeta}{dT}\right\} e^{-sT}dT$$

so that

$$I(T) = 2Re \left\{\frac{\zeta(T)g[\zeta(T)]}{R[\zeta(T)]} \frac{d\zeta}{dT}\right\} H(T - \frac{r}{c_s}) \qquad (2.33)$$

where H denotes the Heaviside unit step function. A simple computational procedure is to assign, sequentially, the values, of ζ on the hyperbola, calculate the various functions of ζ directly, and calculate the value of T from equation (2.31).

When $\theta > \theta_c$, the Cagniard-de Hoop path intersects a branch cut, as shown in Fig. 2.3. Then, in generating a closed contour additonal segments[3], denoted by L_1 and L_2 occur. The associated values of T, found from equation (2.31) are real, positive and physically meaningful, that is, they are greater than r/c_d.

---

[3]The contribution from the circular segment enclosing the branch point $\zeta = -i$ vanishes in the limit.

Equation (2.32) becomes

$$\bar{I}(s) = 2Re\int_{\frac{r}{c_s}}^{\infty} \left\{\frac{\zeta(T)g[\zeta(T)]}{R[\zeta(T)]}\,\frac{d\zeta}{dT}\right\} e^{-sT}dT$$

$$+ 2Re\int_{T_{ds}}^{\frac{r}{c_s}} \left\{\frac{\zeta(T)g[\zeta(T)]}{R[\zeta(T)]}\,\frac{d\zeta}{dT}\right\} e^{-sT}dT \qquad (2.34)$$

Here $T_{ds} = \frac{r}{c_s}[(\frac{c_d^2}{c_s^2} - 1)^{\frac{1}{2}} \cos\theta + \sin\theta]$, the value of T corresponding to $\zeta = -i$. Inversion of equation (2.34) by inspection yields

$$I(T) = 2Re \left\{\frac{\zeta(T)g[\zeta(T)]}{R[\zeta(T)]}\,\frac{d\zeta}{dT}\right\} H(T - r/c_s)$$

$$+ 2Re\left[\frac{\zeta(T)g[\zeta(T)]}{R[\zeta(T)]}\,\frac{d\zeta}{dT}\right][H(T-T_{ds}) - H(T - \frac{r}{c_s})] \qquad (2.35)$$

The attractive features of the method include: (i) the frequent generation of closed form solutions, (ii) their immediate physical interpretation. The last term in equation (2.35) corresponds to a head, or conical, wave, which is shown in Fig. 2.4. It is generated by the P wave emanating from the source interacting with the free boundary. Note that for source-receiver angles $\theta < \theta_c$ no such event is observed. A general discussion of head waves can be found in the text of Cerveny and Ravindra (2.4).

A comment should be made regarding $\theta = \pi/2$, an angle which yields the surface solution. In this case the hyperbolas degenerate to straight lines along the imaginary axis between $-i$ and $-\infty$. Then indentations over the pole at $-ic_d/c_R$ give a contribution--the Rayleigh wave contribution--to the solution.

When point, instead of line, sources arise, cylindrical symmetry takes over, and Hankel transforms replace Fourier transforms. The Cagniard-de Hoop method is now no longer directly applicable, since the exponentials in the integrands--central features--have been replaced by Bessel functions. In addressing

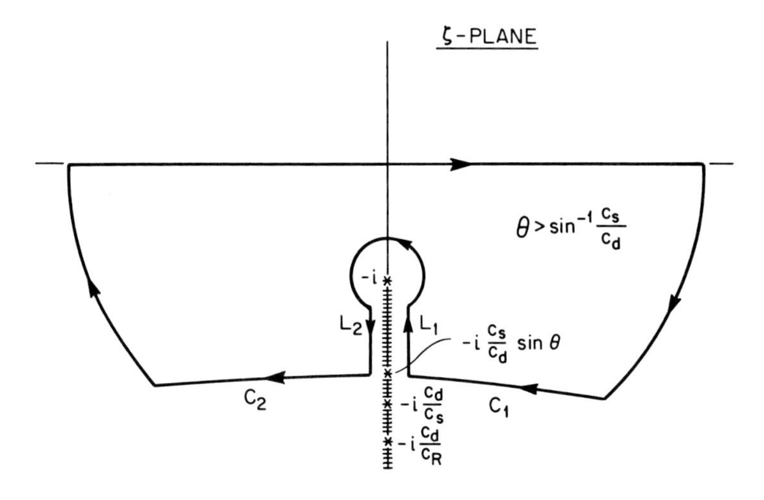

Fig. 2.3. Cagniard-de Hoop paths intersect the branch cut.

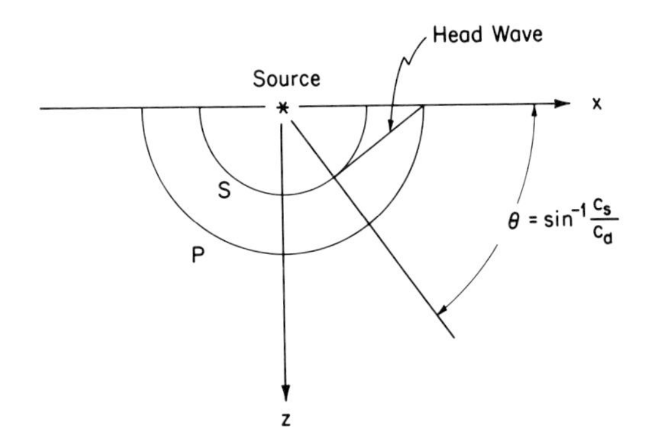

Fig. 2.4. Head wave geometry.

this problem, several different tacks have been taken by elastodynamicists. In one approach, developed by Pekeris, and Aggarwal and Ablow (see Afandi and Scott [2.5] for details), the Bessel functions are expressed in terms of Hankel functions. Then integrals along the real axis are traded for ones along the imaginary axis, thereby generating modified Bessel functions, a profitable step in that they have advantageous Laplace-transform properties. Then a Cagniard-de Hoop like transformation is applied, but in this case the end product is not in closed form. Instead it involves an integral over time. In another approach, used for example, by Eason [2.6], the Bessel functions are replaced by integral representations involving exponentials, but again the result contains integrals. Still another approach uses a far-field approximation. The Bessel functions are replaced by the first term of their large-argument asymptotic expansions (with due caution regarding uniform validity--see Scott and Miklowitz [2.7]). Then expressing the resulting trigonometric functions in terms of exponentials, the technique can be applied immediately.

More complex situations have also been examined. The simple geometrical features of the paths are lost for anisotropic media, but the underlying concepts still carry over (see Scott [2.8]) for details). Layered structures offer a more difficult, but more realistic challenge. They have received considerable attention in both geophysics and mechanics, but will be elaborated upon here, since the literature tends to be incomplete on details and somewhat confusing, a situation not aided by the fact that the approaches in the two fields tend to be different.

The simplest circumstance to consider is a plate subject to the same loads at its top and bottom faces. This can be simulated by considering equations (2.17) and (2.18), and the two equations obtained from them on setting $h = 0$. Solving for the constants and relocating the origin at the plate center for symmetry purposes, one obtains

$$\frac{\tilde{\bar{\psi}}}{\bar{f}(s)} = \frac{\sigma_o}{\sqrt{2\pi}\mu} (2ik\eta_d \operatorname{Sinh} h\eta_d \frac{h}{2} \operatorname{Sinh} h\eta_s z)/D(k,s) \tag{2.36}$$

where (2.37)

$$D(k,s) = (\frac{s^2}{c_s^2} + 2k^2)\operatorname{Cosh} h\eta_d \frac{h}{2} \operatorname{Sinh} h\eta_s \frac{h}{2} - 4k^2\eta_s\eta_d \operatorname{Cosh} h\eta_s \frac{h}{2} \operatorname{Sinh} h\eta_d \frac{h}{2}$$

Following Rosenfeld and Miklowitz [2.9], the hyperbolic functions in equation (2.36) are now written in terms of exponentials, and the denominator is expanded by the Binomial Theorem. Using $z = h/4$ for purposes of illustration one obtains with the aid of equation (2.24)

$$\frac{\tilde{\bar{\psi}}}{\bar{f}(s)} = -\frac{\sigma_o}{\sqrt{2\pi\mu}}\frac{2ik\eta_d}{M_-}\Bigg\{e^{-\frac{3}{4}h\eta_s} + e^{-\frac{1}{4}h\eta_s}$$

$$+ \Gamma\, e^{-\frac{5}{4}\eta_s h} + \ldots - (1+\Gamma)\, e^{-\frac{1}{4}\eta_s h - \eta_d h} + \ldots$$

$$+ (1-2\Gamma^2-\Gamma)\, e^{-\frac{5}{4}\eta_s h - \eta_d h} + \ldots\Bigg\} \tag{2.38}$$

where
$$\Gamma = M_+/M_-$$

On noting that for large s, $\eta_s \sim \frac{s}{c_s}$, $\eta_d \sim \frac{s}{c_d}$, and recalling Tauberian theorems, some of the terms in equation (2.38) have an immediate interpretation in terms of "rays". The first and second terms represent disturbances emanating from the top and bottom surfaces of the plate, respectively, with a wavefront speed of c_s ($\equiv$ shear (S) waves). The third term is a shear wave from the top surface which is reflected from the bottom surface as a shear wave. The fourth term is a P wave[4] from the top reflected from the bottom as an S wave (recall that mode conversion is possible in elastodynamics). Terms like the last one exhibited in equation (2.38) can be confusing. The travel time corresponding to the exponent arises for two different waves, namely (i) S wave from the top, reflecting from the bottom as a P wave, which then reflects from the top as an S wave (ii) P wave from the top reflecting from the bottom as an S wave and then reflecting from the top as an S wave. Reflection (and transmission) coefficients

---

[4] The reader who is uneasy about the presence of longitudinal-type waves in the shear potential ψ, should remember that the current problem involves curved, not plane wavefronts.

provide still more insight. The *plane-wave*, potential reflection coefficient for a P wave reflecting as P and S waves from a stress free boundary is

$$R_{PP}(\alpha,\beta) = \frac{\text{Sin}2\alpha\text{Sin}2\beta - (c_d^2/c_s^2)\text{Cos}^2 2\beta}{\text{Sin}2\alpha\text{Sin}2\beta + (c_d^2/c_s^2)\text{Cos}^2 2\beta} \tag{2.39}$$

where α is the angle of incidence and β is the angle of emergence of the S wave. On making the variable change $\text{Sin}\alpha = \frac{ic_d k}{s}$ and noting Snell's law: $\text{Sin}\beta = (c_d/c_s)\text{Sin}\alpha$, equation (2.39) yields $R_{PP}(k,s) = -\frac{M_+}{M_-} = -\Gamma$. Thus the coefficients in equation (2.38) are seen to be *generalized* reflection and transmission coefficients.

Equation (2.37) can now be rewritten as two series of exponentials, the first representing all rays starting as S events and the second representing all rays beginning as P events

$$\frac{\tilde{\bar{\psi}}}{\bar{f}(s)} = -\frac{\sigma_o}{\sqrt{2\pi\mu}}\frac{2ik\eta_d}{M_-}\left\{e^{-\frac{1}{4}\eta_s h} - R_{SS}\,e^{-\frac{5}{4}\eta_s h} + R_{PS}R_{SP}\,e^{-\frac{5}{4}\eta_s h - \eta_d h} + \ldots\right\} - \frac{\sigma_o}{\sqrt{2\pi\mu}}\frac{\left(\frac{s^2}{c_s^2} + 2k^2\right)}{M_-} \times$$

$$\left\{R_{PS}\,e^{-\frac{1}{4}\eta_s h} - R_{SS}\,R_{PS}\,e^{-\frac{5}{4}\eta_s h - \eta_d h} + \ldots\right\} \tag{2.40}$$

The general pattern is now clear and on inverting the Fourier transforms, one may write

$$\frac{2\pi\mu\bar{\psi}}{\sigma_o\bar{f}(s)} = \sum_S \int_{-\infty}^{\infty} \frac{2ik\eta_d}{M_-} f_n(k)\exp[-sg_n(k,s) - ikx]\,dk$$

$$+ \sum_P \int_{-\infty}^{\infty} \frac{\left(\frac{s^2}{c_s^2} + 2k^2\right)}{M_-} h_n(k)\exp[-sq_n(s,k) - ikx]\,dk \tag{2.41}$$

where g_n and q_n are the ray paths, i.e. expressions such as $-\frac{5}{4}\eta_s h - \eta_d h$, for the rays in question, and f_n and h_n are the appropriate products of generalized reflection and transmission coefficients.

Each term in the series in equation (2.41) can now be inverted by the Cagniard-de Hoop method. It is important to note that in this so-called generalized ray [5] method, solutions can be synthesized without going through the algebraic process outlined here in arriving at equation (2.36), provided the appropriate plane-wave reflection and transmission coefficients are known. A very important issue in the construction of such synthetic seismograms, is the question of which rays contribute at a given station. Hron [2.10] has given a set of rules for the selection of rays for SH waves, a situation where no mode conversion occurs. However no general rules are known for P and SV waves and experience plays a strong role in the selection. For general studies along these lines, the reader is referred to Helmberger et al [2.11- 2.14] and Müller et al [2.15-2.17], and McMechan [2.18].

Finally, it should be noted that problems with spherical symmetry have also been studied on, including work on earth-flattening transformations. See, for example, Helmberger [2.19], Müller [2.20] and Richards [2.21].

§2.3 SINGULARITIES ON "UNPHYSICAL" SHEETS

Current practices in elastodynamics regarding singularities on sheets of a Riemann surface other than the (++) one, will now be described. As should be apparent from the discussion, underlying unifying theories such as those described in part I have yet to be developed, if indeed they exist, and work has been done more or less on a case by case basis.

---

[5] This terminology could be confusing in that the technique does not involve eikonal, or WKB, or related expansions.

As a starting point, the problem considered in section §2.2 will be re-examined using techniques developed by Rosenbaum [2.22] and Phinney [2.23]. Inverting the transforms and specializing for convenience to the case $\bar{f}(s) = 1$, equation (2.23) gives

$$\psi = \frac{i\,\sigma_o}{\pi\mu}\int_{-\infty}^{\infty} ke^{-ikx}\left\{\int_{\gamma-i\infty}^{\gamma+i\infty} \frac{\eta_d}{M_-} e^{-\eta_s z+st}\, ds\right\} dk \qquad (2.42)$$

The singularities of the integrand on the (++) sheet lie on the imaginary axis and so γ in equation (2.42) can be taken to be zero, appropriate indentations around the branch points at $s = \pm ic_s$, $s = \pm ikc_d$ and around the simple poles at $\pm ikc_R$ being understood. Then noting that all the functions involved have the property $f(s^*) = f^*(s)$, the integrals along the negative and positive imaginary axes can be combined so that equation (2.42) becomes

$$\psi = -\frac{2\sigma_o}{\pi\mu}\int_{-\infty}^{\infty} ke^{-ikx}\left\{Im\int_{o}^{i\infty} \frac{\eta_d}{M_-} e^{-\eta_s z+st}\, ds\right\}dk \qquad (2.43)$$

The next step in the procedure is to replace the integral along the imaginary axis in the s-plane by branch line integrals. For clarity, the branch cuts are taken as shown in Fig. 2.5 (one may view BS as collapsing in the limit onto the imaginary axis). Noting that there is no contribution from the integral along the real axis and the arc at infinity, application of the residue theorem to equation (2.42) yields

$$\psi = -\frac{2\sigma_o}{\pi\mu}\int_{-\infty}^{\infty} k\, e^{-ikx}\; Im\left\{\int_{BS} \frac{\eta_d}{M_-} e^{-\eta_s z+st}\, ds\right.$$

$$\left. + \int_{BP} \frac{\eta_d}{M_-} e^{-\eta_s z+st}\, ds - 2\pi i Res\left[\frac{\eta_d\, e^{-\eta_s z+st}}{M_-}\right]_{ikc_R}\right\} dk \qquad (2.44)$$

Here Res is the residue of the integrand at the simple pole

$s = ikc_R$. The branch line integral *BS* can be written as

$$\int_{BS} \frac{\eta_d}{M_-} e^{-\eta_s z+st} ds = \int_{i\infty}^{ikc_s} \frac{\eta_d}{M_-} e^{-\eta_s z+st} ds + \int_{ikc_s}^{i\infty} \frac{\eta_d}{M_-} e^{-\eta_s z+st} ds$$

$$= \int_{ikc_s}^{i\infty} G_S e^{st} ds$$

where

$$G_S \equiv \frac{(\eta_d e^{-\eta_s z})^R M_-^L - (\eta_d e^{-\eta_s z})^L M_-^R}{M_-^L M_-^R} \tag{2.45}$$

Here the superscripts L and R indicate that quantities are to be evaluated at the left and right, respectively, of the branch cut, across which $Im\eta_s$ changes from positive to negative. It is to be understood that the integral in equation (2.45) is along the right side of the branch cut. Making these simplifications, and a similar one for BP, equation (2.44) can be written as

$$-\frac{\pi\mu}{2\sigma_o}\psi = I_S + I_P + I_M \tag{2.46}$$

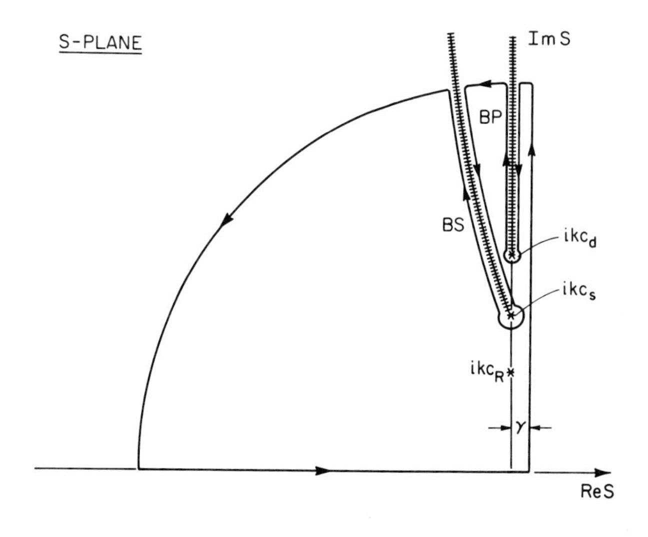

Fig. 2.5 Branch cuts in the s-plane.

where

$$I_S = \int_{-\infty}^{\infty} k\, e^{-ikx}\; Im\left\{\int_{ikc_s}^{i\infty} G_S\, e^{st}\, ds\right\} dk \qquad (2.47)$$

$$I_P = \int_{-\infty}^{\infty} k\, e^{-ikx}\; Im\left\{\int_{ikc_d}^{i\infty} G_P\, e^{st}\, ds\right\} dk \qquad (2.48)$$

and G_P is given by the right side of equation (2.45), with the L and R standing for the left and right side, respectively, of the branch cut running from ikc_d to $i\infty$, and

$$I_M = \int_{-\infty}^{\infty} k\, e^{-ikx} \left\{ -2\pi i Res \left[\frac{\eta_d\, e^{-\eta_s z+st}}{M_-}\right]_{ikc_R}\right\} dk \qquad (2.49)$$

It is at this stage that a key step in the Rosenbaum-Phinney method takes place. Consider the inner integral in I_S, for example. A closed contour is generated which consists of the segment along the right side of the branch cut on the (++) sheet, a circular arc at infinity and a rectilinear segment parallel to the real axis, (denoted by dashed lines in Fig. 2.6), *the latter two lying on the next sheet of the Riemann surface*, i.e., the one for which $Re\ \eta_s < 0$, $Re\ \eta_d > 0$ ($\equiv$(+ −) sheet). In this way another root of the Rayleigh function M_- is exposed. Following Phinney, this "unphysical" root is denoted by PBS2. (general reading of the literature will be facilitated if it is remembered that this root also lies on the (− +) sheet). These poles are usually called leaking mode poles. For steady-state wave propagation they increase exponentially with depth. This apparent "exponential catastrophe" of the time harmonic problem is averted in a time dependent formulation because of the decay in time
for each fixed space point. Note that the occurrence of complex singularities arising from a self-adjoint setting has its genesis in analytic continuation to another sheet, in which the problem

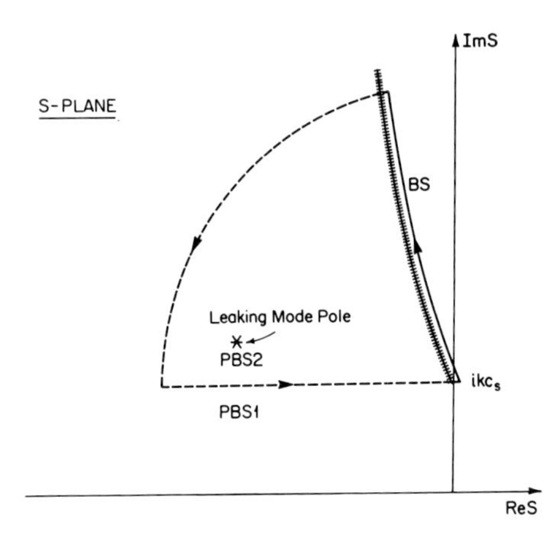

Fig. 2.6. Rosenbaum-Phinney contour modification.

may be viewed as non self-adjoint, a not uncommon procedure (for further discussion of both points, see Dolph [G12]).

The residue theorem can be applied to such "hybrid" contours (see Bliss [2.24], p. 84) and yields

$$I_S = \int_{-\infty}^{\infty} k\, e^{-ikx}\, Im \left\{ \int_{PBS1} G_{SII} e^{st}\, ds \right\} dk$$
$$+ \int_{-\infty}^{\infty} k\, e^{-ikx}\, Im \left\{ 2\pi i Res \left[G_S\, e^{st} \right]_{PBS2} \right\} dk \qquad (2.50)$$

where G_{SII} denotes that the integrand in *PBS1* is to be evaluated on the (+ -) sheet. Note that the range of integration in the second integral is still infinite, since the pole *PBS2* does not migrate to another sheet regardless of the value of k (in contrast to the situation for layered structures).

To investigate integrals such as *PBS1* in equation (2.50), Phinney introduced the transformation

$$q = s - ikc_s \tag{2.51}$$

Using it, and interchanging the order of integration, the first term on the right side of equation (2.50), denoted by I_{S1}, becomes

$$I_{S1} = \int_0^{-\infty} e^{qt} \left\{ \int_{-\infty}^{\infty} G_{SI}\, e^{-ik(x-c_s t)}\, dk \right\} dq \tag{2.52}$$

where, following Phinney, G_{SI} is that function on the (++) sheet which equals G_S at corresponding values of s. Thus the inner integral in equation (2.52) Phinney interpreted as being on the (++) sheet of the k-plane. Next he constructed a closed contour in this latter plane consisting of a circular arc in the upper or lower half-plane, depending on whether $t < x/c_s$ or $t > x/c_s$, respectively. Contributions from the circular arcs are assumed to be zero and consequently the integrals are evaluated in terms of the residues of the poles M_-^L and M_-^R, *both* of which will arise in G_{SI} (see equation (2.44)). The integral *BP* in equation (2.44) can be handled in a similar fashion[6] and the end product in that equation is a complicated interplay between the various poles, with the possibility of their effects canceling out in certain regions. Actually the situation studied by Phinney was more complicated in that since he treated a point load, as opposed to a line load, Bessel functions arose in the place of exponential functions. He replaced them with Hankel functions, discarding the one representing ingoing waves. Also, the limits of the k - integration in equation (2.52) are now 0 to ∞. Closed contours, in either the first or fourth quadrants depending on the value of t, are then constructed by circular arcs at infinity and integrals along the imaginary axis, additional items to be contended with. Phinney approached them by replacing the Hankel functions by their large-

[6] In more complicated cases, this may not be so, since one may want to deform contours onto the (- +) sheet instead of the (+ -) sheet.

argument asymptotic expansions, not always a sound procedure for integrals with limits 0 to ∞. Then using the method of steepest descents, he concluded that for certain times the integrals were negligible compared to the residue terms. Going through the rather complicated "book-keeping" on the poles, Phinney's final conclusion for the half space problem was that retention only of *PBS2* in equation (2.50) gives a good approximation to the complete solution for the time interval $r/c_d < t < r/c_s$, where r is the source - receiver distance. Restriction of the results to a certain time interval, such as between the P and S waves here, seems to be a common feature in attempts to exploit the leaking mode poles.

That Phinney's results are an approximation stems from neglect of the ingoing waves and the treatment of the final integrals along the imaginary axis in the k-plane. The mathematical curiosity of the present authors was first raised by the spectral phenomenon at hand, and they hoped to discern patterns, if they exist, and establish some rigorous results. However, both tasks so far have eluded them. Whether Phinney's result is exact or approximate for the line load is not entirely clear. Chapman [2.25] in discussing a paper of Gupta [2.26], addressed this point. He simply discarded everything but the residue of the pole on the (+ -) sheet, which he then handled by the Cagniard-de Hoop method to get a closed form result. He then made numerical comparisons between this so-called pseudo-wave ($\equiv\bar{P}$wave) and the exact results obtained from the Cagniard-de Hoop technique, getting very good agreement. However the issue is somewhat clouded by the fact that in developing the numerical solutions for the P-pulse, he used an algebraic approximation.

Another interesting numerical exploration of pseudo-waves in the line load problem was given by Gilbert and Laster [2.27]. Using the exact solutions, they plotted surface displacements as function of time for a wide range of the material parameter involved (Poisson's ratio). These plots show that, for certain parameter values, following the P-wave but before the S- wave, there is pulse-like phenomena, fairly poorly defined, that develops with increasing distance from the source (analogous to the Rayleigh wave). They were able to associate this event with the pole on the (+ -) sheet and offered the following explanation: Recall that the point of intersection of the Cagniard-de Hoop path with the branch cuts is a saddle point and corresponds, when η_d arises, to the arrival of the P-wave. Thus thinking in terms of saddle point integration techniques, the proximity of a pole, even if on another sheet of a Riemann surface, is important. If the pole is too close to the saddle point, its effects will merge with the P-wave. As Poisson's ratio increases, the pole migrates away, having less and less effect. In between, the pseudo-wave may be discerned. In another study [2.28] these same authors observed leaking modes in laboratory model experiments.

Considerable complications develop when in place of a half-space, structures consisting of layers on a half space are considered. Even though there are still only four branch points[7], the location of the poles is now determined by transcendental functions of k and s, both of which can be complex variables. See, for example, the case of symmetric waves in a plate, equation (2.37). It is worth noting here that Alsop [2.29] gave a plane wave approach to the leaking mode frequency equation. Tracking down the sheet(s) on which poles lie is a considerably more difficult, but essential, task, in view of the interplay between terms analogous to M_-^L and M_-^R. Also, the poles may not lie on the same sheet for all values of k, so that the range of integration in equations such as (2.50) becomes an issue. For instance, to quote a result of Dainty and Dampney (2.30), a $\bar{P}$ pole starts out at k=0 on the (- +) sheet, but at a certain value of k it crosses both branch points and enters the (+ -) sheet. A further complication in this connection is that only limited areas of the

[7] It is a general rule that layers with a finite thickness do not generate new branch points.

sheets may be exposed in the various contour modifications. The various poles have of course different properties, some leaking S waves and some leaking P and S waves into the substatum. Many investigations into, and uses of, the leaking-mode spectra of layered structures have been made in seismology and mechanics, and these will now be reviewed briefly.

To the seismologist, the ability to correlate observed events on a seismogram with a theory involving material and geometrical parameters is very important. Thus leaking modes, with their sensitivity to P-wave velocity distribution, were a welcome additional tool for the inverse problem, notwithstanding the fact that the numerical effort is considerable, since in inverse problems derivatives with respect to material parameters are required (see Cochran et al [2.31]). A very thorough discussion involving complex frequencies and real wavenumbers was given by Gilbert [2.32] in a review article. He examined (i) two fluid half spaces, (ii) a fluid layer overlying a fluid half space, (iii) a fluid layer on an elastic half space and (iv) a solid layer on an elastic half space. He noted that group velocities, defined as $\frac{d\omega}{dk}$, associated with some of the leaking modes were negative and raised the question, an old and much discussed one, of whether energy could travel with such velocities. Other authors confronting this situation have argued that since the modes are even functions of s and k, there is an image curve in another quadrant for which $\frac{d\omega}{dk}$ is positive and is attainable by a judicious analytic continuation. (See Lloyd and Miklowitz [2.33]). Further early work was done by Laster, Foreman and Linville [2.34] who looked at a layer on a half space and showed the importance of leaking modes for the early part of seismograms, and by Su and Dorman [2.35], who also oonsidered a single layer model. In a series of papers, DeBremaecker et al [2.31,2.36, 2.37, 2.38] presented further studies involving layering. Given in [2.37] is a model consisting of six layers overlying a half space, a more realistic earth model. This type of work was continued by Stalmach and DeBremaecker [2.39], who examined the lattice or terrace like structure exhibited by many of the spectra and the effect of a low P wave velocity channel. Watson [2.40] also contributed to the six layer model, giving real frequency, complex wavenumber data. Dainty [2.41] explored four different earth models, consisting of one or two layers on a half space. Further evidence of the fundamental soundness of the use of leaking modes was given by Dainty and Dampney [2.30], who compared results obtained using them with ones obtained using the generalized ray theory outlined in the previous section and found very good agreement. Another study of the $\bar{P}$ mode on the (+ -) sheet was given by Abramovici [2.42] for a special earth model.

The situation discussed in the preceding paragraphs is actually not the most complicated that can arise in elastodynamics. When two elastic half spaces are considered, with or without intervening layers, eight instead of four branch points arise so that the underlying mathematical structure is an eight sheeted Riemann surface. Looking at just two half spaces, and assuming harmonic time dependency a classical issue has been whether interface waves (called Stoneley waves), i.e., waves propagating without attenuation parallel to the interface and decaying exponentially into both media, exist or not. It was found that such waves exist only for quite restricted classes of solid pairs, but this viewpoint must be tempered in the light of leaking mode poles. Early work on such poles was given by Roever, Vining and Strick [2.43] for a fluid-solid interface and by Podllapollski and Vassillev [2.44] for a solid-solid interface. Evidence that such poles can indeed be correlated with events on seismograms was given by Gilbert and Laster [2.27] who, using exact results obtained by the Cagniard-de Hoop method, exhibited a pulse-like phenomenon that could be related to a root on a sheet other than the (++++) one. Further evidence was supplied by Alterman and Karal [2.45] and Alterman and Loewenthal [2.46], who found such events using finite difference schemes. Recently a very thorough study of these pseudo-Stoneley waves was given by Pilant [2.47], who showed that there were sixteen poles lying on eight Riemann sheets. Considering the density relation $\rho_2 > \rho_1$ he discovered outside the range of classical existence for Stoneley waves the following interesting behavior: (i) if c_{s2} is slightly greater than c_{s1}, then an interface wave attenuated in the direction of propagation arises (ii) if c_{s2} is greater than about three times c_{s1}, energy is propagated along the interface as a Rayleigh wave, and (iii) if $c_{s2} << c_{s1}$ no energy is propagated along the interface. Another recent study is that of Dampney [2.48], who considered a source at the interface between two media. With results obtained using generalized ray theory, he found that the dominant disturbance along the interface was due to the pseudo-Stoneley wave.

Related work in the whole general area is given by Chander et al [2.49], Hirasawa and Berry [2.50], Adler and Sun [2.51], Poupinet and Wright [2.52] and Hill [2.53].

Extensive information on the spectra of plates and other structures has also been available for several years in the mechanics literature, for real frequencies and complex wavenumbers. See for example, the review article of Mindlin [2.54]. To illustrate the complexity of these spectra which could be anticipated for other fields, Fig. 2.7, from [2.54] shows a plot for several modes of real frequency versus real and imaginary wavenumber for equation (2.37). Though in general couched in language different from that of the seismologist[8], mechanicians have also exploited these complex modes. A pioneering work was that of Lloyd and Miklowits [2.33], who considered the excitation of an infinite plate subject to surface loads. On inverting the

---

[8] Analytical continuation is a frequently used device. As general advice the authors can do no better than echo the phrase of Jost "The unphysical region is as bad and as treacherous as any point in the complex plane" (see Nussenzveig, p. 243 [G9]).

Fourier transform first, they were able to write a solution involving integrals over complex branches of the frequency equations. Such branches are important when considering "edge" waves, that is waves whose energy is confined to the vicinity of a boundary. An extremely thorough discussion of the plate spectrum was given by Randles and Miklowitz [2.55]. They found the mappings

$$\chi = (\eta_d - \eta_s) / (\eta_s + \eta_d) , \quad \eta = h(\eta_s + \eta_d)$$

very useful and by means of carefully selected integration paths in the χ-plane were able to obtain good high-frequency representations. This approach was later extended by Randles [2.56] to include anisotropic plates, where the effects of cusped wavefronts were included. Later similar work for a two-layered plate was done by Viano and Miklowitz [2.57]. In quite a different study, Dunkin and Corbin [2.58] considered the role of leaking modes in a problem involving moving loads on a layered elastic half space. Very recent work in the area is that of Stuart [2.59] and [2.60]. He treated the harmonic vibrations of a plate immersed in an acoustic fluid. Even though the plate equation was approximate (Timoshenko-Mindlin equation), poles on several sheets of a Riemann surface still arose. His solution was obtained by saddle-point methods, modified to allow for the close migration of a pole, even if on another sheet. Finally an interesting use of leaking modes should be noted. Alsop et al [2.61] recently used them to obtain approximate reflection and transmission coefficients for a Rayleigh wave encountering a step change in elevation. Similar type of work was done earlier by Wu [2.62] for reflection of harmonic waves from the edge of a semi-infinite plate, and is important in both seismology and in the technology of surface wave devices. Another area that has received considerable attention is the role of complex eigenvalues in elastic surface wave propagation over concave and convex surfaces on which waves of the Rayleigh, Stoneley (and a related type called

Franz waves) and of the Whispering Gallery type can occur. Such a topic warrants a review in itself and the authors can do no more than steer the reader to a recent paper of Frisk and Überall [2.63] as a good reference.

This concludes our survey. It is apparent from it that a rigorous theory of "complex eigenvalues" now exists for the scalar and electromagnetic cases and quantum mechanics, but it is still far from complete. These "complex eigenvalues" admit a semi-group interpretation in the Lax-Phillips theory or alternatively they are poles that arise from analytic continuation of the scattering matrix, an integral operator, or from other mathematical structures having an invariant significance. They lie on the so-called unphysical sheets of the naturally associated Riemann surfaces.

The situation in regard to the influence of branch points and branch cuts is far from being well understood and to date the beginnings of a general theory are yet to emerge. Their extensive use in elastodynamics has been documented above. Also, their use in such different and diverse fields as ocean acoustics (see Labianca [2.64], and Stickler [2.65] who found for his model that the leaky poles did *not* give good approximations) neutron transport theory (See Corngold [2.66]), and Landau poles in plasmas (see Case [2.67]), should not go unnoticed. This conference has also demonstrated their use in electromagnetic theory (see Kuester and Chang [2.68] and Felsen and Marcuvitz [2.69]) and it is hoped that this interdisciplinary review will aid in advancing all areas.

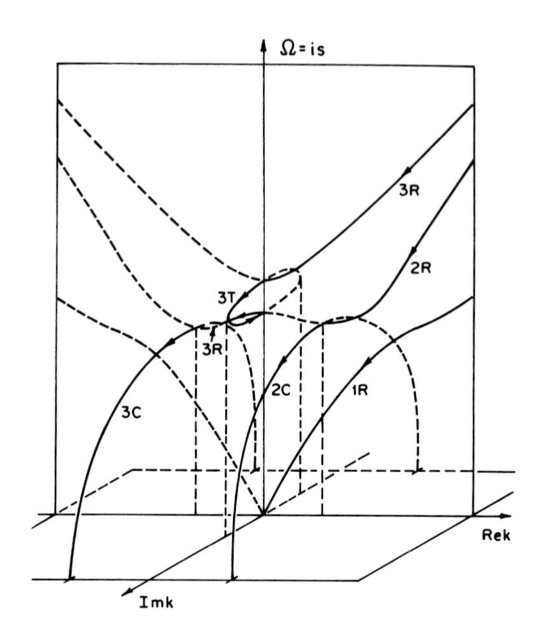

Fig. 2.7. First three symmetric branches of the plate frequency equation.

REFERENCES FOR PART I

A1. VARIATIONAL PRINCIPLES IN THE SCALAR CASE

A1.1. Kerr, D.E., Propagation of Short Radio Waves, Rad. Lab. Ser. MIT, vol. 13, Cambridge, Mass. (1951).

A1.2. Phillips, R.S., Linear Differential Equations of Second Order, Res. Rept. EM 42, New York Univ. (1952).

A1.3. Sims. A.R., *J. Math. Mech. 6*, 247 (1957).

A1.4. MacFarlane, G., *Proc. Camb. Philos. Soc. 43*, 21 (1947).

A1.5. Dolph, C.L., J.E. McLaughlin and I. Marx, *Commun. Pure & Appl. Math. 7*, 621 (1954).

A1.6. Dolph, C.L. and R.K. Ritt, *Math. Zeit. 65*, 309 (1956).

A1.7. Dolph, C.L., *J. Soc. Indust. Appl. Math. 5*, 89 (1957).

A1.8. Weyl, H., *Math. Zeit. 55*, 187 (1952).

A1.9. Müller, C., *Math. Zeit. 56*, 80 (1956).

A1.10. Ikebe, T., *Arch. Ration. Mech. & Anal. 5*, 1 (1960).

A1.11. Morawetz, C.S., *Proc. R. Ir. Acad. A, 72*, 113 (1972).

A1.12. Dolph, C.L., NSF Proposal 1970 and letter to D. Thoe, June 24, 1970.

OTHER REFERENCES

A1.13. Hartree, D.R., et. al., Meteorological Factors in Radio-wave Propagation, *Suppl. Phys. Soc. & R. Meter. Soc.* (1946).
A1.14. Ament, S. and C.L. Pekeris, *Philos. Mag. 38,* 801 (1947).

A.2 RESULTS FROM THE LAX-PHILLIPS THEORY

A2.1 Lax, P.D. and R.S. Phillips, "Scattering Theory", Academic Press, New York (1967).
A2.2. Ralston, J.V., *Commun. Pure & Appl. Math. 22,* 807 (1969).
A2.3. Lax, P.D. and R.S. Phillips, *Commun. Pure & Appl. Math 22,* 737 (1969).
A2.4. Beale, J.T., *Duke Math. J. 41,* 607 (1974).
A2.5. Schmidt, G., "Scattering Theory for Maxwell's Equation in an Exterior Domain", Ph.D. Thesis, Math. Dept., Stanford Univ. (1966).
A2.6. Schmidt, G., A Uniqueness Theorem for the Equation and the Representation of the Potential Scattering Operator, Math. Res. C., Univ. of Wisconsin, Tech. Summary Rept. No. 783 (1967).
A2.7. Schmidt, G., Spectral and Scattering Theory for Maxwell's Equations in an Exterior Domain, Math. Res. C., Univ. of Wisconsin, No. 770 (1967).

OTHER REFERENCES

A2.8. Lax, P.D. and R.S. Phillips, *J. Funct. Anal. 14,* 172 (1973).
A2.9. Lax, P.D. and R.S. Phillips, *Indiana Univ. Math. J. 22,* 101 (1972).
A2.10. R.S. Phillips, On the Exterior Problem for the Reduced Wave Equation, Berkeley Symposium on Partial Differ. Equations, Amer. Math. Soc. (1971).
A2.11. G. Schmidt, *J. Differ. Equations 7,* 389 (1970).
A2.12. Alsholm, P. and G. Schmidt, Spectral and Scattering Theory for Schroedinger Operators, Various Publ. No. 7, Aarhys Univ. (1969).
A2.13. LaVito, J.A., J.R. Schulenberger and C.H. Wilcox, The Scattering Theory of Lax and Phillips and Wave Propagation of Classical Physics, Univ. of Utah Tech. Summary Rept. No. 16 (1971).
A2.14. Beale, J.T., *Commun. Pure & Appl. Math. 26,* 549 (1973).
A2.15. Beale, J.T. and S.I. Rosencrans, *Bull. Am. Math. Soc. 80,* 1276 (1974).
A2.16. Beale, J.T., Spectral Properties of an Acoustic Boundary Condition, preprint, Tulane Univ. (1973).
A2.17. Lax, P.D. and R.S. Phillips *Commun. Pure & Appl. Math. 25,* 85 (1972).

A2.18. Majda, A., *J. Differ. Equations 16*, 515 (1974).
A2.19. Majda, A., *Commun. Pure & Appl. Math. 29*, 261 (1976).

A3. THEOREMS ON THE NATURE AND LOCATION OF POLES

A3.1. Lax, P.D. and R.S. Phillips, *Arch. Ration. Mech. & Anal. 40*, 268 (1971).
A3.2. Ramm. A.G., *Izv. Akad. Nauk. Armjan SSR. Ser. Fiz.-Tekh. (USSR) 3, 6*, 443 (1968).
A3.3. Ramm. A.G., *Soviet Phys. Dokl. 11*, 114 (1966).
A3.4. Beale, J.T., *Commun. Pure & Appl. Math. 26*, 549 (1973).
A3.5. LaVita, J.A., Perturbation of the Poles of the Scattering Matrix, Dept. of Math., Univ. of Denver, MS-R-6903.
A3.6. Steinberg, S., *Arch. Ration. Mech. & Anal. 31*, 372 (1968).
A3.7. Steinberg, S., *Arch. Ration. Mech. & Anal. 38*, 278 (1970).
A3.8. Howland, J.S., *J. Math. Anal. & Appl. 36*, 12 (1971).

OTHER REFERENCES

A3.9. Ladyjenskaya, O.A., *Usp. Mat. Nauk (7.5) 12*, 161 (1957).
A3.10. McLeod, J.B., *Quart. J. Math. 2*, 169 (1967).
A3.11. McLeod, J.B., *Q. J. Appl. Math. 2*, 219 (1967).
A3.12. Haf, H., Zur Theorie Parameterabhangiger Operatorgleichungen, Dissertation, Univ. of Stuttgard, Math. Institut A (1968).
A3.13. Ribaric, M. and I. Vidav., Analytic Properties of the Inverse $A^{-1}(z)$ of an Analytic Linear Operator Function A(z), *Arch. Ration. Mech. & Anal. 32*, 298.

B1. SPECTRAL CONCENTRATION

B1.1. Titchmarsh, E.C., On Some Theorems in Perturbation Theory,
III. *Proc. R. Soc. A. 207*, 321 (1951).
IV. *Proc. R. Soc. A 210*, 30 (1951).
V. *J. Anal. Math. 4*, 187 (1954/1956).
B1.2. Riddell, R.C., *Pac. J. Math. 23*, 377 (1967).
B1.3. Howland, J.S., *Pac. J. Math. 55*, 157 (1974).

OTHER REFERENCES

B1.4. Friedrichs, K.O. and P.A. Rejto, *Commun. Pure & Appl. Math. 15*, 219 (1962).
B1.5. Kato, T., *J. Fac. Sci. Univ. Tokyo 6*, 145 (1951).
B1.6. F. Brownell, *Arch. Ration. Mech. & Anal. 10*, 149 (1962).
B1.7. Howland, J.S., *J. Math. Anal. & Appl. 23*, 575 (1968).
B1.8. Howland, J.S., *Pac. J. Math. 29*, 565 (1969).
B1.9. Howland, J.S., *Am. J. Math. XCI*, 1106 (1969).
B1.10. Howland, J.S., *Trans. Am. Math. Soc. 162*, 141 (1971).
B1.11. Howland, J.S., *Bull. Am. Math. Soc. 78*, 280 (1972).

B2. LOCAL DISTORTION TECHNIQUES

B2.1. Kato, T., Some Results on Potential Scattering, Proc. Int. Conf. Funct. Anal. & Related Topics, Tokyo (1969). Compare also Am. Math. Soc. Summer Inst. on P.D.E. (1971).
B2.2. Ikebe, T., *Pac. J. Math. 15,* 511 (1965).
B2.3. Dolph, C.L., J.B. McLeod and D. Thoe, *J. Math. Anal. & Appl. 16,* 311 (1966).
B2.4. Howland, J.S., *Proc. Am. Math. Soc. 21,* 381 (1969).
B2.5. Shenk, N. and D. Thoe, *J. Math. Anal. & Appl. 37,* 467 (1972).
B2.6. Jensen, A., Local Distortion Technique, Resonances, and Poles of the S. Matrix, to appear in *J. Math. Anal. & Appl.*
B2.7. D. Babbitt and E. Balslev, Local Distortion Techniques and Unitarity of the S-Matrix for the 2 - Body Problem, *J. Math. Anal. & Appl. 54,* 316 (1976).
B2.8. Thomas, L.E., *Helv. Phys. Acta. 45,* 1057 (1972).
B2.9. Combes, J.M., *Commun. Math. & Phys. 12,* 283 (1969).

OTHER REFERENCES

B2.10. Greenstein, D.S., *J. Math. Anal. & Appl. 1,* 355 (1960).
B2.11. Grossman, A., *J. Math. & Phys. 5,* 1025 (1964).
B2.12. Shenk, N. and D. Thoe. See p. 89 of [G5].
B2.13. Nuttall, J., *J. Math. & Phys. 8,* 873 (1967).
See also [G12], [G13], [A1.14] and [A2.12].

B3. GLOBAL DISTORTION TECHNIQUES

B3.1. Bottino, A., A.M. Longoni and T. Regge, *Nuovo Cimento 23,* 954 (1962).
B3.2. Lovelace, C., Three Particle Systems and Unstable Systems, "Scottish Univ. Summer School", (R.G. Moorhouse, ed.) Plenum Press, New York (1963).
B3.3. Combes, J.M., An Algebraic Approach to Quantum Scattering Theory, preprint Marseille.
B3.4. Aguilar, J. and J.M. Combes, *Commun. Math. & Phys. 22,* 269 (1971).
B3.5. Balslev, E. and J.M. Combes, *Commun. Math & Phys., 22,* 280 (1971).
B3.6. Nelson, E., *Ann. Math. 70,* 572 (1959).
B3.7. Combes, J.M. and L. Thomas, *Commun. Math. & Phys. 34,* 251 (1973).
B3.8. Combes, J.M., Spectral Deformation Techniques and Application to N-body Schroedinger Operators, preprint of talk at Int. Cong. Math. 1974, Vancouver, B.C., Canada.
B3.9. Combes, J.M. See p. 243 of [G6].
B3.10. Simon, B., *Ann. Math. 97,* 247 (1973).

B3.11. Babbitt, D. and E. Balslev, *J. Funct. Anal. 18*, 1, (1975).
B3.12. Van Winter, C., The N-Body Problem on a Hilbert Space of Analytic Functions in "Analytic Methods in Mathematical Physics", (R.P. Gilbert and R.G. Newton, eds.) Gordon & Beech (1970).
B3.13. Van Winter, C., *Trans. Am. Math. Soc. 162*, 103 (1971).
B3.14. Van Winter, C., *J. Math. Anal. & Appl. 47*, Part I, p. 633, (1974); *48*, Part II, p. 368 (1974).
B3.15. Van Winter, C., *J. Math. Anal. & Appl. 49*, 88 (1975).

OTHER REFERENCES

B3.16. Simon, B., *Commun. Math. & Phys. 27*, 1 (1972).
B3.17. Babbitt, D. and E. Balslev, *Commun. Math. & Phys. 35*, 173 (1974).
B3.18. Lavine, R.B., *Indiana Univ. Math. J. 21*, 643 (1972).

B4. THE M.S. LIVŠIC MATRIX IN PERTURBATION THEORY

B4.1. Howland, J.S., Regular Perturbations, p. 169 of [G6].
B4.2. Howland, J.S., *J. Math. Anal. & Appl. 50*, 415 (1975).
B4.3. Livšic, M.S., *Transl. Am. Math. Soc. Series 2, 5*, 67 (1957).
B4.4. Livšic, M.S., *Transl. Soviet Phys. Dokl. Am. Phys. Soc. 1*, 620 (1956).
B4.5. Livšic, M.S., *Transl. of J. Exp. Theory Phys. (JETP) 4*, 91 (1957).
B4.6. Livšic, M.S., Operators, Oscillations and Waves, *Am. Math. Soc. Transl. 34* (1973).
B4.7. Brodski, M.S., Triangular and Jordan Representation of Linear Operators, *Am. Math. Soc. Transl. 32* (1971).

S-MATRIX THEORY AND DISPERSION RELATIONS (not discussed)

1. Khuri, N.N., *Phys. Rev. 107*, 1148 (1957).
2. Squires, E.J., An Introduction to Relativistic S-Matrix Theory. See [B3.2].
3. Mandelstam, S., *Nuovo Cimento 30*, 1148 (1963).
4. Eden, R.J., P.V. Landshoff, D.I. Olive and J.C. Polkinghorne, "The Analytic S-Matrix", Cambridge Univ. Press (1966).
5. Gribov, V.N., On the Possibility of Experimental Investigation of Mandelstam Cuts, *Soviet J. Phys.*, 138 (1967).
6. Nussenzveig, H.M., Analytic Properties of Non-Relativistic Scattering Amplitudes, Univ. of Mexico Lecture Notes by J.M. Lozano and L. Sartori.
7. Ciulle, S., C. Pomponius and I. Sabba-Stefanescu, Analytic Extrapolation Techniques and Stability Problems in Dispersion Relation Theory, *Phys. Rep. Lett. 17C* (1975).

8. Saxon, D.S., Abstract Al-11 in *IRE Trans. Antennas & Propag. AP4*, 579 (1959).
See also [G9].

GENERAL REFERENCES

G1. Kato, T., "Perturbation Theory of Linear Operators", Springer-Verlag, Berlin (1966).

G2. Newton, R.G., "Scattering Theory of Waves and Particles", McGraw-Hill (1966).

G3. Wilcox, C.H. (ed.) "Perturbation Theory and its Application in Quantum Mechanics", pub. no. 16, Math Research Center, Univ. of Wisconsin, J. Wiley (1966).

G4. Taylor, J.R., "Scattering Theory", J. Wiley (1972).

G5. NSF 1969 Scattering Conference, Flagstaff, Arizona. Published as: *Rocky Mountain J. Math. 1*, (1971).

G6. LaVita, J.D. and J.P. Marchand (eds.), "Scattering Theory in Mathematical Physics", Proc. NATO Adv. Study Inst. Denver, Colorado, 1973; Reidel Publishing Co., Boston, Ma., (1974).

G7. Wilcox, C.H., Scattering Theory for the D'Alembert Equation in Exterior Domains in "Lecture Notes in Math.", no. 442, (A. Dold and B. Eckmann, eds.), Springer-Verlag, Berlin (1975).

G8. Dolph, C.L., The Integral Equation Method in Scattering Theory, published in "Problems in Analysis", (a symposium in honor of S. Bochner), (R.C. Gunning, ed.), Princeton Univ. Press (1970).

G9. Nussenzveig, H.M., "Causality and Dispersion Relations", Academic Press, New York (1972).

G10. Beck, G.B. and H.M. Nussenzveig, *Nuovo Cimento 10*, 416 (1960).

G11. Dolph, C.L. and F. Penzlin, On the Theory of a Class of Non-Self-Adjoint Operators and its Applications to Quantum Scattering Theory, *Ann. Acad. Sci. Finn. A. 263* (1959).

G12. Dolph, C.L., *Bull. Am. Math. Soc. 67*, 1 (1961).

G13. Dolph, C.L., Positive Real Resolvent and Linear Passive Hilbert Systems, *Ann. Acad. Sci. Finn. A. 336*, (1963).

G14. Krein, M.G., *Am. Math. Soc. Transl. Series 2, 90*, 181 (1970).

G15. Sz-Nagy, B. and C. Foias, "Harmonic Analysis of Operators on Hilbert Space", Am. Elsevier, New York (1970).

G16. Adamjan, V.M. and D.Z. Arov, *Am. Math Soc. Transl. Series 2, 95*, 75 (1970).

G17. Tamarkin, J.D., *Ann. Math. 28*, 127 (1927).

G18. Dunford, N. and G.T. Schwartz, "Linear Operators Part One", Interscience, New York (1958).

G19. Spivak, M. "Calculus on Manifolds", W.A. Benjamin, Inc.,

Menlo Park, California (1972).

REFERENCES FOR PART II

2.1. Cagniard, , L. "Reflection and Refraction of Progressive Seismic Waves", (trans. & rev. by E.A. Flinn & C.H. Dix) McGraw-Hill, New York (1962).
2.2. Achenbach, J.D., "Wave Propagation in Elastic Solids", North-Holland, New York (1973).
2.3. Karlsson, T. & J.F. Hook, *Bull. Seismol. Soc. Am. 53,* 1007 (1963).
2.4. Cerveny, V. and R. Ravindra, "Theory of Seismic Head Waves", Univ. of Toronto Press (1971).
2.5. Afandi, O.F. and R.A. Scott, *Int. J. Solids & Struct. 8,* 1145 (1972).
2.6. Eason, G., *J. Inst. Math. & Appl. 2,* 299 (1966).
2.7. Scott, R.A. and J. Miklowitz, *Int. J. Solids & Struct. 5,* 65 (1969).
2.8. Scott, R.A., Transient Anisotropic Waves in Bounded Elastic Media, in "Wave Propagation in Solids", (J. Miklowitz, ed.) ASME Monograph (1969).
2.9. Rosenfeld, P.L. and J. Miklowitz, Proc. Fourth U.S. Nat. Cong. Appl. Mech., p. 293 (1962).
2.10. Hron, F., *Bull. Seismol. Soc. Am. 61,* 765 (1971).
2.11. Helmberger, D.V., *Bull. Seismol. Soc. Am. 58,* 179 (1968).
2.12. Helmberger, D.V. and D.G. Harkrider, *Geophys. J. R. Astron. Soc. 31,* 45 (1972).
2.13. Wiggins, R.A. and D.V. Helmberger, *Geophys. J. R. Astron. Soc. 37,* 73 (1974).
2.14. Helmberger, D.V. and S.D. Malone, *J. Geophys. Res. 80,* 4881 (1975).
2.15. Müller, G., *Geophys. J. R. Astron. Soc. 21,* 761 (1970).
2.16. Fuchs, K. and G. Müller, *Geophys. J. R. Astron. Soc. 23,* 417 (1971).
2.17. Müller, G., *J. Geophys. Res. 78,* 3468 (1973).
2.18. McMechan, G.A., *Geophys. J. R. Astron. Soc. 37,* 407 (1974).
2.19. Helmberger, D.V. and F. Gilbert, *Geophys. J. R. Astron. Soc. 27,* 57 (1972).
2.20. Müller, G., *Geophys. J. R. Astron. Soc. 23,* 435 (1971).
2.21. Richards, P., *Bull. Seismol. Soc. Am. 66,* 701 (1976).
2.22. Rosenbaum, J.R., *J. Geophys. Res. 65,* 1577 (1960).
2.23. Phinney, R.A., *J. Geophys. Res. 66,* 1445 (1961).
2.24. Bliss, G.A., "Algebraic Functions", Dover Publications, Inc. New York (1966).
2.25. Chapman, C.H., *Pure Appl. Geophys. 94,* 233 (1972).
2.26. Gupta, U., *Pure Appl. Geophys. 80,* 27 (1970).

2.27. Gilbert, F. and S.J. Laster, *Bull. Seismol. Soc. Am. 52,* 299 (1962).
2.28. Gilber, F. and S.J. Laster, *Bull. Seismol. Soc. Am. 52,* 59 (1962).
2.29. Alsop, L.E., *Bull. Seismol. Soc. Am. 60,* 1989 (1970).
2.30. Dainty, A.M. and C.N.G. Dampney, *Geophys. J. R. Astron. Soc. 28,* 147 (1972).
2.31. Cochran, M.D., A.F. Woeber and J.-Cl. De Bremaecker, *Rev. Geophys. & Space Phys. 8,* 321 (1970).
2.32. Gilbert, F., *Rev. Geophys. 2,* 123 (1964).
2.33. Lloyd, J.R. and J. Miklowitz, Proc. Fourth U.S. Nat. Cong. Appl. Mech., p. 255 (1962).
2.34. Laster, S.J., J.G. Foreman and F. Linville, *Geophys 30,* 571 (1965).
2.35. Su, S.S. and J. Dorman, *Bull. Seismol. Soc. Am. 55,* 989 (1965).
2.36. De Bremaecker, J.-Cl., *Bull. Seismol. Soc. Am. 57,* 191 (1967).
2.37. De Bremaecker, J.-Cl., *Nuovo Cimento, Suppl. [1] 6,* 98 (1968).
2.38. Radovich, B. and J.-Cl. De Bremaecker, *Bull. Seismol. Soc. Am. 64,* 301 (1974).
2.39. Stalmach, D.M. and J.-Cl. De Bremaecker, *Bull. Seismol. Soc. Am. 63,* 995 (1973).
2.40. Watson, T.H., *Bull. Seismol. Soc. Am. 62,* 369 (1972).
2.41. Dainty, A.M., *Bull. Seismol. Soc. Am. 61,* 93 (1971).
2.42. Abramovici, F., *Geophys. J. R. Astron. Soc. 16,* 9 (1968).
2.43. Roever, W.L., T.F. Vining and E. Strick, *Philos. Trans. A 251,* 455 (1959).
2.44. Podllapollski, G.S. and Yu. I. Vassillev, A Rayleigh Type Wave at a Non-Free Surface, *Bull. Acad. Sci. (USSR) Geophys. Ser.,* 859 (1961).
2.45. Alterman, Z. and F.C. Karal, *Bull. Seismol. Soc. Am. 58,* 367 (1968).
2.46. Alterman, Z. and D. Loewenthal, Computer Generated Seismograms in "Methods in Computational Physics", vol. 12, (B.A. Bolt, ed.) Academic Press, New York, p. 115 (1972).
2.47. Pilant, W.L., *Bull. Seismol. Soc. Am. 62,* 285 (1972).
2.48. Dampney, C.N.G., *Bull. Seismol Soc. Am. 62,* 1017 (1972).
2.49. Chander, B., L.E. Alsop and J. Oliver, *Bull. Seismol. Soc. Am. 58,* 1849 (1968). See also: Errata, same journal, vol. *59,* ;1475 (1969).
2.50. Hirasawa, T. and M. J. Berry, *Bull. Seismol. Soc. Am. 61,* 1 (1971).
2.51. Adler, E.L. and I.-H. Sun, *IEEE Trans. on Sonics & Ultrason. SU-18,* 184 (1971).
2.52. Poupinet, G. and C. Wright, *Bull. Seismol. Soc. Am. 62,* 1699 (1972).

2.53. D.P. Hill, *Geophys. J. R. Astron. Soc. 34,* 149 (1973).
2.54. Mindlin, R.D., "Waves and Vibrations in Isotropic Elastic Plates", Proc. First Naval Symp. on Struct. Mech. (J.N. Goodier and N.J. Hoff, eds.) Pergamon Press (1958).
2.55. Randles, P.W. and J. Miklowitz, *Int. J. Solids & Struct. 7,* 1031 (1971).
2.56. Randles, P.W., *Int. J. Solids & Struct. 9,* 31 (1973).
2.57. Viano, D.C. and J. Miklowitz, Transient Wave Propagation in a Symmetrically Layered Elastic Plate, *J. Appl. Mech. 41,* 684 (1974).
2.58. Dunkin, J.W. and D.G. Corbin, *Bull. Seismol. Soc. Am. 60,* 167 (1970).
2.59. Stuart, A.D., *J. Acoust. Soc. Am. 59,* 1160 (1976).
2.60. Stuart, A.D., *J. Acoust. Soc. Am. 59,* 1170 (1976).
2.61. Alsop, L.E., A.S. Goodman and S. Gregersen, *Bull. Seismol. Soc. Am. 64,* 1635 (1974).
2.62. Wu, C.H., "Elastic Waves in Semi-Infinite Plates", Ph.D. Thesis, Univ. of Minnesota (1965).
2.63. Frisk, G.V. and H. Überall, *J. Acoust. Soc. Am. 59,* 46 (1976).
2.64. Labianca, F.M., *J. Acoust. Soc. Am. 53,* 1137 (1973).
2.65. Stickler, D.C., *J. Acoust. Soc. Am. 57,* 856 (1975).
2.66. Corngold, N., Quasi-Exponential Decay of Neutron Fields in "Nuclear Science and Technology (E.J. Henley and J. Lewins, eds.) (June, 1972).
2.67. Case, K.M., *Ann. Phys. 7,* 349 (1959).
2.68. Kuester, E.F. and D.C. Chang, Radiadion of a Surface Wave from a Curvature Discontinuity in an Impedance Surface. Part I. Convex Bend, Sci. Rept. No. 16, Electromagnetics Lab, Univ. of Colorado (Jan. 1976).
2.69. Felsen, L.B. and N. Marcuvitz, "Radiation and Scattering of Waves", Prentice-Hall, Inc., Englewood Cliffs, New Jersey (1973).

TOWARD AN ENGINEERING THEORY OF ELECTROMAGNETIC SCATTERING: THE SINGULARITY AND EIGENMODE EXPANSION METHODS

Carl E. Baum
Air Force Weapons Laboratory

*In studying the transient/broad-band response of antennas and scatterers in the resonance region and below, an important way to view the response is the singularity expansion method (SEM) both for efficiency of representation and for physical insight. While the SEM uses the singularities of the response in the complex frequency(*s*) plane, the eigenmode expansion method (EEM) uses the eigenmodes and eigenvalues of integral equations describing the object electromagnetic response. The EEM sheds much insight into, extends the utility of, and is complementary to the SEM. This paper reviews the SEM and EEM with emphasis on some recent developments. These include extensions of pole-related results to second order poles and to contour integral techniques for pole and zero location in the* s *plane. Equivalent circuits are developed at a port of a scatterer or antenna for the object response using SEM and EEM formulas for both short-circuit and open-circuit boundary-value problems. Using contour integral methods, general formulas are developed for the various kinds of terms in the singularity expansion including branch integrals and entire functions. These results point to potential new concepts and new areas of application.*

I. INTRODUCTION

The engineering problem in electromagnetic scattering is not only to be able to calculate or measure an electromagnetic field

ISBN 0-12-709650-7

quantity at some particular position and frequency or time. One would find it useful to be able to have some comprehensive description of the scattering properties over the entire scatterer (or even all space) and over a wide span of frequency and/or time. Such a description might have both quantitative and qualitative features. Besides using such a description to analyze electromagnetic properties one would also like to develop synthesis procedures so that in the true engineering sense one can design the electromagnetic response to have some desirable features. Preferably such synthesis will be based on explicit procedures and not just "hit and miss"; ideally they will result from some direct equations of the form "doing A (in the design) gives B (in the response)".

Another paper in this volume deals with some general properties of complicated scatterers which can be used to simplify the understanding of the object response [6]. One of the concepts consists of dividing the response according to the frequencies of interest in units of the characteristic dimensions of interest on the scatterer; this frequency-region division, leads to a low frequency method (LFM), a singularity expansion method (SEM) for the resonance region, and a high frequency method (HFM) involving asymptotic expansions.

This paper addresses one of these basic frequency division concepts, the singularity expansion method (SEM) together with the eigenmode expansion method (EEM) which sheds important insight into the SEM and extends its utility. There has been a considerable amount of work in this area with some significant success. The SEM was developed for broad-band/transient response dominated by the resonance frequency regime of the scatterer. This has been particularly important in the response of systems (complicated scatterers) to the nuclear electromagnetic pulse (EMP) because of the presence of frequencies with wavelengths ranging from short to long compared with characteristic scatterer dimensions.

Recently this author has written an extensive review of this subject in book form [5] and a review paper on transient electromagnetics containing some additional material of relevance [22]. While reviewing some of this material the present paper emphasizes some concepts of more recent vintage.

For frequency conventions in this paper we use the two-sided Laplace transform (indicated by a tilde ~ above a quantity) with respect to time as

$$\tilde{F}(s) \equiv \int_{-\infty}^{\infty} F(t)e^{-st}dt \ , \ F(t) \equiv \frac{1}{2\pi j}\int_{\Omega_o - j\infty}^{\Omega_o + j\infty} \tilde{F}(s)e^{st}ds \qquad (1.1)$$

where the Laplace transform variable or complex frequency is

$$s \equiv \Omega + j\omega \qquad (1.2)$$

Here $F(t)$ is a scalar, vector, dyadic, or any rank tensor function of time. The two sided Laplace transform integral is assumed to converge in a strip $\Omega_1 < \Omega < \Omega_2$ with definition in the rest of the s plane by analytic continuation from this strip with account taken for singularities; for the inversion Ω_o is in the same strip of convergence so that $\Omega_1 < \Omega_o < \Omega_2$.

The Maxwell equations are summarized as

$$\begin{aligned}
\nabla \times \vec{E}(\vec{r},t) &= -\frac{\partial}{\partial t}\vec{B}(\vec{r},t) - \vec{J}_m(r,t) \\
\nabla \times \vec{H}(\vec{r},t) &= \frac{\partial}{\partial t}\vec{D}(\vec{r},t) + \vec{J}(\vec{r},t) \\
\nabla \cdot \vec{J}(\vec{r},t) &= -\frac{\partial}{\partial t}\ \nabla \cdot \vec{D}(\vec{r},t) = -\frac{\partial}{\partial t}\rho(\vec{r},t) \\
\nabla \cdot \vec{J}_m(\vec{r},t) &= -\frac{\partial}{\partial t}\nabla \cdot \vec{B}(\vec{r},t) = -\frac{\partial}{\partial t}\rho_m(\vec{r},t)
\end{aligned} \qquad (1.3)$$

with magnetic currents and charges included for generality and the consitutive relations with Ohm's law given by

$$\tilde{\vec{D}}(\vec{r},s) = \tilde{\vec{\vec{\varepsilon}}}(\vec{r},s) \cdot \tilde{\vec{E}}(\vec{r},s) \ , \ \tilde{\vec{J}}(\vec{r},s) = \tilde{\vec{\vec{\sigma}}}(\vec{r},s) \cdot \tilde{\vec{E}}(\vec{r},s)$$
$$\tilde{\vec{B}}(\vec{r},s) = \tilde{\vec{\vec{\mu}}}(\vec{r},s) \cdot \tilde{\vec{H}}(\vec{r},s) \ , \ \tilde{\vec{J}}_m(\vec{r},s) = \tilde{\vec{\vec{\sigma}}}_m(\vec{r},s) \cdot \tilde{\vec{H}}(\vec{r},s) \tag{1.4}$$

Note that source terms can also be added to the current densities.

Write an integral equation for a general antenna or scatterer of finite size in free space as

$$\langle \tilde{\vec{\vec{\Gamma}}}(\vec{r},\vec{r}\,';s); \tilde{\vec{J}}(\vec{r}\,',s) \rangle = \tilde{\vec{I}}(\vec{r},s) \ , \ \vec{r} \in S \text{ or } V \tag{1.5}$$

where $\vec{J}$ is the response current density (for volume distributions) or surface current density (for surface distributions) and $\vec{I}$ is the forcing function in the form of some incident or other specified source field. Integration over the object of interest is denoted by $\langle \, , \, \rangle$ with integration over the common spatial coordinates in the terms separated by the comma; a symbol above this comma indicates the type of multiplication used (dot product in (1.5)). The kernel, $\tilde{\vec{\vec{\Gamma}}}$ of the integral equation is related to the dyadic Green's function of free space with the precise form depending on the type of integral equation (E field, H field, etc.) with its corresponding type of incident field. Such integral equations are capable of describing radiation and scattering by a linear stable object in free space as shown in Fig. 1.1. Integral equation concepts also apply to more complicated scattering geometries such as multiple scatterers and the presence of various media by proper definition of the domain of integration and/or the construction of special Green's functions for use in defining the kernel.

For later use we have the dyadic Green's function of free space from [3,21,10]

$$\tilde{G}_0(\vec{r},\vec{r}\,';s) = \gamma \frac{e^{-\zeta}}{4\pi\zeta}$$

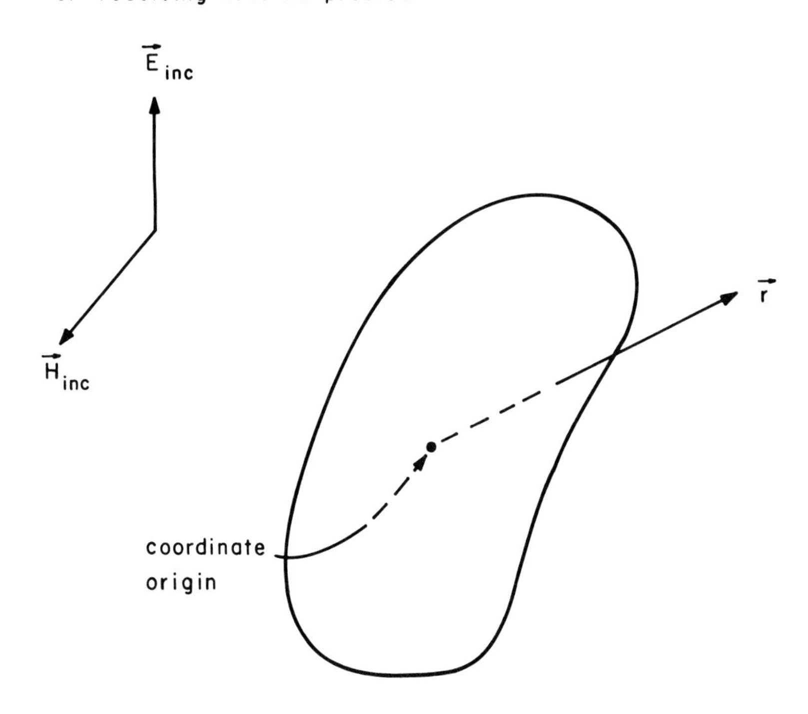

Fig. 1.1. Radiation and scattering of electromagnetic fields by an object.

$$\nabla\tilde{G}_0(\vec{r},\vec{r}\,';s) = \vec{1}_R \frac{\gamma^2}{4\pi}\left[-\zeta^2 - \zeta^{-1}\right]e^{-\zeta}$$

$$\tilde{\overleftrightarrow{G}}_0(\vec{r},\vec{r}\,';s) = \left[\overleftrightarrow{1} - \gamma^{-2}\nabla\nabla\right]\tilde{G}_0(\vec{r},\vec{r}\,';s)$$

$$= \frac{\gamma}{4\pi}\left\{\left[-2\zeta^{-3} - 2\zeta^{-2}\right]e^{-\zeta}\vec{1}_R\vec{1}_R + \left[\zeta^{-3} + \zeta^{-2} + \zeta^{-1}\right]e^{-\zeta}\left[\overleftrightarrow{1} - \vec{1}_R\vec{1}_R\right]\right\} + \frac{1}{3\gamma^2}\delta(\vec{r} - \vec{r}\,')\overleftrightarrow{1} \qquad (1.6)$$

$$R \equiv |\vec{r} - \vec{r}\,'| \ , \quad \zeta \equiv \gamma R$$

$$\gamma \equiv \frac{s}{c} = s\sqrt{\mu_0\varepsilon_0} \ , \quad Z_0 \equiv \sqrt{\frac{\mu_0}{\varepsilon_0}}$$

$$\tilde{\overleftrightarrow{Z}}(\vec{r},\vec{r}\,';s) \equiv s\mu_0\tilde{\overleftrightarrow{G}}_0(\vec{r},\vec{r}\,';s) \quad \text{(impedance kernel)}$$

$$\tilde{\overleftrightarrow{Z}}_m(\vec{r},\vec{r}\,';s) \equiv -Z_0\nabla \times \tilde{\overleftrightarrow{G}}_0(\vec{r},\vec{r}\,';s) = -Z_0\left[\nabla\tilde{G}_0(\vec{r},\vec{r}\,';s)\right] \times \overleftrightarrow{1} \quad \text{(magnetic impedance kernel)}$$

Here a delta function $\delta(\vec{r} - \vec{r}\,')$ is included to handle the integration over a volume with the other terms being treated in such integration as principal value integrals. The radiated or scattered fields may be calculated for an object in free space by an integration over the current densities as

$$\tilde{\vec{E}}(\vec{r},s) = -\langle\tilde{\overleftrightarrow{Z}}(\vec{r},\vec{r}\,';s);\tilde{\vec{J}}(\vec{r}\,',s)\rangle + \langle\tilde{\overleftrightarrow{Z}}_m(\vec{r},\vec{r}\,';s);\frac{1}{Z_0}\tilde{\vec{J}}_m(\vec{r}\,',s)\rangle$$

$$Z_0\tilde{\vec{H}}(\vec{r},s) = -\langle\tilde{\overleftrightarrow{Z}}(\vec{r},\vec{r}\,';s);\frac{1}{Z_0}\tilde{\vec{J}}_m(\vec{r}\,',s)\rangle - \langle\tilde{\overleftrightarrow{Z}}_m(\vec{r},\vec{r}\,';s);\tilde{\vec{J}}(\vec{r}\,',s)\rangle \qquad (1.7)$$

with all polarizability and conductivity effects included in the definitions of the current densities These are summarized in combined field form for the radiated or scattered field as

$$\tilde{\vec{E}}_q(\vec{r},s) = -\langle \tilde{\vec{\vec{Z}}}_q(\vec{r},\vec{r}\,';s);\tilde{\vec{J}}_q(\vec{r}\,',s)\rangle$$

$$\tilde{\vec{E}}_q(\vec{r},s) \equiv \tilde{\vec{E}}(\vec{r},s) + qjZ_0\tilde{\vec{H}}(\vec{r},s) \quad \text{(combined field)} \tag{1.8}$$

$$\tilde{\vec{\vec{Z}}}_q(\vec{r},\vec{r}\,';s) \equiv \tilde{\vec{\vec{Z}}}(\vec{r},\vec{r}\,';s) + qj\tilde{\vec{\vec{Z}}}_m(\vec{r},\vec{r}\,';s) \quad \text{(combined impedance kernel)}$$

$$\tilde{\vec{J}}_q(\vec{r},s) \equiv \tilde{\vec{J}}(\vec{r},s) + \frac{qj}{Z_0}\tilde{\vec{J}}_m(\vec{r},s) \quad \text{(combined current density)}$$

$$q = \pm 1 \quad \text{(separation index)}$$

Related to integral equations as in (1.5) there is the general moment method (MoM) for matricizing the integral equation for numerical solution [2]. By approximating $\tilde{\vec{J}}(\vec{r}\,',s)$ with a set of N "expansion" functions and $\tilde{\vec{I}}(\vec{r},s)$ with a set of N "testing" functions and treating the sets of coefficients as numerical vectors, we have the matricized integral equation as

$$(\tilde{\Gamma}_{n,m}(s)) \cdot (\tilde{J}_n(s)) = (\tilde{I}_n(s)) \tag{1.9}$$

with N component vectors related by an $N \times N$ matrix replacing the integral operator. Corresponding to the different types of kernels related to the dyadic Green's function as in (1.6) there are different MoM matrices. By considering some of the properties of the solution of the matricized integral equation (1.9) one can aid in understanding the properties of the solution of the corresponding integral equations (1.5).

II. SINGULARITY EXPANSION

A. Background

The basic concept of singularity expansion in the context of electromagnetic scatterers (and antennas) was motivated by the observation of typical transient responses of complicated scatterers such as aircraft, missiles, etc. [12]. It was observed that such transient responses seemed to be dominated by one or a few damped sinusoids. Noting that damped sinusoids correspond to pole pairs in the complex frequency plane associated with the Laplace transform, this suggests the general question of the behavior of the response of electromagnetic scatterers in the s plane, and in particular the description of the response in terms of the s plane singularities.

Let us first for generality consider some $f(t)$ which may be any two-sided Laplace-transformable entity that commutes with scalars including such things as:

scalars: $f(t)$

3-vectors (space vectors): $\vec{f}(t)$

N-vectors (numerical vectors): $(f_n(t))$, $n = 1,2,\ldots,N$

3-dyads (space 2-tensors): $\overleftrightarrow{f}(t)$

$N \times N$ matrices (numerical 2-tensors): $(f_{n,m}(t))$, $n,m = 1,2,\ldots,N$

Such things may also be functions of other variables including space coordinates $\vec{r}$. One can also include 4-vectors $\underline{f}_p$ and 4-dyads $\underline{\underline{f}}_p$ for space-time formalisms, except that in Laplace transforming over time the result is not a function of the 4-space coordinates $\underline{r}_p = (x,y,z,pjct)$ with $p = \pm 1$ of Minkowsky space.

Similar comments apply to quaternions $\hat{f}$. Hybrids of the above quantities are also allowed such as:

a. super vectors (for example, numerical vectors of space vectors $(\vec{f}_n(t))$;

b. super matrices (for example, numerical matrices of space dyads $(\overleftrightarrow{f}_{n,m}(t))$.

Our general $f(t)$ may represent any electromagnetic quantity of interest including such items as:

a. the response current density on the scatterer (corresponding to class 1 coupling coefficient);

b. the kernel of the inverse operator of an integral equation (1.5) (corresponding to class 2 coupling coefficients of the response).

In this section generality is maintained in $f(t)$ so that the results may be applied to various types of quantities.

With the two-sided Laplace transform

$$\tilde{f}(s) \equiv \int_{-\infty}^{\infty} f(t)e^{-st}dt, \quad f(t) = \frac{1}{2\pi j}\int_{Br} \tilde{f}(s)e^{st}ds \tag{2.1}$$

and the Bromwich contour defined by

$$\int_{Br} \equiv \int_{\Omega_0 - j\infty}^{\Omega_0 + j\infty} \tag{2.2}$$

in a strip $\Omega_1 < \Omega < \Omega_2$ of convergence of the two-sided transform integral; one can consider the properties of $\tilde{f}(s)$ in the s plane as illustrated in Fig. 2.1. Here we have included some singularities for illustration with conjugate symmetry in the s plane corresponding to the transform of a real valued time function, vector, etc. For convenience we shall adopt a notation for singularity location and type labeling (see Fig. 2.1) as

$$\begin{aligned} p &\equiv \text{pole} \quad (np \equiv \text{nth order pole}) \\ \varepsilon &\equiv \text{essential singularity} \\ b &\equiv \text{branch point} \end{aligned} \tag{2.3}$$

and some other discrete points in the s plane as

$$\begin{aligned} z &\equiv \text{zero} \quad (nz \equiv \text{nth order zero} = \text{-nth order pole}) \\ \Sigma &\equiv \text{saddle point} \quad (n\Sigma \equiv \text{nth order saddle}) \end{aligned} \tag{2.4}$$

We also use

$$e \equiv \text{entire function (singularity at infinity)} \tag{2.5}$$

These and other symbols can be used as subscripts to s and C to denote particular singularities and associated contours. For example, α is often used as an index set to denote which pole is being considered. Other such indices or index sets may be developed for use with sets of other kinds of singularities. As discussed later, α can also be applied to essential singularities.

Note in Fig. 2.1 that singularities are shown only in the left half s plane. This is appropriate for the kernel $\tilde{\Gamma}^{-1}$ of the inverse operator describing the assumed stable scatterer (corresponding to the inversion of (1.5)). Typically the forcing

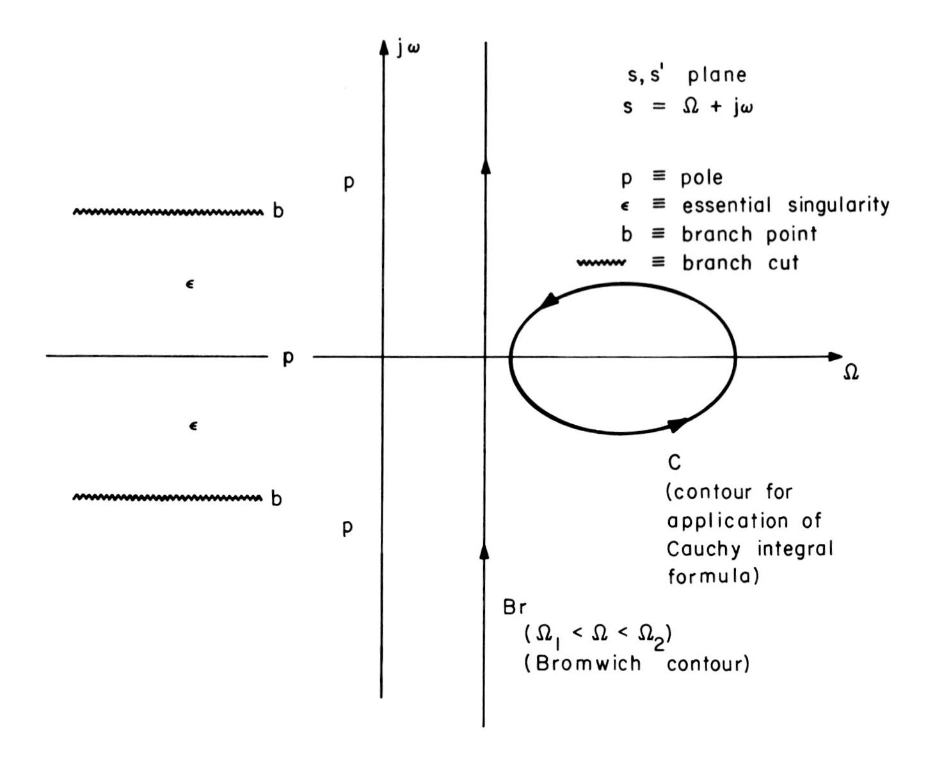

Fig. 2.1. s *plane with some singularities and the Laplace-inversion contour.*

function $\tilde{\vec{I}}$ and the response $\tilde{\vec{J}}$ will also have singularities only for $\Omega \leq 0$, but this need not be the case; it is possible to have physically meaningful forcing functions with singularities in the right half plane and for which the transform integral converges only in a strip, but can be analytically continued both left and right [11]. Such cases of singularities in the right half plane are not considered in this paper, but the concepts are readily extended to these.

It is sometimes convenient to shift the time reference of our functions, vectors, etc. of interest, for which we define

$$g(t) \equiv f(t+t_0)$$

$$\tilde{g}(s) = e^{st_0}\tilde{f}(s) \tag{2.6}$$

$$\tilde{g}(s) \equiv \int_{-\infty}^{\infty} g(t)e^{-st}dt, \quad g(t) = \frac{1}{2\pi j}\int_{Br} \tilde{g}(s)e^{st}ds$$

For cases that $f(t) \equiv 0$ for $t < t_0$, this shift makes $g(t) \equiv 0$ for $t < 0$ so that the two-sided Laplace transform can be made a one-sided Laplace transform; this cannot always be done. Carrying out our analysis in terms of $g(t)$, however, allows us to introduce the concept of a turn-on time t_0 in the singularity expansion in the response $\tilde{\vec{J}}$ (class 1) or inverse operator $\tilde{\vec{\vec{\Gamma}}}^{-1}$ (class 2). For each term in the singularity expansion of $\tilde{g}(s)$, multiplication by e^{-st_0} gives the corresponding term for $\tilde{f}(s)$, or in time domain just replace t by $t - t_0$ to find the corresponding term for $f(t)$. Hence one has a term $e^{-(s-s_\alpha)t_0}$ for discrete singularities (at s_α) and a similar form for more general singularities which can be introduced if desired.

B. Cauchy Integral Formula

The key concpet of this section is the singularity expansion of $\tilde{g}(s)$ through the Cauchy integral formula as

$$\tilde{g}(s) = \frac{1}{2\pi j}\oint_C \frac{\tilde{g}(s')}{s' - s}\, ds' \tag{2.7}$$

or more generally

$$\frac{d^n}{ds^n}\tilde{g}(s) = \frac{1}{2\pi j}\oint_C \frac{\tilde{g}(s')}{(s' - s)^{n+1}}\, ds', \quad n = 0,1,2,... \tag{2.8}$$

for s inside C and $\tilde{g}(s)$ analytic inside and on C. Singularity expansion can be defined in terms of deformation and subsequent decomposition of the coutour C to give representations for each term. For our beginning formula for singularity expansion we use the closed counter-clockwise contour C in the right half of the s plane as in Fig. 2.1, or more generally in the strip of convergence of the transform integral parallel to the $j\omega$ axis.

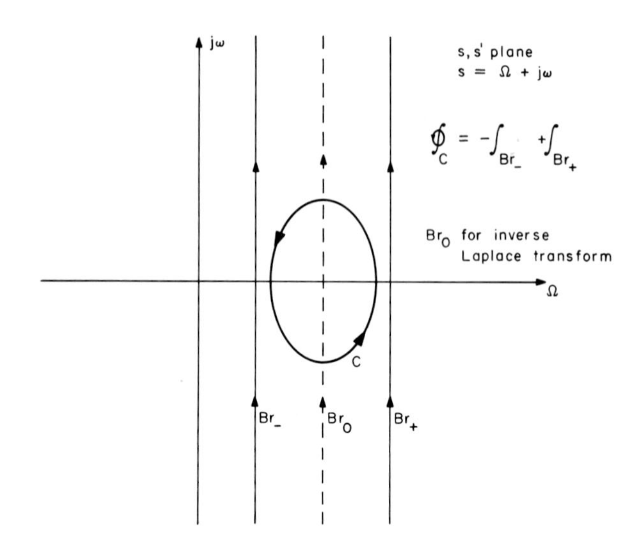

Fig. 2.2. Left-right decomposition of the Cauchy-integral-formula contour into two parallel Bromwich contours in the strip of convergence.

C. Left-Right Decomposition

As our first step in decomposing $\tilde{g}(s)$ according to its singularities in the s plane make a left-right decomposition by deforming C into two parallel Bromwich contours within the strip of convergence as indicated in Fig. 2.2. Let these two contours be labelled Br_- (left) and Br_+ (right) giving the representation for $\tilde{g}(s)$ as

$$\tilde{g}(s) \equiv \tilde{g}_-(s) + \tilde{g}_+(s)$$

$$\tilde{g}_-(s) = -\frac{1}{2\pi j}\int_{Br_-}\frac{\tilde{g}(s')}{s'-s}\,ds', \quad s \text{ to right of } Br_- \tag{2.9}$$

$$\tilde{g}_+(s) = \frac{1}{2\pi j}\int_{Br_+}\frac{\tilde{g}(s')}{s'-s}\,ds', \quad s \text{ to left of } Br_+$$

where $\tilde{g}_-(s)$ can be analytically continued to the left and similarly for $\tilde{g}_+(s)$ to the right. Note that any contribution as $s \to \pm j\infty$ between Br_- and Br_+ must strictly be accounted for. However, for suitably bounded $\tilde{g}(s)$ here this contribution is zero. As required one can make Br_- and Br_+ meet at infinity.

In addition in Fig. 2.2 we have Br_0 lying between Br_- and Br_+ where both $\tilde{g}_-(s)$ and $\tilde{g}_+(s)$ have their integral representations as in (2.9). Let us then represent the time domain form of the left-right decomposition as

$$g(t) = g_-(t) + g_+(t)$$

$$g_-(t) \equiv \frac{1}{2\pi j}\int_{Br_0} \tilde{g}_-(s)e^{st}ds = \frac{1}{2\pi j}\int_{Br_0}\left\{\frac{-1}{2\pi j}\int_{Br_-}\frac{\tilde{g}(s')}{s'-s}\,ds'\right\}e^{st}ds$$

$$= \frac{1}{2\pi j}\int_{Br_-}\tilde{g}(s')e^{s't}u(t)ds' = u(t)g(t) \tag{2.10}$$

$$g_+(t) \equiv \frac{1}{2\pi j}\int_{Br_0} \tilde{g}_+(s)e^{st}ds = \frac{1}{2\pi j}\int_{Br_0}\left\{\frac{1}{2\pi j}\int_{Br_+}\frac{\tilde{g}(s')}{s'-s}\,ds'\right\}e^{st}ds$$

$$= \frac{1}{2\pi j}\int_{Br_+}\tilde{g}(s')e^{s't}u(-t)ds' = u(-t)g(t) = [1-u(t)]g(t)$$

$$u(t) = \begin{cases} 1 & \text{for } t > 0 \\ 0.5 & \text{for } t = 0 \\ 0 & \text{for } t < 0 \end{cases}$$

Hence the left part (left contour in s plane) gives the time domain form for $t > 0$, while the right part gives the time domain form for $t < 0$. One may need to be careful around $t = 0$ if any distributions (delta functions, etc.) are located there. An interpretation of these results of the left-right decomposition is that they show the relation between the two-sided and one-sided Laplace transforms.

D. Right Entire Function

Since we are considering situations with singularities limited to the left half s plane, the right portion of $\tilde{g}(s)$, i.e. $\tilde{g}_+(s)$, is an entire function which we designate as

$$\tilde{g}_{e_+}(s) = \tilde{g}_+(s) \qquad g_{e_+}(t) = g_+(t) \tag{2.11}$$

where the subscripts for "singularity" types in the s plane are carried over to the corresponding time domain terms. This right entire function will be later joined by a left entire function.

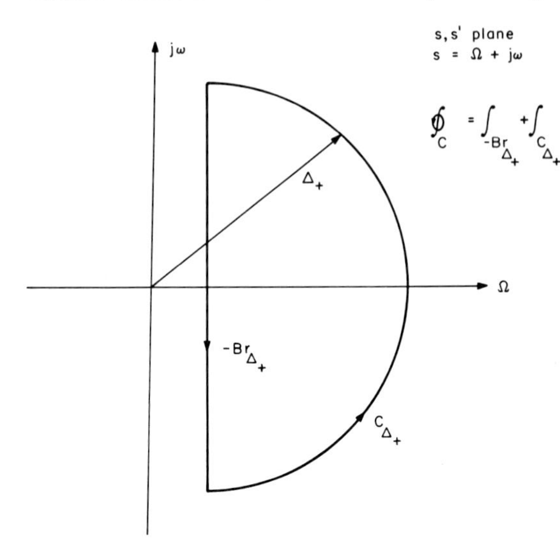

Fig. 2.3. Circular contour for right entire function.

Another way to view the right entire function is to use a circular arc of radius Δ_+ about the origin of the s plane as indicated in Fig. 2.3. Here we have

$$\tilde{g}_{e_+}(s) = \lim_{\Delta_+ \to \infty} \frac{1}{2\pi j} \int_{C_{\Delta_+}} \frac{\tilde{g}(s')}{s' - s}\, ds'$$
$$\tilde{g}_-(s) = \lim_{\Delta_+ \to \infty} \frac{1}{2\pi j} \int_{-Br_{\Delta_+}} \frac{\tilde{g}(s')}{s' - s}\, ds' = \frac{1}{2\pi j} \int_{Br} \frac{\tilde{g}(s')}{s - s'}\, ds' \tag{2.12}$$

provided, of course, that the limits exist. In time domain the right entire function takes the form

$$g_{e_+}(t) = \lim_{\Delta_+ \to \infty} \frac{1}{2\pi j} \int_{C_{\Delta_+}} \tilde{g}(s') e^{s't} u(-t)\, ds' \tag{2.13}$$

while the left part in time domain takes the form as in (2.10).

E. Deformation into Left Half Plane

For decomposing $\tilde{g}_-(s)$ into its singularities and finding the corresponding terms for $g_-(t)$, let us deform Br_- as in Fig. 2.2 into the left half plane. Noting that such deformation can pass only through regions of analycity of the quantity of concern

($\tilde{g}_-(s)$), then as singularities are encountered the contour bends around them. As shown in Fig. 2.4 there are isolated singularities:

a. poles p with closed counterclockwise contours C_α;

b. essential singularities ε with closed counterclockwise contours C_α;

and distributed singularities:

c. branch points b with closed or generally open "counterclockwise" contours C_α, extending to infinity in the left half plane;

d. entire function $\tilde{g}_{e_-}(s)$ associated with contour C_{Δ_-} in the limit as $\Delta_- \to \infty$.

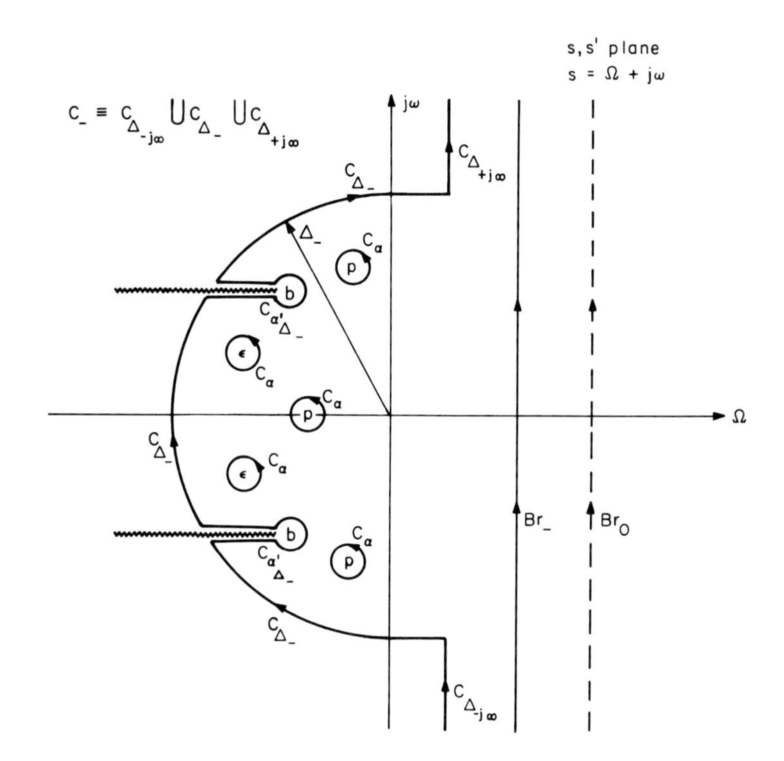

Fig. 2.4. Deformation of the Bromwich contour Br_- *into the left half plane.*

In the deformation Br_- goes into C_- plus the contour associated with the singularities in the left half plane within the semicircle of radius Δ_- about the origin giving

$$\tilde{g}_-(s) = \sum_{\substack{p \\ \text{to right of} \\ C_-}} \frac{1}{2\pi j} \oint_{C_\alpha} \frac{\tilde{g}(s')}{s - s'} ds' + \sum_{\substack{\varepsilon \\ \text{to right of} \\ C_-}} \frac{1}{2\pi j} \oint_{C_\alpha} \frac{\tilde{g}(s')}{s - s'} ds'$$

$$+ \sum_{\substack{b \\ \text{to right of} \\ C_-}} \frac{1}{2\pi j} \oint_{C_{\alpha'_{\Delta_-}}} \frac{\tilde{g}(s')}{s - s'} ds' + \frac{1}{2\pi j} \oint_{C_-} \frac{\tilde{g}(s')}{s - s'} ds' \tag{2.14}$$

$$C_- \equiv C_{\Delta_{-j\infty}} \cup C_{\Delta_-} \cup C_{\Delta_{+j\infty}}$$

Here "to right of C_-" indicates those singularities passed by in the deformation of Br_- to C_-. Portions of Br_- are directly portions of C_-, specifically $C_{\Delta_{-j\infty}}$ and $C_{\Delta_{+j\infty}}$ (the two "ends"). The "middle" portion, C_{Δ_-}, is semicircular of radius Δ_- together with connections to $C_{\Delta_{-j\infty}}$ and $C_{\Delta_{+j\infty}}$. Where branch cuts cross C_Δ, we adopt a convention that C_{Δ_-} has a small break there; this break is closed by one of the $C_{\alpha'_{\Delta_-}}$ contours which can be referred to as a truncated branch cut integral.

In the limit as $\Delta_- \to \infty$, provided the limit exists for each of the terms, we can write

$$\begin{aligned} \tilde{g}_-(s) &= \tilde{g}_p(s) + \tilde{g}_\varepsilon(s) + \tilde{g}_b(s) + \tilde{g}_{e_-}(s) \\ g_-(t) &= g_p(t) + g_\varepsilon(t) + g_b(t) + g_{e_-}(t) \end{aligned} \tag{2.15}$$

for respectively poles, essential singularities, branches, and the left entire function, corresponding to the types of terms in (2.14). Note in Fig. 2.4 the presence of the contour Br_0 for the inverse Laplace transform to convert the singularity terms into time domain. For $t < 0$ Br_0 can be closed to the right giving 0 for the value of each term. For $t > 0$ Br_0 can be deformed to the left and decomposed just like Br_- has been decomposed, thereby giving the time domain form of the terms using the same integration contours as in (2.14).

Another point to note about the deformation of C_- into the left half s plane is that it can be stopped without taking it to infinity to the left. As illustrated in Fig. 2.4, stop the deformation at some finite deformation (not necessarily semicircular). Then use (2.14) to compute $\tilde{g}_-(s)$ (or equivalently in time domain). Alternately one can compute the singularity terms swept by in the finite deformation of C_- as an approximation to $\tilde{g}_-(s)$ or $g_-(t)$. The remaining integral over C_- can then be used as an expression for the error introduced by the truncation of the singularity series associated with the finite deformation of C_-. Convergence of the singularity series in frequency and time domains may be related to the convergence of the integral over C_- to zero or to an entire function as C_- tends to the semicircle at infinity in the left half plane.

F. Poles

The pole part of $\tilde{g}_-(s)$ is a summation over all the poles in the left half plane (assuming convergence of the sum) as

$$\tilde{g}_p(s) = \sum_\alpha \tilde{g}_\alpha(s) \quad , \quad g_p(t) = \sum_\alpha g_\alpha(t)$$

$$\tilde{g}_\alpha(s) = \frac{1}{2\pi j}\oint_{C_\alpha} \frac{g(s')}{s - s'}\, ds' \quad , \quad g_\alpha(t) = \frac{1}{2\pi j}\oint_{C_\alpha} \tilde{g}(s')e^{s't}u(t)ds' \tag{2.16}$$

Suppose near s_α (which is inside C_α) we have an N_αth order pole giving a $\tilde{g}(s)$ near s_α of the form

$$\tilde{g}(s) = \left\{\sum_{n_\alpha=1}^{N_\alpha} \frac{R_{\alpha,n_\alpha}}{(s - s_\alpha)^{n_\alpha}}\right\} + \left\{\begin{matrix}\text{function analytic}\\ \text{at and near } s_\alpha\end{matrix}\right\} \tag{2.17}$$

where R_{α,n_α} are the generalized pole residues which can be a complex scalar, vector, dyad, etc. according to the type of quantity $\tilde{g}(s)$ represents. Having found the presence of a natural frequency (pole of some order) we wish to calculate the form of the associated $\tilde{g}_\alpha(s)$ and $g_\alpha(t)$.

Noting s is assumed to be outside C_α for the integration, the analytic part of $\tilde{g}(s)$ at s_α in (2.17) gives no contribution to

$\tilde{g}_\alpha(s)$ and $g_\alpha(t)$ due to the analytic nature of the integrands in (2.16). Evaluating $\tilde{g}_\alpha(s)$ from the contour integral definition we have

$$\begin{aligned}\tilde{g}_\alpha(s) &= \frac{1}{2\pi j}\oint_{C_\alpha}\left\{\sum_{n_\alpha=1}^{N_\alpha}\frac{1}{(s-s_\alpha)-(s'-s_\alpha)}\frac{R_{\alpha,n_\alpha}}{(s'-s_\alpha)^{n_\alpha}}\right\}ds' \\ &= \sum_{n_\alpha=1}^{N_\alpha}\frac{1}{2\pi j}\oint_{C_\alpha}\frac{1}{s-s_\alpha}\left\{\sum_{n=0}^{\infty}\left(\frac{s'-s_\alpha}{s-s_\alpha}\right)^n\frac{R_{\alpha,n_\alpha}}{(s'-s_\alpha)^{n_\alpha}}\right\}ds' \\ &= \sum_{n_\alpha=1}^{N_\alpha}\frac{R_{\alpha,n_\alpha}}{(s-s_\alpha)^{n_\alpha}}\frac{1}{2\pi j}\oint_{C_\alpha}\frac{ds'}{s'-s_\alpha} = \sum_{n_\alpha=1}^{N_\alpha}\frac{R_{\alpha,n_\alpha}}{(s-s_\alpha)^{n_\alpha}}\end{aligned} \tag{2.18}$$

which uses the fact that only first order poles $(s'-s_\alpha)^{-1}$ give a non-zero contour integral over C_α. So we see that the contour-integral definition of the pole part of the singularity expansion gives s-independent residues R_{α,n_α}, i.e., the poles are all in the simple or "canonical" form. No additional entire function (such as a constant) is included in this form.

In time domain this result is

$$g_\alpha(t) = \sum_{n_\alpha=1}^{N_\alpha} R_{\alpha,n_\alpha}\frac{t^{n_\alpha-1}}{(n_\alpha-1)!}e^{s_\alpha t}u(t) \tag{2.19}$$

Having found a natural frequency s_α of interest, one need only find the set of one or more R_{α,n_α} to have the complete frequency- and time-domain singularity term corresponding to that natural frequency.

Note that for notational convenience one may assign a distinct α to each pole of the set of poles at s_α. In that event $\tilde{g}_\alpha(s)$ and $g_\alpha(t)$ are single terms in the sums in (2.18) and (2.19) respectively. One index in the α index set can correspond to n_α in the above expressions.

G. Essential Singularities

By an essential singularity we mean a point around which the function $\tilde{g}(s)$ is single valued (not a branch point) and analytic, and which is not a pole. Here a pole means a leading term of the form $R_{\alpha,n_\alpha}(s - s_\alpha)^{-n_\alpha}$ as $s \to s_\alpha$. Expanding $\tilde{g}(s)$ around an essential singularity with the constraint of single valuedness, we have a series of powers of $s - s_\alpha$ with the powers extending to $-\infty$. Hence we can regard an essential singularity as an infinite order pole.

Regarding an essential singularity as an infinite order pole, the expressions in (2.16) through (2.19) apply with $N_\alpha = \infty$. Since the essential singularity is a generalized pole we can write

$$\tilde{g}_\varepsilon(s) = \sum_{\substack{\alpha \text{ for} \\ N_\alpha = \infty}} \tilde{g}_\alpha(s) \quad , \quad g_\varepsilon(t) = \sum_{\substack{\alpha \text{ for} \\ N_\alpha = \infty}} g_\alpha(t) \tag{2.20}$$

with the contour-integral definitions for $\tilde{g}_\alpha(s)$ and $g_\alpha(t)$ as in (2.16). One can use the series expansion of $\tilde{g}(s)$ around s_α to define series for $\tilde{g}_\varepsilon(s)$ and $g_\varepsilon(t)$, or one can use the contour integrals in (2.16) directly on some representation of $\tilde{g}(s)$. Using the α index set for both poles and essential singularities, one can modify the sums over α in (2.16) to only include cases of N_α finite. Alternately, one can consider $\tilde{g}_p(s) + \tilde{g}_\varepsilon(s)$ as a generalized pole term, and correspondingly $g_p(t) + g_\varepsilon(t)$ in time domain.

H. Branch Singularities

Now we come to branch points. In going around a branch point, $\tilde{g}(s)$ is by definition not single valued. Hence we cannot close the contour around the branch point, but must deform it to infinity or perhaps around one or more additional branch points that give a single valued $\tilde{g}(s)$ on the total branch contour $C_{\alpha'}$.

Considering the limit as $\Delta_- \to \infty$ and assuming convergence of the individual branch integrals as well as the sum over these, we have

$$\tilde{g}_b(s) = \sum_{\alpha'} \tilde{g}_{\alpha'}(s) \quad , \quad g_b(t) = \sum_{\alpha'} g_{\alpha'}(t)$$

$$\tilde{g}_{\alpha'}(s) = \frac{1}{2\pi j}\int_{C_{\alpha'}} \frac{\tilde{g}(s')}{s - s'}\, ds' \quad , \quad g_{\alpha'}(t) = \frac{1}{2\pi j}\int_{C_{\alpha'}} \tilde{g}(s')e^{s't}u(t)ds'$$

$$C_{\alpha'} = \lim_{\Delta_- \to \infty} C_{\alpha'_{\Delta_-}} \tag{2.21}$$

Divide the branch integral contour as

$$C_{\alpha'_{\Delta_-}} = C_{-\alpha'_{\Delta_-}} \cup C_{+\alpha'_{\Delta_-}} \quad , \quad C_{\alpha'} = C_{-\alpha'} \cup C_{+\alpha'} \tag{2.22}$$

where the subscript - applies to that portion of the contour "below" the branch cut and the subscript + applies to that portion of the contour "above" the contour, with the limit taken as these two pieces of the contour tend toward the branch cut and hence as the two pieces of the contour tend toward each other. As indicated in Fig. 2.5A, "above" or + corresponds to more positive imaginary values of s on $C_{-\alpha'}$.

In taking the limit of $C_{-\alpha'}$ and $C_{+\alpha'}$ as they approach the branch cut, it may be necessary to isolate the branch point $s_{\alpha'}$ (indicated by b) with a small contour (say circular) $C'_{0_{\alpha'}}$, as indicated in Fig. 2.5B. Accordingly $C_{-\alpha'}$ and $C_{+\alpha'}$ are reduced to $C'_{-\alpha'}$ and $C'_{+\alpha'}$ respectively, and we have

$$C'_{\alpha'_{\Delta_-}} = C'_{-\alpha'_{\Delta_-}} \cup C'_{0_{\alpha'}} \cup C'_{+\alpha'_{\Delta_-}} \quad , \quad C'_{\alpha'} = C'_{-\alpha'} \cup C'_{0_{\alpha'}} \cup C_{+\alpha'}$$

$$C'_{\alpha'} = \lim_{\Delta_- \to \infty} C'_{\alpha'_{\Delta_-}} \tag{2.23}$$

If as the radius of $C'_{0_{\alpha'}}$ goes to zero the contribution of the integral over $C'_{0_{\alpha'}}$ goes to zero, this portion need not be considered. If there are some additional positions along the contour where $\tilde{g}(s)$ is nonintegrable, a similar special treatment is required.

For cases that the branch point contribution(s) can be neglected, we can reduce the branch contributions to the forms

$$\tilde{g}_{\alpha'}(s) = \int_{C_{+\alpha'}} \frac{\tilde{R}_{\alpha'}(s')}{s - s'}\, ds' \quad , \quad g_{\alpha'}(t) = \int_{C_{-\alpha'}} \tilde{R}_{\alpha'}(s')e^{s't}u(t)ds'$$

$$\tilde{R}_{\alpha'}(s)\Big|_{s \text{ on cut}} \equiv \frac{\tilde{g}(s_+) - \tilde{g}(s_-)}{2\pi j} \tag{2.24}$$

with s_+ just above and s_- just below the branch cut in the limit as both tend to the branch cut. Here $\tilde{R}_{\alpha'}(s)$ can be considered as a residue distribution for a line of poles; in this sense branch contributions can be considered as generalized poles or "continuous" poles. Similar comments can be made including any special places where the contour is deformed around because of $\tilde{g}(s)$ blowing up; one may even have poles of the usual kind on the branch cut and these may be considered as separate terms.

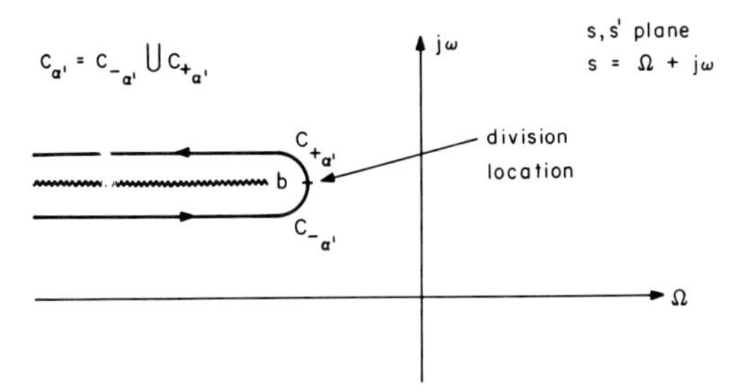

A. Division of contour into lower and upper parts

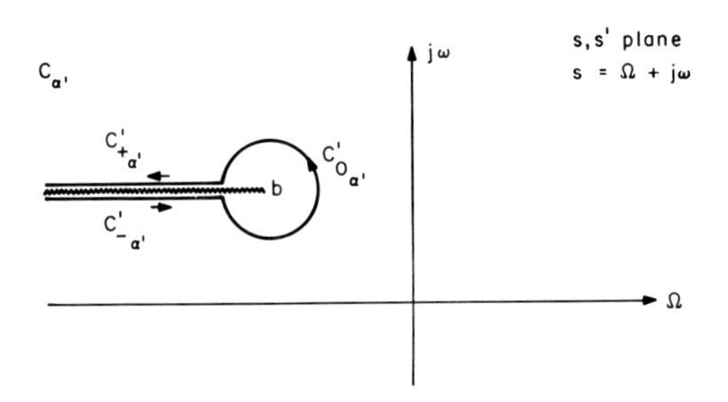

B. Separation of branch point contribution

Fig. 2.5. Branch contours.

I. Left Entire Function

Completing the decomposition of $\tilde{g}_-(s)$ (the left part of $\tilde{g}(s)$), we have the left entire function

$$\tilde{g}_{e_-}(s) = \lim_{\Delta_- \to \infty} \frac{1}{2\pi j} \int_{C_-} \frac{\tilde{g}(s')}{s - s'}\, ds' \tag{2.25}$$

$$g_{e_-}(t) = \lim_{\Delta_- \to \infty} \frac{1}{2\pi j} \int_{C_-} \tilde{g}(s') e^{s't} u(t) ds'$$

As in (2.14) and Fig. 2.4, one can consider three parts to C_-. The portions parallel to the $j\omega$ axis ($C_{\Delta_{-j\infty}}$ and $C_{\Delta_{+j\infty}}$) as $\Delta_- \to \infty$ can be considered by a high frequency method (HFM) such as the geometrical theory of diffraction (GTD) which is valid (asymptotically) along the $j\omega$ axis (and in the right half plane) [4]. For the approximately semicircular contour C_{Δ_-} (radius Δ_- from the origin) the problem is more complicated. As Δ_- is increased, convergence depends on convergence on the sum of the isolated singularities and branch integrals being swept by.

The basic concern is with C_{Δ_-} as $\Delta_- \to \infty$, since the shrinking $C_{\Delta_{-j\infty}}$ and $C_{\Delta_{+j\infty}}$ may typically give zero contribution in the limit. In such a case we would have

$$\tilde{g}_{e_-}(s) = \lim_{\Delta_- \to \infty} \frac{1}{2\pi j} \int_{C_{\Delta_-}} \frac{g(s')}{s - s'} ds'$$

$$g_{e_-}(t) = \lim_{\Delta_- \to \infty} \frac{1}{2\pi j} \int_{C_{\Delta_-}} \tilde{g}(s') e^{s't} u(t) ds' \tag{2.26}$$

J. Entire Function

Section II.D considers the right entire function and Section II.I considers the left entire function. Note that these correspond in time domain to $t < 0$ and $t > 0$ respectively for the entire function contributions.

In deforming the original contour C (Fig. 2.2) used in the Cauchy integral formula, one might keep it circular in shape and center it on the origin as indicated by the contour C_Δ in Fig. 2.6. Let the radius of C_Δ be Δ and let $\Delta \to \infty$ to give the entire function

$$\tilde{g}_e(s) = \tilde{g}_{e_+}(s) + \tilde{g}_{e_-}(s) = \frac{1}{2\pi j} \int_{C_\infty} \frac{\tilde{g}(s')}{s' - s} ds'$$

$$C_\infty \equiv \lim_{\Delta \to \infty} C_\Delta \tag{2.27}$$

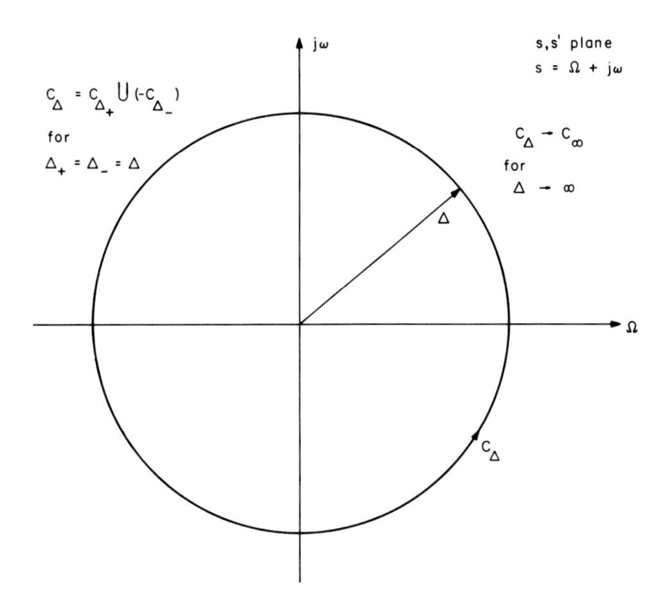

Fig. 2.6. Contour for entire function.

provided the limit exists as the isolated singuarities and branch cuts are swept by. Viewed another way C_Δ can be considered as the combination of two previously used contours ($-C_{\Delta_-}$ meaning integration in the opposite direction) as

$$C_\Delta = C_{\Delta_+} \bigcup (-C_{\Delta_-}) \quad , \quad \Delta_+ = \Delta_- = \Delta \tag{2.28}$$

Of course the contour C_Δ need not be circular as it expands toward ∞.

In time domain we have

$$\begin{aligned} g_e(t) &= g_{e_+}(t) + g_{e_-}(t) = \frac{1}{2\pi j}\int_{Br_0} \tilde{g}_e(s)ds \\ &= \lim_{\Delta\to\infty}\left\{\frac{1}{2\pi j}\int_{C_{\Delta_+}} \tilde{g}(s')e^{s't}u(-t)ds' - \frac{1}{2\pi j}\int_{-C_{\Delta_-}} \tilde{g}(s')e^{s't}u(t)ds'\right\} \\ &= \frac{1}{2\pi j}\oint_{C_\infty} \tilde{g}(s')e^{s't}[u(-t)u(Re[s'] - \Omega_0) - u(t)u(\Omega_0 - Re[s'])]ds' \\ &\qquad \Delta_+ = \Delta_- = \Delta \end{aligned} \tag{2.29}$$

where Ω_0 is chosen within the strip of convergence of the Laplace transform integral (of $g(t)$) and serves as the division between C_{Δ_+} and C_{Δ_-}.

Analogous to (2.24) one can define a residue distribution

$$\tilde{R}(s)\Big|_{s \text{ at } \infty} = \tilde{g}(s)\Big|_{s \text{ at } \infty} \tag{2.30}$$

giving

$$\tilde{g}_e(s) = \frac{1}{2\pi j}\oint_{C_\infty} \frac{\tilde{R}(s')}{s - s'}\, ds'$$

$$\tilde{g}_e(t) = \frac{1}{2\pi j}\oint_{C_\infty} \tilde{R}(s')e^{s't}[u(t)u(\Omega_0 - Re[s']) - u(-t)u(Re[s'] - \Omega_0)]ds' \tag{2.31}$$

so that the entire function can also be considered as a distribution of poles with residue distribution $\tilde{R}(s')$ (at ∞) similar to the branch type of continuous distribution of poles.

K. Some Applications

Looking back over this general procedure of singularity expansion via contour deformation, one can note some general features of the results. Using the Cauchy integral formula and inverse Laplace transform, it is observed that the various types of singularities can be all looked at as generalized poles because of the presence of the $(s - s')^{-1}$ term or $e^{s't}u(t)$ term in the integrand of the integral representation of the singularity term. Essential singularities may be viewed as infinite order poles; branch integral singularities may be viewed as continuous line distributions of poles (with residue distributions); entire functions ("singularities" at infinity) may be viewed as pole distributions along the contour at infinity.

This view of the various types of singularities as generalized poles has application to the characterization of transient waveforms and broad-band frequency-domain responses. Significant recent interest has been generated around Prony's method for characterizing transient waveforms by damped sinusoids because of its relationship to the pole terms in SEM [16]. By generalizing this concept to include higher order poles and continuous distributions of poles corresponding to branch integrals and entire functions, more generally useful representations of transient and broad-band response data may result. The residue distributions corresponding to the distributed poles may be expanded in terms of appropriate series of functions with undetermined coefficients and the integrals over these functions performed analytically; this gives series of terms of the form other than poles (or simple exponentials in time domain). Truncated series of these new kinds of terms can be used along with the usual pole terms in representing responses.

Having explored some general formulas for the verious terms in the singularity expansion, one can apply these formulas to branch contributions as well as to pole contributions (including essential singularities). Pole terms for scattering and antenna problems have received considerable development; other terms need comparable detailed development, and the present integral representations should be helpful.

In the general singularity expansion including branch integrals there is some arbitrariness in the selection of the path of the branch cut, except for the end point(s)(branch point(s)). Furthermore, other representations can be obtained by deforming contours through the branch cuts in the sense of going onto other Riemann sheets.

III. EIGENMODE EXPANSION

Having considered some of the aspects of singularity expansion based on analytic properties of functions, operators, etc. with respect to the complex frequency, let us now look at some of the general properties of the integral equations, specifically from the point of view of diagonalization, i.e. matrix-like properties. The reader is referred to more complete treatments of the subject [15,22] beyond the brief review here. This general method is referred to as the eigenmode expansion method (EEM).

Define eigenvalues and normalized eigenmodes corresponding to the general integral equation (1.5) as

$$
\begin{aligned}
&\langle \tilde{\overleftrightarrow{\Gamma}}(\vec{r},\vec{r}\,';s);\tilde{\vec{j}}_\beta(\vec{r}\,',s)\rangle = \tilde{\lambda}_\beta(s)\tilde{\vec{j}}_\beta(\vec{r}\,',s) \\
&\langle \tilde{\vec{\mu}}_\beta(\vec{r},s);\tilde{\overleftrightarrow{\Gamma}}(\vec{r},\vec{r}\,';s)\rangle = \tilde{\lambda}_\beta(s)\tilde{\vec{\mu}}_\beta(\vec{r},s) \\
&\langle \tilde{\vec{\mu}}_{\beta_1}(\vec{r},s);\tilde{\vec{j}}_{\beta_2}(\vec{r},s)\rangle = \delta_{\beta_1,\beta_2} \equiv \begin{cases} 1 \text{ for } \beta_1 = \beta_2 \\ 0 \text{ for } \beta_1 \neq \beta_2 \end{cases}
\end{aligned}
\tag{3.1}
$$

which have *MoM* approximate forms with N component vectors and $N \times N$ matrices as

$$
\begin{aligned}
&(\tilde{\Gamma}_{n,m}(s)) \cdot (\tilde{j}_n(s))_\beta = \tilde{\lambda}_\beta(s)(\tilde{j}_n(s))_\beta \\
&(\tilde{\mu}_n(s))_\beta \cdot (\tilde{\Gamma}_{n,m}(s)) = \tilde{\lambda}_\beta(s)(\tilde{\mu}_n(s))_\beta \\
&(\tilde{\mu}_n(s))_{\beta_1} \cdot (\tilde{j}_n(s))_{\beta_2} = \delta_{\beta_1,\beta_2} \\
&\det((\tilde{\Gamma}_{n,m}(s)) - \tilde{\lambda}_\beta(s)(\delta_{n,m})) = 0 \qquad \beta = 1,2,\ldots,N
\end{aligned}
\tag{3.2}
$$

For large N the N eigenvalues and corresponding modes can be used to approximate the operator form (3.1). Note that the right eigenmode $\tilde{\vec{j}}_\beta(\vec{r},s)$ can be superscripted with (Γ) to indicate of which operator it is a mode; for an impedance operator one might use (Z) for clarity. Such superscripts can be used with other appropriate quantities as well.

In dyadic form we have

$$
\begin{aligned}
&\tilde{\overleftrightarrow{d}}_\beta(\vec{r},\vec{r}\,';s) \equiv \tilde{\vec{j}}_\beta(\vec{r},s)\tilde{\vec{\mu}}_\beta(\vec{r}\,',s) \\
&\langle \tilde{\overleftrightarrow{\Gamma}}(\vec{r},\vec{r}\,';s);\tilde{\overleftrightarrow{d}}_\beta(\vec{r}\,',\vec{r}\,'';s)\rangle = \langle \tilde{\overleftrightarrow{d}}_\beta(\vec{r},\vec{r}\,';s);\tilde{\overleftrightarrow{\Gamma}}(\vec{r}\,',\vec{r}\,'';s)\rangle \\
&\qquad = \tilde{\lambda}_\beta(s)\tilde{\overleftrightarrow{d}}_\beta(\vec{r},\vec{r}\,'';s) \\
&\tilde{\overleftrightarrow{d}}_\beta^{\,n}(\vec{r},\vec{r}\,';s) \equiv \langle \tilde{\overleftrightarrow{d}}_\beta(\vec{r},\vec{r}\,'';s);\ldots;\tilde{\overleftrightarrow{d}}_\beta(\vec{r}^{\,(n)},\vec{r}\,';s)\rangle \qquad (n \text{ terms}) \\
&\qquad = \tilde{\overleftrightarrow{d}}_\beta(\vec{r},\vec{r}\,';s)
\end{aligned}
\tag{3.3}
$$

$$\tilde{\vec{\vec{\Gamma}}}^{n}(\vec{r},\vec{r}\,';s) = \sum_{\beta} \tilde{\lambda}_{\beta}^{n}(s)\tilde{\vec{\vec{d}}}_{\beta}(\vec{r},\vec{r}\,';s) \qquad n = 0,\pm 1,\pm 2,\ldots$$

The solution of the integral equation takes the form

$$\begin{aligned}\tilde{\vec{J}}(\vec{r},s) &= \sum_{\beta} \tilde{\lambda}_{\beta}^{-1}(s)\langle\tilde{\vec{\vec{d}}}_{\beta}(\vec{r},\vec{r}\,';s);\tilde{\vec{I}}(\vec{r}\,',s)\rangle \\ &= \sum_{\beta} \tilde{\lambda}_{\beta}^{-1}(s)\langle\tilde{\vec{\mu}}_{\beta}(\vec{r}\,',s);\tilde{\vec{I}}(\vec{r}\,',s)\rangle\tilde{\vec{j}}_{\beta}(\vec{r},s)\end{aligned} \tag{3.4}$$

Some special relations are

$$\tilde{\lambda}_{\beta}(s) = \langle\tilde{\vec{\mu}}_{\beta}(\vec{r},s);\tilde{\vec{\vec{\Gamma}}}(\vec{r},\vec{r}\,';s);\tilde{\vec{j}}_{\beta}(\vec{r}\,',s)\rangle$$

$$\frac{d}{ds}\tilde{\lambda}_{\beta}(s) = \langle\tilde{\vec{\mu}}_{\beta}(\vec{r},s);\frac{\partial}{\partial s}\tilde{\vec{\vec{\Gamma}}}(\vec{r},\vec{r}\,';s);\tilde{\vec{j}}_{\beta}(\vec{r}\,',s)\rangle$$

$$\frac{d^{n}}{ds^{n}}\langle\vec{\mu}_{\beta}(\vec{r},s);\vec{j}_{\beta}(\vec{r},s)\rangle = \delta_{\beta_{0}\beta'}\delta_{n,0} \qquad n = 0,1,2,\ldots \tag{3.5}$$

$$\vec{\vec{1}}\delta(\vec{r}-\vec{r}\,') = \vec{\vec{1}}\delta(\vec{r}\,'-\vec{r}) = \vec{\vec{\Gamma}}^{0}(\vec{r},\vec{r}\,';s) = \sum_{\beta}\tilde{\vec{\vec{d}}}_{\beta}(\vec{r},\vec{r}\,';s) = \sum_{\beta}\tilde{\vec{j}}_{\beta}(\vec{r},s)\tilde{\vec{\mu}}_{\beta}(\vec{r}\,',s)$$

$$= \sum_{\beta}\tilde{\vec{\mu}}_{\beta}(\vec{r}\,',s)\tilde{\vec{j}}_{\beta}(\vec{r},s) = \sum_{\beta}\tilde{\vec{j}}_{\beta}(\vec{r}\,',s)\tilde{\vec{\mu}}_{\beta}(\vec{r},s) = \sum_{\beta}\tilde{\vec{\mu}}_{\beta}(\vec{r},s)\tilde{\vec{j}}_{\beta}(\vec{r}\,',s)$$

where $\vec{\vec{1}}\delta(\vec{r}-\vec{r}\,')$ is the identity with respect to the integration domain $\langle\,,\,\rangle$ with $\vec{\vec{1}}$ as the dyadic identity and $\delta(\vec{r}-\vec{r}\,')$ as the three dimensional delta function; for surface-type objects we have $\vec{\vec{T}}\delta_{s}(\vec{r}-\vec{r}\,')$ with the transverse identity dyadic and the two dimensional (surface) delta function. Note the symmetry of the identity with respect to interchange of $\vec{r}$ and $\vec{r}\,'$ and transpose of the dyadic.

For symmetric kernels such as an impedance kernel (as in (1.6) and including symmetric impedance loading), the results simplify in form by setting

$$\tilde{\vec{\mu}}_{\beta}(\vec{r},s) = \tilde{\vec{j}}_{\beta}(\vec{r},s) \tag{3.6}$$

For this case we have

$$\frac{d^{n}}{ds^{n}}\langle\tilde{\vec{j}}_{\beta}(\vec{r},s);\tilde{\vec{j}}_{\beta'}(\vec{r},s)\rangle = \delta_{\beta,\beta'}\delta_{n,0} \tag{3.7}$$

giving the special result

$$< \frac{d}{ds} \tilde{\vec{j}}_\beta(\vec{r},s) ; \tilde{\vec{j}}_\beta(\vec{r},s) > = 0 \tag{3.8}$$

so that the s derivative of the mode (normalized) is orthogonal to the mode.

Corresponding to the normalized eigencurrents $\tilde{\vec{j}}_\beta(\vec{r},s)$ which diagonalize the integral operator, there are normalized eigenfields, both near and far, to describe the scattered or radiated fields. These are calculable using the dyadic Green's function (1.6). Eigencharges can be calculated using the divergence. Eigenpotentials can be calculated using the scalar Green's function (1.6). These are all related via normalized Maxwell equations.

IV. POLE TERMS

Combining the singularity expansion considerations concerning the complex frequency in section II together with the integral equation (1.5) and its eigenmode form in section III, one can understand the behavior of the important SEM pole terms for scatterers and antennas. For a large class of objects such pole terms dominate the response to a broad-band transient excitation containing frequencies spanning the electrically small to electrically large range. It was the experimental observation of the damped sinusoids in typical transient responses that led to the development of the SEM [12]. It was subsequently established that certain types of objects, in particular finite-size perfectly conducting objects in free space, have only pole singularities in the s plane (excluding infinity) for their delta-function responses [12,13,14]. These pole terms can be extended away from the object to radiated or scattered fields as well [7,8].

A. Natural Frequencies and Modes

First consider the natural frequencies which are by definition complex frequencies for which there is a response with no excitation. Such are determined from

$$\langle \tilde{\vec{\vec{\Gamma}}}(\vec{r},\vec{r}\,';s_\alpha);\vec{j}_\alpha(\vec{r}\,')\rangle = \vec{0} \quad , \quad \langle \vec{\mu}_\alpha(\vec{r});\tilde{\vec{\vec{\Gamma}}}(\vec{r},\vec{r}\,';s_\alpha)\rangle = \vec{0}$$

$$(\tilde{\Gamma}_{n,m}(s_\alpha)) \cdot (j_n)_\alpha = (0_n) \quad , \quad (\mu_n)_\alpha \cdot (\tilde{\Gamma}_{n,m}(s_\alpha)) = (0_n) \tag{4.1}$$

$$\det((\tilde{\Gamma}_{n,m}(s_\alpha)) = 0$$

The natural frequencies and modes are labelled with an index set α. Considering the eigenmode equations (3.1), one observes that (4.1) results when at least one eigenvalue is zero. This leads to

$$\alpha \equiv (\beta,\beta') \quad , \quad s_\alpha = s_{\beta,\beta'} \quad , \quad \tilde{\lambda}_\beta(s_{\beta,\beta'}) = 0 \tag{4.2}$$

so that one can consider a particular natural frequency as "belonging" to a particular eigenvalue/eigenmode [15]. β' indicates which zero of the βth eigenvalue is being considered.

The natural modes for the current density $\vec{j}_\alpha(\vec{r})$ (or current or surface current density) are typically normalized to have maximum magnitude of 1 and to have a real value of +1 there to the extent possible (accounting for vector components); if $\vec{j}_\alpha(\vec{r})$ does not have a maximum, a different normalization is required. The natural modes and normalized eigenmodes are then related by

$$\tilde{\vec{j}}_\beta(\vec{r},s_{\beta,\beta'}) = N_{\beta,\beta'}\vec{j}_{\beta,\beta'}(\vec{r})$$

$$\tilde{\vec{\mu}}_\beta(\vec{r},s_{\beta,\beta'}) = M_{\beta,\beta'}\vec{\mu}_{\beta,\beta'}(\vec{r}) \tag{4.3}$$

$$1 = N_{\beta,\beta'}M_{\beta,\beta'}\langle\vec{\mu}_{\beta,\beta'}(\vec{r});\vec{j}_{\beta,\beta'}(\vec{r})\rangle$$

The unnormalized eigenmodes can be set equal to the natural modes at $s_{\beta,\beta'}$ if desired, to consider the eigenmodes as generalizations of the natural modes in the sense of analytic continuation throughout the complex frequency plane. For symmetric kernels so that left and right modes can be set equal to each other, then

$$N_{\beta,\beta'} = M_{\beta,\beta'} \tag{4.4}$$

B. First Order Poles

The kernel of the inverse operator of the integral equation (1.5) gives the formal solution as

$$\tilde{\vec{J}}(\vec{r},s) = \langle \tilde{\vec{\vec{\Gamma}}}^{-1}(\vec{r},\vec{r}\,';s);\tilde{\vec{I}}(\vec{r}\,',s)\rangle \tag{4.5}$$

with the representation

$$\tilde{\vec{\vec{\Gamma}}}^{-1}(\vec{r},\vec{r}\,';s) = \sum_{\alpha} \frac{e^{-(s-s_\alpha)t_0}}{s-s_\alpha} \frac{\vec{j}_\alpha(\vec{r})\vec{\mu}_\alpha(\vec{r}\,')}{\langle \vec{\mu}_\alpha(\vec{r});\frac{\partial}{\partial s}\tilde{\vec{\vec{\Gamma}}}(\vec{r},\vec{r}\,';s)\big|_{s=s_\alpha};\vec{j}_\alpha(\vec{r}\,')\rangle}$$

$$+ \text{ other singularity terms} \tag{4.6}$$

(where t_0 may in general be a function of $\vec{r}$ and $\vec{r}\,'$, but is typically set to zero) for the case of first order poles. The delta-function response has the form

$$\tilde{\vec{U}}^{(J)}(\vec{r},s) = \sum_{\alpha} \tilde{n}_\alpha(\vec{1}_1,s)\vec{j}_\alpha(\vec{r})(s-s_\alpha)^{-n_\alpha}$$
$$+ \text{ other singularity terms} \tag{4.7}$$

where $\vec{1}_1$ (if needed) is the direction of incidence of the excitation with $n_\alpha = 1$ (only) for first order poles. Here we restrict consideration to delta-function excitation; one can add a waveform function to obtain the response

$$\tilde{\vec{J}}(\vec{r},s) = (\text{constant})\tilde{f}(s)\tilde{\vec{U}}^{(J)}(\vec{r},s) \tag{4.8}$$

For present purposes we can regard $\tilde{\vec{I}}$ as a normalized delta function excitation.

To find the class 1 form of the response one expands the terms in the integral equation in series about s_α as

$$\tilde{\vec{\vec{\Gamma}}}(\vec{r},\vec{r}\,';s) = \sum_{\ell=0}^{\infty} (s-s_\alpha)^{\ell}\vec{\vec{\Gamma}}_{\ell_\alpha}(\vec{r},\vec{r}\,')$$

$$\vec{\vec{\Gamma}}_{\ell_\alpha}(\vec{r},\vec{r}\,') = \frac{1}{\ell!}\frac{\partial^\ell}{\partial s^\ell}\tilde{\vec{\vec{\Gamma}}}(\vec{r},\vec{r}\,';s)\big|_{s=s_\alpha}$$

$$\tilde{\vec{I}}(\vec{r},s) = \sum_{\ell=0}^{\infty} (s - s_\alpha)^\ell \vec{I}_{\ell_\alpha}(\vec{r}) \tag{4.9}$$

$$\vec{I}_{\ell_\alpha}(\vec{r}) = \frac{1}{\ell!} \frac{\partial^\ell}{\partial s^\ell} \tilde{\vec{I}}(\vec{r},s)\Big|_{s=s_\alpha}$$

$$\tilde{\vec{U}}^{(J)}(\vec{r},s) = e^{-(s-s_\alpha)t_0} \tilde{\eta}_\alpha(\vec{1}_1,s_\alpha)\vec{j}_\alpha(\vec{r})(s - s_\alpha)^{-1}$$

$$+ \text{ vector function analytic near } s_\alpha$$

with no degeneracy of the $\vec{j}_\alpha$ assumed. Substituting these in the integral equation and collecting according to powers of $s - s_\alpha$, the $(s - s_\alpha)^{-1}$ term gives a result consistent with (4.1). The $(s - s_\alpha)^0$ term gives

$$\begin{aligned} &\langle \vec{\vec{\Gamma}}_{0_\alpha}(\vec{r},\vec{r}\,'); \text{ vector function of } \vec{r}\,'\rangle \\ &+ \langle \vec{\vec{\Gamma}}_{1_\alpha}(\vec{r},\vec{r}\,');\tilde{\eta}_\alpha(\vec{1}_1,s_\alpha)\vec{j}_\alpha(\vec{r}\,')\rangle = \vec{I}_{0_\alpha}(\vec{r}) \end{aligned} \tag{4.10}$$

which with left operation by $\vec{\mu}_\alpha(\vec{r})$ and rearrangement gives the fundamental result

$$\tilde{\eta}_\alpha(\vec{1}_1,s_\alpha) = \frac{\langle \vec{\mu}_\alpha(\vec{r});\vec{I}_{0_\alpha}(\vec{r})\rangle}{\langle \vec{\mu}_\alpha(\vec{r});\vec{\vec{\Gamma}}_{1_\alpha}(\vec{r},\vec{r}\,');\vec{j}_\alpha(\vec{r}\,')\rangle} \tag{4.11}$$

The class 2 form is found by using the inverse operator in (4.6) as

$$\tilde{\eta}_\alpha(\vec{1}_1,s) = \frac{\langle e^{-(s-s_\alpha)t_0}\vec{\mu}_\alpha(\vec{r});\tilde{\vec{I}}(\vec{r},s)\rangle}{\langle \vec{\mu}_\alpha(\vec{r});\vec{\vec{\Gamma}}_{1_\alpha}(\vec{r},\vec{r}\,');\vec{j}_\alpha(\vec{r}\,')\rangle} \tag{4.12}$$

Again note the inclusion of $e^{-(s-s_\alpha)t_0}$ in (4.7) and (4.9) with the turn-on time t_0 (which is included in the integration in case t_0 is taken as a function of $\vec{r},\vec{r}\,'$).

Using the eigenmode representation of the response (3.4) or inverse operator (3.3) and expanding the eigenvalue $\tilde{\lambda}_\beta(s)$ about $s_{\beta,\beta'}$, we have the class 1 coupling coefficient which uses

$$\tilde{\eta}_{\beta,\beta'}(\vec{1}_1,s_{\beta,\beta'}) = \left[\frac{\partial}{\partial s}\tilde{\lambda}_\beta(s)\Big|_{s=s_{\beta,\beta'}}\right]^{-1} \langle \tilde{\vec{\mu}}_\beta(\vec{r},s_{\beta,\beta'});\tilde{\vec{I}}(\vec{r},s_{\beta,\beta'})\rangle \tag{4.13}$$

and the class 2 coupling coefficient has the form

$$\tilde{\eta}_{\beta,\beta'}(\vec{1}_1,s) = \left[\frac{\partial}{\partial s}\tilde{\lambda}_\beta(s)\Big|_{s=s_{\beta,\beta'}}\right]^{-1} \langle e^{-(s-s_{\beta,\beta'})t_0}\tilde{\vec{\mu}}_\beta(\vec{r},s_{\beta,\beta'});\tilde{\vec{I}}(\vec{r},s)\rangle \tag{4.14}$$

(with $\tilde{\vec{j}}_\beta(\vec{r},s_{\beta,\beta'})$ used instead of $\vec{j}_\alpha(\vec{r})$ in this latter case) giving all the SEM quantities for a first order pole from the EEM quantities.

C. Second Order Poles

While less prevalent than first order poles, second order poles are useful for synthesis of antenna or scatterer loading to give what might be referred to as a critically damped antenna or scatterer for some dominant pole. This phenomenon has appeared in the context of resistive loading of an antenna or scatterer [7,22]. Third and higher order poles are in general possible and the present formulas can be extended to apply to such cases.

As our starting point consider the EEM form of the inverse kernel

$$\tilde{\vec{\vec{\Gamma}}}^{-1}(\vec{r},\vec{r}\,';s) = \sum_\beta \tilde{\lambda}_\beta^{-1}(s)\tilde{\vec{\vec{d}}}_\beta(\vec{r},\vec{r}\,';s) = \sum_\beta \tilde{\lambda}_\beta^{-1}(s)\tilde{\vec{j}}_\beta(\vec{r},s)\tilde{\vec{\mu}}_\beta(\vec{r}\,',s) \tag{4.15}$$

The natural frequency of interest is $s_{\beta,\beta'}$, and it is assumed that only one eigenvalue $\tilde{\lambda}_\beta$ is zero there; that zero is second order to proivde a second order pole in the inverse kernel. The eigenmode $\vec{j}_\beta$ (with $\vec{\mu}_\beta$) is assumed nondegenerate. We then have the series expansions around $s_{\beta,\beta'} = s_\alpha$ as

$$\tilde{\lambda}_\beta(s_{\beta,\beta'}) = \sum_{\ell=2}^{\infty} \lambda_{\ell_\beta}(s-s_{\beta,\beta'})^\ell$$

$$\lambda_{\ell_\beta} = \frac{1}{\ell!}\frac{\partial^\ell}{\partial s^\ell}\tilde{\lambda}_\beta(s)\Big|_{s=s_{\beta,\beta'}}$$

$$\lambda_{2_\beta} \neq 0$$

$$\tilde{\vec{j}}_\beta(\vec{r},s) = \sum_{\ell=0}^{\infty} \vec{j}_{\ell_\beta}(\vec{r})(s - s_{\beta,\beta'})^\ell \tag{4.16}$$

$$\vec{j}_{\ell_\beta}(\vec{r}) = \frac{1}{\ell!}\frac{\partial^\ell}{\partial s^\ell}\tilde{\vec{j}}_\beta(\vec{r},s)\Big|_{s=s_{\beta,\beta'}}$$

$$\tilde{\vec{\mu}}_\beta(\vec{r},s) = \sum_{\ell=0}^{\infty} \vec{\mu}_{\ell_\beta}(\vec{r})(s - s_{\beta,\beta'})^\ell$$

$$\vec{\mu}_{\ell_\beta}(\vec{r}) = \frac{1}{\ell!}\frac{\partial^\ell}{\partial s^\ell}\tilde{\vec{\mu}}_\beta(\vec{r},s)\Big|_{s=s_{\beta,\beta'}}$$

Near $s_{\beta,\beta'}$ we have

$$\begin{aligned}
\tilde{\vec{\vec{\Gamma}}}^{-1}(\vec{r},\vec{r}\,';s) &= \tilde{\lambda}_\beta^{-1}(s)\tilde{\vec{j}}_\beta(\vec{r},s)\tilde{\vec{\mu}}_\beta(\vec{r}\,',s) + O(\text{constant}) \\
&= (s - s_{\beta,\beta'})^{-2}\,\lambda_{2_\beta}^{-1}\,\vec{j}_{0_\beta}(\vec{r})\vec{\mu}_{0_\beta}(\vec{r}\,') \\
&\quad + (s - s_{\beta,\beta'})^{-1}\Big\{-\lambda_{2_\beta}^{-2}\,\lambda_{3_\beta}\,\vec{j}_{0_\beta}(\vec{r})\vec{\mu}_{0_\beta}(\vec{r}\,') \\
&\quad + \lambda_{2_\beta}^{-1}\Big[\vec{j}_{1_\beta}(\vec{r})\vec{\mu}_{0_\beta}(\vec{r}\,') + \vec{j}_{0_\beta}(\vec{r})\vec{\mu}_{1_\beta}(\vec{r}\,')\Big]\Big\} \\
&\quad + O(\text{constant})
\end{aligned} \tag{4.17}$$

Note the inclusion of the derivatives of the eigenmodes as well as the third order term in the eigenvalue expansion in the dyadic residue of the first order pole term at $s_{\beta,\beta'}$. Going on to third and higher order poles will introduce more and more higher order derivatives in the residues of the lower order poles at the same natural frequency.

Alternate representations for these terms can be found in the more classical SEM form. The eigenmodes at $s_{\beta,\beta'}$ are

$$\begin{aligned}
\vec{j}_{0_\beta}(\vec{r}) &= N_{\beta,\beta'}\vec{j}_{\beta,\beta'}(\vec{r}) = N_{\beta,\beta'}\vec{j}_\alpha(\vec{r}) \\
\vec{\mu}_{0_\beta}(\vec{r}) &= M_{\beta,\beta'}\vec{\mu}_{\beta,\beta'}(\vec{r}) = M_{\beta,\beta'}\vec{\mu}_\alpha(\vec{r})
\end{aligned} \tag{4.18}$$

$$N_{\beta,\beta'}M_{\beta,\beta'} = \langle\vec{\mu}_{\beta,\beta'}(\vec{r});\vec{j}_{\beta,\beta'}(\vec{r})\rangle^{-1} = \langle\vec{\mu}_\alpha(\vec{r});\vec{j}_\alpha(\vec{r})\rangle^{-1}$$

If we operate by $\tilde{\vec{\vec{\Gamma}}}_{0_\alpha} + (s - s_\alpha)\tilde{\vec{\vec{\Gamma}}}_{\alpha_1} + O((s-s_\alpha)^2)$ on the representation of $\tilde{\vec{\vec{\Gamma}}}^{-1}$ in (4.17) on either left or right, we obtain the identity. Noting that at the natural frequency $s_{\beta,\beta'}$

$$\langle \tilde{\vec{\vec{\Gamma}}}_{0_\alpha}(\vec{r},\vec{r}\,');\vec{j}_{0_\beta}(\vec{r}\,')\rangle = \vec{0} \quad , \quad \langle \vec{\mu}_{0_\beta}(\vec{r});\tilde{\vec{\vec{\Gamma}}}_{0_\alpha}(\vec{r},\vec{r}\,')\rangle = \vec{0} \tag{4.19}$$

and then considering the $(s - s_{\beta,\beta'})^{-2}$ term, we obtain a coefficient of zero. Similarly for the $(s - s_{\beta,\beta'})^{-1}$ term we must have a coefficient of zero, giving by operating on the left

$$\begin{aligned} 0 = {} & \lambda_{2_\beta}^{-1}\langle \tilde{\vec{\vec{\Gamma}}}_{1_\alpha}(\vec{r},\vec{r}\,');\vec{j}_{0_\beta}(\vec{r}\,')\vec{\mu}_{0_\beta}(\vec{r}\,'')\rangle \\ & - \lambda_{2_\beta}^{-2}\lambda_{3_\beta}\langle \vec{\vec{\Gamma}}_{0_\alpha}(\vec{r},\vec{r}\,');\vec{j}_{0_\beta}(\vec{r}\,')\vec{\mu}_{0_\beta}(\vec{r}\,'')\rangle \\ & + \lambda_{2_\beta}\langle \vec{\vec{\Gamma}}_{0_\alpha}(\vec{r},\vec{r}\,');\vec{j}_{1_\beta}(\vec{r}\,')\vec{\mu}_{0_\beta}(\vec{r}\,'')\rangle \\ & + \lambda_{\beta}^{-1}\langle \tilde{\vec{\vec{\Gamma}}}_{0_\alpha}(\vec{r},\vec{r}\,');\tilde{\vec{j}}_{0_\beta}(\vec{r}\,')\tilde{\vec{\mu}}_{1_\beta}(\vec{r}\,'')\rangle \end{aligned} \tag{4.20}$$

where the second and fourth terms on the right are seen to be zero from (4.19). This is consistent with expanding the eigenmode equations (3.1) to $s - s_{\beta,\beta'}$ terms about $s_{\beta,\beta'}$ giving

$$\begin{aligned} &\langle \vec{\vec{\Gamma}}_{1_\alpha}(\vec{r},\vec{r}\,');\vec{j}_{0_\beta}(\vec{r}\,')\rangle + \langle \vec{\vec{\Gamma}}_{0_\alpha}(\vec{r},\vec{r}\,');\vec{j}_{1_\beta}(\vec{r}\,')\rangle = \vec{0} \\ &\langle \vec{\mu}_{0_\beta}(\vec{r});\vec{\vec{\Gamma}}_{1_\alpha}(\vec{r},\vec{r}\,')\rangle + \langle \vec{\mu}_{1_\beta}(\vec{r});\vec{\vec{\Gamma}}_{0_\alpha}(\vec{r},\vec{r}\,')\rangle = \vec{0} \end{aligned} \tag{4.21}$$

Left operation by $\vec{\mu}_{0\beta}$ and right operation by $\vec{j}_{0\beta}$ respectively give in both cases

$$\langle \vec{\mu}_{0_\beta}(\vec{r});\vec{\vec{\Gamma}}_{1_\alpha}(\vec{r},\vec{r}\,');\vec{j}_{0_\beta}(\vec{r}\,')\rangle = 0 \tag{4.22}$$

consistent with the first derivative of the eigenvalue being zero at $s_{\beta,\beta'}$.

Expanding the eigenmode equations to $(s - s_{\beta,\beta'})^2$ terms about

$s_{\beta,\beta'}$ gives

$$\begin{aligned}
&\langle \vec{\vec{\Gamma}}_{2_\alpha}(\vec{r},\vec{r}');\vec{j}_{0_\beta}(\vec{r}')\rangle + \langle \vec{\vec{\Gamma}}_{1_\alpha}(\vec{r},\vec{r}');\vec{j}_{1_\beta}(\vec{r}')\rangle \\
&\quad + \langle \vec{\vec{\Gamma}}_{0_\alpha}(\vec{r},\vec{r}');\vec{j}_{2_\beta}(\vec{r}')\rangle = \lambda_{2_\beta}\vec{j}_{0_\beta}(\vec{r}) \\
&\langle \vec{\mu}_{0_\beta}(\vec{r});\vec{\vec{\Gamma}}_{2_\alpha}(\vec{r},\vec{r}')\rangle + \langle \vec{\mu}_{1_\beta}(\vec{r});\vec{\vec{\Gamma}}_{1_\alpha}(\vec{r},\vec{r}')\rangle \\
&\quad + \langle \vec{\mu}_{2_\beta}(\vec{r});\vec{\vec{\Gamma}}_{0_\alpha}(\vec{r},\vec{r}')\rangle = \lambda_{2_\beta}\vec{\mu}_{0_\beta}(\vec{r}')
\end{aligned} \tag{4.23}$$

Left operation with $\vec{\mu}_{0\beta}$ and right operation with $\vec{j}_{0\beta}$ respectively give in both cases

$$\langle \vec{\mu}_{0_\beta}(\vec{r});\vec{\vec{\Gamma}}_{2_\alpha}(\vec{r},\vec{r}');\vec{j}_{0_\beta}(\vec{r}')\rangle - \langle \vec{\mu}_{1_\beta}(\vec{r});\vec{\vec{\Gamma}}_{0_\alpha}(\vec{r},\vec{r}');\vec{j}_{1_\beta}(\vec{r}')\rangle = \lambda_{2_\beta} \neq 0 \tag{4.24}$$

where (4.21) has been used to convert $\vec{\vec{\Gamma}}_{1_\alpha}$ terms into $\vec{\vec{\Gamma}}_{0_\alpha}$ terms for symmetry of results.

Expanding the eigenmode equations to $(s - s_{\beta,\beta'})^3$ terms about $s_{\beta,\beta'}$ gives

$$\begin{aligned}
&\langle \vec{\vec{\Gamma}}_{3_\alpha}(\vec{r},\vec{r}');\vec{j}_{0_\beta}(\vec{r}')\rangle + \langle \vec{\vec{\Gamma}}_{2_\alpha}(\vec{r},\vec{r}');\vec{j}_{1_\beta}(\vec{r}')\rangle + \langle \vec{\vec{\Gamma}}_{1_\alpha}(\vec{r},\vec{r}');\vec{j}_{2_\beta}(\vec{r}')\rangle \\
&\quad + \langle \vec{\vec{\Gamma}}_{0_\alpha}(\vec{r},\vec{r}');\vec{j}_{3_\beta}(\vec{r}')\rangle = \lambda_{2_\beta}\vec{j}_{1_\beta}(\vec{r}) + \lambda_{3_\beta}\vec{j}_{0_\beta}(\vec{r}) \\
&\langle \vec{\mu}_{0_\beta}(\vec{r});\vec{\vec{\Gamma}}_{3_\alpha}(\vec{r},\vec{r}')\rangle + \langle \vec{\mu}_{1_\beta}(\vec{r});\vec{\vec{\Gamma}}_{2_\alpha}(\vec{r},\vec{r}')\rangle + \langle \vec{\mu}_{2_\beta}(\vec{r});\vec{\vec{\Gamma}}_{1_\alpha}(\vec{r},\vec{r}')\rangle \\
&\quad + \langle \vec{\mu}_{3_\beta}(\vec{r});\vec{\vec{\Gamma}}_{0_\alpha}(\vec{r},\vec{r}')\rangle = \lambda_{2_\beta}\vec{\mu}_{1_\beta}(\vec{r}') + \lambda_{3_\beta}\vec{\mu}_{0_\beta}(\vec{r})
\end{aligned} \tag{4.25}$$

Left operation with $\vec{\mu}_{0\beta}$ and right operation with $\vec{j}_{0\beta}$ respectively gives

$$\begin{aligned}
&\langle \vec{\mu}_{0_\beta}(\vec{r});\vec{\vec{\Gamma}}_{3_\alpha}(\vec{r},\vec{r}');\vec{j}_{0_\beta}(\vec{r}')\rangle + \langle \vec{\mu}_{0_\beta}(\vec{r});\vec{\vec{\Gamma}}_{2_\alpha}(\vec{r},\vec{r}');\vec{j}_{1_\beta}(\vec{r}')\rangle \\
&\quad + \langle \vec{\mu}_{0_\beta}(\vec{r});\vec{\vec{\Gamma}}_{1_\alpha}(\vec{r},\vec{r}');\vec{j}_{2_\beta}(\vec{r}')\rangle = \lambda_{2_\beta}\langle \vec{\mu}_{0_\beta}(\vec{r});\vec{j}_{1_\beta}(\vec{r})\rangle + \lambda_{3_\beta}
\end{aligned} \tag{4.26}$$

$$<\vec{\mu}_{0_\beta}(\vec{r});\vec{\vec{\Gamma}}_{3_\alpha}(\vec{r},\vec{r}');\vec{j}_{0_\beta}(\vec{r}')> + <\vec{\mu}_{1_\beta}(\vec{r});\vec{\vec{\Gamma}}_{2_\alpha}(\vec{r},\vec{r}');\vec{j}_{0_\beta}(\vec{r}')>$$
$$+ <\vec{\mu}_{2_\beta}(\vec{r});\vec{\vec{\Gamma}}_{1_\alpha}(\vec{r},\vec{r}');\vec{j}_{0_\beta}(\vec{r}')> = \lambda_{2_\beta}<\vec{\mu}_{1_\alpha}(\vec{r});\vec{j}_{0_\beta}(\vec{r})> = \lambda_{3_\beta}$$

From (3.5) we have

$$<\vec{\mu}_{0_\beta}(\vec{r});\vec{j}_{1_\beta}(\vec{r})> + <\vec{\mu}_{1_\beta}(\vec{r});\vec{j}_{0_\beta}(\vec{r})> = 0 \tag{4.27}$$

which is used in simplifying half the sum of the two equations (4.26) as

$$\begin{aligned}&<\vec{\mu}_{0_\beta}(\vec{r});\vec{\vec{\Gamma}}_{3_\alpha}(\vec{r},\vec{r}');\vec{j}_{0_\beta}(\vec{r}')>\\ &+\frac{1}{2}\left[<\vec{\mu}_{0_\beta}(\vec{r});\vec{\vec{\Gamma}}_{2_\alpha}(\vec{r},\vec{r}');\vec{j}_{1_\beta}(\vec{r}')> + <\vec{\mu}_{1_\beta}(\vec{r});\vec{\vec{\Gamma}}_{2_\alpha}(\vec{r},\vec{r}');\vec{j}_{0_\beta}(\vec{r}')>\right]\\ &+\frac{1}{2}\left[<\vec{\mu}_{0_\beta}(\vec{r});\vec{\vec{\Gamma}}_{1_\alpha}(\vec{r},\vec{r}');\vec{j}_{2_\beta}(\vec{r}')> + <\vec{\mu}_{2_\beta}(\vec{r});\vec{\vec{\Gamma}}_{1_\alpha}(\vec{r},\vec{r}');\vec{j}_{0_\beta}(\vec{r})>\right] = \lambda_{3_\beta}\end{aligned} \tag{4.28}$$

Having computed the expansion terms of the eigenvalue (λ_{2_β} and λ_{3_β} in (4.24) and (4.28)) we need the terms in the expansion of the eigenmodes in (4.17), noting that $\vec{\vec{\Gamma}}_{0_\alpha}^{-1}$ does not exist because of its definition at $s = s_\alpha$. For this purpose define

$$\vec{\vec{\Gamma}}_{d_\beta}^{-1}(\vec{r},\vec{r}';s) \equiv \sum_{\beta_1 \neq \beta} \tilde{\lambda}_{\beta_1}^{-1}(s)\tilde{\vec{j}}_{\beta_1}(\vec{r},s)\tilde{\vec{\mu}}_{\beta_1}(\vec{r}',s) \tag{4.29}$$

where the subscript d_β indicates that the βth term (the term singular at s_α) is deleted. At the natural frequency we have

$$\vec{\vec{\Gamma}}_{d_{0_\alpha}}^{-1}(\vec{r},\vec{r}') = \sum_{\beta_1 \neq \beta} \lambda_{0_{\beta_1}}^{-1}\vec{j}_{0_{\beta_1}}(\vec{r})\vec{\mu}_{\beta_1}(\vec{r}') \tag{4.30}$$

where the expansion terms around $s = s_\alpha$ for $\beta_1 \neq \beta$ are defined as in (4.16) for the β terms. Note that we have

$$\begin{aligned}<\vec{\vec{\Gamma}}_{d_{0_\alpha}}^{-1}(\vec{r},\vec{r}'');\vec{\vec{\Gamma}}_{0_\alpha}(\vec{r}'',\vec{r}')> &= <\vec{\vec{\Gamma}}_{0_\alpha}(\vec{r},\vec{r}'');\vec{\vec{\Gamma}}_{d_{0_\alpha}}(\vec{r}'',\vec{r}')>\\ &= \vec{\vec{1}}\delta(\vec{r}-\vec{r}') - \vec{j}_{0_\beta}(\vec{r})\vec{\mu}_{0_\beta}(\vec{r}')\end{aligned} \tag{4.31}$$

giving an extension of the concept of inverse at a natural frequency.

Applying (4.30) to (4.21) gives

$$\vec{j}_{1_\beta}(\vec{r}) - \langle \vec{\mu}_{0_\beta}(\vec{r}\,'); \vec{j}_{1_\beta}(\vec{r}\,')\rangle \vec{j}_{0_\beta}(\vec{r}) = -\langle \vec{\vec{\Gamma}}^{-1}_{d_{0_\alpha}}(\vec{r},\vec{r}\,'); \vec{\vec{\Gamma}}_{1_\alpha}(\vec{r}\,',\vec{r}\,''); \vec{j}_{0_\beta}(\vec{r}\,'')\rangle$$

$$\vec{\mu}_{1_\beta}(\vec{r}) - \langle \vec{\mu}_{1_\beta}(\vec{r}\,'); \vec{j}_{0_\beta}(\vec{r}\,')\rangle \vec{\mu}_{0_\beta}(\vec{r}) = -\langle \vec{\mu}_{0_\beta}(\vec{r}\,''); \vec{\vec{\Gamma}}_{1_\alpha}(\vec{r}\,'',\vec{r}\,'); \vec{\vec{\Gamma}}^{-1}_{d_{0_\alpha}}(\vec{r}\,',\vec{r})\rangle \tag{4.32}$$

and to (4.23) gives

$$\vec{j}_{2_\beta}(\vec{r}) - \langle \vec{\mu}_{0_\beta}(\vec{r}\,'); \vec{j}_{2_\beta}(\vec{r}\,')\rangle \vec{j}_{0_\beta}(\vec{r}) = -\langle \vec{\vec{\Gamma}}^{-1}_{d_{0_\alpha}}(\vec{r},\vec{r}\,'); \vec{\vec{\Gamma}}_{1_\alpha}(\vec{r}\,',\vec{r}\,''); \vec{j}_{1_\beta}(\vec{r}\,'')\rangle$$
$$-\langle \vec{\vec{\Gamma}}^{-1}_{d_{0_\alpha}}(\vec{r},\vec{r}\,'); \vec{\vec{\Gamma}}_{2_\alpha}(\vec{r}\,',\vec{r}\,''); \vec{j}_{0_\beta}(\vec{r}\,'')\rangle$$

$$\vec{\mu}_{2_\beta}(\vec{r}) - \langle \vec{\mu}_{2_\beta}(\vec{r}\,'); \vec{j}_{0_\beta}(\vec{r}\,')\rangle \vec{\mu}_{0_\beta}(\vec{r}) = -\langle \vec{\mu}_{1_\beta}(\vec{r}\,''); \vec{\vec{\Gamma}}_{1_\alpha}(\vec{r}\,'',\vec{r}\,'); \vec{\vec{\Gamma}}^{-1}_{d_{0_\alpha}}(\vec{r}\,',\vec{r})\rangle$$
$$-\langle \vec{\mu}_{0_\beta}(\vec{r}\,''); \vec{\vec{\Gamma}}_{2_\alpha}(\vec{r}\,'',\vec{r}\,'); \vec{\vec{\Gamma}}^{-1}_{d_{0_\alpha}}(\vec{r}\,',\vec{r})\rangle \tag{4.33}$$

Substituting (4.32) into (4.33) gives

$$\begin{aligned}
\vec{j}_{2_\beta}(\vec{r}) &- \langle \vec{\mu}_{0_\beta}(\vec{r}\,'); \vec{j}_{2_\beta}(\vec{r}\,')\rangle \vec{j}_{0_\beta}(\vec{r}) \\
&= -\langle \vec{\mu}_{0_\beta}(\vec{r}\,'); \vec{j}_{1_\beta}(\vec{r}\,')\rangle \langle \vec{\vec{\Gamma}}^{-1}_{d_{0_\alpha}}(\vec{r},\vec{r}\,'); \vec{\vec{\Gamma}}_{1_\alpha}(\vec{r}\,',\vec{r}\,''); \vec{j}_{0_\beta}(\vec{r}\,'')\rangle \\
&+ \langle\langle \vec{\vec{\Gamma}}^{-1}_{d_{0_\alpha}}(\vec{r},\vec{r}\,'); \vec{\vec{\Gamma}}_{1_\alpha}(\vec{r}\,',\vec{r}\,'')\rangle^2; \vec{j}_{0_\beta}(\vec{r}\,'')\rangle \\
&+ \langle \vec{\vec{\Gamma}}^{-1}_{d_{0_\alpha}}(\vec{r},\vec{r}\,'); \vec{\vec{\Gamma}}_{2_\alpha}(\vec{r}\,',\vec{r}\,''); \vec{j}_{0_\beta}(\vec{r}\,'')\rangle
\end{aligned} \tag{4.34}$$

$$\begin{aligned}
\vec{\mu}_{2_\beta}(\vec{r}) &- \langle \vec{\mu}_{2_\beta}(\vec{r}\,'); \vec{j}_{0_\beta}(\vec{r}\,')\rangle \vec{\mu}_{0_\beta}(\vec{r}) \\
&= -\langle \vec{\mu}_{1_\beta}(\vec{r}\,'); \vec{j}_{0_\beta}(\vec{r}\,')\rangle \langle \vec{\mu}_{0_\beta}(\vec{r}\,''); \vec{\vec{\Gamma}}_{1_\alpha}(\vec{r}\,'',\vec{r}\,'); \vec{\vec{\Gamma}}^{-1}_{d_{0_\alpha}}(\vec{r}\,',\vec{r})\rangle \\
&+ \langle \vec{\mu}_{0_\beta}(\vec{r}\,''); \langle \vec{\vec{\Gamma}}_{1_\alpha}(\vec{r}\,'',\vec{r}\,'); \vec{\vec{\Gamma}}_{d_{0_\alpha}}(\vec{r}\,',\vec{r})\rangle^2\rangle \\
&- \langle \vec{\mu}_{0_\beta}(\vec{r}\,''); \vec{\vec{\Gamma}}_{2_\alpha}(\vec{r}\,'',\vec{r}\,'); \vec{\vec{\Gamma}}_{d_{0_\alpha}}(\vec{r}\,',\vec{r})\rangle
\end{aligned}$$

From (3.5) we have the relations at $s = s_\alpha$

$$\langle \vec{\mu}_{0_\beta}(\vec{r});\vec{j}_{0_\beta}(\vec{r})\rangle = 1$$

$$\langle \vec{\mu}_{0_\beta}(\vec{r});\vec{j}_{1_\beta}(\vec{r})\rangle + \langle \vec{\mu}_{1_\beta}(\vec{r});\vec{j}_{0_\beta}(\vec{r})\rangle = 0 \tag{4.35}$$

$$\langle \vec{\mu}_{0_\beta}(\vec{r});\vec{j}_{2_\beta}(\vec{r})\rangle + \langle \vec{\mu}_{1_\beta}(\vec{r});\vec{j}_{1_\beta}(\vec{r})\rangle + \langle \vec{\mu}_{2_\beta}(\vec{r});\vec{j}_{0_\beta}(\vec{r})\rangle = 0$$

which for the special case of symmetric kernels reduce to

$$\vec{\mu}_{n_\beta}(\vec{r}) \equiv \vec{j}_{n_\beta}(\vec{r}) \quad , \quad n = 0,1,2,\ldots$$

$$\langle \vec{j}_{0_\beta}(\vec{r});\vec{j}_{0_\beta}(\vec{r})\rangle = 1 \quad , \quad \langle \vec{j}_{0_\beta}(\vec{r});\vec{j}_{1_\beta}(\vec{r})\rangle = 0 \tag{4.36}$$

$$2\langle \vec{j}_{0_\beta}(\vec{r});\vec{j}_{2_\beta}(\vec{r})\rangle + \langle \vec{j}_{1_\beta}(\vec{r});\vec{j}_{1_\beta}(\vec{r})\rangle = 0$$

Substitution from (4.32) into one of the terms in (4.24) gives (using (4.22) and 4.31))

$$\begin{aligned}
&\langle \vec{\mu}_{1_\beta}(\vec{r});\vec{\vec{\Gamma}}_{0_\alpha}(\vec{r},\vec{r}\,');\vec{j}_{1_\beta}(\vec{r}\,')\rangle \\
&= \langle \vec{\mu}_{0_\beta}(\vec{r}^{(1)});\vec{\vec{\Gamma}}_{1_\alpha}(\vec{r}^{(1)},\vec{r}^{(2)});\vec{\vec{\Gamma}}_{d_{0_\alpha}}^{-1}(\vec{r}^{(2)},\vec{r}^{(3)});\vec{\vec{\Gamma}}_{0_\alpha}(\vec{r}^{(3)},\vec{r}^{(4)}); \\
&\qquad \vec{\vec{\Gamma}}_{d_{0_\alpha}}^{-1}(\vec{r}^{(4)},\vec{r}^{(5)});\vec{\vec{\Gamma}}_{1_\alpha}(\vec{r}^{(5)},\vec{r}^{(6)});\vec{j}_{0_\beta}(\vec{r}^{(6)})\rangle \\
&= \langle \vec{\mu}_{0_\beta}(\vec{r}^{(1)});\vec{\vec{\Gamma}}_{1_\alpha}(\vec{r}^{(1)},\vec{r}^{(2)});\vec{\vec{\Gamma}}_{d_{0_\alpha}}^{-1}(\vec{r}^{(2)},\vec{r}^{(3)});\vec{\vec{\Gamma}}_{1_\alpha}(\vec{r}^{(3)},\vec{r}^{(4)}); \\
&\qquad \vec{j}_{0_\beta}(\vec{r}^{(5)})\rangle
\end{aligned} \tag{4.37}$$

giving for (4.24)

$$\begin{aligned}
\lambda_{2_\beta} &= \langle \vec{\mu}_{0_\beta}(\vec{r});\vec{\vec{\Gamma}}_{2_\alpha}(\vec{r},\vec{r}\,');\vec{j}_{0_\beta}(\vec{r}\,')\rangle \\
&\quad - \langle \vec{\mu}_{0_\beta}(\vec{r}^{(1)});\vec{\vec{\Gamma}}_{1_\alpha}(\vec{r}^{(1)},\vec{r}^{(2)});\vec{\vec{\Gamma}}_{d_{0_\alpha}}^{-1}(\vec{r}^{(2)},\vec{r}^{(3)});\vec{\vec{\Gamma}}_{1_\alpha}(\vec{r}^{(3)},\vec{r}^{(4)}); \\
&\qquad \vec{j}_{0_\beta}(\vec{r}^{(5)})\rangle
\end{aligned} \tag{4.38}$$

so that the second order eigenvalue term is expressible in terms of the eigenmodes, eigenvalues and kernel derivatives.

Substituting from (4.32) and (4.34) into (4.28) (using (4.22) and (4.35)) gives

$$\lambda_{3_\beta} = \langle\vec{\mu}_{0_\beta}(\vec{r});\vec{\vec{\Gamma}}_{3_\alpha}(\vec{r},\vec{r}\,');\vec{j}_{0_\beta}(\vec{r}\,')\rangle - \langle\vec{\mu}_{0_\beta}(\vec{r}^{\,(1)});\vec{\vec{\Gamma}}_{2_\alpha}(\vec{r}^{\,(1)},\vec{r}^{\,(2)});$$
$$\vec{\vec{\Gamma}}^{-1}_{d_{0_\alpha}}(\vec{r}^{\,(2)},\vec{r}^{\,(3)});\vec{\vec{\Gamma}}_{1_\alpha}(\vec{r}^{\,(3)},\vec{r}^{\,(4)});\vec{j}_{0_\beta}(\vec{r}^{\,(4)})\rangle - \langle\vec{\mu}_{0_\beta}(\vec{r}^{\,(1)});$$
$$\vec{\vec{\Gamma}}_{1_\alpha}(\vec{r}^{\,(1)},\vec{r}^{\,(2)});\vec{\vec{\Gamma}}^{-1}_{d_{0_\alpha}}(\vec{r}^{\,(2)},\vec{r}^{\,(3)});\vec{\vec{\Gamma}}_{2_\alpha}(\vec{r}^{\,(3)},\vec{r}^{\,(4)});\vec{j}_{0_\beta}(\vec{r}^{\,(4)})\rangle$$
$$+ \langle\vec{\mu}_{0_\beta}(\vec{r}^{\,(1)});\vec{\vec{\Gamma}}_{1_\alpha}(\vec{r}^{\,(1)},\vec{r}^{\,(2)});\langle\vec{\vec{\Gamma}}^{-1}_{d_{0_\alpha}}(\vec{r}^{\,(2)},\vec{r}^{\,(3)});\vec{\vec{\Gamma}}_{1_\alpha}(\vec{r}^{\,(3)},\vec{r}^{\,(4)})\rangle^2;$$
$$\vec{j}_{0_\beta}(\vec{r}^{\,(4)})\rangle \tag{4.39}$$

with a fair bit of combination and cancellation of terms. In addition we need

$$\vec{j}_{1_\beta}(\vec{r})\vec{\mu}_{0_\beta}(\vec{r}\,') + \vec{j}_{0_\beta}(\vec{r})\vec{\mu}_{1_\beta}(\vec{r}\,')$$
$$= -\langle\vec{\vec{\Gamma}}^{-1}_{d_{0_\alpha}}(\vec{r},\vec{r}^{\,(1)});\vec{\vec{\Gamma}}_{1_\alpha}(\vec{r}^{\,(1)},\vec{r}^{\,(2)});\vec{j}_{0_\beta}(\vec{r}^{\,(2)})\vec{\mu}_{0_\beta}(\vec{r}\,')\rangle \tag{4.40}$$
$$-\langle\vec{j}_{0_\beta}(\vec{r})\vec{\mu}_{0_\beta}(\vec{r}^{\,(1)});\vec{\vec{\Gamma}}_{1_\alpha}(\vec{r}^{\,(1)},\vec{r}^{\,(2)});\vec{\vec{\Gamma}}^{-1}_{d_{0_\alpha}}(\vec{r}^{\,(2)},\vec{r}\,')\rangle$$

Let us now write the inverse operator from (4.17) as

$$\tilde{\vec{\vec{\Gamma}}}^{-1}(\vec{r},\vec{r}\,';s) = \vec{\vec{\Gamma}}_{-2_\alpha}(\vec{r},\vec{r}\,')(s-s_{\beta,\beta'})^{-2} + \vec{\vec{\Gamma}}_{-1_\alpha}(\vec{r},\vec{r}\,')(s-s_{\beta,\beta'})^{-1}$$
$$+ O(\text{constant}) \qquad \text{as } s \to s_{\beta\,\beta'} \tag{4.41}$$

The dyadic coefficient of the second order pole term is

$$\vec{\vec{\Gamma}}_{-2_\alpha}(\vec{r},\vec{r}\,') = \lambda^{-1}_{2_\beta}\vec{j}_{0_\beta}(\vec{r})\vec{\mu}_{0_\beta}(\vec{r}\,') \tag{4.42}$$

using the result of (4.38). The dyadic coefficient of the first order pole term is

$$\overset{\leftrightarrow}{\Gamma}_{-1_\alpha}(\vec{r},\vec{r}') = -\lambda_{2_\beta}^{-2}\lambda_{3_\beta}\vec{j}_{0_\beta}(\vec{r})\vec{\mu}_{0_\beta}(\vec{r}') + \lambda_{2_\beta}^{-1}\left[\vec{j}_{1_\beta}(\vec{r})\vec{\mu}_{0_\beta}(\vec{r}') + \vec{j}_{0_\beta}(\vec{r})\vec{\mu}_{1_\beta}(\vec{r}')\right] \tag{4.43}$$

with the terms here obtainable from (4.38) through (4.40) (as well as (3.1) and (4.1)).

If one singularity expands $e^{st_0}\tilde{\overset{\leftrightarrow}{\Gamma}}^{-1}$ and then multiplies by e^{-st_0}, the second order pole at $s_{\beta,\beta'}$ takes the form

$$\begin{aligned}\overset{\leftrightarrow}{\Gamma}^{-1}(\vec{r},\vec{r}';s) &= e^{-(s-s_{\beta,\beta'})t_0}\overset{\leftrightarrow}{\Gamma}_{-2_\alpha}(\vec{r},\vec{r}')(s-s_{\beta,\beta'})^{-2} \\ &+ e^{-(s-s_{\beta,\beta'})t_0}\left[\overset{\leftrightarrow}{\Gamma}_{-1_\alpha}(\vec{r},\vec{r}') + t_0\overset{\leftrightarrow}{\Gamma}_{-2_\alpha}(\vec{r},\vec{r}')\right](s-s_{\beta,\beta'})^{-1} \\ &+ O(\text{constant}) \qquad \text{as} \quad s \to s_{\beta,\beta'}\end{aligned} \tag{4.44}$$

with $\overset{\leftrightarrow}{\Gamma}_{-2_\alpha}$ and $\overset{\leftrightarrow}{\Gamma}_{-1_\alpha}$ as given by (4.42) and (4.43) respectively. The response has the form from (4.7) for class 2

$$\begin{aligned}\tilde{\vec{U}}^{(J)}(\vec{r},s) &= \langle e^{-(s-s_{\beta,\beta'})t_0}\overset{\leftrightarrow}{\Gamma}_{-2_\alpha}(\vec{r},\vec{r}');\tilde{\vec{I}}(\vec{r}',s)\rangle(s-s_{\beta,\beta'})^{-2} \\ &+ \langle e^{-(s-s_{\beta,\beta'})t_0}\left[\overset{\leftrightarrow}{\Gamma}_{-1_\alpha}(\vec{r},\vec{r}') + t_0\overset{\leftrightarrow}{\Gamma}_{-2_\alpha}(\vec{r},\vec{r}')\right];\tilde{\vec{I}}(\vec{r}',s)\rangle \\ &\quad (s-s_{\beta,\beta'})^{-1} + O(\text{constant}) \qquad \text{as} \quad s \to s_{\beta\,\beta'}\end{aligned} \tag{4.45}$$

and for class 1

$$\begin{aligned}\tilde{\vec{U}}^{(J)}(\vec{r},s) &= e^{-(s-s_{\beta,\beta'})t_0}\langle\overset{\leftrightarrow}{\Gamma}_{-2_\alpha}(\vec{r},\vec{r}');\tilde{\vec{I}}(\vec{r}',s_{\beta,\beta'})\rangle(s-s_{\beta,\beta'})^{-2} \\ &+ e^{-(s-s_{\beta,\beta'})t_0}\Big[\langle\overset{\leftrightarrow}{\Gamma}_{-1_\alpha}(\vec{r},\vec{r}') + t_0\overset{\leftrightarrow}{\Gamma}_{-2_\alpha}(\vec{r},\vec{r}');\tilde{\vec{I}}(\vec{r}',s_{\beta,\beta'})\rangle \\ &+ \langle\overset{\leftrightarrow}{\Gamma}_{-2_\alpha}(\vec{r},\vec{r}');\frac{\partial}{\partial s}\tilde{\vec{I}}(\vec{r}',s)\Big|_{s=s_{\beta,\beta'}}\rangle\Big](s-s_{\beta,\beta'})^{-1} \\ &+ O(\text{constant}) \qquad \text{as} \quad s \to s_{\beta,\beta'}\end{aligned} \tag{4.46}$$

For class 2 the second order pole has a coefficient

$$\begin{aligned}&\langle e^{-(s-s_{\beta,\beta'})t_0}\overset{\leftrightarrow}{\Gamma}_{-2_\alpha}(\vec{r},\vec{r}');\tilde{\vec{I}}(\vec{r}',s)\rangle \\ &\quad = \lambda_{2_\beta}^{-1}\langle e^{-(s-s_{\beta,\beta'})t_0}\vec{\mu}_{0_\beta}(\vec{r}');\tilde{\vec{I}}(\vec{r}',s)\rangle\vec{j}_{0_\beta}(\vec{r})\end{aligned} \tag{4.47}$$

and the first order pole has a coefficient

$$<e^{-(s-s_{\beta,\beta'})t_0}\left[\vec{\vec{\Gamma}}_{-1_\alpha}(\vec{r},\vec{r}\,') + t_0\vec{\vec{\Gamma}}_{-2_\alpha}(\vec{r},\vec{r}\,')\right];\tilde{\vec{I}}(\vec{r}\,',s)>$$

$$= \Big\{-\lambda_{2_\beta}^{-2}\lambda_{3_\beta}<e^{-(s-s_{\beta,\beta'})t_0}\vec{\mu}_{0_\beta}(\vec{r}\,');\tilde{\vec{I}}(\vec{r}\,',s)>$$

$$+ \lambda_{2_\beta}^{-1}<e^{-(s-s_{\beta,\beta'})t_0}\vec{\mu}_{1_\beta}(\vec{r}\,');\tilde{\vec{I}}(\vec{r}\,',s)> \tag{4.48}$$

$$+ \lambda_{2_\beta}^{-2}<t_0e^{-(s-s_{\beta,\beta'})t_0}\vec{\mu}_{0_\beta}(\vec{r}\,');\tilde{\vec{I}}(\vec{r}\,',s)>\Big\}\vec{j}_{0_\beta}(\vec{r})$$

$$+ \lambda_{2_\beta}^{-1}<e^{-(s-s_{\beta,\beta'})t_0}\vec{\mu}_{0_\beta}(\vec{r}\,');\tilde{\vec{I}}(\vec{r}\,',s)>\vec{j}_{1_\beta}(\vec{r})$$

Here the second order pole has $\vec{j}_{0_\beta}(\vec{r})$ as its natural mode and its coupling coefficient is the scalar coefficient of this mode in (4.47). The first order pole has two "natural" modes $\vec{j}_{0_\beta}(\vec{r})$ and $\vec{j}_{1_\beta}(\vec{r})$, where the latter is a somewhat different kind of mode, being the derivative of the eigenmode with respect to s at the frequency; each of these modes has its own scalar coupling coefficient as given in (4.48). For class 1 the second order pole has a coefficient

$$e^{-(s-s_{\beta,\beta'})t_0}<\vec{\vec{\Gamma}}_{-2_\alpha}(\vec{r},\vec{r}\,');\tilde{\vec{I}}(\vec{r}\,',s_{\beta,\beta'})>$$

$$= e^{-(s-s_{\beta,\beta'})t_0}\lambda_{2_\beta}^{-1}<\vec{\mu}_{0_\beta}(\vec{r}\,');\tilde{\vec{I}}(\vec{r}\,',s_{\beta,\beta'})>\vec{j}_{0_\beta}(\vec{r}) \tag{4.49}$$

and the first order pole has a coefficient

$$e^{-(s-s_{\beta,\beta'})t_0}\Big[<\vec{\vec{\Gamma}}_{-1_\alpha}(\vec{r},\vec{r}\,') + t_0\vec{\vec{\Gamma}}_{-2_\alpha}(\vec{r},\vec{r}\,');\tilde{\vec{I}}(\vec{r}\,',s_{\beta,\beta'})>$$

$$+ <\vec{\vec{\Gamma}}_{-2_\alpha}(\vec{r},\vec{r}\,');\frac{\partial}{\partial s}\tilde{\vec{I}}(\vec{r}\,',s)\Big|_{s=s_{\beta,\beta'}}>\Big]$$

$$= e^{-(s-s_{\beta,\beta'})t_0}\Big\{-\lambda_{2_\beta}^{-2}\lambda_{3_\beta}<\vec{\mu}_{0_\beta}(\vec{r}\,');\tilde{\vec{I}}(\vec{r}\,',s_{\beta,\beta'})>$$

$$+ \lambda_{2_\beta}^{-1}<\vec{\mu}_{1_\beta}(\vec{r}\,');\tilde{\vec{I}}(\vec{r}\,',s_{\beta,\beta'})> \tag{4.50}$$

$$+ \lambda_{2_\beta}^{-1}t_0<\vec{\mu}_{0_\beta}(\vec{r}\,');\tilde{\vec{I}}(\vec{r}\,',s_{\beta,\beta'})>$$

$$+ \lambda_{2_\beta}^{-1} \langle \vec{\mu}_{0_\beta}(\vec{r}\,'); \frac{\partial}{\partial s} \tilde{\vec{I}}(\vec{r}\,', s)\Big|_{s=s_{\beta,\beta'}} \rangle \Big\} \vec{j}_{0_\beta}(\vec{r})$$
$$+ e^{-(s - s_{\beta,\beta'})t_0} \lambda_{2_\beta}^{-1} \langle \vec{\mu}_{0_\beta}(\vec{r}\,'); \tilde{\vec{I}}(\vec{r}\,', s_{\beta,\beta'}) \rangle \vec{j}_{1_\beta}(\vec{r})$$

Again note the two modes used with the first order pole. There is also a term involving the s derivative of the forcing function. For the class 1 form t_0 in (4.49) and (4.50) may be a function of $\vec{r}$ (but not of $\vec{r}\,'$). The class 1 coupling coefficients are the scalar coefficients of the modes $\vec{j}_{0_\beta}(\vec{r})$ and $\vec{j}_{1_\beta}(\vec{r})$ in (4.49) and (4.50).

In this subsection (IV.C) and the previous (IV.B) we have treated the general forms of pole terms expressible from the integral equation formulation of the problem. Corresponding to each of the formulas for modes and coupling coefficients, there is a *MoM* form using numerical N component vectors and $N \times N$ matrices with dot product multiplication replacing the integration involved in the continuous operator form used here. For brevity only the frequency (s plane) form has been listed, but this is simply interpretable in time domain. For class 1 one simply replaces $e^{-(s - s_\alpha)t_0}(s - s_\alpha)^{-n_\alpha}$ by $(t - t_0)^{n_\alpha - 1} e^{s_\alpha(t - t_0)} u(t - t_0)/(n_\alpha - 1)!$ (with t_0 potentially dependent on $\vec{r}$); for class 2 one convolves with this function (noting that t_0 is potentially dependent on $\vec{r}, \vec{r}\,'$).

D. Determination of Zero and Pole Locations in the s Plane

Associated with the general forms of the pole terms as discussed previously in this section, there is also the problem of computing the locations of these poles in the s plane. There has been considerable attention given to this question and some of the results are summarized here.

Let us assume that we are given some scalar function $\tilde{h}(s)$ which is an analytic function of the complex frequency s with possibly some singularities. This function may be

$$\tilde{h}(s) = \begin{cases} \det(\tilde{\Gamma}_{n,m}(s)) & \text{, with } \tilde{\Gamma}_{n,m}(s) \text{ approximating integral equation operator as in (4.1)} \\ \tilde{\lambda}_\beta(s) & \text{, eigenvalue of scatterer or antenna as in (3.1), (3.2)} \\ \tilde{Y}_a(s) & \text{, antenna input admittance} \\ \tilde{Z}_a(s) & \text{, antenna input impedance} \\ \text{etc.} & \end{cases} \tag{4.51}$$

For determinants and eigenvalues one may wish to find the zeros of $\tilde{h}(s)$ in order to obtain the natural frequencies of, say, some scatterer. For input admittance and impedance one may wish to find both zeros and poles of $\tilde{h}(s)$ in order to obtain both short circuit and open circuit natural frequencies of an antenna at some port.

One of the techniques used for finding function zeros (as natural frequencies) has been the Newton or Muller method [13] based on calculation of $\tilde{h}(s)$ at a few values of s, constructing a polynomial approximation to $\tilde{h}(s)$, finding the appropriate zero of this polynomial as an approximation to $\tilde{h}(s)$, and using this approximate zero as a new starting point in an iterative procedure to home in on the zero. The newer procedures discussed here are based on contour integral formulas of complex variable theory.

Consider a simple, closed, positive (counterclockwise) contour C in the s plane as illustrated in Fig. 4.1a. Suppose that $\tilde{h}(s)$

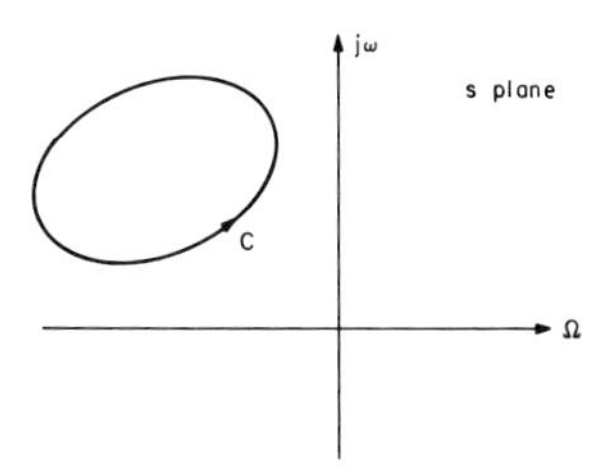

a. Simple closed positive contour.

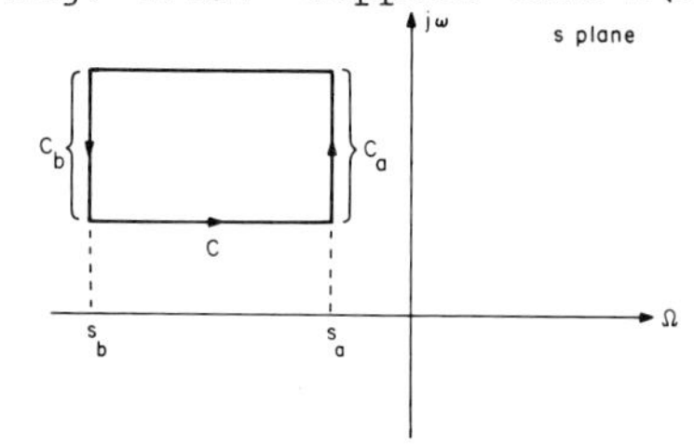

b. Rectangular contour used with exponential normalization.

Fig. 4.1. Contour for zero and pole location in the s plane.

is analytic inside and on C except for possibly poles inside C, and with the restriction of no zeros on C. The residue theorem may then be used to show

$$\frac{1}{2\pi j}\oint_C \tilde{A}(s)\,\frac{1}{\tilde{h}(s)}\,\frac{d}{ds}\,\tilde{h}(s)ds = \sum \tilde{A}(s)\ \text{(at zeros of } \tilde{h}(s) \text{ inside } C) - \sum \tilde{A}(s)\ \text{(at poles of } \tilde{h}(s) \text{ inside } C) \tag{4.52}$$

with multiplicities of zeros and poles counted provided $\tilde{A}(s)$ is analytic within and on C. Choosing C_n as a set of moments as

$$C_n \equiv \frac{1}{2\pi j}\oint_C s^n\,\frac{1}{\tilde{h}(s)}\,\frac{d}{ds}\,\tilde{h}(s)ds \tag{4.53}$$

note that for $n = 0$ we have

$$C_0 = \frac{1}{2\pi j}\oint_C \frac{d}{ds}\,[\ell n(\tilde{h}(s))]ds = \frac{1}{2\pi}\,arg(\tilde{h}(s))\,\Big|_C = N_0 - N_p = N_a$$

$$N_a \equiv \text{argument number} \tag{4.54}$$

$$N_0 \equiv \text{number of zeros within } C$$

$$N_p \equiv \text{number of poles within } C$$

which is just $1/(2\pi)$ times the phase change around C (hence argument number). Provided we restrict $\tilde{h}(s)$ to be analytic inside and on C (and hence no poles inside and on C), C_0 gives the number of zeros inside C. One then knows how many zeros to find, and this gives the basis of this first contour integral technique [19]. By evaluating C_n for $n = 1,2,\ldots,N_a$ we have a set of simultaneous equations of the form

$$C_n = \sum_{m=1}^{N_a} s_m^{\,n} \quad (s_m \text{ at zeros of } \tilde{h}(s)) \tag{4.55}$$

with multiplicity counted. Provided N_a (the number of zeros inside C) is small (say four or less), these equations may be readily solved using analytic formulas for the roots of polynomials of small largest power.

A useful procedure for use in this zero finder is to integrate (4.53) by parts as

$$C_n = N_a s_{ini}^n - \frac{n}{2\pi j} \oint_{C_{ini}} s^{n-1} adj(\ln(\tilde{h}(s)))ds$$

$$= N_a s_{ini}^n - \frac{n}{2\pi j} \oint_{C_{ini}} s^{n-1} [\ln(|\tilde{h}(s)|) + j adj(arg(\tilde{h}(s)))]ds \tag{4.56}$$

s_{ini} ≡ starting and ending point for contour integration (arbitrarily chosen on C)

C_{ini} ≡ contour C with start and end of integration at s_{ini}

adj ≡ adjusted logarithm or adjusted phase (*arg*) to be continuous on C_{ini}, except for the discontinuity (of amount $2\pi N_a$) at the end point or discontinuity point s_{ini}

By pulling out a leading term analytically, the remaining integrand is more smoothly varying and appears to improve the accuracy in a practical way.

If $\tilde{h}(s)$ is, say, the determinant of the *MoM* matrix in a scattering or antenna problem with elements of the matrix containing exponentials of the form $e^{-\frac{s}{c}|\vec{r}_1-\vec{r}_2|}$ corresponding to pairs of positions on the body, then $\tilde{h}(s)$ tends to grow exponentially in the left half plane as $Re[s] \to -\infty$, except very near the zeros. The numerical problems associated with this type of behavior can be reduced by a process of exponential normalization. This is accomplished by defining

$$\tilde{H}(s) \equiv \frac{\tilde{h}(s)}{\tilde{h}_0(s)} \quad , \quad \tilde{h}_0(s) \equiv Be^{As} \quad , \quad Im[A] = 0 \quad , \quad Im[B] = 0 \tag{4.57}$$

so that $\tilde{h}_0(s)$ corresponds to the Laplace transform of a real valued time function (generalized since it is a delta function in time domain corresponding to a time shift of $\tilde{h}(s)$ in time domain). To specify A and B, consider the rectangular contour in Fig. 4.1 (b), where the right boundary C_a is characterized by $Re[s] = s_a$, and the left boundary C_b is characterized by $Re[s] = s_b$. If we constrain that the average $|\tilde{H}(s)|$ on C_a and C_b are both 1.0, we obtain

$$A = \frac{1}{s_a - s_b} \ln\left(\frac{avg\left(|\tilde{h}(s)|\Big|_{C_a}\right)}{avg\left(|\tilde{h}(s)|\Big|_{C_b}\right)}\right) \tag{4.58}$$

$$B = e^{-As_a}\, avg\left(|\tilde{h}(s)|\Big|_{C_a}\right) = e^{-As_b}\, avg\left(|\tilde{h}(s)|\Big|_{C_b}\right)$$

where the average is usually taken as the integral along the contour (C_a or C_b) divided by the length of the contour. This normalization introduces no new zeros into $\tilde{H}(s)$, but makes the magnitude of $\tilde{H}(s)$ have less variation around C than does the magnitude of $\tilde{h}(s)$, thereby improving the accuracy of the integration; this tends to make the phase better behaved as well due to the association of large magnitude variations with large phase variations.

This zero finding procedure ((4.53) through (4.58)) has been used to give significant improvements in the determination of natural frequencies for speed and accuracy as well as excursions into the left half s plane [19].

A new procedure has been developed for treating $\tilde{h}(s)$, which is meromorphic inside C (only pole singularities) and analytic (without zeros) on C, for finding both the zeros and poles inside C. The procedure relies on finding the poles of $\tilde{h}(s)$ inside C and then finding the poles of $\tilde{h}^{-1}(s)$ (zeros of $\tilde{h}(s)$) inside C [20].

Defining a set of moments

$$D_n = \frac{1}{2\pi j}\oint_C s^n \tilde{h}(s)ds \quad , \quad \text{for } n = 0,1,2,\ldots,2N_p \tag{4.59}$$

then in terms of the residues R_m at assumed simple poles s_m of $\tilde{h}(s)$

$$D_n = \sum_{m=1}^{N_p} s_m^n R_m \quad , \quad \text{for } n = 0,1,2,\ldots,2N_p \tag{4.60}$$

where N_p is the number of poles inside C. Writing $\tilde{h}(s)$ as

$$\tilde{h}(s) = \left[\sum_{m=0}^{N_p} a_m s^m\right]^{-1} \tilde{h}_d(s) \tag{4.61}$$

where $\tilde{h}_d(s)$ has no poles inside C, one can construct a matrix equation by eliminating the residues to give

$$\begin{gathered}(A_{m',m}) \cdot (X_m) = (0_{m'}) \quad , \quad \text{for } m',m = 1,2,\ldots,N_p \\ A_{m',m} = D_{m'+m-2} \quad , \quad X_m = a_{m-1}\end{gathered} \tag{4.62}$$

where $(A_{m',m})$ is called a circulant matrix.

By using increasingly larger test values for N_p, starting from zero one can determine a value of N_p for which (4.62) is satisfied. Solving for (X_m), one has the set of coefficients a_m (within a multiplicative constant). The roots of this polymonial are the desired poles. For small values of N_p this procedure is quite accurate, comparing favorably to the previous zero finding procedure. Repeating the process using $\tilde{h}^{-1}(s)$ gives the zeros of $\tilde{h}(s)$ inside C. Finding both zeros and poles can be useful for, say, an antenna input impedance.

V. EIGENIMPEDANCE SYNTHESIS

The concept of eigenimpedance synthesis is one of the new areas of interest related to SEM and EEM. This is discussed in [15,22,9]. Because of its importance in controlling the natural frequencies of antennas and scatterers the concept is briefly summarized here.

The starting point is the integral equation (1.5) together with the eigenmodes of the integral operator as in (3.1). Choosing that form of integral equation using an impedance kernel as in (1.6), we have the impedance integral equation for unloaded (perfectly conducting) objects with the form

$$\langle \tilde{\vec{\vec{Z}}}(\vec{r},\vec{r}\,';s);\tilde{\vec{J}}(\vec{r}\,',s)\rangle = \tilde{\vec{E}}_s(\vec{r},s) \tag{5.1}$$

where $\tilde{\vec{E}}_s$ is some source field such as an indicent field or some specified field at an antenna port (gap). With an assumed source electric field proportional to $\vec{j}_\beta^{(Z)}(\vec{r},s)$, the response current density has the form

$$\tilde{\vec{E}}_s(\vec{r},s) = \tilde{Z}_\beta(s)\vec{J}(\vec{r},s) \tag{5.2}$$

where $\tilde{Z}_\beta(s)$ is the eigenvalue which is now called an eigenimpedance and

$$\tilde{Y}_\beta(s) \equiv \tilde{Z}_\beta^{-1}(s) \quad \text{(eigenadmittance)} \tag{5.3}$$

Now consider impedance loading of the body of the form (over the domain of integration)

$$\begin{gathered}\tilde{\vec{E}}(\vec{r},s) = \tilde{\vec{E}}_s(\vec{r},s) + \tilde{\vec{E}}_{sc}(\vec{r},s) = \tilde{\vec{\vec{Z}}}_\ell(\vec{r},s) \cdot \tilde{\vec{J}}(\vec{r},s) \\ \tilde{\vec{\vec{Z}}}_\ell(\vec{r},s) \equiv \tilde{Z}_\ell(s)\vec{\vec{f}}(\vec{r})\end{gathered} \tag{5.4}$$

where $\vec{\vec{f}}(\vec{r})$ is assumed to be diagonalizable with positive eigenvalues. Constructing the scattered (or radiated) field gives (everywhere)

$$\begin{gathered}\tilde{\vec{E}}_{sc}(\vec{r},s) = -\langle\tilde{\vec{\vec{Z}}}(\vec{r},\vec{r}\,';s);\tilde{\vec{J}}(\vec{r}\,',s)\rangle \\ \langle\tilde{\vec{\vec{Z}}}(\vec{r},\vec{r}\,';s) + \tilde{\vec{\vec{Z}}}_\ell(\vec{r},s)\delta(\vec{r}-\vec{r}\,');\tilde{\vec{J}}(\vec{r}\,',s)\rangle = \tilde{\vec{E}}_s(\vec{r},s)\end{gathered} \tag{5.5}$$

Using the assumed factored form for the loading in (5.4) we have

$$\begin{gathered}\langle\tilde{\vec{\vec{Z}}}'(\vec{r},\vec{r}\,';s) + \tilde{Z}_\ell(s)\vec{\vec{1}}\delta(\vec{r}-\vec{r}\,');\tilde{\vec{J}}(\vec{r}\,',s)\rangle = \tilde{\vec{E}}_s'(\vec{r},s) \\ \tilde{\vec{E}}_s'(\vec{r},s) \equiv \vec{\vec{f}}^{-1}(\vec{r}) \cdot \tilde{\vec{E}}_s(\vec{r},s) \quad , \quad \tilde{\vec{\vec{Z}}}'(\vec{r},\vec{r}\,';s) \equiv \vec{\vec{f}}^{-1}(\vec{r}) \cdot \tilde{\vec{\vec{Z}}}(\vec{r},\vec{r}\,';s)\end{gathered} \tag{5.6}$$

Defining modified eigenimpedances and eigenmodes as

$$\begin{gathered}\langle\tilde{\vec{\vec{Z}}}'(\vec{r},\vec{r}\,';s);\tilde{\vec{j}}_\beta^{(Z')}(\vec{r}\,',s)\rangle = \tilde{Z}_\beta'(s)\vec{j}_\beta^{(Z')}(\vec{r}\,',s) \\ \langle\tilde{\vec{\mu}}_\beta^{(Z')}(\vec{r},s);\tilde{\vec{\vec{Z}}}'(\vec{r},\vec{r}\,';s)\rangle = \tilde{Z}_\beta'(s)\vec{\mu}_\beta^{(Z')}(\vec{r}\,',s)\end{gathered} \tag{5.7}$$

gives the transformations due to impedance loading from (5.6) as

$$\tilde{\vec{j}}_\beta^{(Z')}(\vec{r},s) \xrightarrow[\text{impedance loading}]{} \tilde{\vec{j}}_\beta^{(Z')}(\vec{r},s)$$

$$\tilde{\vec{\mu}}_{\beta}^{(Z')}(\vec{r},s) \xrightarrow[\text{impedance loading}]{} \vec{\mu}_{\beta}^{(Z')}(\vec{r},s) \tag{5.8}$$

$$\tilde{Z}_{\beta}'(s) \xrightarrow[\text{impedance loading}]{} \tilde{Z}_{\beta}'(s) + \tilde{Z}_{\ell}(s)$$

The modified (Z') eigenmodes are unchanged by the loading addition while the load linearity adds to each of the modified eigenimpedances. For $\vec{f}(\vec{r}) = \vec{1}$ (uniform loading) the primes are removed to give the results for the unmodified eigenquantities.

An important thing to note now is that $\tilde{Z}_{\ell}(s)$ is as yet unspecified; one may wish to impose some general properties such as passivity, implying the loading to be a PR function in the circuit sense [1]. Note that the eigenimpedances of a passive unloaded scatterer are also PR functions becuase of the requirement of positive power in a scattered (radiated) wave and the form of (5.2). Given some $\tilde{Z}_{\beta}(s)$ or $\tilde{Z}_{\beta}'(s)$ for an unloaded body with natural frequencies (poles of the scattering problem) as their zeros, by appropriate choice of $\tilde{Z}_{\ell}(s)$ (within constraints of realizability) one can have natural frequencies at some desired set of $s_{\beta,\beta'}$ subject to the synthesis equation for $\tilde{Z}_{\ell}(s)$ as

$$\tilde{Z}_{\ell}(s_{\beta,\beta'}) = -\tilde{Z}_{\beta}'(s_{\beta,\beta'}) \tag{5.9}$$

provided the set of β values and associated β,β' roots selected does not force impossible conditions on $\tilde{Z}_{\ell}(s)$.

This section has concentrated on only the general forms of the equations. For more details and examples the reader should consult the references.

VI. EQUIVALENT CIRCUITS AT ANTENNA OR SCATTERER PORT

One of the promising new applications of SEM and EEM concepts is the construction of equivalent circuit representations of the response of antennas and scatterers. These equivalent circuit representations are currently somewhat formal in nature and the question of realizability using passive lumped elements in any particular circuit form arises. Ideally one can construct such equivalent circuits, or approximations to them, in actual hardware for use in various kinds of testing.

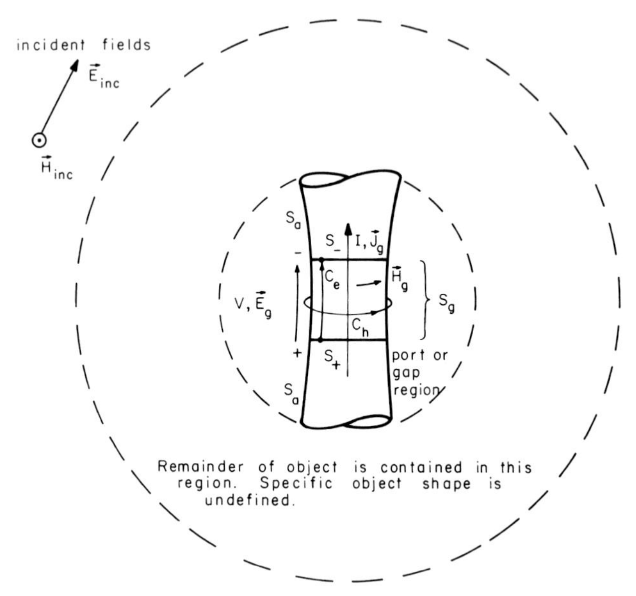

The gap tangential electric field, voltage, and current are shown with a convention to directly give the antenna impedance as $\tilde{V}_{oc} / \tilde{I}_{sc}$.

Fig. 6.1. Antenna or scatterer with single port.

The discussion here centers around the general type of problem as illustrated in Fig. 6.1. We have a gap region (port) which is assumed electrically small and small compared to the rest of the antenna or scatterer. This configuration has a Thevenin equivalent circuit in Fig. 6.2a based on an open circuit voltage V_{oc} at the gap and the antenna (driving point) impedance $\tilde{Z}_a$. Similarly, it has a Norton equivalent circuit in Fig. 6.2b based on the short circuit current I_{sc} and the admittance $\tilde{Y}_a$ (or

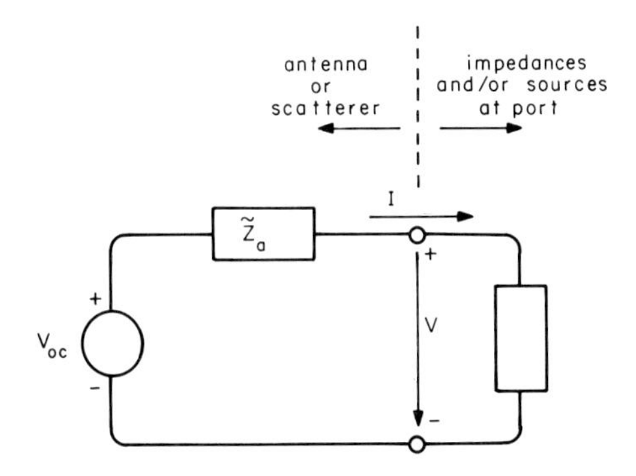

A. Thevenin equivalent circuit

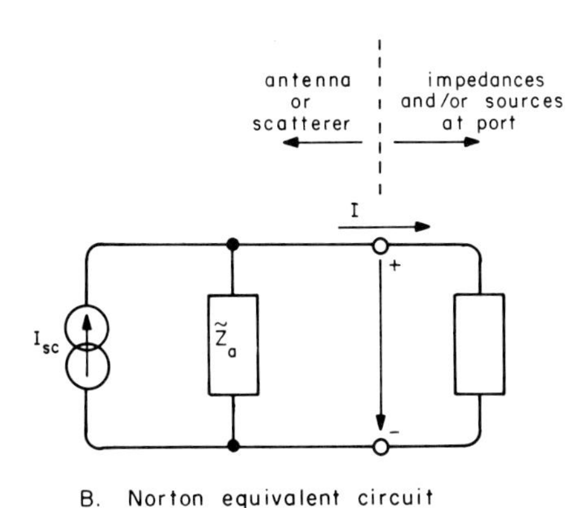

B. Norton equivalent circuit

Fig. 6.2. Thevenin and Norton equivalent circuits of single-port antenna or scatterer.

impedance $\tilde{Z}_a$). This points out the nature of our investigation as the solution of two boundary-value problems and having both solutions contained in a single representation (an equivalent circuit). Some of the simpler forms are based on a term-by-term correspondence of the eigenvalues and natural frequencies of the short-circuit current and admittance, which we include under the general heading of the short-circuit boundary value problem. Other forms are based on the commonality of the eigenvalues and natural frequencies of the open-circuit voltage and impedance, which we include in the open-circuit boundary value problem.

There is much detailed discussion of this topic in a recent reference [17]. This section summarizes some of the results.

Note that throughout this section subscripts will be used on the symmetric product $\langle\, ,\rangle$ for spatial integration. Referring to Fig. 6.1, a subscript "g" will be used to designate the gap region S_g or V_g bounded by S_g, S_+, and S_-. A subscript "a" will be used to designate the remainder of the antenna or scatterer V_a or S_a. A subscript "a+g" designates the entire domain of integration over the antenna or scatterer including the gap region.

For present considerations for equivalent circuits the general SEM form of the response of the object of interest is limited to first order poles plus a possible entire-function contribution. Branch integrals are neglected but can be considered as grouped with the entire-function contribution for the general form of the equivalent circuit, if desired. Certain types of objects (such as finite size perfectly conducting objects in free space) are known to have no branch cuts in the response with appropriately restricted source fields. The waveforms for the incident fields are factored out of the solution before singularity expanding the delta-function response for each wave (with various directions of incidence, polarization, etc.) so that for present purposes only the object singularities are considered in constructing convenient forms of equivalent circuits. Furthermore, class 1 coupling coefficients are used for the poles of the delta-function responses.

A. Parameters for Short-Circuit Form

For calculating the admittance to be used in the short-circuit form we specify a gap electric field of the general quasi-static form

$$\vec{E}_g(\vec{r},t) = -V(t)\vec{e}_g(\vec{r}) \quad , \quad \int_{C_e} \vec{e}_g(\vec{r}) \cdot \vec{1}_e(\vec{r})d\ell = 1 \tag{6.1}$$

where $\vec{e}_g$ is the gradient of a scalar potential function which is consistent on S_+ and on S_- (with a value of 1.0 greater) and therefore has zero curl within the gap region, C_e is any contour crossing the gap as indicated in Fig. 6.1, and $\vec{1}_e$ is a unit vector tangent to this contour in the direction from S_+ to S_-. $V(t)$ is the gap voltage.

For this same problem one needs a corresponding definition of the current $I(t)$ through the gap. For this purpose one might integrate the magnetic field around the gap along the contour C_h as shown in Fig. 6.1, but this has problems with the displacement current in the gap region. A more useful form for our purposes is obtained by averaging the current through the gap region as

$$I(t) \equiv \langle\vec{J}(\vec{r},t);\vec{e}_g(\vec{r})\rangle_g \tag{6.2}$$

This is an average with the gap-electric-field function $\vec{e}_g$ as a weighting function. This formula can also be derived by considering the power developed in the gap in a $\vec{J} \cdot \vec{E}$ sense (integrated) and equating it to the power in a VI sense.

The gap admittance has the form (using the conventions in Fig. 6.1)

$$\tilde{Y}_a(s) = -\frac{\tilde{I}(s)}{\tilde{V}(s)} = -\frac{1}{\tilde{V}(s)} \langle\tilde{\vec{J}}(\vec{r},s);\vec{e}_g(\vec{r})\rangle_g \tag{6.3}$$

This formula for the admittance (when driven by a gap field), together with (6.2) for the short-circuit current when driven by an external incident field, are the basic defining equations used in constructing the short-circuit form of the equivalent circuit.

B. Parameters for Open-Circuit Form

For the impedance to be used in the open-circuit form of the equivalent circuit we specify a gap current density (or surface current density) of the form

$$\vec{J}_g(\vec{r},t) = I(t)\vec{j}_g(\vec{r}) \quad , \quad \int_{S_h} \vec{j}_g(\vec{r}) \cdot \vec{1}_{S_h}(\vec{r})\,dS = 1 \tag{6.4}$$

where $\vec{j}_g$ is the curl of a vector potential function and is therefore divergenceless within the gap region; it does have a divergence where it terminates at S_+ and S_- in Fig. 6.1, but is made to not terminate on S_g. Note that S_h is a surface passing through the gap region and bounded on its edge by the contour C_h, which is any contour circling the gap region as in Fig. 6.1. The unit vector $\vec{1}_{S_h}$ is the surface normal to S_h pointing in the direction to give positive current (from S_+ to S_-). $I(t)$ is the gap current.

The resulting voltage across the gap can be calculated by an integral along C_e of the electric field as in (6.1). However for the open circuit form of the equivalent circuit it is more convenient to use

$$V(t) \equiv -\langle \vec{E}(\vec{r},t);\vec{j}_g(\vec{r})\rangle_g \tag{6.5}$$

which has a form somewhat complementary to (6.2) since it is an average of the gap electric field with the gap-current-density function $\vec{j}_g$ as a weighting function. Again $\vec{J}\cdot\vec{E}$ and VI power considerations lead to the same results.

The gap impedance has the form

$$\tilde{Z}_a(s) = -\frac{\tilde{V}(s)}{\tilde{I}(s)} = \frac{1}{\tilde{I}(s)}\langle \tilde{\vec{E}}(\vec{r},s);\vec{j}_g(\vec{r})\rangle_g \tag{6.6}$$

which is not necessarily the exact reciprocal of $\tilde{Y}_a$ as defined for the short circuit problem due to the two different approximations implicit in the definitions of $\vec{j}_g$ and $\vec{e}_g$ for the two cases. This formula for the impedance, together with (6.5) for the open circuit voltage give us the terms for the open circuit form of the equivalent circuit.

C. Short- and Open-Circuit Boundary Value Problems

Let our general integral equation be of the form of an

impedance integral equation as

$$\begin{aligned}\tilde{\vec{E}}_s(\vec{r},s) &= \langle \tilde{\vec{\vec{\Gamma}}}(\vec{r},\vec{r}\,';s);\tilde{\vec{J}}(\vec{r}\,',s)\rangle \\ \tilde{\vec{\vec{\Gamma}}}(\vec{r},\vec{r}\,';s) &= \tilde{\vec{\vec{Z}}}(\vec{r},\vec{r}\,';s) + \tilde{\vec{\vec{Z}}}_\ell(\vec{r},s)\,\delta(\vec{r}-\vec{r}\,')\end{aligned} \tag{6.7}$$

where the impedance kernel is given by (1.6) and some of the effects of distributed impedance loading have been discussed in section V. Here we are only concerned about the general forms of the equations, and not with the details of their numerical implementation, including the proper handling of the spatial singularities at $\vec{r} = \vec{r}\,'$. Here $\vec{E}_s$ represents a source electric field, either an incident field or a specified field on the gap region. Note that the domain of integration is general but will take on specific portions of the antenna or scatterer as required.

Considering the short circuit form of the boundary value problem, we have

$$\tilde{\vec{E}}_s(\vec{r},s) = \langle \tilde{\vec{\vec{\Gamma}}}(\vec{r},\vec{r}\,';s);\tilde{\vec{J}}(\vec{r}\,',s)\rangle_{a+g} \tag{6.8}$$

For the case of admittance $\vec{E}_s$ is taken as $\vec{E}_g$ which is nonzero only in the gap region. For the case of short-circuit current $\vec{E}_s$ is taken as $\vec{E}_{inc}$, which is in general nonzero over the entire antenna or scatterer.

For the open-circuit form of the boundary value problem, one has first an equation of the general form

$$\tilde{\vec{E}}_s(\vec{r},s) = \langle \tilde{\vec{\vec{\Gamma}}}(\vec{r},\vec{r}\,';s);\tilde{\vec{J}}(\vec{r}\,',s)\rangle_a \tag{6.9}$$

for the current density. For the case of impedance $\vec{E}_s$ is the resulting electric field over the object from the source current density $\vec{J}_g$ in the gap. For the case of open-circuit voltage $\vec{E}_s$ is taken as $\vec{E}_{inc}$, which is in general nonzero over V_a. The resulting electric field in the gap region is

$$\tilde{\vec{E}}(\vec{r},s) = \tilde{\vec{E}}_s(\vec{r},s) - \langle \tilde{\vec{\vec{Z}}}(\vec{r},\vec{r}\,';s);\tilde{\vec{J}}(\vec{r}\,',s)\rangle_a \tag{6.10}$$

which can be used directly for open-circuit voltage. For the case of a source current density in the gap, then for use in impedance calculations the gap electric field becomes

$$\begin{aligned}\tilde{\vec{E}}(\vec{r},s) &= -\langle\tilde{\vec{\vec{Z}}}(\vec{r},\vec{r}';s);\tilde{\vec{J}}_g(\vec{r}',s)\rangle_g - \langle\tilde{\vec{\vec{Z}}}(\vec{r},\vec{r}';s);\tilde{\vec{J}}(\vec{r}',s)\rangle_a \\ &= -\langle\tilde{\vec{\vec{Z}}}(\vec{r},\vec{r}';s);\tilde{\vec{J}}(\vec{r}',s) + \tilde{\vec{J}}_g(\vec{r}',s)\rangle_{a+g}\end{aligned} \tag{6.11}$$

where $\vec{J}$ is zero in V_g and $\vec{J}_g$ is zero in V_a.

For the case that the source electric field is an incident field, one can use a general form

$$\tilde{\vec{E}}_{inc}(\vec{r},s) = E_0 \sum_p \tilde{f}_p(s)\tilde{\vec{\delta}}_p(\vec{r},s) \tag{6.12}$$

Here $\tilde{\vec{\delta}}_p(\vec{r},s)$ is the spatial part of one of the field terms present, basically representing a delta-function response corresponding to one of a set of excitations; the waveforms for each of these terms is given by $\tilde{f}_p(s)$ in frequency domain. In time domain the above form becomes a convolution. For the special case of plane waves $\tilde{\vec{\delta}}_p$ becomes an exponential times a unit vector orthogonal to the direction of propagation. The index p denotes which wave, polarization, etc., is being considered. E_0 is merely a scaling constant with units of volts/meter.

For use in what follows, the subscript sc will be used to apply to all quantities peculiar to the short-circuit boundary value problem, starting with the natural frequencies (poles) $s_{\alpha_{sc}}$ and extending to all the SEM and EEM quantities. Similarly the subscript oc will be used for such quantities peculiar to the open-circuit boundary value problem.

D. Short-Circuit Form of Equivalent Circuit

Define short-circuit EEM quantities from (6.8) as

$$\begin{aligned}\langle\tilde{\vec{\vec{\Gamma}}}(\vec{r},\vec{r}';s);\tilde{\vec{j}}_{\beta_{sc}}(\vec{r}',s)\rangle_{a+g} &= \tilde{\lambda}_{\beta_{sc}}(s)\tilde{\vec{j}}_{\beta_{sc}}(\vec{r},s) \\ \langle\tilde{\vec{\mu}}_{\beta_{sc}}(\vec{r},s);\tilde{\vec{\vec{\Gamma}}}(\vec{r},\vec{r}';s)\rangle_{a+g} &= \tilde{\lambda}_{\beta_{sc}}(s)\tilde{\vec{\mu}}_{\beta_{sc}}(\vec{r}',s) \\ \langle\tilde{\vec{\mu}}_{\beta_{sc_1}}(\vec{r},s);\tilde{\vec{j}}_{\beta_{sc_2}}(\vec{r},s)\rangle_{a+g} &= \delta_{\beta_{sc_1},\beta_{sc_2}}\end{aligned} \tag{6.13}$$

For the admittance we have the source electric field only over the gap region and use (6.3) and (6.1) to give

$$\tilde{Y}_a(s) = -\frac{1}{\tilde{V}(s)} \langle \tilde{\vec{J}}(\vec{r},s);\vec{e}_g(\vec{r})\rangle_g = \sum_{\beta_{sc}} \tilde{Y}_{\beta_{sc}}(s)$$

$$\tilde{\vec{E}}_g(\vec{r},s) = -\tilde{V}(s)\vec{e}_g(\vec{r}) \tag{6.14}$$

$$\tilde{Y}_{\beta_{sc}}(s) = \tilde{\lambda}^{-1}_{\beta_{sc}}(s)\langle \tilde{\vec{\mu}}_{\beta_{sc}}(\vec{r},s);\vec{e}_g(\vec{r})\rangle_g \langle \vec{j}_{\beta_{sc}}(\vec{r},s);\vec{e}_g(\vec{r})\rangle_g$$

The admittance can then be represented as the parallel sum of a set of admittances associated with each short-circuit eigenmode.

For the short-circuit current we have an incident field over the entire structure as in (6.12) which goes with (6.8), (6.2) and (6.14) to give

$$\tilde{\vec{J}}(\vec{r},s) = E_0 \sum_p \tilde{f}_p(s)\left\{\sum_{\beta_{sc}} \lambda^{-1}_{\beta_{sc}}(s)\langle \tilde{\vec{\mu}}_{\beta_{sc}}(\vec{r}\,',s);\tilde{\vec{\delta}}_p(\vec{r}\,',s)\rangle_{a+g}\,\tilde{\vec{j}}_{\beta_{sc}}(\vec{r},s)\right\}$$

$$\tilde{I}_{sc}(s) = \sum_{\beta_{sc}}\left\{\sum_p \tilde{V}_{\beta_{sc},p}(s)\tilde{f}_p(s)\right\}\tilde{Y}_{\beta_{sc}}(s) \tag{6.15}$$

$$\tilde{V}_{\beta_{sc},p}(s) = E_0\,\frac{\langle \tilde{\vec{\mu}}_{\beta_{sc}}(\vec{r}\,',s);\tilde{\vec{\delta}}_p(\vec{r}\,',s)\rangle_{a+g}}{\langle \tilde{\vec{\mu}}_{\beta_{sc}}(\vec{r}\,',s);\vec{e}_g(\vec{r}\,')\rangle_g}$$

Here the short-circuit current $\tilde{I}_{sc}(s)$ is seen also as a sum of terms associated with each short-circuit eigenmode. Furthermore, each term can be written as the product of two terms, one being an individual eigenmode admittance term $\tilde{Y}_{\beta_{sc}}$ as in (6.14), and the other interpretable as a voltage source $\sum_p \tilde{V}_{\beta_{sc},p}(s)\tilde{f}_p(s)$ dependent on a set of voltage source coefficients $\tilde{V}_{\beta_{sc},p}(s)$ for the spatial properties of the incident field and the incident waveforms $\tilde{f}_p(s)$.

Fig. 6.3 combines these results into an equivalent circuit consisting of a parallel set of admittances in series with voltage sources. Each of the pairs of admittance and voltage source corresponds to one value of β_{sc}, the short-circuit eigenmode index. In Fig. 6.3 the impedances and/or sources attached to the gap (port) of the antenna or scatterer are indicated to the right;

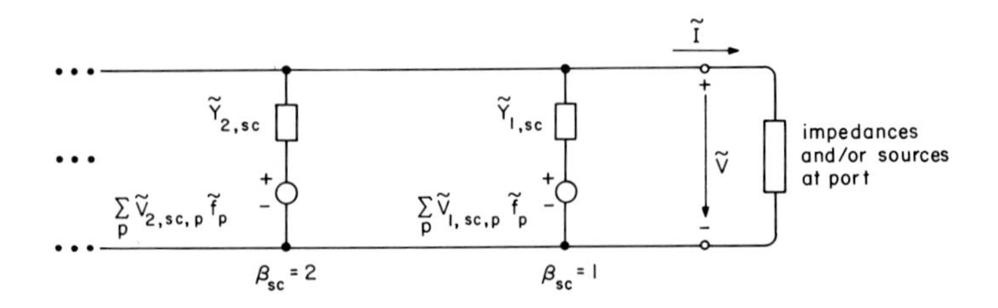

Fig. 6.3. Equivalent circuit for admittance and short-circuit current based on eigenmode expansion of short-circuit boundary-value problem.

for convenience the pairs of admittance and voltage source are indicated progressing to the left starting from $\beta_{sc} = 1$. Since β_{sc} is taking on numerical values (positive integers), then in Fig. 6.3 clarity is added to the admittance and voltage source by immediately following the value of β_{sc} by adding an additional subscript *sc*.

While this form gives a valid equivalent circuit, it leaves something to be desired. The admittances associated with each β_{sc} are quite frequency dependent, being in general more than a single pole pair. More significantly, the voltage source coefficients $\tilde{V}_{\beta_{sc},p}(s)$ are frequency dependent (as well as dependent on angle of incidence, polarization and like parameters), complicating the physical realization of the voltage sources. While the inverse eigenvalue $\tilde{\lambda}^{-1}_{\beta_{sc}}(s)$ in (6.14) is a positive real function (corresponding to an impedance integral equation for a passive object), the additional factor in (6.14) for $\tilde{Y}_{\beta_{sc},p}(s)$ may complicate matters.

An equivalent circuit can be constructed on a pole-by-pole basis using SEM considerations. For this purpose we have a set of short circuit natural frequencies, $s_{\alpha_{sc}}$ or $s_{\beta_{sc},\beta'_{sc}}$, with the basic equations (related to (6.13))

$$\langle \tilde{\vec{\vec{\Gamma}}}(\vec{r},\vec{r}\,';s_{\beta_{sc},\beta'_{sc}});\tilde{\vec{j}}_{\beta_{sc}}(\vec{r}\,',s_{\beta_{sc},\beta'_{sc}})\rangle_{a+g} = \vec{0}$$

$$\langle \tilde{\vec{\mu}}_{\beta_{sc}}(\vec{r}, s_{\beta_{sc},\beta'_{sc}}); \tilde{\vec{\vec{\Gamma}}}(\vec{r}, \vec{r}\,'; s_{\beta_{sc},\beta'_{sc}})\rangle_{a+g} = \vec{0} \tag{6.16}$$

$$\langle \tilde{\vec{\mu}}_{\beta_{sc}}(\vec{r}, s_{\beta_{sc},\beta'_{sc}}); \tilde{\vec{j}}_{\beta_{sc}}(\vec{r}, s_{\beta_{sc},\beta'_{sc}})\rangle_{a+g} = 1$$

These results can be cast in terms of the natural frequency index α_{sc}; however, the combined SEM and EEM designation using β_{sc}, β'_{sc} gives more information. The natural modes with magnitude normalized to 1.0 as in (4.3) are related to the eigenmode form of the natural modes in (6.16) by scaling constants.

In SEM form the admittance is

$$\tilde{Y}_a(s) = \sum_{\beta_{sc}} \tilde{Y}_{\beta_{sc}}(s) = \sum_{\beta_{sc}} \left\{ \left\{ \sum_{\beta'_{sc}} \tilde{Y}_{\beta_{sc},\beta'_{sc}}(s) \right\} + \tilde{Y}_{\beta_{sc},sce}(s) \right\}$$

$$\tilde{Y}_{\beta_{sc},\beta'_{sc}}(s) = \frac{1}{Z_0} a_{\beta_{sc},\beta'_{sc}} (s - s_{\beta_{sc},\beta'_{sc}})^{-1}$$

$$a_{\beta_{sc},\beta'_{sc}} =$$

$$= \frac{Z_0 \langle \tilde{\vec{\mu}}_{\beta_{sc}}(\vec{r}, s_{\beta_{sc},\beta'_{sc}}); \vec{e}_g(\vec{r})\rangle_g \langle \tilde{\vec{j}}_{\beta_{sc}}(\vec{r}, s_{\beta_{sc},\beta'_{sc}}); \vec{e}_g(\vec{r})\rangle_g}{\langle \tilde{\vec{\mu}}_{\beta_{sc}}(\vec{r}, s_{\beta_{sc},\beta'_{sc}}); \frac{\partial}{\partial s} \tilde{\vec{\vec{\Gamma}}}(\vec{r}, \vec{r}\,'; s)\Big|_{s=s_{\beta_{sc},\beta'_{sc}}} ; \tilde{\vec{j}}_{\beta_{sc}}(\vec{r}, s_{\beta_{sc},\beta'_{sc}})\rangle_{a+g}} \tag{6.17}$$

$$= Z_0 \frac{\langle \tilde{\vec{\mu}}_{\beta_{sc}}(\vec{r}, s_{\beta_{sc},\beta'_{sc}}); \vec{e}_g(\vec{r})\rangle_g \langle \tilde{\vec{j}}_{\beta_{sc}}(\vec{r}, s_{\beta_{sc},\beta'_{sc}}); \vec{e}_g(\vec{r})\rangle_g}{\frac{\partial}{\partial s} \tilde{\lambda}_{\beta_{sc}}(s)\Big|_{s=s_{\beta_{sc},\beta'_{sc}}}}$$

where the $\tilde{Y}_{\beta_{sc},\beta'_{sc}}(s)$ can be referred to as pole admittances. For completeness $\tilde{Y}_{\beta_{sc},sce}(s)$ represents the entire function contribution to $\tilde{Y}_{\beta_{sc}}(s)$ (as in (6.14)).

The short circuit current in SEM form expands on (6.15) (using class 1 coupling coefficients after factoring out the $\tilde{f}_p(s)$) as

$$\tilde{I}_{sc}(s) = \sum_p \tilde{f}_p(s)\left\{\sum_{\beta_{sc}} \frac{E_0}{Z_0}\left\{\sum_{\beta'_{sc}} b_{\beta_{sc},\beta'_{sc},p}\,(s - s_{\beta_{sc},\beta'_{sc}})^{-1}\right\} + \tilde{I}_{\beta_{sc},sce,p}(s)\right\}$$

$$b_{\beta_{sc},\beta'_{sc},p} =$$

$$= \frac{Z_0 \langle \tilde{\vec{\mu}}_{\beta_{sc}}(\vec{r}\,', s_{\beta_{sc},\beta'_{sc}});\tilde{\vec{\delta}}_p(\vec{r}\,', s_{\beta_{sc},\beta'_{sc}})\rangle_{a+g} \langle \tilde{\vec{j}}_{\beta_{sc}}(\vec{r}\,', s_{\beta_{sc},\beta'_{sc}});\vec{e}_g(\vec{r}\,')\rangle_g}{\langle \tilde{\vec{\mu}}_{\beta_{sc}}(\vec{r}\,', s_{\beta_{sc},\beta'_{sc}});\frac{\partial}{\partial s}\tilde{\vec{\Gamma}}(\vec{r},\vec{r}\,';s)\Big|_{s=s_{\beta_{sc},\beta'_{sc}}};\tilde{\vec{j}}_{\beta_{sc}}(\vec{r}\,', s_{\beta_{sc},\beta'_{sc}})\rangle_{a+g}}$$

$$= \frac{Z_0 \langle \tilde{\vec{\mu}}_{\beta_{sc}}(\vec{r}\,', s_{\beta_{sc},\beta'_{sc}});\tilde{\vec{\delta}}_p(\vec{r}\,', s_{\beta_{sc},\beta'_{sc}})\rangle_{a+g} \langle \tilde{\vec{j}}_{\beta_{sc}}(\vec{r}\,', s_{\beta_{sc},\beta'_{sc}});\vec{e}_g(\vec{r}\,')\rangle_g}{\frac{\partial}{\partial s}\tilde{\lambda}_{\beta_{sc}}(s)\Big|_{s=s_{\beta_{sc},\beta'_{sc}}}} \tag{6.18}$$

where $\tilde{I}_{\beta_{sc},sce,p}$ is the entire function contribution to each eigenmode term (β_{sc}) in the expansion for the pth incident field term with the waveform $\tilde{f}_p$ factored out. Combining the short circuit current and admittance SEM forms we have

$$\tilde{I}_{sc}(s) = \sum_{\beta_{sc}}\left\{\left\{\sum_{\beta'_{sc}}\left\{\sum_p V_{\beta_{sc},\beta'_{sc},p}\,\tilde{f}_p(s)\right\}\tilde{Y}_{\beta_{sc},\beta'_{sc}}(s)\right\}\right.$$
$$\left. + \left\{\sum_p \tilde{V}_{\beta_{sc},sce,p}(s)\tilde{f}_p(s)\right\}\tilde{Y}_{\beta_{sc},sce}(s)\right\}$$

$$V_{\beta_{sc},\beta'_{sc},p} = \tilde{V}_{\beta_{sc},p}(s_{\beta_{sc},\beta'_{sc}}) = E_0\,\frac{b_{\beta_{sc},\beta'_{sc},p}}{a_{\beta_{sc},\beta'_{sc}}} \tag{6.19}$$

$$= E_0\,\frac{\langle \tilde{\vec{\mu}}_{\beta_{sc}}(\vec{r}\,', s_{\beta_{sc},\beta'_{sc}});\tilde{\vec{\delta}}_p(\vec{r}\,', s_{\beta_{sc},\beta'_{sc}})\rangle_{a+g}}{\langle \tilde{\vec{\mu}}_{\beta_{sc}}(\vec{r}\,', s_{\beta_{sc},\beta'_{sc}});\vec{e}_g(\vec{r}\,')\rangle_g}$$

$$\tilde{V}_{\beta_{sc},sce,p}(s) = \frac{\tilde{I}_{\beta_{sc},sce,p}(s)}{\tilde{Y}_{\beta_{sc},sce}(s)}$$

The voltage-source coefficients $V_{\beta_{sc},\beta'_{sc},p}$ for the poles are conveniently frequency independent. One can further note that the expression for these voltage-source coefficients is simplified by the absence of a term involving the derivative of the kernel (of the integral equation) with respect to the frequency; this term is retained in the admittance for each pole.

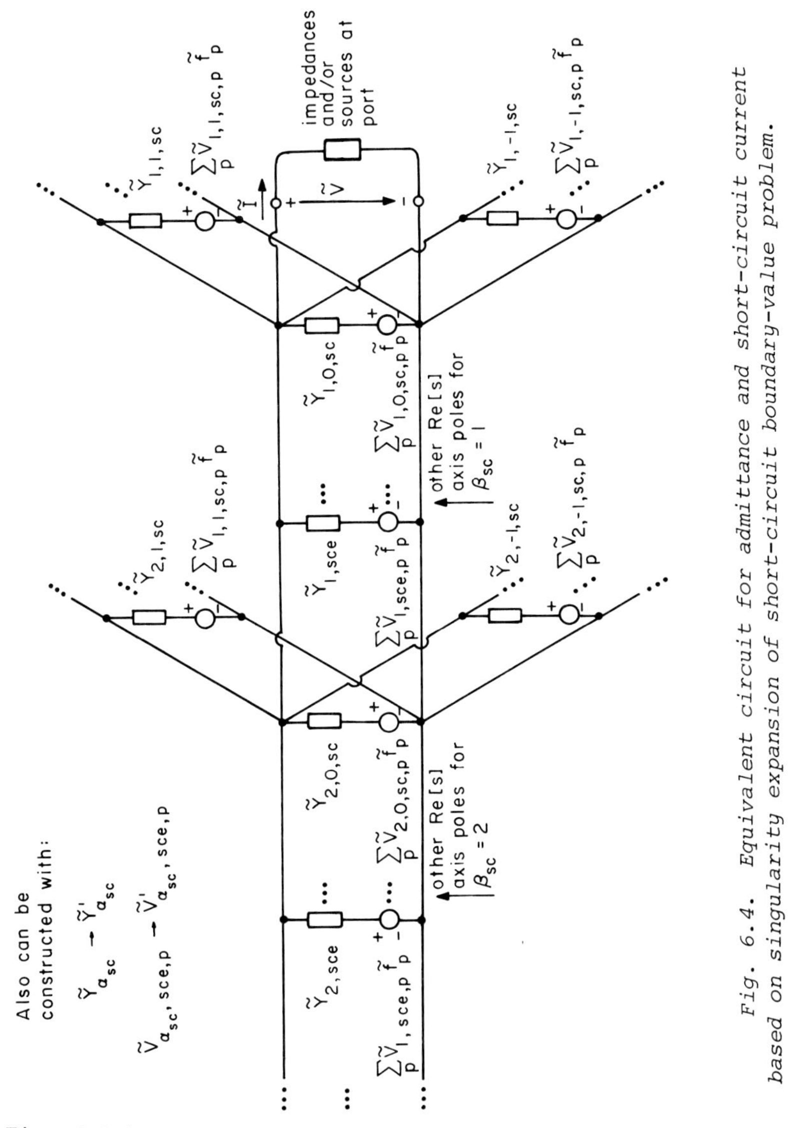

Fig. 6.4. Equivalent circuit for admittance and short-circuit current based on singularity expansion of short-circuit boundary-value problem.

Fig. 6.4 illustrates the equivalent circuit combining the admittance from (6.17) with the short-circuit current from (6.19) as a parallel sum of circuit modules. Each module consists of an admittance (pole admittance or entire-function admittance) in series with a corresponding voltage source. The layout of this equivalent circuit illustrates a few points. Each eigenmode-related term (as in the equivalent circuit in Fig. 6.3) has been expanded into a collection of SEM terms. Fig. 6.4 has these SEM circuit modules grouped according to the eigenmode index β_{sc}. Note that as β_{sc} takes on numerical values an additional subscript sc is added to clearly identify the kind of quantities used. Since pole terms come in conjugate symmetric locations in the s plane (each pair indicated by opposite ($\pm$) values of β'_{sc}), these are arranged in symmetrical locations on the figure. One can envision this diagram superimposed on the s plane with the pole circuit modules symmetrically located in the left half plane corresponding to the pole locations.

The pole admittances used in (6.17) have the property of being non zero at $s = 0$. While for some types of objects this property may be desirable, for other types of objects it is inconvenient. For capacitive (electric) type antennas at low frequencies, the admittance goes to zero. Of course the entire function can compensate for this. However, it is convenient to introduce modified pole admittances which go to zero as $s \to 0$ by subtracting off the zero value, giving for the admittance expansion

$$\tilde{Y}_a(s) = \sum_{\beta_{sc}} \tilde{Y}_{\beta_{sc}}(s) = \sum_{\beta_{sc}} \left\{ \left\{ \sum_{\beta'_{sc}} \tilde{Y}'_{\beta_{sc},\beta'_{sc}}(s) \right\} + \tilde{Y}'_{\beta_{sc},sce}(s) \right\}$$

$$\tilde{Y}'_{\beta_{sc},\beta'_{sc}}(s) = \tilde{Y}_{\beta_{sc},\beta'_{sc}}(s) - \tilde{Y}_{\beta_{sc},\beta'_{sc}}(0)$$

$$= \frac{1}{Z_0} a_{\beta_{sc},\beta'_{sc}} \left[(s - s_{\beta_{sc},\beta'_{sc}})^{-1} + s^{-1}_{\beta_{sc},\beta'_{sc}} \right]$$

$$= \frac{1}{Z_0} \frac{a_{\beta_{sc},\beta'_{sc}}}{s_{\beta_{sc},\beta'_{sc}}} \frac{s}{s - s_{\beta_{sc},\beta'_{sc}}} \tag{6.20}$$

$$\tilde{Y}'_{\beta_{sc},sce}(s) = \tilde{Y}_{\beta_{sc},sce}(s) + \sum_{\beta'_{sc}} \tilde{Y}_{\beta_{sc},\beta'_{sc}}(0)$$

$$= \tilde{Y}_{\beta_{sc},sce}(s) - \frac{1}{Z_0} \sum_{\beta'_{sc}} a_{\beta_{sc},\beta'_{sc}} s^{-1}_{\beta_{sc},\beta'_{sc}}$$

where $\tilde{Y}'_{\beta_{sc},sce}(s)$ is a correspondingly modified entire-function contribution. The short-circuit current can also be written in terms of modified poles as

$$\tilde{I}_{sc}(s) = \sum_{\beta_{sc}} \left\{ \left\{ \sum_{\beta'_{sc}} \sum_{p} V_{\beta_{sc},\beta'_{sc},p} \tilde{f}_p(s) \right\} \tilde{Y}'_{\beta_{sc},\beta'_{sc}}(s) \right\}$$

$$+ \left\{ \sum_{p} \tilde{V}'_{\beta_{sc},sce,p}(s) f_p(s) \right\} \tilde{Y}'_{\beta_{sc},sce}(s) \Bigg\}$$

$$\tilde{I}'_{\beta_{sc},sce,p}(s) \equiv \left\{ \sum_{p} \tilde{V}'_{\beta_{sc},sce,p}(s) \tilde{f}_p(s) \right\} \tilde{Y}'_{\beta_{sc},sce}(s) \tag{6.21}$$

$$= \tilde{I}_{\beta_{sc},sce,p}(s) - \frac{E_0}{Z_0} \sum_{\beta'_{sc}} b_{\beta_{sc},\beta'_{sc},p} s^{-1}_{\beta_{sc},\beta'_{sc}}$$

$$\tilde{V}_{\beta_{sc},sce,p}(s) = \frac{\tilde{I}'_{\beta_{sc},sce,p}(s)}{\tilde{Y}'_{\beta_{sc},sce}(s)}$$

This form using modified poles in the admittance and short-circuit current also gives the same form of equivalent circuit as in Fig. 6.4. Some of the terms are modified by the addition of a "prime" corresponding to the above formulas. Which form is used depends on the type of object under consideration and the associated convergence properties of the two forms of series of circuit modules.

The pole admittances can be written in circuit form (using the shorter α_{sc} notation) as

$$\tilde{Y}_{\alpha_{sc}}(s) = \frac{1}{Z_0} a_{\alpha_{sc}} (s - s_{\alpha_{sc}})^{-1} = \left[sL_{\alpha_{sc}} + R_{\alpha_{sc}} \right]^{-1}$$

$$L_{\alpha_{sc}} = \frac{Z_0}{a_{\alpha_{sc}}}, \quad R_{\alpha_{sc}} = -Z_0 \frac{s_{\alpha_{sc}}}{a_{\alpha_{sc}}} = -s_{\alpha_{sc}} L_{\alpha_{sc}} \tag{6.22}$$

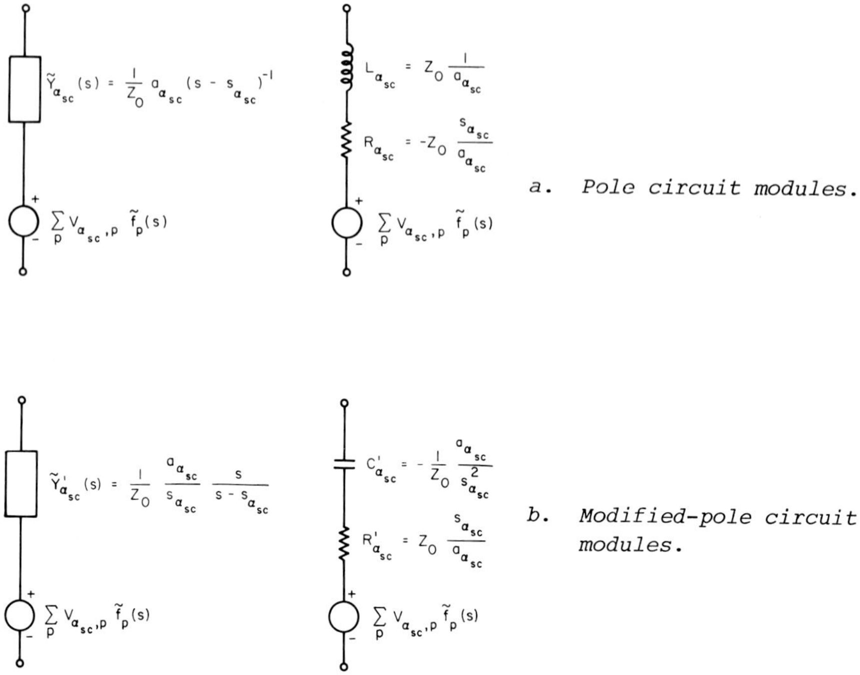

$$\tilde{Y}_{\alpha_{sc}}(s) = \frac{1}{Z_0} a_{\alpha_{sc}} (s - s_{\alpha_{sc}})^{-1}$$

$$L_{\alpha_{sc}} = Z_0 \frac{1}{a_{\alpha_{sc}}}, \quad R_{\alpha_{sc}} = -Z_0 \frac{s_{\alpha_{sc}}}{a_{\alpha_{sc}}}$$

$$\sum_p V_{\alpha_{sc},p} \tilde{f}_p(s)$$

a. Pole circuit modules.

$$\tilde{Y}'_{\alpha_{sc}}(s) = \frac{1}{Z_0} \frac{a_{\alpha_{sc}}}{s_{\alpha_{sc}}} \frac{s}{s - s_{\alpha_{sc}}}$$

$$C'_{\alpha_{sc}} = -\frac{1}{Z_0} \frac{a_{\alpha_{sc}}}{s^2_{\alpha_{sc}}}, \quad R'_{\alpha_{sc}} = Z_0 \frac{s_{\alpha_{sc}}}{a_{\alpha_{sc}}}$$

$$\sum_p V_{\alpha_{sc},p} \tilde{f}_p(s)$$

b. Modified-pole circuit modules.

Fig. 6.5. Elementary circuit modules for pole terms for admittance and short-circuit current for short-circuit boundary-value problem.

as illustrated for a pole circuit module in Fig. 6.5a as a series inductance and resistance, in turn in series with the voltage source. Similarly the modified pole admittance can be written in circuit form as

$$\tilde{Y}'_{\alpha_{sc}} = \frac{1}{Z_0} \frac{a_{\alpha_{sc}}}{s_{\alpha_{sc}}} \frac{s}{s - s_{\alpha_{sc}}} = R'_{\alpha_{sc}} + \frac{1}{s_{\alpha_{sc}} C'_{\alpha_{sc}}}$$

$$C'_{\alpha_{sc}} = -\frac{1}{Z_0} \frac{a_{\alpha_{sc}}}{s^2_{\alpha_{sc}}}, \quad R'_{\alpha_{sc}} = Z_0 \frac{s_{\alpha_{sc}}}{a_{\alpha_{sc}}} = -\frac{1}{s_{\alpha_{sc}} C'_{\alpha_{sc}}} = -R_{\alpha_{sc}} \tag{6.23}$$

as illustrated for a modified-pole circuit module in Fig. 6.5b as a series capacitance and resistance, in turn in series with the voltage source.

For poles off the negative real axis of the s plane, one may

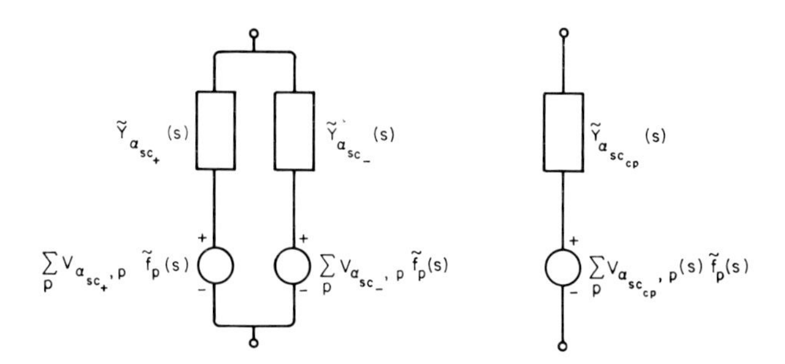

a. Equivalent for pole circuit module conjugate pair.

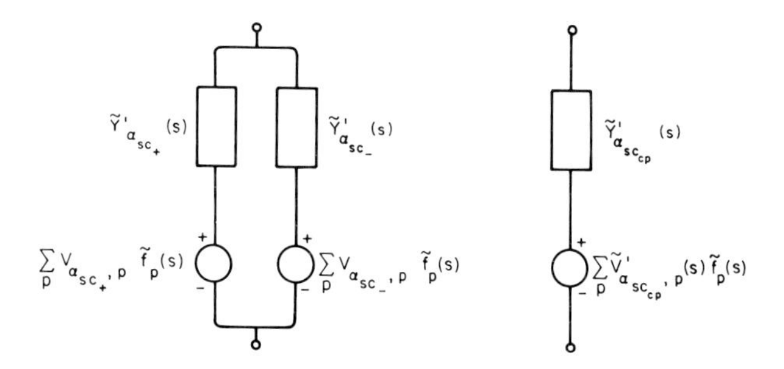

b. Equivalent for modified-pole circuit module conjugate pair.

Fig. 6.6. Thevenin equivalent circuits for pole pairs for short-circuit boundary-value problem.

wish to combine the circuit modules for conjugate pole pairs so that the combination corresponds to real-valued time functions and is perhaps more nearly realizable using passive circuit elements. As illustrated in Fig. 6.6, such pole pair circuits can be constructed in Thevenin form. Denoting the two poles (a conjugate pair) by α_{sc_+} and α_{sc_-} corresponding respectively to upper and lower half s plane, and using a subscript cp (for conjugate pair) in the combined form we have for unmodified pairs as in Fig. 6.6a

$$\tilde{Y}_{\alpha_{sc_{cp}}}(s) = \tilde{Y}_{\alpha_{sc_+}}(s) + \tilde{Y}_{\alpha_{sc_-}}(s)$$

$$\tilde{V}_{\alpha_{sc_{cp}},p}(s) = \frac{V_{\alpha_{sc+},p}\tilde{Y}_{\alpha_{sc+}}(s) + V_{\alpha_{sc-},p}\tilde{Y}_{\alpha_{sc-}}(s)}{\tilde{Y}_{\alpha_{sc+}}(s) + \tilde{Y}_{\alpha_{sc-}}(s)} \tag{6.24}$$

$$= Re\left[V_{\alpha_{sc+,p}}\right] + jIm\left[V_{\alpha_{sc+,p}}\right]\frac{\tilde{Y}_{\alpha_{sc+}}(s) - \tilde{Y}_{\alpha_{sc-}}(s)}{\tilde{Y}_{\alpha_{sc+}}(s) + \tilde{Y}_{\alpha_{sc-}}(s)}$$

For real (or approximately real) pole voltage-source coefficients, the corresponding voltage-source coefficients for the conjugate pair module are frequency independent (or "approximately" frequency independent). For conjugate pairs of modified-pole circuit modules, the formation of the conjugate pair module is shown in Fig. 6.6b. The equations for the terms are as in (6.24) with the addition of primes to give modified-pole quantities.

E. Open-Circuit Form of Equivalent Circuit

Define open-circuit EEM quantities from (6.9) as

$$\begin{aligned}
&\langle\tilde{\vec{\vec{\Gamma}}}(\vec{r},\vec{r}\,';s);\tilde{\vec{j}}_{\beta_{oc}}(\vec{r}\,',s)\rangle_a = \tilde{\lambda}_{\beta_{oc}}(s)\tilde{\vec{j}}_{\beta_{oc}}(\vec{r},s) \\
&\langle\tilde{\vec{\mu}}_{\beta_{oc}}(\vec{r},s);\tilde{\vec{\vec{\Gamma}}}(\vec{r},\vec{r}\,';s)\rangle_a = \tilde{\lambda}_{\beta_{oc}}(s)\tilde{\vec{j}}_{\beta_{oc}}(\vec{r}\,',s) \\
&\langle\tilde{\vec{\mu}}_{\beta_{oc_1}}(\vec{r},s);\tilde{\vec{j}}_{\beta_{oc_2}}(\vec{r},s)\rangle_a = \delta_{\beta_{oc_1},\beta_{oc_2}}
\end{aligned} \tag{6.25}$$

For the impedance we have a source current density over the gap region and use (6.6), (6.4) and (6.11) to give

$$\begin{aligned}
&\tilde{Z}_a(s) = \frac{1}{\tilde{I}(s)}\langle\tilde{\vec{E}}(\vec{r},s);\tilde{\vec{j}}_g(\vec{r})\rangle_g = \sum_{\beta_{oc}}\tilde{Z}_{\beta_{oc}}(s) + \tilde{Z}_g(s) \\
&\tilde{\vec{E}}(\vec{r},s) = -\tilde{I}(s)\langle\tilde{\vec{\vec{Z}}}(\vec{r},\vec{r}\,';s);\tilde{\vec{j}}_g(\vec{r}\,')\rangle_g - \langle\tilde{\vec{\vec{Z}}}(\vec{r},\vec{r}\,';s);\tilde{\vec{J}}(\vec{r}\,',s)\rangle_a \\
&\langle\tilde{\vec{\vec{\Gamma}}}(\vec{r},\vec{r}\,';s);\tilde{\vec{J}}(\vec{r}\,',a)\rangle_a = -\tilde{I}(s)\langle\tilde{\vec{\vec{\mathbf{Z}}}}(\vec{r},\vec{r}\,';s);\tilde{\vec{j}}_g(\vec{r}\,')\rangle_g\,,\ \vec{r}\in V_a
\end{aligned} \tag{6.26}$$

$$\tilde{Z}_{\beta_{oc}}(s) = \tilde{\lambda}_{\beta_{oc}}^{-1}(s)\langle \tilde{\vec{\mu}}_{\beta_{oc}}(\vec{r},s);\langle \tilde{\vec{Z}}(\vec{r},\vec{r}\,';s);\vec{j}_g(\vec{r}\,')\rangle_g\rangle_a \cdot \langle\langle \tilde{\vec{Z}}(\vec{r},\vec{r}\,';s);\tilde{\vec{j}}_{\beta_{oc}}(\vec{r}\,',s)\rangle_a;\vec{j}_g(\vec{r})\rangle_g$$

$$\tilde{Z}_g(s) = -\langle \vec{j}_g(\vec{r});\tilde{\vec{Z}}(\vec{r},\vec{r}\,';s);\vec{j}_g(\vec{r}\,')\rangle_g$$

The impedance is then the series sum of a set of impedances associated with each open-circuit eigenmode, plus an additional "gap" impedance term $\tilde{Z}_g$.

The open-circuit voltage uses an incident field as in (6.12) which we use with (6.9), (6.5) and (6.26) to give

$$\tilde{\vec{J}}(\vec{r},s) = E_0 \sum_p \tilde{f}_p(s)\left\{\sum_{\beta_{oc}} \tilde{\lambda}_{\beta_{oc}}^{-1}(s)\langle \tilde{\vec{\mu}}_{\beta_{oc}}(\vec{r}\,',s);\tilde{\vec{\delta}}_p(\vec{r}\,',s)\rangle_a \tilde{\vec{j}}_{\beta_{oc}}(\vec{r},s)\right\}$$

$$\tilde{\vec{E}}(\vec{r},s) = -\langle \tilde{\vec{Z}}(\vec{r},\vec{r}\,';s);\tilde{\vec{J}}(\vec{r}\,',s)\rangle_a + \tilde{\vec{E}}_{inc}(\vec{r},s)$$

$$\tilde{V}_{oc}(s) = \sum_{\beta_{oc}}\left\{\sum_p \tilde{I}_{\beta_{oc},p}(s)\tilde{f}_p(s)\right\}\tilde{Z}_{\beta_{oc}}(s) + \left\{\sum_p \tilde{I}_{g,p}(s)\tilde{f}_p(s)\right\}\tilde{Z}_g(s) \tag{6.27}$$

$$\tilde{I}_{\beta_{oc},p}(s) = E_0 \frac{\langle \tilde{\vec{\mu}}_{\beta_{oc}}(\vec{r}\,',s);\tilde{\vec{\delta}}_p(\vec{r}\,',s)\rangle_a}{\langle \tilde{\vec{\mu}}_{\beta_{oc}}(\vec{r},s);\langle \tilde{\vec{Z}}(\vec{r},\vec{r}\,';s);\vec{j}_g(\vec{r}\,')\rangle_g\rangle_a}$$

$$\tilde{I}_{g,p}(s) = E_0 \frac{\langle \tilde{\vec{\delta}}_p(\vec{r}\,',s);\vec{j}_g(\vec{r}\,')\rangle_g}{\langle \vec{j}_g(\vec{r});\tilde{\vec{Z}}(\vec{r},\vec{r}\,';s);\vec{j}_g(\vec{r})\rangle_g}$$

The open-circuit voltage is then the sum of terms associated with each open-circuit eigenmode plus a "gap" term. Each term can be written as the product of an impedance term, $\tilde{Z}_{\beta_{oc}}$ or $\tilde{Z}_g$, and a current source, $\sum_p \tilde{I}_{\beta_{oc},p}(s)\tilde{f}_p(s)$ or $\sum_p \tilde{I}_{g,p}(s)\tilde{f}_p(s)$, dependent on a set of current source coefficients, $\tilde{I}_{\beta_{oc},p}(s)$ or $\tilde{I}_{g,p}(s)$, for the spatial properties of the incident field and the incident waveform $\tilde{f}_p(s)$.

The EEM forms for the impedance and open-circuit voltage are configured in an equivalent circuit format in Fig. 6.7. This

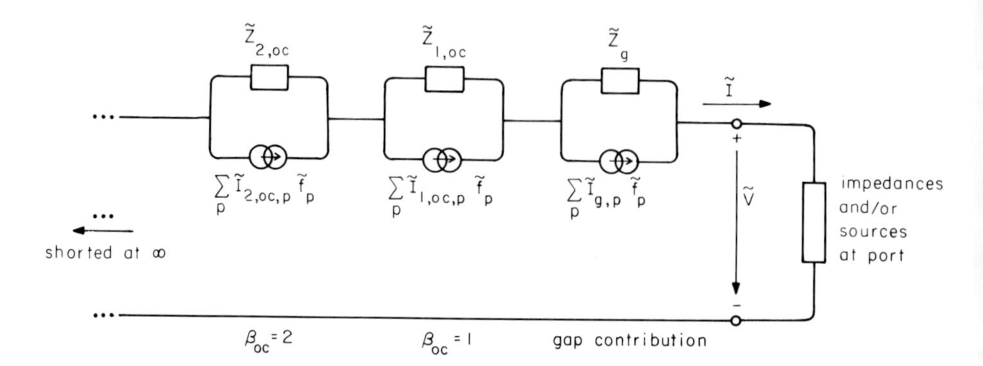

Fig. 6.7. Equivalent circuit for impedance and open-circuit voltage based on eigenmode expansion of open-circuit boundary-value problem.

circuit is configured as the series sum of modules, each module being the parallel combination of an impedance and a current source. With sources and loads at the antenna port indicated to the right, then going to the left, the first module corresponds to the gap term, the second corresponds to $\beta_{oc} = 1$, and the remaining modules correspond to progressively larger values of β_{oc}. For clarity as β_{oc} assumes numerical values, and additional subscript of *oc* is included.

For SEM considerations we have a set of open-circuit natural frequencies, $s_{\alpha_{oc}}$ or $s_{\beta_{oc},\beta'_{oc}}$, with the basic equations (related to (6.25))

$$\begin{aligned}
&\langle \tilde{\vec{\vec{\Gamma}}}(\vec{r},\vec{r}\,';s_{\beta_{oc},\beta'_{oc}});\tilde{\vec{j}}_{\beta_{oc}}(\vec{r}\,',s_{\beta_{oc},\beta'_{oc}})\rangle_a = \vec{0}\\
&\langle \tilde{\vec{\mu}}_{\beta_{oc}}(\vec{r},s_{\beta_{oc},\beta'_{oc}});\tilde{\vec{\vec{\Gamma}}}(\vec{r},\vec{r}\,';s_{\beta_{oc},\beta'_{oc}})\rangle_a = \vec{0}\\
&\langle \tilde{\vec{\mu}}_{\beta_{oc}}(\vec{r},s_{\beta_{oc},\beta'_{oc}});\tilde{\vec{j}}_{\beta_{oc}}(\vec{r},s_{\beta_{oc},\beta'_{oc}})\rangle_a = 1
\end{aligned} \tag{6.28}$$

The natural modes with magnitude normalized to 1.0 as in (4.3) are related to the eigenmode form of the natural modes in (6.28) by scaling constants.

In SEM form the impedance is (using the symmetry of $\tilde{\vec{\vec{Z}}}(\vec{r},\vec{r}\,';s)$)

$$\tilde{Z}_a(s) = \sum_{\beta_{oc}} \tilde{Z}_{\beta_{oc}}(s) + \tilde{Z}_g(s) = \sum_{\beta_{oc}} \left\{ \left\{ \sum_{\beta'_{oc}} \tilde{Z}_{\beta_{oc},\beta'_{oc}}(s) \right\} + Z_{\beta_{oc},oce}(s) \right\} + \tilde{Z}_g(s)$$

$$\tilde{Z}_{\beta_{oc},\beta'_{oc}}(s) = Z_0 a_{\beta_{oc},\beta'_{oc}} (s - s_{\beta_{oc},\beta'_{oc}})^{-1} \tag{6.29}$$

$$a_{\beta_{oc},\beta'_{oc}} = \frac{1}{Z_0} \frac{\left\{ \langle \tilde{\vec{\mu}}_{\beta_{oc}}(\vec{r},s_{\beta_{oc},\beta'_{oc}}) ; \langle \tilde{\vec{\vec{Z}}}(\vec{r},\vec{r}\,';s_{\beta_{oc},\beta'_{oc}}) ; \vec{j}_g(\vec{r})\rangle_g \rangle_a \cdot \langle\langle \tilde{\vec{\vec{Z}}}(\vec{r},\vec{r}\,';s_{\beta_{oc},\beta'_{oc}}) ; \vec{j}_g(\vec{r}\,')\rangle_g ; \tilde{\vec{j}}_{\beta_{oc}}(\vec{r},s_{\beta_{oc},\beta'_{oc}})\rangle_a \right\}}{\langle \tilde{\vec{\mu}}_{\beta_{oc}}(\vec{r},s_{\beta_{oc},\beta'_{oc}}) ; \frac{\partial}{\partial s} \tilde{\vec{\vec{\Gamma}}}(\vec{r},\vec{r}\,';s)\Big|_{s=s_{\beta_{oc},\beta'_{oc}}} ; \tilde{\vec{j}}_{\beta_{oc}}(\vec{r}\,',s_{\beta_{oc},\beta'_{oc}})\rangle_a}$$

$$= \frac{1}{Z_0} \frac{\left\{ \langle \tilde{\vec{\mu}}_{\beta}(\vec{r},s_{\beta_{oc},\beta'_{oc}}) ; \langle \tilde{\vec{\vec{Z}}}(\vec{r},\vec{r}\,';s_{\beta_{oc},\beta'_{oc}}) ; \vec{j}_g(\vec{r})\rangle_g \rangle_a \cdot \langle\langle \tilde{\vec{\vec{Z}}}(\vec{r},\vec{r}\,';s_{\beta_{oc},\beta'_{oc}}) ; \vec{j}_g(\vec{r}\,')\rangle_g ; \tilde{\vec{j}}_{\beta_{oc}}(\vec{r},s_{\beta_{oc},\beta'_{oc}})\rangle_a \right\}}{\frac{\partial}{\partial s} \tilde{\lambda}_{\beta_{oc}}(s)\Big|_{s=s_{\beta_{oc},\beta'_{oc}}}}$$

where the $\tilde{Z}_{\beta_{oc},\beta'_{oc}}(s)$ can be referred to as pole impedances. The entire-function contribution to $\tilde{Z}_{\beta_{oc}}(s)$ is $Z_{\beta_{oc},oce}(s)$. Note that we still have the "gap" term $\tilde{Z}_g(s)$ as in (6.26).

The open-circuit voltage in SEM form expands on (6.27) (using class 1 coupling coefficients after factoring out the $\tilde{f}_p(s)$) as

$$\tilde{V}_{oc}(s) = \sum_p \tilde{f}_p(s) \left\{ \sum_{\beta_{oc}} E_0 \left\{ \sum_{\beta'_{oc}} b_{\beta_{oc},\beta'_{oc},p} (s - s_{\beta_{oc},\beta'_{oc}})^{-1} \right\} + \tilde{V}_{\beta_{oc},oce,p}(s) \right\} + \left\{ \sum_p \tilde{I}_{g,p}(s) \tilde{f}_p(s) \right\} Z_g(s)$$

$$b_{\beta_{oc},\beta'_{oc},p} = \frac{\left\{\langle\tilde{\vec{\mu}}_{\beta_{oc}}(\vec{r}\,',s_{\beta_{oc},\beta'_{oc}});\tilde{\vec{\delta}}_p(\vec{r}\,',s_{\beta_{oc},\beta'_{oc}})\rangle_a \cdot \langle\langle\tilde{\vec{Z}}(\vec{r},\vec{r}\,';s_{\beta_{oc},\beta'_{oc}});\vec{j}_g(\vec{r}\,')\rangle_g;\tilde{\vec{j}}_{\beta_{oc}}(\vec{r},s_{\beta_{oc},\beta'_{oc}})\rangle_a\right\}}{\langle\tilde{\vec{\mu}}_{\beta_{oc}}(\vec{r},s_{\beta_{oc},\beta'_{oc}});\frac{\partial}{\partial s}\tilde{\vec{\Gamma}}(\vec{r},\vec{r}\,';s)\Big|_{s=s_{\beta_{oc},\beta'_{oc}}};\tilde{\vec{j}}_{\beta_{oc}}(\vec{r}\,',s_{\beta_{oc},\beta'_{oc}})\rangle_a}$$

$$= \frac{\left\{\langle\tilde{\vec{\mu}}_{\beta_{oc}}(\vec{r}\,',s_{\beta_{oc},\beta'_{oc}});\tilde{\vec{\delta}}_p(\vec{r}\,',s_{\beta_{oc},\beta'_{oc}})\rangle_a \cdot \langle\langle\tilde{\vec{Z}}(\vec{r},\vec{r}\,';s_{\beta_{oc},\beta'_{oc}});\vec{j}_g(\vec{r}\,')\rangle_g;\tilde{\vec{j}}_{\beta_{oc}}(\vec{r},s_{\beta_{oc},\beta'_{oc}})\rangle_a\right\}}{\frac{\partial}{\partial s}\tilde{\lambda}_{\beta_{oc}}(s)\Big|_{s=s_{\beta_{oc},\beta'_{oc}}}} \tag{6.30}$$

where $\tilde{V}_{\beta_{oc},oce,p}$ is the entire-function contribution to each eigenmode term (β_{oc}) in the expansion for the pth incident field term with the waveform $\tilde{f}_p(s)$ factored out. Combining open-circuit voltage and impedance SEM forms gives

$$\tilde{V}_{oc}(s) = \sum_{\beta_{oc}}\left\{\left\{\sum_{\beta'_{oc}}\left\{\sum_p I_{\beta_{oc},\beta'_{oc},p}\tilde{f}_p(s)\right\}\tilde{Z}_{\beta_{oc},\beta'_{oc}}(s)\right\} + \left\{\sum_p \tilde{I}_{\beta_{oc},oce,p}(s)\tilde{f}_p(s)\right\}\tilde{Z}_{\beta_{oc},oce}(s)\right\} + \left\{\sum_p \tilde{I}_{g,p}(s)\tilde{f}_p(s)\right\}\tilde{Z}_g(s)$$

$$I_{\beta_{oc},\beta'_{oc},p} = \tilde{I}_{\beta_{oc},p}(s_{\beta_{oc},\beta'_{oc}}) = \frac{E_0}{Z_0}\frac{b_{\beta_{oc},\beta'_{oc},p}}{a_{\beta_{oc},\beta'_{oc}}} \tag{6.31}$$

$$= E_0\frac{\langle\tilde{\vec{\mu}}_{\beta_{oc}}(\vec{r}\,',s_{\beta_{oc},\beta'_{oc}});\tilde{\vec{\delta}}_p(\vec{r}\,',s_{\beta_{oc},\beta'_{oc}})\rangle_a}{\langle\tilde{\vec{\mu}}_{\beta_{oc}}(\vec{r},s_{\beta_{oc},\beta'_{oc}});\langle\tilde{\vec{Z}}(\vec{r},\vec{r}\,';s_{\beta_{oc},\beta'_{oc}});\vec{j}_g(\vec{r}\,')\rangle_g\rangle_a}$$

$$\tilde{I}_{\beta_{oc},oce,p}(s) = \tilde{Z}_{\beta_{oc},oce}(s)\tilde{V}_{\beta_{oc},oce,p}(s)$$

The current-source coefficients $I_{\beta_{oc},\beta'_{oc},p}$ for the poles are frequency independent; the expression for these terms is simplified by the cancellation of terms involving the derivative of the kernel of the integral equation with respect to s. Note the presence of the gap term which can be computed from expressions in

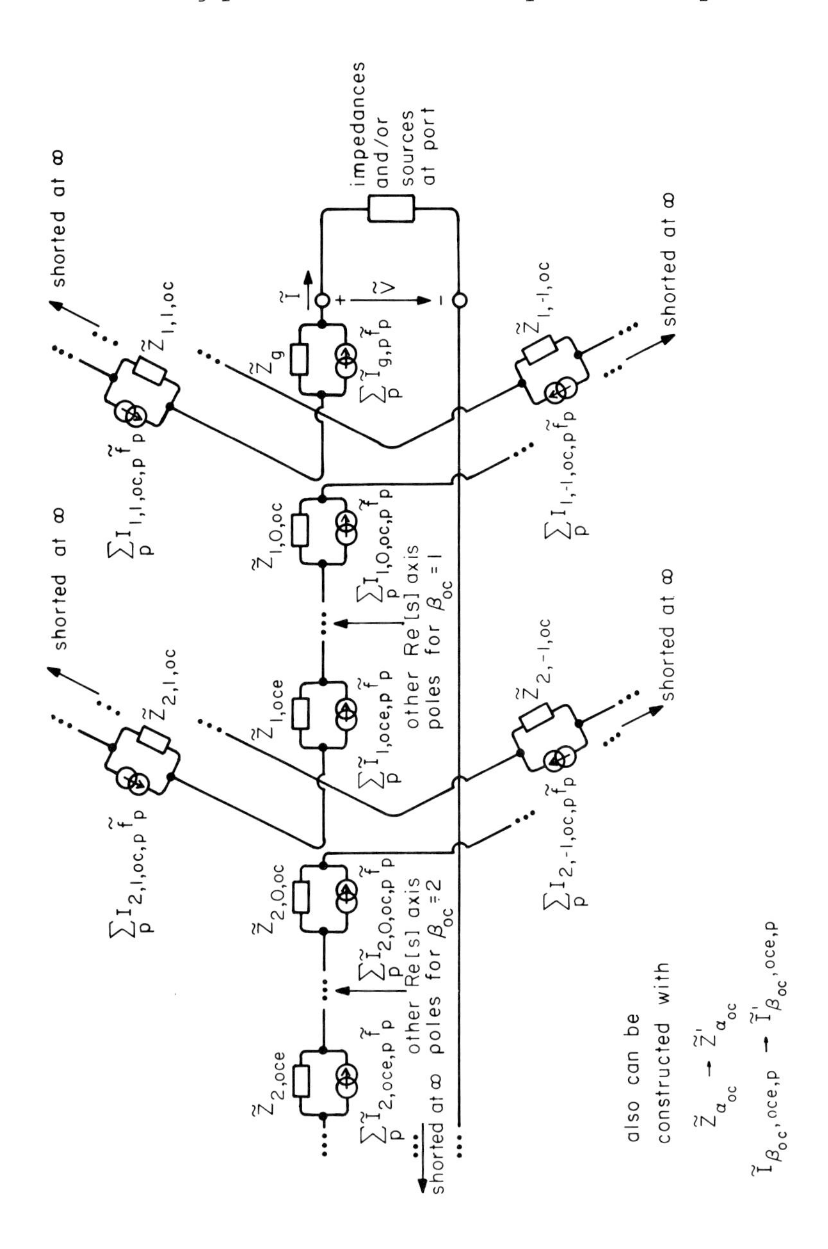

Fig. 6.8. Equivalent circuit for impedance and open-circuit voltage based on singularity expansion of open-circuit boundary-value problem.

(6.26) and (6.27). These terms may also be singularity expanded, but note that as long as the gap region is assumed electrically small and $\tilde{\vec{\delta}}_p(\vec{r},s)$ is chosen so as to have no dependence on s at some coordinate center in the gap region, then the gap terms take on simple dependences involving at most a few powers and inverse powers of s.

Fig. 6.8 gives the equivalent circuit combining the impedance from (6.29) with the open-circuit voltage from (6.31) as a series sum of circuit modules. Each module is the parallel sum of an impedance in parallel with a current source. Starting at the right from the load and sources at the port, we have first the circuit module for the gap followed by the SEM circuit modules grouped according to the eigenmode index β_{oc} in increasing order to the left. As β_{oc} takes on numerical values an additional subscript oc is added for clarity. Within each β_{oc} set of modules the modules are located to correspond to pairs according to their conjugate symmetric locations in the s plane. One can roughly think of Fig. 6.8 as an s-plane diagram with circuit modules symmetrically located in the left half plane corresponding to pole locations.

As with the previous case of the admittance, the pole impedances can be modified by subtraction of the value at $s = 0$. For loop antennas (perfectly conducting) the impedance goes to zero at $s = 0$. For some cases it is then convenient to introduce modified-pole impedances giving for the impedance expansion

$$\begin{aligned}
\tilde{Z}_a(s) &= \sum_{\beta_{oc}} \tilde{Z}_{\beta_{oc}}(s) + \tilde{Z}_g(s) \\
&= \sum_{\beta_{oc}} \left\{ \left\{ \sum_{\beta'_{oc}} \tilde{Z}'_{\beta_{oc},\beta'_{oc}}(s) \right\} + \tilde{Z}'_{\beta_{oc},oce}(s) \right\} + \tilde{Z}_g(s) \\
\tilde{Z}'_{\beta_{oc},\beta'_{oc}}(s) &= \tilde{Z}_{\beta_{oc},\beta'_{oc}}(s) - \tilde{Z}_{\beta_{oc},\beta'_{oc}} \\
&= Z_0 a_{\beta_{oc},\beta'_{oc}} \left[(s - s_{\beta_{oc},\beta'_{oc}})^{-1} + s^{-1}_{\beta_{oc},\beta'_{oc}} \right]
\end{aligned} \tag{6.32}$$

$$= Z_0 \frac{a_{\beta_{oc},\beta'_{oc}}}{s_{\beta_{oc},\beta'_{oc}}} \frac{s}{s - s_{\beta_{oc},\beta'_{oc}}}$$

$$\tilde{Z}'_{\beta_{oc},oce}(s) = \tilde{Z}_{\beta_{oc},oce}(s) + \sum_{\beta'_{oc}} \tilde{Z}_{\beta_{oc},\beta'_{oc}}(0)$$

$$= \tilde{Z}_{\beta_{oc},oce}(s) - Z_0 \sum_{\beta'_{oc}} a_{\beta_{oc},\beta'_{oc}} s^{-1}_{\beta_{oc},\beta'_{oc}}$$

where $Z_{\beta_{oc},oce}(s)$ is a correspondingly modified entire-function contribution. The open-circuit voltage can also be written in terms of modified poles as

$$\tilde{V}_{oc}(s) = \sum_{\beta_{oc}} \left\{ \left\{ \sum_{\beta'_{oc}} \left\{ \sum_p I_{\beta_{oc},\beta'_{oc},p} \tilde{f}_p(s) \right\} \tilde{Z}'_{\beta_{oc},\beta'_{oc}}(s) \right\} \right.$$

$$+ \left\{ \sum_p \tilde{I}'_{\beta_{oc},oce,p}(s) \tilde{f}_p(s) \right\} \tilde{Z}'_{\beta_{oc},oce}(s) \Bigg\}$$

$$+ \left\{ \sum_p \tilde{I}_{g,p}(s) \tilde{f}_p(s) \right\} \tilde{Z}_g(s) \tag{6.33}$$

$$\tilde{V}'_{\beta_{oc},oce,p}(s) = \left\{ \sum_p \tilde{I}'_{\beta_{oc},oce,p}(s) \tilde{f}_p(s) \right\} \tilde{Z}'_{\beta_{oc},oce}(s)$$

$$= \tilde{V}_{\beta_{oc},oce,p}(s) - E_0 \sum_{\beta'_{oc}} b_{\beta_{oc},\beta'_{oc},p} s^{-1}_{\beta_{oc},\beta'_{oc}}$$

$$\tilde{I}'_{\beta_{oc},oce,p}(s) = \tilde{Z}'_{\beta_{oc},oce}(s) \tilde{V}'_{\beta_{oc},oce,p}(s)$$

Fig. 6.8 also describes the equivalent circuit using these modified-pole quantities in the same form as for unmodified poles; primes are added to appropriate quantities.

The pole impedances can be written in circuit form (using the shorter α_{oc} notation) as

$$\tilde{Z}_{\alpha_{oc}}(s) = Z_0 a_{\alpha_{oc}} (s - s_{\alpha_{oc}})^{-1} = \left[sC_{\alpha_{oc}} + G_{\alpha_{oc}} \right]^{-1}$$

$$C_{\alpha_{oc}} = \frac{1}{Z\, a_{\alpha_{oc}}} \quad , \quad G_{\alpha_{oc}} = -\frac{1}{Z_0} \frac{s_{\alpha_{oc}}}{a_{\alpha_{oc}}} = -s_{\alpha_{oc}} C_{\alpha_{oc}} \tag{6.34}$$

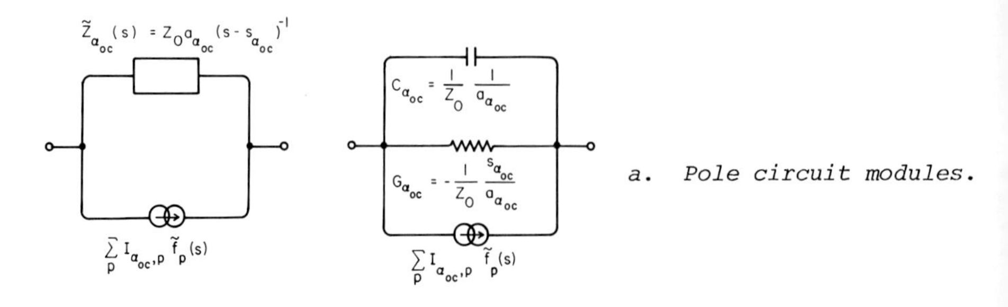

a. Pole circuit modules.

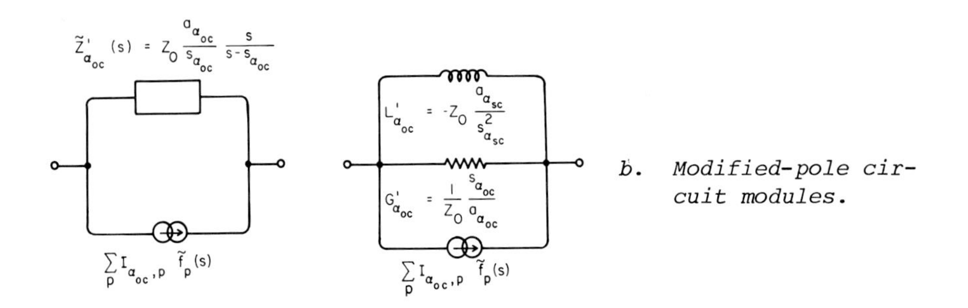

b. Modified-pole circuit modules.

Fig. 6.9. Elementary circuit modules for pole terms for impedance and open-circuit voltage for open-circuit boundary-value problem.

as illustrated for a pole circuit module in Fig. 6.9a as a parallel capacitance and conductance, in turn in parallel with the current source. The modified-pole impedance in circuit form is

$$\tilde{Z}'_{\alpha_{oc}}(s) = Z_0 \frac{a_{\alpha_{oc}}}{s_{\alpha_{oc}}} \frac{s}{s - s_{\alpha_{oc}}} = \left[G'_{\alpha_{oc}} + \frac{1}{sL'_{\alpha_{oc}}} \right]^{-1}$$

$$L'_{\alpha_{oc}} = -Z_0 \frac{a_{\alpha_{oc}}}{s_{\alpha_{oc}}} \quad , \quad G'_{\alpha_{oc}} = \frac{1}{Z_0} \frac{s_{\alpha_{oc}}}{a_{\alpha_{oc}}} = -\frac{1}{s_{\alpha_{oc}} L'_{\alpha_{oc}}} = -G_{\alpha_{oc}} \tag{6.35}$$

as illustrated for a modified-pole circuit module in Fig. 6.9b as a parallel inductance and conductance, in turn in parallel with the current source.

For conjugate pole pairs one may wish to combine the corresponding circuit modules to obtain real valued circuit elements corresponding to real valued time functions, as a step towards

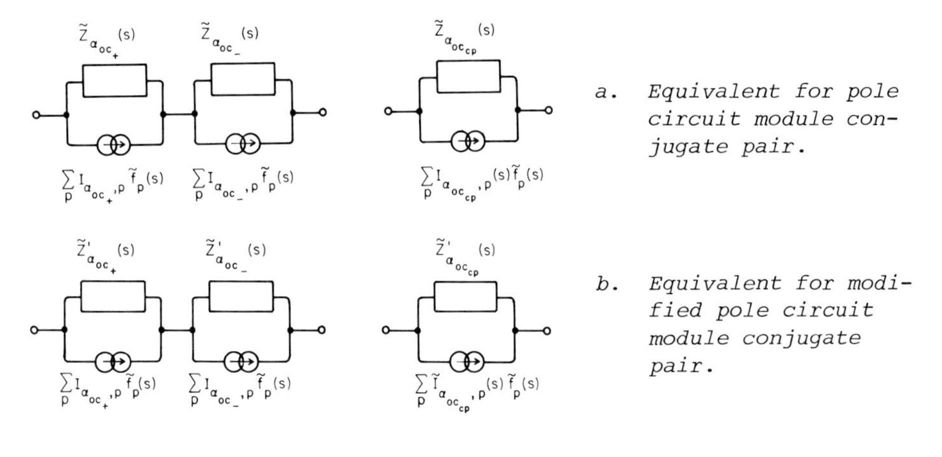

a. Equivalent for pole circuit module conjugate pair.

b. Equivalent for modified pole circuit module conjugate pair.

Fig. 6.10. Norton equivalent circuits for pole pairs for open-circuit boundary-value problem.

more nearly realizable circuits using passive elements. As illustrated in Fig. 6.10, Norton equivalent circuits can be formed. Using α_{oc_+} and α_{oc_-} to designate the two poles (upper and lower half planes respectively) in the conjugate pair (*cp*) of interest, we have for unmodified poles as in fig. 6.10a

$$\tilde{Z}_{\alpha_{oc_{cp}}}(s) = \tilde{Z}_{\alpha_{oc_+}}(s) + \tilde{Z}_{\alpha_{oc_-}}(s)$$

$$\tilde{I}_{\alpha_{oc_{cp}},p}(s) = \frac{I_{\alpha_{oc_+},p}\tilde{Z}_{\alpha_{oc_+}}(s) + I_{\alpha_{oc_-},p}\tilde{Z}_{\alpha_{oc_-}}(s)}{\tilde{Z}_{\alpha_{oc_+}}(s) + \tilde{Z}_{\alpha_{oc_-}}(s)} \tag{6.36}$$

$$= Re\left[I_{\alpha_{oc}},p\right] + jIm\left[I_{\alpha_{oc_+}},p\right] \frac{\tilde{Z}_{\alpha_{oc_+}}(s) - \tilde{Z}_{\alpha_{oc_-}}(s)}{\tilde{Z}_{\alpha_{oc_+}}(s) + \tilde{Z}_{\alpha_{oc_-}}(s)}$$

For real pole current-source coefficients the current-source coefficients for the conjugate pair module are frequency independent, indicating a potential importance for such cases. For conjugate pairs of modified pole circuit modules, as shown in Fig. 6.10b, the equations for the terms are the same as in (6.36) except for the addition of appropriate primes to give modified-pole quantities.

E. Some Comments

This discussion of equivalent circuits illustrates some of the kinds of circuits inferable from the EEM and SEM forms of the response of an object (antenna or scatterer) with an electrical port. The forms of the equations are such as to allow equivalent circuits composed of modules with sources and impedances (admittances) associated with eigenmodes or s-plane singularities. The circuit forms presented here include those based on both short-circuit and open-circuit boundary value problems, and for both poles and modified poles (with the value at $s = 0$ subtracted). Such parallel combinations of Thevenin (series) circuit modules and series combinations of Norton (parallel) circuit modules are not necessarily the only kinds of equivalent circuits that might be developed and prove useful; other forms such as ladder networks might be developed. As discussed in [17], the impedance (admittance) at the port can be described by using both the short-circuit and open-circuit natural frequencies in terms of a ratio of products, giving yet another potential approach.

In the SEM circuit forms discussed here, the pole circuit modules have explicit formulas in terms of the usual SEM quantities describing each pole term and derivable from an appropriate integral equation for the object response. The entire-function circuit modules are another matter. The entire-function contribution can be considered on an eigenmode-by-eigenmode basis (including other contributions, as from the "gap" term), or it can be considered as a single term (and circuit module) for the object response. Other types of contributions (like branch integrals) can also be present for certain classes of objects.

Certain aspects of the entire-function behavior can be determined based on the asymptotic behavior of certain parameters as $s \to 0$ or $s \to \infty$. Certain parameters are positive real functions of s because they represent passive impedance (admittance) functions, implying they behave as As^B for $-1 \leq B \leq 1$ and $A > 0$ as $s \to 0$ and

$s \to \infty$ (with perhaps different values of A and/or B in the two cases). Eigenimpedances (eigenvalues of impedance integral equations) are examples of positive real functions which can be used to aid in establishing the entire-function contribution to each eigenmode part of the equivalent circuit. Additionally, the driving-point impedance $\tilde{Z}_a(s)$ or admittance $\tilde{Y}_a(s)$ is a positive real function of s which limits its behavior as indicated for high and low frequencies by examples in Fig. 6.11; here the distributed gap regions (approximately cylindrical) are assumed to have approximately uniform gap source distributions ($\vec{e}_g(\vec{r})$ and/or $\vec{j}_g(\vec{r})$) as surface distributions over the gap surface [17]. By appropriate choice of arrival time for an incident delta function plane wave, the short circuit current and open circuit voltage will also have restricted high and low frequency behavior for the examples in Fig. 6.11. In combination such restrictions limit the behavior of the entire-function circuit module(s) as $s \to \infty$ (in the right half plane) and as $s \to 0$.

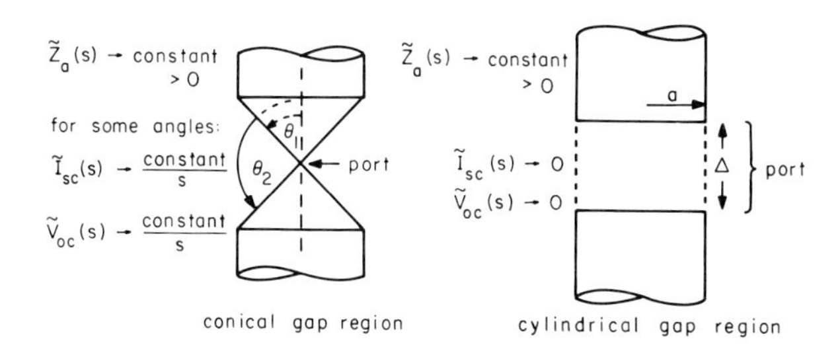

a. High frequencies.

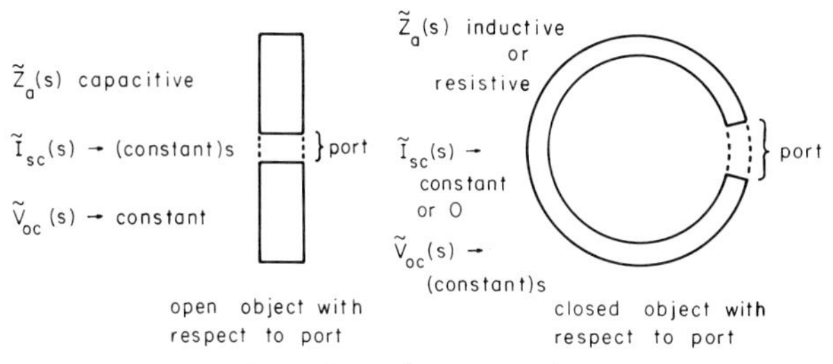

b. Low frequencies.

Fig. 6.11. Some limiting properties of port.

There are various detailed problems associated with the practical evaluation of the various circuit module terms arising from the spatial singularities in the integral-equation kernel(s), especially for the open-circuit boundary value problem. This section has only developed some aspects of the forms of the solutions and some general formulas for the terms. These can be generalized, in principle, to objects with N ports, instead of just a single port. The reader is referred to [17] for more detailed discussion of various of the points in this section. There is, however, a great deal yet to be done on this general topic.

VII. SUMMARY

From its early conception [12] the singularity expansion method has come to considerably expand its scope. Beginning from experimental observations of damped sinusoidal waveforms, inferring poles in the complex frequency plane and generalizing the concept to all singularities in the s plane, the early history of SEM concentrated on the characteristics of pole terms, computing such terms from integral equations for the electromagnetic response of objects (in free space), and working out numerical examples of the transient/broad-band response of some simple scatterers and antennas. This work demonstrated the practical utility and physical insight given by SEM as then understood. Much of this work is summarized in an earlier review [5].

The eigenmode expansion method was then introduced for the purpose of further understanding SEM, extending its utility and tying it into some other perhaps related concepts [15]. This led to alternate formulas for pole terms, ordering of the pole terms according to eigenvalues, introduction of eigenimpedances and connection to circuit concepts of positive real functions and some basic relations for eigenimpedance synthesis by impedance loading (including direct choosing of natural frequency locations in the s plane). These are summarized in another review [22] along with some other approaches to the transient/broad-band scatterer/ antenna problem.

This review has concentrated on some recent developments related to the SEM and EEM concepts. The basic pole terms have been extended to second order poles, important for critically damped objects. Efficient and accurate procedures for finding pole and zero locations using various contour integral formulas of complex variable theory have been summarized. An important use of SEM and EEM is the construction of equivalent circuits of object response at a port; this concept has many ramifications, introducing a variety of possible circuits and providing a way of tying classical circuit concepts (and positive real functions) to antennas and scatterers [17]. By use of certain contour integral formulas, one can also write explicit formulas for the various terms in the singularity expansion of a given function, operator, etc.; this should give some aid and impetus to the study of branch-integral and entire-function contributions.

There are then various new results in SEM and EEM. As one might expect, besides helping resolve some old questions they raise some new questions and possibilities. See also [23].

VIII. REFERENCES

1. Guillemin, E.A., "Synthesis of Passive Networks," Wiley, New York, 1957.
2. Harrington, R.F., "Field Computation by Moment Methods," MacMillan, New York, 1968.
3. Tai, C.T., "Dyadic Green's Functions in Electromagnetic Theory," Intext, New York, 1971.
4. Kouyoumjian, R.G., The Geometrical Theory of Diffraction and Its Application, "Numerical and Asymptotic Techniques in Electromagnetics," R. Mittra, ed., Springer-Verlag, New York, 1975.
5. Baum, C.E., The Singularity Expansion Method, "Transient Electromagnetic Fields," L.B. Felsen, ed., Springer-Verlag, New York, 1975.
6. Baum, C.E., The Role of Scattering Theory in Electromagnetic Interference Problems, National Conference on Electromagnetic Scattering, Chicago, June 1976.

7. Tesche, F.M., Application of the Singularity Expansion Method to the Analysis of Impedance Loaded Linear Antennas, Sensor and Simulation Note 177, May 1973. (Also as Tesche, F.M., The Far-Field Response of a Step-Excited Linear Antenna Using SEM, *IEEE Trans. Antennas & Propag.*, AP-23, 834-848, November 1975.)
8. Baum, C.E., Singularity Expansion of Electromagnetic Fields and Potentials Radiated from Antennas or Scattered from Objects in Free Space, Sensor and Simulation Note 179, May 1973.
9. Blackburn, R.F., Analysis and Synthesis of an Impedance-Loaded Loop Antenna Using the Singularity Expansion Method, Sensor and Simulation Note 214, May 1976.
10. Lee, K.S.H. and L. Marin, Interaction of External System-Generated EMP with Space Systems, Theoretical Note 177, August 1973. (Also as Lee, K.S.H. and L. Marin, Charged Particle Moving Near a Perfectly Conducting Sphere, *IEEE Trans. Antennas & Propag.*, AP-22, 683-689, September 1974.)
11. Baum, C.E., Some Considerations Concerning Analytic EMP Criteria Waveforms, Theoretical Note 285, October 1976.
12. Baum, C.E., On the Singularity Expansion Method for the Solution of Electromagnetic Interaction Problems, *Interaction Note 88*, December 1971.
13. Marin, L. and R. Latham, Analytical Properties of the Field Scattered by a Perfectly Conducting Finite Body, Interaction Note 93, January 1972. (Also as Marin, L. and R. Latham, Representation of Transient Scattered Fields in Terms of Free Oscillations of Bodies, *Proc. IEEE*, 60, 640-641, May 1972; and as Marin, L., Natural Mode Representation of Transient Scattered Fields, *IEEE Trans. Antennas & Propag.* AP-21, 809-818, November 1973.
14. Tesche, F.M., On the Singularity Expansion Method as Applied to Electromagnetic Scattering from Thin Wires, Interaction Note 102, April 1972. (Also as Tesche, F.M., On the Analysis of Scattering and Antenna Problems Using the Singularity Expansion Technique, *IEEE Trans. Antennas & Propag.*, AP-21, 52-63, January 1973.)
15. Baum, C.E., On the Eigenmode Expansion Method for Electro-Magnetic Scattering and Antenna Problems, Part I: Some Basic Relations for Eigenmode Expansions and Their Relation to the Singularity Expansion, Interaction Note 229, January 1975.
16. VanBlaricum, M.L. and R. Mittra, A Technique for Extracting the Poles and Residues of a System Directly from its Transient Response, Interaction Note 245, February 1975.
17. Baum, C.E., Single Port Equivalent Circuits for Antennas and Scatterers, Interaction Note 295, March 1976.
18. Crow, T.T., B.D. Graves and C.D. Taylor, Numerical Techniques Useful in the Singularity Expansion Method as Applied to Electromagnetic Interaction Problems, Mathematics Note 27, December 1972.

19. Singaraju, B.K., D.V. Giri and C.E. Baum, Further Developments in the Application of Contour Integration to the Evaluation of the Zeros of Analytic Functions and Relevant Computer Programs, Mathematics Note 42, March 1976.
20. Giri, D.V. and C.E. Baum, Application of Cauchy's Residue Theorem in Evaluating the Poles and Zeros of Complex Meromorphic Functions and Apposite Computer Programs, Mathematics Note (to be published).
21. Van Bladel, J., Some Remarks on Green's Dyadic for Infinite Space, *IRE Trans. Antennas & Propag.*, AP-9, 563-566, November 1961.
22. Baum, C.E., Emerging Technology for Transient and Broad-Band Analysis and Synthesis of Antennas and Scatterers, *Proc. IEEE*, 64, 1598-1616, November 1976.
23. An updated overview of generation, measurement, simulation, interaction with systems of nuclear electromagnetic pulses may be found in *IEEE Trans. Antennas and Propag. AP-26*, (special issue), 1-187 (1978).

PHENOMENOLOGICAL THEORY OF RADAR TARGETS

J. Richard Huynen
Engineering Consultant
Los Altos Hills, California

I. INTRODUCTION

Radar targets are typically rather complex natural or man made objects. We may distinguish between large single objects such as an airplane or ship, or a distribution of smaller objects such as clouds of chaff, rain, snow or dust, a wheatfield or ripples on an ocean surface. All these targets are potential objects of interest to the radar systems engineer. Because of the variety of possible targets it is customary to study a group of targets separately for radar cross section (RCS) dependence on frequency, polarization, aspect angle etc. One usually embarks on a literature search, an independent measurements program or one may be fortunate enough to apply a well-established theoretical approach which can predict RCS with enough accuracy to satisfy the systems requirements at hand. The advantage of such a selective approach is that specific questions related to the system at hand may obtain a direct answer. The disadvantage is that usually very little of the information often gathered at great expense, is applicable to a new case due to changes in environment, physical targets and systems demands.

In contrast to this selective approach to radar targets, the phenomenological theory concentrates on general properties that are applicable to all targets and hence no a priori knowledge of a particular target is essential for the theory. Instead it considers the target on the basis of the available data that the

ISBN 0-12-709650-7

radar receives from the target. It takes into account the RCS dependence on frequency, polarization, aspect angles etc., under the dynamic conditions of an actually observed radar target in motion. Such a generalized viewpoint obviously is attractive to supplement the selective approach sketched above for target identification or discrimination problems. Other applications are found with dynamic simulation of radar targets [36], inverse scattering, improvement of existing target models, radar altimeter design [50] and glint problems. In general the theory attempts to provide better understanding of radar target behavior such that implementation with a system will lead to meaningful discriminants.

The mathematical framework for the phenomenological theory is provided by the target polarization scattering matrix $T(t)$. The scattering matrix accounts for all information which is contained in the polarized target return; it can be measured by using dual polarized antennas for transmission as well as reception. Even if at present such complete information of targets is lacking, the scattering matrix provides the proper framework for analysis of significant target data. The target scattering matrix accounts for the presence of the target under the conditions of the given radar encounter, it depends on radar frequency, waveform and target observation angle. The target scattering matrix contains a host of interesting and even some remarkable properties, not sufficiently exposed in the current literature, which correspond to the physical properties of general radar targets. Hence properties of symmetry, non-symmetry, convexity and irregularity can in each case be given analytical and numerical support. Examples of most often occurring target matrices are provided in the text.

We thus find, that in general, a time-varying power averaged 'target' (chaff cloud, ocean surface) is described by nine parameters, whereas a fixed single object is given by five parameters. From this observation it follows that the averaged target cannot in general be represented by an equivalent effective single object, since it has four more degrees of freedom. However, it is possible to obtain a decomposition of the average target into an effective single target representing the ensemble, and a 'residue target' or *N*-target which represents target noise at the higher frequencies. Both targets are independent, completely specified, and physically realizable. The *N*-target is a new class of radar targets which cause highly irregular depolarizing scattering behavior; it is used to describe 'residue'. Examples of single *N*-targets are helices, corner-reflectors and troughs or diplanes at selected aspect angles and frequency.

The target decomposition has recently been extended further to apply not only to power averages, but indeed to the instantaneously received signals as well. At each instant the return from a time varying target can be split into components which contribute exclusively to the 'effective target' and components which contribute exclusively to the *N*-target residue, i.e. we can show that $T(t) = a_o(t)T_o + T_N(t)$. Thus the incoming target return is decomposed into a desirable part which is 'signal' and a residue term which is 'noise'. Standard or adaptive signal processing schemes may be used to operate on these signals to optimize radar systems performance. The complexity due to target dependence on polarization is thus turned to advantage. The decomposition theorem makes it possible to focus on the 'effective target' as effectively representing the ensemble. The effective target has the scattering matrix: $T_o(t) = a_o(t)T_o$, where T_o is a *constant* matrix (determined by 5 constant parameters) which represents effective target *structure*, while $a_o(t)$ is a *scalar* time varying signal which has normalized power. This result agrees with the intuitive concept that a complex time varying target may be represented by an 'effective fixed object', T_o, which scattering behavior is determined by a *scalar* $a_o(t)$. The constant structure of the desired part of target scattering provides a powerful tool for adaptive target processing. The residue part $T_N(t)$ may be viewed as clutter which accounts for the approximation.

Thus a match is achieved between electromagnetic behavior of radar targets and practical expediency, which favors the scalar viewpoint. Thus, a radially symmetric chaff cloud has the structure of a sphere: $T_o = I$ = unit matrix, with effective scattering $a_o(t)$ and residue clutter determined by $T_N(t)$, similarly an ocean surface at angles near vertical is structurally a flat surface (also $T_o = I$). Any depolarization for these models is accounted for by the target residue component. These cases are examples of the orientation-independent target model, which is the simplest of all time-varying electromagnetic models [50]. This model has been applied effectively as a model for terrain with the design of a sensitive radio-altimeter [59] and also has been suggested as a model for complex space objects [61]. The scalar function $a_o(t)$ is easily measured for all targets by the (*RC*-*LC*) circularly polarized return signal.

This result puts renewed focus on the use of a circular polarization for target investigations, in contrast to the predominance of linear polarization presently in use. In order to implement the theory discussed in this work we recommend use of dual polarized antennas, which measure $T(t)$, and as second alternative transmit CP, and receive both modes of polarized return.

II. BACKGROUND

The present work spans a period of over 25 years. The literature on all the various aspects of radar target technology has grown into voluminous proportions. It is an impossible task to give credit to all those who contributed to this field in a short review. We have listed in the reference list some of the sources which have relevance to the present work. Several excellent survey books have been published, notably those by Beckmann and Spizzichino [22], Beckmann [51]; Ruck, Barrick, Stuart and Kirchbaum [57], Born and Wolf [39], Skolnik [58], Nathanson [56], Bowman, Senior and Uslenghi [53] and most recently Long [68]. The extensive references quoted in these works should provide the interested reader with ample source material.

The author credits early work by G. Sinclair [1], E. M. Kennaugh [7,8], H. Gent [10] and G.A. Deschamps [2] with inspirational guidance, which started this work in the 1950's [9]. In 1960 the author was led, through consultation by V. H. Rumsey [3] to initiate a measurements program on complete target null-locus plots [17]. Dr. Victor Twersky was instrumental for encouragement to report on this work [20]; some of the results are included in this paper. W. P. Melling, R. H. T. Bates and J. Rheinstein provided necessary editorial support for a 1965 paper [36].

In 1969 the author was provided opportunity and encouragement by Dr. J. P. Schouten from the University of Technology Delft, the Netherlands, to prepare a doctoral thesis on the subject of radar targets. The title of the thesis, which appeared in book form was, *Phenomenological Theory of Radar Targets*, and the present study is an extension of that project. In 1975 the author did private consulting for Nato at SHAPE Technical Center, the Hague, Netherlands. A presentation was given and a report was issued January 1976 [69]. The present work incorporates and updates most of the previous effort in this area.

III. DEPENDENCE ON TARGET ORIENTATION

It is a well-known fact that the radar return from most complex target structures for target illumination with horizontal polarization (H) is different from the return for illumination with vertical polarization (V). Less well understood are the consequences this fact has on the return from a moving target in space. Suppose that the radar antenna is linearly polarized, then because of changes in target orientation angle (see Fig. 1) due to target motion, the effective target illumination, relative to the target axis, has a linear polarization with changes with the target orientation angle. Since the motion of the target

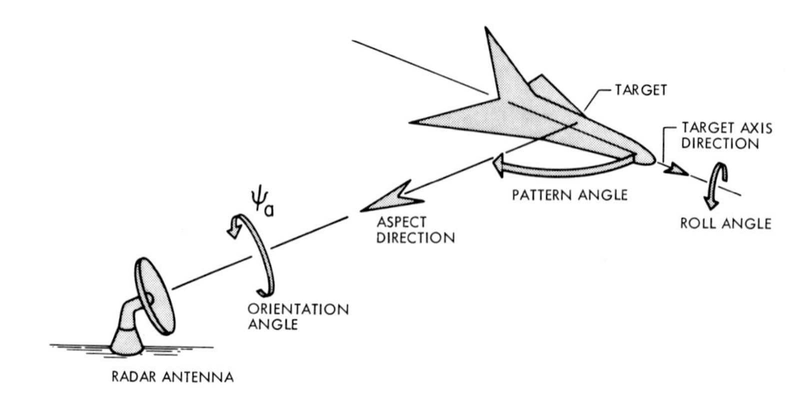

Fig. 1. Target aspect direction and orientation angle.

relative to radar site location may be quite arbitrary, all linear polarization returns at given aspect direction and frequency are necessary to interpret target return for linear polarization. The fact that only one return is actually measured leads to ambiguities in the interpretation of this signature as a credible target parameter. Hence it is not true that the same target exposure, given by a fixed aspect direction, always gives the same radar cross section (RCS), this should be true only for circularly polarized illumination, which is not the most common mode of radar operation. Hence the dependence of the target return at a given aspect direction upon the orientation of linear polarization, which can give differences of 15 db or more, leads to ambiguities: the same target exposure does not always give the same RCS. This behavior of target scattering has caused endless confusion among early investigations aimed at target identification based upon target signature.

These negative results have led to the mistaken belief that target signatures are useless for identification purposes. The mistake was made to take the "signal" received by the linearly polarized radar as a faithful or true target representative without realizing the corrupting influence on true target signature of the orientation angle ψ_a, which is due to incidental target axis orientation relative to radar site position. We will show shortly how to resolve this ambiguity, once the target scattering matrix $T(t)$ is known. It will not be necessary to have a priori knowledge of the target position in space to remove the effect of target orientation from target signature; it can be done solely on the basis of a known $T(t)$ matrix. What remains after orientation-dependence is removed from $T(t)$ is a scattering matrix determined by four so-called orientation-independent parameters:

m = target amplitude (is overall measure of target size)

τ_m = helicity angle (is measure of target non-symmetry)

ν = skip angle (is measure of double-bounce depolarization)

γ = characteristic angle ($\gamma = 0°$ for wire, $\gamma = 45°$ for sphere)

In addition to these pure structure-parameters the target has an "absolute phase" which is a mixed parameter: it can be changed simply by motion of the target along the line of sight, changing by a small amount the target range. The absolute phase is ignored with measurements of power.

We thus find that a target given by scattering matrix $T(t)$ is given by an orientation-angle (ψ) and four orientation-independent structure-parameters, and finally by an absolute phase which is usually ignored. In contrast with this, an averaged distributed target is determined by nine target parameters: A_o, B_o, B_ψ, C_ψ, D_ψ, E_ψ, F, G_ψ, and H_ψ; these are also given physical significance in a later discussion. It should always be understood that above parameters still depend on target exposure (aspect direction: pattern and roll angle, see Fig. 1) and on target frequency and waveform.

The orientation-independent parameters may be used as uncorrupted or true target signatures for conventional or adaptive target discrimination and identification purposes.

IV. THEORY OF TARGET SCATTERING MATRICES

The information contained in the target return is due to interaction of transmitted wave with (fixed) polarization $\underline{a}$ and the illuminated moving target, which is represented by the target scattering matrix $T(t)$. The receiver polarization $\underline{b}$ intercepts part of the returned wave $\underline{r}(t) = T(t)\underline{a}$. Hence the complex voltage measured at the receiver terminals is given by[1]:

$$V(t) = (T(t)\underline{a}) \cdot \underline{b} \tag{4.1}$$

The antenna polarizations are usually linear (H or V) or circular; the general case is shown in Fig. 2. Shown are the four parameters

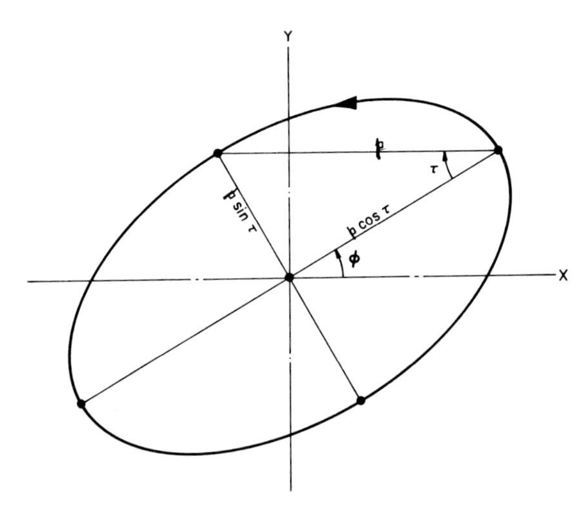

Fig. 2. Left-sensed polarization ellipse in fixed plane.

which define polarization $\underline{a}$: p = amplitude, α = absolute phase angle, ϕ = orientation angle and τ = ellipticity angle. We may write the polarization $\underline{a} = (a_x, a_y)$ as a two-component complex vector, given by four real numbers. The connection between

---

[1] The derivation of this equation which is given in [60] actually is not that simple. We can show that (4.1) satisfies all physical properties that are necessary and sufficient to make (4.1) a valid statement. These are: linearity, reciprocity and the law for maximum reception. In this form $T(t)$ is a complex-symmetric (not hermetian!) matrix; see also Kennaugh [7,8], who first derived this formulation, and also Copeland [16].

coordinate and geometrical parameters is expressed by the following vector identity:

$$\underline{a} = \begin{bmatrix} a_x \\ a_y \end{bmatrix} = pe^{i\alpha} \begin{bmatrix} \cos\phi & -\sin\phi \\ \sin\phi & \cos\phi \end{bmatrix} \begin{bmatrix} \cos\tau \\ i\sin\tau \end{bmatrix} \tag{4.2}$$

The sense is incorporated with the sign of τ. The range of ellipticity angle τ is: $-45° \leq \tau \leq +45°$. For linear polarization $\tau = 0°$, for right circular polarization (RC), $\tau = 45°$; for left-circular polarization (LC), $\tau = -45°$. For horizontal polarization (H), $\phi = 0°$, while for vertical polarization (V), $\phi = 90°$. The range of ϕ is: $-90° \leq \phi \leq +90°$. In all these equations we follow the narrowband convention of suppressing the high frequency wave propagation term: $exp[i(\omega t - kt)]$.

States of polarization are conveniently given by points on a circular polarization chart (Fig. 3) [18]. The circumference

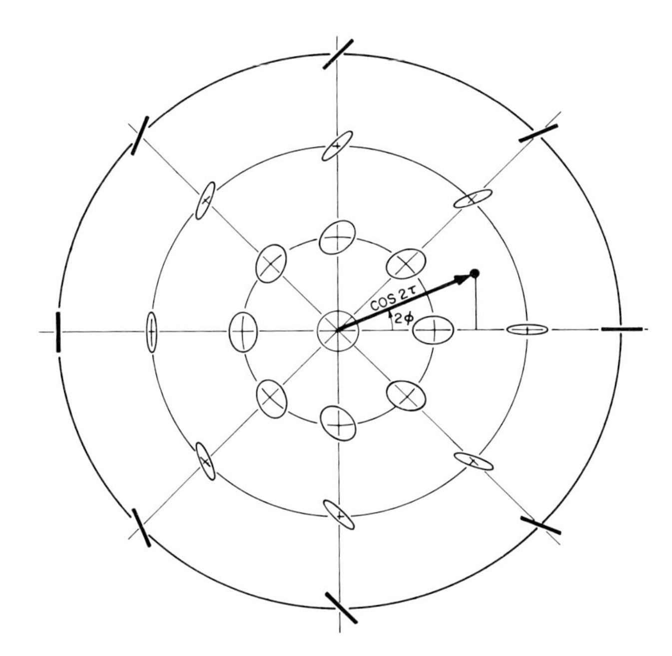

Fig. 3. Polarization chart.

shows the linear polarizations (LP) while the center is circular polarization (CP). Two such charts are necessary to determine all polarizations, one chart for each sense. Both charts are derived as projections from a polarization sphere [2]. Fig. 4 shows the polarization sphere (also called Poincaré sphere). Coordinates of a point on the sphere are called stokes parameters (see section VI). These charts are useful for plotting target parameters as a function of polarization (see section XIII).

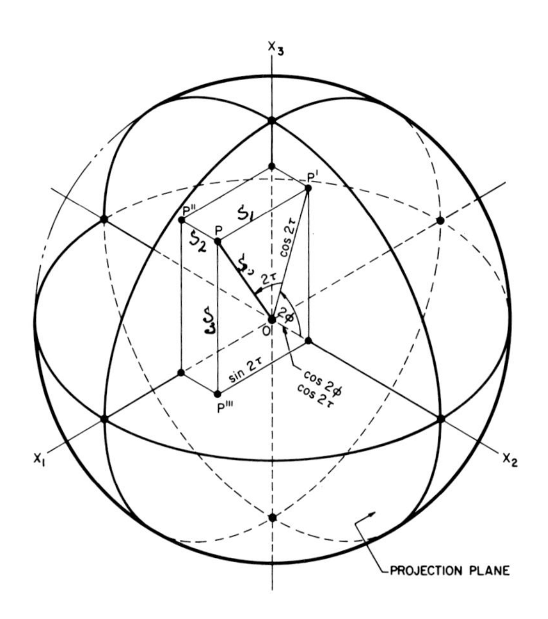

Fig. 4. Polarization sphere.

The target scattering matrix is a 2 × 2 complex valued symmetric matrix. The most convenient base is the H, V polarized coordinate system which was already used in (4.2). In this base the $T(t)$ matrix is given simply by:

$$T(t) = \begin{bmatrix} (H,H) & (V,H) \\ (H,V) & (V,V) \end{bmatrix} = \begin{bmatrix} a(t)+b(t) & c(t) \\ c(t) & a(t)-b(t) \end{bmatrix} \tag{4.3}$$

Here (V, H) is the complex voltage received from the target with vertically polarized transmitter and horizontally polarized receiver and similarly for the other designations. Hence the scattering matrix $T(t)$ is simply a collection of the three voltages (H, H), (V, V) and (V, H). Because of reciprocity $(V, H) = (H, V)$ and hence $T(t)$ is symmetric as shown in (4.3).

The notation with $a(t)$, $b(t)$ and $c(t)$ in (4.3) will be particularly useful for some problems; however a representation in terms of characteristic polarizations (eigenvectors of $T(t)$) will give deeper insight into the target structure. For example in most textbooks the form (4.3) is used and it is shown that a rotation of the coordinates (x, y) by an angle ψ results in a rotated matrix:

$$T_\psi(t) = \begin{bmatrix} a + b \cos 2\psi + c \sin 2\psi & c \cos 2\psi - b \sin 2\psi \\ c \cos 2\psi - b \sin 2\psi & a - b \cos 2\psi - c \sin 2\psi \end{bmatrix} \tag{4.4}$$

This shows the effect of a <u>change</u> ψ of target orientation, but it does not provide information regarding a <u>preferred</u> or intrinsic target orientation. For symmetric targets $T(t)$ becomes diagonal for preferred axis orientation ψ_a; hence: $c' = c \cos 2\psi_a - b \sin 2\psi_a = 0$, which determines ψ_a. In the general case, however, no such criterion is obvious from (4.4).

V. EIGENVALUE PROBLEM FOR $T(t)$

The characteristic eigenvalue problem for $T(t)$ is given in the following form:

$$T\underline{x} = t\underline{x}^* \tag{5.1}$$

In general there are two independent solutions for this equation:

$$\left.\begin{aligned} T\underline{x}_1 &= t_1\underline{x}_1^* \\ T\underline{x}_2 &= t_2\underline{x}_2^* \end{aligned}\right\} \tag{5.2}$$

where the asterisk denotes complex conjugation. The complex scalars t_1 and t_2 are the eigen values; the two normalized vectors $\underline{x}_1$ and $\underline{x}_2$ are the corresponding eigen vectors. Note that (5.1) is not of the form $A\underline{x} = a\underline{x}$ usually associated with hermetian matrices. For those problems, if $\underline{x}_1$ is a solution, $e^{i\alpha}\underline{x}_1$ also is; on the other hand, in (5.2) $\underline{x}_1$ is phase determined with the phase of t_1. We can show that the two eigenvectors in (5.2) are orthogonal, if the eigenvalues are not equal in magnitude (in which case the solution is called degenerate). Since T is symmetric $T^T = T$ (T^T is T transposed):

$$|T\underline{x}_1 \cdot \underline{x}_2| = |\underline{x}_1 \cdot T\underline{x}_2| \tag{5.3}$$

substituting (5.2):

$$|t_1||\underline{x}_1^* \cdot \underline{x}_2| = |t_2||\underline{x}_1 \cdot \underline{x}_2^*| \tag{5.4}$$

Hence if $|t_1| \neq |t_2|$

$$|\underline{x}_1 \cdot \underline{x}_2^*| = 0 \tag{5.5}$$

which is the condition for orthogonality between $\underline{x}_1$ and $\underline{x}_2$. Also $\underline{x}_1$ and $\underline{x}_2$ are normalized; we have:

$$\underline{x}_1 \cdot \underline{x}_2^* = 0; \quad \underline{x}_1 \cdot \underline{x}_1^* = 1; \quad \underline{x}_2 \cdot \underline{x}_2^* = 1 \tag{5.6}$$

Following the normal procedure, it is now possible to construct a unitary transformation $U = [\underline{x}_1, \underline{x}_2]$ for which

$$U^T U^* = \begin{bmatrix} \underline{x}_1 \cdot \underline{x}_1^* & \underline{x}_1 \cdot \underline{x}_2^* \\ \underline{x}_2 \cdot \underline{x}_1^* & \underline{x}_2 \cdot \underline{x}_2^* \end{bmatrix} = I \quad \text{(unit matrix)} \tag{5.7}$$

We can use U to bring T to diagonal form:

$$TU = [T\underline{x}_1, T\underline{x}_2] = [t_1\underline{x}_1^*, t_2\underline{x}_2^*] \tag{5.8}$$

hence:

$$U^T TU = \begin{bmatrix} \underline{x}_1 \cdot t_1\underline{x}_1^* & \underline{x}_1 \cdot t_2\underline{x}_2^* \\ \underline{x}_2 \cdot t_1\underline{x}_1^* & \underline{x}_2 \cdot t_2\underline{x}_2^* \end{bmatrix} = \begin{bmatrix} t_1 & 0 \\ 0 & t_2 \end{bmatrix} = T_d \tag{5.9}$$

Applying the inverse transform, we find T expressed in terms of eigenvalues and eigenvectors (notice $U^{-1} = U^{*T}$):

$$T = U^* T_d U^{*T} \tag{5.10}$$

This relation is fundamental for obtaining insight into target scattering, based on intrinsic properties. We next determine some properties of the total backscattered power:

$$P_{tot} = \underline{E}^S \cdot \underline{E}^{S*} = T\underline{a} \cdot (T\underline{a})^* \tag{5.11}$$

Let us decompose the transmit polarization $\underline{a}$ in terms of orthonormal eigenvectors $\underline{x}_1$ and $\underline{x}_2$:

$$\underline{a} = a_1\underline{x}_1 + a_2\underline{x}_2 \tag{5.12}$$

Substitution into P_{tot} gives:

$$\begin{aligned} P_{tot} &= (a_1 T\underline{x}_1 + a_2 T\underline{x}_2) \cdot (a_1^* T_1^* \underline{x}_1^* + a_2^* T_2^* \underline{x}_2^*) \\ &= |a_1|^2 |t_1|^2 + |a_2|^2 |t_2|^2 = p_A^2 |t_1|^2 - |a_2|^2 (|t_1|^2 - |t_2|^2) \end{aligned} \tag{5.13}$$

where $p_A^2 = |a_1|^2 + |a_2|^2 = g_A$ is total transmit antenna gain. We assume without loss of generality that $|t_1| \geq |t_2|$ and let g_A be fixed. We now wish to determine for with polarization $\underline{a}$, maximum power is returned from target $T(t)$: We find easily that P_{tot} is maximum if $a_2 = 0$ and hence $P_{tot} = g_A |t_1|^2$. To celebrate this result, we assign a target variable $m = |t_1|$, which is called "target-magnitude"; it may be viewed as an overall measure for target size, similarly as wave magnitude p_A gives a measure of antenna size (gain).

Continuing our labelling we will call the principal eigenvector $\underline{x}_1 = \underline{m}$, the "maximum polarization", which is defined by geometric variables ψ and τ_m as follows:

$$\underline{m}(\psi, \tau_m) = \begin{bmatrix} \cos\psi & -\sin\psi \\ \sin\psi & \cos\psi \end{bmatrix} \begin{bmatrix} \cos\tau_m \\ i\sin\tau_m \end{bmatrix} \tag{5.14}$$

The two eigenvalues are written in the following form:

$$
\begin{aligned}
t_1 &= m\, e^{2i(\nu+\rho)} \\
t_2 &= m \tan^2\gamma\, e^{-2i(\nu-\rho)}
\end{aligned}
\tag{5.15}
$$

This notation agrees with $|t_1| = m$ and $|t_1| \geq |t_2|$, if $0 \geq \gamma \leq 45°$, which is the range of γ. The angle γ is called the "characteristic angle"; its role will become clear in later discussion. The angle ν is sometimes called "relative phase", because it assumes that role in the diagonal matrix (5.13), but for general non-symmetric targets this nomenclature loses its significance; we prefer the term "target skip angle" since values of ν in the range $-45° \leq \nu \leq 45°$, have some relationship to depolarization due to the number of bounces of the reflected signal: for single bounce $\nu = 0$, for double bounce $\nu = \pm 45°$. The quantity ρ is called "absolute phase" of the target. It disappears with power measurements, and it changes with the motion of the target along the line-of-right direction.

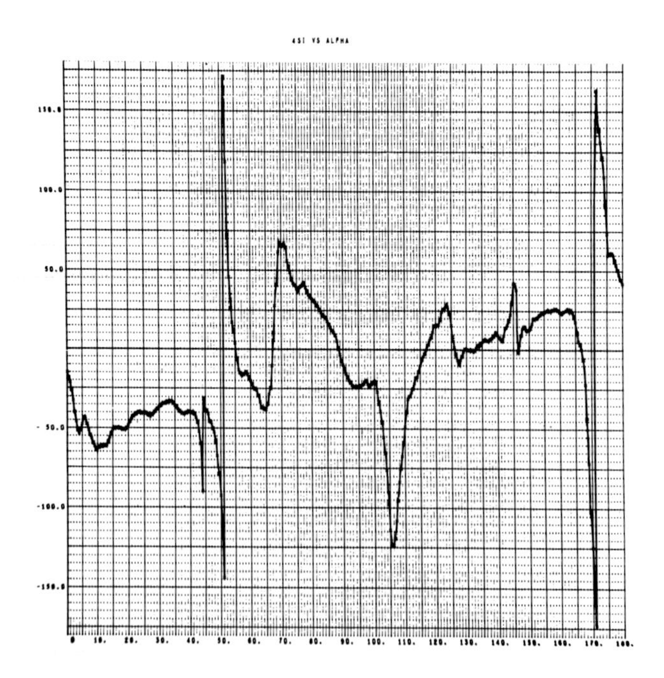

Fig. 5. Target orientation versus aspect angle.

The angle ψ is the "target orientation angle"; it can be made zero simply by rotating the target about the line-of-sight axis, keeping target exposure otherwise unchanged. For roll-symmetric target ψ is closely related to the position of target axis of symmetry, ψ_a, however, for general targets ψ provides a "natural" preferred target orientation. Fig. 5 shows measured values of ψ versus aspect angle for a general non-symmetric target. The angle τ_m is the ellipticity angle of the maximum polarization $\underline{m}$. However, it plays a significant role for determining target symmetry (when $\tau_m = 0$) or nonsymmetry (when $\tau_m \neq 0$). We will call τ_m the target "helicity angle", for reasons which will become clear later. Its range is $-45° \leq \tau_m \leq +45°$. Fig. 6 shows τ_m

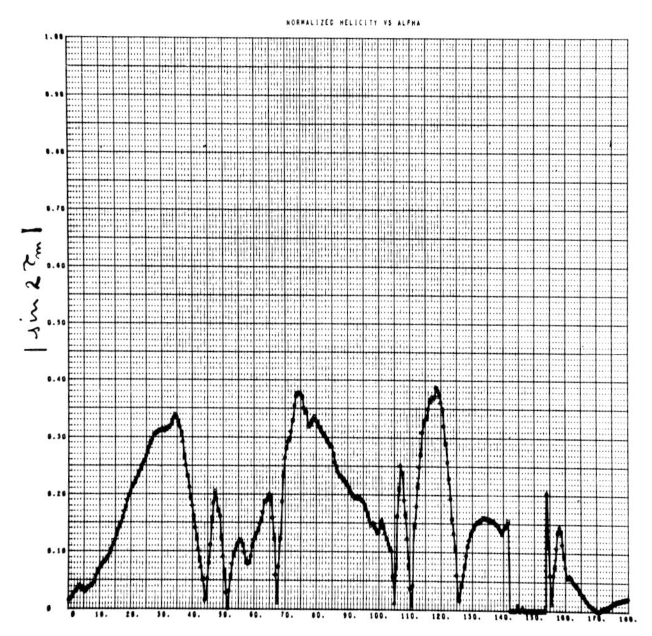

Fig. 6. Normalized helicity versus aspect angle.

versus aspect angle for a nonsymmetric target. With these identifications, we now show how the six target parameters m, ρ, ψ, τ_m, ν and γ determine the scattering matrix T. Since $\underline{m}(\psi, \tau_m)$ is given by (5.14), we find for $\underline{x}_2 = \underline{m}_\perp = \underline{m}(\psi + \pi/2, -\tau_m)$. It is easy to verify that $\underline{m} \cdot \underline{m}_\perp^* = 0$ as required by $\underline{x}_1 \cdot \underline{x}_2^* = 0$. For the unitary transformation we find:

$$U(\psi,\ \tau_m) = \left[\underline{m},\ \underline{m}_\perp\right] = \begin{bmatrix} \cos\psi & -\sin\psi \\ \sin\psi & \cos\psi \end{bmatrix} \begin{bmatrix} \cos\tau_m & i\sin\tau_m \\ i\sin\tau_m & \cos\tau_m \end{bmatrix} \quad (5.16)$$

Substitution of (5.16) into (5.10) gives the result:

$$T(t) = U^* T_d U^{*T} = \begin{bmatrix} \cos\psi & -\sin\psi \\ \sin\psi & \cos\psi \end{bmatrix} \begin{bmatrix} \cos\tau_m & -i\sin\tau_m \\ -i\sin\tau_m & \cos\tau_m \end{bmatrix}$$

$$\cdot \begin{bmatrix} m\, e^{2i(\nu+\rho)} & 0 \\ 0 & m\tan^2\gamma\, e^{-2i(\nu-\rho)} \end{bmatrix} \begin{bmatrix} \cos\tau_m & -i\sin\tau_m \\ -i\sin\tau_m & \cos\tau_m \end{bmatrix}$$

$$\cdot \begin{bmatrix} \cos\psi & \sin\psi \\ -\sin\psi & \cos\psi \end{bmatrix} \quad (5.17)$$

This equation shows explicitly how scattering matrix $T(t)$ depends on six meaningful geometric target parameters. For a target on axis (with principal "axis" horizontal) $\psi = 0$, and what remains are the orientation independent parameters m, ρ, τ_m, ν and γ. We see that if $\tau_m = 0$ (for a roll-symmetric object) and $\psi = 0$ (axis in horizontal x-plane) the scattering matrix is diagonalized; i.e. the returns $(H, V) = (V, H) = 0$. But this is not true for a non-symmetric object ($\tau_m \neq 0$).

VI. RECEIVED BACKSCATTERED POWER

The derivation of received power from $V(t) = (T(t)\underline{a}) \cdot \underline{b}$ and $T(t)$ given by (5.17) is tedious but straightforward. A systematic procedure is given in [60]. The returned power simply is determined by

$$P(t) = |V(t)|^2 = |(T(m,\psi,\nu,\tau_m,\gamma)\underline{a}(p_A,\phi_A,\tau_A)) \cdot \underline{b}(p_B,\phi_B,\tau_B)|^2$$

hence power is given by five target parameters (ρ drops out) and six antenna parameters. We give the result at once:

$$
\begin{aligned}
& P(\phi_A, \tau_A; \phi_B, \tau_B) \\
& = A_0 + B_0 + (-A_0 + B_0) \sin 2\tau_A \sin 2\tau_B \\
& + F(\sin 2\tau_A + \sin 2\tau_B) + A_0 \cos 2(\Phi_A - \Phi_B) \cos 2\tau_A \cos 2\tau_B \\
& + C(\cos 2\Phi_A \cos 2\tau_A + \cos 2\Phi_B \cos 2\tau_B) \\
& + B \cos 2(\Phi_A + \Phi_B) \cos 2\tau_A \cos 2\tau_B \\
& + E \sin 2(\Phi_A + \Phi_B) \cos 2\tau_A \cos 2\tau_B \\
& + D(\sin 2\Phi_A \sin 2\tau_B \cos 2\tau_A + \sin 2\Phi_B \sin 2\tau_A \cos 2\tau_B) \\
& + G(\cos 2\Phi_A \sin 2\tau_B \cos 2\tau_A + \cos 2\Phi_B \sin 2\tau_A \cos 2\tau_B)
\end{aligned}
\tag{6.1}
$$

where

$$
\begin{aligned}
\Phi_A &= \phi_A - \psi \\
\Phi_B &= \phi_B - \psi
\end{aligned}
\tag{6.2}
$$

are the relative orientation angles and

$$
\begin{aligned}
A_0 &= Q f \cos^2 2\tau_m \\
B_0 &= Q(1 + \cos^2 2\gamma - f \cos^2 2\tau_m) \\
B &= Q[1 + \cos^2 2\gamma - f(1 + \sin^2 2\tau_m)] \\
C &= 2Q \cos 2\gamma \cos 2\tau_m \\
D &= Q \sin^2 2\gamma \sin 4\nu \cos 2\tau_m \\
E &= -Q \sin^2 2\gamma \sin 4\nu \sin 2\tau_m \\
F &= 2Q \cos 2\gamma \sin 2\tau_m \\
G &= Q f \sin 4\tau_m
\end{aligned}
\tag{6.3}
$$

$$
Q = \frac{p_A^2 p_B^2 m^2}{8 \cos^4 \gamma} \tag{6.4}
$$

$$
f = 1 - \sin^2 2\gamma \sin^2 2\nu \tag{6.5}
$$

Except for p_A and p_B in Q all these parameters are determined by target parameters only. Notice that target orientation angle ψ

is incorporated with Φ_A, Φ_B, hence all parameters (6.3) are orientation-indpendent; i.e. they depend only on target exposure (aspect angle and roll-angle) frequency and waveform, but not on accidental target orientation (see Fig. 1). Sometimes it is convenient to write expressions (6.4) with Q_0 which does not contain p_A and p_B, i.e. with $Q_0 = m^2/8 \cos^4\gamma$. Since in each context it is clear which form is applicable we will not distinguish notation between these two cases. A rather surprising development is that the long-winded equation (6.1) can be written in elegant matrix form as follows:

$$P(t) = \begin{bmatrix} A_0 + B_0 & F & C & 0 \\ F & -A_0 + B_0 & G & D \\ C & G & A_0 + B & E \\ 0 & D & E & A_0 - B \end{bmatrix} \begin{bmatrix} p_A{}^2 \\ p_A{}^2 \sin 2\tau_A \\ p_A{}^2 \cos 2\tau_A \cos 2\Phi_A \\ p_A{}^2 \cos 2\tau_A \sin 2\Phi_A \end{bmatrix} \bullet$$

$$\bullet \begin{bmatrix} p_B{}^2 \\ p_B{}^2 \sin 2\tau_B \\ p_B{}^2 \cos 2\tau_B \cos 2\Phi_B \\ p_B{}^2 \cos 2\tau_B \sin 2\Phi_B \end{bmatrix} \tag{6.6}$$

(here A_0, B_0, ..., etc. are defined without p_A, p_B; see comments above). Or, in shorthand notation:

$$P(t) = (M(t)g(\underline{a}_\psi)) \bullet h(\underline{b}_\psi) \tag{6.7}$$

The four-component vectors are called <u>stokes vector</u> representations of the antenna polarizations $\underline{a}_\psi$ and $\underline{b}_\psi$ where $\underline{a}_\psi = \underline{a}(\phi_A - \psi, \tau_A)$. The $M(t)$ matrix on the right is called the <u>stokes reflection matrix</u>; it represents the target with measurements of power in a similar way, as the target scattering matrix $T(t)$ represents the target for voltage measurements. Notice that the matrix in the form shown in (6.6) is determined by orientation-independent parameters, since target orientation ψ is incorporated with the antenna orientation. It is an easy exercise to obtain the full target stokes matrix which includes ψ. Since

$$g(\underline{a}_\psi) = \begin{bmatrix} 1 & 0 & 0 & 0 \\ 0 & 1 & 0 & 0 \\ 0 & 0 & \cos 2\psi & \sin 2\psi \\ 0 & 0 & -\sin 2\psi & \cos 2\psi \end{bmatrix} g(\underline{a}) \tag{6.8}$$

and similarly for $h(\underline{b}_\psi)$, we have:

$$M_\psi(t) = \begin{bmatrix} 1 & 0 & 0 & 0 \\ 0 & 1 & 0 & 0 \\ 0 & 0 & \cos 2\psi & -\sin 2\psi \\ 0 & 0 & \sin 2\psi & \cos 2\psi \end{bmatrix} \begin{bmatrix} A_0 + B_0 & F & C & 0 \\ F & -A_0 + B_0 & G & D \\ C & G & A_0 + B & E \\ 0 & D & E & A_0 - B \end{bmatrix} \begin{bmatrix} 1 & 0 & 0 & 0 \\ 0 & 1 & 0 & 0 \\ 0 & 0 & \cos 2\psi & \sin 2\psi \\ 0 & 0 & -\sin 2\psi & \cos 2\psi \end{bmatrix} \tag{6.9}$$

or

$$M_\psi(t) = \begin{bmatrix} A_0 + B_0 & F & C_\psi & H_\psi \\ F & -A_0 + B_0 & G_\psi & D_\psi \\ C_\psi & G_\psi & A_0 + B & E_\psi \\ H_\psi & D_\psi & E_\psi & A_0 - B \end{bmatrix} \tag{6.10}$$

Here:

$$\begin{aligned} H_\psi &= C \sin 2\psi \\ C_\psi &= C \cos 2\psi \end{aligned} \tag{6.11}$$

$$\begin{aligned} G_\psi &= G \cos 2\psi - D \sin 2\psi \\ D_\psi &= G \sin 2\psi + D \cos 2\psi \end{aligned} \tag{6.12}$$

$$\begin{aligned} E_\psi &= E \cos 4\psi + B \sin 4\psi \\ B_\psi &= -E \cos 4\psi + B \sin 4\psi \end{aligned} \tag{6.13}$$

We notice in particular that for a target on axis ($\psi = 0$) $H_\psi = 0$; this fact could not have been anticipated by the conventional form (4.3). The final equation for received power is now written as:

$$P(t) = (M_\psi(t)g(\underline{a})) \cdot h(\underline{b}) \tag{6.14}$$

We notice that M_ψ is a 4 × 4 real, symmetric matrix. Observe the curious trace-rule which $M(t)$ and $M_\psi(t)$ have to obey:

$$\text{trace } M_\psi(t) = 2(A_0 + B_0) \tag{6.15}$$

The three orientation-independent matrix elements are A_0, B_0 and F; each of these parameters has important physical significance.

Some observations are useful at this point: First notice that the form of (6.14) is similar to the equation for voltage (4.1); the complex valued (2 × 2) scattering matrix $T(t)$ has been replaced by the real-valued (4 × 4) stokes reflection matrix $M(t)$, whereas complex valued polarizations $\underline{a}$ and $\underline{b}$ have been replaced by real-valued stokes vectors $g(\underline{a})$ and $h(\underline{b})$. The voltage-expression deals with field-components, whereas (6.14) expresses target scattering in terms of power. All information in (4.1) is retained in (6.14) except for absolute phases for the target and antennas, which disappear with power; however <u>relative</u> phases between matrix components in $T(t)$ are conserved by real quantities in (6.14). This observation leads to many interesting procedures to determine $T(t)$ from power measurements alone [36]. Since $T(t)$ has three complex quantities (a, b and c) and if we discount absolute phase, five real parameters will determine a target at given aspect, frequency and waveform. However, we count nine parameters: A_0, B_0, B_ψ, C_ψ, D_ψ, E_ψ, F, G_ψ and H_ψ with $M_\psi(t)$ matrix. If the two forms are equivalent the nine parameters cannot all be independent. We can show that:

$$\begin{aligned} 2A_0(B_0 + B_\psi) &= C_\psi^2 + D_\psi^2 \\ 2A_0(B_0 - B_\psi) &= G_\psi^2 + H_\psi^2 \\ 2A_0E_\psi &= C_\psi H_\psi - D_\psi G_\psi \\ 2A_0F &= C_\psi G_\psi + D_\psi H_\psi \end{aligned} \tag{6.16}$$

are sufficient conditions to make an independent set.

Some special cases of radar illumination and reception of practical significance can be found from (6.1) or (6.6) which are listed below. For linearly polarized antennas $\tau_A = \tau_B = 0$ and if $\Phi_A = \Phi_B = \Phi = \phi - \psi$ we have for identical linearly polarized radar operation:

$$P_{/\!/}(\phi) = 2A_0 + B_0 + 2C\cos 2\Phi + B\cos 4\Phi + E\sin 4\Phi \quad (6.17)$$

Special cases: $\phi = 0°(H)$, $\phi = 90°(V)$ and $\phi = 45°(D)$.

$$P(H,H) = 2A_0 + B_0 + 2C\cos 2\psi + B\cos 4\psi - E\sin 4\psi \quad (6.18)$$

$$P(V,V) = 2A_0 + B_0 - 2C\cos 2\psi + B\cos 4\psi - E\sin 4\psi \quad (6.19)$$

$$P(D,D) = 2A_0 + B_0 + 2C\sin 2\psi - B\cos 4\psi + E\sin 4\psi \quad (6.20)$$

We find later (Section VIII) for roll-symmetric objects: $E = 0$, $B_0 = B$ and $\psi = 0°$ or $90°$ for target axis in horizontal plane ($\psi_a = 0$).

For linear-orthogonal reception we find: $\tau_A = \tau_B = 0$, $\Phi_A = \Phi = \phi - \psi$, $\Phi_B = \pi/2 + \Phi$.

$$P_{\perp}(\phi) = B_0 - B\cos 4\Phi - E\sin 4\Phi \quad (6.21)$$

with similar special cases. The total backscattered power for linear polarization is the sum of (6.17) and (6.21):

$$P_{tot}(\phi) = 2(A_0 + B_0) + 2C\cos 2\Phi \quad (6.22)$$

For circular polarization, we consider three cases, designated by (RC,RC), (LC,LC) and (RC,LC). Substituting $\tau = 45°$ for RC antennas and $\tau = -45°$ for LC antennas, we find for received power:

$$P(RC,RC) = 2(B_0 + F) \quad (6.23)$$

$$P(LC,LC) = 2(B_0 - F) \quad (6.24)$$

$$P(RC,LC) = 2A_0 \quad (6.25)$$

From (6.23) and (6.24) it follows that $B_0 \geq 0$ and from (6.25) that $A_0 \geq 0$. Disregarding the effect due to orientation, we find from (6.22) that $2(A_0 + B_0)$ is an approximate measure of total returned power. The part $2A_0$ is measured exactly by the (RC,LC) circularly polarized return (6.25). This component is associated with regular, smooth convex type of surfaces which contributes specular returns (for a sphere A_0 is the only non-zero parameter). On the other hand $2B_0$ may be thought of as a measure of total power of the target's irregular, rough, non-convex depolarizing components.

The parameter F in (6.23) and (6.24) shows a bias for RC or LC type of circular polarization. As we shall see shortly, F is characteristic for a right or left wound helix viewed on axis. From (6.3) we know that F is proportional to sin $2\tau_m$. For a roll-symmetric target there can be no preference for LC or RC polarization, hence $F = 0$ and $\tau_m = 0$ are sensitive, orientation-independent indicators of target symmetry. Conversely, $F \neq 0$ and $\tau_m \neq 0$ are indicative of "target helicity." Hence the nomenclature "helicity angle" used to indicate τ_m.

Parameter C is responsible for the difference between (H,H) and (V,V) type of return, equations (6.18) and (6.19). We will see shortly that C is also characteristic for a wire target, whereas $C = 0$ for a sphere; hence C could be called the target, "shape factor." Since C is proportional to cos 2γ, and $\gamma = 0$ for a wire target and $\gamma = 45°$ for a sphere, one is tempted to associate the same "shaping function" with γ, however, for a helix $C = 0$ and $\gamma = 0$ which shows that γ is not directly associated with target shape. Instead the "characteristic angle" γ is closely associated with the theory of target null-polarizations.

We will briefly mention that according to this theory each scattering matrix has two characteristic null-polarizations for which $(T\underline{n}) \cdot \underline{n} = 0$. Thus a sphere has RC and LC as null-polarizations, while for a horizontal wire target the two nulls merge into one, i.e., V polarization. If these two nulls are plotted on the polarization sphere of Fig. 2, the angle between the nulls viewed from the center of the sphere equals 4γ (see Fig. 7). It is easy to understand that knowledge of position of the two nulls N_1 and N_2 on the unit sphere (4 parameters) at each instant almost determines $T(t)$; what is missing is the fifth parameter: maximum target return $m(t)$ which can be plotted separately as a function of time. For complete details on the null polarization technique see [17] and [60].

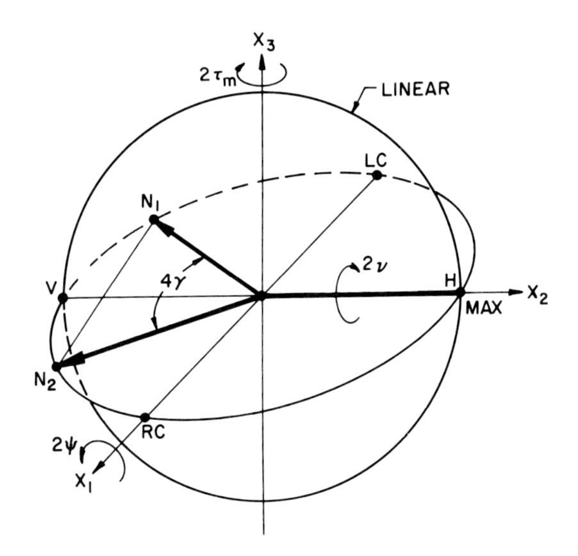

Fig. 7. Polarization nulls.

Equation (6.16) shows that in general $B_0 \geq |B_\psi|$, or for target on axis ($\psi = 0$) $B_0 \geq |B|$, since we know $A_0 \geq 0$. Now for symmetric targets we found $\tau_m = 0$, and from (6.3) it follows that $E = F = G = H = 0$ (if $\psi = 0$, H is always zero), and from the second line of (6.16) we find at once $B_0 = B \geq 0$ for that case. These properties of symmetric targets are summarized in the following section.

VII. THEORY OF TARGET STRUCTURE

We found in the previous section that target return based upon power is determined, for given aspect, frequency and waveform, by the stokes reflection matrix $M_\psi(t)$ which is given by nine parameters: A_0, B_0, B_ψ, C_ψ, D_ψ, E_ψ, F, G_ψ and H_ψ. However, we saw that these parameters, for a single object with scattering matrix $T(t)$, are not independent. We gave conditions (6.16) as sufficient conditions to secure five independent parameters. We now indicate how these conditions and similar relationships obtain their physical significance. They are determined by the degree of coherence of the components of the backscattered wave $s = M_\psi g$. We note from the definition of stokesvector in (6.6) that for a completely polarized (*cp*) or monochromatic wave with stokesvector $s = (s_0, s_1, s_2, s_3)$ the following condition is satisfied:

$$s_0^2 = s_1^2 + s_2^2 + s_3^2 \qquad (7.1)$$

or in matrix form:

$$s_0^2 - s_1^2 - s_2^2 - s_3^2 = \begin{bmatrix} 1 & & & \\ & -1 & & \\ & & -1 & \\ & & & -1 \end{bmatrix} s \cdot s = 0 \qquad (7.2)$$

or, if we substitute $s = M_\psi g$ into (7.2):

$$\left(M_\psi \begin{bmatrix} 1 & & & \\ & -1 & & \\ & & -1 & \\ & & & -1 \end{bmatrix} M_\psi g \right) \cdot g = [Q]g \cdot g = Q_M g \cdot g = 0 \qquad (7.3)$$

The following steps led to the important result (7.3). First the symmetry of M_ψ was used to transpose the M_ψ matrix from the right-hand side of the dot product to the far left; this leads to the matrix $[Q]$ indicated above. Next the following observation was make: if $[Q]$ contains any component in the diagonal proportional to the matrix L, where,

$$L = \begin{bmatrix} 1 & & & \\ & -1 & & \\ & & -1 & \\ & & & -1 \end{bmatrix} \qquad (7.4)$$

then since $Lg \cdot g = g_0^2 - g_1^2 - g_2^2 - g_3^2 = 0$, this component does not contribute to (7.3) and it can be left out. What remains of the $[Q]$ matrix after the L-component is thus removed is the matrix Q_M which has interesting properties: Since it is free from the L-term, it satisfies in general a trace-rule, similar to M_ψ (see 6.15) <u>and in all other aspects, it behaves like an M_ψ stokes</u>

reflection matrix, although it obviously is not physically equivalent to a scattering matrix. For more details on these interesting properties of "higher order matrices of type M" we refer to [60, section 40] and section IX.

Equation (7.3) is satisfied for all values of polarization g, if $Q_M = 0$. This result characterizes a single object (given aspect, frequency and waveform). Hence we obtain nine quadratic conditions, four of which were already displayed in (6.16):

$$Q_M = \begin{bmatrix} Q_{00} & Q_{01} & Q_{02} & Q_{03} \\ Q_{10} & Q_{11} & Q_{12} & Q_{13} \\ Q_{20} & Q_{21} & Q_{22} & Q_{23} \\ Q_{30} & Q_{31} & Q_{32} & Q_{33} \end{bmatrix} = 0 \tag{7.5}$$

Also let:

$$\begin{aligned} Q_{00} &= \frac{Q_1}{2} + \frac{Q_2 + Q_3}{2} \\ Q_{11} &= -\frac{Q_1}{2} + \frac{Q_2 + Q_3}{2} \\ Q_{22} &= \frac{Q_1}{2} + \frac{-Q_2 + Q_3}{2} \\ Q_{33} &= \frac{Q_1}{2} + \frac{Q_2 - Q_3}{2} \end{aligned} \tag{7.6}$$

then it is easily concluded that if $Q_{ii} = 0$, $i = 0, 1, 2, 3$, then $Q_1 = Q_2 = Q_3 = 0$. We can now deduce some important results. Based on the definition of Q_M we can write:

$$\begin{aligned} Q_1 &= (B_0{}^2 - B_\psi{}^2) - (E_\psi{}^2 + F^2) \\ Q_2 &= 2A_0(B_0 + B_\psi) - (C_\psi{}^2 + D_\psi{}^2) \\ Q_3 &= 2A_0(B_0 - B_\psi) - (G_\psi{}^2 + H_\psi{}^2) \end{aligned} \tag{7.7}$$

$$
\begin{aligned}
Q_{01} &= 2A_0F - (C_\psi G_\psi + D_\psi H_\psi) \\
Q_{23} &= -2A_0E_\psi + (C_\psi H_\psi - D_\psi G_\psi) \\
Q_{02} &= C_\psi(B_0 - B_\psi) - (E_\psi H_\psi + FG_\psi) \\
Q_{13} &= -D_\psi(B_0 - B_\psi) - (E_\psi G_\psi - FH_\psi) \\
Q_{03} &= H_\psi(B_0 + B_\psi) - (D_\psi F + C_\psi E_\psi) \\
Q_{12} &= -G_\psi(B_0 + B_\psi) + (C_\psi F - D_\psi E_\psi)
\end{aligned}
\tag{7.8}
$$

Since $Q_{ij} = 0$ all above quantities must be equal to zero. We readily identify the four conditions given by (6.16) as special cases, and the reason for our choice of Q_1, Q_2, Q_3 in (7.6) now becomes apparent. From (7.5) we can verify the trace rule: trace $Q_M = 2Q_{00}$ which would apply even if Q_M were not zero (this case refers to incoherent scattering components which will be discussed in a later section).

Equations (7.7) have particular significance. They show in each term sums of squares of pairs of off-diagonal components. Since equations (7.7) are equally valid if $\psi = 0$ we will consider that case now. For the on-axis case we see that if for instance $A_0 = 0$, then $C = D = G = H = 0$ must follow. Similarly if $B_0 - B = 0$, then $E = F = F = H = 0$ and if $B_0 + B = 0$ then $E = F = C = D = 0$. For this reason the diagonal elements A_0, $B_0 + B$ and $B_0 - B$ are called the generators of the off-diagonal stokes parameters. As we indicated before, $2A_0$ is the generator of target symmetry, $B_0 - B$ is the generator of target non-symmetry and $B_0 + B$ will be called the generator of target irregularity. We notice incidentally that the sum of generating powers equals $2(A_0 + B_0)$, which is in agreement with an earlier observation (see (6.22)). Fig. 8 gives a pictorial representation of the three-fold symmetry of target structure. The three generators are listed in the circles. They interconnect with the off-diagonal parameters which they generate.

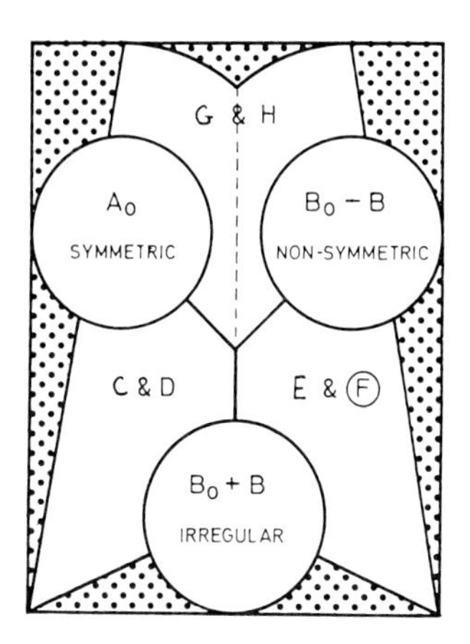

Fig. 8. Target structure diagram.

The symmetric target structure is given by the left half in Fig. 8. If the target is completely symmetric (a roll-symmetric object is symmetric at all aspects and is independent of roll-angle) then $B_0 = B$ and from (25) it follows that $\tau_m = 0$, and hence $E = F = G = H = 0$ (H is always zero for any target on-axis, i.e. for $\psi = 0$). The only distinguishing structural elements, except for generator A_0, are C and D. We have already seen that C plays the role of a shapefactor; the role of D remains obscure, it does not appear in any of the "simple" equations for received power, equations (6.17) ... (6.25). The interested reader is referred to the defining equation (6.3) to examine this question.

Important examples of symmetric objects are: the sphere which has only one term, A_0 (this incidentally is why A_0 is called the generator of symmetry); and any roll-symmetric object with axis either horizontal or vertical: line-targets, ogives, cones, ellipsoids, etc.; also a trough, two plates intersecting at 90°, when viewed as a corner reflector at normal incidence. The important feature with all these targets is the existence of a plane of mirror symmetry with respect to a horizontal or vertical axis

(x- or y-axis).

The non-symmetric part of a general object is given by the right-hand part of Fig. 8. A completely non-symmetric target is determined by $A_0 = 0$, and since this condition is valid for all target orientations ψ, these targets are not restricted to any particular "on-axis" orientation ($\psi = 0$), as was the case for symmetric objects, and for this reason it is better to reintroduce ψ as part of the discussion below. The class of non-symmetric targets are called *N*-targets. *N*-targets play an important role in the theory of distributed targets, to be discussed later. There they play the role of representing "residue" or clutter-noise at the higher frequencies. In addition to the noise aspect, they represent the most asymmetric type of depolarization. Examples of *N*-targets are left- and right-wound helices, for which $\gamma = 0°$ and $\tau_m = \pm 45°$, and hence from (6.3) and (6.13): $A_0 = 0$, $E_\psi = 0$, $B_\psi = 0$ and $B_0 = |F|$, i.e. all the generator-power is represented by $|F|$. This property of helices is responsible for the nomenclature "target helicity" to indicate F. Since F is proportional to $\sin 2\tau_m$, the angle τ_m is named the "helicity-angle" of the target, as was indicated before.

Other *N*-targets are troughs, viewed at normal incidence, at any orientation-angle. For troughs $A_0 = 0$ ($\gamma = 45°$, $\nu = 45°$, $\tau_m = 0$) and $F = 0$. Hence a trough is both a symmetric target on axis ($\psi = 0$) and an *N*-target for any target orientation ψ. The troughs are characterized by $\nu = \pm 45°$ and for this reason ν is called the degree of double-bounce or "skip-angle". No special property has been coined yet to target parameter E, except that it represents non-symmetric depolarization.

Finally we have parameters G_ψ and H_ψ; they represent target "coupling" terms. For a purely non-symmetric target $A_0 = 0$, as we have seen, and $G_\psi = H_\psi = 0$ for any orientation. On the other hand, for a purely symmetric target ($\tau_m = 0$) at arbitrary orientation we have $G_\psi = D \sin 2\psi$ and $H_\psi = C \sin 2\psi$; and both terms become zero if $\psi = 0$. In the general case G_ψ is non-zero even if

$\psi = 0$, while $H_\psi = H = 0$ for $\psi = 0$. H_ψ determines coupling due to target orientation, whereas G couples the symmetric and non-symmetric parts of the target; G_ψ, however, is not a pure indicator. This discussion points out again the importance of orientation ψ also for the general case, for the understanding of target behavior: First H_ψ is to be made zero by a rotation ψ of the target about the line of sight direction, then with target on-axis G represents the coupling term between symmetric and non-symmetric parts. A complete listing of matrices T and M for elementary targets is provided in the next section.

VIII. TARGET MATRIX REPRESENTATIONS

In this section we give matrix representations for some of the most common types of targets. Shown are the target-scattering matrix T and the corresponding stokes matrix M.

A. Sphere Target

The sphere target is one of the most common and useful, since it represents all types of "specular" returns from targets which have local "flare-spots". It also represents rough surfaces or chaff clouds, which on the average have an effective-sphere structure. We notice that the T-matrix is simply a unit matrix, but the corresponding M-matrix is not the unit matrix!

Sphere target: $\gamma = 45°$, $\nu = 0°$, $\tau_m = 0°$, $A_0 = 1/2$, $B_0 = B = 0$.

$$T_0 = I = \begin{bmatrix} 1 & 0 \\ 0 & 1 \end{bmatrix}; \quad M_0 = \frac{1}{2}\begin{bmatrix} 1 & 0 & 0 & 0 \\ 0 & -1 & 0 & 0 \\ 0 & 0 & 1 & 0 \\ 0 & 0 & 0 & 1 \end{bmatrix}$$

B. Horizontal or Vertical Trough

The next two target models represent troughs, i.e. a diplane with its face viewed at normal incidence. The cross-symbol represents troughs which axes are <u>either</u> horizontal or vertical, since both positions have the same matrix representation.

<u>Horizontal or vertical trough</u>: $\gamma = 45°$, $\nu = 45°$, $\tau_m = 0°$, $\psi = 0°$ or $90°$, $A_0 = 0$, $B_0 = B = 1/2$.

$$T_T = \begin{bmatrix} 1 & 0 \\ 0 & -1 \end{bmatrix}; \quad M_T = \frac{1}{2}\begin{bmatrix} 1 & 0 & 0 & 0 \\ 0 & 1 & 0 & 0 \\ 0 & 0 & 1 & 0 \\ 0 & 0 & 0 & -1 \end{bmatrix}$$

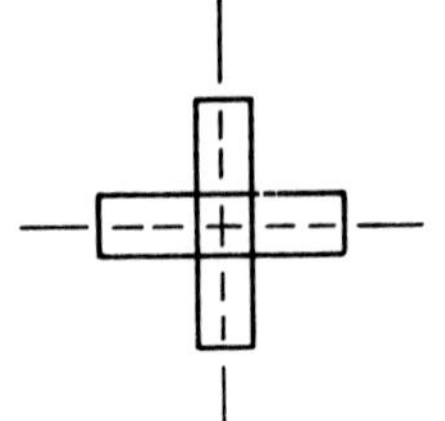

C. Troughs Oriented at ±45°

This figure shows the same trough structure rotated by 45°. The troughs are important in the theory of distributed targets. There they represent target clutter. For troughs $A_0 = 0$:

<u>Troughs oriented at ±45°</u>: $\gamma = 45°$, $\nu = 45°$, $\tau_m = 0°$, $\psi = \pm 45°$, $A_0 = 0$, $B_0 = 1/2$, $B = -1/2$.

$$T_T(45) = \begin{bmatrix} 0 & 1 \\ 1 & 0 \end{bmatrix}; \quad M_T(45) = \frac{1}{2}\begin{bmatrix} 1 & 0 & 0 & 0 \\ 0 & 1 & 0 & 0 \\ 0 & 0 & -1 & 0 \\ 0 & 0 & 0 & 1 \end{bmatrix}$$

D. Long Horizontal Wire

A long horizontal wire target is often used as a model for depolarization. We notice C, which was called the shapefactor. The wire return is always linearly polarized in the plane containing the axis.

<u>Long horizontal wire</u>: $\gamma = 0°$, $\tau_m = 0°$, $\psi = 0$, $A_0 = 1/8$, $B_0 = B = 1/8$, $C = 1/4$.

$$T_L = \begin{bmatrix} 1 & 0 \\ 0 & 0 \end{bmatrix}; \quad M_L = \frac{1}{4}\begin{bmatrix} 1 & 0 & 1 & 0 \\ 0 & 0 & 0 & 0 \\ 1 & 0 & 1 & 0 \\ 0 & 0 & 0 & 0 \end{bmatrix}$$

E. Long Wire Oriented at ψ_a

The wire oriented at angle ψ_a is characterized by $F = 0$, $A_0 = B_0$, $G_\psi = D_\psi = 0$. Hence these properties are also true on the average for an arbitrary distribution of dipoles.

<u>Long wire oriented at ψ_a</u>: $\gamma = 0°$, $\tau_m = 0°$, $\psi = \psi_a$.

$$T_L(\psi_a) = \begin{bmatrix} \cos^2\psi_a & \frac{1}{2}\sin 2\psi_a \\ \frac{1}{2}\sin 2\psi_a & \sin^2\psi_a \end{bmatrix};$$

$$M_L(\psi_a) = \frac{1}{4}\begin{bmatrix} 1 & 0 & \cos 2\psi_a & \sin 2\psi_a \\ 0 & 0 & 0 & 0 \\ \cos 2\psi_a & 0 & \cos^2 2\psi_a & -\frac{1}{2}\sin 4\psi_a \\ \sin 2\psi_a & 0 & -\frac{1}{2}\sin 4\psi_a & \sin^2 2\psi_a \end{bmatrix}$$

F. Symmetric Target, $\psi_a = 0°$

The symmetric target on-axis has $\tau_m = 0$ and $E = F = G = H = 0$. We notice that ψ is either 0° or 90°, <u>which means that either (V,V) or (H,H) reception gives the maximum return for all polarizations</u>. Examples: ogives, cones, any roll-symmetric object.

<u>Symmetric target, $\psi_a = 0°$</u>: $\tau_m = 0°$, $\psi = \phi_m = 0°$ or $90°$, $B_0 = B$, $a + b = e^{+2i\nu}$, $a - b = e^{-2i\nu}\tan^2\gamma$, $E = F = G = H = 0$, $4A_0B_0 = C^2 + D^2$.

$$T_S(0) = \begin{bmatrix} a+b & 0 \\ 0 & a-b \end{bmatrix}; \quad M_S(0) = \begin{bmatrix} A_0 + B_0 & 0 & \pm C & 0 \\ 0 & -A_0 + B_0 & 0 & \pm D \\ \pm C & 0 & A_0 + B_0 & 0 \\ 0 & \pm D & 0 & A_0 - B_0 \end{bmatrix}$$

G. Symmetric Target Oriented at ψ_a

A symmetric target at arbitrary axis orientation ψ_a has $F = 0$. By averaging over ψ, all terms containing ψ_a vanish, which results in the orientation-independent target model.

Symmetric target oriented at ψ_a: $\tau_m = 0$, $\psi = \psi_a + \phi_m$, $a + b = e^{+2i\nu}$, $a - b = e^{-2i\nu}\tan^2\gamma$.

$$T_S(\psi_a) = \begin{bmatrix} a + b\cos 2\psi & b\sin 2\psi \\ b\sin 2\psi & a - b\cos 2\psi \end{bmatrix}$$

$$M_S(\psi_a) = \begin{bmatrix} A_0 + B_0 & 0 & C\cos 2\psi & C\sin 2\psi \\ 0 & -A_0 + B_0 & -D\sin 2\psi & D\cos 2\psi \\ C\cos 2\psi & -D\sin 2\psi & A_0 + B\cos 4\psi & -B\sin 4\psi \\ C\sin 2\psi & D\cos 2\psi & -B\sin 4\psi & A_0 - B\cos 4\psi \end{bmatrix}$$

H. Helix with Right Screw Sense

The helix has $\gamma = 0°$ and $\tau_m = \pm 45°$, also $A_0 = 0$. Typical for helices is the structure parameter F, called target helicity.

Helix with right screw sense: $\gamma = 0°$, $\tau_m = 45°$, ψ = arbitrary.

$$T_{HR} = \frac{1}{2}\begin{bmatrix} 1 & -i \\ -i & -1 \end{bmatrix}; \quad M_{HR} = \frac{1}{4}\begin{bmatrix} 1 & 1 & 0 & 0 \\ 1 & 1 & 0 & 0 \\ 0 & 0 & 0 & 0 \\ 0 & 0 & 0 & 0 \end{bmatrix}$$

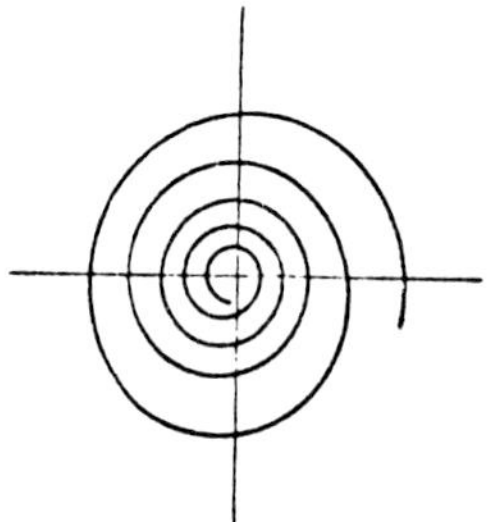

I. Helix with Left Screw Sense

This helix also has $\gamma = 0°$, $\tau_m = -45°$, and $A_0 = 0$.

Helix with left screw sense: $\gamma = 0°$, $\tau_m = -45°$, ψ = arbitrary.

$$T_{HL} = \frac{1}{2}\begin{bmatrix} 1 & i \\ i & -1 \end{bmatrix}; \quad M_{HL} = \frac{1}{4}\begin{bmatrix} 1 & -1 & 0 & 0 \\ -1 & 1 & 0 & 0 \\ 0 & 0 & 0 & 0 \\ 0 & 0 & 0 & 0 \end{bmatrix}$$

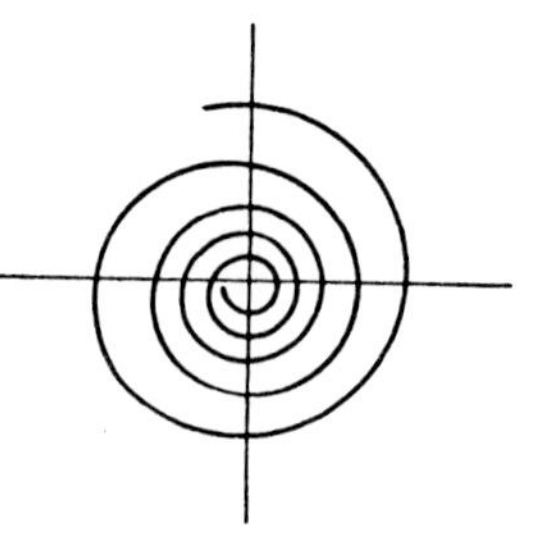

J. *N*-Target

The *N*-targets play an important role in the theory of target decompositions. They have $A_0 = 0$ and $C_\psi = D_\psi = G_\psi = H_\psi = 0$ for all orientations of the *N*-target.

N-target: $\tau_m = \pm 45°$, $|b^2 + c^2| = \tan^2\gamma$, ψ = arbitrary, $A_0 = 0$, $a = 0$.

$$T_N = \begin{bmatrix} b & c \\ c & -b \end{bmatrix} \qquad M_N = \begin{bmatrix} B_0 & F & 0 & 0 \\ F & B_0 & 0 & 0 \\ 0 & 0 & B & +E \\ 0 & 0 & E & -B \end{bmatrix}$$

Between B_0, B, E, and F we have: $B_0^2 = B^2 + E^2 + F^2$. The N-target plays a significant role in the theory of distributed nonsymmetric radar targets.

IX. THEORY OF DISTRIBUTED RADAR TARGETS

The instantaneous target return from a time-varying object was characterized either by the (2×2) complex valued target-scattering matrix $T(t)$, which relates to received voltage $V(t) = (T(t)\underline{a}) \cdot \underline{b}$, where $\underline{a}$ and $\underline{b}$ were the transmit and receiver polarizations; or the (4×4) real valued stokes reflection matrix $M(t)$, which relates to received power (in this discussion we omit index ψ) $P(t) = (M(t)g(\underline{a})) \cdot h(\underline{b})$, where $g(\underline{a})$ and $h(\underline{b})$ are stokes vector representations of the polarizations above. A target which consists of a large number of smaller objects, i.e. a chaff or weather cloud, terrain or sea state, will have a return whose amplitudes and phases are random variables.

It is customary to take the expected value of power as measure of average RCS (expected values of voltage are not as useful since in most cases these will be zero). In our formulation of power, with constant antannas, this leads at once to consideration of the expected value of stokes matrix $R = \langle M(t) \rangle$ as representing the averaged distributed target, such that $\langle P(t) \rangle = (Rg) \cdot h$. Likewise, the stokes return vector $s(t) = M(t)g$ will no longer be completely polarized (cp); instead it will be a partially polarized (pp) wave. A partially polarized wave has a lesser degree of coherence between x- and y-components, while a cp wave has complete coherence between these components. For cp waves we found previously, if $s = (s_0, s_1, s_2, s_3)$ is the stokes vector, then $s_0^2 = s_1^2 + s_2^2 + s_3^2$. However, for pp waves we have instead: $\langle s_0\rangle^2 \geq \langle s_1\rangle^2 + \langle s_2\rangle^2 + \langle s_3\rangle^2$. In many cases when no confusion is possible we omit the brackets and write for a pp wave:

$$s_0^2 \geq s_1^2 + s_2^2 + s_3^2 \tag{9.1}$$

The proof of this fact, which is simple, will be left to the reader (see [60]). We now can substitute $s = Rg$ into (9.1):

$$\begin{aligned} s_0^2 - s_1^2 - s_2^2 - s_3^2 &= (Ls) \cdot s = (LRg) \cdot (Rg) = [RLR]g \cdot g \\ &= [Q]g \cdot g = Q_R g \cdot g \geq 0 \end{aligned} \tag{9.2}$$

First we transpose R to the left of the dot-product sign; this is permitted since R is a symmetric matrix. L was defined as in (7.4), also $[Q] = [RLR]$. In the last step we subtracted from $[Q]$ a term αL such that the trace rule for Q_R holds:

$$\text{trace } Q_R = Q_1 + Q_2 + Q_3 \tag{9.3}$$

where Q_1, Q_2 and Q_3 are defined in (7.7). The last step in (9.2) was permitted since $Lg \cdot g = g_0^2 - g_1^2 - g_2^2 - g_3^2 = 0$, which expresses the fact that the transmit polarization g is completely polarized. If R is a single (i.e., non-distributed) target, $R = M$; we found before that $Q_R = 0$, and this in fact was a necessary and sufficient condition for R to represent a single object.

However, if R is to represent a distributed target or a sum of multiple targets, what are the necessary and sufficient conditions imposed on any R-matrix to represent such a case? A detailed proof of this problem is given in [60]. We simply state the result: R is a single realizable distributed target iff trace $R = 2(\langle A_0\rangle + \langle B_0\rangle)$, $\langle A_0\rangle \geq 0$, $\langle B_0\rangle - \langle B_\psi\rangle \geq 0$, $\langle B_0\rangle + \langle B_\psi\rangle \geq 0$ (the usual conditions also true for a single object), and if $Q_1 \geq 0$, $Q_2 \geq 0$ and $Q_3 \geq 0$!

We will now proceed to show that these conditions indeed seem plausible. Some very remarkable properties about the structure of R and Q_R matrices will emerge. We repeat that from $s = Rg$, if R is a realizable stokes matrix, then the backscattered wave s must be a realizable stokes vector such that (9.1) is satisfied. Substitution of $s = Rg$ into (9.1) leads to (9.2): $Q_R g \cdot g \geqq 0$. What conditions should be imposed on Q_R such that this is satisfied for any cp polarization g? This is easily answered as a special case of $Q_R g \cdot h \geqq 0$ for any cp stokes vectors g and h. But $Q_R g \cdot h \geqq 0$ is exactly the problem we started with: i.e., when is $Rg \cdot h \geqq 0$? Any matrix which has this property will be called: 'of type R.' Hence R is of type R if Q_R is of type R! But if R is of type R we found previously $2 \langle A_0 \rangle \geqq 0$, $\langle B_0 \rangle - \langle B_\Psi \rangle \geqq 0$ and $\langle B_0 \rangle + \langle B_\Psi \rangle \geqq 0$ are necessary conditions. Now the corresponding elements of Q_R are:

$$
\begin{aligned}
Q_1 &= (\langle B_0 \rangle^2 - \langle B_\Psi \rangle^2) - (\langle E_\Psi \rangle^2 + \langle F \rangle^2) \geq 0 \\
Q_2 &= 2 \langle A_0 \rangle (\langle B_0 + B_\Psi \rangle) - (\langle C_\Psi \rangle^2 + \langle D_\Psi \rangle^2) \geq 0 \\
Q_3 &= 2 \langle A_0 \rangle (\langle B_0 - B_\Psi \rangle) - (\langle G_\Psi \rangle^2 + \langle H_\Psi \rangle^2) \geq 0
\end{aligned}
\tag{9.4}
$$

The trace rule (9.3) is satisfied for Q_R, by the formation of Q_R, and thus all conditions previously listed are satisfied.

The next question is whether these conditions are sufficient, or are there perhaps some other conditions, not yet discovered? We can show by a simple and most remarkable property of higher order matrices of type R that the conditions listed above indeed are sufficient. What is at stake is to show that Q_R is indeed of type R, such that $s = Q_R g$ represents a p.p. realizable stokes vector which has property (9.1). To show this, we proceed as in (9.2) with R replaced by Q_R. We can now define $Q_R^{(2)}$ such that $[Q_R L Q_R] g \cdot g = Q_R^{(2)} g \cdot g \geqq 0$ with the usual argument that trace $Q_R^{(2)} = Q_1^{(2)} + Q_2^{(2)} + Q_3^{(2)}$, analogous with (9.3), after a term αL was subtracted from $[Q_R L Q_R]$. We now have a higher order matrix of type R: $Q_R^{(2)}$ which must satisfy $Q_R^{(2)} g \cdot h \geqq 0$, just as $Rg \cdot h \geqq 0$! This matrix must now satisfy: $Q_1^{(2)} \geqq 0$, $Q_2^{(2)} \geqq 0$, and $Q_3^{(2)} \geqq 0$ which leads to complicated fourth-order conditions on elements of R similar to the second-order relations (9.4). This process could be repeated indefinitely, each time leading to new and more complicated higher-order term conditions.

Fortunately, it turns out we do not have to go beyond R and Q_R matrices because the following remarkable relationships between higher order matrices of type R were found to be true [60]:

$$Q_R^{(2)} = \chi \quad R \tag{9.5}$$

$$Q_R^{(3)} = \chi^2 \quad Q_R \tag{9.6}$$

$$Q_R^{(4)} = \chi^5 \quad R \tag{9.7}$$

$$Q_R^{(5)} = \chi^{10} \, Q_R \tag{9.8}$$

----

where

$$\chi = -2\langle A_0\rangle Q_1 + \langle B_0 - B_\psi\rangle Q_2 + \langle B_o + B_\psi\rangle Q_3 + 2\langle E_\psi\rangle Q_{23} - 2\langle F\rangle Q_{01} \tag{9.9}$$

Since χ is a scalar of third order in R-terms and $Q_R^{(n)}$ a matrix with terms of order 2^n, it is easy to find the general rule; for any integer n:

$$Q_R^{(2n)} = \chi^{\frac{1}{3}\left(\begin{smallmatrix} 2n \\ 2-1 \end{smallmatrix}\right)} R \tag{9.10}$$

$$Q_R^{(2n+1)} = \chi^{\frac{2}{3}\left(\begin{smallmatrix} 2n \\ 2-1 \end{smallmatrix}\right)} Q_R \tag{9.11}$$

From (9.5) we have the special case:

$$Q_1^{(2)} = Q_2 Q_3 - (Q_{23}^2 + Q_{01}^2) = 2A_0 \, \chi \geq 0 \tag{9.12}$$

and from this follows, since $\langle A_0 \rangle \geqq 0$: $\chi \geqq 0$. The reader is invited to prove this property directly from (9.9)!

The case $\chi = 0$ is of special significance, it indicates that the distributed target R consists of the sum of two incoherent single targets: $R = M_1 + M_2$. The proof of this interesting theorem is given in [60]. We recall that for one single target: $R = M$ the condition was: $Q_R = 0$ (zero-matrix).

We may now ask ourselves; is it possible that R represents both a single object and the sum of two objects, i.e., can it be that $M = M_I + M_{II}$? We will show that this is impossible, unless the matrix M_I is proportional to M_{II}. The proof is simple: we compute Q_1, Q_2 and Q_3 which are defined by (9.4).

$$\begin{aligned} Q_1 &= \langle B_0^I + B_0^{II}\rangle^2 - \langle B_\Psi^I + B_\Psi^{II}\rangle^2 - \langle E_\Psi^I + E_\Psi^{II}\rangle^2 - \langle F^I + F^{II}\rangle^2 = \\ &= Q_1^I + Q_1^{II} + 2\langle B_0^I\rangle\langle B_0^{II}\rangle - 2(\langle B_\Psi\rangle^I \langle B_\Psi\rangle^{II} + \langle E_\Psi\rangle^I \langle E_\Psi\rangle^{II} + \langle F\rangle^I \langle F\rangle^{II}) \\ &= 2\langle B_0^I\rangle\langle B_0^{II}\rangle(1 - \cos\delta) \geq 0 \end{aligned} \tag{9.13}$$

Where we used $Q_1^I = \langle B_0^I\rangle^2 - \langle B_\Psi^I\rangle^2 - \langle E_\Psi^I\rangle^2 - \langle F^I\rangle^2 = 0$ and $Q_1^{II} = 0$ for a single target, and $\langle B_0^I\rangle\langle B_0^{II}\rangle \cos\delta$ is the scalar product between brackets. The same type of result would be found from Q_2 and Q_3. From (9.13) it is clear that equality is possible only if $\delta = 0$, i.e., the vectors in the scalar product are parallel, or in other words, corresponding terms of matrices M_I and M_{II} must be proportional. Since this applies to all terms (resulting from Q_2 and Q_3) matrices M_I and M_{II} must then be the same, except for a constant factor.

The physical meaning of this theorem has profound significance to electromagnetic scattering problems: A single radar target is irreducible, i.e., it cannot be decomposed into two or more independent radar targets. This theorem points to a fundamental limitation of traditional attempts at 'sectionalizing' an object of complex shape, such as an airplane, into independent simpler shapes. Although at higher frequencies when size to wavelength ratio is large such methods for computing RCS have had some success, the theorem shows the futility of such attempts at the lower frequencies.

X. TARGET DECOMPOSITION THEOREMS

We found previously that given aspect, frequency and waveform, a general distributed target has a stokes matrix with nine independent parameters, whereas a single composite target is determined by five parameters. It thus seems natural to consider the possibility of decomposing the nine-parameter target structure R into an average single effective object, M_0 (given by 5 parameters), and a residue part N which contains the 4 remaining degrees of freedom.

$$R = \langle M(t) \rangle = M_0 + N \tag{10.1}$$

We choose the N-target (section VII) for residue because of all targets, it has the most non-symmetry and irregularity and it is determined by four parameters: $\langle B_0^H \rangle$, $\langle B_\Psi^N \rangle$, $\langle E_\Psi^N \rangle$ and $\langle F^N \rangle$ and all these properties are valid for any orientation Ψ. The proof of physical realizability of (10.1) constitutes the main result of the target decomposition theorem. To simplify notation, we omit brackets indicating averages and orientation symbol Ψ: For effective target we use superscript T. Let:

$$B_0 = B_0^T + B_0^N \quad , \quad B = B^T + B^N \quad , \quad E = E^T + E^N \quad , \quad F = F^T + F^N \tag{10.2}$$

For M_0 to be a single target, we have the four conditions (6.16):

$$\begin{aligned} 2A_0(B_0^T + B^T) &= C^2 + D^2 \quad , \quad 2A_0(B_0^T - B^T) = G^2 + H^2 \\ 2A_0E^T &= CH - DG \quad , \quad 2A_0F^T = CG + DH \end{aligned} \tag{10.3}$$

These relationships determine B_0^T, B^T, E^T and F^T of the M_0 target uniquely and hence also B_0^N, B^N, E^N and F^N are determined (we assume that R is known). Thus the decomposition $R = M_0 + N$ is unique, if we can prove the physical realizability of the N-target. For this to be true we have to show that $B_0^N \geqq 0$ and $Q_1^N \geqq 0$ (see section IX) since for the N-target all other conditions are satisfied. The condition $B_0^N \geqq 0$ is easily shown:

$$\begin{aligned} 4A_0\ B_0^N &= 4A_0\ B_0 - 4A_0\ B_0^T = 4A_0\ B_0 - (C^2 + D^2 + G^2 + H^2) \\ &= [2A_0(B_0+B) - (C^2+D^2)] + [2A_0(B_0-B) - (G^2+H^2)] = Q_2+Q_3 \geq 0 \end{aligned} \tag{10.4}$$

where we use definitions of Q_2 and Q_3 (9.4) and relations (10.2) and (10.3). Now since $A_0 \geqq 0$, $B_0^N \geqq 0$ follows.

Next our main task is to prove $Q_1^N \geqq 0$. This is done by obtaining the following relationships:

$$Q_1^N = B_0^{N^2} - B^{N^2} - E^{N^2} - F^{N^2} =$$

$$= (B_0 - B_0^T)^2 - (B - B^T)^2 - (E - E^T)^2 - (F - F^T)^2 =$$

$$= (B_0^2 - B^2 - E^2 - F^2) + (B_0^{T^2} - B^{T^2} - E^{T^2} - F^{T^2})$$

$$- 2(B_0 B_0^T - B B^T - E E^T - F F^T) \tag{10.5}$$

The second term in brackets is zero: $Q_1^T = 0$ for a single object.

$$Q_1^N = Q_1 - 2(B_0 B_0^T - B B^T - E E^T - F F^T) \tag{10.6}$$

After multiplication with $2A_0^2$ and using (10.3) to eliminate index T variables we arrive after some algebra at the following interesting result:

$$2A_0^2 Q_1^N = Q_2 Q_3 - [Q_{01}^2 + Q_{23}^2] = Q_1^{(2)} = 2A_0 \chi \geq 0 \tag{10.7}$$

by (9.12). Since $A_0 \geqq 0$ it follows that $Q_1^N \geqq 0$ as was required. Thus we proved, the N-target is physically realizable and the decomposition (10.1) is unique and feasible.

This was the status in 1970 as presented in "Phenomenological theory of radar targets" [60]. Subsequently, a considerable improvement of the target decomposition theorem was found which applies not only to power averages (represented by stokes matrices) but to the instantaneously received signals from the target as well. The decomposition of average powers, expressed by (10.1) suggests two incoherent voltage sources, since corresponding average powers are additive, which contribute to the total return. Since received voltage was given by $V(t) = (T(t)\underline{a}) \cdot \underline{b}$ (see section IV) where $\underline{a}$ and $\underline{b}$ are fixed antenna polarizations and $T(t)$ represents the time varying target, we anticipate a decomposition:

$$T(t) = T_0(t) + T_N(t) \tag{10.8}$$

such that $T_0(t)$ contributes exclusively to M_0 above and $T_N(t)$ contributes only to the N-target in (10.1). The N-target presents no problem whatever, since the sum of single N-targets is a distributed N-target. However, $T_0(t)$ has to be of special form, because M_0 is a single average target. We found in the previous section that a single radar target is irreducible , i.e., it cannot be written as the sum of simpler, independent objects. To see how this theorem is applied we first compute the instantaneous power in terms of stokes matrices:

$$M(t) = M_0(t) + M_N(t) + 2Re\ W(T_0(t),\ T_N^*(t)) \tag{10.9}$$

where the first two terms on the left indicate power contributions due to $T_0(t)$ and $T_N(t)$ and the third term represents target interaction. Taking the expected value results in:

$$R = \langle M(t) \rangle = \langle M_0(t) \rangle + \langle M_N(t) \rangle = M_0 + N \tag{10.10}$$

since the last term in (10.9) vanishes due to the fact that targets $T_0(t)$ and $T_N(t)$ are uncorrelated. Now $\langle M_N(t) \rangle = N$, and $\langle M_0(t) \rangle = M_0$. According to the irreducibility of single target M_0, the only form $M_0(t)$ can take is:

$$M_0(t) = \hat{p}_0(t)\ M_0 \tag{10.11}$$

where M_0 is a constant matrix and $\hat{p}_0(t)$ is a real non-negative time varying scalar with $\langle \hat{p}_0(t) \rangle = 1$. This has as a consequence that now $T_0(t)$ also is restricted in form:

$$T_0(t) = \hat{a}_0(t)\ T_0 \tag{10.12}$$

where T_0 is a constant scattering matrix which corresponds to M_0. For examples of such correspondences we refer to section VIII. The complex time varying scalar $\hat{a}_0(t)$ has normalized average power: $\langle a_0(t) a_0^*(t) \rangle = 1$.

We found the following result: Corresponding to the decomposition of average powers (10.1), the return voltage from a radar target at each instant, can be decomposed into a part $\hat{a}_0(t)T_0$ which contributes exclusively to the effective target and a part

$T_N(t)$ which contributes only to residue. The two parts are mutually uncorrelated. Hence:

$$T_0(t) = \hat{a}_0(t)\ T_0 + T_N(t) \tag{10.13}$$

The physical significance of $T_0(t) = \hat{a}_0(t)T_0$ is particularly interesting. This result confirms the intuitive concept of an 'effective target' as a target which has constant structure given by T_0, which scattering behavior is described by the scalar $\hat{a}_0(t)$. We may think of the scalar Swerling models to illustrate this fact [70]. On the other hand the 'residue' part $T_N(t)$ may be viewed as 'clutter' which is induced as residue to account for the approximation. Thus a match is achieved between electromagnetic behavior of radar targets and practical expediency, which favors the scalar concept.

Of particular interest, is to give a physical account for the term $\hat{a}_0(t)$. We know that the (RC,LC) type of circularly polarized return from a radar target gives $2A_0$ (equ. 6.25). For an N-target, since $A_0^N = 0$ (section VIII) this return will be zero, and hence <u>the (RC, LC) type of return is a direct measure of $\hat{a}_0(t)$</u> (normalized to average power equal unity).

These results will lead to investigation concerning target discrimination and identification based on invariant target structure. The equation (10.13) gives a method by which the total incoming signal is decomposed into a desirable part which is 'signal' and a clutter term which is 'noise.' For most cases (hi-frequency, large D/λ ratio) the target noise may be considered Gaussian and independent of the 'signal' component [70]. Standard or adaptive signal processing schemes may be used for improved radar performance. The effective structure matrix T_0 may be measured in real time to optimize discrimination between wanted from unwanted targets. In order to implement such a system, it will be necessary to measure the complete target scattering matrix by using dual polarized antennas for transmission as well as for reception.

Other areas of application are with the improvement of current models for fluctuating targets. These Swerling models are basically scalar in nature and there is no method by which one can account for complete polarization response of the target. One way to implement Swerling models with the present theory is to assign the scalar $\hat{a}_0(t)$ to the conventional Swerling distribution. For the noise part $T_N(t)$ we choose a Gaussian model. The combined approach leads to an improved model which includes target response to polarization [70].

XI. ORIENTATION-INDEPENDENT TARGETS

A class of targets of particular practical interest will be the next topic for discussion. These targets have the desirable property that there is no preferred orientation bias (see fig. 1, for definition of orientation as rotation about the line of sight). For terrain this represents plains and fields viewed at close to normal direction, but not a trench, ridge or river bed which would have a preferred direction of orientation. The orientation-independent target model is useful even to describe physical targets which in fact have a preferred orientation because in that case, since there is usually uncertainty about its direction, one can treat orientation Ψ of the target as a random variable which has equal probability in all directions. With that provision the target then becomes orientation-independent. Obviously not all target models have these properties (see in fact the dipole cloud model published by Poelman [73], which has a preferred orientation).

Examples of orientation independent targets (not to be confused with orientation independent target parameters) are homogeneous terrain and sea state surfaces at close to vertical incidence. Also homogeneous clouds of rain, dust particles of chaff. It turns out that scattering from orientation independent targets is uniquely determined by three target parameters: $\langle A_0 \rangle$, $\langle B_0 \rangle$ and $\langle F \rangle$, where $\langle F \rangle = 0$ for targets without helicity. The conditions are easily derived from the general equation (6.1) by assuming the orientation angle Ψ to be an independent random variable with uniform distribution. Then all terms of the stokes matrix containing Ψ will vanish, such that only $<A_0>$, $<B_0>$ and $<F>$ remains. If we require also symmetry between (LC, LC) and (RC, RC) type of target return, then $<F> = 0$, i.e., the target in the average has no helicity (a bedspring surface would have helicity). The received power from an orientation independent target will thus be:

$$<P(\phi_A, \tau_A; \phi_B, \tau_B)> = <A_0> + <B_0> + (-<A_0> + <B_0>) \cdot \sin 2\tau_A \sin 2\tau_B + <A_0> \cos 2\tau_A \cos 2\tau_B \cos 2(\phi_A - \phi_B) \quad (11.1)$$

We seek a decomposition of the target scattering into an 'effective single object' and 'residue' components. The only orientation-independent symmetric single target is given by a sphere or flat facet at normal incidence. Hence $<A_0>$ is associated with the sphere structure and $<B_0>$ with residue. We may write $B_0^N = <B_0> = B_1 + B_2$ and since also $B^N = B_1 - B_2 = 0$ we find $B_1 = B_2 \neq 0$. The B_1 and B_2 are associated with 'trough noise' type of return.

$$R = \begin{bmatrix} <A_0> + <B_0> & 0 & 0 & 0 \\ 0 & -<A_0> + <B_0> & 0 & 0 \\ 0 & 0 & <A_0> & 0 \\ 0 & 0 & 0 & <A_0> \end{bmatrix} =$$

$$R = <A_0> \begin{bmatrix} 1 & & & \\ & -1 & & \\ & & 1 & \\ & & & 1 \end{bmatrix} + B_1 \begin{bmatrix} 1 & & & \\ & 1 & & \\ & & 1 & \\ & & & -1 \end{bmatrix} + B_2 \begin{bmatrix} 1 & & & \\ & 1 & & \\ & & -1 & \\ & & & 1 \end{bmatrix} \quad (11.2)$$

We refer to section (VIII) for the matrix identifications. At each instant in time the corresponding scattering-matrix decomposition will be:

$$T(t) = a(t) \begin{bmatrix} 1 & 0 \\ 0 & 1 \end{bmatrix} + b(t) \begin{bmatrix} 1 & 0 \\ 0 & -1 \end{bmatrix} + c(t) \begin{bmatrix} 0 & 1 \\ 1 & 0 \end{bmatrix} \quad (11.3)$$

where $2<A_0> = <|a(t)|^2>$, $2B_1 = <|b(t)|^2>$ and $2B_2 = <|c(t)|^2>$

and $2B_1 = 2B_2 = \langle B_0 \rangle$. We find that for this case the matrix elements $a(t)$, $b(t)$, $c(t)$ are mutually uncorrelated. Pictorially (11.3) is reproduced by:

$$T(t) = a(t) \bigoplus b(t) \boxplus c(t) \boxtimes \qquad (11.4)$$

where circled additions are incoherent and $\langle |b|^2 \rangle = \langle |c|^2 \rangle$. The orientation-independent target has many interesting properties. The return from the 'effective' part is that from a sphere. The 'residue' part is given by the diagonal N-target matrix:

$$N = \begin{bmatrix} B_0 & & & \\ & B_0 & & \\ & & 0 & \\ & & & 0 \end{bmatrix} \qquad (11.5)$$

For linearly polarized illumination, the backscattered return-vector $s_N = Ng(\underline{a})$ is found simply:

$$s_N = \begin{bmatrix} B_0 & & & \\ & B_0 & & \\ & & 0 & \\ & & & 0 \end{bmatrix} \begin{bmatrix} p_A^2 \\ 0 \\ p_A^2 \cos 2\phi_A \\ p_A^2 \sin 2\phi_A \end{bmatrix} = p_A^2 B_0 \begin{bmatrix} 1 \\ 0 \\ 0 \\ 0 \end{bmatrix} \qquad (11.6)$$

hence s_N is a completely unpolarized stokes vector. Fig. 9 shows the return from the orientation independent target for vertical illumination.

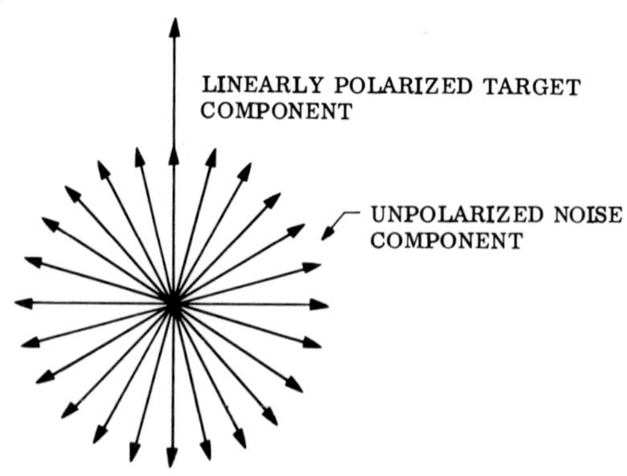

Fig. 9. Return from orientation independent target.

Special cases of practical interest are worth mentioning: For identical antennas we derive from (11.2) with $P_{//} = Rg \cdot g$: and $P_{\perp} = Rg \cdot g_{\perp}$:

$$P_{//} = (2\langle A_0\rangle + \langle B_0\rangle)g_0^2 + (-2\langle A_0\rangle + \langle B_0\rangle)g_1^2 \qquad (11.7)$$

$$P_{\perp} = \langle B_0\rangle g_0^2 - (-2\langle A_0\rangle + \langle B_0\rangle)g_1^2 \qquad (11.8)$$

where we used: $g_0^2 = g_1^2 + g_2^2 + g_3^2$. The total power return is the sum of (11.7) and (11.8) which gives $P_{tot} = 2(\langle A_0\rangle + \langle B_0\rangle)g_0^2$, and this agrees with our previous results.

For linear polarization $g_1 = 0$, $P_{//} = (2\langle A_0\rangle + \langle B_0\rangle)g_0^2 = \sigma g_0^2$ and $P_{\perp} = \langle B_0\rangle g_0^2$. The ratio:

$$X = \frac{P_{\perp}}{P_{//}} = \frac{\langle B_0\rangle}{2\langle A_0\rangle + \langle B_0\rangle} \qquad (11.9)$$

is sometimes used as measure for depolarization, whereas σ is used as 'RSC.' In terms of σ and x, equation (11.7) obtains an interesting form:

$$P_{//} = \sigma[g_0^2 + (2x - 1)g_1^2] \qquad (11.10)$$

For $x = \frac{1}{2}$, (11.10) reduces to particularly simple form: $P_{//} = \sigma g_0^2$.

Orientation-independent target models were proposed by the author to represent terrain with the design of a sensitive radar altimeter [50] and have also been proposed to represent complex space-objects [61].

XII. APPLICATIONS

A. Average RCS for Systems Design

To get started with an illustrative example, we refer to Fig. 10. A missile body is observed at nose-on aspect angles. The measured radar cross-section patterns are shown for horizontal and vertical polarizations. Two different 'average RSC' can be obtained, and since there is no a priori information on which polarization is preferred, it is difficult to determine a unique 'figure of merit' which is used for system design (i.e., some systems wish to achieve a minimum none-on RSC in order to prevent detection). With the new technique, it is possible to provide a unique criterion for the systems RCS for any polarization.

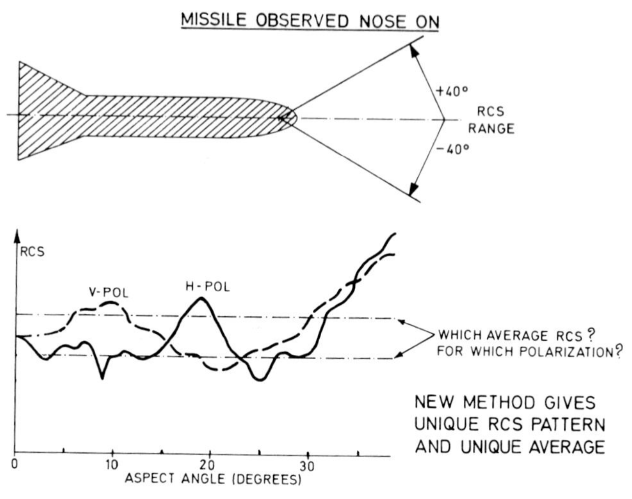

Fig. 10. Missile nose on RCS patterns.

As 'average' will be chosen the effective target average which is determined by the (RC, LC) circularly polarized return. Since the object at nose-on is mostly convex (we assume large D/λ, or high-frequency case) the effective structure matrix T_0 is the unit-matrix and hence the (RC, LC) pattern suffices to determine effective RCS both instantaneously as function of aspect angle and in the average over a range of desired aspects. Notice that these data which were obtained by using the circular polarization pattern are now in fact criteria, usable for system design at all polarizations! The effect of polarization enters merely as a multiplicative constant (db reference for the pattern). All deviations of actual RCS from the adjusted circular pattern are treated in this case (for convex surface) as 'residue' or target clutter. For more complicated surfaces the circular (RC, LC) pattern still is used as describing instantaneous effective target return for all polarizations, however, the structure matrix T_0 may take more complicated form which has to be determined separately by measuring complete scattering matrix $T(t)$ for a range of aspects. All that will change in the end, is the level of db reference for the CP pattern of the effective target, in each range of aspect angles.

B. Dynamic Simulation of Radar Targets

The figure gives another illustration. A reentry vehicle (RV) is aimed at a target objective and one wishes to know how well the RV can be observed by a tracking radar close to the mission objective.

One can estimate RV vulnerability through computer simulation of the observed RCS time history. The problem is to determine what kind of RCS information based on range measurements (so-called static data) is required to provide accurate simulation. This looks like a simple problem: because of motion of the RV one can measure theoretically the static RCS at all aspect angles, target orientation angles and if the RV is non-symmetric (due to protruding fins etc.) for various roll-positions of the target. This would mount to a large and costly measurements program, and these data have to be acquired at various frequencies.

The problem is to determine the minimum amount of measurements one would need for dynamic simulation. A solution based upon equation (6.17) for received power was found for symmetric targets: one uses at each frequency three linear polarizations: horizontal, 45° and vertical, for transmission and reception. Thus, three static patterns of the RV are obtained as input to the dynamic simulation program. For non-symmetric objects one obtains at each frequency and for each roll angle, 5-linear polarization patterns at 0°, 30°, 60°, 90° and 120° orientations [36]. This procedure results in considerable reduction of expenses. These two applications illustrate the use of the phenomenological approach, since no a priori model of the target or its trajectory were used to solve the general problem. Only actual measurements provided inputs to data processing applicable to any target situation.

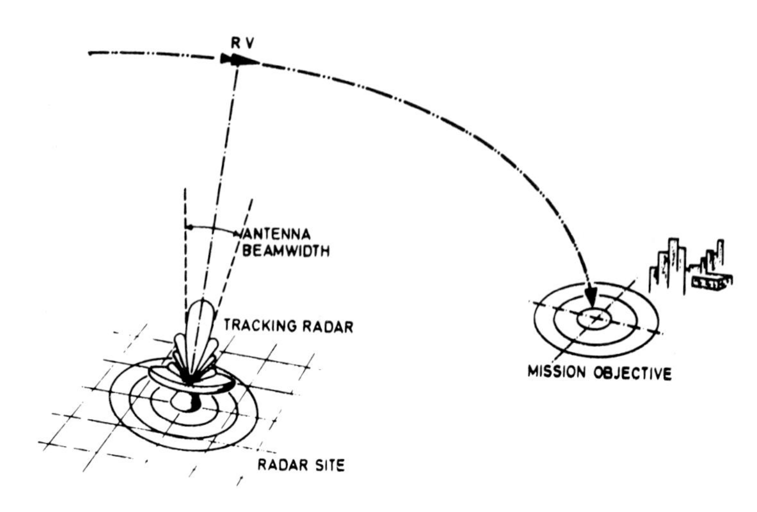

Fig. 11. Dynamic simulation of radartargets.

In Fig. 12 the five measured patterns which were used as input to the target simulation program are shown. The target was highly non-symmetric and one can see changes greater than 20 dB at certain aspect angles between one linear polarization and another. At the pattern-null the change is even greater than 30 dB.

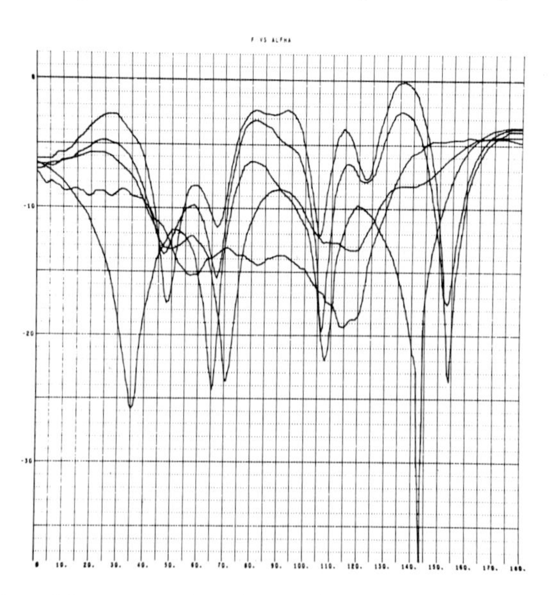

Fig. 12. Dynamic simulation input patterns.

C. Target Model for Radar Altimeter Design

Another area of application is with the improvement of existing target models. One practical example is with the design of a very sensitive radio altimeter, which will measure the height above ground with great accuracy. It is clear that an improved model for terrain should give better altimeter performance. The so-called orientation independent target model which includes polarization dependence was suggested by this writer, and was used for the radar altimeter design. The theory was successfully implemented in the design of the system and is now in actual operation for about three years [59]. See also section XI.

D. Improved Fluctuating Target Models

Still another area is the improvement of current models for fluctuating targets. These so-called Swerling models are basically scalar in nature, and there is no consideration given to polarization response of the target. The way to implement Swerling models with our present theory would be to assign to the scalar $\hat{a}_0(t)$ the conventional Swerling distribution. For the noise part: $T_N(t)$ we choose a Gaussian model. The combined approach leads to an improved model which includes target responses to polarization [70].

We now summarize the philosophy of the phenomenological approach. The idea is to depend less on specific target models, because in most cases radar targets can be very complicated structures (air planes, terrain types, sea-state, etc.). Even if one has a well-analyzed model, it is quite another problem to apply this model to an actual case. One has to verify that the conditions, upon which the model was based are applicable to the actual target.

In contrast to this approach, the idea here is to use all the information contained in the scattering matrix and to use concepts of symmetry, non-symmetry, irregularity and statistical properties of the target. For example, for any roll-symmetric object there is a universal parameter $F = 0$. Also we use only general electromagnetic properties such as reciprocity and if targets are mostly convex, they may be identified by specular reflections, (scattering centers or flare-spots) or if a target is mostly nonconvex, with protruding edges, etc., it will cause depolarization. These properties are of course a function of size/wavelength dependence of the target. Only recently it was realized that a dipole cloud of almost any configuration can be characterized by having four parameters of the Stokes matrix equal to zero ($A_0 = B_0$, $D = 0$, $F = 0$, $G = 0$), see section VIII, case (E).

XIII. SOME EXPERIMENTAL RESULTS

Although many experimental results are available on the subject of RCS (radar cross section), see for instance "Radar Cross-section Handbook" [57], it will soon become apparent, that complete data about radar targets are scarce or non-existent. Even for simple-shaped objects, such as cones, sphere capped cones, cylinders, etc., there is rarely complete information on RCS polarization dependence, and phase measurements are scarce [31]. In other words, the complete scattering matrix of the object is not known and hence information is incomplete.

A good size manual could probably be written about the reasons for this state of affairs. Radar measurements are expensive to obtain and target models have to be constructed with great precision in shape and according to specified material properties. Scaling presents serious credibility problems for non-perfectly conducting objects and because of the required tolerances. Radar test equipment is expensive, and requires periodic calibration and experienced personnel to operate and maintain.

Radar range facilities suffer from notorious background interference problems and are susceptible to weather conditions. For economic reasons, usually only a minimal amount of information about the target is acquired. For complex shapes the minimal RCS data are not nearly adequate to fully determine the radar target response. A step in the direction of improving RCS prediction quality with less expense was made by this writer. This method, which was outlined previously, uses the 5-linear polarization technique for simulation of dynamic RCS of non-symmetric objects and 3-linear polarizations for roll-symmetric targets [36].

A. Fixed Targets

Some early attempts at acquiring complete data were measured during the years 1951-1955 [9]. A radar set was pointed at a 'fixed' target such as a bridge, raincloud, etc. By continuously varying polarization and recording intensity, a so-called polarization map of the target was obtained at fixed aspect angle. In order to describe the display technique which was used, we refer to the polarization sphere Fig. 4. Fig. 13 shows the polarization sphere labelled with contours of constant received intensity. Each point on the sphere corresponds to a state of polarization for identical radar transmitter and receiver antennas. The radar remains pointed at the target (at fixed aspect angle) until all polarizations are scanned.

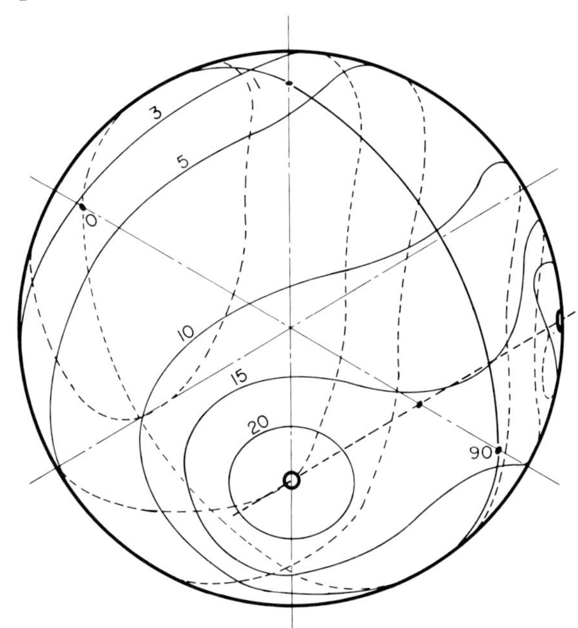

Fig. 13. Polarization sphere with contours of construct received intensity (db).

We notice a curious behaviour at two points on the sphere. Here the contours converge into a single point. At these two locations, the target return is polarized orthogonal to the antenna receiver polarization and hence a 'null' is registered. We can show that each target at fixed aspect angle exhibits this kind of behaviour. Each target is represented by a scattering matrix, which can be found from the unique location of the two nulls on the sphere (4 parameters) and an intensity associated with 0 dB (maximum target return). We thus find the required 5 parameters which determine the scattering matrix. For these reasons the target data obtained is called 'complete,' however we have to keep in mind the restriction placed on aspect angle and frequency behaviour implied by this method.

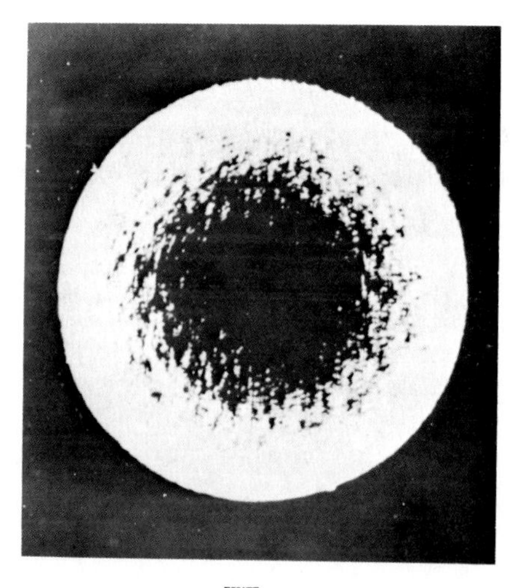

FIGURE
Subject: Cloud - Stratus layer at about 2000 ft. ceiling
Range: 1 mile Frequency band: K
Weather: Windy, misty Sense: R. C.

Fig. 14. Polarization map for a cloud.

Some actual measurements are shown in Figs. 14 and 15. Fig. 14 shows one polarization map for a cloud (the other map for opposite sense is similar). As expected, the center of the chart (CP) shows a deep 'null' area (the other null is for opposite sense CP).

This would be typical for specular targets for which P(LC, LC) = P(RC, RC) = 0. Hence the 'average effective' cloud target is represented by a sphere. Due to the distributed nature of the cloud and the fact that time changes between measured points on the chart, we notice the temporal clutter variations. We have here an example of an orientation-independent target with scattering matrix decomposition given as in equation (11.4).

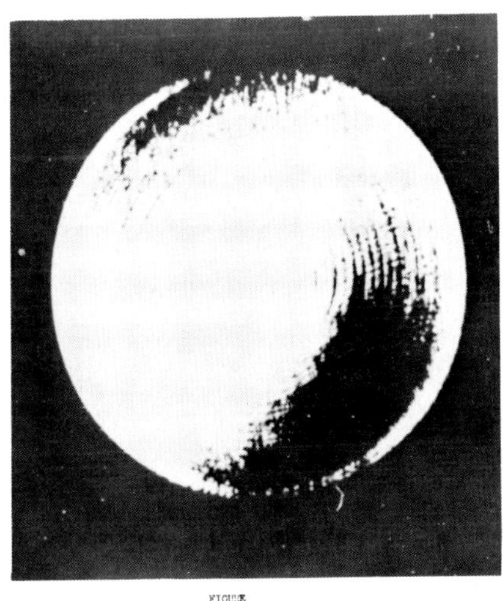

Fig. 15. Polarization map for a bridge (RS).

Fig. 15 shows the polarization map which was measured for a bridge [9]. The bridge is viewed at a slant from a top of a hill (fixed aspect). The two polarization-nulls are clearly indicated. The upper null-area near 45° LP extend over both maps (only one for the right sense is shown). This target obviously is highly non-symmetric. The general decomposition (10.13) now applies, and the constant average target structure matrix, instead of being the unit matrix as in the previous case now has the general form.

Although the bridge was considered 'fixed,' in fact, it will be time varying due to changing wind-conditions ground movements, temperature variations, etc., this is indicated by the cluttered fringes around the null areas.

B. Symmetric Targets as Function of Aspect Angle

The previous examples show measurements of time varying extended targets (a cloud, a large bridge) at fixed angles of observation. The next two examples will show measurements on single targets at varying angles of observation. The method uses the fact that for a single target the scattering matrix is determined (except for an amplitude factor) by the location of the two characteristic nulls on the polarization chart. By using a judiciously chosen set of measurements and a so-called 'polarization-null-locator' chart, it was possible to measure in fact the complete scattering matrix as a function of aspect angle by obtaining a characteristic null-locus for each target. (For details we refer to [60, p. 101] and [17].)

Fig. 16 shows the null-locus (polarization-null as a function of aspect angle displayed on the polarization chart) for a diplane (two plates at 90° angle) which is being rotated along the vertical axis. Fig. 17 shows the null-locus for a convex object which is being rotated about the vertical axis. Due to symmetry of both targets, it was not necessary to measure the locus for the <u>second</u> null at each aspect separately, since this is already determined by the locus of the first null. In order to interpret the charts it is useful to recall that the chart center is CP which is a characteristic null-polarization for a specular type of reflection. On the other hand double-bounce reflections such as would occur with the diplane have a null at ±45° LP (only the locus ending in -45° LP is shown in Fig. 16). The interesting 'question-mark' locus for the diplane is generated. We start with aspect facing one plate at normal incidence (specular case, null at center) after which the null moves to the 'double-bounce' (-45° LP) at 45° degrees incidence.

Observe the striking contrast between the measured loci of the two targets. Because a 'convex' object exhibits mostly specular type reflections, the characteristic nulls at any aspect

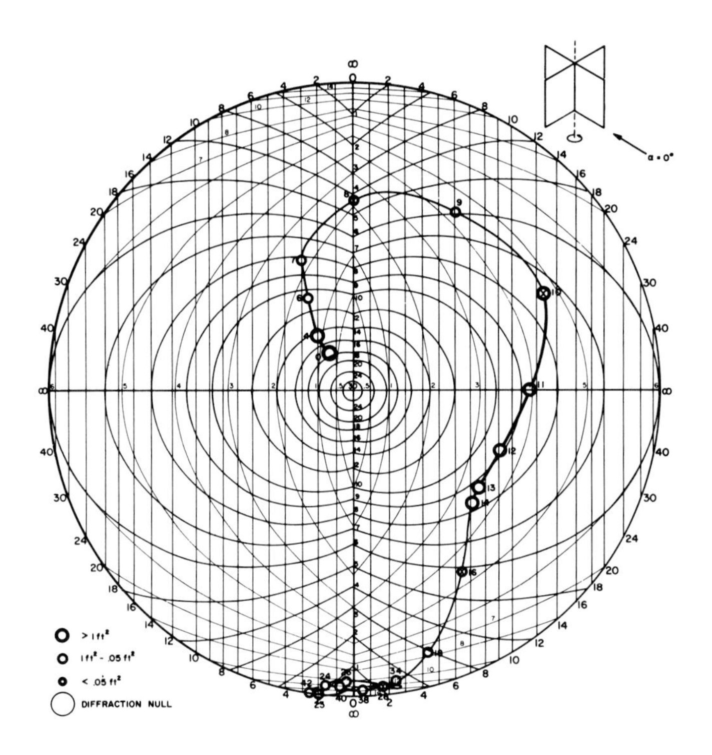

Fig. 16. Null locus for dihedral.

can be expected to occur at the center of the chart (CP). However, the measured deviations from the center indicate that the target is only truly specular in an average sense. These deviations could be treated on a statistical basis which again would lead to the orientation-independent-target model as a credible representation of convex-type objects in space [61]. The statistical basis is derived from the fact that usually the aspect angle of the target is unknown. The diplane model is often projected as a credible decoy for smooth convex-objects, since they give large RCS/size ratios. We see, however, that due to different polarization characteristics it can easily be discriminated against by using CP instead of the usual LP mode of radar operation.

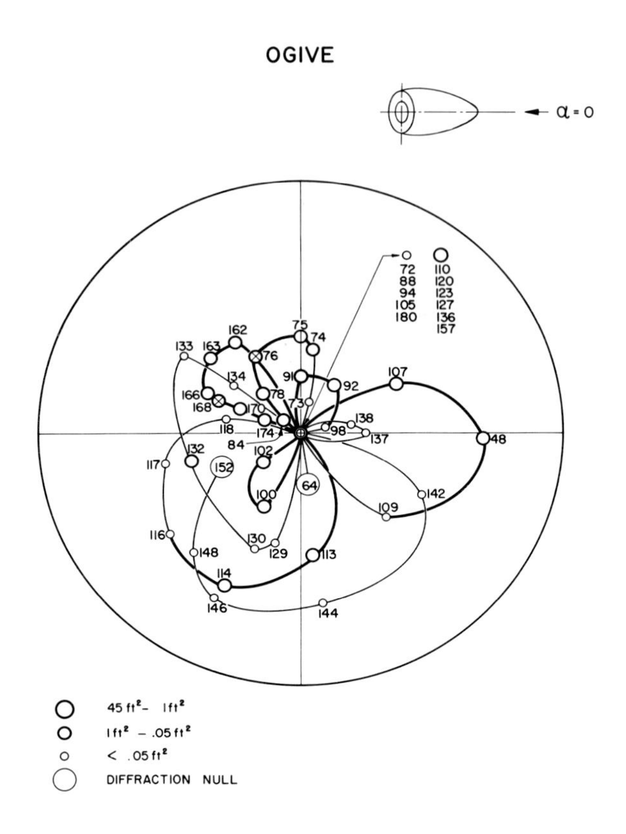

Fig. 17. Null locus for ogive.

XIV. REFERENCES

1. Sinclair, G., The Transmission and Reception of Elliptically Polzarized Waves, *Proc. IRE 38*, 148-151 (1950).
2. Deschamps, G.A., Part II - Geometrical Representation of the Polarization of a Plane Electromagnetic Wave, *Proc. IRE 39*, 540-544 (1951).
3. Rumsey, V.H., Part I - Transmission Between Elliptically Polarized Antennas, *Proc. IRE 39*, 535-540 (1951).
4. Kales, M.L., Part III - Elliptically Polarized Waves and Antennas, *Proc. IRE 39*, 544-549 (1951).
5. Bohnert, J.I., Part IV - Measurements on Elliptically Polarized Antennas, *Proc. IRE*, 549-552 (1951).
6. Booker, H.G., Techniques for Handling Elliptically Polarized Waves with Special Reference to Antennas, *Proc. IRE 39*, 533-534 (1951).
7. Kennaugh, E.M., Polarization Properties of Radar Reflections, Antenna Lab., The Ohio State University Research Foundation,

Columbus, RADC Contract AF28(099)-90, Project Report 389-12 (AD 2494) (March 1, 1952).

8. Kennaugh, E.M., A.F. Butler, and L.S. Taylor, Effects of Type of Polarization on Echo Characteristics, Antenna Lab., The Ohio State University Research Foundation, Columbus, Contract AF28(099)-90, Progress Resports OSURF 389-4, 9, 12, 14, 19 (June 1950 to July 1953).
9. Huynen, J.R., R.W. Thille, and W.H. Thormahlen, Interim Engineering Report on Effect of Polarization on the Radar Return from Ground Targets and Rain, Dalmo Victory Company, Belmont, Calif., Contract AF33(039)-20926 (June 1951 to May 1952) and Contract AF33(600)-22711 (January 1953 to May 1954) (declassified).
10. Gent, H., Elliptically Polarized Waves and Their Reflection from Radar Targets: a Theoretical Analysis, Telecommunications Research Establishment, TRE Memo. 584, March 1954.
11. Graves, C.D., Radar Polarization Power Scattering Matrix, *Proc. IRE 44*, 248-252 (1956).
12. Van de Hulst, H.C., "Light Scattering by Small Particles," Wiley, New York (1957).
13. Wiltse, J.C., S.P. Schlesinger, and C.M. Johnson, Backscattering Characterisitcs of the Sea in the Region from 10 to 50 k Mc, *Proc. IRE 45(2)* (1957).
14. Cosgriff, R.L., W.H. Peake, and R.C. Taylor, Terrain Scattering Properties for Sensor Systems Design (Terrain Handbook II), Engineering Experiment Station Bulletin, Vol. 29, No. 3, The Ohio State University (1960).
15. Crispin, J.W., Jr., et al., The Measurement and use of Scattering Matrices, Radiation Lab., University of Michigan, Ann Arbor, Report 2500-3-T (May 1960).
16. Copeland, J.R., Radar Target Classification by Polarization Properties, *Proc. IRE 48*, 1290-1296 (1960).
17. Huynen, J.R., Study on Ballistic-Missile Sorting Based on Radar Cross-section Data: Report No. 4, Radar Target Sorting Based upon Polarization Signature Analysis, AFCRC-TN-60-588, LMSD-288216 (May 1960).
18. Huynen, J.R., Radar Measurements on Scattering Operators, *Proc. 6th Symp. on Ballistic Missile and Aerospace Technology II* (1961).
19. Bechtel, M.E., and R.A. Ross, Interim Report on Polarization Discrimination, Cornell Aeronautical Lab., Inc., Ithaca, N.Y., Report UB-1376-S-105 (September 1962).
20. Huynen, J.R., A New Approach to Radar Cross-Section Measurement, *1962 IRE Intern. Conv. Rec. Pt. 5, 10*, 3-11.
21. Huynen, J.R., Dynamic Radar Cross-Section Predictions, *1963 Proc. 7th Mil-E-Con.*
22. Beckmann, P., and A. Spizzichino, "The Scattering of Electromagnetic Waves from Rough Surfaces," Macmillan, New York (1963).
23. Huynen, J.R., Complete Radar Cross-Section Measurements, 1964 Radar Reflectivity Measurements Symp., Vol. 1, p. 93; Report RADC-TDR-64-25-Vol. 1.

24. Webb, J.A., and W.P. Allen, Precision Measurements of the Radar Scattering Matrix, 1964 Radar Reflectivity Measurement Symp., Vol. 1, p. 233. See also Lockheed Space & Missiles Company, Ga., Report RADC-TDR-64-25, Vol. 1.
25. Olin, I.D., and F.D. Queen, Dynamic Measurement of Radar Cross Section, *Proc. IEEE 53*, 954-961 (1965).
26. Blacksmith, P., Jr., R.E. Hiatt, and R.B. Mack, Introduction to Radar Cross-Section Measurements, *Proc. IEEE 53*, 901-920 (1965).
27. Kouyoumjian, R.G., and L. Peters, Jr., Range Requirements in Radar Cross-Section Measurements, *Proc. IEEE 53* (1965).
28. Freeny, C.C., Target Support Parameters Associated with Radar Reflectivity, *Proc. IEEE 53* (1965).
29. Beckmann, P., Scattering by Composite Rough Surfaces, *Proc. IEEE 53* (1965).
30. Blacksmith, P., Jr., and R.B. Mack, On Measuring the Radar Cross Sections of Ducks and Chickens, *Proc. IEEE 53* (1965).
31. Bahret, W.F., and C.J. Sletten, A Look into the Future of Radar Scattering Research and Development, *Proc. IEEE 53* (1965).
32. Bickel, S.H., and R.H.T. Bates, Effects of Magneto-ionic Propagation on the Polarization Scattering Matrix, *Proc. IEEE 53*, 1089 (1965).
33. Kell, R.E., and N.E. Pedersen, Comparison of Experimental Radar Cross-Section Measurements, *Proc. IEEE 53* (1965).
34. Bickel, S.H., Some Invariant Properties of the Polarization Scattering Matrix, *Proc. IEEE 53* (1965).
35. Lowenschuss, O., Scattering Matrix Application, *Proc. IEEE 53*, 988 (1965).
36. Huynen, J.R., Measurement of Target Scattering Matrix, *Proc. IEEE 53*, 936-946 (1965).
37. Freeny, C.C., Experimental and Analytical Investigation of Target Scattering Matrices, Rome Air Development Center Report RADC-TR-65-298 (December 1965).
38. Papas, C.H., "Theory of Electromagnetic Wave Propagation," pp. 118-148. McGraw-Hill, New York (1965).
39. Born, M., and E. Wolf, "Principles of Optics," 3rd Ed., Pergamon, London-New York (1965, 1966).
40. Long, M.W., On the Polarization and the Wavelength Dependence of Sea Echo, *IEEE Trans. Antennas & Propag.*, 749 (1965).
41. Kanareykin, D.B., N.F. Pavlov, and V.A. Potckhin, "The Polarization of Radar Signals," Sovyetskoye Radio, Moscow (1966) (in Russian).
42. Krishen, K., W.W. Koepsel, and S.H. Durrani, Cross-Polarization Measurements and their Relation to Target Surface Properties, *IEEE Trans. Antennas & Propag. 14,* 629-635 (1966). See also Tech. Report EE-TK-2, Elec. Eng. Dept., Kansas State University, Manhattan, Kansas (1966).
43. Huynen, J.R., Invariant Radar Target Parameters, *IEEE International Antennas and Propagation Symposium*, Palo Alto, Calif. (December 1966).

44. Fung, A.K., Scattering and Depolarization of Electromagnetic Waves from a Rough Surface, *Proc. IEEE (Letters) 54,* 395-396 (1966).
45. Fung, A.K., Vector Scatter Theory Applied to Moon and Venus Radar Return, *Proc. IEEE 54(7),* 996-998 (1966).
46. Knittel, G.H., The Polarization Sphere as a Graphical Aid in Determining the Polarization of an Antenna by Amplitude Measurements Only, *IEEE Trans. Antennas & Propag. AP-15,* 217-221 (1967).
47. Hagfors, T., A Study of the Depolarization of Lunar Radar Echoes, *Radio Sci. 2(5) (1967).* See also "Radar Astronomy" (J.V. Evans and T. Hagfors, Eds.), McGraw-Hill, New York (1968).
48. Special Issue on Partial Coherence, *IEEE Trans. Antennas & Propag. AP-15,* 2-199 (1967).
49. Bolinder, E.F., Geometric Analysis of Partially Polarized Waves, *IEEE Trans. Antennas & Propag. AP-15(H01),* 37 (1967).
50. Huynen, J.R., An Orientation-Independent Terrain Scattering Model for Vertical Incidence Radar, Lockheed Missiles & Space Company, Sunnyvale, California, Report B-70-68-4 (January 1968).
51. Beckmann, P., "The Depolarization of Electromagnetic Waves," The Golem Press, Boulder, Colorado (1968).
52. Crispin, J.W., and K.M. Siegel, "Methods of Radar Cross Section Analysis," Academic Press (1968).
53. Bowman, J.J., T.B.A. Senior, and P.L.E. Uslenghi, "Electromagnetic and Acoustic Scattering by Simple Shapes," North Holland Publishing Co. (1969).
54. Moore, R.K., Radar Return from the Ground, Bulletin of Engineering No. 59, University of Kansas (1969).
55. Fung, A.K., and H.I. Chan, Backscattering of Waves by Composite Rough Surfaces, *IEEE Trans. Antennas & Propag. AP-17(5)* (1969).
56. Nathanson, F.E., "Radar Design Principles," McGraw-Hill (1969).
57. Ruck, G.T., D.E. Barrick, W.D. Stuart, and C.K. Kirchbaum, "Radar Cross-Section Handbook, Vol. I and II," Plenum Press, New York (1970).
58. Skolnik, J., "Radar Handbook," McGraw-Hill (1970).
59. Williams, C.S., J.A. Cooper, and J.R. Huynen, Antenna-Polarization and Terrain Depolarization Effects on Pulse-Radar Return from Extended Areas at the Near Vertical, *Proc. IEEE 58(9),* 1322-1328 (1970).
60. Huynen, J.R., Phenomenological Theory of Radar Targets, Doctoral Thesis, Technical University, Delft, The Netherlands (1970) (obtainable from the author).
61. Varshauchuk, M.L., and V.O. Kobak, Cross-Correlation of Orthogonally Polarized Components of Electromagnetic Field, Scattered by an Extended Object, *Radio Eng. & Electron. Phys. 16(2),* 201-205 (1971).
62. Poelman, A.J., and J.P. van der Voort, The Polarization

Dependence of Received Back-scattered Power from a Random Dipole Cloud, SHAPE Technical Centre, The Hague, The Netherlands, Technical Memorandum TM-276 (AD 520 934) (May 1972) (NATO unclassified; limited distribution).

63. Lin, Y.T., and A. Ksienski, Utilization of Phase Information in Radar Target Identification, in 1973 G-AP International Symposium.
64. Parashar, S.K., A.W. Biggs, A.K. Fung, and R.K. Moore, Investigation of Radar Discrimination of Sea Ice, *Proc. of the Ninth International Symposium on Remote Sensing of Environment I* (1974).
65. Goggins, W.B., Jr., P.H. Blacksmith, and C.J. Sletten, Phase Signature Radars, *IEEE Trans. Antennas & Propag. AP-22(6)* (1974).
66. Mitchell, R.L., Models of Extended Targets and Their Coherent Radar Images, *Proc. IEEE 62* (1974).
67. Bahret, W.F., The Use of RCS Data, NAECON 1974, Record, 169-176.
68. Long, Maurice, "Radar Reflectivity of Land and Sea," Lexington Press (1975).
69. Huynen, J.R., Phenomenological Theory of Radar Targets, Report for SHAPE Technical Center (December 1975) (obtainable from SHAPE or from the author).
70. Huynen, J.R., F. McNolty, and E. Hanson, Component Distributions for Fluctuating Radar Targets, *IEEE Trans. Aerosp. & Electron. Syst.*, 1316-1332 (1975).
71. Ksienski, A.A., Y.T. Lin, and L.J. White, Low Frequency Approach to Target Identification, *Proc. IEEE 63(12)* (1975).
72. Poelman, A.J., On Using Orthogonally Polarized Noncoherent Receiving Channels to Detect Target Echoes in Gaussian Noise, *IEEE Trans. Aerosp. & Electron. Syst.*, 660-662 (1975).
73. Poelman, A.J., Cross-correlation of Orthogonally Polarized Back Scatter Components, *IEEE Trans. Aerosp. & Electron. Syst.*, 674-681 (1976).

TRENDS IN ARRAY ANTENNA RESEARCH

Robert J. Mailloux
Deputy for Electronic Technology (RADC)
Hanscom Air Force Base, Massachusetts

This paper describes a number of analytical developments in the history of phased array research and analyzes the present state of maturity of that field. The main conclusion of this study is that the technology is evolving so rapidly, and the number of different array types and requirements growing so swiftly that past analytical developments are vastly inadequate to handle the problems posed by present day array systems. The paper highlights those areas where intensified research is necessary.

I. INTRODUCTION

The electromagnetic theory of antennas has long been an area of fruitful research with obvious application to the mission-oriented goals of the Air Force. Phased array research is a newer discipline but the emergence of this technology, based upon the apparently simple combination of antenna elements has been a strong impetus for research on some extremely subtle and intriguing diffraction phenomena. This flurry of activity began in the mid-1960's with the discovery of anomalous scanning behavior in array radiators, and resulted in substantial advances in the theory and measurement of element interaction and its effects.

Most significant is that the stimulus came from a technological advance within a mature field of research, and that these new discoveries required yet more detailed research. At

ISBN 0-12-709650-7

present the study of array phenomena is itself reaching a state of maturity and many of the canonical problems are now understood, but again a vast number of important research areas are being uncovered because of the accelerating pace of innovation.

This paper reflects the thesis stated above, and expresses the belief that the study of phased arrays, far from the stage of merely tying down loose ends, is emerging as an even more fruitful, productive and increasingly relevant area for Air Force sponsored research. The paper is intended to highlight the technical developments and requirements that provide stimulus, and the most obvious areas requiring intensified research.

Electronically scanned (phased) arrays have found practical use in applications requiring a rapid change in antenna pattern as a function of time, and fixed beam array antennas are used to produce certain specialized beam patterns that cannot be adequately reproduced by lens or reflector geometries. The most important application to date has been to large ground based array radars for surveillance and air traffic control, and this application is primarily responsible for most of the development that has taken place. Other important applications have been to multi-function aircraft arrays and various smaller communications arrays, but progress in these fields is limited by the weight, complexity and primarily the cost of array systems.

The major factors influencing the future of array antennas are the weight of these past developments, the accelerating pace of technology, cost, and the burden of meeting new requirements imposed by systems that are currently being planned. As noted above, the most important factor to date has been the development of large ground based arrays like the FPS-85, Hapdar, Cobra Dane and others. These major efforts have stimulated research into array element coupling, space and corporate feeds and microwave circuit components like diode and ferrite phase shifters.

Future trends in array research may not be so closely aligned to the needs of ground based radar, but instead the more fruitful paths will originate from the requirements of a growing list of special purpose arrays; arrays designed for the single application that is their intended use. Many new system specifications require arrays with such unique characteristics that the only economical solution is to design the array tailored to the task at hand. Costs can be reduced by production methods but in certain cases they are reduced far more dramatically by choice of array type. In addition to cost, new array systems will be required to meet increasingly difficult performance specifications. Most important of these are the low sidelobe characteristics required for defense against anti-radiation missiles, and the null steering requirements of broadband anti-jam communication links. New system types place their own demands upon the antenna circuits, and the growth of satellite communications requirements has become a stimulus for both satellite and aircraft antenna technology. Similarly the rate of growth of microwave technology itself is a stimulus to array development. Examples will be cited later to show that the fact of an advancing technology with new transmission media and with solid state microwave transmitters or receivers available at each element has become an increasingly strong driving force for array research. Conventional array elements are not well suited to couple into new strip-line and microstrip transmission circuits, and thus a number of different elements have recently been developed and many more will soon be developed for application to scanning arrays. This fact, coupled with the radiating and reflecting properties of active and adaptive antenna circuits provide a collection of new and very difficult phased array analysis problems that will challenge the technology and chart the course of research for many years to come.

This survey attempts to deal with these historical forces and influences that affect the future of array research. Section II describes the basic analytical formulation for a typical array problem. The presentation is tutorial in style; scalar equations are used wherever possible in order to avoid the added complexity of vectorial solutions. In general there has been no attempt to survey all of the possible kinds of analytical solution to any one problem; the analysis is included because it aids in explaining some of the physical phenomena observed in phased array systems, and because it serves to illustrate the magnitude of the analytical problem for the case of finite arrays. Section III describes several new array geometries categorized as "Special Purpose Arrays"; these are typical responses to specific system problems that require arrays subject to external constraints. The special purpose arrays considered are conformal arrays, arrays for hemispherical coverage and null steering, and array techniques for limited sector coverage and multiple frequency arrays. Section IV describes certain aspects of new technology that will serve to force the development of arrays with novel kinds of strip line and microstrip elements. Radomes, polarizers and spatial filters are also described in this section; these components are undergoing an intense period of change and their design is becoming integral with the associated array or antenna design.

In addition to these areas there are many other topics that can be expected to affect the future of arrays and array research. Some of these which have not been discussed here are problems associated with antennas over the earth, the science of HF ground screen development, transient analysis of arrays and the impact and technology of the various active and adaptive array techniques. These were omitted because their proper consideration is beyond the scope of the paper.

These are but a few of the requirements, the technology and the trends. Subject to the author's personal biases and limited perspective, these describe the present state of the technology. Each of the contributing factors is discussed to present a cohesive exposition of this one view of the future of array antennas.

II. THE ARRAY AS A BOUNDARY VALUE PROBLEM

A. Introduction

An analytical study of phased array radiation follows the conventional approach from diffraction theory of obtaining solutions of Maxwell's equations for two spatial regions; the external region is free space and the internal region is the inside of the various transmission lines or waveguide exciting the radiating elements. The solution in the exterior region must satisfy the boundary conditions that apply on the surface supporting the array, and this gives rise to the major problems in the analysis of arrays conformal to specific structures like aircraft or spacecraft antennnas, or arrays mounted over the earth.

Most often the exterior region is considered to be unbounded or bounded by a half space; in these circumstances the Greens function is derived from combinations of retarded potential type terms.

The dyadic form of Green's symmetrical theorem gives the free space fields in terms of integrals over all currents, charges and aperture fields in the exterior space or on its boundary [1].

Although arrays are as commonly comprised of wire elements as aperture elements, this review will treat only the aperture case. The dual situation involving wire elements leads to field expressions derivable from vector potential integrals over the currents in the wires and their images, and the resulting boundary value problem arising at the array face is a series of integral equations on the surfaces of the wires. In these cases the interior field solution is usually idealized to the extent that the fields are replaced by a delta function voltage source as in the Halleen's equation [2]. More recent work has removed some of these assumptions about the idealized nature of the source and has considered the implications of the use of an approximate kernel for the Greens function for wires [3].

The free space field in the half space bounded by a perfectly conducting half plane with an array of apertures as shown in Fig. 1, can be written in terms of integrals over the aperture fields as shown below [1].

For $z \geq 0$

$$\overline{B}(\overline{r}) = j2\omega\mu\varepsilon \sum_m \int_{S_m} \underline{\Gamma}^{o}(\overline{r},\overline{r}') \cdot (\hat{z} \times \overline{E})\, dS'_m$$

$$\overline{E}(\overline{r}) = 2 \sum_m \int_{S_m} \nabla \overline{G}(\overline{r},\overline{r}') \times (\hat{z} \times E)\, dS'_m \tag{1}$$

where

$$\underline{\Gamma}^{o}(r,r') = (\underline{U} + \frac{1}{k_o^2} \underline{\nabla\nabla}) G(\overline{r},\overline{r}') \tag{2}$$

and

$$G(r,r') = \frac{e^{-jk_o |\overline{r} - \overline{r}'|}}{4\pi |\overline{r} - \overline{r}'|}$$

and

$$|\overline{r} - \overline{r}'| = \sqrt{(x - x')^2 + (y - y')^2 + z^2}$$

The index "m" corresponds to the "m'th" - apertures in the array, and $\hat{z} \times E$ is evaluated at the m'th aperture.

An $\exp(+j\omega t)$ time dependence is assumed and has been suppressed. Vectors are denoted by a bar above the expression and dyadics by a bar below. $\underline{U}$ is the unity dyad and $\underline{\Gamma}^{o}$ is the conventional free space dyadic Green's function. Equation (2) is used to express the radiation fields, and also as the basis of the electromagnetic boundary value problem at the junction of the fields determined from (2) satisfy the appropriate boundary conditions on the perfectly conducting plane at $Z = 0$ and assure continuity of the tangential $\overline{E}$ field at the apertures. One can obtain a set of integro-differential equations at each aperture by

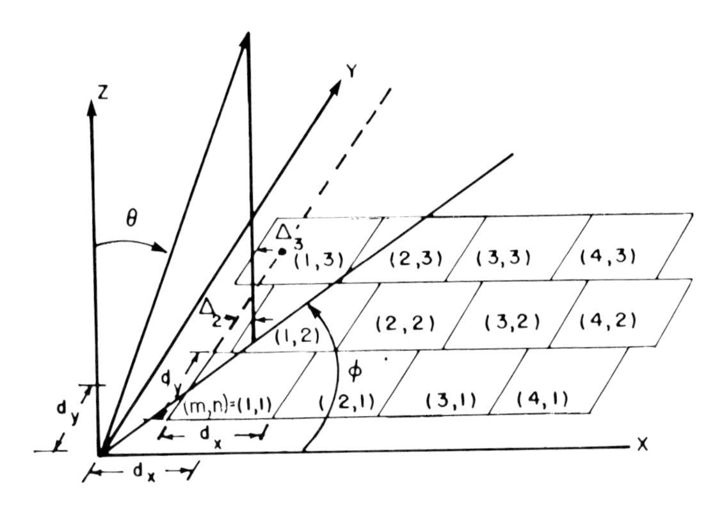

Fig. 1. Array coordinates.

expanding the fields in the feed waveguides in terms of *TE* and *TM* modes, using the internal $\bar{E}$ field as the tangential $(\hat{z} \times \bar{E})$ aperture field in each aperture and then equating the magnetic fields of the internal and external expansions across each aperture. This procedure is extremely cumbersome and has not been carried out in such generality except for several distinct canonical geometries.

An expression which is entirely equivalent to (2) is obtained by defining the magnetic Hertzian potential $\bar{\pi}_m(r)$ as:

$$\bar{\pi}_m(\bar{r}) = \underline{j2} \sum_m \int_{S_m} (\hat{z} \times \bar{E}) G(r,r') dS_m' \tag{3}$$

The corresponding fields are written

$$\bar{B}(\bar{r}) = \nabla(\nabla \cdot \bar{\pi}_m) + k_o^2 \bar{\pi}_m \tag{4}$$

$$\bar{E}(\bar{r}) = - j\omega\nabla \times \bar{\pi}_m .$$

For the case of rectangular waveguides exciting rectangular

apertures one can expand the waveguide fields in terms of magnetic potential functions by defining two scalar Hertzian potentials π'_{mx} and π'_{my} such that:

$$\bar{\pi}'_m = x\pi'_{mx} + y\pi'_{my} \text{ and } \pi'_{m_z} = 0 \,. \quad (5)$$

Equating tangential fields in the apertures leads to the following equations for the difference between internal and external Hertzian potential components.

$$\left(\frac{\partial^2}{\partial_x^{\;2}} + \frac{\partial^2}{\partial_y^{\;2}} + k_o^{\;2}\right)(\pi'_{mx} - \pi_{mx}) = 0$$
$$\left(\frac{\partial^2}{\partial_x^{\;2}} + \frac{\partial^2}{\partial_y^{\;2}} + k_o^{\;2}\right)(\pi'_{my} - \pi_{my}) = 0 \quad (6)$$

$$\frac{\partial}{\partial_x}(\pi'_{my} - \pi_{my}) = -\frac{\partial}{\partial}(\pi'_{mx} - \pi_{mx}) \quad (7)$$

These three integrodifferential equations, repeated at each aperture, define all of the radiation and inter-element coupling for the array of apertures. They are similar to those obtained for a number of classical diffraction problems [4], and clearly show the vector nature of the solution unless ∂/∂_x or ∂/∂_y are zero. Arrays scanned in a single plane and with translational invarience in the second plane can have scalar field solutions. In addition, it is often convenient and appropriate to neglect the cross-polarized component of radiation or coupling when that neglect can be shown to have no adverse effect upon the critical aspects of the solution [5,6].

B. Solution for an Infinite Array

The kernal of equations (2) or (3) involves a summation of retarded field integrals over the elements of an array. The special case of an equally spaced infinite array provides particularly simple form of kernal that has solutions with the form of Floquet's spatial harmonic series.

For a two dimensional array with dimensions shown in Fig. 1, the main beam of the array is scanned to an angle (θ_o, ϕ_o) by application of incident fields in each waveguide (p,q) having the form:

$$E_{inc} = E_o e^{-jk_o(u_o md_x + v_o ndy)} \quad (8)$$

where

$$u_o = \sin \theta_o \cos \phi_o$$
$$v_o = \sin \theta_0 \sin \phi_o \quad (9)$$
$$k_o = 2\pi/\lambda_o$$

and (u_o, v_o) are the direction cosines of the main beam position vector.

This periodic incident field results in the same periodicity in the aperture field and the accompanying simplification of the summations.

For the case of an array scanned in one plane, summations of the form

$$\int_{x'} \int_{y'} \sum_{m=-\infty}^{\infty} \frac{e^{-jk_o \sqrt{(x-x'_m)^2 + (y-y',^2 + z^2}}}{\sqrt{(x-x'_m)^2 + (y-y')^2 + z^2}} E(x'_m, y') \quad (10)$$

become, using

$$E(x_m, y) = E(x_o, y) e^{-jk_o u_o md_x}$$

and using

$$x'_p = x'_o + md_x ,$$

the above yields:

$$\int_{x'}\int_{y'} \frac{-j\pi}{d} E(x_o',y') \sum_{m=-\infty}^{\infty} e^{-jk_o u_m (x-x_o')} H_o^{(2)}(k_o r\sqrt{1-u_m^2}) \tag{11}$$

where

$$r = \sqrt{(y-y')^2 + z^2}$$

and

$$u_m = u_o + m/d_x$$

This form shows that the series is now summed over the spatial parameter u_m and has the characteristics of a spatial harmonic[7] series in this parameter. To complete the evaluation of equation (3), this equation must be integrated over the y' parameter, and the near fields thus assume a relatively complex form in general. In the special case of infinite slots in the y' dimension with $\gamma_y = 0$ the series takes on an extremely simple form even in the near field. After performing this integration, expression (11) becomes:

$$= \frac{j2\pi}{d_x} \sum_{m=-\infty}^{m=\infty} \frac{e^{-j(K_m|z| + k_o u_m x)}}{K_m} F(u_m)$$

where

$$K_m = k_o\sqrt{1 - u_m^2} \quad \text{and} \quad F(u_m) = \int_{x'} E(x')e^{jk_o u_m x'} dx' \tag{12}$$

This expression is now clearly the sum over a series of waves that propagate or decay outside of the array depending upon whether K_m is real or imaginary. The sum is called a grating lobe series and the spatial angles at which $\exp[K_m|z| + k_o u_m x]$ is unity are grating lobe angles. The field in space is thus represented as an infinite series of waves with excitation coefficients $F(u_m)$.

For an array scanned in both planes, the summations become:

$$\int_{x'}\int_{y'}\sum_{n=-\infty}^{\infty}\sum_{m=-\infty}^{\infty}\frac{e^{-jk_o\sqrt{(x-x'_m)^2+(y-y'_n)^2+z^2}}}{\sqrt{(x-x'_m)^2+(y-y'_n)^2+z^2}}E(x'_m,y'_n)$$

$$= -\frac{1}{d_x d_y}\sum_m\sum_n\frac{e^{-j(k_o u_m x + k_o v_n y + K_{mn}|z|)}}{K_{mn}}F(u_m,v_n) \qquad (13)$$

where here

$$K_{mn} = k_o\sqrt{1 - u_m^2 - v_n^2}$$

and

$$F(u,v) = \int_{y'}\int_{x'} E(x',y')e^{jk_o(ux' + vy')}$$

For an array scanned in one plane and under certain special circumstances, the array electromagnetic field can be scalar. Examples of such scalar problems are the E-plane scan of a parallel plane array with TEM incident modes, and H-plane scan of the array shown in Figure 2. This array geometry is a novel design and uses the properties of dielectric slab loaded waveguides to support efficient radiation at two frequencies that are separated by about an octave. Since $\gamma_1 = 0$ the array is equivalent to a parallel plane structure for H-plane scan. This equivalence is shown by removal of the horizontal metal separators at $y = \frac{dy}{2} \cdot (2n - 1)$.

The solution proceeds by expanding the waveguide fields in terms of an infinite series of waveguide modes ($LSE_{p,o}$) and using this field expansion in equation (6).

The interior potential function for a mode with transverse incident field distribution $e_p(x)$ is:

$$\pi_p = e^{-\gamma_p z}e_p(x) - \sum_{q=1}^{\infty}\Gamma_q e^{\gamma_q z}e_q(x) \qquad (14)$$

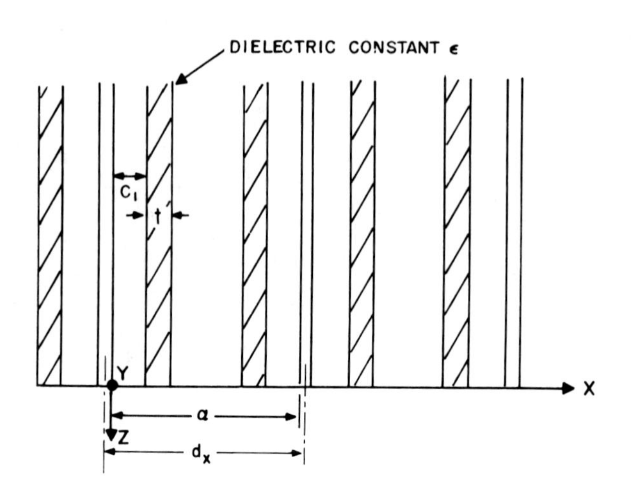

Fig. 2. Array Geometry - H-plane scan.

where γ_p and γ_q are the modal propagation constants for the slab loaded waveguide [8]. The waveguide eigenfunctions $e_p(x)$ are orthogonal [9] and are normalized so that

$$\int_{-a/2}^{a/2} e_p(x) e_q d_x = \delta_{pq} = \begin{cases} 1 & p = q \\ 0 & p \neq q \end{cases}$$

The coefficients Γ_q give the amplitude and phase of the waves reflected from the aperture face ($z = 0$) and include propagating and non-propagating modes. The aperture field for the incident mode P is (at $z = 0$)

$$E_{y_p} = jw \frac{\partial \pi_p}{\partial z} = jw[\gamma_p e_p(x) + \sum_{q=1}^{\infty} \Gamma_q \gamma_q e_q(x)] \tag{15}$$

Within the waveguides the magnetic field is given by

$$B = \frac{-\partial^2 \pi}{\partial x^2} \tag{16}$$

and in the exterior region it is obtained from equation (4) as:

$$Bx_p = -\frac{1}{d_x}[\gamma_p \int_{-a/2}^{a/2} e_p(x') \sum_{m=-\infty}^{\infty} e^{-j\beta_m(x-x')} \xi m dx'$$
$$+ \sum_{q=1}^{\infty} \Gamma_q \gamma_q \int_{-a/2}^{a/2} e_q(x') \sum_{m=-\infty}^{\infty} e^{-j\beta_m(x-x')} \xi_m dx'] \quad (17)$$

where

$$\xi_m = \begin{cases} j\sqrt{k_o^2 - \beta_m^2} & \text{for } \beta^2 > k_o^2 \\ \sqrt{\beta_m^2 - k_o^2} & \text{for } \beta^2 < k_o^2 \end{cases}$$

and

$$\beta_m = k_o[u_o + m/d_x]$$

Equating these magnetic field expressions at $z = 0$, truncating the series at $q = Q$, multiplying by $e_\ell(x')$, using orthogonality and defining the integral

$$\text{Int}_q(\beta_m) = \int_{-a/2}^{a/2} e_q(x')\, e^{+j\beta_m x'} dx' \quad (18)$$

one obtains a set of Q equations for the Q coefficients of Γ_q for the incident p' mode excitation:

$$[\delta_{p\ell}\gamma_p^2 - \frac{1}{d_x}\gamma_p S_{p\ell}] = \sum_{q=1}^{Q} \Gamma_q [\gamma_q^2 \delta_{q\ell} + \frac{1}{d_x}\gamma_q S_{q\ell}] \quad (19)$$

$$\text{for } \ell = 1 \text{ to } Q$$

and where

$$S_{q\ell} = \sum_{m=-M}^{M} \xi_m \text{Int}_q(-\beta_m)\, \text{Int}_\ell(\beta_m)$$

Solution of the above matrix equation gives the waveguide field distribution at each aperture, and includes all of the mutual coupling effects for the infinite array. The particular array studied here uses two incident modes ($p = 1,2$) at the high frequency, and so the set of equations above must be solved twice to obtain a solution for the combined two mode excitation. The series over m is truncated at $\pm M$ (usually between 40 and several hundred terms) as required for convergence.

C. Analysis for a Finite Array

Solutions like the above have been extremely useful for the analysis and design of large arrays such as those used for ground based radar. Smaller arrays with ten or fewer elements in each plane have behavior dominated by edge effects and for these the infinite array analysis has little meaning. There have been analytical treatments [10,11] of semi-infinite arrays that give insight into the phenomena of edge effects in large arrays without including higher order modal effects. The vast majority of finite array studies have been performed using a scattering matrix that includes only a single waveguide mode; a procedure that can be highly inaccurate when the array is operated at a frequency or scan angle near which an array resonance can occur. These resonances or "blind-spots" have been the subject of substantial controversy over the past decade and will be described in more detail later.

Equation (19) can be re-written as an infinite set of simultaneous equations and then truncated to yield a solution of any desired accuracy. This is accomplished by expanding the field in each mth waveguide in terms of a sum over all incident and reflected modes. In general this involves both components of the vector solution, but again it is more convenient for the purposes of illustrations to restrict the analysis to a finite array of "M" of the infinite columns of Fig. 2 for H-plane scan.

The potential function for the mth waveguide is written:

$$\Pi(m) = a_m e^{-\gamma_1 z} e_1(x') - \sum_{q=1}^{\infty} b_m^q e^{-\gamma_q z} e_q(x') \tag{20}$$

Here it is assumed that only the single dominant mode is incident in each waveguide, but that all modes are reflected. The notation b_m^q is the coefficient of the q'th reflected mode in waveguide m. After obtaining the aperture electric field

$$E(m) = j\omega[a_m \gamma_1 e_1(x') e^{-\gamma_1 z} + \sum_{q=1}^{\infty} b_m^q \gamma_q e^{-\gamma_q z} e_q(x')] \tag{21}$$

and inserting this field in the integro-differential equation (7) into the equivalent equation obtained by equating the tangential magnetic fields at both sides of the aperture as in equation (19). The resulting equation, multiplied by the sequence of $e_1(x)$ for $\ell = 1$ to Q and integrated over x yields a series of "Q" algebraic equations at each aperture and can be written on the form of a multi-modal scattering matrix. The ℓ'th equation at the mth waveguide is:

$$\left.\begin{aligned} &\sum_{q=1}^{Q} [b_m^q \gamma_q^2 \delta_{q,\ell} - \sum_{n=1}^{N} b_n^q \gamma_q P_{mn}^{q\ell}] \\ &= [a_m \delta_{q,\ell} \gamma_1^2 + \sum_{n=1}^{N} a_n \gamma_1 P_{mn}^{1\ell}] \end{aligned}\right\} \quad \ell = 1, Q \tag{22}$$

where

$$P_{mn}^{q\ell} = \frac{1}{2\pi} \int dx\, e_\ell(x) K_{mn}^{q\ell}(x)$$

and

$$K_{mn}^{q\ell}(x) = (\frac{\partial^2}{\partial_x^2} + k_o^2) \int_{x'} e_q(x') \int_{-\infty}^{\infty} \frac{e^{-jk_o|r-r'|}}{|r-r'|} dx' dy'$$

$$= -j\pi \quad (\frac{\partial^2}{\partial^2} + k_o^2) \int_{x'} e_q(x') H_o^{(2)}(k_o|x_m - x_n'|) d'x$$

for

$$|r-r'| = \sqrt{(x_m - x_n')^2 + (y')^2}$$

N such sets of equations are required, one set at each aperture; thus leading to a set of $N \times Q$ equations to be solved to complete the array solution.

The numerical evaluation of solutions like the one above are indeed formidable and the solution is most often approximated using only one or two terms of the series.

Although the most common analytical practice is to assume a set of incident fields $\{a\}$ and solve the set (22) for the reflected signals $\{b\}^q$ for all modes q, one could obviously assume a sequence of independent incident modes and solve the set for each incident a_m. This solution is the scattering matrix for each mode q of the array.

$$b_m^q = \sum_{n=1}^{N} S_{mn}^q a_n \text{ or } b^q = S^q a \tag{23}$$

Written in this form there are Q such scattering matrices required to describe the Q waveguide modes reflected from the apertures (for a single mode incident on each).

It is important to observe that the whole set of higher order modes enters into the equation (23), and so the scattering coefficients S_{mn}^q include the mutual coupling of these higher order modes.

Arrays with more than one incident mode (like that of Fig. 2) can be analyzed by repeating the above procedure for the several incident modes and superimposing the solutions. Although this formulation gives a complete solution of the multi-element array radiation and inter-element problem, the amount and complexity of the required numerical analysis often makes such a solution impractical. Suitable approximations include using only one or several modes in each waveguide, neglecting cross-polazied interactions, utilizing asymptotic approximations of the scattering coefficients for the widely spaced elements and neglecting the interaction between the higher order modes in the evaluation of scattering coefficients. The implications of several of these approximations will be discussed in subsequent sections.

D. Array Radiation and the Concept of an Element Pattern

Equation (4) gives the complete radiated field for an array of apertures in a perfectly conducting ground plane. Determination of the tangential E fields in these apertures is achieved by solving the boundary value problem at the waveguide aperture interface by the methods outlined in the previous section, or by other techniques to be mentioned later.

The far field approximation to equation (2) is obtained by using

$$|\bar{r}-\bar{r}'| \doteq R_o - \bar{r}' \cdot \hat{\rho} \tag{24}$$

where R_o is measured from the coordinate origin in the aperture to the given point in space at R_o, θ, ϕ and

$$\bar{r}' = \hat{x}x' + \hat{y}y'$$

and

$$\hat{\rho} = \hat{x}u + \hat{y}v + \hat{z}\cos\theta \tag{25}$$

Using this approximation it is customary to write

$$G(\bar{r},\bar{r}') = \frac{e^{-jk_oR_o}}{4\pi R_o} e^{jk_o(\bar{r}'\cdot\hat{\rho})}$$

Evaluation of equation (2) for apertures in the plane $z = 0$ yields:

$$E(\bar{r}) = \frac{jk_o}{2\pi} \frac{e^{-jk_oR_o}}{R_o} \sum_m \int_{S_m} dS_m [\cos\theta\, \bar{E}_T(x'_m, y'_m) - \hat{z}\hat{\rho}\cdot E_T(x'_m, y'_m)]$$

where $\bar{E}_T$ is the tangential field in the aperture.

This relationship is also given in the text by Amitay et al 12 .

The tangential field $\bar{E}_T$ is a two component vector in general, but for the array of Fig. 2 and (approximately) for the case of thin slots, the aperture field can be described by a single component. For tutorial purposes the remainder of this description will treat the scalar case in which the cross polarized radiation is neglected or identically zero and the waveguide polarization is in the $\hat{y}$ direction. In this case the aperture fields are written using equation (4). The field in the m'th waveguide (at = 0) is:

$$E_{y_m} = j\omega[a_m \gamma_1 e_1(x') + \sum_{q=1}^{Q} b_m^q \gamma_q e_q(x')] \tag{26}$$

or, using the scattering matrix representation of equation (23)

$$E_{y_m} = j\omega[a_m \gamma_1 e_1(x') + \sum_{n=1}^{M} a_n \sum_{q=1}^{Q} e_q(x') S_{mn}^q \gamma_q] \tag{27}$$

Unlike the infinite array, equation (27) shows that a finite array with periodic incident fields has non-periodic aperture fields because of the lack of symmetry in the element scattering matrices.

Defining the aperture integrals

$$I_q(u,v) = \int_{-a/2}^{a/2} dx' \int_{-b/2}^{b/2} dy' e_q(x',y') e^{jk_o(x'u + y'v)} \tag{28}$$

one obtains (neglecting constants) the following expression for the array far field

$$F(\theta,\phi) = \sum_{m=1}^{M} e^{+jk_o(ux_m + vy_m)} \cdot \{a_m \gamma_1 I_1(u,v) + \sum_{n=1}^{M} a_n \sum_{q=1}^{Q} S_{mn}^q \gamma_q I_q(u,v)\} \tag{29}$$

where x_m and y_m are the position coordinates for the m'th waveguide.

This expression can be re-written in the following form

$$F(\theta,\phi) = \sum_{m=1}^{M} e^{jk_o(ux_m + vy_m)} j\omega a_m[\gamma_1 I_1(u,v) + \sum_{n=1}^{M} \sum_{q=1}^{Q} S_{mn}\gamma_q I_q(u,v)]$$

$$= \sum_{m=1}^{M} e^{jk_o(ux_m + vy_m)} a_m f_m(\theta,\phi) \qquad (30)$$

which makes it evident that the far field is a superposition of fields due to each element located at position x_m, y_m, excited by a coefficient a_m and having a spatial variation $f_m(\theta,\phi)$. For a large planar array forming a single pencil beam, one can show that the main beam gain is related to the square of the magnitude of this sum times cos θ except for angles very near to end-fire. Thus, for a large, two dimensional array

$$f_m(\theta,\phi) \sqrt{\cos\theta}$$

is the element pattern of the m'th element. The cos θ term is not present for a column array.

Like the aperture distribution, this element pattern differs for each element of a finite array. Furthermore, the element pattern has in it all of the effects of mutual coupling and so can be an extremely complex function of the space coordinates (θ,ϕ). Proper element pattern control is the prime requisite of array design, and the formidable task of element pattern evaluation is not a choice to be taken lightly. Unfortunately the history of phased array development reveals the closeted skeletons of arrays that were build using single mode approximations for mutual coupling. These and other details will be described in subsequent sections, but it is important to note here that the pattern $f_m\sqrt{\cos\theta}$ of the m'th element is exactly what one measures in the far field when only that element is excited. Because of reciprocity it is also the signal received at that element from a distant transmitter, and so its measured value includes, for any given array, all of the coupling and higher order modal effects that will be observable when the array is excited as a whole. Element pattern measurement is thus an extremely powerful tool of array design, because it is possible to record this single mode parameter and still account for all of the subleties that occur at the array face.

E. Historical Perspective and the Blindness Phenomena

The previous sections have shown one method of analyzing waveguide arrays including the mutual coupling between all array elements. Waveguide elements were chosen for these examples because they have been the subject of extensive research over the past ten years and because they conveniently illustrate many of the phenomena that will be described later. Early studies of mutual coupling were performed mainly for arrays of dipole elements [13] with assumed sinusoidal current distributions and later [14, 15] for current distributions that contained several higher order terms to approximate the exact distribution. These analyses were based upon various forms of Hallen's integral equation and the discovery that higher order modes were important came about mainly through the diligence of researchers working in the field. These theoretical efforts were accompanied by extensive experimental programs and the use of higher order current approximations was motivated primarily by a concern that any analytical solution for current and charge distributions be adequate to allow an accurate description of the near-field. Despite the fact that these earlier dipole array studies were performed many years ago, the dipole has remained a subject of continued interest. Recent analytical studies have been based primarily on moment-method approaches [16,17] which are applicable to a wide variety of wire antenna shapes and orientations, and for which there are now a number of available computer programs of very great generality.

Air Force sponsorship in this area has been a factor of major importance. Starting with the basic studies of Carter [13] and King [14,15] and continuing to the present day, the Air Force 6.1 effort has funded many of the major analytical developments in dipole antenna arrays.

The recent concern with waveguide arrays reflects the fact that by the mid 1950's the analytical background for this technology lagged far behind that of dipole arrays. Customary waveguide array solutions dealt almost exclusively with single mode approximations to the waveguide field, but did properly account for the full spatial harmonic series (grating lobe series) in the free space half space. Some earlier studies of single radiating waveguides used stationary solutions of the aperture integral equation in order to obtain variational formulas for input impedance [18,19], but until the mid-60's there were no published multi-model solutions of even this basic radiating geometry.

If little effort had been devoted to the single radiatior problem, even less has been done to describe the couling between waveguides. One of the first studies of this sort was performed by G.W. Wheeler [20] who assumed the coupled radiators were in the far field of one another. In 1956 Levis [21] derived general equations for a variational formulation to obtain the coupling between a number of generally cylindrical waveguides radiating through a common ground plane. He applied the method to a set of coupled annular slots. Galejs [22] applied a stationary formulation due to richmond [23] to solve the problem of two parallel slots in a ground plane, with both slots backed by waveguides. His method yielded usable and convenient formulas; but includes the implicit assumption that the tangential magnetic field at the coupled waveguide aperture is the same as the magnetic field which would be present on the ground plane if the coupled aperture were not present. In this manner Galejs avoided the problem of solving an integral equation.

Other researches that evolved from the point of view of antenna element coupling were the study by Lyon [24] et. al., to determine the power coupling between various structures including arbitrarily oriented open ended waveguide and two studies by Mailloux [25,26] that dealt with the multiple mode solution of collinear radiating waveguides, and the induction of cross polarized fields in mutually coupled waveguides, and the induction of cross polarized fields in mutually coupled waveguides with arbitrary orientation. This last paper described some approximate procedures to account for coupling in large arrays where the numerical evaluation of all the higher order terms would otherwise become unwieldly. The use of such inter-element coupling approaches to array theory has not been popular until recnetly [27,28] because the coupling integrals are two dimensional with singular kernals and the resulting matrices are often so large that it seemed unreasonable to consider including higher order effects unless there was an extremely good reason to do so. In recent years this approach has gained some favor because of the availability of large computers and because of an increased awareness of the need for accurate array calculations.

The stimulus that intensified research into array mutual coupling phenomena was called array "blindness", and went undiscovered by university or government sponsored research programs. Its discovery occurred when several array systems exhibited poor scanning performance and so to these investigators "array blindness" was not an interesting phenomenon but a plague; once uncovered it was found in numerous systems and proposed systems. Blindness is evidenced by a null well within the normal scan sector of an array. It is maily a problem for large arrays and so was not found in tests of arrays that consisted of only a few elements in each plane. Before describing and commenting further on the history of this important development, I would stress that this was an area that should have been uncovered by researchers before it became a crisis to be discovered by system manufacturers.

Given the cost and importance of such systems, there was clearly not an adequate concern for fundamental studies at a time when they could have averted the serious problems that followed.

The phenomenon of array blindness became a factor of extreme confusion for a number of years. Examples of this confusion abound throughout the early literature where, for example, one auther stresses the importance of including waveguide higher-order modes in any analysis for predicting blind spots, and another author uses a single-mode theory for a different structure to accurately predict an occurrence.

The reasons for this confusion, as explained by Knittel et. al. [29], is that, depending upon the array structure, there are two basic types of cancellation resonances: those that occur external to and those that occur within the array waveguide apertures.

The waveguide higher-order modes play a dominant role for the internal-type resonance, but are relatively unimportant for an external resonance. This is because the external resonance occurs only for arrays that have a structure of some kind beyond the array face, and the resonance is caused by the interaction between the radiating mode and a higher-order external mode supported by this structure. An internal resonance can be viewed as a cancellation effect between the dominant and a higher-order waveguide mode radiation. An awareness of this distinction is useful for categorizing the various reports of array blind spots.

The first convincing demonstration of the existence of an array null was obtained experimentally by Lechtreck [30] using an array of circularly polarized coaxial horns with separate hemispherically shaped radomes for each element. The null occurred for the electric field perpendicular to the ground plane, and was called an external resonance by Oliner [31].

Farrell and Kuhn [32,33] presented the first theoretical evidence of internal resonance nulls in all planes of a triangular grid array (Figs. 3, 4) and in the E-plane of a rectangular

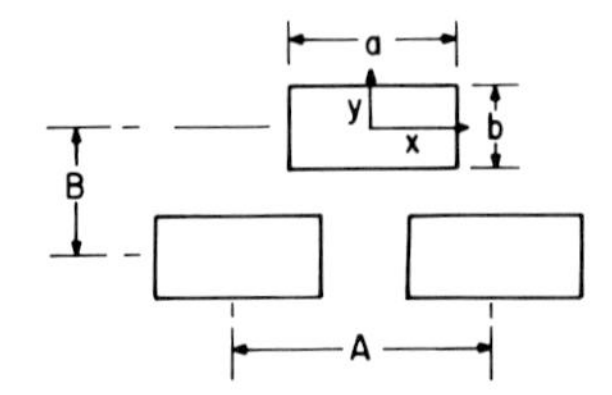

TRIANGULAR GRID ARRAY

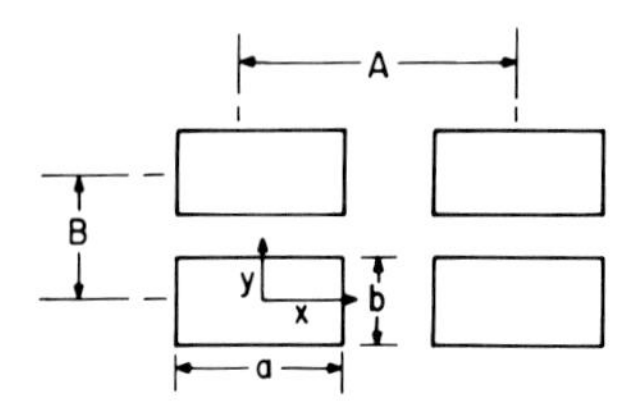

RECTANGULAR GRID ARRAY

Fig. 3. Triangular and rectangular grid lattices.

grid array. They also presented experimental verification of the *E*-plane rectangular grid null, but they were able to verify the existence of nulls only in the *H*-plane and inter-cardinal planes of the triangular grid array.

Amitay [34] and Galindo analyzed circular waveguide phased arrays in rectangular grid orientations and observed that incomplete nulls occur for intercardinal planes of scan.

DuFort [35,36] found nulls for a TE mode parallel plate array and a triangular grid array of rectangular waveguies on an *H*-plane corrugated surface, and Mailloux [37] found blind spots for the *E*-plane scan of an array of TEM mode parallel plane waveguides with conducting fences between adjacent radiators. To the extent that these effects occur because of the external structure, they are external resonances.

External resonances associated with the use of dielectric layers were observed experimentally by Bates [38] and Byron and Frank [39], experimentally in a phased array waveguide simulator

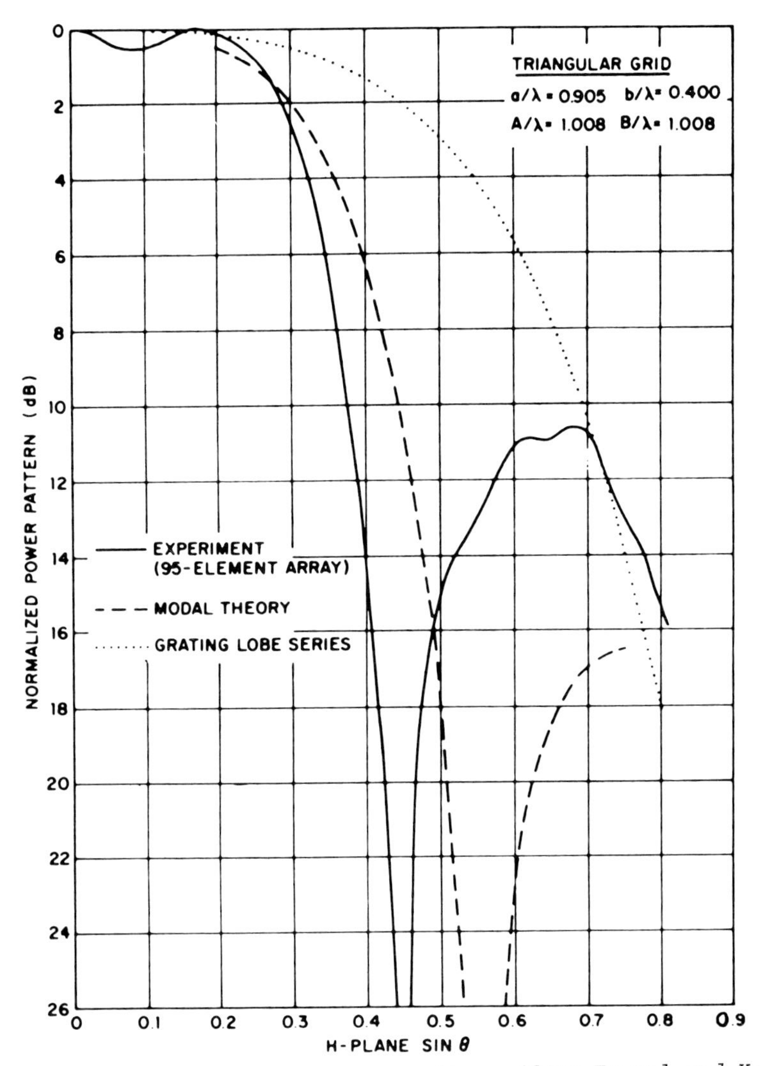

Fig. 4. Array element power pattern (from Farrel and Kuhn [49]) showing array blindness.

by Hannan [40], Byron and Frank [39], and Gregorwich et. al. [41]; and predicted theoretically by Frazita [42], Knittel et. al. [29], and Parad [43] using one-mode approximations (the grating lobe series), and by Galindo and Wu [44], Wu and Galindo [45], and Borgiotti [46] using higher-order modal analyses.

In addition to the growing list of blind-spot occurrences,

the nature of the phenomenon has become relatively well understood, and some techniques for avoiding or eliminating the difficulties are available.

The initial impression of Allen [47] and Lechtreck [30,48], that the null was due to coupling into a surface wave, gave an incomplete picture because the array elements are not reactively terminated and the elements are placed more than one-half wavelength apart, thus eliminating any conventional surface wave propagation. Farrell and Kuhn [32,33] performed the first rigorous analysis of an array with a blind spot, and they were the first to observe that certain waveguide higher-order modes play a dominant role in achieving the cancellation necessary for a null. Diamond [49] and later Borgiotti [46] confirmed all of these findings for waveguide arrays.

Oliner and Malech [50] suggested what is now generally accepted as true, that the blind spot is associated with the normal mode solution of an equivalent reactively loaded passive array, and that the condition for a complete null on the real array occurs when the elements are phased to satisfy the boundary conditions for the equivalent passive array. Knittel [28] et. al. developed this theory and showed that in the vicinity of the null the solution corresponds to a leaky wave of the passive structure, but that a surface-wave-like field exists immediately at the null. This is consistent with the results of an analysis made earlier by Wu and Galindo [45], who demonstrated that the only radiating (fast) wave of the periodic structure spatial harmonic spectrum is identically zero at a null, and that, for this reason, a structure with a period greater than one-half wavelength can have a normal mode. Along with these contributions to the understanding of the physics of a phased array null, a number of authors showed that both the waveguide aperture and lattice dimensions are critical in determining the likelihoood of a blind spot [51,52,53,54]. Figures 5 and 6 illustrate the use of a graphical technique used by Knittel [55] bo reveal a direct relation

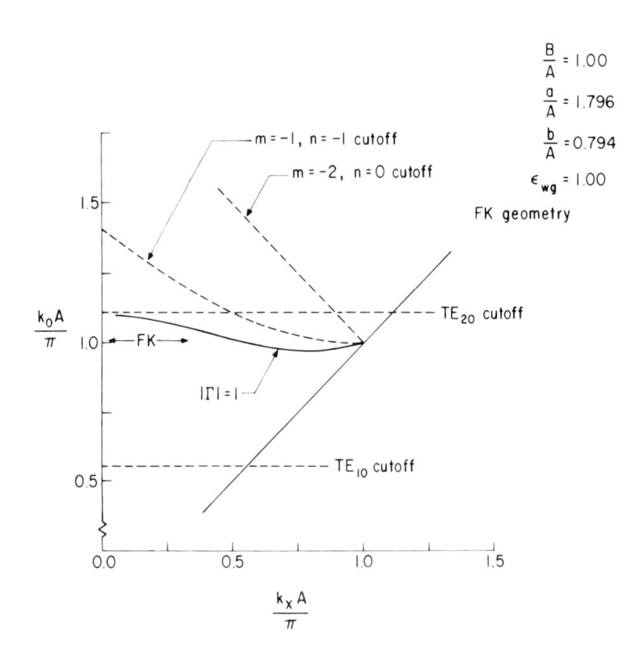

Fig. 5. K-β Diagram showing null locus ($^a/_A = 1.796$), courtesy Dr. George Knittel.

between the blindness effect and the cutoff conditions of the next-higher waveguide mode and lattice mode (grating lobe). Fig. 5 shows the locus on a *K*-β diagram of the blind spot for the array studies by Farrell and Kuhn [32] (denoted *F-K* on the figure). It is significant that the curve begins at the TE_{20} cutoff for a null at bradside and ends at the intersection of the $m = -1$, $n = -1$, $m = -2$, $n = 0$ higher order grating lobe cutoff loci at maximum scan. The curve never crosses any of these mode cutoff loci, because crossing the TE_{20} cutoff would allow energy to leak back into the waveguides, and crossing the grating lobe cutoff line would allow energy to radiate by means of a grating lobe. In neither case could the passive equivalent array sustain an unattenuated normal mode (unless the odd mode too were reactively terminated). Fig. 6 shows that if the waveguide size is reduced

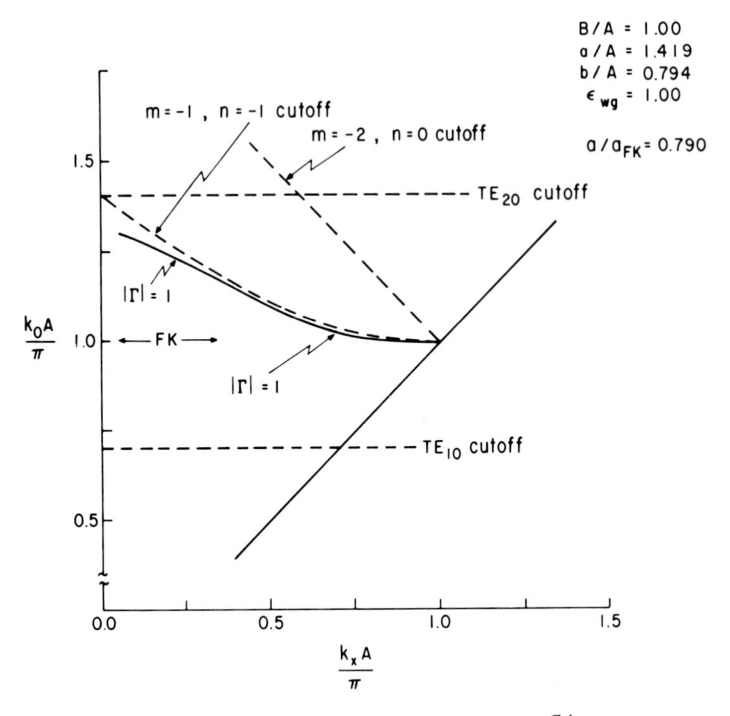

Fig. 6. K-β diagram showing null locus ($^{a}/A = 1.419$), courtesy Dr. George Knittel.

and no other dimensions are changed, the TE_{20} cutoff become unimportant and the null curve is nearly asymptotic to the grating lobe locus.

These two figures were included to demonstrate the power of this graphical technique for predicting the onset of blindness difficulties. In all other cases shown by Knittel the blindness locus remained nearly asymptotic to the waveguide or grating lobe cutoff loci, whichever occurred at lower frequency. The implication for design is obviously that the null can be avoided by choosing dimensions sufficiently smaller than those for the cutoff conditions.

Certain exceptions to the above conditions can occur; for example it is possible [56] to have blindness occurring after the onset of a grating lobe if both the main beam and the grating lobe lie in the same element pattern null, but the insight provided by the graphical approach remains the best available guide for design.

Recently a number of design techniques have been proposed [57] that make use of the data uncovered by these earlier studies and the blindness phenomenon is now considered much less threatening as long as the basic limitations of grating and element size are respected. Studies of waveguide array interaction have become less fashionable and now very few basic efforts are being conducted along these lines.

Almost all of the studies described above used analytical techniques valid for infinite arrays. Recently there have also been a number of studies that have included the effects of edges in semi-infinite arrays [11] and finite arrays [10,27,28, 58]. These were conducted by assuming single mode aperture fields.

These and other studies too numerous to mention, have brought planar waveguide array theory to an advanced state of development that parallels the present state of dipole array theory. There is a peril in assuming that these works mark a reasonable end to analytical studies in waveguide arrays. There remain a multiplicity of problems relating to waveguides conformal to various structures, to wideband and multi-frequency waveguides, and to the synthesis of low sidelobe distributions for finite arrays including the mutually coupled terms. These remaining problems are not simple variations of available solutions; they are as varied and general and as fit subjects for research as any other problems in electromagnetic diffraction, and they are more important than most.

Before leaving this generalized history of progress on array boundary value problems, there are several other issues which should be raised. First, the early solutions of the dipole antenna were performed by expanding the dipole current using several judiciously selected distributions [15], and then forcing the integral equation to be satisfied at the appropriate finite number of points along the dipole. This method is now called "point matching", and has also been used to solve waveguide aperture problems, where the chosen aperture distributions are the waveguide modal fields. More general forms of the method of moments [59] have since achieved substantial success in dealing with wire antenna problems and these have been adapted to aperture problems as well [60.61].

Finally, in concluding this section on analysis I should point out that most of the work in the last ten years has dealt with dipoles and waveguides, and so these subjects were highlighted here. We appear to be entering an era of much more generalized radiators, fed by stripline and microstrip and offering a vast, indeed staggering list of boundary value problems that will have crucial impact upon a number of military systems. Some of the beginnings of this technology will be described in succeeding sections of the paper, but as was the case for the waveguide work described here, the technology will be the forcing function and the research efforts will be highly directed toward specific problems. Present research funding is not adequate to uncover all of the anomalous behavior with all of the geometries and so the research in this important area will continue to always be relevant, and in many cases will be performed in a state of crisis.

III. SPECIAL PURPOSE ARRAYS

A. Conformal Arrays and Arrays for Hemispherical Coverage

Aerodynamic requirements for spacecraft and high performance aircraft have stimulated an increasing concern for the design of low profile and conformal antennas. The technological problems of these applications differ, and the technology of conformal arrays is really several technologies. Aircraft fuselage mounted arrays, whether conformal or planar, are expected to provide nearly hemispherical scanning. Spacecraft arrays are sometimes wrapped entirely around the vehicle, and the major design requirement becomes the study of a commutating matrix for steering the beam [62-65] . Arrays on cones have many special problems [66-68]; their radiated polarization is strongly angle dependent, there is little room for the array feed near the tip, and finally, their steering control is necessarily very complex. The specialized problems of an array on a concave surface are discussed in a paper by Tsandoulas and Willwerth [69].

Common to all of these structures is the underlying fact that they are mounted on non-planar surfaces, and this alters their radiation and mutual interaction. Analytical treatments have progressed to the rigorous solutions of coupling in infinite arrays or slits on cylinders [70,71] and on conical surfaces [72]. Simpler formulations have been developed using extensions of the geometrical theory of diffraction 73-75 , and with these it is now possible to perform analytical studies of finite arrays on generalized conformal structures. Figures 7 and 8 show results obtained by Steyskal [75] for the reflection coefficient of the center element in an array of 156 dielectric loaded circular waveguides mounted on a cylinder of 11.6λ diameter for the two principals polarizations. Analytical results of this sort can be applied for cylinders with radii of 2λ or greater.

Studies of arrays on cylinders and designed for nearly hemispherical scan coverage [71,76] have emphasized the difficulty in using conventional array approaches for such wide angle scan. By matching the array near the horizon (say 80° from the zenith), the array can be made to have gain variation of only about 6dB over the hemisphere but the array matched at this angle is mismatched at other scan angles, and can suffer a gain reduction of up to 4dB at broadside [76]. Recent studies sponsored by RADC have demonstrated coupling into a surface-wave mode or operation for near-endfire radiation. These efforts [77,78] have included the use of dielectric structures over or in the vicinity of the array, and have shown that these means also improve gain

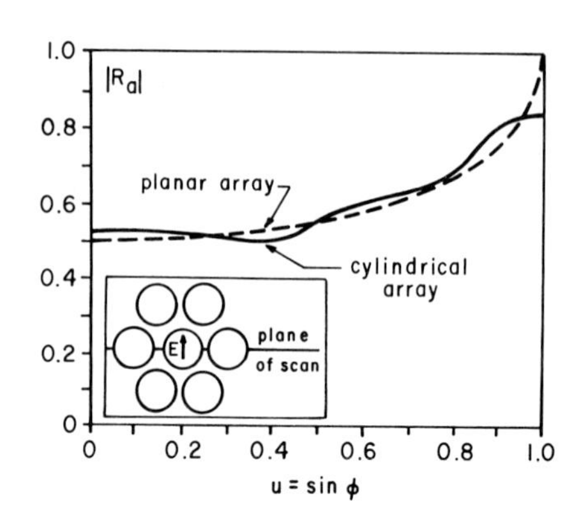

Fig. 7. Conformal array active reflection coefficient H-plane scan.

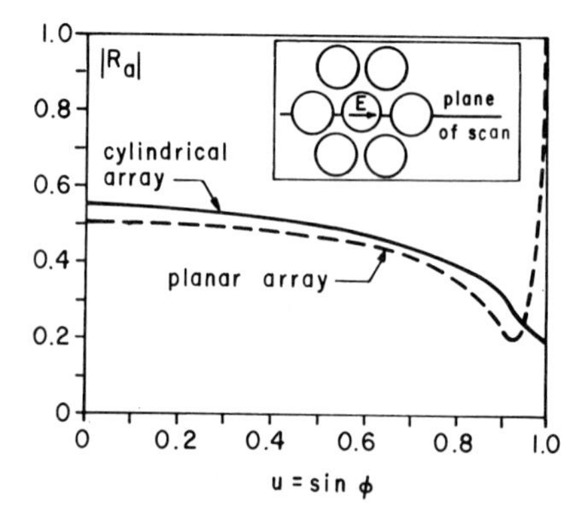

Fig. 8. Conformal array active reflection coefficient E-plane scan.

coverage within the hemisphere so that the envelope of peak radiation gain is always within about 6dB of the maximum radiation over a narrow frequency range. Computations [78] have shown that the coverage obtainable from an array with 23dB nominal gain presents maximum oscillations of 8dB over a 10-percent bandwidth. This data was obtained for an array covered with a layer of $C_r = 4$ material 0.075λ thick extending over and beyond the array. Such wide angle scanning does not seem feasible at present for arrays with 30dB gain.

Inhouse studies at RADC/ET have used the array in a conventional manner except at endfire where coverage is provided by short circuiting the array elements to form a corrugated surface that can support a surface wave for endfire radiation. This technique can provide highly efficient radiation over a hemisphere for one plane of scan, but is also gain limited at about 20dB for a square array.

Figure 9 shows an array of sixty four waveguide elements excited by an eight element feed. Although not shown in the figure, the array is also excited by a sixty-four way power divider and 8-phase shifters to form a beam scanned in the elevation plane. In practice, the waveguides would be short circuited by diodes or mechanical shorting switches to form the corrugated surface for endfire radiation, but the experiment was conducted using fixed short circuits.

The groundplane, partially shown in Fig. 9, measured six feet wide and had a four foot curved surface with 84 inch radius extending in front of the antenna structure.

Figure 10 shows the measured array gain at 9.5GHz for a number of beams within the sector including a beam scanned to the horizon and one formed by the excited corrugated structure. A cosine envelope distribution is also included for reference. The data show that the surface wave beam provides approximately 6dB gain increase at the horizon as compared with the scanned endfire beam, and that in fact the achievable gain at the horizon is 17dB; only 4dB below the maximum, and the peak at 80° is nearly equal to the broadside gain. The minimum of the pattern gain envelope occurs at about 69°, and shows a dip down to approximately 15dB which is within about 1dB of the array projection factor (cos θ).

These studies are but the beginning of the necessary research to develop flush mounted aircraft antennas that can be scanned over the entire hemisphere. This research is crucial because of

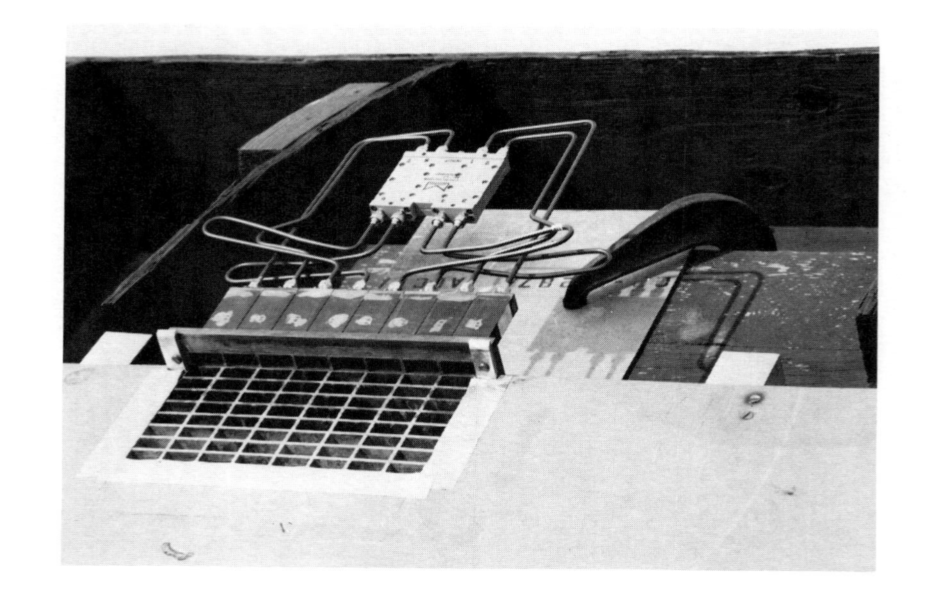

Fig. 9. Waveguide array used in hemispherical scan experiments.

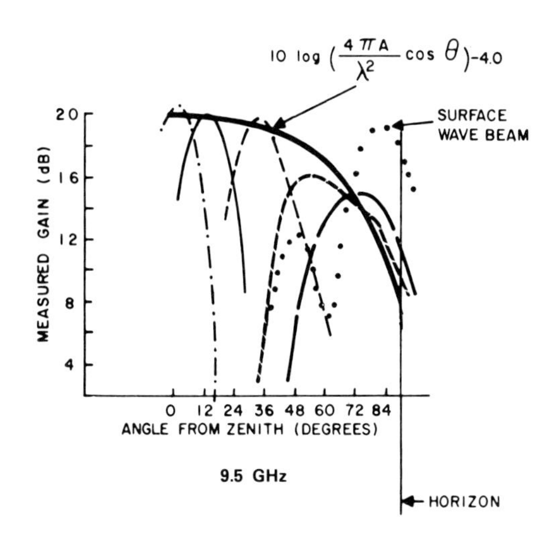

Fig. 10. Scan data for hemispherical scan array at 9.5GHz.

the potential for substantially reduced antenna size and lower cost. Its importance is understood by realizing that there are a great number of aircraft intended to have SHF communications links by the mid-1980's, and present antenna technology requires over-design by nearly 10dB. It is difficult to overestimate the importance of research in this extremely difficult technical area.

For a number of applications a flat array offers no advantage over a cylindrical or spherical array. In such situations it is better to avoid the natural disadvantage of the flat array and use the vertical projection of a structure with curved front face to achieve some increased gain at the horizon. This is done to an extent in some of the cylindrical array studies but the Dome [79,80] antenna capitalizes upon this projection in a way that no other system does. This basic antenna, shown in Fig. 11, uses a passive, spherical lens to extend the scan coverage of a conventional planar array to hemispheric coverage or greater. Each dome module consists of a collector element, a fixed delay and a radiator element. The dome assembly is radiated by the feed array,

Fig. 11. The Dome Antenna: a technique for hemispherical scan.

a conventional electronically scanned space fed lens with an F/D of 0.75. The array generates a non-linear pahse front to steer the dome from zenith to and somewhat below the horizon. Although some sacrifice in array efficiency is traded for wide angle scanning, the dome scan characteristics can be tailored to optimize the radiation over desired sub-sectors. Examples quoted in the literature show experimental results for a dome with phase shift modules chosen to form two different gain scan contours; one with a peak at 60° from zenith, and one at 120° from zenith. Such flexibility of selection allows this technique to become a reasonable choice for a number of system applications; it will continue to be of importance as a subject for research and development in order to improve efficiency, bandwidth and sidelobe levels and so achieve optimized designs for a number of specialized requirements.

B. Low Sidelobe and Null Steered Arrays

Among the areas of prime interest in radar and special purpose arrays are the requirements of providing low sidelobe and null steered radiation patterns. These two concerns have grown, because of the military threats presented by ARM (Anti-Radiation Missiles) and the increased use of jammers. Obviously the solution is just to use the well-known aperture distributions that have low sidelobe Chebyshev or Taylor pattern functions and so reduce sidelobes to the theoretical limits; but this solution seldom is applicable except to certain broadside arrays or slot arrays with fixed beam positions. Studies 81-84 of random phase and amplitude effects and of pattern distortion due to phase quantization have led to statistical predictions of sidelobe levels for fixed beam and scanned arrays.

Another problem that limits the sidelobe ratio maintainable by an array is that the element pattern $f_m(\theta,\phi)$ differs for each element "m" in an array. Equation (30) gives a general expression for this complex function and shows that it depends upon all of the mutual coupling terms from everywhere in the array, and so only the elments near the array center have the same element patterns. The patterns for elements near the array edges are not only different in amplitude, but can have different spatial dependence than those for central elements. This means that any amplitude weighting specified for sidelobe suppression must vary with scan angle to achieve the lowest possible sidelobes. Problems of this sort have caused little concern in the past because edge effects are not dominant in large radar arrays, and because it has become common practice to leave a number of unexcited elements near the edges of these arrays so that the excited elements have more similar element patterns. Specifying extremely low sidelobes for small arrays may cause this problem area to grow until it poses a fundamental limitation on array performance. As yet there has been little research expended on this potentially troublesome area, but as better phase shifters and more accurate power division schemes become available, element pattern distortion will remain as the dominant limitation on sidelobe reduction for small arrays.

Among the more important areas of array research is the topic of pattern null steering to elminate jamming interference. Most recent contributions to this subject [85,86] have included consideration of mutual coupling effects and proceed from an equation similar to (30). Aduming only one mode in each aperture ($Q = 1$), equation (30) is written:

$$F(\theta,\phi) = \gamma_1 I_1(\theta,\phi) j\omega \sum_{m=1} e^{jk_o(ux_m + vy_m)} a_m [1 + \sum_{n=1} S_{mn}]$$

$$= \gamma_1 I_1(\theta,\phi) j\omega C^T e \tag{31}$$

where $C = (S + I)a$, a is a column vector of excitations, and I is the $M \times M$ element identity matrix.

$$a = \begin{pmatrix} a_1 \\ a_2 \\ \vdots \\ a_m \end{pmatrix} \qquad C = \begin{pmatrix} C_1 \\ C_2 \\ \vdots \\ C_m \end{pmatrix}$$

and (32)

$$e = \begin{bmatrix} e^{jk_o(ux_1 + vy_1)} \\ \vdots \\ e^{jk_o(ux_m + vy_m)} \end{bmatrix}$$

and

$$C^T = a^T (S + I)^T$$

The power pattern has the form

$$P(\theta,\phi) = F(\theta,\phi) \cdot F^*(\theta,\phi)$$

$$= |\gamma_1^2||I_1^2(\theta,\phi)|\omega^2 [C^\dagger (e \cdot e^\dagger) C] \tag{33}$$

$$= |\gamma_1^2||I_1^2(\theta,\phi)|\omega^2 a^\dagger [(I + S)^\dagger e e^\dagger (I + S)] a$$

where the sumbol † denotes the combined transpose and complex conjugate operations.

Thus:

$$P(\theta,\phi) = |\gamma_1^2||I_1(\theta,\phi)|\omega^2 (a^\dagger \bar{A}\ a) \tag{34}$$

where

$$A = (I + S)^\dagger\ e e^\dagger\ (I + S)$$

is a one term dyad.

The directive gain is defined as the ratio of this $P(\theta,\phi)$ to the total radiated power P_R, which is given by:

$$P_R = a^\dagger\ (I - S^\dagger S) a \tag{35}$$

Thus a normalized quadratic form for directive gain, in terms of the scattering matrix $\bar{S}$, assuming single mode excitation and only single mode contributions present in the far field radiation expansion is:

$$G(\theta,\phi) \quad \left| I^2(\theta,\phi) \right| \frac{a^\dagger A\ a}{a^\dagger B\ a} \tag{36}$$

where

$$B = I - S^\dagger S$$

This expression for gain is a quadratic form. It could include the computed or measured scattering matrix data, and is a convenient form for optimizing subject to various constrints. Some details of an appropriate optimization are given in the literature [85,86] and a recent report illustrates [86] and details the specific procedure used for optimizing the directive gain of the single mode waveguide array problem described above. The procedure has been applied in a number of situations for dipole and waveguide arrays with finite numbers of elements and including mutual coupling. It has never been applied to situations that included important higher order mutual coupling expressions. Of particular importance is that the numerics of the problem become simpler as the number of constraints are increased. Thus this sort of optimization has been used to place a number of nulls close together within a sector of a pattern, and so produce a trough that would eliminate the effects of narrowband jammers over a relatively wide spatial sector. The recent study [86] addresses the effects of random errors in phase and amplitude control on null and trough formation, and concludes that 0.1 db of amplitude and 1^0 rms of phase control are necessary to maintain a -40dB trough. When phase shifters, power dividers and engineering practices allowing such required tolerances can be obtained; then again the fundamental technical problem of higher order radiating fields will remain and assume its dominance as the ultimate limit to radiation suppression.

Techniques like the one mentioned above imply that good amplitude control is available in addition to the required phase control. This is seldom the case, and has been the major factor hindering the advance of technology based upon such optimization. The increasing importance of antijamming protection will however make these systems the subject of intensified research interest as pressures increase to simplify feed networks and broaden system bandwidth. Several other options for antijam array techniques will be discussed in subsequent sections of the paper.

C. Array Techniques for Limited Sector Coverage

One of the most important classes of special purpose array techniques are those which trade scan capability for decreased cost. These are called "limited scan arrays", and they exist because there are many military and civilian requirements for high gain, electronically scanned antennas that need scan only some restricted sector of space. Military requirements include weapons locators, antennas for synchronous satellites and for air traffic control. Civilian requirements are mainly for air traffic control.

Two general classes of arrays are used for limited scan systems; the first class, which is historically the earliest and the most successful, is an array that is placed in or near the focal region of a reflector or lens antenna to scan its beam. The second class consists of large aperture elements and a beam forming network that includes some means of suppressing the system grating lobes. In either case there exists a minimum number of control elements that are required for beam scanning over any given sector and this serves as one measure of the efficiency of the scanning system.

One measure of this minimum number is the number of orthogonal beams within the scan sector. Another is the theorem of J. Stangel [87] which states that the minimum number of elements is:

$$N = \frac{1}{4\pi} \oiint G_o(\theta,\phi)\, d\Omega \tag{37}$$

where $G_o(\theta,\phi)$ is the maximum gain achievable by the antenna in the (θ,ϕ) direction, and $d\Omega$ is the increment of solid angle.

Another measure of the minimum number of array elements is contained in the definition of a parameter introduced by Patton [88] and called the "element use factor". This parameter will be used in a somewhat generalized form to compare the number of phase shifters in competing systems with unequal principal plane beamwidths. The factor N/N_{min} where N is the actual number of phase shifters in the control array, and N_{min} is a reasonable number of control elements as defined below

$$N_{min} = \left(\frac{\sin \theta^1_{max}}{\sin \frac{\theta^1_3}{2}} \right) \left(\frac{\sin \theta^2_{max}}{\sin \frac{\theta^2_3}{2}} \right) \qquad (38)$$

θ^1 and θ^2 max are the maximum scan angles in the two orthogonal principal planes measured to the peak of each beam, and θ^1_3 and θ^2_3 are the half power beamwidths in these planes. Thus N - min is approximately four times the product of the number of beamwidths scanned in each principal plane, and as will be shown later, is also approximately equal to the number of orthogonal beam positions for a rectangular array with beams filling a rectangular sector in direction cosine space. Although more general, Stangel's formula reduces to Patton's in the limiting case of small scan angle and using the approximate formula

$$G_o = \frac{4\pi}{\theta^1_3 \theta^2_3} \qquad (39)$$

and integrating over a rectangular sector.

In the case of a periodic array with a square or rectangular grid, the condition of minimum phase controls can be shown to require that the following relation be satisfied in each plane.

$$\left(\frac{d}{2}\right) \sin \theta_{max} = 0.5 \qquad (40)$$

A more recent study by Borgiotti [89] describes a similar bound and presents a technique for synthesizing patterns with various sidelobe levels that satisfy the criterion of requiring a minimum number of controls.

Given that there is a minimum number of required controls for any given sidelobe level, beamwidth and scan or multiple beam coverage sector, the remaining issue is to investigate techniques for optimizing scanning systems, subject to given sidelobe requirements, so that their characteristics approach those of an ideal scanner.

Optical techniques that combine single or dual reflectors or lenses with phased arrays to achieve sector scanning have the advantage that they have grown out of techniques for large apertures and so naturally provide high gain. Alternatively, the array techniques that have existed in the past lend themselves to much greater control of aperture distribution for sidelobes, but are not so readily suited to the high gain requirements of present systems for limited scan coverage. Cost factors and the availability of technology have brought about an intense period of creative engineering that has resulted in the current state of optical feed limited scan systems.

Design of this clss of systems is dominated by optical considerations, and problems of spill-over, aperture blockage, off acis focusing and induced cross polarization often effect the design more than the ratio of beam positions to control elements. Aperture efficiency is another parameter of importance for many applications. Many of the reflector or lens geometries require oversize apertures because the feed structure illuminates only a spot on the main aperture, and that spot moves with scan angle. For these devices, aperture efficiencies can be of the order of 25% instead of the usual 55 to 60% for non-scanning reflectors, and so scanning reflectors or lenses often require double the aperture of the more efficient fixed beam structures. The type of beam steering required is also a factor of great importance; the simplest being row and column steering with progressive phases in both planes. Certain antennas however require complex steering functions for off axis scans, and these result in slower beam steering and larger computer data storage.

Optical limited scan techniques have their origin in the development of mechanically scanned reflector and lens geometries using feed tilt [90] and displacement [91] and in the development of feeds to correct the wide angle performance of beam shaping reflectors [92]. Present devices include single and dual reflection of lens geometries in combination with a phased array, that produce an electronically scanned beam using relatively few phase controls.

Design principles for such a wide ranging collection are themselves so varied that they cannot be developed from basic principles in a text of this length. Instead, this paper will outline some of the more important contributions to the technology and to the analytical and conceptual tools that made the technology possible. Several recent sources include comprehensive surveys of developments in these areas [93,94].

Many of the early studies on mechanical scanners were performed using geometrical optics, and even today this method receives wide usage for computing required feed locations, focusing conditions and phase shifter controls, and for investigating general design parameters. Recent studies [94] have emphasized the deficiencies of the geometrical optics approach for obtaining intensity information about focal region fields and have demonstrated the use of physical optics for design. Analytical methods used in design are noted in the descriptions that follow.

Early studies of scanning parabolas have uncovered a number of useful design concepts. Using ray optical techniques, Sletten et. al. [92,93], have investigated the location of focal (or caustic) surfaces for paraboloidal reflectors receiving off axis plane waves, and have shown how these characteristics can be used to develop midpoint correctors for elevation beam shaping while maintaining a narrow focused beam in the azimuthal plane, and ridge line correctors for forming several pencil beams in elevation without destroying azimuthal focus. Ruze [91] has used a scalar plane wave theory to analyze the scanning characteristics of a parabola with a laterally displaced feed located at the Petzval surface. This analysis was used to obtain scanning patterns, to evaluate coma lobe contributions and to derive equations for the number of beamwidths scanned by such displacement for a -10.5dB coma lobe at the scan limit. This number is given below.

$$N = 0.44 + 22(f/D)^2$$

Recent work by Rusch and Ludwig [94] has included a numerical evaluation of focal region fields for a paraboloid receiving an off axis plane wave. Results of this study show that the maximum focal field locus, and therefore the position of optimum feed location, does not coincide with the Petzval surface, but remains relatively close to it for low f/D values. Reflectors with f/D greater than 0.5 have their optimum feed location closer to the focal point than the Petzval surface, but this location tends toward that line for large scale angles.

Imbriale et.al. [95] have also considered parabolic reflectors with large lateral feed displacements, and have compared the results of Ruze's scalar theory with the complete vector theory solution and experimental data for various feed displacements. This study demonstrated that the coma-lobe level is sometimes vastly underestimated by the scalar theory when used for feeds with large displacement.

Another recent work of significant import has been reported by Rudge [96] who has demonstrated the spatial fourier transform relationship between the aperture fields of a parabolic reflector and its focal plane fields. In an extension of this work, Rudge and Withers have also shown [97,98] that the fields in a specified off axis focal plane bear the same transform relationship to the aperture field under the excitation of an inclined wave.

Not surprisingly, the first viable limited scan antennas consisted of a parabolic reflector [100] with an array placed between the reflector and the focal point as shown in Fig. 12. The array is then required to produce the complex conjugate of the field that it would receive from a distant point source at a given angle, and the extent to which it can do this determines the quality of the antenna radiation pattern. Because of the

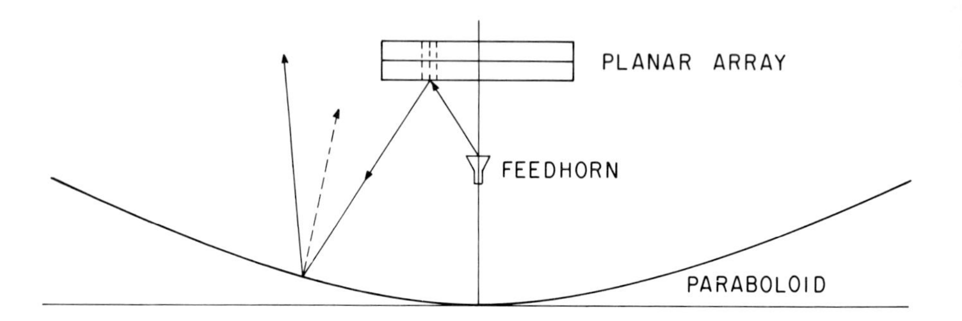

Fig. 12. Reflector/array combination for limited sector coverage.

complexity of the converging field, White and DeSize [101] placed an array of feeds on a spherical surface concentric with the parabola focal point, and demonstrated scanning for that case. More recent structures do use parabolas with non-linear phase controls in the array, and achieve sidelobes at the -18 to -20 dB level. Examples of these structures are the AGILTRAC antenna [102,103] and the AN/TPN-19 Precision Approach Radar Antenna (Fig. 13). One final example of a single reflector or lens geometry scanned by a phased array is shown in Fig. 14. This antenna differs fundamentally from the other optical schemes because the main reflector or lens is not restricted by a focusing condition. This new concept in limited scan antennas, proposed by A. C. Schell [104] uses an array disposed around a cylinder to scan a reflector or lens surface that is contoured according to an opti-

Fig. 13. Precision Approach Radar Antenna AN/TPN-19.

mum scan condition, rather than a focusing condition. The reflector is then stepped or the lens phase corrected to achieve focusing. Preliminary design results [105] show that in one plane of scan the technique achieves an element use factor of about unity, while using an oversize final aperture to again allow motion of the illuminated spot.

Figure 14 shows a schematic view of the array-lens combination and Figure 15 demonstrates its scanning properties. The array element currents are equal in amplitude and have a progressive phase given by $\beta n \Delta\theta$. The reflector surface (or lens back face) is chosen to transform this phase variation into a linear

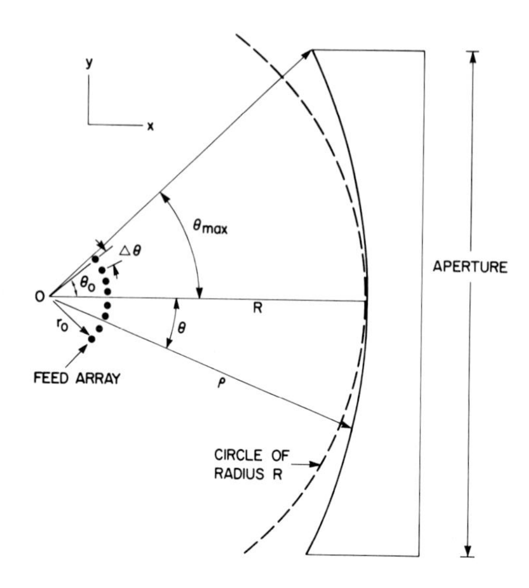

Fig. 14. Scan corrected lens antenna.

wavefront normal to the beam direction. The condition for determining the curvature of the reflector (or lens) is that a constant incremental phase change in θ along the circular arc tangent to the center of the back face of the lens produces a constant incremental phase change in the "y" coordinate along the aperture.

$$\frac{dy}{d\theta} = \text{Constant} = R \tag{41}$$

and, since $y = \rho \sin\theta$

$$\rho = \frac{R\theta}{\sin\theta} \tag{42}$$

This curvature statisfies the scan requirement, but does not guarantee that the wave will focus. Focusing is achieved for the reflector through the use of confocal parabolic sections stepped so that their centers lie along the scan surface, and for the lens by adjusting the path lengths so that they are equal at some angle. The array represented by the data of Fig. 15 consists of

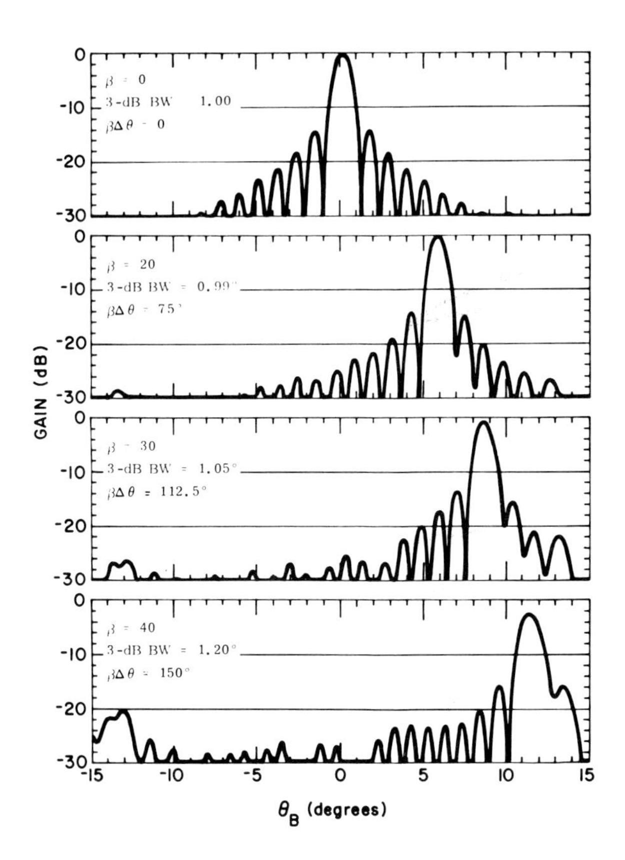

Fig. 15. Pattern characteristics of scan corrected lens.

25 elements. The half angle θ_o subtended by the array is 45°, $k_oR = 197$, $k_oa = 48$ and k_o times the final aperture width is approximately 526 ($D/\lambda = 83.7$).

The far field beam angle is given by

$$\sin \theta_B = \beta/kR, \tag{43}$$

and for the case shown in the figure the maximum θ_B is about 11.5°. At this scan angle the gain is reduced about 2.5dB with respect to broadside and a far-sidelobe has risen to the -20dB level. The aperture illumination is nearly uniform, and the -13 dB near sidelobe ratio is maintained throughout the scan sector.

These computations by McGahan [105] have been confirmed experimentally for the lens geometry scanning in one plane. An element use factor of 0.95 would result if the same economy of phase controls can be maintained for the lens scanning in two dimensions. Since the amplitude distribution on the array is transferred very simply onto the inner lens surface, it is possible to produce very low sidelobe patterns with this geometry. Preliminary theoretical data indicate that with perfect phase and amplitude control this structure can have sidelobes below -40dB.

Single reflector or lens structures with a phased array feed are simple but, with the exception of the technique described by Schell and McGahan, they require a relatively large number of phase controls (Element use factors of 2.5 - 3.25). All of these antennas require oversize main apertures because the illumination moves with scan (typical aperture efficiencies are 20-25%). Thus large aperture size coupled with weight and cost limitations, usually restricts the choice of final aperture to that of a reflector, and the resulting blockage can cause sidelobe problems and the need for offset feeds.

Studies of dual reflector or lens combinations illuminated by a phased array have followed two distinct paths. One class has used relatively small sub-reflectors, linear progressive phase control but comparatively large element use factors because the array cannot be used optimally. The small sub-reflector requires that the array scan sector be limited, and if the array is not designed to take advantage of this fact, then the element use factor will be larger than the theoretical minimum. Examples of this first category are the near field cassegrain geometry and the offset fed gregorian geometry of Fitzgerald [106-108], The second category of dual reflector or lens scanning antennas uses a much larger secondary aperture and an array that scans over wide angles. This type of antenna can have element use factors close to unity, but the required secondary aperture sizes make the structure bulky. Comparison of two antennas in this category

[108,109], with the Fitzgerald studies indicate that element use factors of 1.4 can be maintained using secondary apertures of approximately 0.7 the diameter of the main reflector, but restricting the subreflector size to 0.35 to 0.25 of the main reflector diameter led to element use factors of between 2.5 and 4. Larger sub-reflectors also allow more accurate control of the main reflector illumination and in one case [109] resulted in approximately -20dB sidelobes over the scan sector.

In addition to these combinations of array and optical structures, there is a growing class of antennas that scan efficiently over limited sectors using novel array techniques. Each achieves its relatively low cost by using large array elements or subarrays and so reducing the number of required phase controls for a given size final aperture. This use of oversize elements in a periodic array results in grating lobes, which are suppressed by careful control of the subarray element pattern, or by scanning the element patterns to null certain of the lobes. Alternatively, other approaches have used pseudo-random array grids to reduce the peak levels of the grating lobes by redistributing their energy over a wider sector of space.

The radiated field of the array of aperture elements shown in Fig. 1 given in direction -cosine space for a beam at (u_o, v_o) as:

$$E(u,v) = \frac{e(u,v)}{N} \sum_{m=1}^{N_x} \sum_{n=1}^{N_y} A_{mn} e^{j\frac{2\pi}{\lambda}(umd_x + \Delta_n u + vnd_y)} \tag{44}$$

where

$$u = \sin\theta \cos\phi$$
$$v = \sin\theta \sin\phi$$

and

$$A_{mn} = I_{mn} e^{-j\frac{2\pi}{\lambda}(u_o md_x + u_o \Delta_n + v_o nd_y)}$$

Here A_{mn} is the signal at the element (m,n) located at the position $x = md_x + \Delta_n$, $y = nd_y$ as required to form the beam at (u_o, v_o). I_{mn} is the signal amplitude and $e(u,v)$ is the element pattern of each aperture in the array. $N = N_x N_y$ is the number of elements in the array.

Assuming that the array is excited by a separable distribution so that $I_{mn} = I_m^H I_n^E$, and assuming that the element pattern is also separable, then $e(u,v) = e_x(u)e_y(v)$ and this field distribution can be written in the following form:

$$E(u,v) = \frac{1}{N_x}\left\{\sum_{m=1}^{N_x} I_m^H e_x(u) e^{j\frac{2\pi}{\lambda}(u-u_o)md_x}\right\}$$
$$\cdot \frac{1}{N_y}\left\{\sum_{n=1}^{N_y} I_n^E e_y(v) e^{j\frac{2\pi}{\lambda}[(v-v_o)nd_y+\Delta_n(u-u_o)]}\right\} = E_x(u)E_y(u,v) \qquad (45)$$

This pattern is not separable in general, but may be so for certain choices of (Δ_n). The choice of optimum (Δ_n) for grating lobe reduction will be considered later.

If the rows are not displaced, then the field is fully separable and the exponential forms in the above summations have principal maxima at the grating lobe locations:

$$u_p = u_o + \frac{p\lambda}{d_x}$$
$$v_q = v_o + \frac{q\lambda}{d_y} \qquad (46)$$

for all $p.q$ bounded by the inequality

$$K_{pq} = \frac{2\pi}{\lambda}\sqrt{1 - u_m^2 - v_n^2} \leq \frac{2\pi}{\lambda}$$

These points are shown in (u,v) space as a regularly spaced grating lobe lattice about the main beam location (u_o, v_o) in Fig. 16. The circle with unity radius represents the bounds of the above inequality; all grating lobes within the circle represent

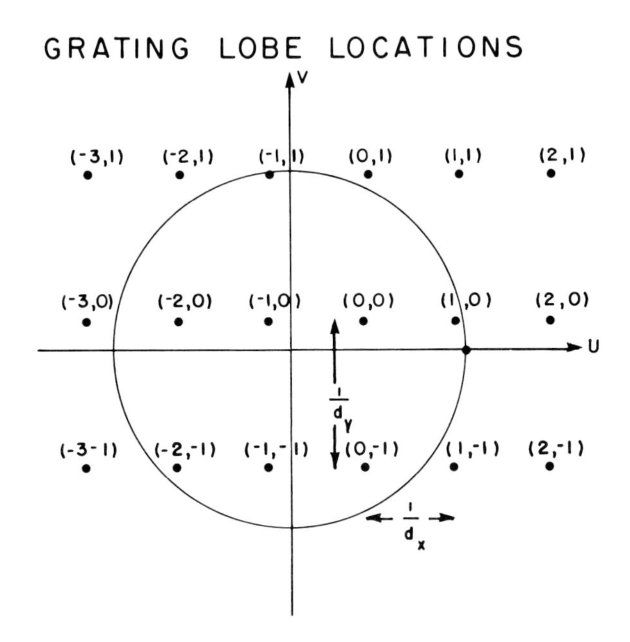

Fig. 16. Periodic array grating lobe lattice.

those radiating into real space, and those outside do not radiate.

Figure 17 shows how the array factor and element pattern combine to produce the resulting radiated distribution in one principal plane. This figure illustrates how the effects of squinting or narrowing the element pattern or of destroying the periodicity can serve to reduce resulting grating lobes by altering either of the two factors in this product.

Efforts to maintain non-periodic grids for grating lobe reduction have centered mostly about use of circularly disposed arrays with an aperiodic arrangement of elements of one or several sizes.

An example is the array investigated by W. Patton [111]. This structure, shown schematically in Fig. 18, consists of a circular array of dipole subarrays arranged in an aperiodic fashion. This array is locally periodic, and does have vestigal grating lobes, but these are considerably suppressed for a large array. Patton describes a 30 ft. diameter array and a 10 ft.

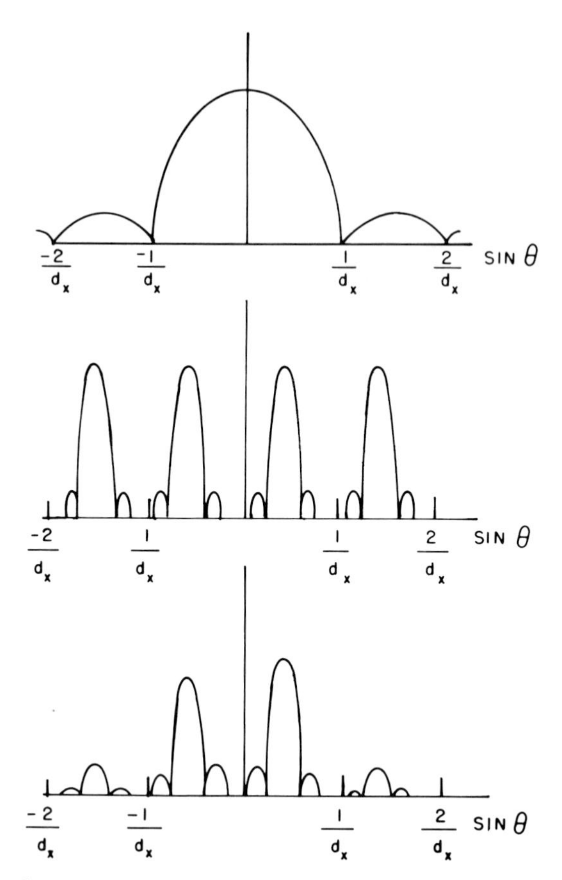

Fig. 17. The array pattern, element factor product.

diameter array at C-band. The 30 foot array consists of one thousand elements that scan a 0.36° beam approximately 5° with an element use factor of 1.3. The system has high average sidelobes at its maximum scan and losses which add to 5.94dB for the 10 ft. model and a projected 4.21dB for the 30 ft. array. The transmission line interconnections may also make an X-band design somewhat less practical. Peak sidelobes were measured at the -15dB level for the 10 ft. diameter array, and are projected at -20.9dB for the 30 ft. array, but the item of primary importance is the achievement of this extremely low element use factor and the low generalized *f/D* ratio achievable with aperiodic array technology.

Fig. 18. Element location diagram for the REST array; a technique for limited sector coverage.

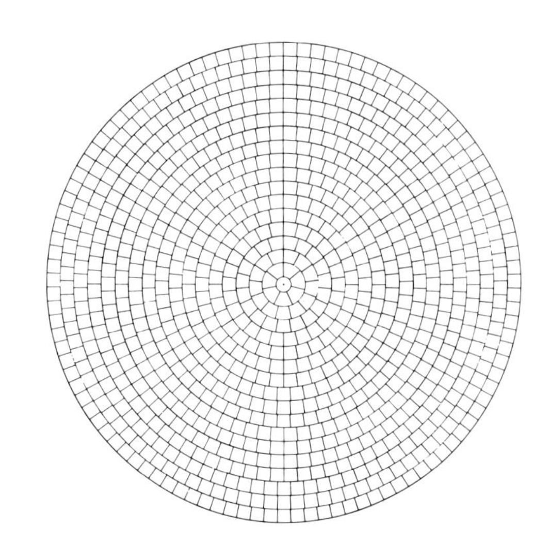

A similar antenna, but using unequal size elements has recently been described by Manwarren and Minuti [112]. This antenna has been designed to provide a 1° pencil beam at 1300 MHz to scan a conical section with 8° half angle with 20dB grating lobes. An S-band model has also been configured. The antenna consists of 412 elements of three different sizes to make up the array surface. The elements are arranged in concentric rings to produce a pseudobrandom grid as done in the REST program, but with additional randomness introduced by the unequal size elements. Computed patterns show graceful gain degradation with scan and grating lobes at the desired levels.

A recent effort at RADC/ET has revealed that relatively large aperture horns can be used as elements of a limited scan array if the higher order mode amplitude is actively controlled [113]. The technique is called multi-mode scanning and consists of choosing odd mode amplitudes and phases so that the combined element radiation pattern from any horn has a zero at the angle of the grating lobe nearest to broadside.

The required odd mode amplitude and phases are obtained from a knowledge of the element patterns for even and odd modes. An array for E-plane scan has its field pattern given by: (for $(\Delta_n) = 0$)

$$E(u,v) = E_x(u) \sum_{n=1}^{N_y} I_n^E e_y(v) e^{j\frac{2\pi}{\lambda}(v-v_o)nd_y} \qquad (47)$$

element pattern $e_y(v)$ is zero at the grating lobe positions $v_q = \pm q\lambda/d_y$ corresponding to broadside main beam position for uniformally illuminated element. The growth of the $q = -1$ grating lobe as a function of scan can be nearly eliminated by actively controlling $e_y(v)$ to place a zero at this grating lobe for all scan angles. In a waveguide circuit such control is accomplished by exciting the aperture with two modes (the LSE_{10} and LSE_{11}) instead of just the dominant LSE_{10} mode, so that $e_y(v)$ becomes the sum of two terms, with a zero at $v_o = v_o - \lambda/d_y = v_{-1}$. Choosing the ratio of odd mode to even mode as R_{11}, the combined element pattern is:

$$e_y(v) = e_{yo}(v) + R_{11} e_{y1}(v) \qquad (48)$$

choosing $e_y(v)$ to be zero at the position of the $q = -1$ grating lobe one obtains for R_{11}

$$R_{11} = \frac{-e_{yo}(v_{-1})}{e_{y1}(v_{-1})} \,. \qquad (49)$$

Since the various waveguide modes have constant phase aperture distribution e_{yo} is a real function and e_{y1} is a pure imaginary, so the R_{11} is pure imaginary and increases with scan in order to maintain the null position coincident with the center of the $q = -1$ grating lobe. The relative odd mode phase is thus fixed at $\pm$ 90° with respect to the even mode phase depending upon the sense of the scan angle. The allowable element spacing for E-plane horns is

$$(d_y/\lambda) \sin \theta_{max} = 0.6 \qquad (50)$$

The laboratory model shown in Fig. 19 is an array designed for

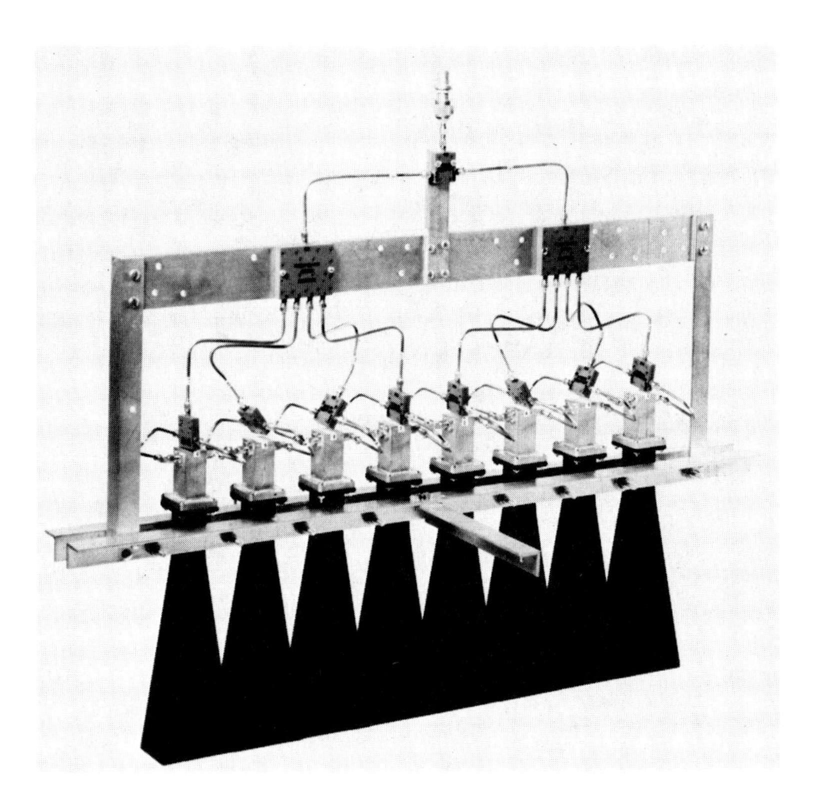

Fig. 19. Laboratory model multimode scanning array.

E-plane scan (±12°).

Fig. 20 shows the E-plane pattern for the array phased at broadside and the elements excited with the central four at uniform amplitude, the second element in from each end of the array at -3dB amplitude, and the outer elements at -6dB amplitude. This taper should have first sidelobes at about -19dB, but due to phase errors the level is approximately -17dB. The grating lobes at +19° (-16dB) and +40° (-26dB) can be reduced by using a dielectric lens in each horn.

Figure 21 shows two cases at the maximum scan angle +12°. The dashed curve is the horn array radiation pattern without odd modes and shows that the main beam gain is reduced more than 5dB with respect to the broadside array and the grating lobe at -7°

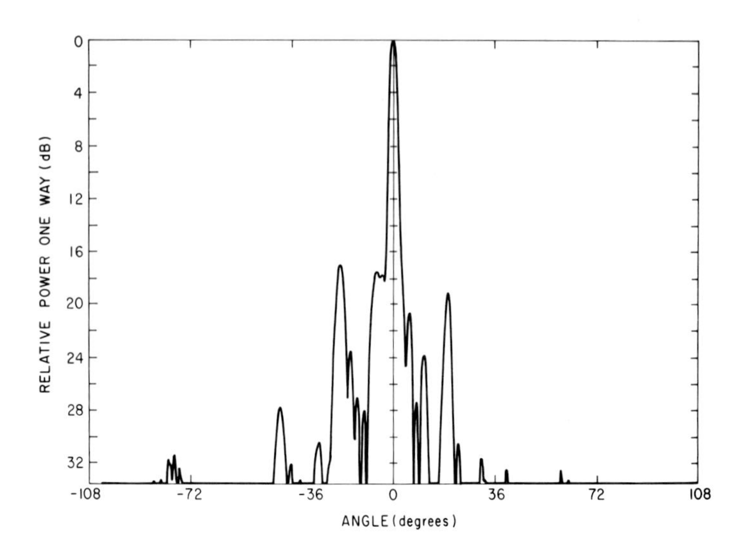

Fig. 20. Broadside pattern data (8 element array).

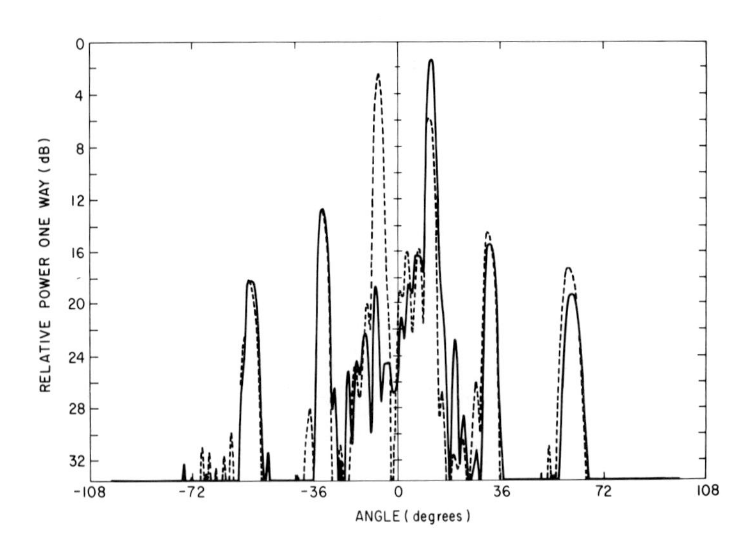

Fig. 21. End of scan pattern data (8 element array).

is larger than the main beam by 1.8dB. Other grating lobes are at tolerable levels. The solid curve shows that when the element is excited by two modes the offending grating lobe is reduced to approximately the -20dB level, and the main beam increased to -1.2dB with respect to broadside because the new element pattern has its peak tilted toward the main beam. The second grating lobe (q = -2) is approximately the -12dB level.

Conventional aperture tapering procedures can be used to reduce near sidelobe levels to -30dB or less. The nulled grating lobe is suppressed 20 to 25 dB at center frequency for a small array (8 elements), but substantially more for larger arrays. Residual grating lobes at wider angles are unaffected by array tapering and remain the major limitation of the technique.

Full two dimensional scanning requires the suppression of three grating lobes, however, and so a total of three higher order modes must be controlled as a function of scan. The dominant grating lobes to be cancelled are those nearest broadside (p,q) = (-1,0), (-1,-1), (0,-1) for general scan angles, and this control is achieved using four phase shifters for each multi-mode horn to form an element pattern that is separable in u-v space and positions the three nulls properly. In practice, it is also sometimes appropriate to narrow the horn H-plane patterns by dielectrically loading them [114]. This correction is added to minimize broadside H-plane grating lobes.

Bandwidth and far sidelobe levels are the most important limitations of the technique. Good performance has been achieved over narrow bandwidths (~3%), and bandwidths of up to 10% appear feasible. Far sidelobe (grating lobe levels) of -20dB can now be obtained using various aperiodic row displacements $[\Delta_n]$ and spatial filtering combinations as will be described later, but it is unlikely that sidelobes can be reduced much below that level. The main advantages of the technique in comparison with most of the reflector or lens schemes are the availability of extremely low near sidelobes, the naturally high aperture efficiency and small antenna volume for any desired gain, and the use of row-column steering commands.

A final limited-scan antenna type is described as having "overlapped sub-arrays". This concept is an outgrowth of the realization that the ideal element for a limited scan system would have a flat top and no sidelobes. An element pattern like that of Fig. 22 would allow the beam to scan out to some maximum scan angle ($\sin \theta_m$) while suppressing all grating lobes as long as they did not occur within the range $-\sin \theta_m \leq \sin \theta \leq \sin \theta_m$. Since the grating lobes occur at positions given by equation (46) then for a very large array one can optimize the inter-element spacing for a given maximum scan angle by choosing

$$(d_x/\lambda) \sin \theta_m = 0.5 \tag{51}$$

where d_x is the inter-subarray spacing. This condition was derived earlier (Eq. 40) from the basis of satisfying the criteria given for minimizing the number of phase controls, but here its results fromm choosing the widest possible flat-topped subarray pattern consistent with good grating lobe suppression. In this case a large array with main beam at $\sin \theta = (0.5\lambda/d_x)$ for some arbitrarily small value will have its nearest grating lobe at $\sin \theta = -(0.5\lambda/d_x)$, and all grating lobes will be completely suppressed. Such an array is characterized as a limited scan design because it can take advantage of limitations imposed upon the scan sector in order to increase aperture size d_x/λ. In principle an array with $\sin \theta_m = 0.1$ can use a 5λ inter-element spacing while for $\sin \theta_m = 0.05$, a 10λ spacing can be used. These size increases and associated reductions in the number of required phase controls for restricted coverage illustrate the goal of limited scan antenna designs.

The aperture field corresponding to this far-field distribution is of the form:

$$i(x) = \frac{\sin\left(\frac{2\pi}{\lambda} x \sin \theta_m\right)}{\left(\frac{2\pi}{\lambda} x \sin \quad _m\right)}$$

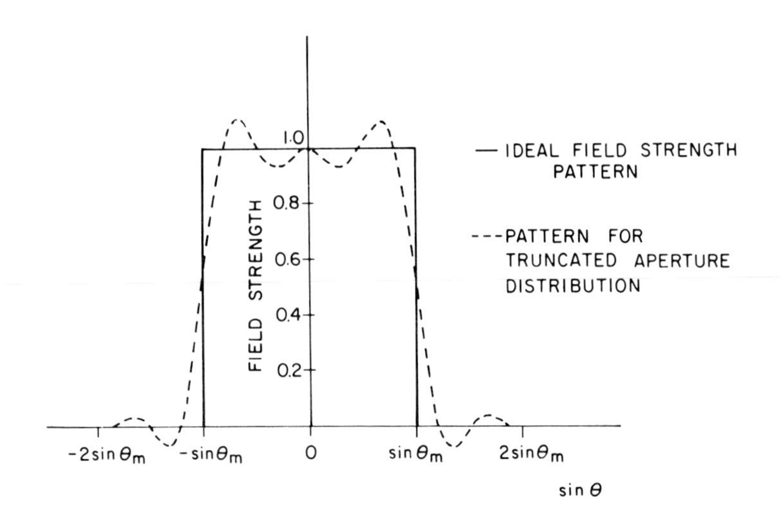

Fig. 22. Ideal and approximate subarray patterns for overlapped subarray.

where x is the distance measured from the center of the subarray. If the maximum ideal spacing $d_x/\lambda = (0.5/\sin\theta_m)$ is used, then this aperture distribution has zeros at $x = \pm nd_x$ excluding $n = 0$, and one must include a number of elements in order to reproduce the $i(x)$ distribution faithfully. Thus, each phase shifter must feed a multiplicity of sub-arrays and the sub-arrays can be said to be overlapped. Obviously, the ideal aperture field can only be approximated; it must be truncated and then approximated by realizable distributions at each element. The dashed curve shows the flat-topped sub-array pattern achievable if the $i(x)$ is truncated at $x = \pm 3d_x$. In this case, 20dB grating lobe suppression can be obtained for scan out to the angle

$$(d_x/\lambda)\sin\theta_m = 0.43 \tag{54}$$

The required overlapped distribution implies the interconnection of a number of array elements and so is extremely difficult to fabricate in microwave circuitry. Consequently the circuit approach has received only limited attention. Alternatively, space feed systems can quite naturally achieve overlapped sub-array distributions that have proven very practical. These systems use feed-through lenses or reflectors that can faithfully reproduce a substantial part of the $f(x)$ distribution as compared with microwave circuit systems.

Examples of such schemes for producing optically overlapped sub-arrays includes the HIPSAF antenna [115] and the dual reflector-array design of Tang et al [109]. Figure 23 shows schematically that exciting two adjacent feed horns results in two overlapped aperture illuminations at the main reflector. These "sub-array" aperture distributions have approximately $(\sin x)/x$ fields and so have rectangular shaped radiation patterns as appropriate for good grating lobe suppression. Figure 24 shows a calculated and a measured pattern from the central sub-array of the experimental reflector. Details of this extensive analytical and experimental study are included in the reference [109] but in general the program demonstrated that such optical techniques can produce low sidelobe (<-20dB) scanned patterns over limited spatial sectors using only about 1.4 times the theoretical minimum number of phase controls.

Apart from these configurations using quasi optical techniques, the ultimate in overlapped circuitry for low sidelobe arrays will, of necessity, be synthesized using constrained feed distribution networks. This is a new area of technology and there has been relatively little work in this area. Several studies of overlapped sub-arrays are reported by Tang and modifications of these have recently been implemented for fire control radars. In addition, the sub-array distribution that produces an approximate flat topped pattern can be approximated by higher order mode distributions [116] in horn apertures so that

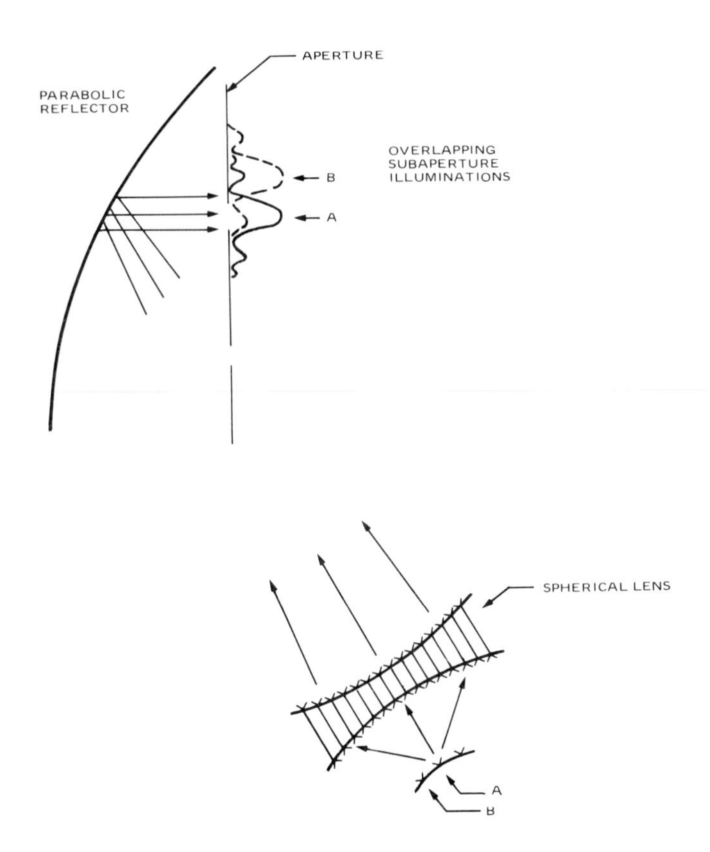

Fig. 23. Aperture illumination from optically overlapped feed.

the element spacings can be made equal to the distance d between sub-arrays. This work is an outgrowth of studies on the active odd-mode control of element radiation patterns, but the developments in overlapped sub-arrays differ in concept, in means and in results from the multimode scanning technique. The study describes a passive interconnecting network to synthesize a flat topped, symmetric, sub-array pattern, while the multimode scanning technique requires active control of odd mode amplitude and achieves much greater scan per element (although slightly less per phase shifter).

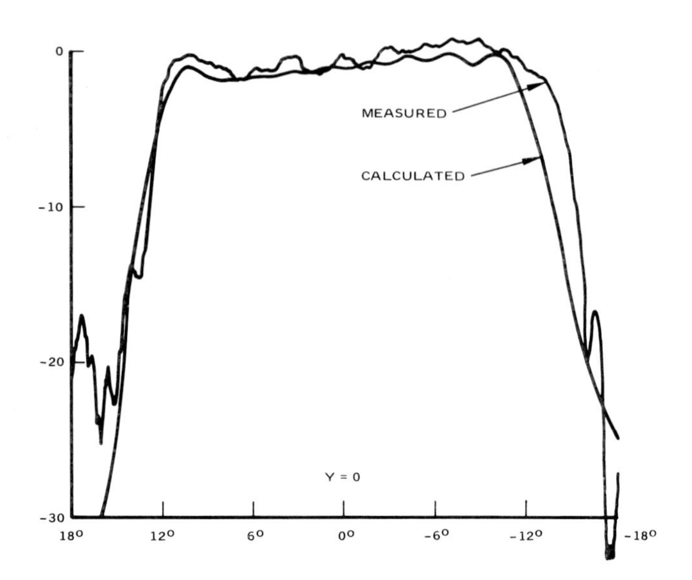

Fig. 24. Subaperture far field pattern for central subaperture.

The basic circuit allows an element size times scan angle product in the E-plane of approximately

$$(d_y/\lambda)\sin\theta_m = 0.33 \tag{55}$$

The largest array grating lobes are less than -16dB for maximum scan. Circuits have been devised to provide similar overlapped behavior for two planes of scan, but there is not yet sufficient data to compare the relative advantage gained by using the second plane.

The discussion of limited scan arrays has dealt mainly with a description of methods devised to reduce array costs; and these methods form the basis of an evolving technology. The most significant change forthcoming in this area is the development of techniques for extremely low sidelobe control. These techniques will be aided by some established methods of sidelobe suppression (random element positions, and tunnel structures) and by the spatial filtering technique to be described later, but the basic antenna structures themselves must be substantially improved in order to achieve sidelobe levels between -35 and -45dB. The only limited scan antenna with evidence showing that such sidelobe levels are achievable is the array-lens concept of Schell. These data are not yet published and consist at present of analytical calculations that neglect coupling and near field effects; but they confirm that the possibility of such extra-low sidelobe control exists.

Other quasi-optical approaches can conceivably produce extremely low sidelobes; in particular those schemes based upon overlapped sub-arraying approaches should produce very low sidelobe distributions, although not out to the scan limits given in this description.

None of the array techniques discussed here can produce patterns with such low sidelobes except through the use of spatial filtering; but techniques based upon constrained feed circuits for overlapped distributions can ultimately produce the lowest sidelobe limited scan systems. As yet there has been relatively little effort directed toward synthesizing such networks, and this remains an area where much work is needed.

D. Broadband and Multiple Frequency Arrays

Though considered together, broadband and multiple frequency arrays call for fundamentally different technology. Wideband arrays have one beam formed by a feed network and a set of phase shifters, but multi-band technology has developed by interleaving relatively narrow band elements with different center frequencies, and with separate beamformers for each frequency.

The maximum theoretical bandwidth of linearly polarized rectangular waveguide phased arrays is about 60% [117]. Studies [118,119] of such elements have indeed shown that these bandwidths can be achieved with low VSWR and wide scan coverage.

Recent efforts sponsored by AFCRL (now RADC/ET) have developed double ridged waveguides [120] and novel strip line radiators [121] (see Fig. 25) that can provide good performance over an octave bandwidth (67%). Arrays with circular polarization have much narrower bandwidths, with 25% seen as a reasonable outer limit [122,123].

Most of the development in dual band arrays has concerned inter-leaved arrays with ingenious brickwork patterns of various size elements [124-127]; with each frequency occupying a portion of the total aperture. A new RADC effort has led to the structure shown in Fig. 26 as an array for two frequencies, one roughly double the other. An analysis of this structure was included in Section II for tutorial purposes. The advantages of this geometry are that both frequencies occupy the whole array aperture and that separate terminals are provided for independent steering of the two beams. Figure 27 shows the H-plane scan characteristics of this array at two distinct frequencies, and indicates that the array has bood scan characteristics with no blind spots within the scan sector at either frequency. The proper use of array matching techniques should improve these characteristics and so make the technique viable for high power radiation at both frequencies.

Other multiple frequency arrays have been proposed and developed for distinct applications, and this area of technology is evidently destined to play an expanding role in the future of array antennas as the number of aircraft terminals grows to meet the needs of satellite communication systems.

IV. NEW TECHNOLOGY

A. New Technology as a Forcing Function

The techniques that have been discussed thus far represent major subject areas for research; the methods described represent present day solutions and may not correspond to ultimate

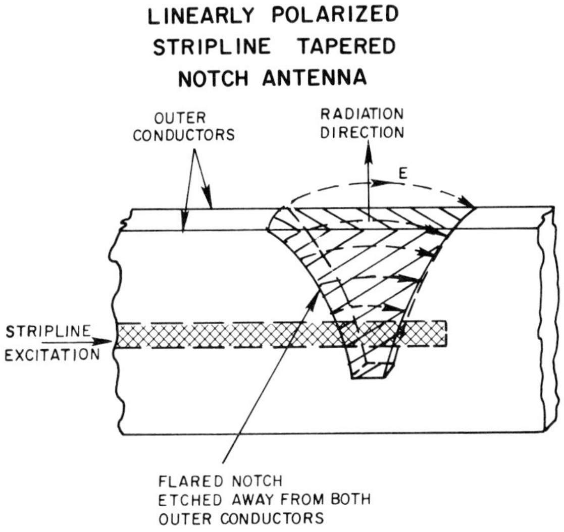

LINEARLY POLARIZED
STRIPLINE TAPERED
NOTCH ANTENNA
OUTER
CONDUCTORS
RADIATION
DIRECTION
E
STRIPLINE
EXCITATION
FLARED NOTCH
ETCHED AWAY FROM BOTH
OUTER CONDUCTORS

Fig. 25. Wide band stripline flared notch element.

LOW FREQUENCY COAXIAL LINE INPUT
4 HIGH
FREQUENCY WAVEGUIDE INPUTS
END OF
BIFURCATION
L
4 DIELECTRIC SLABS

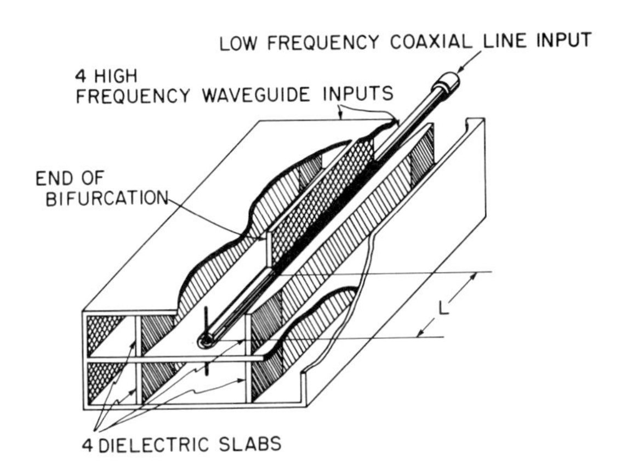

Fig. 26. Dual frequency array element.

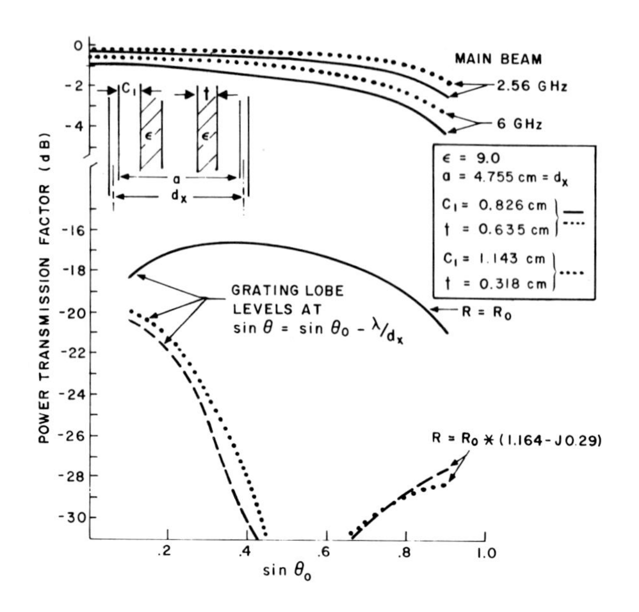

Fig. 27. H-plane scanning characteristics of dual frequency array element.

solutions. The categorization "Special Purpose Arrays" thus defines an area that will be of major importance for many years. This section is addressed to a different kind of stimulus for array research; one based primarily upon the wide variety of transmission media. The thesis proposed is that the very rapid change in this technology can be a strong force that guides and propels a major part of the future of array antennas. The newer elements of technology include improved phase shifters, the emergence of microwave integrated circuit technology and developments in strip line and microstrip transmission circuits. These new developments compliment existing array technology, but in addition they act as a stimulus to further advances in array techniques.

Fig. 28 shows a waveguide phase shifter developed by Raytheon Company. The phase shifter developed by Raytheon itself is a three bit analog non-reciprocal device that handles 3.5Kw peak and has 1dB loss. The photograph shows the driven circuit incorporated into the body of the phase shifter. Both ends of the phase shifter are matched to the environment they occupy. The front face is allinearly polarized C-band waveguide loaded with dielectric that has been matched to provide good properties over the design scan sector, while the back face is matched to optimize pickup from the space feed network. The main reason for showing this illustration is to indicate that indeed such scan matching has become practice; array behavior is calculated or measured in simulators, phase shifters are incorporated to achieve compact units that plug into the array and can be conveniently replaced.

A second item of technology that further illustrates some of the above is shown in Fig. 29. This laboratory prototype developed by Hughes Corporation is a 3-bit resistive gate diode phase shifter operating at S-band. Total loss is approximately 1.5dB, the device can control 300 watts of peak power with five watts average. Its size is approximately 1 inch by 1 inch by 2 inches, ant it switches in 10 microseconds. The chief advantage claimed for this device is that the phase shifter does not require any forward bias current; the only bias current flowing through the device is a forward leakage current of several microamps. Alternatively, the commonly used PIN diode phase shifters require 50-200MA current at one volt forward and 100 volts at 1 microamp reverse. Total power required for phase control per phase shifter and driver combination may thus be on the order of 0.3 to 0.6 watts. The total power for a 2000 element array would thus be nearly one kilowatt for the PIN diode array, but less than .2

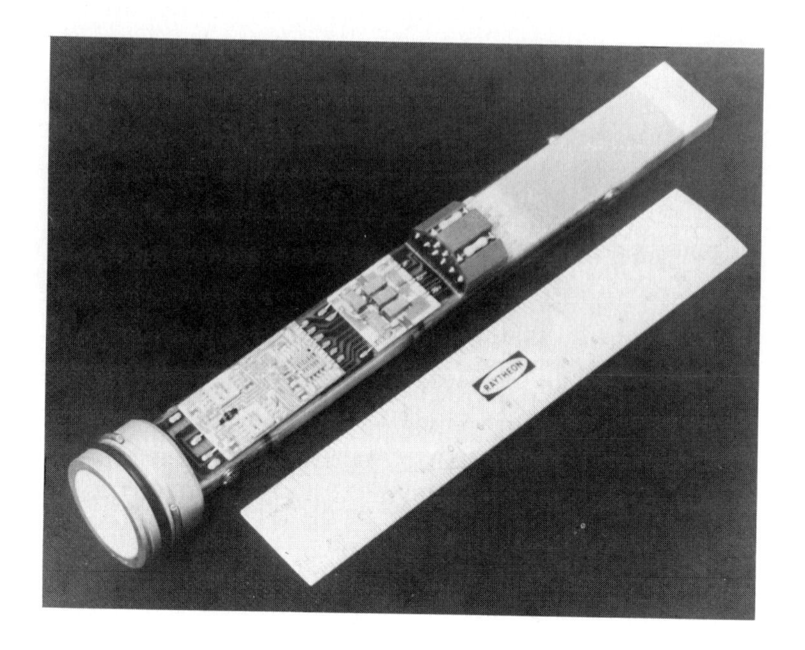

Fig. 28. Exciter, phase shifter and array element (courtesy Dr. Lewis, Raytheon Corp.)

watts for a resistive gate diode array. This extremely low driver power requirement makes it possible to control the array steering from the beam steering computer without an additional driver and high power supply.

Apart from the obvious fact that diode phase shifters have come a long way, the second issue raised by this technology is that the emergence of microstrip transmission circuits has not carried through to microstrip scanned antennas. There have been a number of developments in microstrip devices with fixed beams [128-131], but most of these early devices were not well suited to electronic scanning. Figure 30 shows a microstrip array of spiral elements developed by Raytheon Corporation. The array structure combines a corporate feed, power dividers, baluns and phase shifting network on one printed circuit board.

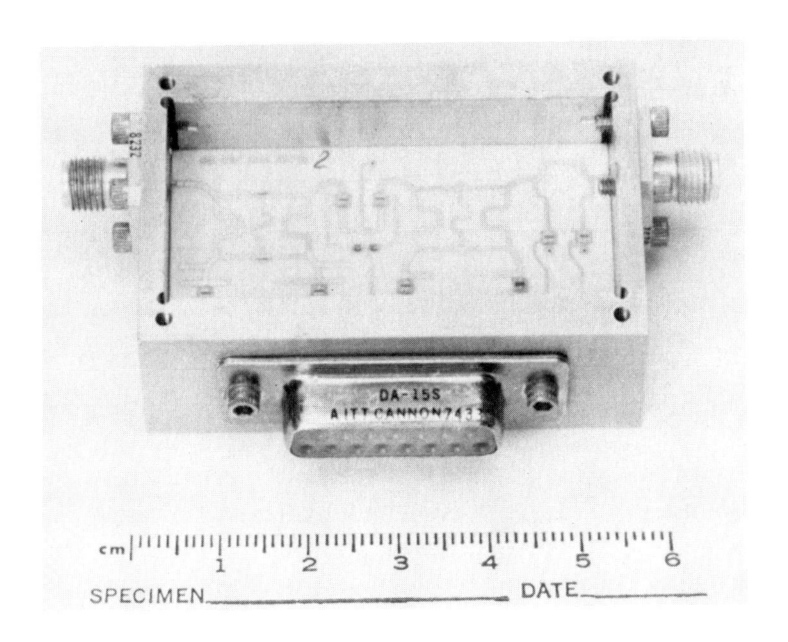

Fig. 29. Resistive gate phase shifter (courtesy of Mr. R. Tang, Hughes Aircraft Corp.).

To date, there are no comprehensive theoretical treatments of even single microstrip patch antennas. Nevertheless, the technology itself has advanced to such a degree that many of the larger corporations are developing numerous variations of the original designs, and the day of full scanned arrays is clearly very near. Research is needed in this important area in order to avoid some of the pitfalls that led to the problems of array blindness. Air Force applications for this type of lightweight, inexpensive, and comfortable antenna array are many, and extend from man-pack designs to flush-mounted air-craft antennas. Studies of microstrip and the newer types of strip-line antennas should be undertaken to assure that a valid technological base exists, and that

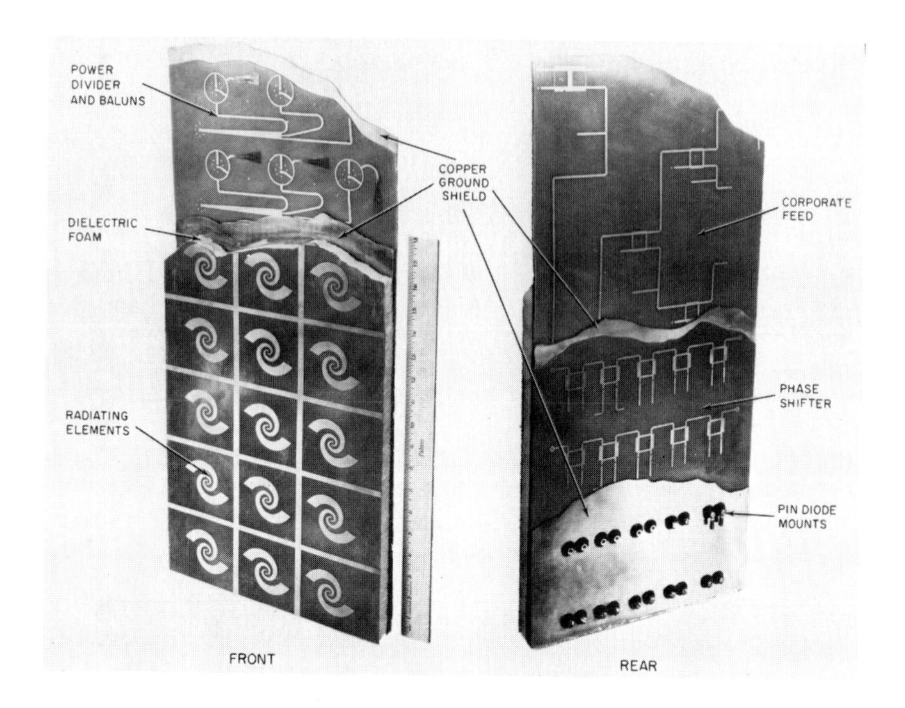

Fig. 30. Microstrip spiral array elements and constrained feed network (courtesy of Dr. R. Lewis, Raytheon Corp.).

its depth is sufficient to sustain the rapid technological growth that lies ahead.

A second technological area that has been a stimulus and could become a much more important factor in array design is the growth of active microwave integrated circuitry. Mixers, oscillators and microwave amplifiers are now available throughout most of the microwave range, and are cost competitive for a growing number of array system applications.

At frequencies up to 4 GHz transistors offer viable alternatives to the use of microwave tubes for many array applications. At higher frequencies it is convenient to combine transistors with varactor multipliers. This procedure can, for example, yield 10 watts at 4 GHz with better than 40% efficiency [132]. For frequencies up to X-band IMPATT diodes can provide several watts of power and nearly 10 watts is available from varactor multipliers. Gunn diodes can provide microwave signals at up to 70 GHz, but with relatively low signal levels at the high frequencies. A stimulating and timely survey of the current state of this art is given in reference [132].

At present these devices tend to be too expensive for many applications, and the market is so small as to preclude the use of truly inexpensive production methods. With time the use of solid state transmitters and receivers at each array element will become commonplace. This use will provide further stimulus for development of microstrip antenna types, and may also foster new developments in non-uniform array synthesis. The reasons for this additional concern is that such amplifiers are usually operated in a saturated mode, and it is difficult to amplitude weight the array elements as would be required for sidelobe suppression. The alternative is to allow uniform illumination and use non-uniform spacing for the purposes of tapering. This practice is not new; it is implemented in several military systems and numerous prototype design programs, but further exploration of these techniques for sidelobe suppression without undue complication of beam steering control requirements could bring about important advances in solid state radar arrays.

The continuing need to produce lower cost arrays has also led to the concept of an integrated sub-array module approach. The array of Fig. 30 is one early example of the technology required for such an approach, but the concept could be carried substantially further. Studies presently being undertaken by Hughes Aircraft Company are directed toward advancing such technology. In this approach a large number of radiators, phase shifters and a feed network are combined into an integrated subarray module which is used as the basic building block of an antenna. These components are combined on one common substrate of a high dielectric constant material such as alumina using thick film printing techniques. This printing technique can produce not only the conductor pattern of the circuits, but also the microwave capacitors and resistors as well. Radiators such as metallic discs are attached to the other side of the substrate (the ground plane side of the phase shifter circuit). This approach eliminates most of the interconnections such as coaxial cables and connectors, thereby reducing manufacturing and assembly labor as well as improving reliability. RF testing is performed at the sub-array module level instead of the individual component level, hence, minimizing the testing cost. The ultimate sub-array would have a continuous scanning aperture, i.e., the phase shift across the radiating aperture is varied continuously for beam scanning. For example, a ferrite slab can be used as a radiating aperture. The index of refraction across the ferrite slab can be varied continuously by external magnetization for beam scanning. Preliminary results with a scanned aperture of this type have been demonstrated at Lincoln Laboratory [133].

This description brings us to a limiting case, but emphasizes one of the main issues raised earlier. Array elements of the future may be very different from the waveguides and dipoles of the present. This technology must be supported by the same level of intense research activity that was necessary during the 60's because mistakes will be even more costly in the future.

B. Radomes, Polarizers and Spatial Filters

1. *Metallic Grid Structures for Radomes, Dichroic Reflectors and Polarizers.*

The present state-of-the-art in dielectric radomes is summarized in reference [134]. There is a growing use of metallic gratings for radomes, polarizers, dichroic sub-reflectors, and, now possibly for spatial filters. These devices represent an area of research that is strongly influenced by, and can itself influence, phased array research.

The survey by J.R. Wait [135] compares various theories of wire grid and mesh structures that are the basis of this new technology. Much of the basic analysis was performed in the interest of developing improved ground plane surfaces and not for radomes or polarizers. Studies of artificial dielectrics as summarized by Collin [136,137] are also directly applicable to the radome problem as are the work of Kieburtz and Ishimaru [138], Chen [139], Pelton and Monk [140] and the report by R. J. Lubbers [141], which includes an extensive bibliography and presents a cataloging of the various periodic slot array geometries analyzed using modal matching techniques. There appear to be only several references that describe multiple layers of metallic gratings, and these are restricted to identical gratings [142].

Apart from these analytical concerns, there has emerged an entirely new area of technology that offers metallic grid radomes or combinations of dielectric layers and metallic grids. The structures have been shown to have satisfactory wide-angle transmission characteristics over moderate frequency ranges, and to incorporate the advantages of rigid, lightweight metallic structures with the desired electromagnetic qualities. A logical extension of the radome studies, the use of such grids for dichroic sub-reflectors has become common in recent years. Although there appears to be no single reference that summarizes this work, the references given in the Luebbers report [141] serve as a good introduction to the subject.

The related subject of wave polarizers for use with reflectors or array antennas is described in references [143,144].

2. *Spatial Filters for Sidelobe Suppression*

Low antenna sidelobe levels are a desirable attribute of ECM-resistant radar and communications systems. For many applications one of the best ECCM features is a sidelobe level substantially below the range that is common to current systems. In order to reduce the vulnerability of existing systems that do not have very low sidelobes, it is often necessary to completely redesign the antenna. However, a new technology has been developed to provide an option that in certain cases can upgrade the ECCM capability of a radar or communications system without requiring antenna replacement. This technology is called spatial filtering.

Spatial filters are structures that are placed in front of an existing antenna to provide minimally attenuated transmission in the angular region near the main beam while suppressing radiation in other directions. They consist of several parallel layers of uniform dielectric or metallic gratings with reflection coefficients of the layers and interlayer spacings chosen to produce the desired angular filter characteristic. Tradeoffs can be made among the frequency bandwidth, angular range of transmission and filter characteristics by varying the physical parameters of the filter

To date, only one example of a microwave spatial filter is found in the literature [145]. This filter was designed using dielectric layers and synthesized to have Chebyshev characteristics in space.

The principles of layered dielectric frequency domain filters are well established [146], and insofar as possible the techniques for analysis and synthesis have been extended to the spatial domain. The fundamental difference between synthesis in the frequency domain and in the spatial domain arises because the transmission coefficients of layers that have a high dielectric constant are strongly frequency dependent but relatively invariant with the angle of incidence. If a wave from a medium having a low dielectric constant is incident on a medium of high dielectric constant, then for any angle of incidence the wave propagation direction in the latter is almost perpendicular to the interface.

This property necessitates a fundamental change in filter design from the frequency domain transformers synthesized by

Collin [139] and others [146,149], which consist of various dielectric layers sandwiched together. The spatial domain filters synthesized to date consist of quarter-wave sections of dielectric separated by half-wave or full-wave air spaces to produce a Chebyshev bandpass characteristic for the transmission response over an angular region.

Collin used the wave matrix formalism to derive convenient expressions for the transmission properties of layered impedance sections [137] and to describe the spatial properties of abutting dielectric layers [136]. The same formalism will be used here to derive properties of the stratefied dielectric filter. Consider the basic filter section shown in Fig. 31. The incident and reflected waves in medium 1 are a_1 and b_1 respectively. The incident electric fields are assumed to be either parallel-polarized or perpendicularly polarized, with no cross-polarized components excited. The input-output parameters of the section are related by

$$\begin{pmatrix} a_1 \\ b_1 \end{pmatrix} = \begin{pmatrix} A_{11} & A_{12} \\ A_{21} & A_{22} \end{pmatrix} \begin{pmatrix} a_2 \\ b_2 \end{pmatrix} \tag{56}$$

The parameters A_{mn} can be related in terms of the conventional scattering matrix. The wave matrix of a cascade of networks is the product of the wave matrices of each network. A parameter of particular importance is

$$A_{11} = \left.\frac{a_1}{a_2}\right|_{b_2 = 0} \tag{57}$$

which is the inverse of the filter transmission coefficient. For the case of near-broadside incidence on a quarter-wave dielectric slab and air space of width S, the wave matrix is approximately

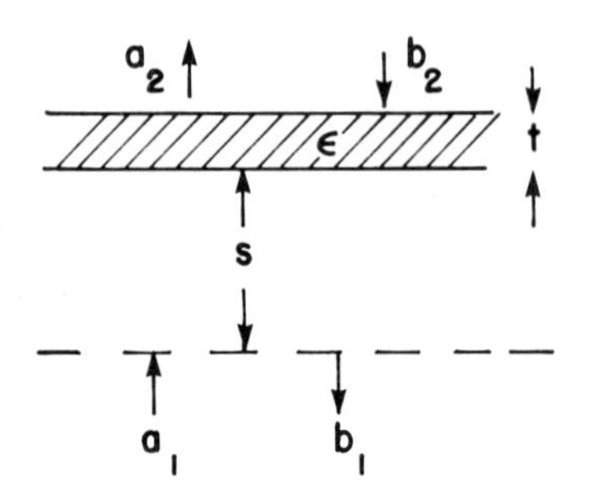

Fig. 31. Spatial filter element.

$$[A] = \begin{pmatrix} A_{11} & A_{12} \\ A_{21} & A_{22} \end{pmatrix} = \frac{1}{S_{12}} \begin{pmatrix} e^{jk_z S} & -S_{11} e^{jk_z S} \\ S_{11} e^{-jk_z S} & -e^{-jk_z S} \end{pmatrix} \quad (58)$$

where $S_{11} = \frac{\varepsilon - 1}{\varepsilon + 1}$, $S_{12} = \frac{-2j\sqrt{\varepsilon}}{\varepsilon + 1}$, and $k_z = k_o \cos\theta$.

The angle θ is the polar angle from broadside in air.

The wave matrix of a filter comprised of a number of such sections is obtained by multiplying in sequential order the matrices of the sections, as

$$[A]_{\text{filter}} = \prod_{n=1}^{N} [A(S_n, \varepsilon_n)] \quad (59)$$

This procedure is used with exact, angle dependent values of S_{11} and S_{12} to derive the filter properties for parallel and perpendicular polarizations and arbitrary angles of incidence [145], but the synthesis procedure is accomplished using the broadside values of these parameters.

The procedure for filter synthesis depends on the properties of the polynomial expression for the power loss ratio, defined by Collin [137] as the power ratio associated with the inverse of the filter transmission coefficient:

$$P = A_{11}A_{11}^{*} \tag{60}$$

The power loss ratio of a filter consisting of quarter wave sections of transmission line can be expressed as an even polynomial of degree $2n$, where n is the number of sections in the filter. Synthesis is accomplished by equating the power loss ratio of the n layer filter equal to unity plus the square of a Chebyshev polynomial of order n.

The difference between the design for spatial filters and the work of Collin and others is the inclusion of the air spaces S_j between the dielectric layers. The electrical path length through the dielectric and the air-dielectric interface characteristics do not vary appreciably with the angle of incidence. It is the variation with angle of the paths through the air spaces that causes the filter properties.

A representative result of the synthesis procedure is shown in Fig. 32. The filter consists of four layers with an interlayer separation of 0.5λ. The pass band extends to ±11.5° from broadside. The two inner layers of dielectric have a permittivity of $\varepsilon_2 = 15.14$, while the outer dielectric layers have an $\varepsilon_1 = 3.08$. The transmission characteristics for this filter have been calculated using the accurate formulation of the wave matrix, and the results are shown in Figures 33a and 33b. The filter reflection for the transmitted polarization and the cross polarization are given. The u-axis ($u = \sin\theta \cos\phi$) is the H-plane, while the v-axis ($v = \sin\theta \sin\phi$) is the E-plane.

A four-section dielectric layer spatial filter has been constructed. The filter consists of two outer layers of $\varepsilon = 3$ dielectric and two inner layers of $\varepsilon = 15$ dielectric material, each layer having a thickness of a quarter wavelength in the material at 11 GHz. The spacing between layers is a free-space wavelength at 11 GHz. Each layer is approximately 25 × 75 cm in

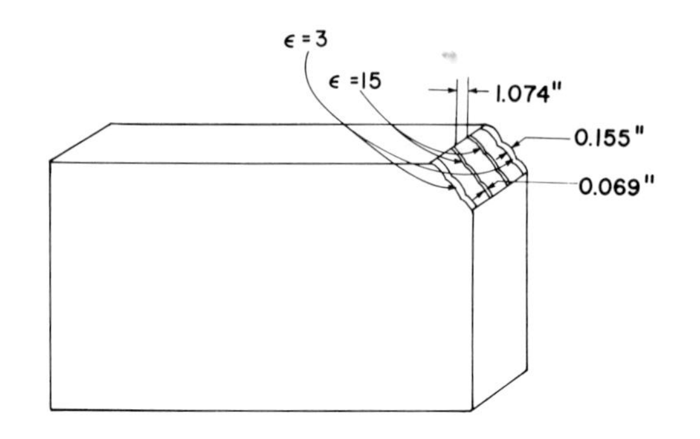

Fig. 32. Experimental model spatial filter.

planar extent, and styrofoam is used between layers for stability.

Preliminary tests of this filer have been made using an array source. In Fig. 34 are shown the radiation patterns of the array alone and the array with the filter. This array has a set of grating lobes that can be easily distinguished. Note that the filter attenuated those lobes falling within the stop-band of the filter, but did not attenuate the grating lobe at $\theta = 60°$, which is within the second pass-band of the filter.

Although this preliminary study has been conducted using dielectric layers, the use of metallic grid structures has obvious advantages in both weight and cost as compared with dielectric layer filters. Under the assumption that mutual coupling can be neglected, the grid structures become shunt susceptances, and the synthesis techniques described by Matthai and Young [150] can be used directly. To date there are no rigorous analytical results available that treat the problem of combining several unequal wire grids as required for spatial filter analysis.

Aside from the work described here, there is relatively little known about the characteristics of such spatial filters

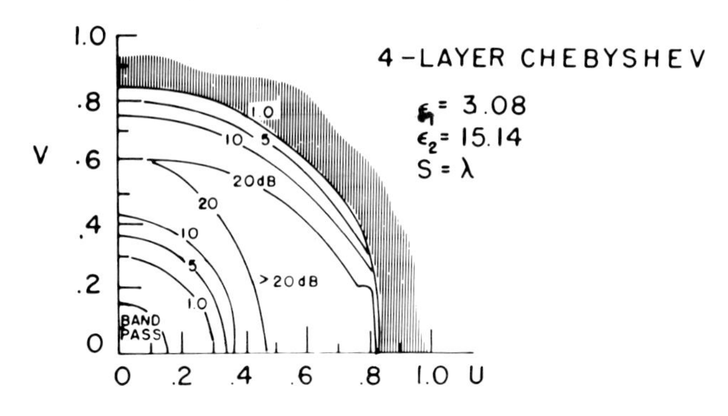

ISOLATION BETWEEN
TRANSMITTED AND
CROSSED POLARIZATION

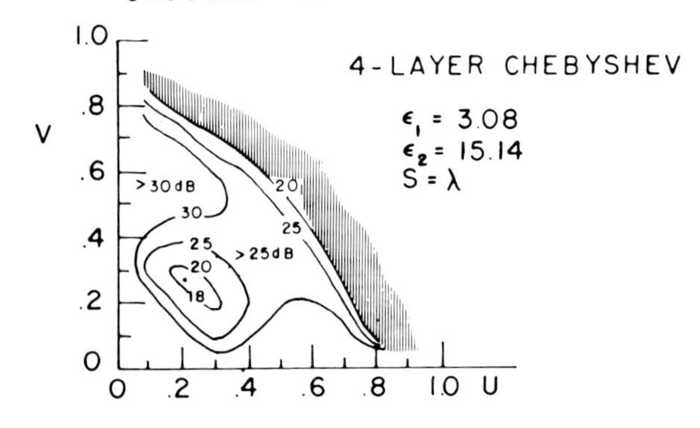

Fig. 33. Characteristics of experimental filter.

and their sidelobe suppression qualities. The example concerned the special case of a limited scan array with grating lobes, and even near the array the fields can be characterized by well defined plane waves. The array also has near fields that are characterized as non-propagating waves (reactive fields), but these did not appear to have any significant influence on the experimental results. Remaining questions include what benefit the

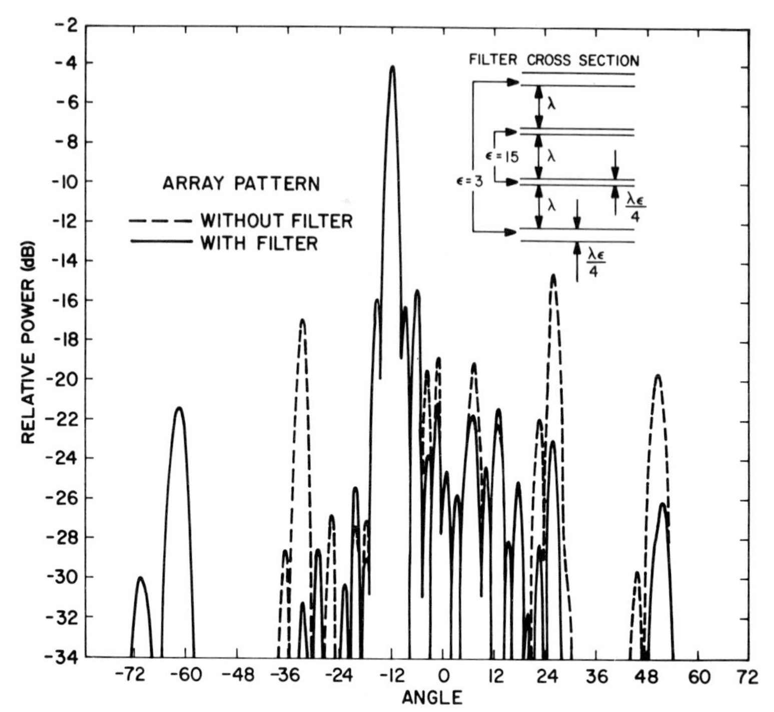

Fig. 34. Grating lobe suppresion using the experimental filter.

spatial filter can offer for far sidelobes that are due to random phase shifter errors or for sidelobes due to aperture blockage.

Preliminary results indicate that although the filter will obviously not improve the gain already reduced by blockage, it can reduce the resulting far sidelobe structure. In addition to studies of metallic grating structures for use as filter elements, there is thus a need for studying near field effects such as the use of a filter near small diffracting obstacles, and in the presence of fields with pseudo-random phase variations. The potential advantages of the use of such filters is very great, but substantial research is required before this potential can be realized.

V. CONCLUSION

The purpose of this paper has been to provide some data that can be useful in predicting general trends in phased array technology over the next few years, and to identify pertinent research areas that will support this technology. The method chosen for developing these conclusions has been to describe some of the history of phased array research and then to show evidence of the acceleration pace of technological innovation. I believe this changing technology will uncover even more fundamental topics for array research than have been studied in prior years.

Stimulus for this growth is provided by military requirements for radar and communication; in particular by the need for rapid scanning, wide or multiple bandwidths, very low sidelobes, null steering and the constraints imposed by cost, size and in some cases the physical environment near the array. Born of this increased activity and these new stimuli is the "special purpose array"; a collection of many different array types that are each designed to satisfy only one set of requirements, and are economically viable vecause they trade off other capability not related to any primary requirement. The growth of this collection of arrays has become so rapid as to provide a major focal point for activity in electromagnetic research. Added to this variety of new topics there is antenna technology growing in response to the availability of solid state devices; this has caused development of a variety microstrip and strip-line antennas which are quite fundamental new radiators with properties not yet fully investigated.

Finally, there is an area of growth that is occuring in response to uniquely military requirements, wide band antennas for ECM, multiple frequency antennas for satellite communication, extra low sidelobe and null steered arrays for ECCM radiation defense. Taken together, these forces have produced a major change in the direction of array research, and a need for increased research activity on a widening variety of topics within the general subject of array antennas.

VI. REFERENCES

1. Levine, H. and J. Schwinger, *Commun. Pure & Applied Math. 44,* 355-391 (1950-51).
2. Hallen, E., Theoretical Investigations into Transmitting and Receiving Antennae, *Nova Acta Regiae Soc. Sci. Upsaliensis (4) 11, 1* (1938).
3. King, R.W.P. and C.W. Harrison, "Antennas and Waves: A Modern Approach", see section 3.10, MIT Press, Cambridge, Massachusetts (1969).
4. Bouwkamp, C.J., *Rep. Prog. Phys. 17,* 35-100 (1954).
5. Mailloux, R.J., *IEEE Trans. AP-17, 1,* 49-55 (1969).
6. Lewin, L., *IEEE Trans. MTT-18, 7,* 364-372 (1970).
7. Brouillion, L., "Wave Propagation in Periodic Structures", Dover Publications, Inc. (1953).
8. Seckelmann, R., *IEEE Trans. MTT-14,* 518-527 (1966).
9. Collin, R.E., "Field Theory of Guided Waves", McGraw-Hill Book Co., Inc., New York (1960).
10. Borgiotti, G.V., *IEEE Trans. AP-19, 5,* 593-599 (1971).
11. Wasylkiwskyj, W., *IEEE Trans. AP-21, 3,* 277-285 (1973).
12. Amitay, N., V. Galindo and C.P. Wu, "Theory and Analysis of Phased Array Antennas, Wiley Interscience, New York (1972).
13. Carter, P.S. Jr., *IRE Trans. AP-8,* 276-285 (1960).
14. King, R.W.P., "The Theory of Linear Antennas", Harvard Univ. Press, Cambridge, Massachusetts (1956).
15. King, R.W.P. and S.S. Sandler, *IEEE Trans. Antennas & Propag. AP-12,* 269-275 & 276-280 (1964).
16. Harrington, R.F., "Field Computation by Moment Methods", Mc Millan Co., New York (1968).
17. Harrington, R.F. and J.R. Mautz, *IEEE Trans. AP-15, 4,* 502-515 (1967).
18. Lewin, L. "Advanced Theory of Waveguides", Iliffe and Sons Ltd., London (1951). Chapter 6.
19. Cohen, M.H., T.H. Crowley and C.A. Levis, The Aperture of a Rectangular Waveguide Radiating into a Half Space, (ATI-133 707) Antenna Laboratory, Ohio State Univ. Research Foundation, Rept. 339-22 (Nov. 1951).
20. Wheeler, G.W., "Coupled Slot Antennas", Ph.D. Thesis, Harvard University, Cambridge, Massachusetts (May 1950).
21. Levis, C.A., Variational Calculations of the Impedance Parameters of Coupled Antennas, Ohio State Univ. Research Foundation, Rept. 667-16, Contract AF33(616)3353 (Aug. 1956).
22. Galejs, J., *Radio Sci. 69D, 2,* 179-189 (1965).
23. Richmond, J.H., *IRE Trans. AP, AP-8,* 515-520 (1961).
24. Lyon, J.A.M. R.M. Kalafus, Y.K. Kwon, C.J. Diegenis, M.A.H. Ibrahim, and C.C. Chen, Derivation of Aerospace Antenna

Coupling--Factor Interference Predication Techniques, Tech. Report AFAL-TR-66-57, The Univ. of Michigan, Radiation Lab. (April 1966).

25. Mailloux, R.J., *IEEE Trans. Antennas & Propag. AP-17, 1,* 49-55 (1969).
26. Mailloux, R.J., *IEEE Trans. AP-17,* 740-746 (1969).
27. Bailey, M.C,, *IEEE Trans. AP-22,* 178-184 (1974).
28. Steyskal, H., *IEEE Trans. AP-22,* 594-597 (1974).
29. Knittel, G.H., A. Hessel and A.A. Oliner, *Proc. IEEE 56,* 1822-1836 (1968).
30. Lechtreck, L.W., Cumulative Coupling in Antenna Arrays, IEEE G-AP International Symposium Digest, 144-149 (1965).
31. Oliner, A.A. and R.G. Malech, "Speculation on the Role of Surface Waves, Microwave Scanning Antennas", Vol. 2, 308-322, Academic Press, New York (1966).
32. Farrell, G.F. Jr. and D.H. Kuhn, *IEEE Trans. AP-14,* 652-654 (1966).
33. Farrell, G.F., Jr. and D. H. Kuhn, *IEEE Trans. AP-16,* 405-414 (1968).
34. Amitay, N. and V. Galindo, *Bell Sys. Tech. J.,* 1903-1932 (1968).
35. DuFort, E.C., *IEEE Trans. AP-16,* 37-46 (1968a).
36. DuFort, E.C., *Proc. IEEE 56,* 1851-1860 (1968a).
37. Mailloux, R.J., *IEEE Trans. AP-20,* 160-166 (1972).
38. Bates, R.H.T., *IEEE Trans. Antennas & Propag. AP-13,* 321-322 (1965).
39. Byron, E.V. and J. Frank, *IEEE Trans. AP-16,* 496-499 (1968a).
40. Hannan, P.W., *IEEE Trans. AP-15,* 574-576 (1967).
41. Gregorowich, W.S., A. Hessel, G.H. Knittel and A.A. Oliner, A Waveguide Simulator for the Determination of a Phased-Array Resonance, IEEE G-AP International Symposium Digest, 134-141 (1968).
42. Frazita, R.F., *IEEE Trans. AP_15,* 823-824 (1967).
43. Parad, L.I., *IEEE Trans. AP-15,* 302-304 (1967).
44. Galindo, V. and C.P. Wu, *Bell Sys. Tech. J. 47,* 93-116 (1968).
45. Wu, C.P. and V. Galindo, *Bell Sys. Tech. J. 47,* 117-142 (1968).
46. Borgiotti, G.V., *Proc. IEEE 56,* 1881-1892 (1968).
47. Allen, J.L., *IEEE Trans. AP-13,* 638-639 (1965).
48. Lechtreck, L.W., *IEEE Trans. AP-16,* 31-37 (1968).
49. Diamond, B.L., *Proc. IEEE 56, 11,* 1837-1851 (1968).
50. Oliner, A.A. and R.G. Malech, Speculation on the Role of Surface Waves in "Microwave Scanning Antennas", vol. 2, 308-322, Academic Press, New York.
51 Ehlenberger, A.G., L. Schwartzman and L. Topper, *Proc. IEEE 56, 11,* 1861-1872 (1968).
52. Byron, E.V. and J. Frank, *IEEE Trans. AP-16,* 601-603 (1968b).
53. Hessel, A. and G.H. Knittel, A Loaded Groundplane for the Elimination of Blindness in a Phased-Array Antenna, IEEE G-AP International Symposium Digest, 163-169 (1969).
54. Knittel, G.H., The Choice of Unit-Cell Size for a Waveguide

Phased Array and its Relation to the Blindness Phenomenon, presented at Boston Chapter Antennas and Propagation Group Meeting (1970).

55. Hessel, A. and G.H. Knittel, *IEEE Trans. Antennas & Propag. AP-18,* 121-123 (1970).
56. Mailloux, R.J., Blind Spot Occurrence in Phased Arrays - When to Expect It and How to Cure It, AFCRL-71-0428, Physical Sci. Res. Papers, no. 462, Air Force Cambridge Res. Labs. (Aug. 1971).
57. Lee, S.W. and W.R. Jone, *IEEE Trans. AP-19,* 41-51 (1971).
58. Lee, S.W. *IEEE Trans. AP-15, 5,* 598-606 (1967).
59. Harrington, R.F., *Proc. IEEE 55, 2,* 136-149 (1967).
60. Amitay, N. V. Galindo and C.P. Wu, "Theory and Analysis of Phased Array Antennas", Wiley-Interscience, New York (1972).
61. Harrington, R.F. and J.R. Mautz, A Generalized Network Formulation for Aperture Problems, AFCRL-TR-75-0589, Sci. Rept. no. 8, Contract F19628-73-C-0047 (Nov. 1975).
62. Shelley, B., *Proc. IEEE 56, 11,* 2016-2027 (1968).
63. Holley, A.E., et al., *IEEE Trans. AP-22,* 3-12 (1974).
64. Bogner, B.F., *IEEE Trans. AP-22,* 78-81 (1974).
65. Boyns, J.E., et al., *IEEE Trans. AP-18, 5,* 590-595 (1970).
66. Munger, A.D., et al., *IEEE Trans. AP-22,* 35-43 (19740).
67. Hsiao, J.K. and A.G. Cha, *IEEE Trans. AP-22,* 81-84 (1974).
68. Gobert, W.B. and R.F.H. Young, *IEEE Trans. AP-22,* 87-91 (1974).
69. Tsandoulas, G.S. and F.G. Willwerth, *Microwave J. 16, 10,* 29-32 (1973).
70. Borgiotti, G.V. and Q. Balzano, *IEEE Trans. AP-18,* 55-63 (1970).
71. Borgiotti, G.V. and Q. Balzano, *IEEE Trans. AP-20,* 547-553 (1972).
72. Balzano, Q. and T.B. Dowling, *IEEE Trans. AP-22,* 92-97 (1974).
73. Golden, K.E., et al., *IEEE Trans. AP-22,* 44-48 (1974).
74. Shapira, J., et al., *IEEE Trans. AP-22,* 49-63 (1974).
75. Steyskal, H., Mutual Coupling Analysis of Cylindrical and Conical Arrays, IEEE/AP-S Int. Symp. Record, 293-294, (June 1974).
76. Maune, J.J., An SHF Airborne Receiving Antenna, Twenty-Second Annual Symposium on USAF Antenna Res. and Development (Oct. 1972).
77. Villeneuve, A.T., M.C. Behnke and W.H. Kummer, Hemispherically Scanned Arrays, AFCRL-TR-74-0084, Contract No. F19628-72-C-0145, Sci. Rept. no. 2 (Dec. 1973).
78. Balzano, Q., L.R. Lewis and K. Siwiak, Analysis of Dielectric Slab-Covered Waveguide Arrays on Large Cylinders, AFCRL-TR-73-0587, Contract No. F19628-72-C-0202. Sci. Rept. no. 1 (Aug. 1973).
79. L. Schwartzman and J. Stangel, *Microwave J. 18, 10,* 31-34 (1975).
80. Esposito, F.J., L. Schwartzman and J.J. Stangel, The Dome

Antenna-Experimental Implementation, URSI/USNC Meeting Digest Commission 6, (June 1975).
81. Ruze, J., Physical Limitations on Antennas, MIT Research Lab Electronics Tech. Rept. 248 (Oct. 1952).
82. Miller, C.J., Minimizing the Effects of Phase Quantization Errors in an Electronically Scanned Array, Proc. Symp. Electronically Scanned Array Techniques and Applications, RADC-TDR-64-225, vol. 1, 17-38 (AD448421) (1964).
83. Allen, J.L., Some Extensions of the Theory of Random Error Effects on Array Patterns, in J.L. Allen et al., "Phased Array Radar Studies", Tech. Rept. No. 236, Lincoln Lab. MIT, July 1960-1961).
84. Elliott, R.S., *IRE Trans. AP-6,* 114-120 (1958).
85. McIlvenna, J.F. and C.J. Drane, *Radio & Electron. Eng. 41, 12,* 569-572 (1971).
86. McIlvenna, J., et al., The Effects of Excitation Errors in Null Steering Antenna Arrays, RADC-TR=76-183, Rome Air Development Center (June 1976).
87. Stangel, J.J., A Basic Theorem Concerning the Electronic Scanning Capabilities of Antennas, URSI Commission VI, Spring Meeting (June 1974).
88. Patton, W.T., Limited Scan Arrays in "Phased Array Antennas; Proceedings of the 1970 Phased Array Antenna Symposium", (A. A. Oliner and G.H. Knittel Eds.), 332-343, Artech House, Inc. Dedham, Massachusetts (1972).
89. Borgiotti, G. Design Criteria and Numerical Simulation of an Antenna System for One-Dimensional Limited Scan, AFCRL-TR-75-0616 (Dec. 1975).
90. Silver, S. and C.S. Pao, Paraboloid Antenna Characteristics as a Function of Feed Tilt, MIT Radiation Lab, Cambridge, Massachusetts, Rept. 479 (1944).
91. Ruze, J., *IEEE Trans. Antennas & Propag. AP-13,* 660-665 (1965).
92. Sletten, C.J., et al., *IEEE Trans. AP-6, 3,* 239-251 (1958).
93. Collin, R.E. and F.J. Zucker, "Antennas Theory Part 2", Chapter 17, McGraw-Hill Book Co., New York (1969).
94. Rusch, W.V.T. and A.C. Ludwig, *IEEE Trans. Antennas & Propag. AP-21,* 141-147 (1973).
95. Imbraile, W.A., et al., *IEEE Trans. AP-22, 6,* 742-745 (1974).
96. Rudge, A.W., *Electron. Lett. 5,* 610-612 (1969).
97. Rudge, A.W. and M.J. Withers, *Proc. IEEE 118, 7,* 857-863 (1971).
98. Rudge, A.W. and M.J. Withers, *Electron. Lett. 5,* 39-41 (1969).
99. Rudge, A.W. and D.E.N. Davies, *Proc. IEE, 117,* 351-358 (1970).
100. Winter, C., *Proc. IEEE 56, 11,* 1984-1999 (1968).
101. White, W.D. and L.K. DeSize, Scanning Characteristics of Two-Reflector Antenna Systems, IRE International Convention Rec., Part 1, 44-70 (1962).
102. Tang, C.H., Application of Limited Scan Design for the AGIL-TRAC-16 Antenna, 20th Annual USAF Antenna Research and

Development Symposium, Univ. of Illinois (Oct. 1970).
103. Howell, J.M., Limited Scan Antennas, IEEE/AP-S International Symposium (1974).
104. Schell, A.C., A Limited Sector Scanning Antenna, IEEE G-AP International Symposium (Dec. 1972).
105. McGahan, R., A Limited Scan Antenna Using a Microwave Lens and a Phased Array Feed, M.S. Thesis, MIT (1975).
106. Fitzgerald, W.D., Limited Electronic Scanning with a Near Field Cassegrainian System, ESD-TR=71-271, Tech. Rept. 484, Lincoln Laboratory, (Sept. 1971).
107. Fitzgerald, W.D., Limited Electronic Scanning with an Offset Feed Near-Field Gregorian System, ESD-TR-71-272, Tech. Rept. 486, Lincoln Laboratory (Sept. 1971).
108. Tang, C.H. and C.F. Winter, Study of the Use of a Phased Array to Achieve Pencil Beam over Limited Sector Scan, AFCRL-TR=73-0482, ER 73-4292, Raytheon Co. Final Rept. Contract F19628-72-C0213 (July 1973).
109. Tang, R. et al., Limited Scan Antenna Technique Study, Final Rept. AFCRL-TR-75-0448, Contract No. F19628-73-C-0129 (Aug. 1975).
110. Miller, C.J. and D. Davis, LFOV Optimization Study, Final Rept. 77-0231, Westinghouse Defense and Electronic Systems Center, System development Div. Baltimore, Md., (Jan. 1972).
111. Patton, W.T. Limited Scan Arrays in "Phased Array Antennas", Proceedings of Phased Array Antenna Symposium (A.A. Oliner and G.H. Knittel, Eds.) Artech House, Inc. (1972).
112. Manwarren, T.A. and A.R. Minuti, Zoom Feed Technique Study, RADC-TR-74-56, Final Tech. Rept. (March 1974).
113. Mailloux, R.J. and G.R. Forbes, *IEEE Trans. Antennas & Propag. AP-21, 5,* 597-602 (1973).
114. Tsandoulas, G.N. and W.D. Fitzgerald, *IEEE Trans. AP-20, 1,* 69-74 (1972).
115. Tang, R., Survey of Time-Delay Beam Steering Techniques in "Phased Array Antennas", Proceedings of the Phased Array Antenna Symposium (A.A. Oliner and G.H. Knittel, Eds.) Artech House, Inc. (1972).
116. Mailloux, R.J., An Overlapped Subarray for Limited Scan Applications, *IEEE Trans. AP-22* (May 1974).
117. Tsandoulas, G.N., *Microwave J. 15, 9,* 49-56 (1972).
118. Chen, C.C., *IEEE Trans. AP-21,* 298-302 (1973).
119. Laughlin, G.J., et al., *IEEE Trans. AP-20,* 699-704 (1972).
120. Chen, C.C., Octave Band Waveguide Radiators for Wide-Angle Scan Phased Arrays, IEEE AP-S Int. Symposium Record, 376-377 (1972).
121. Lewis, L.R., M. Fassett and J. Hunt, A Broadband Stripline Array Element, IEEE AP-S Int. Symposium Record (June 1974).
122. Chen, M.H. and G.N. Tsandoulas, Bandwidth Properties of Quadruple-Ridged Circular and Square Waveguide Radiators, IEEE AP-S Int. Symposium Record, 391=394 (Aug. 1973).
123. Tsandoulas, G.N. and G.H. Knittel, *IEEE Trans. AP-21,* 796-

808 (1973).
124. Hsiao, J.K., *IEEE Trans. AP-19,* 729-735 (1971).
125. Boyns, J.E. and J.H. Provencher, *IEEE Trans. AP-20,* 106-107 (1972).
126. Hsiao, J.K., *IEEE Trans. AP-20,* 505-506 (1972).
127. Harper, W.H., et al., NRL Report No. 7369, Naval Research Laboratory (Feb. 1972).
128. Munson, R.E., *IEEE Trans. AP-22,* 74-78 (1974).
129. Howell, J.Q., *IEEE Trans. AP-23,* 90-93 (1975).
130. Kaloi, C., Asymmetrically Fed Electric Microstrip Dipole Antenna, TR-75-03, Naval Missile Center (Jan. 1975).
131. Derneryd. A., *IEEE Trans. AP-24,* 846 (1976).
132. *Microwave J. 20, 2,* Special Issue on Solid State Power (Feb. 1977).
133. Stern, E. and G.N. Tsandoulas, *IEEE Trans. AP-23, 1,* 15-20 (1975).
134. Walton, J.D., ed., "Radome Engineering Handbook", Georgia Tech. (1970).
135. Wait, J.R., Theories of Scattering from Wire Grid and Mesh Structures, Proc. of National Conf. on EM Scattering, Univ. of Illinois (June 1976).
136. Collin, R.E., *Proc. IRE 43, 2,* 179-185 (1955).
137. Collin, R.E., "Field Theory of Guided Waves", McGraw-Hill Book Co., 79-93 (1960).
138. Kieburtz, R.B. and A. Ishimaru, *IRE Trans. AP-9,* 663-671 (1962).
139. Chen, C.C., *IEEE Trans. MTT-19, 5,* 475-481 (1971).
140. Pelton, E.L. and B.A. Monk, *IEEE Trans. AP-22, 6,* 799-804 (1974).
141. Luebbers, R.J., Analysis of Various Periodic Slot Array Geometries Using Modal Matching, Report AFAL-TR-75-119, Ohio State Univ. (Feb. 1976).
142. Monk, B.A., et al., *IEEE Trans. AP-22,* 804-809 (1974).
143. Young, L., L. Robinson and C. Hacking, *IEEE Trans. AP-21,* 376-378 (1973).
144. D.S. Lerner, *IEEE Trans. AP-13,* 3-7 (1965).
145. Mailloux, R.J., *IEEE Trans. Antennas & Propag. AP-24, 2,* 174-181 (1976).
146. Cohn, S.B., *IRE Trans. MTT-3, 2,* 16-21 (1955).
147. Riblet, H.H., *IRE Trans. MTT-5, 1,* 36-43 (1957).
148. Young, L., *IRE Trans. MTT-7, 2,* 233-244 (1959).
149. Young, L., *IRE Trans. MTT-10, 5,* 339-359 1962).
150. Mathai, G., L. Young and E.M.T. Jones, "Microwave Filter Impedance Matching Networks and Coupling Structures", Mc Graw-Hill Book Co., New York (1964). Ch. 9.

A
B
C 8
D 9
E 0
F 1
G 2
H 3
I 4
J 5